Soil Hydrology, Land Use and Agriculture

Measurement and Modelling

Soil Hydrology, Land Use and Agriculture

Measurement and Modelling

Edited by

Manoj K. Shukla

New Mexico State University
USA

www.cabi.org

CABI is a trading name of CAB International

CABI Head Office
Nosworthy Way
Wallingford
Oxfordshire OX10 8DE
UK

Tel: +44 (0)1491 832111
Fax: +44 (0)1491 833508
E-mail: cabi@cabi.org
Website: www.cabi.org

CABI North American Office
875 Massachusetts Avenue
7th Floor
Cambridge, MA 02139
USA

Tel: +1 617 395 4056
Fax: +1 617 354 6875
E-mail: cabi-nao@cabi.org

A catalogue record for this book is available from the British Library, London, UK.

Library of Congress Cataloging-in-Publication Data

Soil hydrology, land use and agriculture : measurement and modelling/edited by Manoj K. Shukla.
p. cm.
Includes bibliographical references and index.
ISBN 978-1-84593-797-3 (alk. paper)
1. Soil moisture--Measurement--Computer simulation. 2. Hydrologic models. 3. Hydrology. 4. Land use--Environmental aspects. I. Shukla, Manoj. II. Title.

S594.S665 2011
631.4′32--dc22

2011003332

ISBN-13: 978 1 84593 797 3

Commissioning editor: Claire Parfitt
Production editor: Shankari Wilford

Typeset by SPi, Pondicherry, India.
Printed and bound in the UK by MPG Books Group.

Contents

The colour plate section can be found following p. 212

Contributors

Gabriele Antolini, Environmental Protection Agency of Emilia-Romagna, HydroMeteoClimate Service (ARPA-SIMC), Bologna, Italy. E-mail: gantolini@arpa.emr.it

Allah Bakhsh, Department of Irrigation and Drainage, University of Agriculture, Faisalabad 38040, Pakistan. E-mail: bakhsh@iastate.edu

Satish Bastola, Irish Climate Analysis and Research Units, Department of Geography, National University of Ireland, Maynooth, Ireland. E-mail: Satish.Bastola@nuim.ie

A. Salim Bawazir, Civil and Geological Engineering Department, New Mexico State University, PO Box 30001, MSC 3CE, Las Cruces, NM 88003-8001, USA. E-mail: abawazir@nmsu.edu

Marco Bittelli, Department of AgroEnvironmental Science and Technology, University of Bologna, Italy. E-mail: marco.bittelli@unibo.it

Max P. Bleiweiss, Department of Entomology, Plant Pathology and Weed Science, New Mexico State University, PO Box 30003, MSC 3BE, Las Cruces, NM 88003-8003, USA. E-mail: mbleiwei@taipan.nmsu.edu

Sanjit K. Deb, Department of Plant and Environmental Sciences, New Mexico State University, MSC 3Q, PO Box 30003 Las Cruces, NM 88003, USA. E-mail: Sanjit@nmsu.edu

Sjoerd W. Duiker, Department of Crop and Soil Sciences, The Pennsylvania State University, 116 ASI Building, University Park, PA 16802-3504. E-mail: sduiker@psu.edu

David Favis-Mortlock, Environmental Change Institute, School of Geography and the Environment, University of Oxford, South Parks Road, Oxford, OX1 3QY, UK. E-mail: david.favis-mortlock@ouce.ox.ac.uk

Alexander G. Fernald, Department of Animal and Range Sciences, New Mexico State University, PO Box 30003, MSC 3I, Las Cruces, NM 88003, USA. E-mail: fernald@nmsu.edu

Dennis C. Flanagan, USDA-ARS NSERL, 275 S. Russell Street, West Lafayette, IN 47907-2077, USA. E-mail: Dennis.Flanagan@ars.usda.gov

Markus Flury, Department of Crop and Soil Sciences, Washington State University, Pullman, WA 99164-6420, USA. E-mail: flury@wsu.edu

Nicola Fohrer, Ecology Centre, Institute for the Conservation of Natural Resources, Department of Hydrology and Water Resources Management, CAU (Christian-Albrechts Universität) Kiel, Olshausenstr. 75, D-24118 Kiel, Germany. E-mail: nfohrer@hydrology.uni-kiel.de

A.K. Gosain, Department of Civil Engineering, Indian Institute of Technology (IIT Delhi), Haus Khas, New Delhi 110 016, India. E-mail: gosain@civil.iitd.ac.in

Steven J. Guldan, Alcalde Sustainable Agriculture Science Center, PO Box 159, Alcalde, NM 87511, USA. E-mail: sguldan@nmsu.edu

Khaled Hardan, Institute of Water Management, Hydrology and Hydraulic Engineering, Department of Water, Atmosphere and Environment, BOKU (University of Natural Resources and Applied Life Sciences, Vienna), Muthgasse 18, A-1190 Vienna, Austria. E-mail: khaled.hardan@boku.ac.at

Manon Janssen, Institute of Agricultural Climate Research, Johann Heinrich von Thünen-Institut (vTI), Federal Research Institute for Rural Areas, Forestry and Fisheries, Braunschweig, Germany. E-mail: manon.janssen@uni-rostock.de

Ramesh S. Kanwar, Department of Agricultural and Biosystems Engineering, Iowa State University, 3013 Agronomy Hall, Ames, IA 50011, USA. E-mail: rskanwar@iastate.edu

Andreas Klik, Institute of Water Management, Hydrology and Hydraulic Engineering, Department of Water, Atmosphere and Environment, BOKU (University of Natural Resources and Applied Life Sciences, Vienna), Muthgasse 18A, A-1190 Vienna, Austria. E-mail: Andreas.Klik@boku.ac.at

Suresh Kumar, Irish Climate Analysis and Research Units, Department of Geography, National University of Ireland, Maynooth, Ireland. E-mail: jogis@tcd.ie

Bernd Lennartz, Institute for Land Use, Rostock University, 18051 Rostock, Germany. E-mail: bernd.lennartz@uni-rostock.de

A. Mani, Department of Agricultural Engineering, College of Agriculture, Rajendranagar, Hyderabad – 500030, India. E-mail: manidigumarthi@rediffmail.com

Biswajit Mukhopadhyay, Water Infrastructure Program, Jacobs Engineering Group, Inc., 7950 Elmbrook Drive, Dallas, TX 75247-4951, USA. E-mail: Biswajit.Mukhopadhyay@jacobs.com

Donal Mullan, School of Geography, Archaeology and Palaeoecology, Queen's University Belfast, Belfast BT7 1NN, Northern Ireland. E-mail: dmullan15@qub.ac.uk

Conor Murphy, Irish Climate Analysis and Research Units, Department of Geography, National University of Ireland, Maynooth, Ireland. E-mail: Conor.Murphy@nuim.ie

Hans-Peter Nachtnebel, Institute of Water Management, Hydrology and Hydraulic Engineering, Department of Water, Atmosphere and Environment, BOKU (University of Natural Resources and Applied Life Sciences, Vienna), Muthgasse 18, A-1190 Vienna, Austria. E-mail: hans_peter.nachtnebel@boku.ac.at

João Pedro Nunes, CESAM and Department of Environment and Planning, Universidade de Aveiro, Campus Universitário de Santiago, 3810-193 Aveiro, Portugal. E-mail: jpcn@ua.pt

Carlos G. Ochoa, Department of Animal and Range Sciences, New Mexico State University, Las Cruces, NM 88003, USA. E-mail: carochoa@nmsu.edu

Roberto Orsini, Department of Environmental and Crop Sciences, Polytechnic University of Marche, Ancona, Italy. E-mail: r.orsini@univpm.it

Naresh Pai, Department of Biological and Agricultural Engineering, University of Arkansas, Fayetteville, AR 72701, USA. E-mail: npai@uark.edu

Alberto Pistocchi, Institute for Environment and Sustainability (IES), European Community Joint Research Centre, Ispra, Italy. E-mail: alberto.pistocchi@eurac.edu

Sandhya Rao, INRM Consultants Pvt Ltd, New Delhi 110 075, India. E-mail: sandhya.delhi@gmail.com

Pier Paolo Roggero, Desertification Research Group, Department of Agronomy and Plant Breeding, University of Sassari, Sassari, Italy. E-mail: pproggero@uniss.it

Dharmendra Saraswat, Department of Biological and Agricultural Engineering, Division of Agriculture Cooperative Extension Service, University of Arkansas, Little Rock, AR 72204, USA. E-mail: dsaraswat@uaex.edu

Britta Schmalz, Ecology Centre, Institute for the Conservation of Natural Resources, Department of Hydrology and Water Resources Management, CAU (Christian-Albrechts Universität) Kiel, Olshausenstrasse 75, D-24118 Kiel, Germany. E-mail: bschmalz@hydrology.uni-kiel.de

Thomas Schmugge, Physical Sciences Laboratory, New Mexico State University, PO Box 30002, Las Cruces, NM 88003, USA. E-mail: tschmugge@psl.nmsu.edu

Júlia Seixas, CENSE (Centre for Environmental and Sustainability Research), Faculdade de Ciências e Tecnologia, Universidade Nova de Lisboa, Lisbon, Portugal. E-mail: mjs@fct.unl.pt
Manoj K. Shukla, Plant and Environmental Sciences Department, New Mexico State University, MSC 3Q, PO Box 30003, 945 College Avenue, Las Cruces, NM 88003, USA. E-mail: shuklamk@nmsu.edu
Vijay P. Singh, Department of Biological and Agricultural Engineering, Texas A&M University, Scoates Hall, 2117 TAMU, College Station, TX 77843-2117, USA. E-mail: vsingh@tamu.edu
John Sweeney, Irish Climate Analysis and Research Units, Department of Geography, National University of Ireland, Maynooth, Ireland. E-mail: John.Sweeney@nuim.ie
Bärbel Tiemeyer, Institute of Agricultural Climate Research, Johann Heinrich von Thünen-Institut (vTI), Federal Research Institute for Rural Areas, Forestry and Fisheries, Braunschweig, Germany. E-mail: baerbel.tiemeyer@vti.bund.de
Marco Toderi, Department of Environmental and Crop Sciences, Polytechnic University of Marche, Ancona, Italy. E-mail: m.toderi@univpm.it
Fausto Tomei, Environmental Protection Agency of Emilia-Romagna, HydroMeteoClimate Service (ARPA-SIMC), Bologna, Italy. E-mail: ftomei@arpa.emr.it
Pushpa Tuppad, Blackland Research and Extension Center, Texas AgriLIFE Research, 720 East Blackland Road, Temple, TX 76502, USA. E-mail: ptuppad@brc.tamus.edu
Antje Ullrich, Environmental Agency (UBA), Wörlitzer Platz 1, D-06844 Dessau-Roßlau, Germany. E-mail: antje.ullrich@yahoo.com
Martin Volk, Department of Computational Landscape Ecology, Helmholtz Centre for Environmental Research, UFZ (Umweltforschungszentrum), Permoserstrasse 15, D-04318 Leipzig, Germany. E-mail: martin.volk@ufz.de
Xiuying Wang, Blackland Research and Extension Center, Texas AgriLIFE Research, 720 East Blackland Road, Temple, TX 76502, USA. E-mail: swang@brc.tamus.edu
Jimmy R. Williams, Blackland Research and Extension Center, Texas AgriLIFE Research, 720 East Blackland Road, Temple, TX 76502, USA. E-mail: jwilliams@brc.tamus.edu

Preface

The understanding of the complex interaction between soil hydrology, agricultural land use and management is critically important for sustaining soil and water resources and agricultural production. Agriculture is strongly affected by changes in soil hydrology as well as by changes in land use and management practices. *Soil Hydrology, Land Use and Agriculture: Measurement and Modelling* addresses past, present and future issues and challenges related to the measurement and modelling of hydrological properties and processes. Soil hydrological properties and processes vary at multiple scales – starting from the scale of a single pore, up to the scale of a pedon, to multiple pedons and to landscape to the pedosphere scales – and are also influenced at each of these scales by human activities, land use and management practices, and natural or management-related perturbations. Soil hydrological properties and processes are usually linked non-linearly, with complex interactions driven by the soil–plant–atmosphere continuum.

This book presents some of the complex interactions between soil hydrology and land use management changes on a watershed scale, and determines the influence of these changes on soil, water and solute dynamics within the vadose zone. The book synthesizes information on several existing soil hydrological models, their capabilities, theories and input requirements, addresses the consequences of land use and management changes for agriculture and presents research results including those from field measurements and modelling. The book also attempts to present results on the possible impacts of climatic change on soil hydrological processes and, to a limited degree, on its impacts on agriculture.

Over the past decade, several attempts have been made to understand the impact of changes in land use management and soil hydrology on agriculture. These efforts include extensive field measurements and the use of physically (or semi-physically) based distributed models to quantify interactions between soil hydrology and agriculture. This book is designed to bring these two aspects together in such a way that it presents the state of the art on these issues. The book covers the application of physically based distributed hydrological models under various ecological and climatic conditions and scales in much greater detail than is available in any peer-reviewed journal article.

This book comprises 19 chapters which start with an introduction to soil hydrology and the application of hydrological models on a mesoscale, while listing the variability of hydrological properties and some of the past, present and future challenges associated with soil hydrology (Chapters 1–2). As there are a number of soil hydrological models that are in the public domain, the book presents a detailed overview of some of these physically/semi-physically based models (Chapter 3), followed by case studies on the application of some of the models to determine the impact of land use and management on various soil hydrological parameters under different climates and ecosystems (Chapters 4–11). The book presents case studies relating to soil water and nutrient management for the sustainable use of agricultural sources (Chapters 12–14), lists different climate data sets for soil hydrological modelling (Chapter 15) and discusses the influence of climate change on soil hydrology, soil erosion, and agriculture (Chapters 16–19). The book provides the state of the art on hydrological models and is a useful reference for graduate students and scientists working on the interface among soil physics, soil hydrology, land use management, agricultural engineering, agronomy, natural resources and climate change. It also advances our understanding of complex, linked non-linear interactions among soil hydrological properties and processes, and familiarizes the scientific community with the diverse applications of these models.

The editor would like to thank CAB International for its efforts in publishing this book. Special thanks are due to Gwenan Spearing, Claire Parfitt, Sarah Mellor, Meredith Carroll, Shankari Wilford and Everild Haynes from CAB International for support in getting the book published on time. The editor also wants to thank New Mexico State University Agricultural Experiment Station. Thanks are due also to several of my faculty colleagues from various institutions across the USA for their time and suggestions as reviewers of the manuscripts, and to students for their help in putting together some of the material for the book. I would also like to thank my parents, my wife Neeta and sons Utkarsh and Amogh for their support, understanding and constant encouragement.

Manoj K. Shukla

1 Introduction to Soil Hydrology: Processes and Variability of Hydrological Properties

Manoj K. Shukla*

Introduction

The term hydrology is derived from the Greek words 'hudor' and 'logos', which mean 'water' and 'study', respectively. Thus, the simplest definition of hydrology can be that it is a branch of science pertaining to the study of water. A much more comprehensive definition of hydrology could be that it is a branch of science that encompasses the study of the occurrence, distribution and movement of, and of changes in the quantity, quality and state of atmospheric, surface, soil-borne, plant-borne and subsurface water on earth.

Water is one of the most dynamic entities on the earth. It can stay in all three forms, liquid, solid and vapour, at the same time and at room temperature. The path that water takes through the environment in all these three forms is represented by the hydrological cycle, which describes the continuous movement of liquid, vapour and/or solid (ice) water in the atmosphere, on the soil and on plant surfaces and through the soil and plant surfaces. Different components of the hydrological cycle are grouped into precipitation, interception, snow melt, surface runoff, infiltration, percolation, deep percolation, subsurface flow, evaporation, transpiration (or evapotranspiration), condensation, return flow, etc. Sun or solar energy is the driving force for the hydrological cycle, and the overall total mass of water on earth stays fairly constant over time. However, the resident time for water in these different storage components varies from a few days (water vapour) to several thousand years (deep groundwater, ice on the poles) (Table 1). Various components of the hydrological cycle are strongly influenced by human activities – such as changing land use from natural forest to agriculture, from agriculture or forest land to urban land, by the construction of reservoirs and dams, by deforestation and afforestation, and by changing from mechanical tillage to no-tillage or from tillage or no-tillage to conservation tillage, etc.

Soil, the interface between the lithosphere and the atmosphere, is the most basic resource that interacts with the biosphere and the hydrosphere to support life on earth. The interaction of the soil with the various components of hydrology is important for the hydrological cycle and is known as soil hydrology. Soil hydrology has a strong influence on the water uptake and release by plants during photosynthesis. Thus, soil hydrology can also be defined as study pertaining to agricultural water management. Soil hydrology takes into account all of the components of water related

* Corresponding author: shuklamk@nmsu.edu

Table 1.1. Average residence time of water in different stores (Source: http://www.PhysicalGeography.net. Chapter 8: Introduction to the Hydrosphere. Accessed: 8 February 2010).

Storage	Residence time (years unless stated otherwise)
Antarctica	20,000
Deep aquifers (may be confined)	10,000
Oceans	3,200
Shallow aquifers (unconfined)	100–200
Lakes	50–100
Glaciers	20–100
Snow cover (mostly seasonal)	2–6 months
Rivers	2–6 months
Soil moisture	1–2 months
Atmosphere	9 days

to irrigation and drainage, percolation and recharge to groundwater, capillary rise, root and plant water uptake and release, evaporation from soil and plants, and transpiration. The fate of the total amount of water applied to the soil as irrigation or natural rainfall is usually determined by taking into account all the different components of the hydrological cycle and conducting a water balance study.

Water Balance Components

In hydrology or soil hydrology, we usually use a water balance equation that takes into account all of the water that goes in or out of the system. The general water balance equation is expressed as:

Water in = Water out + Change in water storage

Water balance can be subdivided into several categories, such as surface water balance, groundwater balance or root-zone water balance, depending upon the purpose of the study as well as on the user – namely, hydrologist, soil scientist, agronomist, groundwater hydrologist, etc. Usually, components related to soil, plant and atmosphere are present in most water balance equations.

The surface water balance equation involves components of the hydrological cycle and components of soil and plants, and is expressed as follows (Fig. 1.1):

Precipitation + Irrigation + Snow melt = Evaporation + Transpiration + Surface runoff + Infiltration + Change in storage of surface water above ground surface

The root zone is the maximum depth at which the roots of a plant of interest can be found. The root-zone water balance involves all the water that can move in and out of the root zone and can be expressed as follows (Fig. 1.1):

Infiltration + Capillary rise from below root zone = Evaporation + Transpiration + Percolation below root zone + Change of water storage of root zone

The transition zone can be defined as the part of the vadose zone immediately below the root zone that extends to the water table. The water balance equation for the transition zone can be written as (Fig. 1.1):

Percolation (below root zone) + Seepage (into transition zone) + Capillary rise from water table = Capillary rise into root zone + Subsurface drainage + Deep drainage into groundwater table + Change of water storage of transition zone

Similarly, water balance equations can be written for the water that moves in and out of aquifers, both confined and unconfined. Aquifer water balance will involve part of the applied water that joins the aquifer as well as that which leaves via either pumping or seepage.

Importance of Soil Hydrology

Soil hydrology can also be called vadose zone hydrology, soil physics or environmental soil physics. Soil hydrology can be defined as a systematic study of the physical properties of the soil environment, also known as the vadose zone, and the associated physical processes taking place within the vadose zone from the air–soil interface to the soil–groundwater interface. It includes the measurement and prediction, as well as the modelling, of vadose zone properties and processes for a variety of

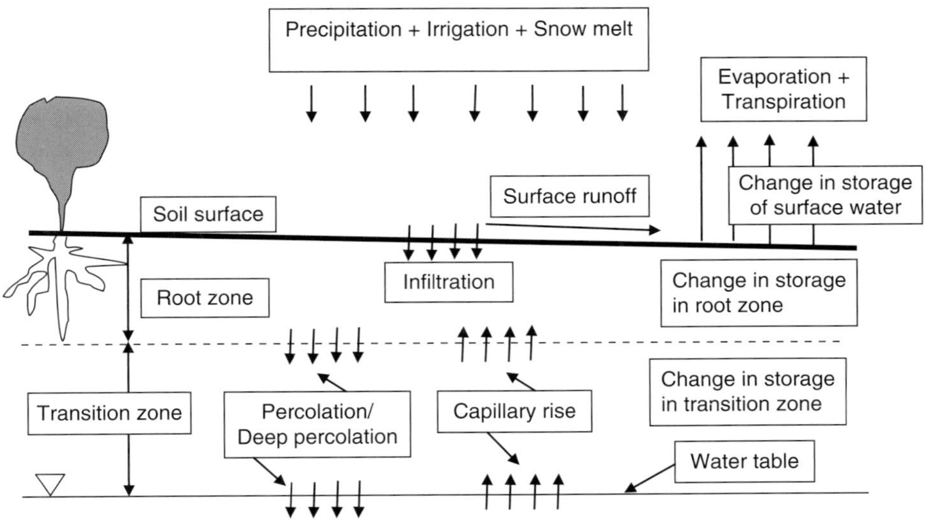

Fig. 1.1. Components of water balance on the soil surface and in the root zone and the transition zone.

ecosystems and land-use and management systems, in relation to the functions of value to the sustenance of life on earth.

Important soil hydrological processes are: infiltration, drainage, water redistribution within the vadose zone, evaporation, transpiration and deep percolation (Fig. 1.1). All of these processes occur at the microscopic or pore scale within the vadose zone and are primarily governed by the amount, orientation, size, distribution and connectivity of pores, but not just by the total micro- or macroporosities. The geometry and connectivity of pores have a profound influence on water retention and transport through the vadose zone, which is a complex network of irregular interconnected and tortuous flow channels. The pores can be formed as a result of the natural arrangement of particles, micro- or macro-aggregates, swelling and shrinking of soils, meniscus pressures, physical and chemical reactions, and biological activities, including root growth, migration and decay, earthworm and other soil-borne animal activities, and human activities related to management and land use. Soil water storage and transport can also be influenced by the type of vegetation, geology and chemistry of the vadose zone. Thus, soil hydrology is an interdisciplinary field that interacts closely with hydrology, physics, chemistry, engineering, environmental science, pedology, soil science, mathematics, geostatistics and plant science, among others (Lal and Shukla, 2004) (Fig. 1.2). Soil hydrological processes are important components of the water budget and directly influence plant growth and sustenance. Soil hydrology addresses practical problems encountered by practitioners, researchers and farmers in real-life situations, and in collaboration with other disciplines plays a pivotal role in human endeavour to sustain agricultural productivity while maintaining soil, water and environmental quality. With the world population growing at a rapid rate, the importance as well as the role of soil hydrology is becoming critical for not only sustaining or increasing total grain production but also maintaining water and environmental quality, thus maintaining life on earth.

An understanding of soil hydrology principles and processes is important for maintaining environmental quality (Fig. 1.3). Surface soil physical properties control the infiltration of water into the soil and thus are important for causing surface runoff and soil erosion. Runoff water may contain dissolved or suspended sediments as well as chemicals and nutrients. These pollutants can be transferred to surface water resources such as lakes, ponds

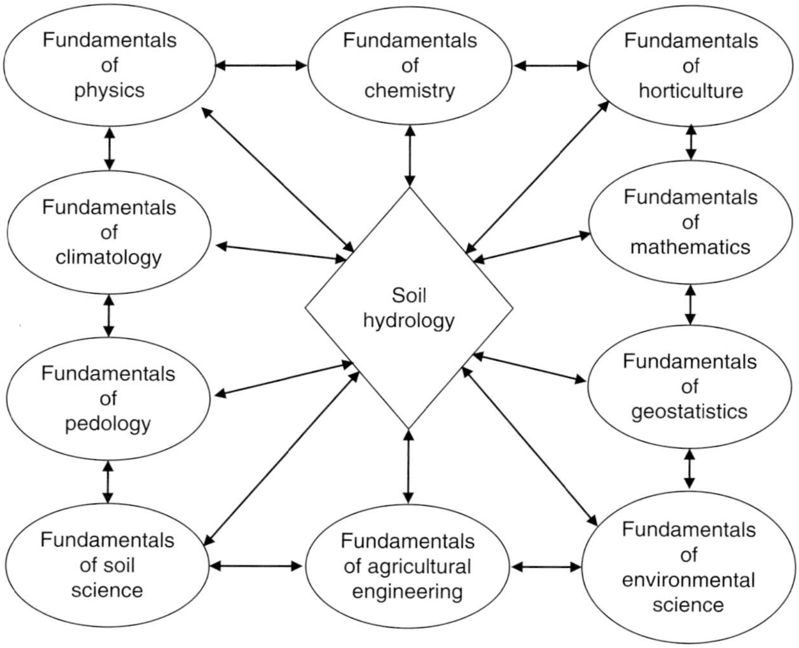

Fig. 1.2. Interactions of soil hydrology with other disciplines. Note: several disciplines included on this figure can also interact directly with each other.

or rivers, so soil hydrology is important for maintaining the water quality of surface water bodies. Vadose zone soil physical properties control the storage and movement of water and dissolved chemicals or nutrients through the profile, and control the migration of contaminants towards the groundwater. Thus, knowledge of soil hydrology is important for preventing groundwater contamination. Soil physical properties and surface soil moisture contents also control the migration of fine dust particles into the atmosphere. The high specific surface area associated with these fine dust particles means that they can also carry with them other chemicals that are sorbed on the surface of these particles. Consequently, in controlling the concentration of airborne particulate matter, soil hydrological properties are an important component of air quality.

Soil hydrological properties are greatly influenced by land use and management. For example, intense tillage can cause the breakdown of aggregates, the creation of a plough layer, an increase in soil bulk density with attendant lowering of porosity, reductions in soil water storage and transport through pores, and the exposure of organic matter to degradation, thus causing an overall decline in soil quality which can lead to a decline in agricultural productivity. In contrast, no-tillage or conservation tillage can have the exactly opposite outcome, and can improve soil quality and increase agricultural productivity. Land management practices can also influence greenhouse gas emissions from agricultural fields. The applications of nitrogenous fertilizers, along with excess water application, can increase N_2O emissions from the soil. Similarly, depending upon the quality of biomass and on management practices, CO_2 and CH_4 emissions can increase and degrade environmental quality. The preservation of the resource base and of environmental quality is immensely important for sustaining life on earth. The degradation or aggradation of groundwater quality, surface water quality, air or environmental quality and soil quality has a profound influence on the productivity and

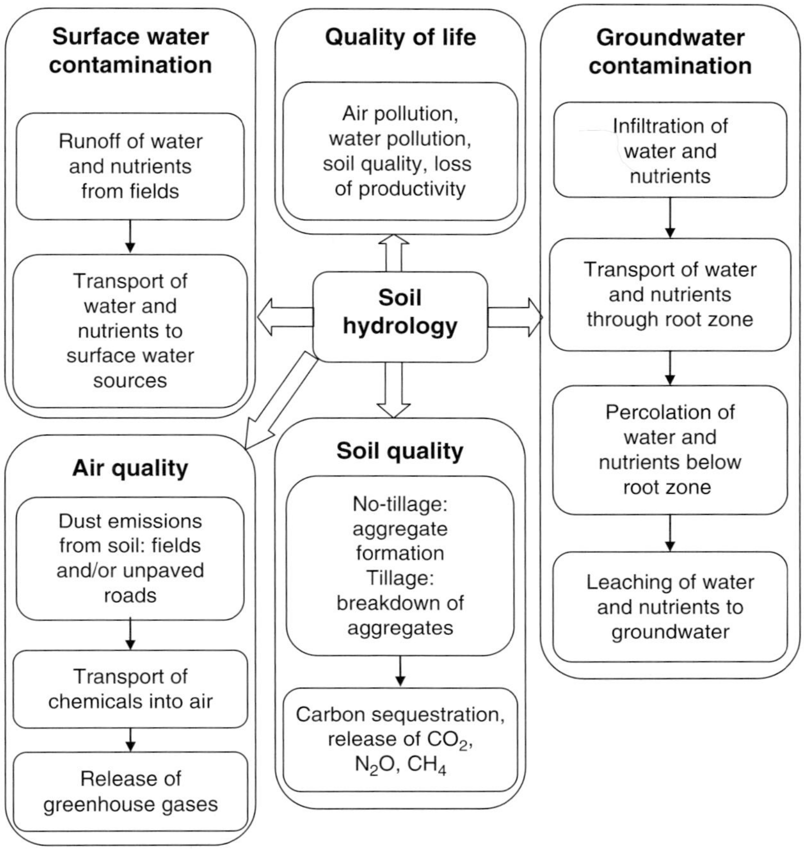

Fig. 1.3. Influence of soil hydrology on groundwater, surface water, air and soil quality, and overall quality of life on earth.

sustainability of agriculture, and on human and animal health, and thus have a profound influence on the overall quality of life on earth.

Variability of Soil Hydrological Properties

Soil hydrological properties show both short- and long-range variability, and are multivariate in nature (Nielsen *et al.*, 1973). Soil hydrological properties are greatly influenced by intrinsic factors of soil formation as well as by extrinsic factors associated with land use and management, and vary both in time and space (van Es, 2002). Intrinsic variability is caused by pedogenesis and usually takes place at large timescales. The variability caused by the pedogenetic processes is described as regionalized, with nearby areas considered to be more similar than areas that are further away (van Es, 2002). In contrast, the variability caused by extrinsic factors can take effect relatively quickly and cannot be treated as regionalized.

In addition to these known sources of variability, decisions and choices made by investigators during sampling, sampling designs, availability of resources, number of investigators involved in sampling and analysis, skill level of investigators, type and quality of tools and equipment used to collect samples and analyses, scale of the domain, availability of time, accessibility of sites, criteria of success and assumptions made during sampling and analysis have profound influences on variability. Methods of measurement also strongly affect variability; for example, infiltration rates measured using a single ring may produce significantly different mean and standard errors from those measured using a double ring. Sample support can also influence variability; for example, increasing or decreasing the size of the infiltrometer rings can change the mean, variability and amount of the infiltration rate. Similarly, hydraulic conductivity measured in the field could show a much larger variability than that measured in the laboratory.

Vadose zone hydrological processes are usually represented by state variables representing the state of the soil at that location and time, and can be explained on the basis of various hydrological properties. For example (see Table 1.2), the process of water redistribution through a soil profile can be represented by state variables – matric potential and soil water content, and by hydrological properties – hydraulic conductivity and soil water diffusivity or capacity (van Es, 2002). Similarly, the process of infiltration can be described by state variables – matric potential, soil water content, and by hydrological properties – hydraulic conductivity and texture.

Soil properties and state variables influence the variability of soil hydrological properties in many different ways. Depending upon whether a state variable or a soil property is measured or estimated, and upon the method used for measurement (laboratory, *in situ* or field) or estimation (inverse modelling, pedotransfer functions), there can be profound influences on variability (Nielsen and Wendroth, 2003). A process can be influenced by a single state variable and by multiple hydrological properties simultaneously. The variability of the associated individual hydrological properties can also be highly dissimilar, and differences can range from very low to several orders of magnitude. In addition, some of the properties could have low spatial variability but high temporal variability, while others could display the exactly opposite behaviour.

Indices for Expressing Variability

The variability of a soil hydrological property can be expressed by range, interquartile range, variance, standard deviation, coefficient of variation, skewness and kurtosis. Range, the difference between the largest and the smallest value, is commonly used to express variability, but it does not truly describe the characteristics of the data sets, especially if there are outliers. Interquartile range provides

Table 1.2. Soil processes, state variables and associated soil hydrological properties.

Process	State variables	Properties
Aeration	Porosity, matric potential, water content	Air permeability, air content
Infiltration	Matric potential, water content	Hydraulic conductivity, texture
Evaporation	Matric potential	Isothermal and thermal conductivity
Water redistribution	Matric potential, water content	Unsaturated hydraulic conductivity, soil water diffusivity
Water retention	Porosity, water content	Texture, aggregation, pore size distribution
Heat flow	Soil temperature	Thermal conductivity, specific heat capacity, thermal diffusivity
Chemical transport	Water content, particle size	Clay content, apparent dispersion coefficient, retardation factor, decay/production coefficients, pH
Adsorption	Osmotic potential	Specific surface area, specific charge, pH

a measure of variability of the middle 50% of values and is slightly better than range at eliminating the outliers at both ends of the values. Variance, the second moment about the mean, is the average of the square of deviations of a value from the mean (also known as the first moment), standard deviation is the square root of variance, and the coefficient of variation is the ratio of standard deviation and mean. Standard deviation is also known as the measure of absolute variability, whereas coefficient of variation is a measure of relative variability. Skewness is calculated as the ratio of the third moment about the mean and standard deviation, and is a measure of the asymmetry of the probability distribution of a random variable. Kurtosis, the ratio of the fourth moment about the mean and standard deviation, is a measure of the peakedness of the probability distribution of a random variable. A higher peakedness means more variance, probably as a result of outliers (Isaaks and Srivastava, 1978; Nielsen and Wendroth, 2003).

Indices of statistical variability of some soil hydrological properties for soils under agricultural cropping systems under different land-use and management systems for some states situated in the mid-west, east, south and south-west USA are presented in Tables 1.3–1.9. In general, data presented in these tables show that saturated hydraulic conductivity (K_s) displayed the greatest variability expressed as coefficient of variation (CV) across sites, although some of the important hydraulic properties affecting K_s, such as total porosity (TP) and field capacity water content (FC), showed much smaller variability across these sites. Such behaviour supported earlier observations that pore size, shape and connectivity are more important than TP. The data in these tables also show that for some individual soil properties, such as bulk density (BD), and silt and clay contents, CVs were linearly related to their range across different fields; however, for some others, including available water content (AWC), FC, K_s and TP across different fields, CVs were not related to their range at all. Similarly, no stochastic correlation was seen between the CV and the size of the field. This could be the result of small sample size, but could also be the result of the multi-scale variability of individual soil properties across these domains.

Influence of Sample Support on Saturated Hydraulic Conductivity

Saturated hydraulic conductivity is one of the most important parameters for soil–water–plant interactions, and for water and solute movement and retention through the soil profile. It is a critically important parameter for estimation of various other soil hydrological parameters necessary for modelling flow through the naturally unsaturated vadose zone. Among different soil hydrological properties, K_s is reported to have the greatest statistical variability by several authors (Biggar and Nielsen, 1976; Jury, 1989; Webb *et al.*, 2000; Shukla *et al.*, 2004a,b; Iqbal *et al.*, 2005). The variability of K_s is associated with soil types, land use, position on the landscape, depth, instruments and methods of measurement, and

Table 1.3. Indices for statistical variability of soil hydrological properties from a 6 ha field under agriculture in Central Iowa (Cambardella *et al.*, 1994).

Property[a]	Mean	Median	Var.[b]	SD[b]	CV[b]	Min.[b]	Max.[b]	Range
Sand	33.0	33.0	5476.0	74.0	2.2	3.0	58.0	55.0
Silt	34.0	31.0	144.0	12.0	0.4	12.0	72.0	60.0
Clay	33.0	31.0	169.0	13.0	0.4	6.0	66.0	60.0
BD	1.3	1.3	0.0	0.2	0.1	0.9	1.6	0.8
WSA	59.0	59.0	81.0	9.0	0.2	26.0	82.0	56.0

[a]BD, bulk density (g cm^{-3}); Sand, silt and clay given as %; WSA, water stability of aggregates (%).
[b]CV, coefficient of variation; Max., maximum value; Min., minimum value; SD, standard deviation; Var., variance.

Table 1.4. Indices for statistical variability of soil hydrological properties from a 5 ha field under agriculture in Columbus, Ohio (Source: Shukla *et al.*, 2003b).

Property[a]	Mean	Median	Var.[b]	SD[b]	CV[b]	Min.[b]	Max.[b]	Range	Skewness	Kurtosis
Sand	13.0	13.3	3.1	1.8	0.1	8.6	15.3	6.7	−1.3	3.4
Silt	44.4	44.0	8.5	2.9	0.1	41.3	50.0	8.7	0.8	−0.4
Clay	42.6	42.7	5.0	2.2	0.1	38.7	45.4	6.7	−0.2	−0.7
BD	1.3	1.4	0.0	0.1	0.1	1.1	1.5	0.5	−0.5	0.1
K_s	84.0	72.1	7377	85.9	1.0	0.6	327	327	2.3	6.4
TP	0.5	0.5	0.0	0.0	0.1	0.4	0.6	0.2	0.5	0.1
FC	0.3	0.3	0.0	0.0	0.1	0.3	0.3	0.1	0.5	−1.2
AWC	1.8	1.6	0.2	0.4	0.2	1.1	2.5	1.5	0.3	−0.1
WSA	91.8	92.9	13.1	3.6	0.0	82.3	95.4	13.1	−1.8	3.7
GMD	1.8	1.8	0.0	0.2	0.1	1.4	2.1	0.6	−0.3	−0.5

[a]AWC, available water content (cm); BD, bulk density (g cm^{-3}); FC, field capacity water content (cm^3 cm^{-3}); K_s, hydraulic conductivity (cm h^{-1}); GMD, geometric mean weight diameter of aggregates (mm); Sand, silt and clay given as %; TP, total porosity; WSA, water stability of aggregates (%).
[b]CV, coefficient of variation; Max., maximum value; Min., minimum value; SD, standard deviation; Var., variance.

Table 1.5. Indices for statistical variability of soil hydrological properties from a 250 ha field under agriculture in Coshocton, Ohio (Source: Shukla *et al.*, 2003a).

Property[a]	Mean	Median	Var.[b]	SD[b]	CV[b]	Min.[b]	Max.[b]	Range	Skewness	Kurtosis
Sand	25.8	25.1	38.3	6.2	0.2	15.6	41.6	26.0	0.0	0.6
Silt	59.2	59.9	35.3	5.9	0.1	46.0	68.7	22.7	−0.3	−0.5
Clay	15.0	14.9	4.5	2.1	0.1	10.7	19.4	8.7	−0.5	0.1
BD	1.6	1.6	0.0	0.2	0.1	1.3	1.9	0.6	−0.8	−0.4
K_s	13.9	7.3	281.5	16.78	1.2	0.1	86.9	86.8	3.1	12.1
TP	0.4	0.4	0.0	0.1	0.2	0.3	0.5	0.2	−0.8	0.4
FC	0.3	0.3	0.0	0.1	0.2	0.2	0.4	0.2	−0.7	0.4
AWC	1.8	1.9	0.3	0.6	0.3	0.6	3.0	2.5	0.5	−0.4
WSA	74.9	75.3	263.6	16.2	0.2	31.1	95.8	64.7	1.9	−1.4
GMD	1.4	1.3	0.0	0.2	0.1	1.2	1.7	0.6	−1.0	0.5

[a]AWC, available water content (cm); BD, bulk density (g cm^{-3}); FC, field capacity water content (cm^3 cm^{-3}); K_s, hydraulic conductivity (cm h^{-1}); GMD, geometric mean weight diameter of aggregates (mm); Sand, silt and clay given as %; TP, total porosity; WSA, water stability of aggregates (%).
[b]CV, coefficient of variation; Max., maximum value; Min., minimum value; SD, standard deviation; Var., variance.

Table 1.6. Indices for statistical variability of soil hydrological properties from a 200 ha field under agriculture in South Charleston, Ohio (Source: Shukla and Lal, 2005).

Property[a]	Mean	Median	Var.[b]	SD[b]	CV[b]	Min.[b]	Max.[b]	Range	Skewness	Kurtosis
Sand	23.9	25.7	20.3	4.5	0.2	14.4	29.1	14.7	−1.1	0.1
Silt	59.6	57.3	25.7	5.1	0.1	54.3	69.0	14.7	1.0	−0.5
Clay	16.5	16.6	2.6	1.6	0.1	14.3	20.3	6.0	0.9	1.0
BD	1.2	1.2	0.0	0.1	0.1	1.1	1.3	0.2	0.1	−1.1
K_s	23.7	18.2	446.4	21.1	0.9	1.3	64.2	62.9	1.0	0.0
TP	0.6	0.6	0.0	0.0	0.0	0.5	0.6	0.1	−0.1	−1.1
FC	0.3	0.3	0.0	0.0	0.2	0.2	0.3	0.1	0.7	0.3
AWC	2.2	2.2	0.1	0.3	0.1	1.8	2.6	0.8	−0.5	−0.5
WSA	54.5	58.8	364.8	19.1	0.4	23.3	79.8	56.4	−0.4	−1.2
GMD	1.2	1.1	0.0	0.2	0.2	0.9	1.5	0.6	0.3	−1.5

[a]AWC, available water content (cm); BD, bulk density (g cm^{-3}); FC, field capacity water content (cm^3 cm^{-3}); K_s, hydraulic conductivity (cm h^{-1}); GMD, geometric mean weight diameter of aggregates (mm); Sand, silt and clay given as %; TP, total porosity; WSA, water stability of aggregates (%).
[b]CV, coefficient of variation; Max., maximum value; Min., minimum value; SD, standard deviation; Var., variance.

Table 1.7. Indices for statistical variability of soil hydrological properties from a 162 ha field under agriculture in Mississippi (modified from Iqbal *et al.*, 2005).

Property[a]	Horizon	Mean	Median	Var.[b]	SD[b]	CV[b]	Min.[b]	Max.[b]	Range	Skewness
Sand	Surface	27.7	25.5	366.7	19.2	0.7	1.1	74.6	73.5	0.5
Sand	Subsurface	24.4	19.6	427.7	20.7	0.8	0.8	85.9	85.1	0.8
Sand	Deep	28.3	24.5	538.7	23.2	0.8	0.3	91.4	91.1	0.5
Silt	Surface	61.0	64.9	276.2	16.6	0.3	16.0	95.1	79.1	0.5
Silt	Subsurface	60.7	63.9	279.9	16.7	0.3	12.9	96.0	83.1	0.5
Silt	Deep	58.5	59.1	404.4	20.1	0.3	7.4	97.2	89.8	0.1
Clay	Surface	11.3	10.1	53.9	7.3	0.6	5.0	33.9	28.9	1.1
Clay	Subsurface	14.9	12.8	91.6	9.6	0.6	0.6	46.0	45.3	0.9
Clay	Deep	13.2	10.9	104.2	10.2	0.8	0.6	53.3	52.7	1.2
BD	Surface	1.2	1.2	0.0	0.1	0.1	0.9	1.4	0.5	0.2
BD	Subsurface	1.3	1.3	0.0	0.1	0.1	1.0	1.5	0.5	0.4
BD	Deep	1.2	1.2	0.0	0.1	0.1	0.8	1.5	0.7	0.5
K_s	Surface	1.0	0.4	2.7	1.6	1.6	0.0	11.8	11.8	3.5
K_s	Subsurface	0.3	0.1	0.2	0.4	1.5	0.0	2.6	2.6	3.0
K_s	Deep	0.5	0.2	0.8	0.9	1.7	0.0	6.4	6.4	3.6
FC	Surface	2.8	2.9	0.4	0.7	0.2	1.2	4.4	3.2	0.3
FC	Subsurface	3.0	3.1	0.5	0.7	0.2	1.3	4.9	3.7	0.2
FC	Deep	3.2	3.3	0.8	0.9	0.3	1.0	5.7	4.6	0.3
WP	Surface	1.7	1.7	0.4	0.6	0.4	0.5	3.5	3.0	0.3
WP	Subsurface	2.0	2.1	0.6	0.8	0.4	0.6	4.4	3.8	0.2
WP	Deep	2.3	2.3	0.8	0.9	0.4	0.4	4.3	3.9	0.1
AWC	Surface	1.2	1.1	0.1	0.4	0.3	0.5	2.7	2.2	1.0
AWC	Subsurface	1.0	0.9	0.1	0.3	0.3	0.4	1.9	1.4	0.8
AWC	Deep	0.9	0.9	0.1	0.3	0.3	0.4	2.5	2.1	1.4

[a]AWC, available water content (cm); BD, bulk density (g cm^{-3}); K_s, hydraulic conductivity (cm h^{-1}); FC, field capacity water content (cm^3 cm^{-3}); Sand, silt and clay given as %; WP, wilting point (cm^3 cm^{-3}).
[b]CV, coefficient of variation; Max., maximum value; Min., minimum value; SD, standard deviation; Var., variance.

Table 1.8. Indices for statistical variability of soil hydrological properties from a 31 ha field under agriculture in Anthony, New Mexico (Source: Ikemura *et al.*, 2008).

Property[a]	Mean	Median	Var.[b]	SD[b]	CV[b]	Min.[b]	Max.[b]	Range	Skewness	Kurtosis
Sand	29.1	29.0	131.0	11.4	0.4	11.0	48.4	37.4	−0.1	−0.6
Silt	49.6	48.8	82.6	9.1	0.2	37.0	64.4	27.4	0.6	−0.7
Clay	21.2	23.3	13.8	3.7	0.2	14.6	25.0	10.4	−0.9	−0.8
BD	1.4	1.4	0.0	0.1	0.1	1.2	1.6	0.4	−0.6	−0.7
K_s	0.1	0.0	0.0	0.1	1.8	0.0	0.4	0.4	2.0	2.3
TP	0.5	0.5	0.0	0.0	0.1	0.4	0.6	0.1	0.6	−0.7
FC	0.4	0.4	0.0	0.0	0.1	0.3	0.4	0.2	0.1	−0.4
AWC	2.4	2.4	0.1	0.2	0.1	2.0	2.9	0.8	0.7	1.0

[a]AWC, available water content (cm); BD, bulk density (g cm^{-3}); FC, field capacity water content (cm^3 cm^{-3}); K_s, hydraulic conductivity (cm h^{-1}); Sand, silt and clay given as %; TP, total porosity.
[b]CV, coefficient of variation; Max., maximum value; Min., minimum value; SD, standard deviation; Var., variance.

also with experimental errors (Stockton and Warrick, 1971). It has been suggested that more studies are needed on the variability of K_s across different landscapes (Bouma, 1973). The variability of K_s has a profound influence on the overall hydrology of the soil system. Therefore, the focus in the next few sections will be centred on the variability of

Table 1.9. Indices for statistical variability of soil hydrological properties from a 12 ha agricultural field in North California (modified from Duffera *et al.*, 2007).

Property[a]	Mean	Median	Var.[b]	SD[b]	CV[b]	Min.[b]	Max.[b]	Range	Skewness	Kurtosis
Sand	55.5	55.2	68.9	8.3	0.1	25.3	73.7	48.4	−0.2	0.3
Silt	25.6	24.6	31.4	5.6	0.2	5.0	52.9	47.9	1.5	5.6
Clay	18.9	19.2	59.3	7.7	0.4	6.3	51.9	45.6	0.4	0.2
BD	1.6	1.6	0.0	0.1	0.1	1.2	1.9	0.7	−0.7	0.9
K_s	5.3	2.9	29.2	5.4	1.0	0.0	22.9	22.9	1.1	0.0
TP	0.4	0.4	0.0	0.0	0.0	0.3	0.6	0.3	0.7	0.9
FC	0.2	0.2	0.3	0.1	0.3	0.1	0.4	0.3	0.4	−0.4
AWC	1.3	1.2	0.1	0.3	0.2	0.7	2.4	1.7	1.1	2.4
WP	0.1	0.1	0.1	0.0	0.4	0.0	0.2	0.2	0.6	0.4

[a]AWC, available water content (cm); BD, bulk density (g cm^{-3}); FC, field capacity water content (cm^3 cm^{-3}); K_s, hydraulic conductivity (cm h^{-1}); Sand, silt and clay given as %; TP, total porosity; WP, wilting point (cm^3 cm^{-3}).
[b]CV, coefficient of variation; Max., maximum value; Min., minimum value; SD, standard deviation; Var., variance.

saturated hydraulic conductivity – or steady-state infiltration rate, while some data and discussion will be on unsaturated hydraulic conductivity.

Data on the effect of sample support on saturated and unsaturated hydraulic conductivity are presented in Table 1.10, which presents data from infiltration tests conducted under six different suction values (matric potentials) using different sample supports (disc sizes) (Das Gupta *et al.*, 2006). The experiment was conducted near College Station, Texas using disc infiltrometers with five different disc sizes (10, 15, 17, 20 and 24 cm). Looking at the range of the data, and at other indices of statistical variability, only minor differences were observed for unsaturated hydraulic conductivity at a given suction. Similarly, only minor differences were observed for near-saturation hydraulic conductivity for disc sizes of 10, 20 and 24 cm. The CV for unsaturated hydraulic conductivity – as well as for near-saturation hydraulic conductivity – generally increased with disc size for most suction values. Although there were some inconsistencies observed in the disc sizes of 10 and 24 cm, the increase in CV with increasing disc size could be due to the inclusion of larger sample volumes, larger heterogeneity and the possible contribution of macropore channels to water transport.

Influence of Measurement Devices on Saturated Hydraulic Conductivity

The data presented in Table 1.11 consist of K_s values determined on 7.8 cm long and 7.8 cm diameter soil cores using a constant head method in the laboratory, and steady-state infiltration rate (i_c) measured at the same (within 0.5 m radius) field location before the core sampling – or adjacent to the core sampling location – by conducting long-duration (>2.5 h) double-ring infiltrometer tests using 27 cm diameter outer rings and 15 cm diameter inner rings. These limited data for soils in Ohio under different land-use and management practices showed that variance or standard deviations were always lower for i_c than for K_s. However, an exactly opposite result is also possible from a similar study. There can be a variety of reasons for this trend, including smaller support for the laboratory (core) measurement than in the field experiment, spaces between the core and the soil, hitting or missing macropores in the soil, etc.

Unlike the data in Table 1.11, the data presented in Table 1.12 indicate that i_c values measured using a double-ring infiltrometer were always smaller than K_s values measured

Table 1.10. Indices for statistical variability for hydraulic conductivity ($K(h)$, K_s, as m s^{-1}) under different suctions (matric potentials) for five infiltration disc sizes (data modified from Das Gupta *et al*., 2006).

$K(h)/K_s^a$	Mean	SD[b]	CV[b]	Max.[b]	Min.[b]	Range
			0.10 m diameter disc			
$K(-0.20)$	–	–	–	–	–	–
$K(-0.15)$	–	–	–	–	–	–
$K(-0.10)$	2.3×10^{-6}	7.0×10^{-7}	0.3	3.0×10^{-6}	1.6×10^{-6}	1.4×10^{-6}
$K(-0.05)$	3.0×10^{-6}	2.7×10^{-7}	0.1	3.3×10^{-6}	2.8×10^{-6}	5.3×10^{-7}
$K(-0.02)$	3.4×10^{-5}	2.0×10^{-7}	0.6	5.4×10^{-5}	1.4×10^{-5}	4.0×10^{-5}
$K(0.00)$	3.7×10^{-5}	1.9×10^{-5}	0.5	5.5×10^{-5}	1.8×10^{-5}	3.7×10^{-5}
			0.15 m diameter disc			
$K(-0.20)$	–	–	–	–	–	–
$K(-0.15)$	–	–	–	–	–	–
$K(-0.10)$	6.7×10^{-7}	9.1×10^{-7}	1.4	1.8×10^{-6}	–	–
$K(-0.05)$	3.5×10^{-6}	1.4×10^{-6}	0.4	5.4×10^{-6}	2.0×10^{-6}	3.4×10^{-6}
$K(-0.02)$	1.9×10^{-5}	8.0×10^{-6}	0.4	2.7×10^{-5}	8.0×10^{-6}	1.9×10^{-5}
$K(0.00)$	5.9×10^{-5}	1.6×10^{-5}	0.3	8.2×10^{-5}	4.6×10^{-5}	3.6×10^{-5}
			0.17 m diameter disc			
$K(-0.20)$	–	–	–	–	–	–
$K(-0.15)$	4.2×10^{-8}	4.6×10^{-7}	10.9	4.0×10^{-7}	–	–
$K(-0.10)$	8.3×10^{-7}	2.4×10^{-7}	0.3	1.1×10^{-6}	5.3×10^{-7}	5.7×10^{-7}
$K(-0.05)$	4.4×10^{-6}	2.4×10^{-7}	0.5	7.6×10^{-6}	2.1×10^{-6}	5.5×10^{-6}
$K(-0.02)$	2.9×10^{-5}	1.0×10^{-5}	0.4	4.3×10^{-5}	2×10^{-5}	2.3×10^{-5}
$K(0.00)$	6.5×10^{-5}	2.7×10^{-5}	0.4	1.0×10^{-4}	4.3×10^{-5}	6.0×10^{-5}
			0.20 m diameter disc			
$K(-0.20)$	4.1×10^{-7}	2.2×10^{-7}	0.5	8.3×10^{-7}	1.5×10^{-7}	6.8×10^{-7}
$K(-0.15)$	3.9×10^{-7}	1.2×10^{-7}	0.3	6.3×10^{-7}	2.4×10^{-7}	3.9×10^{-7}
$K(-0.10)$	9.8×10^{-7}	2.3×10^{-7}	0.2	1.4×10^{-7}	5.4×10^{-7}	0.9×10^{-6}
$K(-0.05)$	8.2×10^{-6}	6.9×10^{-6}	0.9	2.1×10^{-5}	1.4×10^{-6}	1.9×10^{-5}
$K(-0.02)$	8.0×10^{-6}	1.0×10^{-5}	1.3	3.3×10^{-5}	5.6×10^{-7}	3.2×10^{-5}
$K(0.00)$	2.2×10^{-5}	1.9×10^{-5}	0.9	5.7×10^{-5}	8.2×10^{-7}	5.6×10^{-5}
			0.24 m diameter disc			
$K(-0.20)$	4.0×10^{-7}	2.3×10^{-7}	0.6	7.6×10^{-7}	1.1×10^{-7}	6.5×10^{-7}
$K(-0.15)$	4.7×10^{-7}	2.6×10^{-7}	0.6	9.5×10^{-7}	1.5×10^{-7}	8.0×10^{-7}
$K(-0.10)$	9.2×10^{-7}	5.1×10^{-7}	0.6	1.8×10^{-6}	2.7×10^{-7}	1.5×10^{-6}
$K(-0.05)$	7.2×10^{-6}	7.6×10^{-6}	1.0	2.5×10^{-5}	6.4×10^{-7}	2.4×10^{-5}
$K(-0.02)$	7.3×10^{-6}	1.0×10^{-5}	1.4	3.3×10^{-5}	2.0×10^{-7}	3.3×10^{-5}
$K(0.00)$	2.1×10^{-5}	1.8×10^{-5}	0.9	5.5×10^{-5}	5.7×10^{-7}	5.4×10^{-5}

[a]$K(h)$, unsaturated conductivity, $K(-0.20)$ to $K(-0.02)$; K_s, saturated conductivity, $K(0.00)$.
[b]CV, coefficient of variation; Max., maximum value; Min., minimum value; SD, standard deviation.

on intact soil cores under each of the three management practices investigated in Ohio. Similarly, indices of statistical variability, including variance, standard deviation and range, were also much smaller for i_c than for K_s for locations under annual chisel tillage, no-tillage and woodland.

In further experiments, K_s was determined *in situ* using two different field measurement devices – a tension infiltrometer and a pressure infiltrometer, and one laboratory measurement device – the soil core method (Tables 1.13–1.15). The field methods almost always have a disadvantage of soil being field saturated and not fully saturated, and measurements being made under a quasi steady-state condition, usually ascertained from the measured equilibrium time. Here, K_s values were

Table 1.11. Indices for statistical variability for saturated hydraulic conductivity (K_s) in the laboratory using soil core measurements and for steady-state infiltration (i_c) measured in the field using a double-ring infiltrometer for three locations in Ohio (Source: Shukla *et al.*, 2003a,b).

	Location 1		Location 2		Location 3	
Index[a]	K_s^b	i_c^b	K_s	i_c	K_s	i_c
Mean	84.0	14.9	13.9	16.0	23.7	2.5
Median	72.1	11.2	7.3	15.7	18.2	1.2
Var.	7376.8	150.7	281.5	77.6	446.4	8.7
SD	85.9	12.3	16.8	8.8	21.1	2.9
CV	1.0	0.8	1.2	0.5	0.9	1.2
Min.	0.6	2.0	0.1	3.8	1.3	0.6
Max.	327.4	43.7	86.9	41.6	64.2	12.0
Range	326.8	41.8	86.8	37.8	62.9	11.4
Skewness	2.3	1.5	3.1	0.6	1.0	2.7
Kurtosis	6.4	1.8	12.1	0.7	0.0	8.3

[a]CV, coefficient of variation; Max., maximum value; Min., minimum value; SD, standard deviation; Var., variance.
[b]K_s and i_c as cm h^{-1}.

Table 1.12. Indices for statistical variability for saturated hydraulic conductivity (K_s) in the laboratory using soil cores and steady-state infiltration (i_c) measured in the field using a double-ring infiltrometer for some fields under annual tillage (using chisel or mouldboard ploughs), no-tillage (6–15 years) and woodland in Ohio (Source: Shukla and Lal, 2005).

	Annual tillage		No-tillage		Woodland	
Index[a]	K_s^b	i_c^b	K_s	i_c	K_s	i_c
Mean	48.1	7.9	26.7	14.6	85.7	24.8
Median	44.3	5.8	10.9	12.9	73.3	22.4
Var.	1349.9	37.4	2956.0	117.7	4743.0	204.9
SD	36.7	6.1	54.4	10.8	68.9	14.3
CV	0.76	0.78	2.03	0.74	0.80	0.58
Min.	0.6	1.8	0.1	0.6	10.1	6.0
Max.	101.7	23.2	327.4	43.7	178.0	45.3
Range	101.0	21.4	327.3	43.1	167.9	39.3
Skewness	0.2	1.3	4.7	0.8	0.3	0.3
Kurtosis	−1.6	1.7	25.0	0.4	−2.1	−0.8

[a]CV, coefficient of variation; Max., maximum value; Min., minimum value; SD, standard deviation; Var., variance.
[b]K_s and i_c as cm h^{-1}.

determined for positive pressures of 1 and 2 cm in the field using the tension infiltrometer, and pressures of 0.11 ± 0.03 cm and 0.31 ± 0.10 cm for the pressure infiltrometer. The laboratory tests were conducted by the constant or falling head method using large hydraulic head gradients (up to ten). In general, the variability (expressed as CV) was large irrespective of the method of measurement used. However, looking at Tables 1.13–1.15, one can conclude that there were no definite trends and absolutely no correlation among these three methods for measuring K_s (Reynolds *et al.*, 2000). The possible explanation could be the differences in flow domains or sample sizes and flow geometries. For example, flow is three-dimensional when using a tension infiltrometer; however, it is closer to being a one-dimensional flow when using either a pressure infiltrometer or the soil core method. The surface area over which the

Table 1.13. Indices for statistical variability for saturated hydraulic conductivity ($K_s \times 10^{-5}$ m s^{-1}) measured in a conventionally tilled field using a tension infiltrometer or a pressure infiltrometer, and measured in the laboratory using the soil core method (Source: Reynolds *et al.*, 2000).

Infiltrometer type/Soil core	Index[a]/Soil type					
	GM	CV	Max.	Min.	Range	Soil
Tension	3.1	59.6	6.8	1.3	5.6	Sand
Pressure	9.5	51.1	17.9	3.1	14.8	Sand
Soil core	8.0	48.6	16.5	3.8	12.7	Sand
Tension	1.6	163.9	7.7	0.2	7.6	Loam
Pressure	1.5	101.8	4.4	0.3	4.1	Loam
Soil core	1.2	218.6	6.6	0.2	6.4	Loam
Tension	1.0	44.5	2.1	0.6	1.6	Clay loam
Pressure	0.1	362.4	1.1	0.0	1.1	Clay loam
Soil core	0.0	140,000	4.3	0.0	4.3	Clay loam

[a]CV = coefficient of variation; GM = geometric mean; Max. = maximum value; Min. = minimum value.

Table 1.14. Indices for statistical variability for saturated hydraulic conductivity ($K_s \times 10^{-5}$ m s^{-1}) measured in a no-tillage field using a tension infiltrometer or a pressure infiltrometer, and measured in the laboratory using the soil core method (Source: Reynolds *et al.*, 2000).

Infiltrometer type/Soil core	Index[a]/Soil type					
	GM	CV	Max.	Min.	Range	Soil
Tension	2.6	47.3	5.3	1.2	4.1	Sand
Pressure	5.4	58.1	9.9	1.6	8.3	Sand
Soil core	8.1	73.7	38.7	3.3	35.4	Sand
Tension	4.2	68.2	16.0	2.4	13.6	Loam
Pressure	6.9	79.5	15.7	1.7	14.0	Loam
Soil core	3.4	344.9	34.3	0.2	34.1	Loam
Tension	2.3	62.8	5.1	1.0	4.2	Clay loam
Pressure	1.9	5058.2	126.3	0.0	126.2	Clay loam
Soil core	13.6	206.6	68.7	1.5	67.2	Clay loam

[a]CV = coefficient of variation; GM = geometric mean; Max. = maximum value; Min. = minimum value.

Table 1.15. Indices for statistical variability for saturated hydraulic conductivity ($K_s \times 10^{-5}$ m s^{-1}) measured in a native woodland using a tension infiltrometer or a pressure infiltrometer, and measured in the laboratory using the soil core method (Source: Reynolds *et al.*, 2000).

Infiltrometer type/Soil core	Index[a]/Soil type					
	GM	CV	Max.	Min.	Range	Soil
Tension	2.1	53.1	6.2	1.0	5.2	Sand
Pressure	21.7	57.8	39.0	7.1	31.8	Sand
Soil core	21.6	95.5	59.3	4.5	54.8	Sand
Tension	4.5	97.4	10.0	0.8	9.2	Loam
Pressure	23.8	63.8	81.6	12.2	69.4	Loam
Soil core	32.4	84.3	88.2	8.6	79.6	Loam
Tension	6.3	83.5	12.3	2.0	10.3	Clay loam
Pressure	7.2	53.3	18.3	3.6	14.7	Clay loam
Soil core	25.2	81.0	78.2	10.8	67.4	Clay loam

[a]CV = coefficient of variation; GM = geometric mean; Max. = maximum value; Min. = minimum value.

infiltration was taking place was also much higher (491 cm^2) for the tension infiltrometer than for the other methods (79 cm^2).

An experiment was conducted in a hay field of Purdue University, Indiana, where a 1 × 1 m square infiltrometer (or box) was installed in the soil and a constant head of 8 cm was maintained for 1 h; after that, the infiltration rate was measured using the falling head method (Haws et al., 2004). Once all the water had infiltrated into the soil, the infiltrometer was refilled and the steady-state infiltration rate (near asymptotic value) was determined using the falling head method (Table 1.16). On the following day, the test area was divided into one hundred 10 ×10 cm cells, and cylindrical cores of 7.2 cm diameter were inserted; steady-state infiltration rate was again measured over a 2-day period. Most steady-state infiltration rate values were lower for the core than for the box, for both different soils and horizons (Table 1.16), indicating that, as the support or infiltration area decreases from 1 m^2 to 0.41 m^2, in general the steady-state infiltration rate also decreases. However, no significant linear correlation was observed between support area and steady-state infiltration rate. There could be several explanations for these results, including the likely blockage of macropores by core walls, and the possibility that experimental artefacts could, on the whole, be greater for core than for box infiltration tests.

For a randomly distributed domain, measuring K_s over a larger volume of soil can be equivalent to pooling the measurements from within the smaller volumes (Parkin and Robinson, 1992). In this case the small and large supports are centred on the same mean because they are sampling the same population. Additional infiltration experiments conducted by Haws *et al.* (2004) using three different concentric square infiltrometers (100 × 100, 60 × 60 and 20 × 20 cm^2) at hill-slope and landscape transects showed that steady-state infiltration rate increased with increasing support area (Table 1.17). However, CV for steady-state infiltration rate was highest for the infiltrometers with the smallest sample support (infiltration) area.

Table 1.16. The steady-state infiltration rate (cm h^{-1}) measured in the field using a 1 × 1 m box and 7.2 cm diameter cores installed within the same area for three different soil types (Drummer, Brenton, Dana) and horizons (modified from Haws *et al.*, 2004).

	Drummer		Brenton		Dana	
Horizon	Box	Core	Box	Core	Box	Core
Ap and A	4.8	2.07	9.25	9.55	18.0	8.74
Btg1	9.42	7.72	10.8	10.5	12.6	8.16
Btg2	10.74	8.73	7.2	4.92	8.4	5.23

Table 1.17. The influence of sample support on steady-state infiltration rate (cm h^{-1}) and its statistical variability measured in the field using three concentric square infiltrometers of sizes 100 × 100, 60 × 60 and 20 × 20 cm^2 at hill-slope and landscape positions/scales (modified from Haws *et al.*, 2004).

Infiltrometer size (cm^2)	Mean	SD[a]	Var.[a]	CV[a]	Min.[a]	Max.[a]	Range
		Hill-slope position/scale					
100 × 100	3.03	1.6	2.56	0.53	0.79	10.85	10.06
60 × 60	2.25	1.0	1.04	0.45	0.43	4.61	4.18
20 × 20	2.16	1.8	3.35	0.85	0.3	10.14	9.84
		Landscape position/scale					
100 × 100	4.26	2.9	8.29	0.68	1.03	19.00	17.97
60 × 60	3.37	3.7	13.91	1.11	0.91	29.74	28.83
20 × 20	2.64	3.0	8.94	1.13	0.16	18.18	18.02

[a]CV, coefficient of variation; Max., maximum value; Min., minimum value; SD, standard deviation; Var., variance.

Table 1.18. The influence of sample support on saturated hydraulic conductivity (K_s, cm min^{-1}) and its statistical variability measured in the field using concentric square infiltrometers with six different inner ring diameters (cm) (source: Lai and Ren, 2007).

Infiltrometer inner ring diameter (cm)	Mean	Median	Var.[a]	SD[a]	CV[a]	Min.[a]	Max.[a]	Range	Skewness
10	0.02	0.01	0.0006	0.02	1.54	0.0001	0.26	0.26	5.89
20	0.01	0.01	0.0004	0.02	1.39	0.0000	0.18	0.18	4.87
40	0.01	0.01	0.0002	0.01	0.95	0.0001	0.10	0.10	2.30
80	0.01	0.01	0.0003	0.02	1.16	0.0002	0.14	0.14	4.25
120	0.01	0.01	0.0002	0.02	1.04	0.0002	0.11	0.11	3.22
200	0.01	0.01	0.0001	0.01	0.74	0.0003	0.07	0.07	2.46

[a]CV, coefficient of variation; Max., maximum value; Min., minimum value; SD, standard deviation; Var., variance.

In a study in China, infiltration tests were conducted by using six different diameters of inner ring of the infiltrometer, and maintaining a constant head of 5 cm (Lai and Ren, 2007). The results showed that increasing the size of the inner ring did not change the saturated hydraulic conductivity of the soil (Table 1.18). However, except for the inner ring with 40 cm diameter, K_s values tended to become more normal, with decrease in both range and skewness with increasing inner ring diameter. These results were slightly counter-intuitive because, as the support area increases, the chances of open macropores getting involved in the infiltration also increase. However, no appreciable differences in steady-state K_s values do point to a lack of open macropores at the soil surface.

Influence of Land Use on Saturated Hydraulic Conductivity

In the study whose data are presented in Table 1.12, steady-state hydraulic conductivity was measured in the field using a double-ring infiltrometer, and in the laboratory by the soil core method, under three different land uses: annual tillage by chisel or mouldboard ploughing, no-tillage (6 to 15 years) and woodland. Both K_s and steady-state infiltration rate were higher in woodland than in agricultural fields, and steady-state infiltration rate values varied in the order: woodland > no-tillage > annual tillage. However, average values of K_s did not follow the conventional wisdom, and were higher for fields under annual tillage than under no-tillage. This finding could be due to a number of factors, including the larger sample size used for determining the K_s from no-tillage fields than from fields under annual tillage, measurement errors in the field and laboratory while collecting and preparing the core samples, the timing of tillage operations and errors during sample analyses.

Tables 1.13–1.15 also present the K_s values under three different types of land use and management, including conventionally tilled soils, no-tilled soils and soils of natural woodlands. Most of the time, the K_s values follow the order, woodland > no-tillage > conventional tillage, for all three soils and all three methods of measurement. Such a trend is not surprising because of the higher macroporosity of the soils of the natural woodland than of the soils under no-tillage or under a conventional tillage system. The tension infiltrometer seems to underestimate the K_s values for sand under all three land-management systems. A possible explanation could be the restriction of flow from tubes, or air in the Mariotte bottle used to supply water. Other possible reasons could be the arrangement of macropores, three-dimensional infiltration, and restrictions to flow by the membrane (Reynolds *et al.*, 2000).

Temporal Variability of Saturated Hydraulic Conductivity

The temporal variability of K_s has been determined by several researchers. One such study estimated K_s from four adjacent fields located in Arkansas; one of the fields was under prairie while the other fields were under different amounts of time under cropping (Scott *et al.*, 1994). The cropped fields had been under a rice–soybean rotation for 3, 14 and 32 years, respectively. K_s was measured on intact soil cores in the laboratory using the constant-head method, and data on the temporal variability found are presented in Fig. 1.4. Different measurements, presented as symbols, are joined together with a line that may or may not represent the true variation. The prairie and the fields cropped for 14 and 32 years showed a good quadratic relationship with time, with K_s values being highest in the prairie and lowest in the field cropped for 24 years. Although not consistent, in general a pattern emerged, and K_s values started to increase from spring until early summer, stayed at similar levels until early winter and then decreased. These variations can be partly explained based on biological influences and types of management practices, including tillage and harvesting, root development and possibly earthworm activities.

The temporal variability of unsaturated hydraulic conductivity ($K(h)$) was determined by Das Gupta *et al.* (2006) for three different matric potentials (−0.2, −0.15 and −0.1 m) using 0.2 and 0.24 m infiltration discs. The observations were made during May 2003 and January 2005 on an abandoned agricultural field near Texas A&M University field station near College Station, Texas. The average values from these two discs showed that $K(h)$ varied with time quite remarkably (Fig. 1.5).

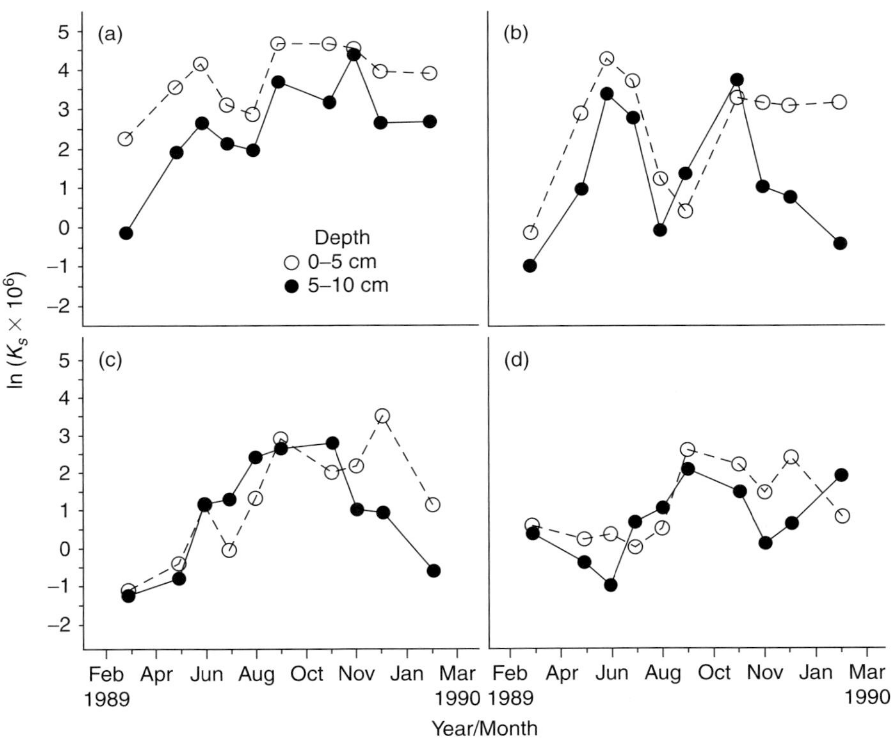

Fig. 1.4. Temporal variability of saturated hydraulic conductivity (K_s) measured during March 1989 and March 1999 at two depths in four different fields located in Arkansas: (a) prairie, (b) 3 years of cropping, (c) 14 years of cropping, and (d) 32 years of cropping (Scott *et al.*, 1994). Reprinted with permission from Soil Science Society of America.

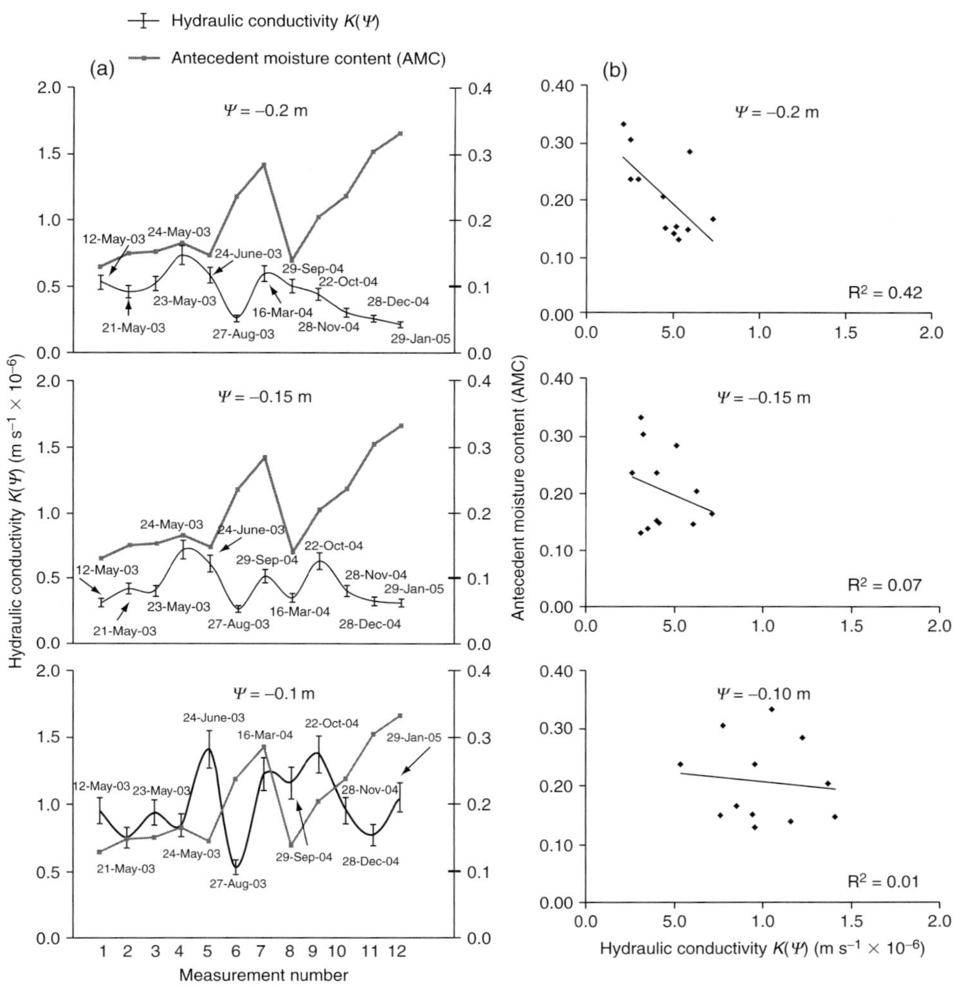

Fig. 1.5. The temporal variability of unsaturated hydraulic conductivity ($K(\Psi) = K(h)$) and antecedent soil moisture (AMC) content at –20, –15 and –10 cm matric potential for a field located near College Station, Texas (Das Gupta *et al.*, 2006). Reprinted with permission from the Soil Science Society of America.

Spatial Variability of Soil Hydrological Parameters

It is commonly known that most soil hydrological properties exhibit both short- and long-range variability (Nielsen *et al.*, 1973). It has also been generally accepted that soil samples collected close to each other are more similar than those collected at greater distances, and the similarity decreases as the separation distance between samples increases – up to a certain separation distance beyond which samples are known as spatially uncorrelated or independent. Spatial dependence is reported to occur at scales ranging from a few metres to several kilometres (Trangmar *et al.*, 1987; Ovalles and Collins, 1988; Gaston *et al.*, 2001). Geostatistical analysis is usually carried out to understand the spatial structure and spatial variability of soil hydrological properties better. Geospatial analysis can also provide more insight on the spatial variability of a property, whether it is structured, unstructured or directional. A detailed overview of these methods and of their application to field data sets can be found in Warrick and Nielsen (1980), Nielsen and Wendroth (2003) and Webster (1985).

The classification that expresses the spatial dependence of soil properties as strong, moderate or weak was suggested by Cambardella *et al.* (1994); it is very popular among researchers and has been extensively cited by researchers. This classification, known as nugget to sill ratio (NSR), or nugget ratio, uses the ratio of nugget variance and total sill. Spatial dependence is classified into three categories: strong for NSR < 0.25, moderate for 0.25 < NSR < 0.75 and weak for NSR > 0.75. As an example, we discuss the data shown in Table 1.19 (Iqbal *et al.*, 2005). These data suggest that most soil physical properties have a nugget ratio ranging from 0.25 to 0.75 and, using the criteria suggested by Cambardella *et al.* (1994), Iqbal *et al.* (2005) classified them as moderately spatial dependent. In spite of the similarity of NSR values, various soil physical properties displayed wide variations in their ranges of spatial dependence. Among the exponential model results reported in Table 1.19, strong spatial dependence occurred only twice, and the corresponding ranges of spatial dependence were 99 m and 421 m; moderate spatial dependence occurred 13 times and ranges varied from 78 m to 238 m. Among the spherical model results, strong spatial dependence also occurred twice, and the corresponding ranges of spatial dependence were 425 m and 861 m, respectively; moderate spatial dependence was reported four times, and the range of spatial dependence varied from 153 m to 997 m. Similarly, the range of spatial dependence for a soil property also varied greatly for different horizons. This research provided very useful information on the structure of the variability and the spatial dependence of a soil property, although the question 'What could be the best sampling strategy for collecting samples

Table 1.19. Spatial variability of soil properties presented using semivariogram parameters for a 162 ha cotton field in Perthshire, Massachusetts (Source: Iqbal *et al.*, 2005).

Variable[a]	Horizon	Model	Nugget	Sill	NSR[b] (%)	Range (m)	Spatial class[c]
Sand	Surface	Exponential	78	427	18	421	S
	Subsurface	Exponential	155	452	34	238	M
	Deep	Exponential	151	520	29	137	M
Clay	Surface	Exponential	16	65	25	218	M
	Subsurface	Exponential	32	95	34	139	M
	Deep	Exponential	54	108	50	144	M
BD	Surface	Exponential	0.002	0.007	29	106	M
	Subsurface	Exponential	0.003	0.009	33	107	M
	Deep	Exponential	0.006	0.017	35	132	M
K_s	Surface	Exponential	0.46	1.51	31	94	M
	Subsurface	Exponential	0.46	0.92	50	110	M
	Deep	Exponential	0.59	2.19	27	111	M
FC	Surface	Spherical	16	60	27	741	M
	Subsurface	Spherical	18	76	23	861	S
	Deep	Spherical	33	102	32	997	M
WP	Surface	Spherical	12.0	70	18	425	S
	Subsurface	Spherical	22.0	70	31	425	M
	Deep	Spherical	21	77	28	153	M
AWC	Surface	Exponential	4	12	31	93	M
	Subsurface	Exponential	2	7	22	99	S
	Deep	Exponential	2	7	30	78	M

[a]AWC, available water content ($cm^3 cm^{-3}$); BD, bulk density ($g\ cm^{-3}$); FC, field capacity water content ($cm^3\ cm^{-3}$); K_s, hydraulic conductivity ($cm\ h^{-1}$); WP, wilting point water content ($cm^3\ cm^{-3}$).
[b]NSR, nugget to sill ratio.
[c]M, moderate spatial dependence (25% < NSR < 75%); S, strong spatial dependence (NSR < 25%).

for analysing various soil properties that are spatially independent or uncorrelated?' was not definitively answered. The differences in spatial class and range of spatial dependence among different horizons also indicated the large inherent spatial variability of soil properties in general.

Spatial variability was similarly, assessed for a 12 ha field in Kingston, North Carolina and spatial dependence was again described using the classification suggested by Cambardella *et al.* (1994). The important difference between the data sets presented in Tables 1.19 and Table 1.20 is that the spatial variability of K_s was

Table 1.20. Spatial variability of soil properties presented using semivariogram parameters for a 12 ha field in Kingston, North Carolina (Source: Duffera *et al.*, 2007).

Variable[a]	Depth (cm)	Model	Nugget	Sill	NSR[b] (%)	Range (m)	Spatial class[c]
Sand	4–12	Spherical	0.1	102	0.1	81	S
	19–27	Spherical	0.1	85	0.1	79	S
	34–42	Spherical	12	27	43	112	M
	49–57	Gaussian	18	37	49	168	M
	64–72	Spherical	19	38	50	390	M
Silt	4–12	Spherical	0.1	56	0.2	76	S
	19–27	Spherical	0.1	42	0.2	64	S
	34–42	Linear	24	25	95	–	W
	49–57	Linear	24	25	96	–	W
	64–72	Exponential	0.01	16	0.1	–	W
Clay	4–12	Spherical	0.01	10	0.1	86	S
	19–27	Spherical	0.6	23	3	75	S
	34–42	Exponential	0.01	25	0.04	63	S
	49–57	Spherical	19	63	31	411	M
	64–72	Exponential	21	43	50	–	W
K_s	4–12	Linear	24	24	99	–	W
	19–27	Linear	2	2	98	–	W
	34–42	Linear	31	31	100	–	W
	49–57	Linear	30	30	100	–	W
	64–72	Exponential	26	52	50	–	W
BD	4–12	Linear	0.01	0.011	91	–	W
	19–27	Spherical	0	0.02	0.1	70	S
	34–42	Spherical	0.01	0.02	47	–	W
	49–57	Linear	0.01	0.012	94	–	W
	64–72	Exponential	0.001	0.01	15	92	S
TP	4–12	Linear	15	16	95	–	W
	19–27	Spherical	0.01	21	0.1	70	S
	34–42	Spherical	13	26	50	–	W
	49–57	Linear	16	17	94	–	W
	64–72	Exponential	2	14	15	94	S
FC	4–12	Spherical	0.01	19	0.1	81	S
	19–27	Spherical	0.01	16	0.1	94	S
	34–42	Spherical	6	22	25	365	S
	49–57	Gaussian	8	35	21	496	S
	64–72	Spherical	9	17	49	–	W
WP	4–12	Spherical	0.001	3	0.04	83	S
	19–27	Spherical	0.01	4	0.3	77	S
	34–42	Exponential	0.01	4	0.2	75	S
	49–57	Spherical	3	6	50	224	M
	64–72	Exponential	5	10	50	–	W

[a]BD, bulk density (g cm^{-3}); FC, field capacity water content (cm^3 cm^{-3}); K_s hydraulic conductivity (cm h^{-1}); TP, total porosity; WP, wilting point water content (cm^3 cm^{-3}).

[b]NSR, nugget to sill ratio.

[c]M, moderate spatial dependence (25% < NSR < 75%); S, strong spatial dependence (NSR < 25%) W, weak spatial dependence (NSR > 75%).

reported as moderate in the former, but weak in the latter. On looking at the CV, in both of these studies K_s was reported as highly variable (Duffera *et al.*, 2007). As K_s is an important parameter for water and solute application efficiencies and for triggering greenhouse gas (GHG) emissions from an ecosystem, knowledge of spatial structure and spatial variability on a landscape scale is a prerequisite for designing site-specific management. In order to conserve water (both surface and groundwater) resources, and make efficient use of the available water – without polluting the water resources, and also by preventing or minimizing GHG emissions from agricultural fields, there is a need to increase overall knowledge on farm water application and water use efficiency. Here, an accurate knowledge of the variability of K_s is a prerequisite for initiating an efficient water management scheme.

Soil Hydrology and Climate Change

Increasing global population has put pressure on land-use systems in many parts of the world, and changing land use has resulted in changes in the quality of the land, especially agricultural land. The quality of a land is essentially controlled by its soil hydrological properties, and drastic change in land use can cause drastic changes in soil hydrological properties. The need to feed the increasing number of people on the earth and to increase per capita grain production has led to green revolutions in several parts of the world. Growing more crops per year could have also caused an enormous stress on the resistance and resilience of soils. Intensive field cultivation can seriously affect soil hydrological and soil chemical properties, and in certain parts of the world has caused a build-up of salinity, loss of organic matter from soil, and increased use of fertilizers and other chemicals to control weeds and pests. Loss of organic matter and intensive tillage could have also altered soil hydrological properties. Barring a few industrialized nations, in general, an increase in the use of water is reported as primarily due to population growth and economic development (Bates *et al.*, 2008). It has been reported that the irrigated area under agriculture globally has increased at a rate of 2% a year since 1960 (Bruinsma, 2003; Bates *et al.*, 2008); however, the quality of ground- and surface water has shown a general decline (UNESCO, 2006).

So far, only a few accounts, if any, are available on change in land use due to the climate change. Climate change predictions can have large uncertainties, but, if climate predictions are accurate, they can cause substantial changes in the amounts and frequency of annual precipitation, resulting in drastic changes in soil moisture, temperature and length of growing seasons. These changes will be different in different parts of the world, and will be confounded by fundamental complex agricultural systems and crop response to the changed frequency of precipitation, moisture, temperature and carbon dioxide regimes (Rosenzweig and Hillel, 1995). Predicted changes in water availability (the median from 12 climate models) at the end of the 21st century (2090–2099 versus 1980–1999) are shown in Fig. 1.6 (Bates *et al.*, 2008), which shows that there is a likelihood of increased water stress in the Mediterranean basin, Central America and the subtropical regions of Africa and Australia (Bates *et al.*, 2008). This could cause substantial changes in land use around the world. For a more elaborate discussion of the effect of climate on land use, refer to the full IPCC report of Bates *et al.* (2008): *Climate Change and Water*. Changes in water availability and water stress will force soil hydrologists to develop new methods and technologies to manage soil and water resources more efficiently.

Summary

The design of management practices for sustaining soil and environmental quality requires an in-depth understanding of soil hydrological properties and processes. The design of best management practices involves increasing application efficiencies of water and nutrients to prevent deep percolation of water and the leaching of solutes eventually leading to groundwater pollution. The understanding of

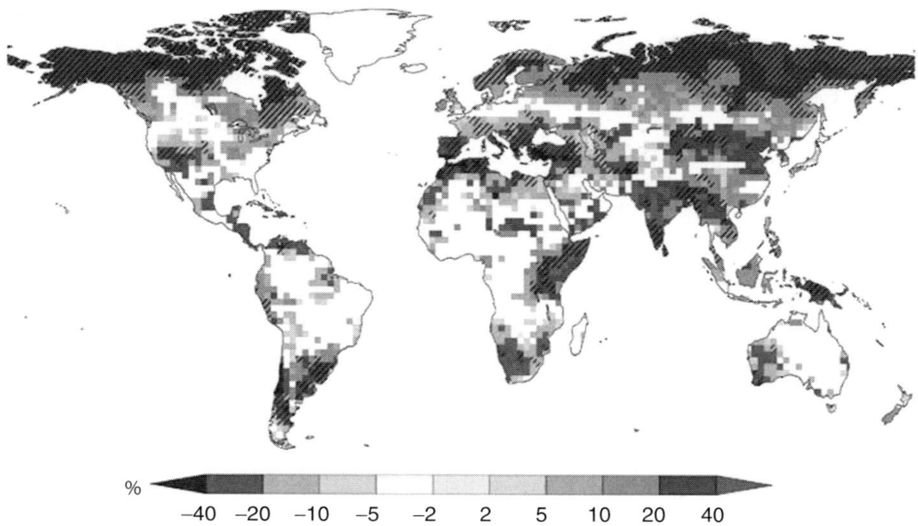

Fig. 1.6. Predicted changes in water availability at the end of the 21st century. The percentage scale refers to large-scale relative changes in annual runoff for the period 2090–2099, relative to the period 1980–1999. White and (cross) hatched areas indicate disagreement/agreement among different models (Bates *et al.*, 2008). Reprinted with permission from the original source: IPCC, 2008: *Climate Change and Water.* Technical Paper of the Intergovernmental Panel on Climate Change, Figure 2.10. IPCC Secretariat, Geneva, Switzerland.

the migration of water and solutes requires an understanding of soil hydrological properties and their variations in space and time. Water in the soil pores can influence other soil physical, chemical and biological properties. The fluctuations of water in the soil can cause disease propagation, increase in the salinity or sodicity of the topsoil, and increase in emissions of N_2O and other potent GHGs.

Water management is one of the key issues for the sustainability of natural resources, global food security, contamination of groundwater and surface water, and soil conservation. It is also an important component in the understanding of vegetation–soil–human interactions. Thus, soil hydrology is important for several curricula centred on sustainability issues related to the soil–water–plant continuum. Some of the disciplines which include the fundamentals of soil hydrology in their curricula are agricultural engineering, civil engineering, crop science, environmental science, forestry, horticulture, range science, soil science, vadose zone hydrology, watershed management, etc. The principles of soil hydrology are taught through a series of classes included in the curricula of several agricultural, engineering, environmental science and soils classes. As a stand-alone course, soil hydrology is a required or optional class for several undergraduate and graduate degrees, and is taught under different names, such as soil physics, advanced soil physics, contaminant transport modelling, soil–plant interactions, vadose zone hydrology, physics of porous media, etc.

Increasing global population has increased the pressure on natural resources, and an understanding of soil hydrology principles is becoming increasingly important to develop sustainable techniques for ensuring global food security, and maintaining the quality of water resources, environmental quality and the overall quality of life on earth. There is a need to continue to calibrate existing semi-physically based hydrological models under various climatic scenarios to model the measured soil, water and yield scenarios across the globe. There is also an increasing need to use these semi-physically based models to forecast what can happen under changed scenarios across continents by conducting collaborative research covering soil, water, plant and atmospheric interactions.

References

Bates, B.C., Kundzewicz, Z.W., Wu, S. and Palutikof, J.P. (eds) (2008) *Climate Change and Water.* IPPC Technical Paper VI. IPCC (Intergovernmental Panel on Climate Change) Secretariat, Geneva, Switzerland.

Biggar, J.W. and Nielsen, D.R. (1976) Spatial variability of leaching characteristics of a field soil. *Water Resources Research* 12, 78–84.

Bouma, J. (1973) Use of physical methods to expand soil survey interpretations of soil drainage conditions. *Soil Science Society of America Proceedings* 37, 413–421.

Bruinsma, J. (2003) *World Agriculture: Towards 2015/2030. An FAO Perspective.* Earthscan, London.

Cambardella, C.A., Moorman, T.B., Parkin, T.B., Karlen, D.L., Turco, R.F. and Konopka, A.E. (1994) Field scale variability of soil properties in Central Iowa soils. *Soil Science Society of America Journal* 58, 1501–1511.

Das Gupta, S., Mohanty, B. and Köhne, J.M. (2006) Soil hydraulic conductivities and their spatial and temporal variations in a vertisol. *Soil Science Society of America Journal* 70, 1872–1881.

Duffera, M., White, J.G. and Wesz, R. (2007) Spatial variability of southern U.S. coastal plain soil physical properties: implications for site specific management. *Geoderma* 137, 327–339.

Gaston, L.A., Locke, M.A., Zablotowicz, R.M. and Reddy, K.N. (2001) Spatial variability of soil properties and weed populations in the Mississippi delta. *Soil Science Society of America Journal* 65, 449–459.

Haws, N.W., Liu, B., Boast, C.W., Rao, P.S.C., Kladivko, E.J. and Franzmeier, D.P. (2004) Spatial variability and measurement scale of infiltration rate on an agricultural landscape. *Soil Science Society of America Journal* 68, 1818–1826.

Ikemura Y., Shukla, M. K., Tahboub, M. and Leinauer, B. (2008) Physical and chemical properties of soil in an age chronosequence of organic farms for a semi-arid ecosystem of New Mexico. *Journal of Sustainable Agriculture* 31, 149–170.

Iqbal J., Thomasson, J.A., Jenkins, J.N., Owens, P.R. and Whisler, F.D. (2005) Spatial variability analysis of soil physical properties of alluvial soils. *Soil Science Society of America Journal* 69, 1338–1350.

Isaaks, E.H. and Srivastava, R.M. (1978) *Applied Geostatistics.* Oxford University Press, New York.

Jury, W.A. (1989) Spatial variability of soil properties. In: Hern, S.C. and Melancon, S.M. (eds) *Vadose Zone Modeling of Organic Pollutants.* Lewis Publishers, Boca Raton, Florida, pp. 245–269.

Lai, J. and Ren, L. (2007) Assessing the size dependency of measured hydraulic conductivity using double ring infiltrometers and numerical simulations. *Soil Science Society of America Journal* 71, 1667–1675.

Lal, R., and Shukla, M.K. (2004) *Principles of Soil Physics.* Marcel Dekker, New York.

Nielsen, D.R. and Wendroth, O. (2003) *Spatial and Temporal Statistics – Sampling Field Soils and their Vegetation.* Catena Verlag, Reiskirchen, Germany.

Nielsen, D.R., Biggar, J.W. and Erh, K.T. (1973) Spatial variability of field measured soil water properties. *Hilgardia* 42, 215–259.

Ovalles, F.A. and Collins, M.E. (1988) Evaluation of soil variability in northwest Florida using geostatistics. *Soil Science Society of America Journal* 52, 1702–1708.

Parkin, T.B. and Robinson, J.A. (1992) Analysis of lognormal data. *Advances in Soil Science* 20, 193–235.

Reynolds, W.D., Bowman, B.T., Brunke, R.R., Drury, C.F. and Tan, C.S. (2000) Comparison of tension infiltrometer, pressure infiltrometer, and soil core estimates of saturated hydraulic conductivity. *Soil Science Society of America Journal* 64, 478–484.

Rosenzweig, S. and Hillel, D. (1995) Potential impacts of climate change on agriculture and food supply. Consequences 1(2), Summer 1995. Available at: http://www.gcrio.org/CONSEQUENCES/summer95/agriculture.html (accessed 3 February 2011).

Scott, H.D., Mauromoustakos, A., Handayani, I.P. and Miller, D.M. (1994) Temporal variability of selected properties of loessial soil as affected by cropping. *Soil Science Society of America Journal* 58, 1531–1538.

Shukla, M.K. and Lal, R. (2005) Erosional effects on soil properties in an on-farm study on alfisols in west central Ohio. *Soil Science* 170, 445–456.

Shukla, M.K., Lal, R., Owens, L.B. and Unkefer, P. (2003a). Land use and management impacts on structure and infiltration characteristics of soils in the North Appalachian Region of Ohio. *Soil Science* 168, 167–177.

Shukla, M.K., Lal, R. and Ebinger, M. (2003b) Tillage effects on physical and hydrological properties of a typic Argiaquoll in central Ohio. *Soil Science* 168, 802–811.

Shukla, M.K., Lal, R. and Ebinger, M. (2004a) Principal component analysis for predicting corn biomass and grain yields. *Soil Science* 169, 215–224.

Shukla, M.K., Slater, B., Lal, R. and Cepuder, P. (2004b) Spatial variability of soil properties and potential management classification of a chernozemic field in Lower Austria. *Soil Science* 169, 852–860.

Stockton, J.G. and Warrick, A. (1971) Spatial variability of unsaturated hydraulic conductivity. *Soil Science Society of America Proceedings* 35, 847–848.

Trangmar, B.B., Yost, R.S. and Uehara, G. (1987) Spatial variation of soil properties and rice yield on recently cleared land. *Soil Science Society of America Journal* 51, 668–674.

UNESCO (2006) *Water: A Shared Responsibility. The United Nations World Water Development Report 2(WWDR 2).* No. UN-WATER/WWAP/2006/3. World Water Assessment Programme (WWAP), UNESCO (United Nations Educational, Scientific and Cultural Organization), Paris. Available at: http://unesdoc.unesco.org/images/0014/001444/144409e.pdf (accessed 3 February 2011).

van Es, H.M. (2002) Soil variability. In: Dane, J. and Topp, C. (eds) *Methods of Soil Analysis, Part 4: Physical Properties.* SSSA Publication 5. Soil Science Society of America, Madison, Wisconsin, pp. 1–13.

Warrick, A.W. and Nielsen, D.R. (1980) Spatial variability of soil physical properties in the field. In: Hill, D. (ed.) *Applications of Soil Physics.* Academic Press, New York, pp. 319–344.

Webb, T.H., Claydon, J.J., and Harris, S.R. (2000) Quantifying variability of soil physical properties within soil series to address modern land-use issues on the Canterbury Plains, New Zealand. *Australian Journal of Soil Research* 38, 1115–1129.

Webster, R. (1985) Quantitative spatial analysis of soil in the field. *Advancement in Soil Science* 3, 1–70.

2 Hydrological Modelling at Mesoscopic Scales Using Global Data Sets to Derive Stream Water Availability Models of River Basins

Biswajit Mukhopadhyay* and Vijay P. Singh

Introduction[1]

The development of reliable hydrological models for regional river basins has been a long-standing goal of various researchers on hydrology and water resources management professionals because such models have significance in a broad range of global studies, such as quantifying the effect of climate change, water resources planning and management, design of hydraulic works, ecosystem management, energy production and urban development. Several attempts in the past have met with varied degrees of success (see Singh and Woolhiser, 2002, for a review). Some of the impediments include lack of availability of the reliable data that are necessary to develop such models, establishment of sound procedures for building the models, and lack of widely and easily available data for validation of the models. Nonetheless, this is an important area of hydrology, where the future holds much promise due to increasing availability of high-quality global data sets and the increasing power and sophistication that can be introduced into modelling land surface hydrological processes with the aid of geographical information system (GIS) technology.

The objective of this chapter is to present an approach that can be applied to develop hydrological models of regional river basins, using global data sets that have become available recently and are expected to evolve further in the future, to provide fair approximations of natural streamflows on a monthly basis. Such a model can be called a 'stream water availability model' (SWAM), and it can be used in a more generalized water availability model of a river basin. A water availability model of a river basin has great importance in engineering, as well as in water resources management for meeting demands from multiple sectors – such as drinking water supply, irrigation and agriculture, hydropower generation, and waste water disposal.

The chapter is organized as follows. The scales at which various hydrological models are developed and river basin models fitted in, the purpose of and need for such models, and the problems and prospects in the development of these models are discussed first. Subsequently, the data that are necessary for building such models, currently available as global databases – which are expected to be improved in quality and coverage through continuous refinements, are discussed. The modelling approach that has been adopted and proposed in the chapter is described next. This is followed by a discussion on the essential hydrological processes in the generation

* Corresponding author: Biswajit.Mukhopadhyay@jacobs.com

of streamflows. The final section of the work contains an application of this model to a major river basin with sparse data and great geopolitical significance. The impact of climate change is also included in the discussion.

River basin modelling at mesoscopic scales

Spatial and temporal scales of hydrological models

Hydrology can be studied at various spatial and temporal scales. Generally speaking, a macroscopic scale typifies a large drainage basin covering a continental land surface at an order of a million square kilometres. An example is the entire basin drained by the Mississippi–Missouri River System with a drainage area of 3,250,000 km^2 on the North American continent. A mesoscopic scale represents a spatial domain that extends over several thousands of square kilometres regionally to encompass the drainage area of a major river or a segment of that river and its tributaries. An example is the 103,600 km^2 area drained by the Tennessee River, one of the major tributaries of the Ohio River which, in turn, is a major tributary of the Mississippi River. A local-scale model in hydrology refers to a drainage basin with limited areal extent, such as hundreds or tens of square kilometres, or even less, to cover an area drained by a stream and its tributaries which are parts of a regional river system.

The scale at which a particular hydrological analysis is performed depends on the purpose of the undertaking. For example, for small-scale engineering applications, such as localized flood or erosion control projects and diversion work for irrigation and drainage, hydrological models are required at local scales. For a local-scale model, the data on topographic, meteorological and land surface characteristics are used on a spatial scale that is finely resolved within a few metres, or less than a metre, and on a temporal scale with a resolution of an hour or minutes. For water resources management, such as water availability models for agriculture, community water supply, hydroelectric potential and regional river engineering, the spatial domain for a hydrological model usually extends to the mesoscale, and the time unit is usually a month or a day. For certain broader purposes, such as for an understanding of how global climate change affects hydrology, the spatial domain is at a macro-scale, with temporal scales ranging from annual to decadal. Such scale dependence of hydrological analysis requires a formal demarcation between local and mesoscale and between mesoscale and macro-scale. Operational definitions of these scales call for a formalization of certain common terms loosely used in hydrology to designate drainage units.

In the hydrological literature, terms such as drainage area, catchment, watershed and basin have been used interchangeably. In the USA, watershed is the standard term for a drainage area, whereas, in Europe, Australia and certain Asian countries, the standard term is catchment. Hence, it is necessary to establish a convention in a logical fashion for the use of these terms (e.g. see Olivera *et al.*, 2002a). In this chapter, the following convention is adopted.

A catchment is considered to be the smallest drainage unit defined by the minimum or threshold contributing area that is used to define a stream on a digital elevation model (DEM). This convention is adopted because the currently common practice (which is expected to continue in the future) for hydrological modelling is the delineation of drainage units with the aid of DEMs, and the size of the threshold contributing area in a DEM analysis can vary with the objective of the analysis and also with its scale. Thus, the freedom of choosing the size of threshold contributing area offers the flexibility of not having a rigid upper limit of the spatial extent for a drainage unit to be designated as a catchment, and the choice can vary with the scale and scope of an investigation. A river basin, in contrast, is the largest drainage area formed by integrating all of the catchments in the study area. Other terms, such as sub-basins, watersheds and subwatersheds, are reserved to denote specific drainage units within a river basin. The term watershed is used to denote a subsystem of a river basin, e.g. the drainage area of a particular tributary of a main river system. A sub-watershed can be defined by a set of selected points on the channel network, such as gauging stations or water-quality monitoring stations on a stream, to serve certain particular operational purposes. A sub-watershed based on a recording station of a stream encompasses all of the catchments from where flow originates upstream of that point.

With the adoption of the convention of using different terms with specific connotations to denote various drainage units, the limits for local, mesoscopic and macroscopic scales in hydrology can be defined for specific river basins. An example is provided in Fig. 2.1. Individual river basins of regional extents are natural land surface units for mesoscale studies because they represent a self-contained hydrological system for the transportation, distribution and accumulation of water and its constituents at the land surface. Mesoscopic or river-basin scale models represent a synoptic view of earth's land surface hydrology at a scale larger than individual watersheds. Watershed-scale hydrological studies are conducted at a local scale with physically or statistically based models, using hydrological data at fine spatial resolutions and at short time steps. Typical engineering applications are drawn from such local-scale hydrological analyses. Macro-scale hydrological models can be considered as a hierarchical, nested system of mesoscale models (Vörösmarty *et al.*, 1993); such macro-scale models are intended to simulate water movement on a continental scale and should consider those movements in all three spatial dimensions.

Importance of hydrological models of river basins at mesoscopic scales

A principal goal of hydrological modelling is the quantitative prediction of streamflows. At local scale, rainfall-runoff processes, based on either physical or empirical models of land surface exchanges and finely resolved data sets, are simulated to quantify temporal and spatial distributions of runoff generation. These models find various applications in the engineering of flood-hazard mitigation and water-control structures. However, for management of water resources with the aim of equitable allocation of water to meet the demands arising from various sectors – such as municipal, agricultural and power generation – it is vitally important to have reliable water availability models and decision-support systems of river basins at mesoscopic scales. An availability model provides the quantitative measures of supplies, demands and storages in a river basin at various control points.

Several water availability models have already been developed and are being successfully used. MODSIM (Labadie, 2005) and WAM (Wurbs, 2005) are two examples of such models. These models derive additional importance from the severe pressures that have been placed on water managers worldwide, as many river basins have been experiencing stresses originating from extreme hydrological conditions, such as floods and droughts, rising water demand owing to population growth, urbanization accompanied by degradation of ecosystems, industrialization with a sharp increase in electrical power demand, sustainability of agriculture and the environment, and vagaries arising from global warming and subsequent climate change.

One of the key inputs for a water availability model is accurate estimates of naturalized monthly streamflows at critical control points on the channel network of a river basin. Naturalized streamflows are typical streamflows at various locations of a river network that occur naturally without any anthropogenic effects. Usually, these are estimated at gauging station locations with several years of historical flow records. Naturalized streamflows at gauged control points are then distributed to other ungauged control points (Wurbs, 2006). In other words, SWAMs form an input as well as a subset of larger availability models which take into account groundwater and other water diversions and withdrawals.

Even though naturalized streamflow estimates, or SWAMs, are developed from historical records where dependable stream gauge data are available, the task becomes problematic for ungauged or poorly gauged basins. As Sivapalan (2003) has noted: 'Yet, over 3000 years after river gauging began

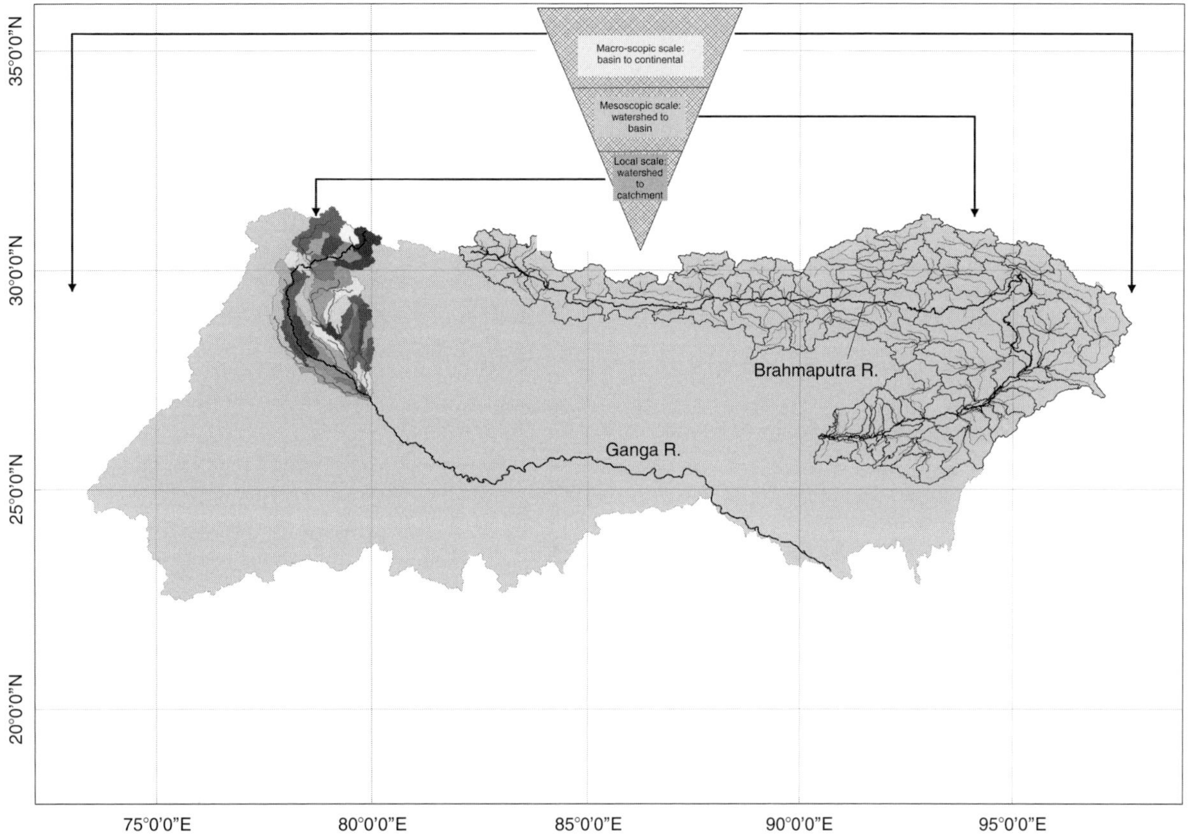

Fig. 2.1. Macroscopic to local-scale systems for multi-level hydrological analysis. Shown in grey shade is the Ganga–Brahmaputra basin, extending over the northern Indian subcontinent, eastern Himalayan range and southern Tibetan plateau (macroscopic scale). A mesoscopic scale is illustrated with the river basin drained by the upper to middle section of the Brahmaputra River, with outlines of the constituent catchments and major streams. A local watershed scale is shown by the catchments drained by the mountainous section of the Ganga River.

on the Nile, the Earth's land surface remains ungauged or poorly gauged, especially in many developing countries, and in some cases the measurement networks have been actually declining in recent decades', there are a great number of important river basins that are mostly ungauged. Predictions in ungauged basins (PUB) are the grand challenges facing present and future hydrology (e.g. Littlewood *et al.*, 2003; Sivapalan *et al.*, 2003; Wagener *et al.*, 2004).

Reliable estimates of the spatial and seasonal distributions of stream water availability within a river basin are very important for authorization of new water projects in the face of sustainable management and the operation of existing water projects and facilities as these face mounting pressures from political, economic and environmental fronts. Thus, the necessity of the development of SWAMs for important river basins is becoming essential as the field of integrated water resources management (IWRM) burgeons. A SWAM also owes its significance to the assessment and prediction of the effects of climate change on river-basin hydrology. Hydrological models at river-basin scales are crucial for improved understanding of the impacts of the factors contributing to global climate change on the environment and society.

Review of the literature

The basic objective of this chapter is not to provide an exhaustive literature review but to present a practical approach towards mesoscopic hydrological modelling, which can be used to approximate monthly streamflows in large river basins from globally gridded data sets on topography, climate, land and snow cover, land use and soil types. Thus, an overview of all pertinent models developed previously is beyond the scope of the present chapter. Instead, in this section an emphasis is placed on the distinction between the mesoscopic model presented in this work and some of the other hydrological models applicable at a similar spatial scale. It should also be noted at this point that these models, including the one that is presented here, are not geared towards detailed modelling of flood routing from individual storm events. Mesoscopic hydrological models require time series inputs of meteorological variables to output time series of streamflows, and the time base for both input and output time series can be monthly, daily or sub-daily.

One of the earlier models applied to large geographical domains is the VIC-2L (Variable Infiltration Capacity – Two soil Layers) model (Wood *et al.*, 1992; Liang *et al.*, 1994). Other models, which emerged subsequently, include the ARNO model (Todini, 1996), SWAT (Soil and Water Assessment Tool) model (Arnold *et al.*, 1998, 1999; Arnold and Fohrer, 2005), Macro-PDM (Arnell, 1999), TOPKAPI (topographic kinematic approximations and integration) model (Liu and Todini, 2002), and BTOP model (Takeuchi *et al.*, 2007), just to mention a few of the commonly known ones. Some of these models have been refined in a continued fashion and have evolved through various versions. For example, the current version of the VIC model is VIC-4.1.1, and that of the SWAT model is SWAT2009.93.3. Some of the models are in the public domain.

All of the models mentioned above are meant to operate on a daily time step, run for many years and can be applied to large river basins at mesoscopic scales. Some of these models, such as VIC-2L and Macro-PDM, can also be adapted for continental-scale (macroscopic) modelling. Nevertheless, all of the models have their merits and limitations. For example, BTOP and SWAT models are fully distributed models, whereas VIC-2L and ARNO models are semi-distributed. In the ARNO model, hydrological input and processes are lumped at catchment scales. In the case of VIC-L2, the basin can be divided into grids with coarse spatial resolutions such as 1° or $1/8^\circ$ (for smaller basins), and both inputs and outputs are lumped over those coarse grids. Even though all of these models incorporate major hydrological processes such as precipitation, infiltration, evapotranspiration, snow melting and base flow, and utilize the principles of water balance and channel flow routing in the simulation of streamflow characteristics, differences among the models lie in the way the hydrological processes are modelled. For example, infiltration in both VIC-2L and ARNO models has been modelled in a similar fashion by taking into account initial soil moisture content, infiltration capacity and variation of the saturated fraction of the soil layer. Such models are often referred to as Explicit Soil Moisture Accounting (ESMA) models. In contrast, models like SWAT take an empirical approach in modelling infiltration, such as a curve number approach, which is essentially like using a runoff coefficient. Other hydrological processes, such as snow-melt generation, base flow production and evapotranspiration, have been modelled with varying degrees of details and data requirements.

One of the differences between the model presented here and other mesoscopic scale models is the time base for model inputs and outputs. As pointed out by Singh and Woolhiser (2002), this aspect of a model greatly influences the type of the model or the details to be included in the model. A watershed model meant to provide long-term monthly averages of river discharge is quite different in its architecture and construct from an hourly or daily streamflow model. A second feature that distinguishes the present model from some of the other models is the spatial resolution of the input fields. The model presented here is a fully distributed model, in which each catchment is constituted of a number of cells with high spatial resolution (~100 m) and each grid cell of the model domain receives its own input parameters for simulation of the hydrological processes at this scale. This is in contrast to the models where input values for specific model parameters are lumped over catchments or coarse grids. Thirdly, the models mentioned above require time series data from station records, at least for precipitation and temperature inputs. In addition to precipitation and temperature, some models require time series data of other climatic variables, such as wind speed. In these models, certain inputs, like temperature, are used to internally calculate other energy inputs, such as short-wave and long-wave radiation. In the case of semi-distributed models, the time series inputs are spatially lumped over a catchment or a coarse grid. The model presented here has been developed with an aim to take advantage of the evolving data sets on monthly averages of climatic variables, such as temperature, precipitation and short-wave and long-wave radiation. These data sets have been developed on global geographical grids and cover several decades of the past. Thus, for this model, daily time series data from station records are not necessary. For this reason, this model can be applied to any river basin, including those where station records are either meagre or difficult to obtain, or even absent. The availability of station records indeed provides an added opportunity for this model to validate global databases or to assess database reliability. An example in this regard will be provided in connection with the application of the model to a particular river basin.

One of the difficulties in the detailed or process-based models, such as the ESMA models, is that a large number of parameters are introduced into the models. Several authors have recognized this problem of over-parameterization, and have even conceded that requirement of a large number of parameters in a model does not necessarily improve the physical representation of the processes, and the performance of the model in its capability to predict streamflows from observed data on precipitation, temperature, etc. Such model parameters must be determined for each catchment from available spatial data; these data are often not readily available or may require extensive field-based experiments and observations. Then again, as pointed out by Jakeman and Hornberger (1993), even though the physics involved in the transformation of precipitation to streamflow may be highly complex, the observed correlation between precipitation and runoff warrants models of only very limited complexity.

The model presented in this chapter operates on a monthly timescale. Hence, it is assumed that over the timescale of a month, the complexities of the processes involved in infiltration, and thereby the transformation of actual rainfall to potential runoff, can be simplified to a water balance type approach. In this approach, soil moisture is calculated on a daily time step and in this way estimation is made of the monthly storage of water in the ground. The process of snow-melt generation has been modelled with a physical basis by keeping the number of parameters required to account for loss of meltwater to a minimum of one or two. Similarly, evapotranspiration has been simulated with a process-oriented approach, and only one parameter is required to relate potential and actual evapotranspiration in the process of soil moisture accounting.

The work described here is certainly not the first attempt to include global- or continental-scale data sets covering long-term monthly averages of meteorological variables to predict long-term monthly mean river discharges for regional river basins. Vörösmarty *et al.* (1989) developed a monthly water balance model (WBM), relying on the techniques developed by Thornthwaite and Mather (1957), to predict monthly soil moisture, evapotranspiration and runoff for the Amazon River basin, encompassing a 5.8×10^6 km^2 area. The model was derived from gridded data sets covering precipitation, temperature, potential evapotranspiration, soil types, vegetation and elevation, with the grids having a spatial resolution of $0.5° \times 0.5°$ (latitude × longitude). Vörösmarty and Moore (1991) applied the same model to the Zambezi River basin, a 1.4×10^6 km^2 area. In this application, the authors used data on temperature, irradiance and cloud cover to calculate monthly potential evapotranspiration. In subsequent work, Vörösmarty *et al.* (1996) updated the previous work on the Amazon River basin to change the time steps from monthly to daily, and compared the calculated river discharges with those derived from satellite data. The daily values of precipitation were derived from the long-term monthly averages used in the previous developments. Several other workers also applied a WBM to river basins. For example, Mishra and Hata (2006) applied a slightly modified version of the WBM of Thornthwaite and Mather (1957) to generate monthly water surplus for 0.5° grid cells of the Upper Blue Nile Basin (an area of 175, 000 km^2). Similarly, there are other variations of the WBM (e.g. Vandewiele *et al.*, 1992; Zhao, 1992). Jiang *et al.* (2007) applied six such WBMs to the Dongjian Basin of South China to evaluate the hydrological impacts of climate change.

Of the hydrological models applicable to a river basin with large geographical extent, no single model has emerged as the preferred tool or as a standard model for analysis or hydrological predictions. This is an evolving field of hydrology where contributions are continuing to come from various sources. Unlike the field of engineering hydrology, where only a limited number of models have become industry standards, there may never be a set of mesoscopic hydrological models that will prove to be standards for analysis and for use as predictive tools for regional river basins.

Problems and prospects in the development of river basin models at mesoscopic scales

The availability and quality of topographic, meteorological and observational hydrological data significantly constrain the form and output of a hydrological model. In the case of mesoscopic hydrology, quality of data not only denotes accuracy but also spatial resolutions and temporal coverage. For this reason, one of the essential requirements in this endeavour is to employ the input data that have best overall accuracy, highest spatial resolution and greatest temporal coverage. In certain parts of the world, such as the USA, both meteorological and hydrological data with sufficient spatial and temporal coverage are readily available and can be used to develop reliable models of regional river basins. In contrast, there are vast land surfaces on the globe, such as various geographical regions on the continents of Asia, Africa and South America, where such data are either sparse or non-existent. In some areas, even if a certain volume of data exists, access to those data is either prohibitive or extremely restricted; a good example of such an area is the Indian subcontinent, which has several trans-boundary river basins.

The challenges noted above are being partly met with the recent developments of various global databases and observational networks that have direct significance in hydrological modelling at mesoscopic scales. These global databases and observational networks relevant to the hydrology of regional river basins started to emerge in recent decades. Refinements and expansions of these data sets are expected to continue in the future, leading to highly evolved databases. Furthermore, applications of remote sensing technology in the field of hydrology and, in particular, the marriage of the global databases with GIS technology for the processing and applications of the data will enable the hydrologists of today and tomorrow to develop reasonable mesoscale models in a way that was not possible in the past.

The following section deals with the global data sets that can be used to develop regional river-basin models to predict monthly stream water availability at various control points of a basin.

Materials and Methods

Global data sets for mesoscopic river-basin modelling

Topography

Launched in February 2000, the space shuttle Endeavour, on its 11-day mission remotely sensed the earth's surface using single-pass, across-track, Interferometric Synthetic Aperture Radar (IFSAR) to provide the first near-global high-resolution elevation data (Farr and Kobrick, 2000). The project, known as the Shuttle Radar Topography Mission (SRTM) project, has created an unparalleled data set of global elevations for mesoscale hydrological modelling. The SRTM project has been a collaborative effort of the National Aeronautics and Space Administration (NASA) of the United States, the National Geospatial-Intelligence Agency (NGA) of the US Department of Defense, the German Aerospace Center and the Italian Space Agency. The data from the original mission underwent several corrections and edits during post-processing of the original data (Gesch *et al*., 2006). NASA's Jet Propulsion Laboratory (JPL) managed the mission and processed the original data set, whereas the Earth Resources Observation and Science (EROS) Data Center of the U.S. Geological Survey (USGS) has the responsibility of hosting, distributing and archiving the processed SRTM data products (SRTM, 2011).

The data set, commonly referred to as the SRTM-3 data set, as processed and edited by NASA/JPL, is available at a three-arc second (approximately 90 m at the equator) resolution for the global land masses between the parallels 60° N and 56° S (for the USA and its territories the spatial resolution is one-arc second or nearly 30 m). However, this data set has certain drawbacks which inhibit its direct use in various applications. The data exhibit typical IFSAR artefacts, such as scattered voids due to shadow and layover effects, poor signal returns over some terrains and occasional phase unwrapping errors. The terrains where radar-specific problems especially prevented the production of high-quality elevation data are the mountainous regions, such as the Himalayas and Andes, and certain land surfaces, such as the Sahara Desert.

The NGA further conducted quality-assurance checks and then carried out several additional edits of the data to comply with the standards of the Digital Terrain Elevation Data (DTED) format of NASA. One of the major problems that still remained with the data, known as the DTED Level 1 data, was the presence of voids (no elevation data at certain points). At present, there are three reliable sources from where void-filled seamless SRTM data are available. One source is the digital elevation model (DEM) data set produced by Lehner *et al*. (2006) as part of the development of the HydroSHEDS (Hydrological data and maps based on SHuttle Elevation Derivatives at multiple Scales) database. Another void-filled SRTM database has been produced by the Environmental Systems Research Institute (ESRI) as part of the Data and Maps data set that has come with ESRI's ArcGIS software package since 2006 (ESRI, 2006). The third void-filled SRTM data set has been produced by the Consultative Group for International Agriculture Research-Consortium for Spatial Information (CGIAR-CSI) (Jarvis, 2004; CGIAR-CSI, 2008).

There are distinct differences in the methods that have been employed in the production of void-filled SRTM data by the three sources mentioned above. The differences arise from the different algorithms used in void filling, as well as from the starting and auxiliary data sets that are used in the void filling. Figure 2.2 is a flow chart that shows the different levels at which SRTM data with certain characteristics are currently available

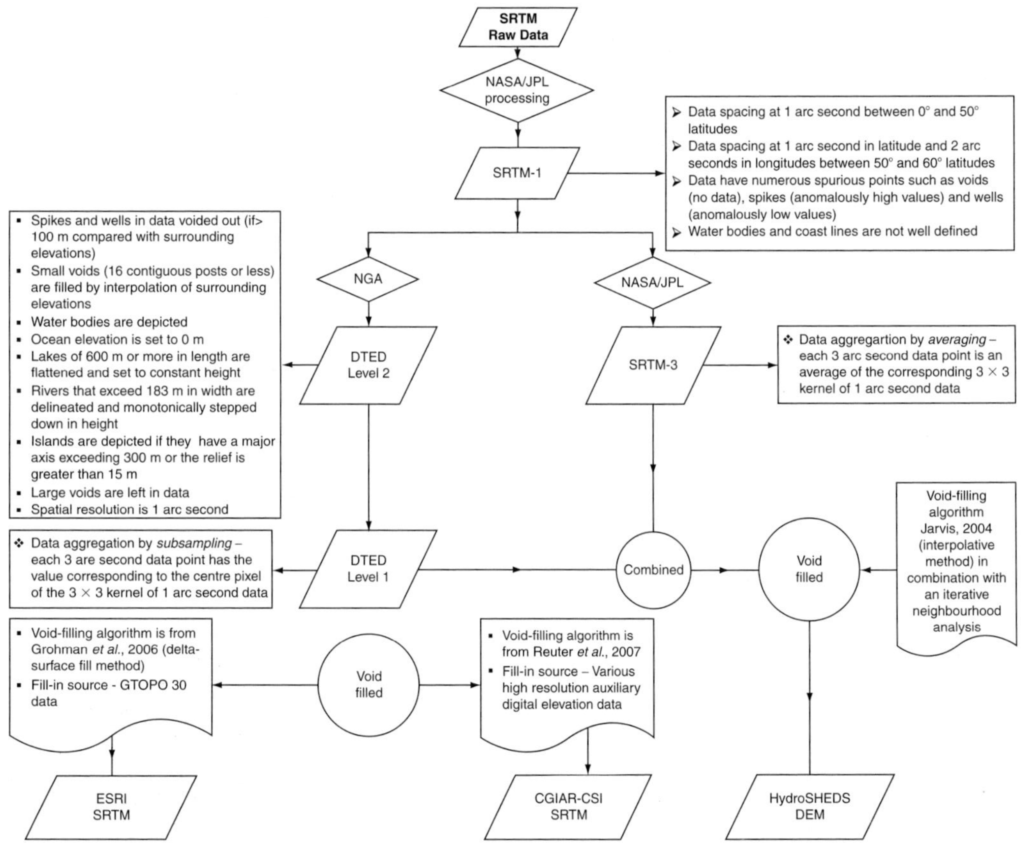

Fig. 2.2. Availability of SRTM (Shuttle Radar Topography Mission) data products at various levels and data characteristics at various levels of SRTM data. CGIAR-CSI, Consultative Group for International Agriculture Research Consortium for Spatial Information; DTED, Digital Terrain Elevation Data; ESRI, Environmental Systems Research Institute; HydroSHEDS, Hydrological data and maps based on SHuttle Elevation Derivatives at multiple Scales; NASA-JPL, National Aeronautics and Space Administration of the United States-Jet Propulsion Laboratory; NGA, National Geospatial-Intelligence Agency (NGA) of the US Department of Defense.

for hydrological modelling. The void-filled SRTM data distributed by CGIAR-CSI are perhaps the best-quality data set because the various auxiliary sources that are used during void filling not only have made the data set more reliable for mountainous regions but also have reduced elevation errors that exist with the DTED Level 1 or SRTM-3 data sets. As noted by Falorni *et al.* (2005), in high-relief terrains the original SRTM data have notable vertical inaccuracies or errors in elevation values.

Nevertheless, the global coverage of the digital elevation data on three-arc second resolution offers almost ten times finer spatial resolution than the previous GTOPO30 global elevation data (Gesch *et al.*, 1999) (GTOPO30 is a global DEM developed by the USGS). The GTOPO30 data set has 30 arc seconds spatial resolution (approximately 1 km^2 grid sizes). In spite of such coarse spatial resolution, the GTOPO30 data sets had also been used to derive some large spatial scale (i.e. small cartographic-scale ratios on maps) river networks (e.g. Olivera *et al.*, 2002b; Olivera and Raina, 2003) and land surface hydrological models (e.g. Poveda *et al.*, 2007). However, the GTOPO30 data sets are not particularly suitable for resolving the finer scale landscape features necessary for hydrological analysis of regional river networks. The three-arc second data are still coarse for small spatial-scale engineering applications at local levels, but are sufficiently good for regional analysis extending to several thousand square kilometres. As a matter of fact, for regional analysis on a small cartographic scale, further finer resolution of cell sizes can produce such a large number (in the order of billions) of cells that their practicability may become infeasible.

Temperature and precipitation

In the past two decades, several global precipitation and temperature data sets have been developed using different input sources, such as ground observations, satellite estimates and climate model simulation (Fekete *et al.*, 2004). The development of mesoscopic hydrological models of regional river basins requires climatic variables, such as long-term monthly mean precipitation and temperature data on geographical grids. The gridded data sets, by virtue of having long and uninterrupted time series for all the grid points, are extremely useful in the quantitative description of spatial and temporal patterns of regional climates, and are particularly suitable for areas with large spatial coverage but with only a set of limited individual recording stations. The extent of spatial resolution necessary for such a grid depends on the application, but production of the same depends on the density of available observational stations and the quality of data that can be obtained from those stations. The nature and quality of station records also control the temporal coverage that can be maintained by such data sets. There exist different data sets that cover the global land surface at medium spatial resolutions and are built on long periods of recorded observational data. Some of the globally gridded climatic data sets have been developed on the basis of observed data at existing weather stations with long periods of records, and by subsequent applications of certain spatial interpolation techniques. Thus, the principal controls on the quality of a global climatic data set are the lengths of periods of records, number and spatial coverage of stations, and the interpolation techniques used in the generation of gridded data sets. The selection of a data set for an application must be governed by the accuracy present in the data sets. Three global precipitation and temperature data sets that can potentially be used in the development of regional river-basin scale water availability models are described below.

The Climatic Research Unit (CRU) of the University of East Anglia in the UK produced a global database of monthly values of nine climatological variables for more than 100 years covering the global land surface (excluding Antarctica) from weather station records and interpolated at longitude–latitude grids with a spatial resolution of 0.5° × 0.5° (Mitchell and Jones, 2005). This data set is an expanded and improved version of the data set developed earlier by New *et al.* (2000). The updated data set, known as CRU TS 2.1, comprises 1224 monthly time series of nine climatic variables for the period 1901–2002. Recently the data set (now CRU TS 3.0) has been expanded to extend the end date to June 2006 (BADC, 2008). The climatic variables in the data set are temperature, diurnal temperature range, daily maximum and minimum temperatures, precipitation, wet-day frequency, frost day frequency, vapour pressure and cloud cover. The database was developed from raw station data derived from seven sources (Table 2.1). However, the raw station data were extensively processed to correct for inhomogeneities, and for the development of reference series, selection of neighbouring stations influencing the reference series, combination of the neighbours and development of station anomalies. The details of the methods used in the processing of the raw station data have been described by Mitchell and Jones (2005). In the anomaly approach, the processed station data are interpolated with the thin-plate smoothing spline technique, considering latitude, longitude and elevation as parameters (New *et al.*, 2000). New *et al.* (2002) also updated their earlier data sets through further interpolations to create the grids at a spatial resolution of 10 arc minute latitude/longitude, covering the period from 1961 to 1990.

Matsuura and Willmott (2009a,b) from the Center for Climate and Land Surface Change of the University of Delaware, USA have produced another comprehensive database of global monthly precipitation and temperature values for each of the years from 1900 to 2008. The station data sources used in the development of this database are also listed in Table 2.1. This data set is a much improved and enhanced version of the global precipitation and temperature database originally developed by Legates and Willmott (1990) and Peterson and Vose (1997). In the Matsuura and

Table 2.1. Sources of station records in two global precipitation and temperature data sets.

Sources of station records – CRU TS 2.1[a]		
Source	Reference	Variables[b]
Jones	Jones and Moberg (2003)	T
Hulme	Mike Hulme, unpublished records	Ppt
GHCN V. 2	Peterson *et al.* (1998)	Ppt, T
Mark New	New *et al.* (2000)	T
Hahn	Hahn and Warren (1999)	T
MCDW	William Angel, unpublished records	Ppt, T
CLIMAT	UK Met Office, unpublished records	Ppt, T

Sources of station records – Matsuura and Willmott (2009a,b)		
Station/Organization	Source	Variables
GHCN2[c]	NCDC[d], USA	Ppt, T
Atmospheric Environment Service	Environment Canada	Ppt, T
State Hydrometeorological Institute	St Petersburg, Russia	Ppt, T
GC-Net Data	Steffen *et al.* (1996)	Ppt, T
Automatic Weather Station Project	University of Wisconsin-Madison	Ppt, T
National Center for Atmospheric Research	University of Colorado	Ppt
African Precipitation data	Sharon Nicholson (2001)	Ppt
South American Monthly Precipitation Station Records	Webber and Willmott (1998)	Ppt
Global Surface Summary of Day	NCDC, USA	Ppt, T
Global Synoptic Climatology Network	NCDC, USA	T

[a]CRU TS 2.1 Climate Database, based on global climate data produced by the Climatic Research Unit (CRU) of the University of East Anglia, UK. Available at: http://www.cgiar-csi.org/data/climate/item/52-cru-ts-21-climate-database.
[b]Ppt, Precipitation (mm); T, Temperature (°C).
[c]Global Historical Climatology Network.
[d]National Climatic Data Center.

Willmott data set, station values of monthly total precipitation (mm) and monthly mean air temperature (°C) are interpolated to a 0.5° × 0.5° degree of latitude–longitude grid, where the grid nodes are centred on 0.25°. The processing of the station data used in the development of this data set is fundamentally different from that employed in the development of the CRU TS 2.1 data set. The Matsuura and Willmott database is based on a background climatology developed from the station data and a climatologically aided interpolation method (Willmott and Robeson, 1995). For temperature, interpolation is also assisted by a DEM (Willmott and Matsuura, 1995). Elevations at grid nodes are obtained from the GTOPO30 data set (K. Matsuura, Delaware, 2009, personal communication). Each average monthly station air temperature is first brought down to sea level at an average environmental lapse rate of 6.0 °C km^{-1}. An interpolation scheme is then applied to the adjusted-to-sea-level average monthly station air temperatures and the gridded sea-level temperature values are brought up to the DEM grid height using the same environmental lapse rate. Interpolation into the grid nodes is accomplished with the spherical version of Shepard's algorithm, which employs an enhanced distance-weighting method (Shepard, 1968; Willmott *et al.*, 1985). Furthermore, a more robust neighbour-finding algorithm, based on spherical distance, is also used. The average number of nearby stations that influence a grid-node estimate is 20. Matsuura and Willmott (2009a,b) also determined the cross-validation errors; one station at a time was dropped during interpolation and the interpolated values at the dropped station location were compared with the station data; in general, very low cross-validation errors were obtained. Further details about the data processing and methodologies employed in the development of this data set can be found at Willmott, Matsuura and Collaborators' Global Climate Resource Pages (http://climate.geog.udel.edu/~climate/).

Another global precipitation and temperature database covering the period from 1950 to 2000 at a spatial resolution of 30 arc second has been developed by Hijmans *et al.* (2005) and is available from World Clim (http://www.worldclim.org). Compared with the two other databases noted above, this database has a relatively shorter temporal coverage but a higher spatial resolution. This data set also used various station records and the thin-plate smoothing spline algorithm (Hijmans *et al.*, 2005). However, the station records were not processed before application of the interpolation method.

Surface radiation

Global surface radiation data are available from three sources originating from three different NASA programmes: the Global Energy and Water Cycle Experiment – Surface Radiation Budget programme (GEWEX-SRB); the International Satellite Cloud Climatology Project (ISCCP); and the Clouds and the Earth's Radiant Energy System (CERES) experiments carried on by the Terra and Aqua satellites of the Earth Observing System (EOS) programme (details available at: http://eospso.gsfc.nasa.gov/). With these varied sets of radiation data, a sound decision is necessary to use the data set that is in line with the objective of a particular study. For the purposes of a mesoscale hydrological investigation, the minimum requirement is to obtain monthly average values of surface radiation fluxes that are quantified with high spatial resolution and long temporal coverage.

The CERES project, involving multi-satellite missions, has started to produce surface and top-of-atmosphere (TOA) radiative fluxes on 1° × 1° scales with a timespan beginning in 2001 (Wielicki *et al.*, 1996; Doelling *et al.*, 2006). The temporal coverage for this data set (SRBAVG) is considerably shorter than that of the GEWEX-SRB and ISCCP data sets.

The GEWEX-SRB project has recently released (version 3.0) a set of high-quality monthly average values of short-wave (SW) and long-wave (LW) surface radiation under both clear-sky and all-sky conditions. The data are given at a spatial scale of 1° × 1° grid, covering the period July 1983 to December 2007. The GEWEX-SRB project is a major component of NASA's radiation research, with the objective of determining surface, TOA and atmospheric SW and LW radiative fluxes with the precision needed to predict transient climate variations and decadal-to-centennial climate trends (e.g. Stackhouse *et al.*, 2001, 2004; Gupta *et al.*, 2006). The SW and LW data sets are derived with two sets of algorithms, known as primary and quality-check algorithms, and a variety of data sources. The primary SW algorithm is adapted from Pinker and Laszlo (1992) and the primary LW algorithm is an adaptation from Fu *et al.* (1997). The quality-check SW and LW algorithms are developed after Gupta *et al.* (2001) and Gupta *et al.* (1992), respectively. The primary data sources are the visible and infrared radiances, the ISCCP DX data set for deriving surface and cloud parameters, and the Goddard Earth Observing System reanalysis products (GEOS-4.0.3; Bloom *et al.*, 2005) for temperature and moisture profiles. Additionally, column ozone data from the Total Ozone Mapping Spectrometer (TOMS) and clear-sky TOA albedo data from the Earth Radiation Budget Experiment (ERBE) are used as inputs to the SW models. The project also makes use of global observations from CERES and moderate-resolution imaging spectroradiometer instruments on the Aqua and Terra satellites for surface albedo (for a quality check of SW radiation) and emissivity.

The ISCCP Project has also produced a new 25-year (1983–2007) global radiative flux data product called ISCCP FD. In general, these data sets provide physically consistent surface and TOA radiative fluxes. The most obvious difference between the ISCCP-FD and GEWEX-SRB data sets is the spatial resolution. The spatial resolution in the ISCCP-FD is a 280 km × 280 km equal area grid (approximately 2.5° × 2.5°) and hence this data set is appropriate for GCM (global climate model) or macro-scale modelling. This grid system is completely different from the 1° × 1° grid in the GEWEX-SRB data sets. An overview of the ISCCP-FD data products is given by Zhang *et al.* (2004). In summary, the ISCCP-FD method uses satellite retrievals at specific visible wavelength channels and then uses radiative transfer to compute SW fluxes and integrates overall wavelengths. The GEWEX-SRB method converts the specific visible channel radiances to broadband (integrated over all solar wavelengths) albedo and then uses radiative transfer to optimize the surface albedo and fluxes to agree with the estimated TOA albedo as much as possible. Both methods give about the same results with regard to bias over the continents of the world, but have larger differences in the polar daytime latitudes. Sorting out those differences is the subject of ongoing research at NASA Langley Research Center (P.W. Stackhouse, Virginia, 2009, personal communication). The ISCCP-FD tends to have a slightly better RMS (root mean square) for monthly averaged fluxes, because the GEWEX-SRB uses a smaller number of satellite pixels to determine the fluxes due to higher spatial resolution. However, the GEWEX-SRB seems to capture the horizontal variability due to higher spatial resolution better. The ISCCP-FD tends to be a little more consistent when the position of geosynchronous satellites changes, so the month-to-month variability is slightly less, but both data sets contain some artefacts from changes in these observing systems. Comparison between these data sets is continuing as the two teams work together to improve the products (P.W. Stackhouse, Virginia, 2009, personal communication).

Extensive validations of the GEWEX-SRB flux values have included ground measurements obtained from the Baseline Surface Radiation Network (BSRN), the Swiss Federal Institute of Technology's Global Energy Balance Archive (GEBA), the Climate Monitoring and Diagnostics Laboratory of the National Oceanic and Atmospheric Agency (NOAA), and other national and international networks. Results of monthly averaged SW and LW radiative fluxes show that generally flux errors are within 10 W m^{-2}. Larger errors are observed where there are larger uncertainties in the input and where the site data do not represent the entire grid box. The most difficult portion of the net radiation is its validation over large areas. Downwelling SW and LW radiation is measurable and is used to assess the downward radiation fluxes. However, upwelling SW and LW fluxes tend to be very locally dependent on the surface type and not very applicable to 1° × 1° scales. Thus, the true uncertainty with the monthly average at 1° × 1° spatial resolution is estimated to be in the range of ±10 W m^{-2} with higher errors at snow/ice-covered surfaces

(Koster *et al.*, 2006; Zhang *et al.*, 2007; Lin *et al.*, 2008). For net fluxes, it is important to note that the surface type (i.e. vegetation type and thickness, soil types) has an important role at small scales, even though the 1° × 1° uncertainty possibly represents the region as a whole and not the sub-grid processes contained within the region.

The Version 3 GEWEX-SRB data are substantially improved from the previous versions through refinements of algorithms and input data sets used in the development of data products. Thus, currently this data set stands as the best global radiation data set, with highest spatial resolution and longest temporal coverage.

Land use and land cover

There are at least two sets of global land cover (GLC) data available in the public domain. The first set of GLC data is produced by the USGS EROS National Center, the University of Nebraska at Lincoln and the Joint Research Centre (JRC) of the European Commission (EC), and is available from EROS. The first version (Version 1.2) of the GLC database was released to the public in November 1997 and underwent several improvements to reach its present version (Version 2.0; 2009). These data sets are derived from the data collected by the Advanced Very High Resolution Radiometer (AVHRR) instruments (sensors) on board the polar-orbiting satellites of NOAA, and span a 12-month time period from April 1992 to March 1993. The data sets are available in geographical as well as projected coordinate systems, in certain raster formats with pixel resolutions of 1 km × 1 km (approximately 30 arc seconds).

The second set of GLC data is available from the Global Environment Monitoring (GEM) unit of the Institute for Environment and Sustainability (IES) of the JRC of the EC. The GEM produced a GLC raster map with 1 km spatial resolution from the data acquired by the VEGETATION instrument on board the SPOT 4 satellite, launched on 24 March 1998. The GEM GLC 2000 data are considered to be the internationally standardized land cover data, with 1 km × 1 km spatial resolution and producing the land cover information of the earth for the year 2000 using the land cover classification system proposed by the Food and Agriculture Organization (FAO) of the United Nations (UN).

Snow cover

The Earth Science Division of the Science Mission Directorate of NASA launched a coordinated series of polar-orbiting and low inclination satellites for long-term global observations of the land surface, biosphere, solid earth, atmosphere and oceans as part of the EOS umbrella programme.

The Moderate Resolution Imaging Spectroradiometer (MODIS) refers to two instruments which are 36-channel visible to thermal-infrared sensors. They are currently collecting data as part of NASA's EOS programme. The first MODIS instrument was launched on board the Terra satellite on 18 December 1999, and the second was launched on board the Aqua satellite on 4 May 2002. MODIS is a key instrument aboard the Terra and Aqua satellites in the EOS mission. Terra's orbit around the earth is timed so that it passes from north to south across the equator in the morning, while Aqua passes south to north over the equator in the afternoon. Terra MODIS and Aqua MODIS view the entire earth's surface every 1 to 2 days, acquiring data in 36 spectral bands, or groups of wavelengths.

The MODIS instrument provides high radiometric sensitivity (12 bit) in 36 spectral bands ranging in wavelengths from 0.4 μm to 14.4 μm. MODIS obtains measurements with spatial resolutions of 250 m (bands 1 and 2), 500 m (bands 3–7) and 1000 m (bands 8–36) using a continuously rotating double-sided scan mirror. In short, the MODIS instruments provide calibrated, geo-referenced radiance data from individual bands, and a series of geophysical products from land, ocean and atmosphere disciplines that can be used for studies of processes and trends on local to global scales.

The National Snow and Ice Data Center (NSIDC) in Boulder, Colorado, USA, archives and distributes snow cover data derived from MODIS instruments from both Terra and Aqua satellites, with overlapping dates for certain time periods (NSIDC, 2011). A variety of snow and ice products are produced from the MODIS sensors, and products are available at a variety of spatial and temporal resolutions. Hall and Riggs (2007) provide a detailed account of the MODIS products.

In summary, the MODIS snow cover data are based on a snow mapping algorithm that employs a normalized difference snow index (NDSI) and other test criteria. The NDSI is a measure of the difference between the infrared reflectance of snow in visible and SW wavelengths. For Terra data, the algorithm uses MODIS bands 4 (0.55 μm) and 6 (1.6 μm) to calculate the NDSI. MODIS band 6 detectors failed on Aqua shortly after launch, so band 7 (2.1 μm) is used to calculate the NDSI for Aqua. Also with Aqua data, the NDSI/NDVI (normalized difference vegetation index) test for snow in vegetated areas was disabled because the use of band 7 resulted in too much false snow detection. The MODIS snow product suite begins with a 500 m resolution, 2330 km swath snow cover map which is then girded to a sinusoidal grid. The sequence proceeds to climate-modelling grid (CMG) products on a latitude/longitude (cylindrical equidistant projection) grid with a spatial resolution of 0.05°× 0.05°. Monthly average snow cover is calculated from daily global products for the month.

Two sets of MODIS-derived snow cover data, derived from both Terra and Aqua, are obtainable from NSIDC. The MODIS/Aqua Snow Cover Monthly L3 Global 0.05° CMG (MYD10CM) data set, new for Version 5 (V005), contains snow cover and quality assessment (QA) data in a Hierarchical Data Format-Earth Observing System (HDF-EOS) format, and corresponding metadata. This data set consists of 7200 column by 3600 row global arrays of snow cover in a 0.05° CMG. These data, stored in HDF-EOS format, are available for the period ranging from 4 July 2002 to the present. The MODIS/Terra Snow Cover Monthly L3 Global 0.05° CMG (MOD10CM) data set, new for Version 5 (V005), also contains snow cover and QA data in HDF-EOS format, and corresponding metadata. This data set further consists of 7200 columns and 3600 rows of global arrays of snow cover in a 0.05° CMG. Data are available from 24 February 2000 to the present. However, owing to insufficient data received, the MOD10CM time series does not contain the granules for the months of June 2001, March 2002 and December 2003. The granule for the month of December 2003 is present in the MYD10CM time series. However, for the two missing months, 8-day composite data can be used.

Soil cover

A key piece of information, necessary for the development of a hydrological model at any scale, is the soil characteristics of the land surface being modelled. FAO has produced the vector data sets of the soil map of the world at a scale of 1:25,000,000, which are obtainable from FAO (FAO/UNESCO, 2003). The soil classification follows the world reference base for soil resources (FAO, 1998).

Nachtergaele *et al.* (2009) have updated the FAO-UNESCO Soil Map of the World to produce the Harmonized World Soil Database (HWSD). Over 16,000 different soil mapping units are recognized in the HWSD, and selected soil parameters are included in the database to characterize each of the soil units. These soil parameters include available water storage capacity, textural class, granulometry (gravel, sand silt and clay fractions), soil depth, organic carbon content, pH, cation exchange capacity of the soil and clay fraction, total exchangeable nutrients, lime and gypsum contents, sodium exchange percentage and salinity. The database, developed in Microsoft Access, is linked with a raster data set that has a grid resolution of 30 arc seconds by 30 arc seconds. Thus, the raster data set consists of 21,600 rows and 43,200 columns, of which 221 million grid cells cover the globe's land surface. The HWSD is particularly useful in hydrological modelling at mesoscopic scale owing to the information about available water capacity (*AWC*), expressed as mm of water per m of soil column), and the textural classification used by the US Department of Agriculture (USDA), which are contained in the database. These two soil parameters, namely *AWC* and USDA textural class, are needed for soil moisture accounting in water balance calculations and for assessment of infiltration capacities of various soil horizons present in a study area.

River flows

River-basin scale hydrological models require actual river discharge data collected through long-term monitoring at several gauging stations for validation, calibration and continuous refinements. However, from a global perspective, the current observational data concerning surface water are not adequate. Several authors have already noted this limitation (e.g. Shiklomanov *et al.*, 2002; Alsdorf *et al.*, 2003). The problem has arisen for two major reasons. First, within many land areas, there are inadequate data collection stations. Moreover, as noted above, in the recent decades, there has been a further widespread decline in hydrological monitoring stations. Secondly, access to existing data is neither easy nor unrestricted everywhere. Brakenridge *et al.* (2005) quoted the Global Runoff Data Center (http://grdc.bafg.de/) as stating in 2004 that 'Currently, only a few national hydrological services distribute their data in accordance with World Meteorological Organization resolutions which call for free and unrestricted exchange.' While the source for the first problem can be simply economic, that of the second problem is often political. For example, a nation's position within a river basin can be such that sharing of hydrological data for trans-boundary rivers can cause political conflicts. An issue, for example, can arise from the operation of existing dams and reservoirs or the construction of new dams and reservoirs which can conflict with the interests of water demand downstream.

In the face of the current problem of unavailability of widespread river discharge data, based on gauging station records, a recent development to directly measure river discharge using orbital satellite passive microwave sensors holds great potential without regard to political boundaries. The Advanced Microwave Scanning Radiometer Earth Observing System (AMSR-E) on NASA's Aqua satellite provides global coverage on a near-daily basis without severe interference from cloud. Brakenridge *et al.* (2007) have developed a methodology, using the AMSR-E band at 36.5 GHz, descending orbit, horizontal polarization daily global data product to measure daily river discharge. The methods have been tested along US rivers monitored by *in situ* gauging stations and have been calibrated for several sites worldwide even if only fragmentary monthly mean discharge data were available. River flow data from numerous stations covering all major rivers are now available online from River Watch, a cooperative project between the Dartmouth Flood Observatory, Dartmouth College, Hanover, New Hampshire, USA and GDACS-GFDS (Global Disaster Alert Coordination System-Global Flood Detection System) from the JRC of the EC, Ispra, Italy (details available at: http://www.dartmouth.edu/~floods/AMSR-E%20Gaging%20Reaches/IndexMap.htm). For many rivers, the data are available from 2002 as monthly mean discharges. In addition, other important hydrological statistics for a station are also available. These data will prove to be quite useful in river-basin scale hydrological modelling, in addition to addressing a wide variety of applications, such as the GFDS from the JRC of the EC (details available at: www.gdacs.org/floods). In closing this section, it needs to be mentioned that wherever gauging station-based discharge data are available, such as the gauge data provided by the USGS, these should also be used.

Digital maps

For GIS-based hydrological modelling, it is often necessary to use digital maps and charts, prepared by electronic digitization of hard copy maps and atlases, for comparison with DEM-based delineations of channel networks and drainage units, or for other analytical purposes. Three of the best sources of such data are as follows.

Using operational navigation charts, also known as aeronautical charts, ESRI (1993) produced the Digital Chart of the World (DCW) for the US Defense Mapping Agency (DMA) as vector maps at a scale of 1:1,000,000. The DCW is generally considered to provide the most comprehensive and consistent global river network data currently available. ESRI (1992, and updated subsequently) also produced a data set known as Arc World which included a global vector map of surface water bodies at a scale of 1:3,000,000. Thus, the Arc World data are digitized at a coarser spatial scale than the DCW data. However, the Arc World data include some corrections and updates as compared with the DCW, and provide a consistent focus on major rivers and lakes of the world.

Lehner *et al*. (2006), as part of the Hydro-SHEDS project and database development, produced global river networks at a spatial resolution of 15 arc seconds. In the process of the development of this channel network data set, the Arc World data were used during conditioning of the digital elevation data. In the end, the HydroSHEDS river network shows significantly better accuracy than the river network of Arc World. Generally, the HydroSHEDs channel network also exhibits better accuracy than the DCW. However, the accuracy of both data sets varies with the physiographic set-up. In some regions, where HydroSHEDS is particularly susceptible to errors, such as vegetated flood plains, the quality of DCW is superior. The HydroSHEDS data products also include digital delineations of river basins, although those are not so useful for specific purposes. For work with a specific river basin or its watersheds, delineations of catchments and other drainage units should be carried out on a case-by-case basis.

The modelling approach

A hydrological model, irrespective of scale, is essentially a mathematical abstraction of the hydrological processes involved in the transformation of inputs into the drainage basin system, such as rainfall and snow melt, to the outputs from the system, such as evapotranspiration, runoff and subsurface storage. In essence, all such models are based on the fundamental principle of mass balance. Variations in various models chiefly lie in the manner in which the hydrological processes are accounted for. In some models, the representation of hydrological processes is based on physical principles, whereas in others it can be empirical or statistical, or even a combination thereof.

The model presented in the current work has been developed with the principal goal that it will be based on spatially distributed physical data with a minimum number of calibration parameters, will be a practical tool for estimation of streamflows at river-basin scales, will be capable of assessing the effect of climate change on streamflows, and will follow the topology of a river basin.

River-basin topology and abstraction of the channel network

In a topological network, a catchment, the smallest drainage unit, is abstracted as a source node that drains into a junction node. The drainage of the source node to a junction node is abstracted as an overland link. Two junction nodes are connected by a stream link, which is a vector directed towards the downstream junction node. Thus, the topological model (TM) of a basin, the largest drain unit, is a network of nodes and links with connectivity and adjacency rules. This network serves as a framework of the SWAM of a basin. In engineering hydrology, such networks are routinely used for event-based rainfall-runoff modelling. For river-basin scale hydrology, the event-based modelling is inconsequential. Thus, for a river-basin scale, a TM can be used for the prediction of monthly river discharge and sediment load at various nodes using either deterministic or stochastic methods.

Delineation of drainage areas with different spatial scales can lead to the development of topological networks with varying spatial resolution. The topology of the network obeys the law of gravity and enters into the solution of the discrete equation of continuity within the network. This equation is presented as follows. Let $q(l,t)$ denote the river discharge at the downstream junction node of a link l of a network with $\mathfrak{L}$ stream links; $r(h,l,t)$ the surface runoff at time t draining through link l and originating from the h source nodes; and $s(l,t)$ the storage volume (s) within link l with time.

Then, the continuity equation can be compactly expressed as:

$$\frac{\Delta s(l,t)}{\Delta t} = -q(l,t) + \sum_h r(h,l,t_{i-1}-t_i) + \sum_u q(u,t_{i-1}-t_i) + g(l,t_{i-1}-t_i) \quad (2.1)$$

where $q(u,t)$ denotes the discharge from all those links whose downstream node, u, is the same as the upstream node of l and $g(l,t)$ is the base flow contribution to l ($l \in \mathfrak{L}$; $u \in \mathfrak{L}$; $l \neq u$). In this convention, runoff from source nodes h, $r(h,l)$, is routed through the upstream node of l.

The SWAM presented here is a distributed model in terms of input and output. However, a discussion is necessary about the parameterization of hydrological processes that transform an input to an output. In principle, parameters $\Phi(x,t)$[1] can vary both spatially and temporally. Thus, irrespective of the spatial and temporal variation of an input $\mathrm{I}(x,t)$ to which a process Γ is subjected, the output $\mathrm{o}(x,t)$ is spatially and temporally varying. The output $\mathrm{o}(x,t)$ is calculated on a cell-by-cell basis and then the aggregated output is computed on a catchment scale. Thus the model can be represented as:

(See Equation 2.2 at the bottom of the page.)

in which symbol Θ is an operator for a process on an input field, and A_c is the area of the catchment under consideration. Finally, we compare the lumped observation O at a basin or sub-basin or watershed scale in order to test and calibrate the model according to the approximation given below:

$$\mathrm{O} \approx \iint_{x \in A} \mathrm{o}(x,t)\,dA(x) \quad (2.3)$$

For practical reasons, we consider the spatial and temporal variation of $\Phi(x,t)$ in a limited manner. This is because a consideration of cell-by-cell variations of model parameters would require millions or billions of values of the parameters for a mesoscale river basin. Furthermore, it should be realized that most parameters of the processes are conceptual representations or approximations of non-measurable watershed characteristics (runoff coefficient is an example). For these reasons, in the model presented here, while $\mathrm{I}(x,t)$ varies from cell to cell of the DEM representing a river basin, the model parameters are spatially lumped over either a subspace or over the entire space domain, depending on its degree of variability over the entire basin.

Flow generation

In most of the previous mesoscopic hydrological models that adopted a water balance approach, land cover characteristics have largely been ignored by assuming that precipitation and evapotranspiration are uniform over an area such as a grid cell or a catchment. In some studies, land cover characteristics were considered only in ascertaining *AWC* of the soil cover. In the present study, land cover characteristics are explicitly used to distinguish between precipitation that can infiltrate the ground and precipitation that can be intercepted. Accordingly, both precipitation and potential evapotranspiration (E_P) are allocated proportionately to a grid cell (i,j), where E_p represents the value for the entire cell regardless of land cover. Furthermore, the present model incorporates snow melt that results from both perennial snow and ice-covered areas (SCA_p) and seasonal snowfall. Seasonal snow-covered area (SCA_s) is obtained from classification of precipitation into rain and snowfall, whereas SCA_p is derived from land cover data. However, if MODIS snow cover data are used, then total snow-covered area (SCA) can be derived directly (e.g. Mukhopadhyay and Dutta, 2010b; Mukhopadhyay, 2011).

Total precipitation (mm month^{-1}) is classified as rain or snowfall according to the following criterion:

$$\mathrm{P}(i,j) = \begin{cases} P_r(i,j) \text{ if } T_a(i,j) \geq -1\,^{\circ}\mathrm{C} \\ P_s(i,j) \text{ if } T_a(i,j) < -1\,^{\circ}\mathrm{C} \end{cases} \quad (2.4)$$

$$\int_{A_c} \Gamma[\Phi(x,t)\Theta \mathrm{I}(x,t)]\,dA(x) = \int_{x \in A_c} \mathrm{o}(x,t)\,dA(x) \quad (2.2)$$

where $P(i,j)$ is the total amount of precipitation over a grid cell, $T_a(i,j)$ is the monthly mean temperature in the grid cell, and subscripts r and s denote the rain and the snow, respectively.

Let $\Delta_a(i,j)$ be the fraction of the area of a grid cell that is not available for direct infiltration but is available to interception (e.g. canopy cover, bare rocks or an impervious area of an urbanized section of a watershed). Then, the total moisture available in that cell for infiltration, $P_n(i,j)$, is given as:

(See Equation 2.5 at the bottom of the page.)

where *SM* denotes snow melt originating from both perennial snow and ice-covered area and monthly snowfall.

Similarly, E_P', the potential amount of evapotranspiration (mm month^{-1}) that affects the infiltrated water in a grid cell after consideration of land cover-characteristics is given as:

(See Equation 2.6 at the bottom of the page.)

Once the intercepted fraction of the available moisture is separated from the total moisture, the daily soil moisture content (W_s) is calculated as a function of evapotranspiration and maximum possible soil moisture content (W_{max}) or soil moisture storage capacity. In general, W_{max} = *AWC*, which is the available water capacity or field capacity of a soil column. It should be noted here that, for areas under perennial snow and glacial covers, barren rocks or lithosols, W_{max} is virtually zero. However, for these areas, a value of 1 mm m^{-1} is assigned to soil *AWC* to avoid division by zero during the evaluation of soil moisture extraction (discussed below).

The governing equations for soil moisture storage calculated at a daily time step (t) are given as:

$$\frac{dW_s(i,j)}{dt} = 0 \text{ for } W_s(i,j) = W_{\max} \text{ and } P_n(i,j) > E_P' \quad (2.7a)$$

$$\frac{dW_s(i,j)}{dt} = P_n(i,j) - E_P' \text{ for } W_s(i,j) < W_{\max} \text{ and } P_n(i,j) > E_P' \quad (2.7b)$$

$$\frac{dW_s(i,j)}{dt} = \beta[P_n(i,j) - E_P'] \text{ for } W_s(i,j) < W_{\max}; W_s \neq 0 \text{ and } P_n(i,j) \leq E_P' \quad (2.7c)$$

where β is the soil moisture extraction function given as the ratio of the actual evapotranspiration (E_A') and E_P'. This ratio can be determined as a function of W_s/W_{max} and by assuming a certain nature of soil drying under moisture-stressed conditions. In the accounting of soil moisture, daily values of precipitation are obtained by simply dividing a monthly average by the number of days in a month. Evapotranspiration is usually calculated as a daily value (mm day^{-1}).

The monthly (m) change in storage of water in the ground (SG) is given by:

$$\frac{\Delta SG(i,j)}{\Delta m} = SG_m - SG_{m-1} \quad (2.8)$$

where the final ground storage for a month is taken as the value of W_s obtained for the last day of the month.

Monthly, actual evapotranspiration from rain from bare land surface, E_A', is given as:

$$E_A' = \begin{cases} E_P' \text{ for } P_n(i,j) \geq E_P'(i,j) \\ |\Delta S_m(i,j)| + P_n(i,j) \text{ for } P_n(i,j) < E_P'(i,j) \end{cases} \quad (2.9a)$$

Evapotranspiration of intercepted rain (E_A'') is given as:

(See Equation 2.9b at the bottom of the page.)

Total actual evapotranspiration is given by:

$$E_A(i,j) = E_A'(i,j) + E_A''(i,j) \quad (2.9c)$$

$$P_n(i,j) = \begin{cases} [1-\Delta_a(i,j)]\, P_r(i,j) \text{ for } (i,j) \notin SCA \\ [1-\Delta_a(i,j)]\, SM(i,j) \text{ for } (i,j) \in SCA_s \\ SM(i,j) \text{ for } (i,j) \in SCA_p \end{cases} \quad (2.5)$$

$$E_P'(i,j) = \begin{cases} [1-\Delta_a(i,j)]E_P \text{ for } (i,j) \notin SCA_p \\ E_P \text{ for } (i,j) \in SCA_p \end{cases} \quad (2.6)$$

$$E_A'' = \begin{cases} \Delta_a(i,j)E_P(i,j) \text{ for } \Delta_a(i,j)P_r(i,j) \geq \Delta_a(i,j)E_P(i,j) \\ \Delta_a(i,j)P_r(i,j) \text{ for } \Delta_a(i,j)P_r(i,j) < \Delta_a(i,j)E_P(i,j) \end{cases} \quad (2.9b)$$

Monthly runoff (RO_m) that can result from surplus moisture is then calculated as: (See Equation 2.10 at the bottom of the page.)

It should be pointed out here that RO_m includes both surface and subsurface runoff. In other words, in the accounting process described above there is no need to separately evaluate the base flow component. The contribution to streamflow from this fraction (surplus moisture) has often been referred to as *slow runoff*.

Another fraction originating from intercepted rain and melted snowfall also contributes to streamflow; this fraction that contributes to streamflow is sometimes referred to as *quick runoff*. The loss of abstracted rain mostly occurs through evapotranspiration. Thus, the quick flow component can be reasonably estimated by subtracting the amount of evapotranspiration affecting the parts that intercept rain from the intercepted volume of rain. As the fraction of intercepted snow that contributes to streamflow through melting and other losses cannot be easily estimated, that fraction is accounted for by introducing a model parameter (f_s) which can be determined through model verification and calibration. Thus, the total amount of water available for runoff to a river in a given month (PR_m) is:

$$PR_m(i,j) = RO_m(i,j) + [\Delta_a(i,j)P_r(i,j) - E''_A] + f_s\Delta_a(i,j)P_s(i,j) \quad (2.11)$$

However, due to the time lag between runoff generation on the land surface and flow through a stream, monthly river discharge (SR_m) is given as:

$$SR_m(i,j) = K_c PR_m(i,j) + (1 - K_c)PR_m - 1(i,j) \quad (2.12)$$

where K_c is defined as the catchment storage constant or the ratio of discharge to storage in the catchment. The value of K_c depends on the area of the catchment, topography, land cover characteristics, etc. Typically, K_c ranges between 0.50 and 0.60, but it can be considered as a calibration parameter. Catchments with steep slopes and less impervious areas can assume K_c values close to unity. Thus, in summary, the model has two parameters, namely f_s and K_c. The variable $\Delta_a(i,j)$ can also be used as a calibration parameter but it can be estimated quite reasonably from land cover data and hence is less susceptible to significant adjustments.

In general, the volumetric flow rate (R_m) through a catchment outlet is obtained from point estimation of runoff as:
(See Equation 2.13 at the bottom of the page.)

Numerical approximation of Eqn 2.13 over a DEM to compute monthly total runoff ($r(h,l)$) from a catchment h contributing to a link l is given as:
(See Equation 2.14 at the bottom of the page.)

where R_m represents surface runoff from the current month, R_{m-1} represents surface runoff from the previous month and a_{ij} denotes the area of the $(i,j)^{th}$ cell in the DEM. Note that the point-wise average, such as $PR_m(x,y)$ in Eqn 2.13, is replaced by the cell-wise lumped values for the DEM in Eqn 2.14.

Flow routing

For the routing of monthly streamflow through a link according to Eqn 2.1, i.e. for the evaluation of $s(l,t)$, a linear reservoir model is most appropriate. The continuity equation in that event is expressed as a first-order linear differential equation given as:

$$K\frac{dq(l,t)}{dt} + q(l,t) = [r(h,l,t) + q(u,t)] \quad (2.15)$$

In Eqn 2.15, K is the time lag between the input centroid and the output centroid, which is simply the travel time along link l.

$$RO_m(i,j) = \begin{cases} P_n(i,j) - E'_A(i,j) - \Delta SG_m(i,j) \text{ for } W_s(i,j) = W_{\max} \\ 0 \text{ for } W_s(i,j) < W_{\max} \end{cases} \quad (2.10)$$

$$R_m = \iint_{A_c} [K_c PR_m(x,y) + (1 - K_c)PR_{m-1}(x,y)]dxdy \quad (2.13)$$

$$r(h,l) = \sum_{i,j \in A_c} [K_c PR_m(i,j) + (1 - K_c)PR_{m-1}(i,j)]a_{ij} \quad (2.14)$$

It should be noted here that the storage constant, K, in the Muskingum method is also approximately equal to the travel time through the reach, and in the absence of any other data it is often estimated in this way.

The solution to Eqn 2.15 with the boundary condition $q(l,t) = 0$ at $t = 0$ is given by Eqn 2.16:

$$q(l)=I\left[1-e^{-\left(\frac{t}{K}\right)}\right] \tag{2.16}$$

where:

$$I=\sum_h r(h,l)+\sum_u q(u) \tag{2.17}$$

It follows from the continuity principle that $q(l)$ of the nth month of a year is given by $[e^{(-t/K)}]q(l)_{n-1} + q(l)_n$, where n denotes a month.

Estimation of the values of K for various links requires knowledge of link length and flow velocity, v. For continental-scale river flow routing, Miller *et al.* (1994) and Coe (2000) estimated v from an equation given as:

$$v=c\sqrt{\frac{i}{i_0}} \tag{2.18}$$

where i is the mean slope of the channel reach (link), which can be estimated from the DEM, i_0 is the reference topographic index set at 0.5×10^{-4} and c is the constant coefficient ($c = 0.35$). This formulation is similar to that provided by Clark (1945) for the estimation of K (in hours), and is given as:

$$K=\frac{cL}{\sqrt{i}} \tag{2.19}$$

where L is the length of the channel reach (in km) and the value of c ranges from 0.5 to 1.4. Eqn 2.18 and Eqn 2.19 yield similar results. However, the use of Eqn 2.19 results in a slightly better accounting of channel storage, and hence greater hydrograph attenuation.

Hydrological processes

Snow melt

The total energy flux, Q_M (E L^{-2} T^{-1}) (E, L and T are energy, length and time in any consistent set of units), at a point on the snow surface is given by:

$$Q_M = Q_{NR} + Q_A + Q_P + Q_G - Q_I \tag{2.20}$$

where subscripts NR, A, P, G and I denote the fluxes from net radiation, overlying atmosphere, rain on snow, underlying ground and change in internal energy, respectively. In this energy balance process, if all the terms on the right-hand side of Eqn 2.20 are correctly quantified, then surface melt can be determined correctly. All of these terms are rarely measured, but their values can be reasonably estimated provided that certain sets of observational data are available (e.g. Walter *et al.*, 2005). An alternative to this process-based snow-melt model is the empirical degree-day method (Martinec, 1975; Martinec *et al.*, 1994, 2008). The degree-day method has been widely used throughout the world to model snow-melt runoff. However, one of the most severe limitations of this method is that it ignores solar heating, the most important source of energy in causing snow melt, especially in high-altitude terrains. Consequently, several workers take a hybrid approach in which the radiation component is added to the degree-day component (e.g. Bengtsson, 1986; Kustas *et al.*, 1994). Mukhopadhyay and Dutta (2010a) adopted such an approach for modelling snow-melt runoff in the Upper Indus Basin (UIB). In their approach, they also included the convective heat transfer from liquid precipitation falling on snow. Furthermore, they accounted for the loss of snow melt in runoff due to refreezing and storage within the ice pack by introducing a coefficient similar to the runoff coefficient used in the degree-day method. The previous approach is further refined with a more mechanistic approach in the present study, so that snow melt from both seasonal and perennial snows can be calculated with greater accuracy and on a better physical basis.

In the present model, the heat fluxes ($W m^{-2}$) from each source except from the atmosphere (Q_A) are calculated as follows:

$$Q_{NR} = Short\ wave\downarrow - Short\ wave\uparrow + Long\ wave\downarrow - Long\ wave\uparrow \quad (2.21)$$

$$Q_P = \rho_w\, c_{pw}\, P_r (T_r - T_m) \quad (2.22)$$

$$Q_G = 2.0 \quad (2.23)$$

$$Q_I = (\rho_i c_{pi} - P_w c) z_f (T_s - T_m) \quad (2.24)$$

where the symbols ↓ and ↑, respectively, denote the downwelling and upwelling short-wave and long-wave surface radiation; ρ_i and ρ_w denote the densities of ice and water, respectively; c_{pi} and c_{pw} denote the specific heat of ice and water, respectively; z_f is the depth of freezing within the snow/ice cover, T_r is the temperature of rain, T_s is the temperature of snow/ice, and T_m is the melting temperature for ice. Note that the terrestrial heat flux is assumed to be constant following the suggestions given by the US Army Corps of Engineers (1998).

Melting rate (cm day^{-1}) from the radiative, convective and conductive heat transfer is then calculated as:

$$M_1 = \frac{Q_{NR} + Q_P + Q_G}{\rho_w \lambda_f} \quad (2.25)$$

where λ_f is the latent heat of fusion of ice.

The melting rate (cm day^{-1}) from the turbulent heat exchange of sensible and latent heat involving convection and condensation is calculated from the restricted degree-day method (Brubaker *et al.*, 1996) as:

$$M_2 = \kappa T_d \quad (2.26)$$

where κ is the restricted degree-day factor, which, unlike a degree-day factor, does not vary considerably with seasons and typically assumes a value of 0.2–0.25 cm $°C^{-1}$ day^{-1} (Martinec, 1989). In Eqn 2.26, T_d is the average day temperature, which is taken as the average monthly temperature (T_a) for a grid cell.

The total snow melt (*SM*) is then calculated as:

$$SM(i,j) = \frac{M_1(i,j) + M_2(i,j)}{B} - \frac{Q_I(i,j)}{\rho_w \lambda_f} \quad (2.27)$$

where *B* is the thermal quality of the snow/ice.

Refreezing resulting from the change in internal energy is computed from Eqn 2.24, which is a modified version of the equation given by Gray and Prowse (1992). In this modification, instead of the entire depth of the snowpack, only the depth through which refreezing takes place is used in the calculation of internal energy. Snow melting occurs when the temperature of the snowpack is at 0 °C. When the temperature of the overlying air is below 0 °C, then heat is transported from the snowpack to the atmosphere and refreezing occurs from the surface of the snowpack. The depth of the refreezing front penetrates further into the snowpack with time, as long as the temperature of the atmosphere remains below freezing. Bengtsson (1982) analysed this process mathematically, and showed that the depth of freezing is a function of time and overlying air temperature. The exact value of the depth of refreezing depends on certain physical characteristics of the snowpack, such as its thermal diffusivity, liquid-water-holding capacity and density. As these properties of the snowpack are not known, the depth of freezing (z_f) is used as a calibration parameter in the present snow-melt model. Another point to note regarding the use of Eqn 2.24 is snow/ice temperature. Several studies have shown that, when the overlying atmospheric temperature is below freezing, the temperature of the top layer of snow is either close to the air temperature or 2–3 °C below the air temperature. Following the suggestion of Brubaker *et al.* (1996), we set $T_s = T_a - 2.5$ °C when $T_a < 0$ °C. In both Eqn 2.22 and 2.24, we set $T_m = 0$ °C, and in Eqn 2.22 we assume $T_r = T_a$ when $T_a > 0$ °C.

Another calibration parameter in the model is the thermal quality of snow. If the snowpack contains no interstitial water, then $B = 1.0$. After melting begins, some interstitial water develops and $B < 1.0$. For example, the US Army Corps of Engineers (1998) suggests that, for a 'ripe' snowpack, $B = 0.95$–0.97. In other words, a thermal quality less than unity indicates that the snowpack is at 0 °C and contains liquid water. If $B > 1.0$, then the snowpack is at a temperature less than 0 °C and contains no liquid water, a condition that typically persists during winter or over perennial snow when the only energy flux that can induce surface melting is the net radiation.

The temperature at a computation or elevation grid (T_d) is calculated from the temperature at the centre of a climatic grid using:

$$T_d = T_{i,j}^{st} + \lambda_a \left[h_{i,j}^{st} - h_{i,j}^{gr} \right] \quad (2.28)$$

where $T_{i,j}^{st}$ is the temperature in the (*i,j*) climatic grid cell, λ_a is the adiabatic lapse rate, $h_{i,j}^{st}$ is the elevation of the (*i,j*) climatic grid, and $h_{i,j}^{gr}$ is the actual elevation of the DEM cell (*i,j*).

Depending on humidity, the adiabatic lapse rate in a terrain can vary between the dry adiabatic lapse rate of 0.0098 °C m^{-1} and the saturated adiabatic lapse rate of 0.0060 °C m^{-1}. For most snow-melt modelling, the global mean lapse rate of 0.00649 °C m^{-1} is used. This is known as the environmental lapse rate, set by the Civil Aviation Organization for the International Standard Atmosphere (ISA) from the sea level up to 11 km in altitude. However, in mountainous terrains, λ_a can show great diurnal and seasonal variations and can depart from the standard environmental lapse rate. For example, Archer (2003) noted that, within the north-western part of the UIB, the value of λ_a ranges from 0.0065 to 0.0075 °C m^{-1}. Jain *et al.* (2008) observed that λ_a varied from 0.0060 to 0.0074 °C m^{-1} in the Sutlej River basin of the Western Himalayas. Li and Williams (2008) used lapse rate as a calibration parameter for snow-melt runoff modelling in the Tarim River basin, located to the north of the Karakoram Mountains in south-western China. Reliable local lapse rates and their temporal dynamics can only be obtained if multiple weather stations at representative elevations in a river basin are available.

The computational algorithm used in the implementation of the snow-melt module of the model is schematically shown in Fig. 2.3. In this model, the entire snow-covered area is divided into two zones separated by the freezing-point line. This temperature divide is not the same as the equilibrium-line elevation, although there is a relationship between these two lines. Furthermore, this 0 °C isotherm is not stationary throughout the year. An idealized summer situation is shown in Fig. 2.3, where the sub-zero zone persists over perennial snow, and glacial covers and seasonal snow covers can persist in the zone where the air temperature is above freezing. Conversely, during winter, the line can move well over seasonal snow covers. Snow melting principally occurs at interfaces of the atmosphere and the ground (Male and Gray, 1981). As the ground heat flux is negligible, the bulk of the melting occurs at the surface, although meltwater does percolate through the snowpack and a basal saturated zone is developed. For perennial snow covers, this basal saturated zone does not exist owing to the great thickness of the pack. In this case, refreezing usually occurs before meltwater reaches the basal zone. Thus, in this zone, the principal process by which snow melt is lost from runoff is refreezing. However, during the summer, meltwater can be stored in the pack as pools or lakes, and this loss to runoff cannot be easily accounted for in a process-based model. Table 2.2 lists the values of physical constants and model parameters that are involved.

Infiltration and soil moisture

One critical element present in a model that simulates the infiltration process by soil moisture accounting is how the nature of soil moisture extraction is simulated. Eqns 2.7a–2.7c provide the guiding principles. However, further discussion on the nature of soil moisture extraction function β is necessary. Mather (1974) presented a series of graphs to show the various types of relationships that can exist between the ratio of daily actual evapotranspiration (E_A) to potential evapotranspiration (E_P) and available soil moisture. Those graphical relationships can be generalized with the aid of the following mathematical function, which can be termed a soil moisture extraction function, and is a modified version of such a function proposed by Minhas *et al.* (1974):

$$\beta = \frac{E_A}{E_P} = \frac{1 - \exp\left[-\phi \left(\frac{W_s}{AWC} \right) \right]}{1 - \exp(-\phi)} \quad (2.29)$$

As defined above, W_s is the daily soil moisture (mm d^{-1}) and *AWC* is the available water

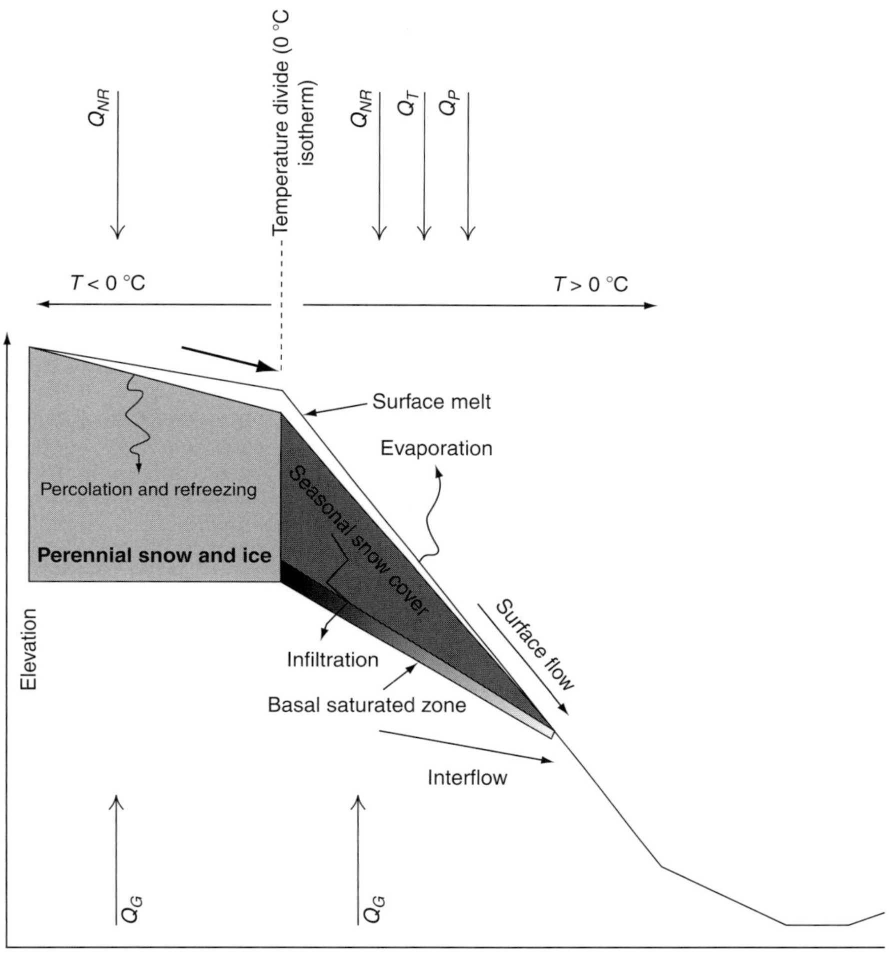

Fig. 2.3. A physical representation of the snow-melt model is shown schematically. Note that the temperature divide is non-stationary in both space and time. The condition depicted in this diagram represents an idealized summer month. During a cooler season, the 0 °C isotherm can shift over seasonal snow cover. Q_G, Q_{NR}, Q_P, Q_T, are the total energy fluxes on the snow surface from the underlying ground, net radiation, rain on snow and the overlying atmosphere (atmospheric heat), respectively.

capacity of soil (mm). Thus, the ratio W_s/AWC provides a measure of the available soil moisture. The nature of the daily moisture extraction can be simulated with various values of parameter ϕ. The graphs shown in Fig. 2.4 represent three models for three different values of ϕ. For a value of 0.5, the relationship is almost linear, which can be simply expressed as $(E_A/E_P) = \phi(W_s/AWC)$. Mather (1974) also showed three such linear relationships for which ϕ = 1, 1.44 and 2.0. Such linear relationships are expected for normal conditions, where the rate of evapotranspiration is directly proportional to the available moisture. However, the non-linear behaviour shown in Model 1 is expected under dry conditions or in arid regions, whereas the nature of soil drying shown by Model 3 indicates humid conditions.

Evapotranspiration

Evapotranspiration can significantly affect the spatial and temporal characteristics of streamflows at river-basin scales. Numerous

Table 2.2. Physical constants and model parameters.

Physical constant or parameter	Symbol	Value	Unit
Physical constants			
Density of ice	ρ_i	917	kg m^{-3}
Density of water	ρ_w	1000	kg m^{-3}
Specific heat of ice	c_{pi}	2102	J kg^{-1} °C^{-1}
Specific heat of water	c_{pw}	4186	J kg^{-1} °C^{-1}
Specific heat of air	c_{pa}	1004.6	J kg^{-1} K^{-1}
Latent heat of fusion of ice	λ_f	334.9	kJ kg^{-1}
Adiabatic lapse rate	λ_a	0.0060–0.0075	°C m^{-1}
Model parameters			
Restricted degree-day factor	κ	0.20–0.25	cm °C^{-1} day^{-1}
Depth of refreeze	z_f	1.0–50	cm
Thermal quality of snow	B	0.95–1.2	Dimensionless
Critical temperature	T_c	0.0–3.0	°C
Aerodynamic factor	α_{PT}	1.25–1.75	Dimensionless
Soil moisture extraction parameter	ϕ	–5.0 – +5.0	Dimensionless
Snowfall coefficient	f_s	Variable	Dimensionless
Catchment storage coefficient	K_c	0.50–0.75	Dimensionless

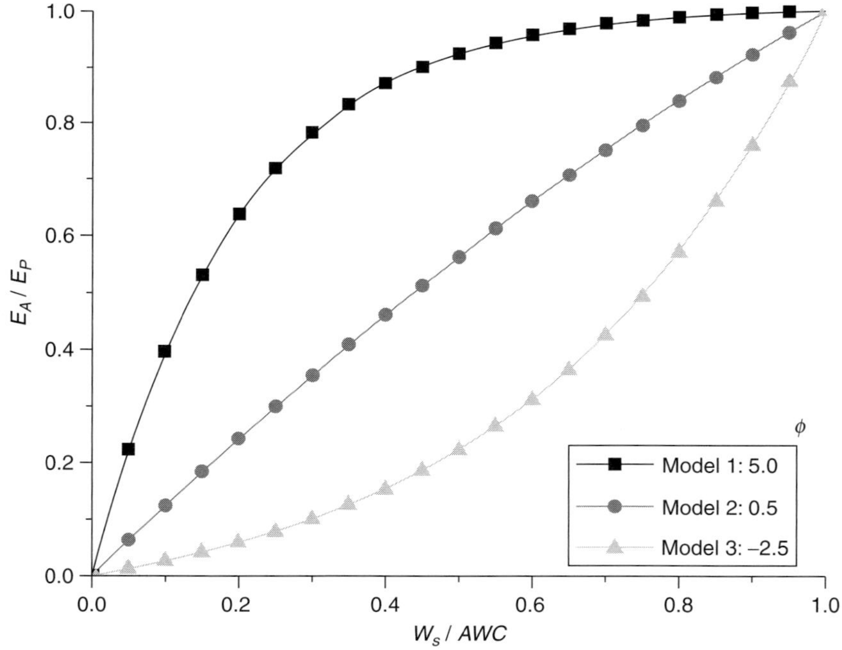

Fig. 2.4. Variation of the soil moisture extraction function (β) with the parameter ϕ, as shown in Eqn 2.29, where E_A/E_P is the ratio of actual to potential daily evapotranspiration (= β) and W_S/AWC is the ratio of the daily soil moisture to the available water capacity. For the value of ϕ = 0.5 (Model 2) the relationship is almost linear and can be simply expressed as $E_A/E_P = \phi(W_S/AWC)$; for the other two values of ϕ (Models 1 and 3) the relationships are distinctly non-linear.

empirical equations are available for the estimation of evaporation and evapotranspiration, based on either climatic or radiation data or a combination thereof. Those equations were developed from studies in various parts of the world that can be climatically different from the river basin under consideration. Nonetheless, several studies have evaluated and established the global applicability of these methods. For example, Weiß and Menzel (2008) compared four potential evapotranspiration equations on a global basis to demonstrate their differences and assess their impact on the calculation of streamflows. From this study, they concluded that the radiation based Priestley–Taylor equation proved to be most suitable for global applications. Xu and Singh (2000) also observed that the Priestley–Taylor method of computation of monthly evaporation (Priestley and Taylor, 1972) resulted in values that agreed most closely with pan evaporation in their study region in Switzerland.

The equation for calculating potential evaporation according to the Priestley–Taylor method can be given by grid-by-grid accounting of the variations in net radiation, temperature and psychrometric constant as:

$$E_P(i,j) = \alpha_{PT} \frac{\Delta(i, j)}{\Delta(i, j) + \gamma_p(i, j)} E_r(i, j) \quad (2.30)$$

where α_{PT} is a constant coefficient, Δ is the gradient of the saturated vapour pressure curve at air temperature (kPa °C^{-1}), γ_p is the psychrometric constant (kPa °C^{-1}), and E_r is the evaporation rate (mm d^{-1}) from open water due to net radiation. Calculations of the variables in Eqn 2.30 follow the following equations.

The conversion of net radiation (W m^{-2}) to equivalent evaporation rate (mm d^{-1}) is given by:

$$E_r(i, j) = \frac{Q_{NR}(i, j)}{\lambda_v(i, j)\rho_w}(86.4 \times 10^6) \quad (2.31)$$

where λ_v is the latent heat of vaporization (J kg^{-1}) and ρ_w is the density of water (1000 kg m^{-3}), and λ_v is given as a function of temperature by:

$$\lambda_v(i, j) = 2501.0 \times 10^3 - 2361.0T_d(i, j) \quad (2.32)$$

The psychrometric constant γ_p (kPa °C^{-1}) is calculated as a function of pressure by:

$$\gamma_p(i, j)=\frac{c_{pa}P(i, j)}{0.622\lambda_v(i, j)} \quad (2.33)$$

where c_{pa} is specific heat capacity of air, and P is the pressure (kPa) calculated from elevation (h) as given by:

$$P(i,j)=101.3-0.01055\times h(i,j) \quad (2.34)$$

The slope of the saturated vapour pressure curve at air temperature is calculated as:

$$\Delta(i,j)=\frac{4098e_s(i,j)}{\left[237.3+T_d(i,j)\right]^2} \quad (2.35)$$

where e_s is the saturation vapour pressure (kPa) given as a function of temperature by:

$$e_s(i,j)=0.6108\exp\left[\frac{17.27T_d(i,j)}{237.3+T_d(i,j)}\right] \quad (2.36)$$

The value of the factor α_{PT} (see Eqn 2.30), which accounts for the, aerodynamic component, is typically taken as 1.26–1.30. However, Jensen *et al.* (1990) showed that these values are valid for humid areas only. For arid areas, a value of 1.70–1.75 is more appropriate in order to account for advection. Weiß and Menzel (2008) also lent credence to this distinction in their studies separating humid and arid regions.

Another method of estimation of monthly evapotranspiration that has found wide applicability is the Turc method (Turc, 1961), which can be described as:

$$E_P(i,j)=0.013\left[\frac{T_a(i,j)}{T_a+15}\right]\left[Q_{SR}(i,j)+50\right] \quad (2.37)$$

where Q_{SR} is the solar or short-wave radiation (given in cal cm^{-2}). However, the Turc method provides estimation of potential evapotranspiration more accurately in humid regions than in arid regions.

Results

Application of the model to the Upper Indus Basin (UIB)

An introduction to the UIB

The UIB is one of the largest drainage basins in the world. The Indus is a noteworthy river for various reasons. It is a trans-boundary river, and it is one of the mightiest rivers in Asia in terms of its volumetric flow and sediment load on an annual basis. The river is extremely important for irrigation of vital crop-producing areas and hydroelectric power generation. As well as these roles, it carries significant geopolitical weight. For example, the Indus Water Treaty, signed between India and Pakistan in 1960, aimed to provide for equity of the water resources between the two neighbouring nations. Yet dispute persists over water use in this and associated basins (e.g. Lafitte, 2007). In sharp contrast to the importance of this river, meagre data and study reports exist in the literature on this river and its drainage basin. Several factors might have contributed to this paucity of knowledge. First, for example, the UIB straddles politically sensitive territories and transects international borders. So no single authority collects basin-wide data and thereby conducts systematic studies. Secondly, the extremely rugged and remote nature of the terrain makes it inhospitable for land-based study. Thirdly, like many major river basins of the world that remain ungauged and are facing declining gauging station networks, the UIB is largely an ungauged or poorly gauged basin. For all these various reasons, the UIB is a classic case of a major river basin, with sparse data, for which practical hydrological modelling can be accomplished using the approach presented above to augment a decision-support system.

The Indus (~3000 km long) originates at an elevation of about 5166 m in remote western Tibet on the westward slopes approaching Lima la Pass, within the Kailash Range. In its mountainous course, the Indus flows north-west between the Ladakh Range and the Great Karakoram mountains to the north, and the Zanskar Range and Great Himalayan Range to the south, until it reaches the Hindu Kush Mountains. There it makes a sharp bend to continue its southward journey through expansive tracts in the fertile plains where the Indus Valley civilization flourished about 2500 BC.

The entire Indus River basin is a large basin with several major tributaries that originate in the western Himalayas; confluence of these with the main stem of the Indus occurs upstream of its mouth in the Arabian Sea. In the following discussion, the focus will be only on the UIB, which covers the vast and extremely rugged mountainous terrain stretching from western Tibet and the Ladakh region of Kashmir to the foothills of the Himalayas. This part of the basin is drained by the Indus and its tributaries upstream of the impoundment reservoir of the Tarbela Dam in Pakistan (Fig. 2.5).

A few studies concerning the general hydrological character of the north-western portion of the UIB had been conducted in the past (see Archer, 2003; Archer and Fowler, 2004; Ali and De Boer, 2007). Mukhopadhyay and Dutta (2010a) used an approach similar to the one presented above to derive a spatially distributed SWAM of the UIB. Subsequently, Mukhopadhyay and Dutta (2010b) and Mukhopadhyay (2011) extended the approach using MODIS snow cover data for the period 2000 to 2009 to show that there has been a significant decrease in SCA_p in the UIB since 1992, with a hydrological effect that includes a significant decrease in summer discharges and the shifting of the peak flow from the middle of summer to late spring. The following presentation is an extension of these two previous investigations.

Development of the mesoscopic hydrological model of the Upper Indus Basin

GENERAL METHODOLOGIES. The principal tool used in the model development is GIS technology. The GIS software system used for this purpose is ArcGIS-ArcInfo, developed by ESRI. The topological model (TM) of the river basin is developed by using the raster and vector processing functions known as

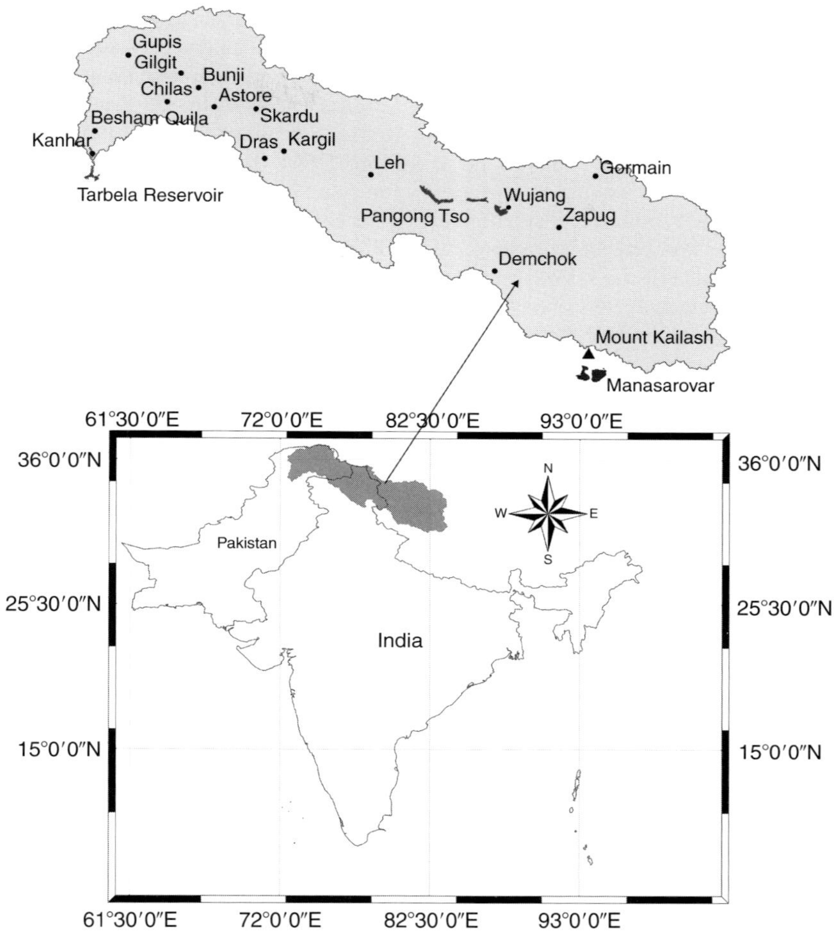

Fig. 2.5. Location of the Upper Indus Basin (UIB, top) in relation to the political boundaries of India and Pakistan (bottom). Within the UIB, the locations of 15 major townships (•) and Pangong Tso, a series of three lakes in the middle of the UIB, are shown. The Tarbela Reservoir is at the western outlet of the basin. Outside the UIB to the south-east, shown for reference, are the famous Lake Manasarovar (the eastern-most of two lakes shown, the western-most is Lake Rakshatal) and Mount Kailash (▲).

the Arc Hydro tools, variously described in Maidment (2002). The SWAM is developed using model builder tools employing various map algebra and spatial analyst functions and Visual Basics for Applications programming language that are available within the ArcGIS software suite.

MAP PROJECTION. All of the data described above are available in various formats that are initially stored in a database with a geographic coordinate system (GCS), which is a system of spherical coordinates. When represented on an ellipsoidal earth, these data must be projected on to a planar coordinate system so that linear and planar features such as stream lengths and catchment areas can be quantified. Correct estimates of such parameters are necessary for better accuracy of a SWAM which uses this information at various steps in the computation.

Finlayson and Montgomery (2003) showed that a careful selection of DEM projection can minimize length and area distortion when analysing large portions of the earth in the two-dimensional plane of a DEM. Thus, it is important to establish a standard projection scheme for model development. We have selected the Lambert Equal Area Conformal Conic projection with the parameters given in Table 2.3. This is one of the best projection systems for middle latitudes and is known to have minimal distortion in area. The distance is known to be correct along the standard parallels but the scale is reduced between the parallels and increased beyond them. A quantitative measure of the length distortion factor is the scale factor *SF* (McDonnell, 1979). An $SF \approx 1$ indicates minimal distortion in length when data are projected from the GCS on to a flat paper with a certain projection scheme. As shown by the *SF*s in Table 2.3, the maps used in the analysis possibly have a 2% error in linear and planar measurements.

After projecting the original data sets into the planar coordinate system, the input data sets are stored in a geo-database. The spatial domain of all input and output data sets in this geo-database is the boundary of the UIB.

A TOPOLOGICAL MODEL. The SRTM elevation data were obtained as $1° \times 1°$ raster data set tiles in a GCS. Several such data tiles extending from longitudes 70° E to 85° E and latitudes 30° N to 40° N were meshed together to form a large DEM. This DEM was processed with the Arc Hydro tools to obtain a preliminary delineation of the UIB. The process included filling and cutting of the spurious sinks and sources in the original DEM, and the creation of a series of flow direction and flow accumulation grids. In the next step, the DEM representing the UIB was used for the development of a TM consisting of refined channel networks and catchments with connectivity and adjacency rules established on the basis of the principle of surface water flow under gravity.

Two sets of channel network data within the limits of the UIB were first created from the HydroSHEDS and DCW databases. In general, good agreements were found between these two data sets. As far as the major streams were concerned, the digital

Table 2.3. Parameters and scale factor of the Lambert Conformal Conic projection used in the map analysis of the Upper Indus Basin (UIB) described in the text.

Datum	WGS[a] 1984			
Central meridian	105° 0 00" E			
Latitude of origin	0° 0' 00"			
First standard parallel	30° 0' 00" N			
Second standard parallel	62° 0' 00" N			
	Point 1	Point 2	Point 3	Point 4
Coordinates of points[b] to determine scale factor (*SF*)[c]	31° 4' 00" N 81° 18' 45" E	34° 10' 12" N 77° 34' 48" E	35° 18' 00" N 75° 37' 00" E	36° 8' 32" N 74° 29' 25" E
Distance along great circle (sphere length, *d*)[d]	–	1–2: 491.67	2–3: 219.27	3–4: 138.39
Map distance	–	1–2: 485.58	2–3: 215.08	3–4: 135.25
Scale factor (*SF*)	–	0.9876	0.9809	0.9773

[a]WGS, World Geodetic System.

[b]Four points are selected to cover the distance from south to north and east to west (point 1 and point 2 represent the Mount Kailash and Rakaposhi peaks, respectively, whereas point 3 and point 4 represent the townships of Leh and Skardu, respectively.

[c] $SF = \frac{\text{Map length}}{\text{Sphere length}}$

[d] $d = r\cos^{-1}\left[\sin\delta_1 \sin\delta_2 + \cos\delta_1 \cos\delta_2 \sin(\phi_1 - \phi_2)\right]$

where δ and ϕ are latitude and longitude of a place measured in radians, with subscripts 1 and 2 denoting point 1 and point 2, respectively, and r is the radius of the earth (6378.1 km).

channel networks matched quite well with hard copy maps and atlases. However, in the areas to the south-east of the basin (the Tibetan plateau area), some adjustments (editing) to the stream lines were necessary to establish proper connectivity from upstream to downstream. The final channel network was used to correct and adjust the raw DEM by the stream-burning process, whereby the stream courses were enforced into the elevation surface. This DEM has 22,976,072 cells, each with a spatial resolution of 107.52 m. The highest and lowest elevations are 8566 m and 423 m, respectively, with a mean of 4708 m.

Subsequent to the conditioning of the DEM, catchments were delineated by assigning 1600 km^2 as the minimum contributing area for the DEM cells to be classified as major streams. This area was experimentally determined so that small catchments draining first-order streams are all integrated into one drainage unit. In the final analysis, 91 catchments constitute the basin drained by the Upper Indus and all of its tributaries. In the TM, each catchment is designated with a number, called HydroID. For the catchments of the UIB, HydroID numbers are 1–91, and these increase from north to south. The relative positions of the catchments in the north–south direction are determined by the locations of the centroids of the catchments. Each stream link representing a major stream has a reach number that increases from the upstream to the downstream direction. In other words, the number that designates a link has all of the upstream links with lower number designations. Junction nodes are also numbered in the same fashion. The junction number also increases from upstream to downstream. Thus, the number designation for a link is the same as it is for its upstream junction node. Each junction node has a number which is greater than the numbers that designate all of its upstream nodes. Some of these nodes, or hydrological junctions, can be recognized as the confluence of two streams with known geographical names, whereas the confluence of two unnamed streams has only a number designation. Similarly, some of the stream links represent major streams with known names, whereas others have only numerical identifiers.

It should be noted here that we used the void-filled SRTM data sets from three sources (Fig. 2.2) to build the TM of the UIB and to test the relative merits of the data sets. In terms of topology, virtually identical results were obtained, but the one presented here is based on the data set obtained from CGIAR-CSI, because we found that, compared with the other two data sets (from ESRI and HydroSHEDS), this one had the fewest elevation errors.

RE-SAMPLING AND AVERAGING OF CLIMATIC GRIDS. As noted above, the grid resolutions for the climatic data are coarser than those of the DEM. Hence, the climatic grids are re-sampled into newer grids with cell sizes equalling those of the DEM grid. This does not necessarily improve the resolution of the spatial variability of the climatic variables, but the procedure not only is necessary for grid-based computation but also avoids uncertainties that can arise from statistical downscaling. The approach is reasonable because the scale of climatic variability is always far greater than the scale at which topographic variability occurs.

We tested both the CRU TS 2.1 and the Matsuura and Willmott databases to determine their relative accuracies. We found that predicted values from the Matsuura and Willmott database match quite well with the monthly values at several locations within the UIB where station records spanning several decades are available (e.g. data given in Ahmed and Joyia, 2003; Ali and De Boer, 2007). At least for the UIB, the accuracy of the CRU TS 2.1 database was found to be very poor. Consequently, we selected the Matsuura and Willmott database for use in our model.

We have used the SRB Version 3 SW and LW data to compute monthly averages of net radiation using the following formula and the 'surface' data for 'all sky' conditions:

$$\text{Net radiation} = SW_{(\text{downward})} - SW_{(\text{upward})} + LW_{(\text{downward})} - LW_{(\text{upward})}$$

In addition to re-sampling, monthly averages were derived from the year-by-year monthly grids. Thus, for each of the 12 months, the precipitation and temperature grids represent averages of 109 years and the net radiation grids represent averages of 25 years.

BASIN CHARACTERISTICS. The total drainage area of the UIB is 265,598 km^2. Of this, 132,547 km^2 (49.91%) lies in Tibet, whereas 58,860 km^2 (22.16%) and 73,921 km^2 (27.83%) lie within the jurisdictions of India and Pakistan, respectively. The remaining 269 km^2 (0.10%) of the basin area is within the boundary of Afghanistan. These estimates are based on the political boundaries obtained from DCW and are not meant to attest any claim or controversy that exists in this region for the demarcation of international borders.

Tables 2.4 and 2.5 list the land-cover characteristics of the UIB according to the 1992 and 2000 global land cover data. Table 2.6 provides data on the soil characteristics of the basin. Figure 2.6 (a,b) shows the USDA textural classes (soil types) and *AWC* of the soil cover of the UIB. It should also be noted that, for the 1992 and 2000 land cover characteristics, parameter Δ_a has been evaluated separately.

DRAINAGE CHARACTERISTICS. The complete channel network of the UIB is shown in Fig. 2.7, with the major streams and certain prominent lakes marked. The length of the Upper Indus River is 1468 km, and that of its main tributary, the Shyok River, is 494 km. The drainage density of the basin is 0.18 km km^{-2}, and stream

Table 2.4. Land cover characteristics of the Upper Indus Basin (UIB) according to 1992 global land cover data. Source: Mukhopadhyay and Dutta (2010a).

Land cover type	Percentage	Δ_a^a
Open shrublands	61.70	0.55
Barren or sparsely vegetated	19.34	0.00
Snow and ice	8.16	1.00
Grasslands	7.47	0.40
Water bodies	1.26	1.00
Crop/natural vegetation mosaic	0.93	0.25
Croplands	0.68	0.20
Permanent wetlands	0.28	1.00
Closed shrublands	0.10	0.60
Woody savannahs	0.05	0.50
Evergreen needle leaf forest	0.01	0.45
Deciduous broadleaf forest	0.01	0.45
Mixed forest	0.01	0.50
Savannahs	0.01	0.45
Urban and built up	0.001	0.90

[a]Δ_a, fraction of the area not available for direct infiltration but available to interception.

Table 2.5. Land cover characteristics of the Upper Indus Basin (UIB) according to 2000 global land cover data.

Land cover type	Percentage	Δ_a
Sparse herbaceous/shrub	29.342	0.45
Shrubs	16.291	0.60
Herbaceous with sparse tree/shrub	12.488	0.40
Herbaceous, single layer	12.363	0.35
Snow and Ice	6.962	1.00
Bare soil/other unconsolidated materials	6.241	0.00
Bare rock	3.876	1.00
Cropland	3.209	0.15
Water	2.941	1.00
Consolidated	2.865	0.85
Wetland	1.455	1.00
Water (60–70%) and small sand and silt islands	0.797	0.90
Needle leaf deciduous forest	0.518	0.40
Cropland/natural vegetation mosaic	0.279	0.25
Needle leaf evergreen forest	0.185	0.45
Unconsolidated	0.095	0.60
Urban	0.025	0.90
Rice paddy	0.022	0.10
Bare	0.020	0.00
Gravels, stones and boulders	0.017	0.65
Loose and shifting sands	0.005	0.00
Broadleaf deciduous forest	0.003	0.45
Broadleaf evergreen forest	0.001	0.50

[a]Δ_a, fraction of the area not available for direct infiltration but available to interception.

Table 2.6. Soil characteristics of the Upper Indus Basin (UIB).[a]

Soil type	Percentage	*K(h)* (mm/hour)
Loam	49.0	3.40
Sandy loam	35.5	10.90
Sandy clay loam	2.7	1.50
Silty clay	0.3	0.50
Clay	0.3	0.30
Ice covered (no exposed soil)	12.2	0.00

[a]*K(h)* values are adopted from Rawls *et al.* (1983). Other data are developed from the Harmonized World Soil Database from the work of Nachtergaele *et al.* (2009).

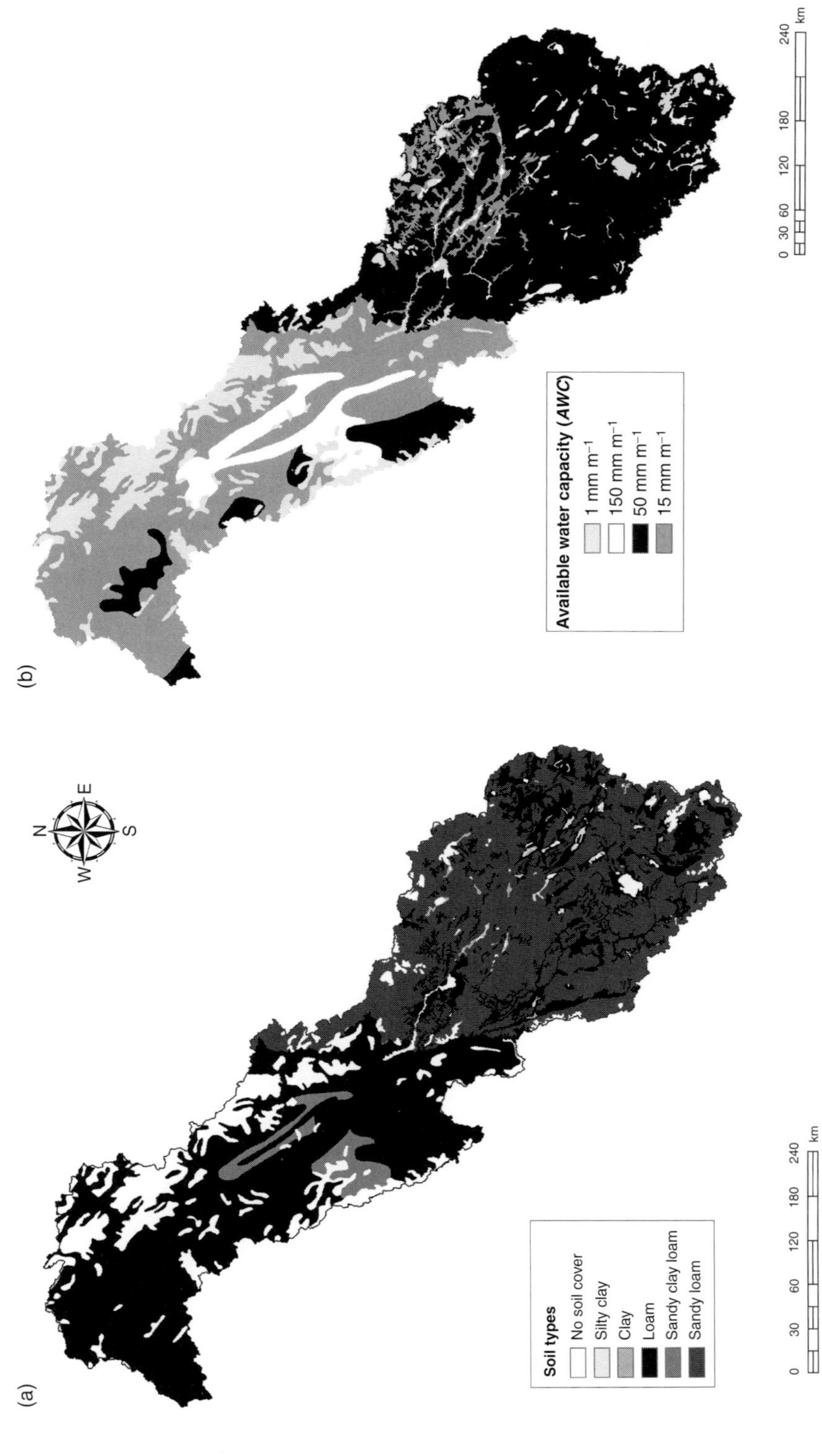

Fig. 2 6. Soil map of the Upper Indus Basin (UIB) showing USDA soil textural classes/soil types (a) and available water capacity (*AWC*) (b).

Fig. 2.7. The channel network of the Upper Indus Basin (UIB) derived from the digital elevation model (DEM). The major streams are marked with bold lines. A few lakes are also shown.

frequency (or stream density) is 3.2 for every 100 km². For a river basin of this size, such low drainage density and hence such low stream frequency are due to the arid climatic conditions. The low values of these drainage parameters also indicate the long overland flow paths, moderate runoff and high permeability of the terrain.

DRAINAGE BASIN TOPOLOGY. Figure 2.8 shows the catchments or the drainage units that constitute the UIB. The reaches through which flow-routing calculations are performed are shown in Fig. 2.9. Table 2.7 provides the geographical references for the numerical designations of the reaches and stream junctions that form the topological network of the UIB. The schematic topological network with overland and channel flow paths are depicted in Fig. 2.10. For reference, known geographical names of the reaches and the junctions are presented in Table 2.7.

SNOW-COVERED AREAS. In the present study, the perennial snow (and ice)-covered areas (SCA_ps) were obtained from the 1992 and 2000 global land cover (GLC) data. These SCA_p values (Figs 2.11 and 2.12) are combined with seasonal snow-covered area (SCA_s) values derived from monthly average precipitation and temperature (Eqn 2.4, where temperature values are corrected for elevation according

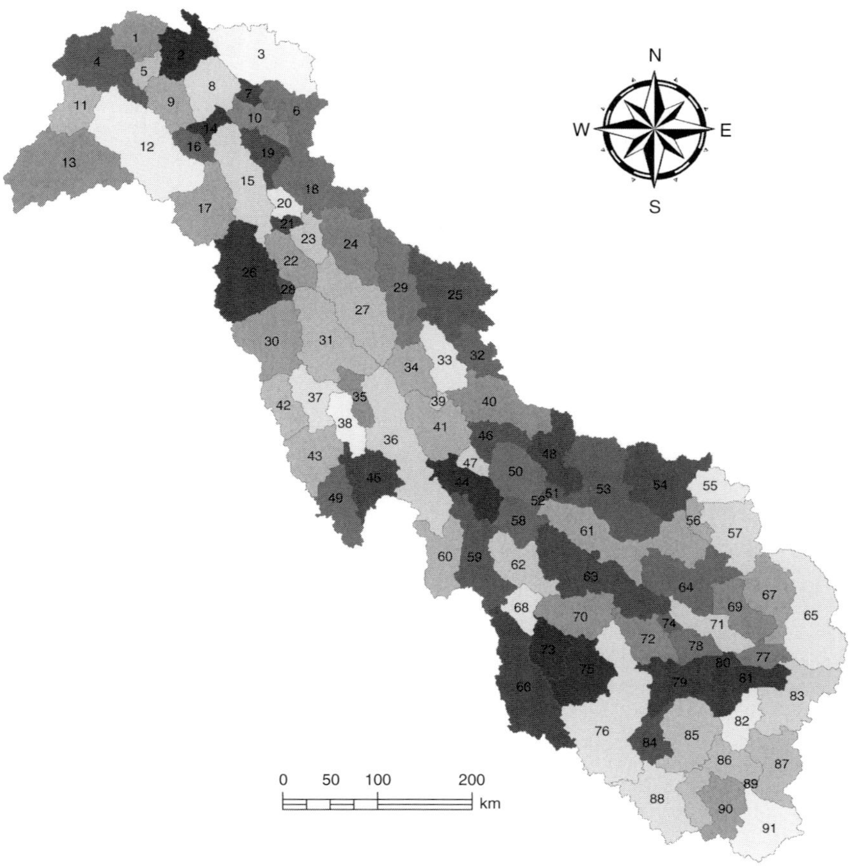

Fig. 2.8. The catchments of the Upper Indus Basin (UIB) as defined in this study. A number (the HydroID) is assigned to each catchment.

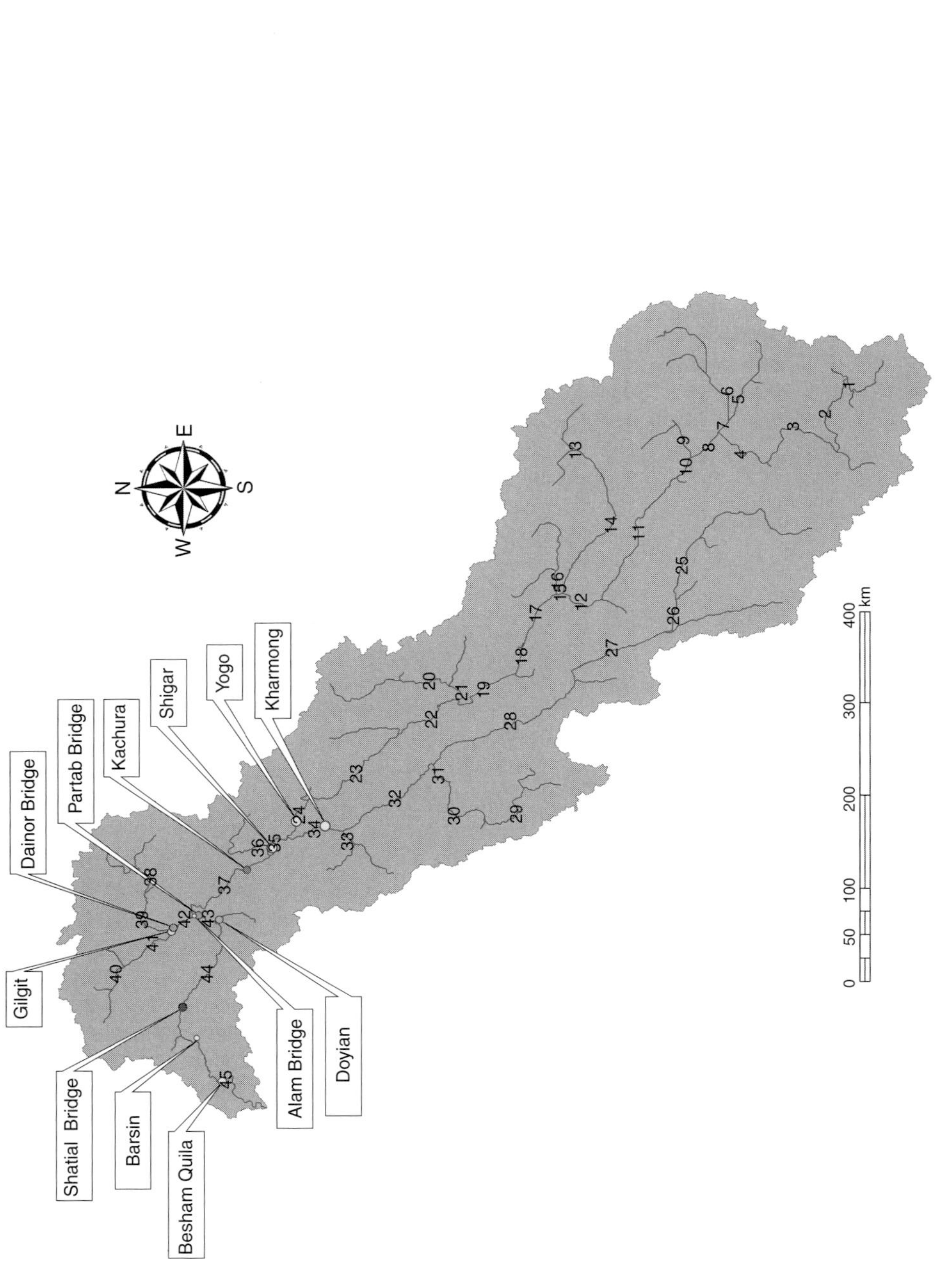

Fig. 2.9. The stream reaches used in hydrological routing calculations in the Upper Indus Basin (UIB) are shown with the reach numbers. Locations and names of past or present gauging stations are also shown. Note that all of the known gauging stations are in the north-western part of the basin.

Table 2.7. Reach number and corresponding geographical name of the river (if known); junction number and the corresponding confluence of rivers with known names in the Upper Indus Basin (UIB).

Reach number	River name	Junction number	Confluence	Nearest gauging station
19	Chusul Nala	22	Shyok River – Chusul Nala	
20–24	Shyok River	23	Shyok River – Nubra River	
25–28	Indus River	24	Hushe River – Indus River	Yugo (downstream)
29–31	Zanskar River	32	Zanskar River – Indus River	
32	Indus River	33	Shingo River – Dras Nala	
33	Shingo – Dras	34	Dras Nala – Indus River	Kharmong (downstream)
34–35	Indus River	35	Shyok River – Indus River	
36	Shigar River	37	Shigar River – Indus River	Kachura (downstream)
37	Indus River	38	Shimshal River – Hunza River	
38–39	Hunza River	42	Gilgit River – Hunza River	Alam Bridge (downstream)
40–41	Gilgit River	43	Gilgit River – Indus River	Partab Bridge (downstream)
42	Gilgit – Hunza or Gilgit River	44	Astore River – Indus River	Shatial Bridge (downstream)
43–45	Indus River	46	Upstream of Tarbela Reservoir (near Kanhar)	Besham Quila (upstream)

to Eqn 2.28) for the computation of snow melt according to Eqn 2.27. Thus, two different sets of computations are obtained for the snow-covered areas. In the first set, the perennial snow and ice-covered areas represent 1992 conditions, while in the second set they represent 2000 conditions. It is interesting to note that according to the 1992 land cover data, the snow- and ice-covered area constitutes 8.16% of the total basin area, whereas according to the 2000 land cover map 6.96% of the total area of the UIB is snow and ice covered.

The MODIS data products can also be used to delineate the snow-covered areas (Mukhopadhyay and Dutta, 2010b; Mukhopadhyay, 2011). Figure 2.13 shows the time series plot of total *SCA* ($SCA_p + SCA_s$) in the UIB from March 2000 to December 2009. From this plot, SCA_p can be derived from the annual minima of *SCA*. The 2000 MODIS data show that the SCA_p in the UIB is close to 6.78% of the basin area. This is materially identical with the estimate (6.96%) obtained from the 2000 GLC data, as noted above. The average SCA_p from 2001 to 2008 is 5.65 ± 0.33% (1 σ or SD), which indicates a 2.51% decrease from the year 1992. However, the 2009 data indicate a slight (0.84%) increase in SCA_p compared with the period 2001 to 2008. Nevertheless, if the years 2000 and 2009 are included in the averaging process, then the average SCA_p in UIB during the last 10-year period is 5.85 ± 0.51% (1 σ) of the total basin area. Thus, the SCA_p in the UIB during the period 2000–2009 is 2.31% less than that in the year 1992 and 4.22% less than the SCA_p derived from the DCW (Mukhopadhyay and Dutta, 2010b; Mukhopadhyay, 2011).

As the basic purpose of this case study is to demonstrate the applicability of the mesoscopic hydrological model of a river basin presented in this chapter, the subsequent discussions are based on calculations made from the two sets of *SCA* data discussed above.

STREAM WATER AVAILABILITY. Figure 2.14 (a–e) shows the annual hydrographs at some of the key hydrological junctions of the UIB. In the present streamflow models of the UIB, a value of 0.75 is assumed for the catchment storage constant, K_c. The flows at the nearest downstream gauging stations are also plotted in the respective graphs. Most of the gauging stations are in the north-western part of the basin. The Pakistan Water and Power Development Authority (WPDA) collects streamflow data from these gauging stations (see Table 2.7), but it is not easy to obtain these data (D. Hashmi, Lahore, Pakistan, 2010, personal communication). However, Archer (2003) and Ali and De Boer (2007) obtained monthly flow data from the WPDA that were averages of nearly 40 years of records (1960–1998). These are the best

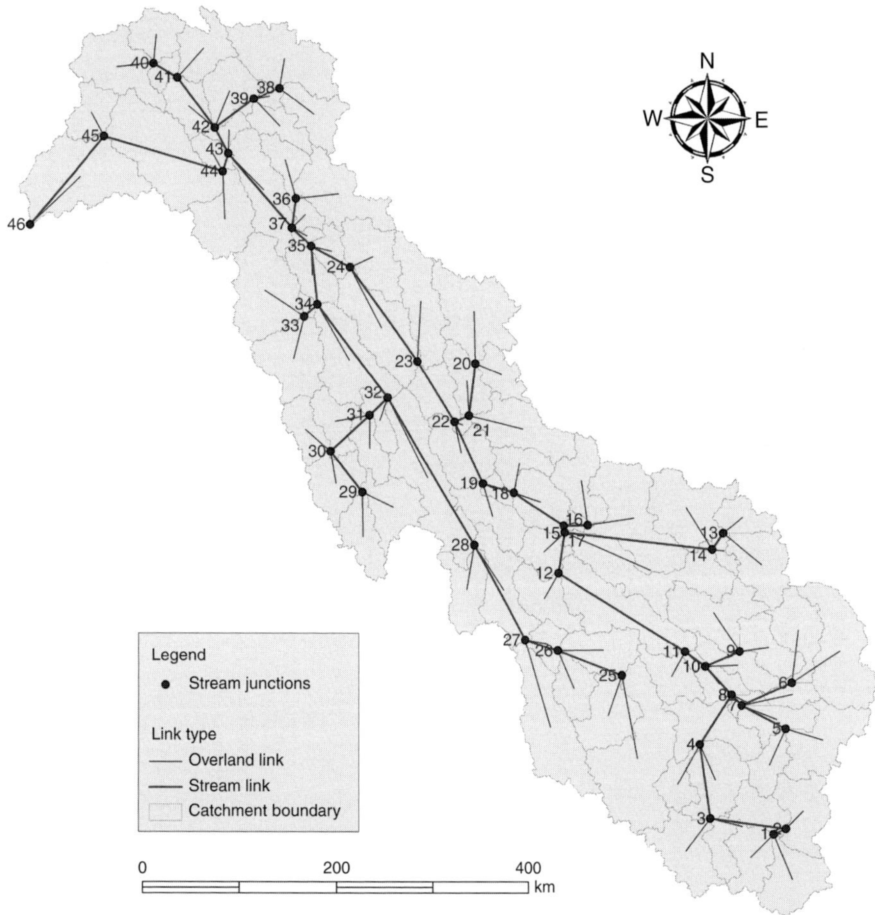

Fig. 2.10. The topological data model of the Upper Indus Basin (UIB) developed from the digital elevation model (DEM) and the channel and catchment delineations presented in this study. The junction numbers are marked.

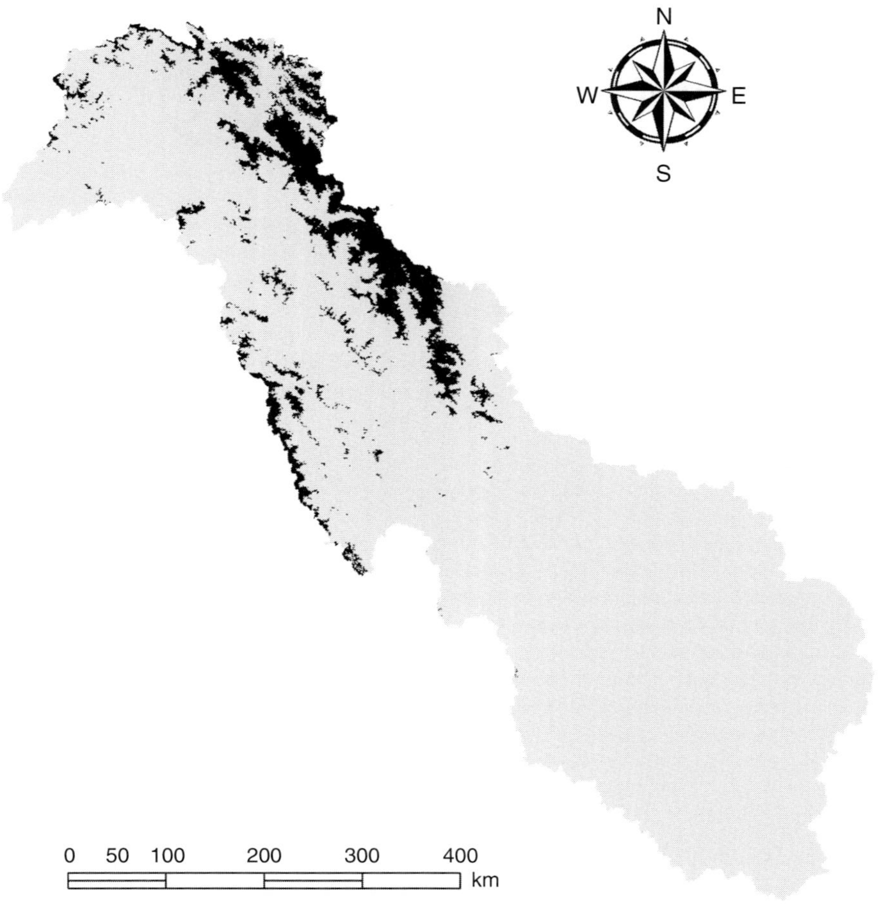

Fig. 2.11. Perennial snow- and ice-covered areas (shown in black) within the Upper Indus Basin (UIB) according to 1992 land cover data.

readily available data that show the general characteristics of monthly streamflows in the Upper Indus River within the region that is monitored by the WPDA. Archer (2003) cautioned that these data are at best of moderate quality owing to the difficulties in gauging mountainous streams and the procedures employed therein. Young and Hewitt (1990) also noted the doubtful quality of the gauge data owing to the practice that is in place for data collection, such as infrequent measurements during the low-flow seasons, manual measurements only during the daytime, etc. In spite of these inadequacies, the available data are used first to validate a model based on the land cover characteristics of 1992. The aim was not to achieve absolute accuracy for any particular hydrological junction for a particular month, but to ensure that reasonably realistic estimates of flows and the general pattern were generated by the model. From this exercise, two model parameters, namely z_f and B, were established for each of the months. The calculated flow patterns match with the reported patterns that show that the streamflows are low from the months of January to April, start to pick up in May with the advent of summer and reach very high flow regimes during July and August, which persist up to September, and then start to decline from October before the beginning of winter, and again become

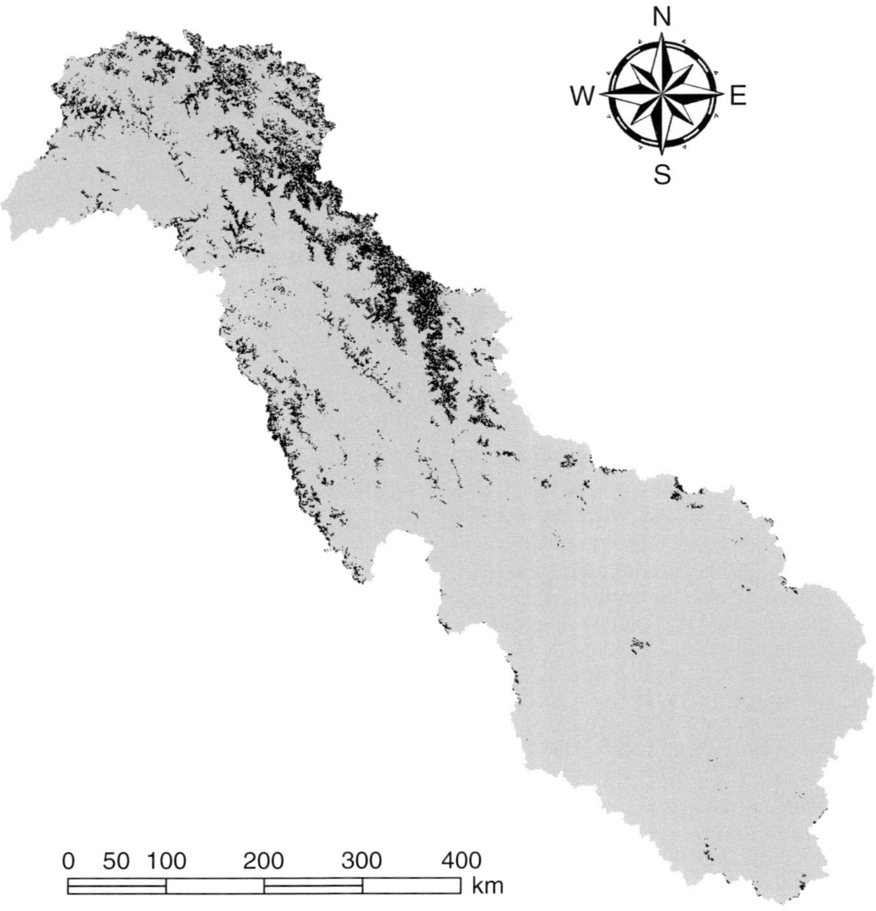

Fig. 2.12. Perennial snow-and ice-covered areas (shown in black) within the Upper Indus Basin (UIB) according to 2000 land cover data.

very low for the months of November and December. The available gauge data and the 1992 model consistently show that at all hydrological junctions, the highest volumetric flow rate occurs in July, followed by August, and that volumetric flow rates in June in most junctions are lower than those in August. Raina (2009) also noted that all Himalayan river basins exhibit peak river runoff from mid-July to mid-August. The close match between the observed average flows and the calculated monthly means attests to the overall validity of the model.

In the channel network, for the stem that originates to the north in the eastern part of the UIB and becomes Chusul Nala to the east of Leh and then joins the Shyok River at junction 22 (J22) where the latter makes a sharp turn in its flow directions from south-eastward to north-westward (Fig. 2.10), the river discharge is low throughout the year (maximum $642\,m^3\,s^{-1}$ in July). However, discharge sharply increases from junction 23 (J23) to junction 24 (J24) as shown in Fig. 2.14a. This is because of the large volumes of meltwater that a river like the Nubra carries from the Siachen Glacier[2] and adjoining snowfields in eastern Karakoram. The model indicates that streamflow occurs in the Shyok River through J24 at rates of 1236, 2094 and

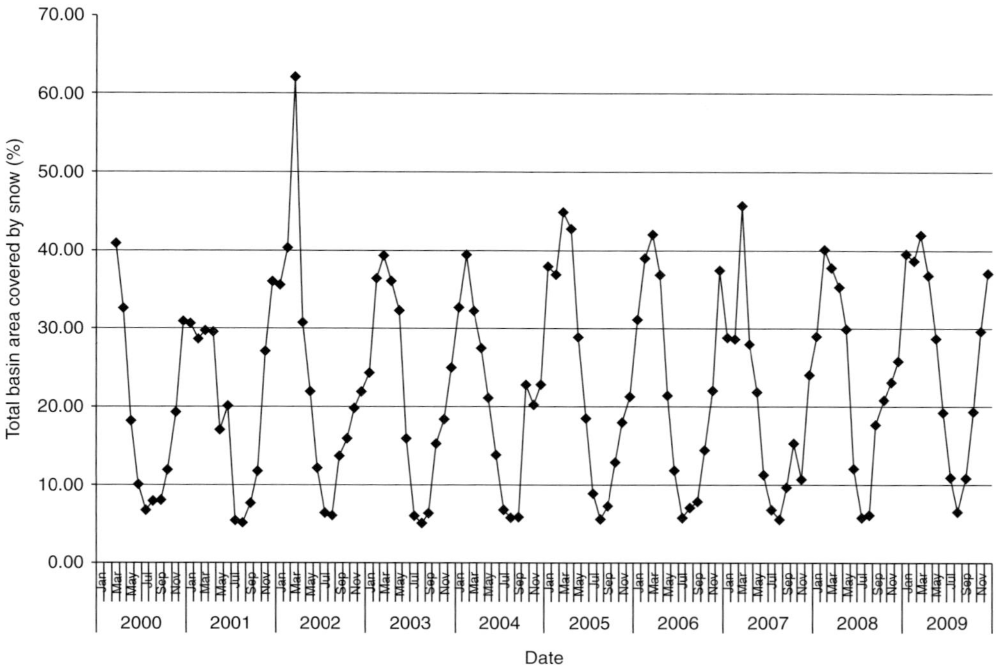

Fig. 2.13. Time series plots showing monthly total snow-and ice-covered area in the Upper Indus Basin (UIB) for the period March 2000 to December 2009, as derived from MODIS (Moderate Resolution Imaging Spectroradiometer) snow cover data products.

1988 m^3 s^{-1} during June, July and August, respectively.

For the main stem of the Upper Indus River, the streamflow at its confluence with Zanskar River at junction 32 (J32) reaches a maximum of 663 m^3 s^{-1} during July (Fig. 2.14b). However, at junction 34 (J34), upstream of the confluence of the Indus and the Shyok Rivers, the flow increases to 995, 1197 and 1102 m^3 s^{-1} during the months of June, July and August, respectively. At the confluence of the Indus and Shyok at junction 35 (J35), the flow reaches 1005 m^3 s^{-1} in May and attains a maximum of 3360 m^3 s^{-1} in July; it persists at up to 1252 m^3 s^{-1} in September (Fig. 2.14c).

Downstream of the confluence of the Indus and Shyok Rivers, the streamflow steadily increases (Fig. 2.14d–e). It reaches 1942 m^3 s^{-1} in May at the confluence of the Indus and Gilgit Rivers at junction 43 (J43) and peaks in July to volumetric flow rates of 6980 m^3 s^{-1}; it declines to 715 m^3 s^{-1} in October (Fig. 2.14d). At the outlet of the basin at junction 46 (J46), the calculated flow rates from May to October are 2083, 4871, 7273, 6852, 2740 and 761 m^3 s^{-1}, respectively (Fig. 2.14e).

DISCUSSION OF RESULTS OF THE UIB MODEL. The hydrological characteristics of the UIB were described by Young and Hewitt (1990, 1993) and Archer (2003). Archer's characterization of the hydrological regimes of the UIB was primarily derived from an investigation conducted on three watersheds drained by the Hunza, Astore and Khan Khwar Rivers (the catchments numbered from 1 to 16 and part of 17 shown in Fig. 2.8). He analysed historical records of rainfall, temperature and stream discharge spanning 12–39 years and came to the conclusion that runoff generated during a summer in the UIB can be classified into three contrasting regimes: (i) meltwater of glaciers and permanent snow, where summer runoff is predominantly controlled by the energy input during that summer; (ii) meltwater of seasonal snow, controlled by the volume of precipitation that occurred during

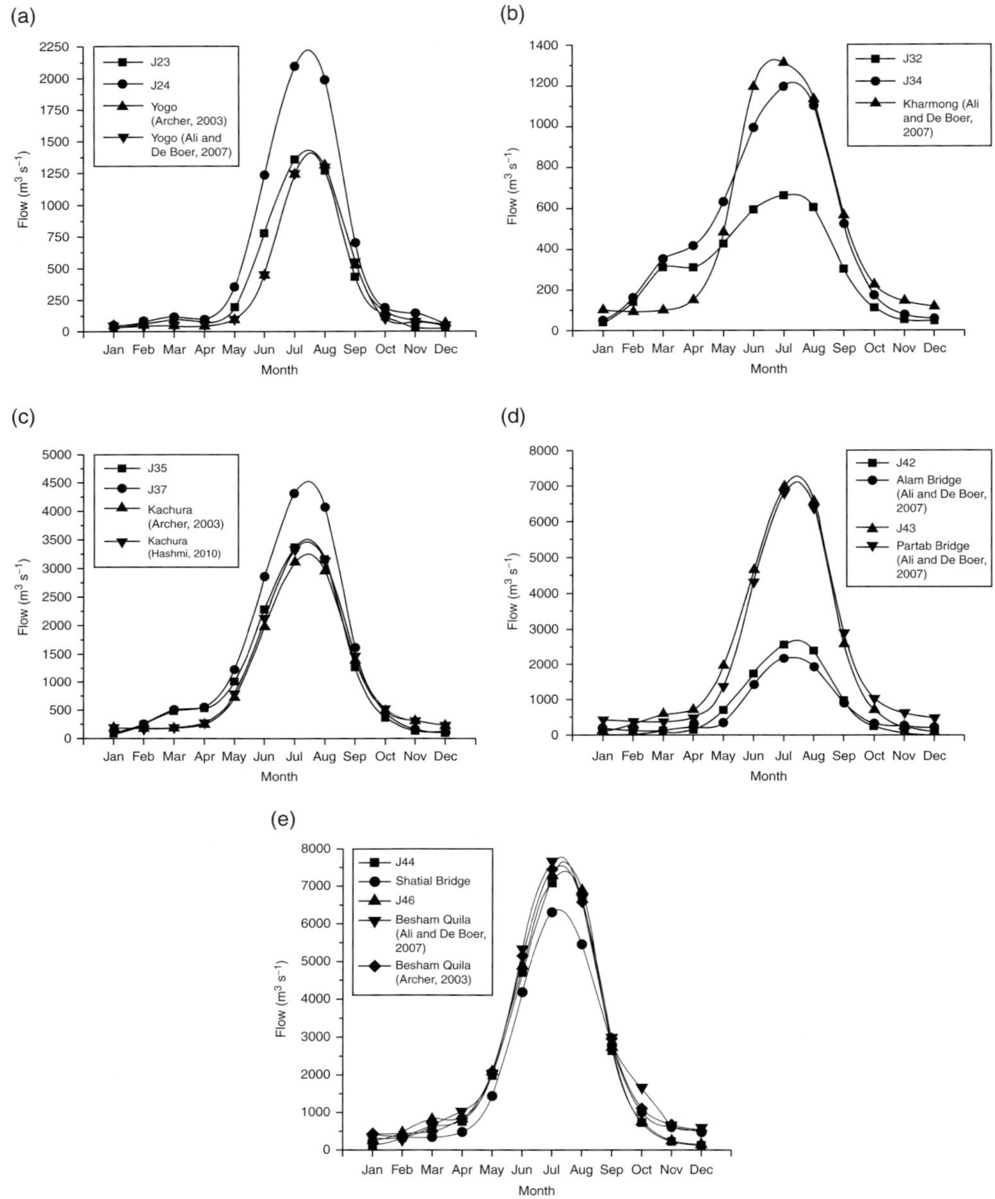

Fig. 2.14. Annual hydrographs at key hydrological junctions of the Upper Indus Basin (UIB) with land cover characteristics of the year 1992. The land cover characteristics include perennial snow and glacial covers. The reported flows at the gauging stations (named, some with data sources) close to the numbered junctions (J) are plotted to show the close match between the calculated flows and the observed average flows. (a) Streamflow characteristics along the Shyok River; (b) Streamflow characteristics along the Indus River before its confluence with the Shyok River; (c) Streamflow characteristics of the Indus River from the point of its confluence with the Shyok River to the point of its confluence with the Shigar River; (d) Streamflow characteristics of the Indus River from its confluence with the Shigar River to its confluence with the Gilgit River as well as the flow characteristics of the Gilgit River at its confluence with the Hunza River; (e) Streamflow characteristics of the Indus River for its confluence with the Astore River to upstream of Tarbela Reservoir.

the preceding winter and spring; and (iii) rainwater, controlled by precipitation in the current season.

In contrast, Young and Hewitt (1990, 1993) looked at the broader basin and concluded that essentially there were two hydrological regimes in the UIB, namely, the glaciated and non-glaciated contributions to streamflows. According to these authors, the non-glaciated component constitutes the liquid precipitation and seasonal snow melt at elevations below 3000 m, and the volumetric contribution of this component is negligible as a result of small amounts of rainfall, a very high evaporation rate, low moisture availability to the ground, and the dry and hot atmosphere that prevails in a semi-desert condition at elevations below 2500 m. However, from elevations 2500–3000 m to 4800–5000 m, a zone of ablation of glaciated terrain exists and this glaciated component is the principal contributor to streamflows.

The results obtained from the model runs presented above generally validate the observations made by Young and Hewitt (1990, 1993). We also note that the streamflows during the high flow season (May to September, with peak flows in July) chiefly originate from snowmelt and that most of the precipitation as rainfall is lost by evaporation. Contributions to streamflows from seasonal rainfall principally come in March and are mostly from the western Himalayan ranges near the downstream sections of the basin. Thus the SWAM, as presented in this work, reflects two hydrological regimes, namely, the meltwater from the permanent glacial and snow fields as the principal component, and rainfall as the component contributing negligible flows to the streams. The third regime of Archer (2003), originating from the seasonal snow melt, is lost quickly to evaporation and infiltration (ground storage).

IMPACT OF CLIMATE CHANGE. The SWAM presented here allows for assessment of the impact of reduction in snow-covered areas in the UIB on streamflow rates. As noted above, flow in the Indus River significantly increases after its confluence with the Shyok River. This is also due to the fact that Nubra River, the main tributary of Shyok River, originates at the Siachen Glacier and drains the snowfields around this glacial field. Downstream of the confluence of the Indus and the Shyok Rivers the principal contributors to the increase in the river discharge are rivers such as the Shigar, Shimshal and Hunza, which originate from the Karakoram Range to the north and the Gilgit River from Hindu Kush to the north-west. This clearly shows the importance of the glacial melt and snow melt in the flow of the Indus River. Thus, any effect of global or regional climate change on the SCA_p in the Karakoram will affect the flows in the Indus and the stream water availability within the UIB.

Figure 2.15 (a–e) shows the annual hydrographs at the hydrological junctions along the three stems of the major river system in the UIB, assuming year 2000 snow and land cover characteristics. The 2000 land cover data indicate that there has been a 1.2% decrease in SCA_p in the UIB since 1992 (Tables 2.4 and 2.5). The calculated hydrographs show that only a 1.2% decrease in SCA_p translates into reductions of flows by various proportions for all months at all of the hydrological junctions. Compared with the 1992 estimates, peak (July) junction flows at J23 and J24 originating from eastern Karakoram are 8–20% less during the high-flow season in 2000. Flow reductions for the melting season are in the order of 16–23% at J32 and J34 on the main stem of the Indus River draining the northern slopes of the Greater Himalayas, before its confluence with Shyok River. In the central part of the basin, at J35 and J37, summer peak flow reductions are in the order of 16–21%. At J42, where the flows originate from western Karakoram and Hindu Kush, the estimated flows have decreased by 11–20% in 2000 compared with those estimates made for the year 1992. In the downstream sections of the basin, at J43 and J44, such decreases are in the order of 10–16%. At the outlet of the basin (J46), the maximum decrease in peak (July) flow is 11%. Comparisons of these various proportions of decrease in flows during the high-flow or melting season indicate that the loss of SCA_p has been greater in the Himalayan ranges than that in the Karakoram and Hindu Kush ranges.

The decline in streamflows due to reduction in perennial snow- and ice-covered areas within the UIB can potentially have adverse effects on various sectors, such as

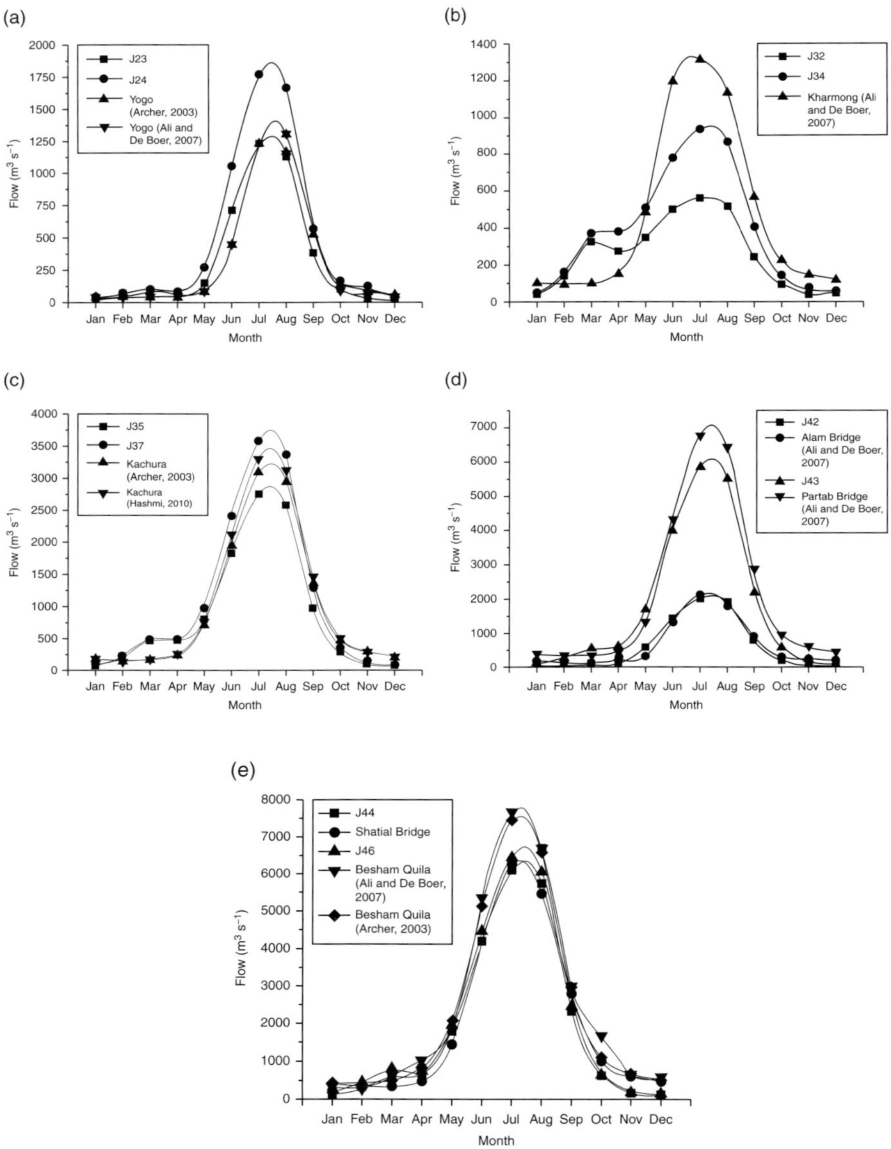

Fig. 2.15. Annual hydrographs at key hydrological junctions of the Upper Indus Basin (UIB) with land cover characteristics of the year 2000. The land cover characteristics include perennial snow and glacial covers. The reported flows at the gauging stations (named, some with data sources) close to the numbered junctions (J) are plotted to show the close match between the calculated flows and the observed average flows. Compare the hydrographs with those shown in Fig. 2.14 and note the decrease in discharge rates, especially during high-flow seasons, at various hydrological junctions. (a) Streamflow characteristics along the Shyok River; (b) Streamflow characteristics along the Indus River before its confluence with the Shyok River; (c) Streamflow characteristics of Indus River from the point of its confluence with the Shyok River to the point of its confluence with the Shigar River; (d) Streamflow characteristics of the Indus River from its confluence with the Shigar River to its confluence with the Gilgit River as well as the flow characteristics of the Gilgit River at its confluence with the Hunza River; and (e) Streamflow characteristics of the Indus River for its confluence with Astore River to upstream of the Tarbela reservoir.

agriculture, water supply and hydropower generation. Mukhopadhyay and Dutta (2010b) and Mukhopadhyay (2011) presented further details on the changes in monthly streamflow characteristics within the UIB that had occurred during the period of 2000–2009 from a baseline condition of 1992, as a result of decrease in *SCA* within the basin. In this analysis, monthly snow- and ice-covered areas for the period 2000–2009 were estimated from MODIS-derived snow cover data products described above.

Conclusions

Concluding remarks on the model

Before the SRTM project, for almost a decade, analyses of river networks on a small cartographic scale were mostly conducted with the GTOPO30 digital elevation data sets. However, the 1 km spatial resolution was a serious limitation for its use in the analysis and modelling of detailed hydrology and geomorphology of river basins on a regional scale. Researchers in these areas relied on local maps for topography. Digitization or photogrammetry is not only a time-consuming and costly procedure to produce high-resolution DEMs, it is also difficult to develop systematic and manageable topological models (TMs) of large river basins from these DEMs. This perhaps has been a great impediment to the development of models for important river basins outside developed countries. Hydrological models, such as seasonal water availability and long-term water balance models, are needed for all of the major river basins of the world for the purposes of integrated water resources and environmental management (IWREM), understanding the consequences of global climate change on the hydrological cycle, and economic developments in parity with environmental preservation and sustainability. The current availability of DEMs with high spatial resolutions is likely to change the way in which hydrological research can be performed and applied, bringing local catchment- and watershed-scale modelling into the realm of global applicability. At the same time, it should be recognized that there are elevation errors in the present format of the SRTM data, and continued efforts should be undertaken to refine and filter the data to minimize the effects of these errors. Nonetheless, at present, the SRTM data provide the best means to systematize the topological models of large river basins, where other sources of digital elevation data are absent.

With high-resolution DEMs, river-basin scale hydrological models also must take advantage of the present generation of global climatic and other data sets that are available in point, polygonal and gridded format. In this respect, the other valuable data sets for the climatic variables of regional river basins with sparse data are the Matsuura and Willmott database of global temperature and precipitation from 1900 to 2008 on a year-by-year, month-by-month basis, and the SRB database of short-wave and long-wave radiation from 1983 to 2007 on a year-by-year and month-by-month basis. It is expected that these databases will evolve over time to go through further refinements and additional temporal coverage. In addition, the global land cover and soil cover data are important for deriving several model parameters. When a TM derived from the SRTM project data is combined with the climatic data from these data sets and other global data sets containing land and soil cover information, SWAMs of large regional river basins can be developed. The hierarchical framework in the TM identifies the points on the streams where gauging stations should be placed; which streams contribute to and to what extent they contribute to the flow through the main stems; and the flow through which overland links should be managed for proper water distribution, diversion, irrigation, flood control and hydroelectric power generation.

Another great hindrance hitherto faced by hydrological researchers engaged in river-basin scale modelling is the lack of river discharge data for model validation and calibration. However, this obstacle is gradually being removed as global river discharge data are obtained through satellite observations and are disseminated over the Internet (e.g. Dartmouth Flood Observatory).

Finally, it should be also noted that most of the predictions in ungauged basins (PUB) have been based on statistical regression of observed peak flow statistics in a geographical region without any knowledge of the physical processes that transform rainfall to runoff. In contrast, the approach presented in this work is largely physically based and can easily be applied for PUB.

Future outlook

Although much progress has been achieved in hydrological modelling, there is a greater road ahead. A basic question is: What modelling technology is better? Because of the confusion surrounding models and their applicability, the modelling technology developed decades ago is still in use in many parts of the world. This state of affairs is partly the result of lack of consensus as to the superiority of one type of technology over another. Also, it has not been possible to develop physically based models in a true sense and define their limitations. Thus, it is not always clear when and where to use which type of model.

Hydrological models will have to be described in simple terms such that the interpretation of their results would not tax the ability of the user. They must be designed to serve a practical end, and their constituency is one of users. After all, hydrological models are to be used, not to be confined to academic shelves. Thus, model building will have to gravitate around the central theme of the eventual practical use of models in IWREM.

If hydrological models are to become practical tools, then they must be built on a common platform, e.g. a widely used GIS software. Furthermore, mesoscopic models should require a minimal number of parameters for calibration. They will need to assess errors and determine how they propagate, define the reliability with which they accomplish their intended functions, and require the end user to possess only a minimal amount of hydrological training. Moreover, the models will have to 'learn' from the user as well as from empirical experience. Many of these functions can be performed by the use of expert systems in hydrological modelling. Usually, the user is interested in what a model yields and its accuracy, and how easy it is to use, and not in the biology, chemistry, physics, geology and hydrology that it is based on. Although much progress has been made in hydrological modelling at a small cartographic scale, there is still a long way to go before the models become 'household' tools for decision and policy makers and lending or funding authorities.

Distributed models require large quantities of data which can be stored, retrieved, managed and manipulated with the use of GIS and a database management system (DBMS). With the use of remote sensing, radar and satellite technology, our ability to observe data over large and inaccessible areas and to map these areas spatially is vastly improved, making it possible to develop truly distributed models for both gauged and ungauged watersheds. This is possible because of the literally unlimited computing capability available these days, which will be even more available in the future.

Applications of hydrological models to IWREM will grow in the future. The models will be required to be practical tools – readily usable in planning and decision making. They will have to be interfaced with economic, social, political, administrative and judicial models. Thus, hydrological models will become a component in the larger management strategy. Furthermore, these models will become more global, not only in the sense of spatial scale but also in the sense of hydrological details.

The future of hydrological models will be shaped by various factors: increasing societal demand for IWREM; growing need for globalization by the incorporation of geological, biological, chemical and physical aspects of the hydrological cycle; assessment of the impact of climate change; rapid advances in remote sensing and satellite technology, GIS, DBMS and expert systems; an enhanced role of models in planning and decision making; mounting pressure on the transformation of models to user-friendly forms; and clearer statements of the reliability and risk associated with model results.

With growing technologies triggered by the information revolution, remote sensing, satellite technology, GIS, expert systems, advanced statistical and mathematical analysis packages (for error analysis, sensitivity analysis and reliability analysis), data-mining technologies, visual graphics and DBMS, there is real potential to take hydrological models to the next level of sophistication and integrate them with environmental and ecological management and other process models. Hydrological models will have to embrace the rapid advances occurring in these technologies.

Acknowledgements

The authors are grateful to Dr Paul Stackhouse of the Langley Research Center of NASA for providing the short-wave and long-wave radiation data prior to their release to the public, making the data sets available in requested formats, and for providing valuable information about this very important and high-quality data set. Dr Kenji Matsuura of the University of Delaware supplied important information on the University of Delaware data sets and some key references. Dr Jim Dyer of Ohio University extended much help in generation of some of the codes for the soil moisture accounting procedure. The senior author completed most of the work presented in this article during his tenure as the Program Manager for Water Resources Engineering Program at HDR Engineering, Inc. (HDR). He is very thankful to HDR for providing the necessary support and logistics to continue research and developmental activities alongside consulting engineering.

Notes

[1] As is common practice, physical variables (e.g. temperature, amount of snow melt, etc.) that are actually used in the model calculations are presented in italics in the text and in equations. Abstract representations of the operators (hydrological processes) and inputs/outputs to/from the models (parameters and variables) are presented as non-italic (roman) characters; these include, ϕ, Γ, o and O.

[2] The Siachen Glacier is the second longest glacier in the world outside the polar region. The entire glacier is controlled by India but claimed by Pakistan (Wikipedia, 2009). Available from: http://en-wikipedia.org/wiki/Siachen_Glacier (accessed July 2011).

References

Ahmed, S. and Joyia, M.F. (2003) *Northern Areas Sustainable Development Strategy* (NASDS)., NASDS Background Paper, Water IUCN (International Union for Conservation of Nature and Natural Resources) Pakistan, Northern Areas Programme, Gilgit, Pakistan.

Ali, K. and De Boer, D.H. (2007) Spatial patterns and variation of suspended sediment yield in the upper Indus River basin, northern Pakistan. *Journal of Hydrology* 334, 368–387.

Alsdorf, D., Lettenmaier, C., Vörösmarty, C. and NASA Surface Water Working Group (2003) The need for global satellite-based observations of terrestrial surface waters. *EOS, Transactions, American Geophysical Union* 84, 269, 275–276.

Archer, D. (2003) Contrasting hydrological regimes in the upper Indus basin. *Journal of Hydrology* 274, 198–210.

Archer, D. and Fowler, H.J. (2004) Spatial and temporal variations in precipitation in the Upper Indus Basin, global teleconnections and hydrological implications. *Hydrology and Earth System Sciences* 8, 47–61.

Arnell, N.W. (1999) A simple water balance model for the simulation of streamflow over a large geographic domain. *Journal of Hydrology* 217, 314–335.

Arnold, J.G. and Fohrer, N. (2005) SWAT2000: current capabilities and research opportunities in applied watershed modeling. *Hydrological Processes* 19, 563–572.

Arnold, J.G., Srinivasan, R., Muttiah, R.S. and Williams, J.R. (1998) Large area hydrologic modeling and assessment Part I: model development. *Journal of the American Water Resources Association* 34, 73–89.

Arnold, J.G., Srinivasan, R., Muttiah, R.S. and Allen, P.M. (1999) Continental scale simulation of the hydrologic balance. *Journal of the American Water Resources Association* 35, 1037–1051.

BADC (British Atmospheric Data Centre) (2008) University of East Anglia Climate Research Unit (CRU) Datasets. Available from http://badc.nerc.ac.uk/data/cru (accessed 7 February 2011).

Bengtsson, L. (1982) The importance of refreezing on the diurnal snowmelt cycle with application to a northern Swedish catchment. *Nordic Hydrology* 13, 1–12.

Bengtsson, L. (1986) Snowmelt simulation models in relation to space and time. *IAHS Publication* No. 155, 115–123.

Bloom, S., Silva, A. da, Dee, D., Bosilovich, M., Chern, J.-D., Pawson, S., Schubert, S., Sienkiewicz, M., Stajner, I., Tan, W.-W. and Wu, M.-L. (2005) *Documentation and Validation of the Goddard Earth Observing System (GEOS) Data Assimilation System – Version 4.* Technical Report Series on Global Modeling and Data Assimilation NASA/TM-104606, Vol. 26. NASA (National Aeronautics and Space Administration) Global Modeling and Assimilation Office, Maryland.

Brakenridge, G.R., Nghiem, S.V., Anderson, E. and Chen, S. (2005) Space-based measurement of river runoff. *EOS, Transactions, American Geophysical Union* 86, 185, 188.

Brakenridge, G.R., Nghiem, S.V., Anderson, E. and Mic, R. (2007) Orbital microwave measurement of river discharge and ice status. *Water Resources Research* 43, W04404, doi: 10.1029/2006WR005238.

Brubaker, K., Rangbo, A. and Kustas, W. (1996) Incorporating radiation inputs into the snowmelt runoff model. *Hydrological Processes* 10, 1329–1343.

CGIAR-CSI (Consultative Group for International Agriculture Research Consortium for Spatial Information) (2008) The CGIAR Consortium for Spatial Information: Applying GeoSpatial Science for a Sustainable Future. Available at: http://srtm.csi.cgiar.org/ (accessed 4 February 2011).

Clark, C.O. (1945) Storage and the unit hydrograph. *Transactions of the American Society of Civil Engineers* 110, 1419–1488.

Coe, M.T. (2000) Modeling terrestrial hydrological systems at the continental scale: testing the accuracy of an atmospheric GCM. *Journal of Climate* 13, 686–604.

Doelling, D. R., Young, D.F., Wielicki, B.A., Wong, T. and Keyes, D.F. (2006) The newly released 5-year Terra-based monthly CERES radiative flux and cloud product. *12th Conference on Atmospheric Radiation*, 10–14 July 2006, Madison, Wisconsin, American Meteorological Society, Boston, Massachusetts.

ESRI (Environmental Systems Research Institute) (1992) *Arc World 1:3 Mio. Continental Coverage.* ESRI, Redlands, California.

ESRI (1993) *Digital Chart of the World 1:1 Mio.* ESRI, Redlands, California.

ESRI (2006) *ESRI Data and Maps 2006: An ESRI White Paper.* ESRI, Redlands, California.

Falorni, G., Teles, V., Vivoni, E.R., Bras, R.L. and Amaratunga, K. (2005) Analysis and characterization of the vertical accuracy of digital elevation models from the shuttle radar topography mission. *Journal of Geophysical Research – Earth Surface* 110(F2), F02005. doi:10.1029/2003JF000113.

FAO (Food and Agriculture Organization of the United Nations) (1998) *World Reference Base for Soil Resources.* World Soil Resources Report 84. FAO, Rome.

FAO/UNESCO (2003) WRB (World Reference Base for Soil Resources) Map of World Soil Resources, 1:25,000,000 – January 2003. Available from: http://www.fao.org/AG/agL/agll/wrb/soilres.stm (accessed 7 February 2011).

Farr, T.G. and Kobrick, M. (2000) Shuttle radar topography mission produces a wealth of data. *EOS, Transactions American Geophysical Union* 81, 583–585.

Fekete, B.M., Vörösmarty, C.J., Roads, H.O. and Willmott, C.J. (2004) Uncertainties in precipitation and their impacts on runoff estimates. *Journal of Climate* 17, 294–304.

Finlayson, D.P. and Montgomery, D.R. (2003) Modeling large-scale fluvial erosion in geographic information systems. *Geomorphology* 53, 147–164.

Fu, Q., Liou, K.N., Cribb, M.C., Charlock, T.P. and Grossman, A. (1997) Multiple scattering parameterization in thermal infrared radiative transfer. *Journal of Atmospheric Sciences* 54, 2799–2812.

Gesch, D., Verdin, K. and Greenlee, S. (1999) New land surface digital elevation model covers the Earth, *EOS, Transactions, American Geophysical Union* 80, 69–70.

Gesch, D., Farr, T., Slater, J., Muller, J.P. and Cook, S. (2006) New products from the Shuttle Radar Topography Mission. *EOS, Transactions, American Geophysical Union* 87, 174.

Gray, D.M. and Prowse, T.D. (1992) Snow and floating ice. In: Maidment, D.R. (ed.) *Handbook of Hydrology.* McGraw-Hill, New York, pp. 7.1–7.58.

Grohman, G., Kroenung, G. and Strebeck, J. (2006) Filling SRTM voids: the delta surface fill model. *Photogrammetric Engineering and Remote Sensing* 72, 213–216.

Gupta, S.K., Darnell, W.L. and Wilber, A.C. (1992) A parameterization for longwave surface radiation from satellite data: recent improvements. *Journal of Applied Meteorology* 31, 1361–1367.

Gupta, S.K., Kratz, D.P., Stackhouse, P.W. Jr and Wilber, A.C. (2001) *The Langley Parameterized Shortwave Algorithm (LPSA) for Surface Radiation Budget Studies, Version 1.0.* Report NASA/TP-2001-211272, NASA, Langley Research Center, Hampton, Virginia.

Gupta, S.K., Stackhouse, P.W. Jr, Cox, S.J., Mikovitz, J.C. and Zhang, T. (2006) 22-year surface radiation budget data set. *GEWEX News* 16(4, November), 12–13.

Hahn, C.J. and Warren, S.G. (1999) *Extended Edited Synoptic Cloud Reports from Ships and Land Stations over the Globe, 1952–1996.* ORNL/CDIAC-123, NDP-026C, CDIAC, ORNL, US Department of Energy, Oak Ridge, Tennessee.

Hall, D.K. and Riggs, G.A. (2007) Accuracy assessment of the MODIS snow products. *Hydrological Processes* 21, 1534–1547.

Hashmi, D. (2010) *Influence of Climate Change on Upper Indus Basin Rivers.* Available at: http://www.tpe.ac.cn/Resources/PPT/03.pdf (accessed 14 May 2010).

Hijmans, R.J., Cameron, S.E., Parra, J.L., Jones, P.G. and Jarvis, A. (2005) Very high resolution interpolated climate surfaces for global land areas. *International Journal of Climatology* 25, 1965–1978.

Jain, S.K., Goswami, A. and Saraf, A.K. (2008) Determination of land surface temperature and its lapse rate in the Satluj River basin using NOAA data. *International Journal of Remote Sensing* 29, 3091–3103.

Jakeman, A.J. and Hornberger, G.M. (1993) How much complexity is warranted in a rainfall-runoff model? *Water Resources Research* 29, 2637–2649.

Jarvis, A. (2004) SRTM Data Processing Methodology. Available at: http://srtm.csi.cgiar.org/SRTMdataProcessingMethodology.asp (accessed 4 February 2011).

Jensen, M.E., Burman, R.D. and Allen, R.G. (eds) (1990) *Evapotranspiration and Irrigation Water Requirements.* American Society of Civil Engineers, New York.

Jiang, T., Chen, Y.D., Xu, C.-Y., Chen, X., Chen, X. and Singh, V.P. (2007) Comparison of hydrological impacts of climate change simulated by six hydrological models in the Dongjiang Basin, South China. *Journal of Hydrology* 336, 316–333.

Jones, P.D. and Moberg, A. (2003) Hemispheric and large-scale surface air temperature variations; an extensive revision and an update to 2001. *Journal of Climate* 16, 206–223.

Koster, R.D., Fekete, B.M., Huffman, G.J. and Stackhouse, P.W. Jr (2006) Revisiting a hydrological analysis framework with International Satellite Land Surface Climatology Project Initiative 2 rainfall, net radiation, and runoff fields. *Journal of Geophysical Research* 111, D22S05, doi:10.1029/2006JD007182.

Kustas, W.P., Rango, A. and Uijlenhoet, R. (1994) A simple energy budget algorithm for the snowmelt runoff model. *Water Resources Research* 30, 1515–1527.

Labadie, J. (2005) MODSIM: river basin management decision support system. In: Singh, V.P. and Frevert, D.K. (eds) *Watershed Models.* CRC Press, Boca Raton, Florida, pp. 569–591.

Lafitte, R. (2007) The Indus Waters Treaty 1960. Baglihar Hydroelectric Plant. Expert Determination on Points of Difference Referred by the Government of Pakistan under the Provisions of the Indus Waters Treaty. Governments of Pakistan and India. Available at: http://siteresources.worldbank.org/SOUTHASIAEXT/Resources/223546-1171996340255/BagliharSummary.pdf (accessed 4 February 2011).

Legates, D.R. and Willmott, C.J. (1990) Mean seasonal and spatial variability in gauge-corrected, global precipitation. *International Journal of Climatology* 10, 111–127.

Lehner, B., Verdin, K. and Jarvis, A. (2006) HydroSHEDS, Technical Documentation, Version 1.0. World Wildlife Fund US, Washington, DC. Available at: http://hydrosheds.cr.usgs.gov/HydroSHEDS_TechDoc_v10.doc (accessed 4 February 2011).

Li, X. and Williams, M. (2008) Snowmelt runoff modeling in an arid mountain watershed, Tarim Basin, China. *Hydrological Processes* 22, 3931–3940.

Liang, X., Lettenmaier, D.P., Wood, E. and Burges, S. (1994) A simple hydrologically based model of land surface water and energy fluxes for general circulation models. *Journal of Geophysical Research* 99 (D7), 14415–14428.

Lin, B., Stackhouse, P.W. Jr, Minnis, P., Wielicki, B.A., Hu, Y., Sun, W., Fan, T.-F. and Hinkelman, L.M. (2008) Assessment of global annual atmospheric energy balance from satellite observations. *Journal of Geophysical Research* 113, D16114, doi:10.1029/2008JD009869.

Littlewood, I.G., Croke, B.F.W., Jakeman, A.J. and Sivapalan, M. (2003) The role of 'top-down' modeling for Prediction in Ungauged Basins (PUB). *Hydrological Processes* 17, 1673–1679.

Liu, Z. and Todini, E. (2002) Towards a comprehensive physically-based rainfall-runoff model. *Hydrology and Earth System Sciences* 6, 859–881.

Maidment, D. (2002) *Arc hydro GIS for Water Resources*. ESRI, Redlands, California
Male, D.H. and Gray, D.M. (1981) Snow cover ablation and runoff. In: Gray, D.M. and Male, D H. (eds) *Handbook of Snow*. Pergamon Press, Elmsford, New York.
Martinec, J. (1975) Snowmelt-runoff model for stream flow forecasts. *Nordic Hydrology* 6, 145–154.
Martinec, J (1989) Hour-to-hour snowmelt rates and lysimeter outflow during an entire ablation period. In: *Snow Cover and Glacier Variation, Proceedings of the Baltimore Symposium, Maryland, May 1989. IAHS Publication* No.193, 19–28.
Martinec, J., Rango, A. and Roberts, R (1994) *Snowmelt Runoff Model (SRM) User's Manual, Updated Version 3.2*. Geographica Bernensia P29 (Geographie für die Praxis), Department of Geography, University of Berne.
Martinec, J., Rango, A. and Roberts, R. (2008) *Snowmelt Runoff Model (SRM) User's Manual, Updated Edition for Windows, WinSRM Version 1.11* (Gómez-Landesa, E. and Bleiweiss, M.P., eds). USDA (United States Department of Agriculture) Jornada Experimental Range, New Mexico State University, Las Cruces, New Mexico. Available at: http://aces.nmsu.edu/pubs/research/weather_climate/SRMSpecRep100.pdf (accessed 4 February 2011).
Mather, J.R. (1974) *Climatology: Fundamentals and Applications*. McGraw-Hill, New York.
Matsuura, K. and Willmott, C.J. (2009a) Terrestrial air temperature: 1900–2008 gridded monthly time series. Available at: http://climate.geog.udel.edu/~climate/ (accessed 4 February 2011).
Matsuura, K. and Willmott, C.J (2009b) Terrestrial precipitation: 1900–2008 gridded monthly time series. Available at: http://climate.geog.udel.edu/~climate/ (accessed 4 February 2011).
McDonnell, P.W. Jr (1979) *Introduction to Map Projections*. Marcel Dekker, New York.
Miller, J., Russell, G. and Caliri, G. (1994) Continental scale river flow in climate models. *Journal of Climate* 7, 914–928.
Minhas, B.S., Parikh, K.W. and Srinivasan, T.N. (1974) Toward the structure of a production function for wheat yields with dated input of irrigation water. *Water Resources Research* 10, 383–393.
Mishra, A. and Hata, T. (2006) A grid-based runoff generation and flow routing model for the Upper Blue Nile basin. *Hydrological Sciences Journal* 51, 191–206.
Mitchell, T.D. and Jones, P.D. (2005) An improved method of constructing a database of monthly climate observations and associated high-resolution grids. *International Journal of Climatology* 25, 693–712.
Mukhopadhyay, B (2011) Detection of dual effects of degradation of perennial snow and ice covers on the hydrologic regime of a Himalayan river basin by stream water availability modeling. *Journal of Hydrology* (in press).
Mukhopadhyay, B. and Dutta, A. (2010a) A stream water availability model of Upper Indus Basin based on a topologic model and global climatic datasets. *Water Resources Management* 24, 4403–4443.
Mukhopadhyay, B. and Dutta, A. (2010b) Signal of climate change within the cryosphere of Upper Indus Basin and its hydrologic outcome. Presentation at *Hydrology Conference 2010: The Changing Physical and Social Environment: Hydrologic Impacts and Feedbacks*, 11–13 October 2010, San Diego, California (available from first author upon request).
Nachtergaele, F., van Velthuizen, H. and Verelst, L. (Coordinators) (2009) *Harmonized World Soil Database, Version 1.1, March 2009*. FAO/IIASA/ISRIC/ISSCAS/JRC. FAO, Rome and IIASA (International Institute for Applied Systems Analysis), Laxenburg, Austria Available at: http://www.iiasa.ac.at/Research/LUC/External-World-soil-database/HWSD_Documentation.pdf (accessed 4 February 2011)
New, M., Hulme, M. and Jones, P.D. (2000) Representing twentieth-century space-time climate variability. Part II: Development of 1901–1996 monthly grids of terrestrial surface climate. *Journal of Climate* 13, 2217–2238.
New, M., Lister, D., Hulme, M. and Makin, I. (2002) A high-resolution data set of surface climate over global land areas. *Climate Research* 21, 1–25.
Nicholson, S.E. (2001) A semi-quantitative, regional precipitation data set for studying African climates of the nineteenth century, Part I. Overview of the data set. *Climatic Change* 50, 317–353.
NSIDC (National Snow and Ice Data Center) (2011) MODIS Data. Available at: http://nsidc.org/data/modis/index.html (accessed 7 February 2011).
Olivera, F. and Raina, R. (2003) Development of large scale gridded river networks from vector stream data. *Journal of American Water Resources Association* 39, 1235–1248.
Olivera, F., Furnans, J., Maidment, D., Djokic, D. and Ye, Z. (2002a) Drainage systems. In: Maidment, D. (ed.) *Arc hydro GIS for Water Resources*. ESRI. Redlands, California, pp. 55–86.

Olivera, F., Lear, M.S., Famiglietti, J.S. and Asante, K. (2002b) Extracting low-resolution river networks from high-resolution digital elevation model. *Water Resources Research* 38, 1231–1238.

Peterson, T.C. and Vose, R.S. (1997) An overview of the Global Historical Climatology Network temperature database. *Bulletin of the American Meteorological Society* 78, 2837–2849.

Peterson, T.C., Vose, R., Schmoyer, R. and Razuvaev, V. (1998) Global Historical Climatology Network (GHCN) quality control of monthly temperature data. *International Journal of Climatology* 18, 1169–1179.

Pinker, R.T. and Laszlo, I. (1992) Modeling surface solar irradiance for satellite applications on a global scale. *Journal of Applied Meteorology* 31, 194–211.

Poveda, G., Velez, J., Mesa, O., Cuartas, A., Barco, J., Mantilla, R., Mejia, J.F., Hoyos, C.D., Ramirez, J.M., Ceballos, L.I., Zuluaga, M.D., Arias, P.A., Botero, B.A., Montoya, M.I., Giraldo, J.D. and Quevedo, D.I. (2007) Linking long-term water balances and statistical scaling to estimate river flows along the drainage network of Columbia. *Journal of Hydrologic Engineering* 12, 4–13.

Priestley, C.B.H. and Taylor, R.J. (1972) On the assessment of surface heat flux and evaporation using large scale parameters. *Monthly Weather Review* 100, 81–92.

Raina, V.K. (2009) *Himalayan Glaciers: A State-of-Art Review of Glacial Studies, Glacial Retreat, and Climate Change*. MoEF Discussion Paper, Ministry of Environment and Forests, Government of India, New Delhi.

Rawls, W.J., Brakensiek, D.L. and Miller, N. (1983) Green-Ampt infiltration parameters from soil data. *Journal of Hydraulic Engineering* 109, 62–70.

Reuter, H.I., Nelson, A. and Jarvis, A. (2007) An evaluation of void filling interpolation methods for SRTM data. *International Journal of Geographic Information Science* 219, 983–1008.

Shepard, D. (1968) A two-dimensional interpolation function for irregularly-spaced data. *Proceedings of the 1968 ACM National Conference*. Association for Computing Machinery, New York, pp. 517–523.

Shiklomanov, A.I., Lammers, R.B. and Vörösmarty, C. (2002) Widespread decline in hydrological monitoring threatens Pan-Arctic research. *EOS, Transactions, American Geophysical Union* 83, 13, 16–17.

Singh, V.P. and Woolhiser, D.A. (2002) Mathematical modeling of watershed hydrology. *Journal of Hydrologic Engineering* 7, 270–292.

Sivapalan, M. (2003) Prediction in ungauged basins: a grand challenge for theoretical hydrology. *Hydrological Processes* 17, 3163–3170.

Sivapalan, M., Takeuchi, K., Franks, S.W., Gupta, V.K., Karambiri, H., Lakshmi, V., Liang. X., McDonnell, J.J., Mendiondo, E.M., O'Connell, P.E., Oki, T., Pomeroy, J.W., Schertzer, D., Uhlenbrook, S. and Zehe, E. (2003) IAHS decade on Predictions in Ungauged Basins (PUB), 2003–2012: shaping an exciting future for the hydrological sciences. *Hydrological Sciences* 48, 857–880.

SRTM (2011) Shuttle Radar Topography Mission: The Mission to Map the World. Available at: http://www2.jpl.nasa.gov/srtm/ (accessed 7 February 2011).

Stackhouse, P.W. Jr, Cox, S.J., Gupta, S.K., Chiacchio, M. and Mikovitz, J.C. (2001) The WCRP/GEWEX surface radiation budget project release 2: an assessment of surface fluxes at 1 degree resolution. In: Smith, W.L. and Timofeyev, Y. (eds) *IRS 2000: Current Problems in Atmospheric Radiation*. International Radiation Symposium, St Petersburg, Russia, 24–29 July 2000. A. Deepak Publishing, Hampton, Virginia, p. 147.

Stackhouse, P.W. Jr, Gupta, S.K., Cox, S.J., Mikovitz, J.C., Zhang, T. and Chiacchio, M. (2004) 12-Year surface radiation budget data set. *GEWEX News* 14 (No. 4, November), 10–12.

Steffen, K., Box, J.E. and Abdalati, W. (1996) Greenland Climate Network: GC-Net. In: Colbeck, S.C. (ed.) *Special Report on Glaciers, Ice Sheets and Volcanoes. Tribute to M. Meier*. CRREL 96–27, 98–103. US Army Cold Regions Research and Engineering Laboratory, Hanover, New Hampshire.

Takeuchi, K., Hapuarachchi, P., Zhou, M., Ishidaira, H. and Magome, J. (2007) A BTOP model to extend TOPMODEL for distributed hydrological simulation of large basins. *Hydrological Processes* 22, 3236–3251.

Thornthwaite, C.W. and Mather, R.J. (1957) *Instructions and Tables for Computing Potential Evapotranspiration and Water Balance*. Publication No. 10, Laboratory of Climatology, Drexel Institute of Technology, Centerton, New Jersey.

Todini, E. (1996) The ARNO rainfall-runoff model. *Journal of Hydrology* 175, 339–382.

Turc, L. (1961) Evaluation des besoins en eau d'irrigation, évapotranspiration potentielle, formule climatique simplifiée et mise à jour. *Annals of Agronomy* 12, 13–49.

US Army Corps of Engineers (1998) *Runoff from Snowmelt*. Engineering Manual (EM) No. 1110–2–1406. Washington, DC.

Vandewiele, G.L., Xu, C.Y. and Ni-Lar-Win (1992) Methodology and comparative study of monthly water balance models in Belgium, China and Burma. *Journal of Hydrology* 134, 315–347.

Vörösmarty, C.J. and Moore, B. III (1991) Modeling basin-scale hydrology in support of physical climate and global biogeochemical studies: an example using the Zambezi River. *Studies in Geophysics* 12, 271–311.
Vörösmarty, C.J., Moore, B. III, Grace, A.L., Gildea, M.P., Melillo, J.M., Peterson, B.J., Rastetter, E.B. and Steudler, P.A. (1989) Continental-scale models of water balance and fluvial transport: an application to South America, *Global Biogeochemical Cycles* 3, 241–265.
Vörösmarty, C.J., Gutowski, W.J., Person, M., Chen, T.-C. and Case, D. (1993) Linked atmosphere–hydrology models at the macroscale. In: *Macroscale Modeling of the Hydrosphere, Proceedings of the Yokohama Symposium, July 1993. IAHS Publication* No. 214, 3–27.
Vörösmarty, C.J., Willmott, C.J., Choudhury, B.J., Schloss, A.L., Stearns, T.K., Robeson, S.M. and Dorman, T.J. (1996) Analyzing the discharge regime of a large tropical river through remote sensing, ground-based climatic data, and modelling. *Water Resources Research* 32, 3137–3150.
Wagener, T., Sivapalan, M., McDonnell, J., Hooper, R., Lakshmi, V., Liang, X. and Kumar, P. (2004) Predictions in ungauged basins as a catalyst for multidisciplinary hydrology. *EOS, Transactions, American Geophysical Union* 85, 451, 457.
Walter, M.T., Brooks, E.S., McCool, D.K., King, L.G., Molnau, M. and Boll, J. (2005) Process-based snowmelt modeling: does it require more input data than a temperature-index modeling? *Journal of Hydrology* 300, 65–75.
Webber, S.R. and Willmott, C.J. (1998) *South American Precipitation: 1960–1990 Gridded Monthly Time Series (Version 1.02)*. Center for Climatic Research, Department of Geography, University of Delaware, Newark, Delaware.
Weiß, M. and Menzel, L. (2008) A global comparison of four potential evapotranspiration equations and their relevance to stream flow modeling in semi-arid environments. *Advances in Geosciences* 18, 15–23.
Wielicki, B.A., Barkstrom, B., Harrison, E.F., Lee, R., Smith, G. and Cooper, J. (1996) Clouds and the Earth's Radiant Energy System (CERES): an earth observing system experiment. *Bulletin of the American Meteorological Society* 77, 853–868.
Willmott, C.J. and Matsuura, K. (1995) Smart interpolation of annually averaged air temperature in the United States. *Journal of Applied Meteorology* 34, 2577–2586.
Willmott, C.J. and Robeson, S.M. (1995) Climatologically aided interpolation (CAI) of terrestrial air temperature. *International Journal of Climatology* 15, 221–229.
Willmott, C.J., Rowe, C.M. and Philpot, W.D. (1985) Small-scale climate maps: a sensitivity analysis of some common assumptions associated with grid-point interpolation and contouring. *American Cartographer* 12, 5–16.
Wood, E.P., Lettenmaier, D.P. and Zartarian, V.G. (1992) A land-surface hydrology parameterization with subgrid variability for general circulation models. *Journal of Geophysical Research* 97 (D3), 2717–2728.
Wurbs, R.A. (2005) Texas Water Availability Modeling System. *Journal of Water Resources Planning and Management* 131, 270–279.
Wurbs, R.A. (2006) Methods for developing naturalized monthly flows at gauged and ungauged sites. *Journal of Hydrologic Engineering* 11, 55–64.
Xu, C.-Y. and Singh, V.P. (2000) Evaluation and generalization of radiation-based methods for calculating evaporation. *Hydrological Processes* 14, 339–349.
Young, G.J. and Hewitt, K. (1990) Hydrology research in the Upper Indus Basin, Karakoram Himalaya, Pakistan. In: *Hydrology of Mountainous Areas, Proceedings of the Strbské Pleso Workshop*, Czechoslovakia, June 1988. *IAHS Publication* No. 190, 139–152.
Young, G.J. and Hewitt, K. (1993) Glaciohydrological features of the Karakoram Himalaya: measurement possibilities and constraints. *IAHS Publication* No. 218, 273–283.
Zhang, Y., Rossow, W.B., Lacis, A.A., Oinas, V. and Mishchenko, M.I. (2004) Calculation of radiative fluxes from the surface to top of atmosphere based on ISCCP and other global data sets: refinements of the radiative transfer model and the input data. *Journal of Geophysical Research* 109, D19105.
Zhang, Y., Rossow, W.B. and Stackhouse, P.W. Jr (2007) Comparison of different information sources used in surface radiative flux calculation: radiative properties of the surface. *Journal of Geophysical Research* 112, D01102.
Zhao, R.J. (1992) The Xinanjiang model applied to China. *Journal of Hydrology* 135, 371–381.

3 An Overview of Some Soil Hydrological Watershed Models

Sanjit K. Deb* and Manoj K. Shukla

Introduction

Soil hydrological processes including the storage, distribution and movement of soil water through the vadose zone involve complex and dynamic interactions among climate, crop and soils. Soil processes are an integral part of the hydrological cycle that exerts short- and long-term spatio-temporal controls over water and energy balances and controls interrelationships between different components of the hydrological cycle, such as precipitation, infiltration, surface runoff, drainage, evapotranspiration (ET) and groundwater recharge. Over the past several decades, environmental concerns related to the transport and transformation of agrochemicals, the degradation of soil and water resources, erosion and sedimentation, non-point source pollution, total maximum daily loads (TMDLs) for the pollutants of concern, climate change, etc., have helped to heighten interest in intricate soil processes in mathematical models of watershed hydrology. Integration of the agricultural system into computer simulation methods by assembling numerous and complex variables and interacting hydrological processes involved within the soil–vegetation–atmosphere system has been one of the most advanced and critical challenges during the last two decades. The popularity of mathematical models of watershed hydrology is no surprise owing to their ability to provide holistic representations of hydrological processes in agricultural and mixed land-use watersheds, and assist in understanding the dynamic interactions between climate and hydrological systems, and in managing soil and water resources around the world. All of these aspects of hydrological models have created major breakthroughs in management and simulation technology for agricultural systems.

In the 1950s, as the concept of mathematical models of watershed hydrology emerged, most modelling efforts focused primarily on the rainfall–runoff component of the hydrological cycle (Singh and Frevert, 2002a). In 1966, the development of updated versions of the Stanford Watershed Model (SWM) (Crawford and Linsley, 1966), currently known as the Hydrological Simulation Program–FORTRAN (HSPF) model, exemplified the modelling concept of integrating different components of the hydrological cycle. Modelling technology has since undergone a digital revolution, especially with regard to computing capability, data acquisition, data processing, database management, graphical design and display, etc. During the 1970s

* Corresponding author: sanjit@nmsu.edu

and 1980s, a number of important mathematical models were formulated for simulating watershed hydrology, environmental and ecosystem impacts, and agricultural management scenarios. Water-quality components have also been incorporated into some of these models as the importance of non-point sources of pollution was gradually recognized. Since the early 1990s, most hydrological modelling activities have been focused on the development of model graphical user interfaces (GUIs) and integration with geographic information systems (GIS) and remote sensing data. The emphasis has been placed on the integration of hydrological (terrestrial, pedological and lithological), atmospheric and hydrospheric systems, especially in the large-scale physically based models. Advances in computer technology, computational efficiency, integration of GIS and remote sensing, employment of multiple databases and development of GUIs, as well as the development of new mathematical representations of physical processes, have outdated many models. In this light, the development of new models and/or improvement of already existing models are ongoing process, which are never going to stop.

The SWM, developed by Linsley and Crawford in 1960 (Crawford and Linsley, 1966), has been commonly considered to be the first watershed hydrological model. Since then, numerous watershed hydrological and water-quality models have been developed around the world. Some of the commonly used agricultural watershed-scale hydrological models include the following: Agricultural Policy Environmental eXtender (APEX) (Williams, 1995; Williams *et al.*, 2008a,b), Water Erosion Prediction Project (WEPP) model (Flanagan and Nearing, 1995; Flanagan *et al.*, 2001), Soil and Water Assessment Tool (SWAT) (Arnold *et al.*, 1998; Neitsch *et al.*, 2005), Annualized Agricultural NonPoint Source (AnnAGNPS) model (Bingner *et al.*, 2009), Dynamic Watershed Simulation Model (DWSM) (Borah *et al.*, 2002, 2004), Distributed Hydrology Soil Vegetation Model (DHSVM) (Wigmosta *et al.*, 1994, 2002), continuous Areal Nonpoint Source Watershed Environment Response Simulation (ANSWERS–2000 or ANSWERS–continuous) model (Bouraoui and Dillaha, 1996, 2000), Système Hydrologique Européen (SHE) TRANsport (SHETRAN) model (Ewen, 1995; Ewen *et al.*, 2000), and MIKE SHE (Refsgaard and Storm, 1995). An overview of the current status of these existing watershed hydrological models is of great importance for modellers, developers and users, including hydrologists, agriculturists, engineers, soil scientists, ecologists, biologists, climatologists, geologists, watershed stakeholders, policy makers and students.

The aim of this chapter is to provide an overview of some of the commonly used watershed-scale models, and of a field-scale Root Zone Water Quality Model (RZWQM) (RZWQM Team, 1992; Ahuja *et al.*, 2000). The RZWQM was chosen because of its comprehensive representation and integration of agricultural system processes, and future efforts and directions centred on watershed modelling (Ma *et al.*, 2007a). Broadly, the models described in the chapter were selected based on several criteria. Some of these were the use of models by the scientific community in agricultural and mixed land-use watersheds, publication in refereed journals, availability of organized model support, inclusion of best management practices (BMPs), and provision of event-based and/or continuous simulation. The chapter describes the major components, capabilities, processes involved, model inputs, modelling methods and approaches, and enhancements or adaptations of the APEX, SWAT, WEPP and RZWQM models. Major components, capabilities, processes and modelling methods are also presented for six other models, namely ANSWERS–2000, AnnAGNPS, DHSVM, DWSM, MIKE SHE and SHETRAN. For an overview of additional field-scale and/or watershed-scale soil hydrological models, readers are referred to Singh (1995) and Singh and Frevert (2002a,b). Earlier reviews and a compilation of watershed models by Borah and Bera (2003, 2004) also provide some background on the mathematical bases of some of the currently available models and some application outcomes of those models.

Model Overviews: APEX, SWAT, WEPP and RZWQM

Model development history

APEX

The Soil and Water Resources Conservation Act (RCA) of 1977 charged the US Department of Agriculture (USDA) with the responsibility of assessing the status of the nation's soil and water resources on a regular basis. The first RCA appraisal, conducted in 1980, revealed a significant need for improved technology for evaluating the impacts of soil erosion on soil productivity. In the early 1980s, the Erosion Productivity Impact Calculator (EPIC) model was developed by a USDA modelling team to support assessments of soil erosion impacts on soil productivity for soil, climate and cropping conditions representative of a broad spectrum of US agricultural production regions (Williams *et al.*, 1984). The first major application of the EPIC model was a national analysis performed in support of the 1985 RCA assessment. Various components from the Chemicals, Runoff, and Erosion from Agricultural Management Systems (CREAMS) (Knisel, 1980) and Simulator for Water Resources in Rural Basins (SWRRB) (Williams *et al.*, 1985) models were combined to develop EPIC and GLEAMS (Groundwater Loading Effects on Agricultural Management Systems) (Leonard *et al.*, 1987). A pesticide component was added later.

EPIC now stands for Environmental Policy Integrated Climate, which reflects the greater diversity of processes to which the model is currently applied. The drainage area considered by EPIC is up to about 100 ha, where weather, soils and management systems are assumed to be homogeneous. The shortcomings of the EPIC model in simulating key landscape processes at farm or small watershed scales were revealed at the onset of the US Environmental Protection Agency (EPA)-funded project 'Livestock and the Environment: A National Pilot Project (NPP)', which was initiated in the early 1990s to address water quality and other environmental problems associated with intensive livestock production (Gassman *et al.*, 2009). The NPP served as a catalyst for the development of the initial versions of the APEX model (Williams, 1995; Williams and Izaurralde, 2006; Williams *et al.*, 2006, 2008a,b), a multi-field version of the predecessor EPIC model for addressing environmental problems associated with livestock and other agricultural production systems on a whole-farm or small watershed basis. The qualitative description of the EPIC model was provided by Williams (1995), while the expanded qualitative description of the APEX was reported by Williams *et al.* (2006). Williams and Izaurralde (2006) reported an exhaustive qualitative description of the APEX coupled with mathematical theory for several components. Williams *et al.* (2008b) provide an in-depth theoretical description for the APEX model (Version 0604; available at http://epicapex.brc.tamus.edu/) and a comprehensive review of the current version of APEX is given by Gassman *et al.* (2009).

SWAT

The SWAT model was developed at the USDA Agricultural Research Service (ARS) Grassland, Soil and Water Research Laboratory in Temple, Texas (Arnold *et al.*, 1998; Neitsch *et al.*, 2005). The development of SWAT is a continuation of ARS modelling experience that spans roughly a 30-year period, and a detailed history of SWAT's development can be found in Neitsch *et al.* (2005), Gassman *et al.* (2007) and Williams *et al.* (2008a). In the mid-1970s the ARS invited a team of interdisciplinary scientists to develop a process-based, non-point source simulation model for the field scale, and from that effort the CREAMS model was developed. In the 1980s, several new models were developed based on the CREAMS, including EPIC and GLEAMS. The SWAT model emerged mainly from the SWRRB model and contains features from CREAMS, GLEAMS and EPIC. Development of SWRRB began in the early 1980s with modification of the daily rainfall hydrology model from CREAMS. A major modification was the expansion of surface runoff and other computations for up to ten sub-basins, as opposed to a single field, to predict basin water yield. Other enhancements included an improved peak runoff rate method, calculation of transmission losses, and the addition of several new components, including groundwater return flow, reservoir storage, the EPIC crop growth sub-model, a weather generator and sediment transport. Further modifications of the SWRRB in the late 1980s included the incorporation of the GLEAMS pesticide fate component, optional USDA Soil Conservation Service (SCS) technology for estimating peak runoff rates and developed sediment yield equations. The SWRRB and ROTO (Routing Outputs to Outlet) (Arnold *et al.*, 1995) models were merged into the single SWAT model to overcome the SWRRB's limitation of allowing only ten sub-basins. As part of the HUMUS (Hydrologic Unit Model for the United States) project, which was designed to provide the technical basis for conducting the appraisal of water resources for the 1997 RCA Appraisal Report, the SWAT model was used across the entire continental USA.

The key enhancements for SWAT versions 94.2, 96.2, 98.1, 99.2, and 2000 are described by Neitsch *et al.* (2002a,b) and Arnold and Fohrer (2005), and include the incorporation of in-stream kinetic routines from the Enhanced Stream Water Quality Model (QUAL2E) (Brown and Barnwell, 1987). Four versions of the SWAT model (SWAT98.1, SWAT99.2, SWAT2000 and SWAT2005 are currently being distributed at the SWAT website (http://swatmodel.tamu.edu/). A description and review of the current version SWAT2005 can be found in Neitsch *et al.* (2004, 2005) and Gassman *et al.* (2007), respectively. As reported by Arnold *et al.* (2010), several new enhancements are included in the SWAT2009, e.g. an enhanced C (carbon) dynamics sub-model, improved sediment routing routine, plant competition, improved irrigation, improved filter strip representation, the ability to simulate grassed waterways at the Hydrological Response Units (HRU) level and temporal scheduling of management practices. Several other SWAT2009 enhancements, which are currently being tested and refined, include the fate and transport of heavy metals and pharmaceuticals (hormones, antibiotics, etc.), landscape routing of water, sediment and chemicals, septic tanks, urban management practices, etc. (Arnold *et al.*, 2010). ArcSWAT has been developed as an ArcGIS-ArcView extension of SWAT and a GUI interface for SWAT; its current version, ArcSWAT 2009.93.4, is compatible with the SWAT2009.

WEPP

The WEPP soil erosion model was developed by an interagency group of scientists working for the ARS, the USDA Natural Resources Conservation Service (NRCS), the USDA Forest Service, US Department of Interior's Bureau of Land Management (USDI BLM), and the US Geological Survey (USGS). A detailed history of WEPP development can be found in Flanagan *et al.* (2007). Scientists from the aforementioned agencies throughout the USA have been working since 1985 to develop new and improved erosion prediction technology, which was intended to be process oriented and, conceptually, a significant enhancement over the Universal Soil Loss Equation (USLE) (Wischmeier and Smith, 1978). In the late 1989, WEPP model development became largely the responsibility of personnel at ARS National Soil Erosion Research Laboratory (NSERL) in West Lafayette, Indiana. The cooperating action agencies, including the BLM, Forest Service and SCS, also provided suggestions.

The basic WEPP hillslope model components were weather generation, surface hydrology, hydraulics of overland flow, hillslope erosion, sediment transport and deposition, water balance, plant growth, residue management and decomposition, soil disturbance by tillage and irrigation (Foster and Lane, 1987). The prototype hillslope profile model version 89 was developed in 1989 (Lane and Nearing, 1989). An important aspect of WEPP technology was to separate the erosion processes into rill detachment as a function of excess flow shear stress and inter-rill detachment as a function of rainfall intensity. A substantial number of modifications have been made in WEPP 89, including: the addition of process-based components for sprinkler and furrow irrigation; spatially varying non-uniform overland flow hydrology; winter routines for snow accumulation and density, snow melt, soil frost and thaw; and subsurface lateral drainage. The plant growth and residue decomposition components have also undergone major revisions.

The 1995 release of WEPP 95.7 included a functional user interface on an MS DOS platform and allowed both hillslope profile and watershed simulations (Flanagan and Nearing, 1995). In early 1996, the ARS began efforts to create a graphical Windows interface. An initial prototype of common interface program known as MOSES (Modular Soil Erosion System) that allowed the use of WEPP and other models (e.g. the Revised USLE or RUSLE) through common screens and databases was distributed in 2000 (Flanagan *et al.*, 2001). Initial Windows interface work was completed in 1999 (WEPP 99.5) to produce a user-friendly software program still widely used today (the current version is 2010.1). A beta version of the WEPP watershed model, an extension of the WEPP hillslope model, was completed late in 1990 (Stone *et al.*, 1990). The watershed model continued to evolve as watershed channel and impoundment routines were added, linkages to the hillslope model became fully operational, channel hydrology was updated using hillslope model hydrology components, and the channel erosion equations were tested and improved (2001.3). Additional work was initiated at NSERL in 1996 to link the WEPP model with GIS and utilize digital elevation data to automatically delineate watersheds, channels, hillslopes, and representative hillslope profiles (Cochrane and Flanagan, 1999). The ArcView 3.× extension software for use with the WEPP was distributed in 2001. The Web-based interface for cropland and rangeland applications (WEPP Version 2004.7) and enhanced interfaces for forest hydrology or subsurface lateral flow (2006.5) were released in 2004 and 2006, respectively. The winter hydrology component of the WEPP was modified in Version 2008.907. The current version of WEPP (2010.1) is distributed at the WEPP website (http://www.ars.usda.gov/Research/docs.htm?docid=10621).

RZWQM

In 1985, a workshop organized by ARS scientists in cooperation with other federal agencies and private industries reviewed state water quality modelling and identified the need for a process-based water quality model to quantify root-zone processes as affected by agricultural management practices (Malone *et al.*, 2004; Ma *et al.*, 2007a). A team from several ARS research units nationwide was assigned responsibility for building and developing the RZWQM (Ahuja *et al.*, 2000; Ma *et al.*, 2007a). The task was to utilize the capability of existing models, e.g. the Pesticide Root Zone Model (PRZM) (Carsel *et al.*, 1985), Nitrogen Tillage Residue Management (NTRM) (Shaffer and Larson, 1987), CREAMS and GLEAMS, and incorporate additional features needed for simulating advanced soil chemistry and nutrient transformations, improved pesticide dynamics, plant growth, chemical transport via macropores, tile drainage, and soil–water–plant system management.

The RZWQM version 1.0 was completed in 1992 (RZWQM Team, 1992). Since then, it has undergone extensive verification, evaluation and refinement, particularly in cooperation with the MSEA (Management Systems Evaluation Areas) water-quality projects in the five US Midwestern States. The RZWQM model was re-released in 1998 (RZWQM98) for general users with a Windows-compatible interface, and has received continuous updates and improvements (RZWQM Development Team, 1998; Ahuja *et al.*, 2000; Malone *et al.*, 2004; Ma *et al.*, 2007a). The RZWQM model was linked with the Decision Support System for Agrotechnology Transfer (DSSAT3.5) (Jones *et al.*, 2003) in 2005 (Ma *et al.*, 2005, 2006); this was later updated using the DSSAT4.0 package and released as the RZWQM2 model (Ma *et al.*, 2008). The current version of RZWQM2 (version 1.80.2009) is distributed at the RZWQM website (http://www.ars.usda.gov/Main/docs.htm?docid=17740).

Model components

The major components or capabilities, processes, and modelling methods or approaches of the APEX, SWAT, WEPP and RZWQM models are summarized in Table 3.1.

Table 3.1. Summary of APEX (Agricultural Policy Environmental eXtender), SWAT (Soil and Water Assessment Tool), WEPP (Water Erosion Prediction Project) and RZWQM (Root Zone Water Quality Model) model components.

Component/capability/process/criterion	APEX	SWAT	WEPP	RZWQM/RZWQM2
Components/processes	Climate; infiltration, surface runoff, ET (evapotranspiration); subsurface flow; water and wind erosion, sediment; manure erosion; nutrient and C cycling; pesticides; soil temperature; crop growth; tillage; economics; channel/reservoir routing (water, sediment, nutrients/pesticides); groundwater; grazing and manure management.	Climate; infiltration; surface runoff; ET; subsurface flow; water erosion and sediment; nutrients/pesticides/bacteria; soil temperature; land cover/plants; agricultural and urban management; channel and reservoir routing (water, sediment, nutrients/pesticides/bacteria).	Climate; infiltration; surface runoff; winter processes; ET; subsurface flow; water erosion and sediment; irrigation; soil erodibility; plant growth; residue decomposition and management; tillage; channel and impoundment routing (water and sediment).	Climate; infiltration; macropore flow; infiltration–excess runoff in absence of macropores; lateral infiltration into soil surrounding macropores; ET; snow accumulation and melt; heat transport; nutrients/pesticides or agrochemicals over, within and below crop root zone; management practices.
Watershed representation/spatial scale	Watershed is subdivided into many homogeneous sub-areas (fields, soil, land use, management, weather, etc.); sub-areas are linked to channels.	Sub-basins or sub-watersheds are grouped based on climate, HRUs[a] (lumped land areas with land cover, soil and management), ponds, groundwater, and main channels.	Small watershed is subdivided into hillslopes that are linked to channels/impoundments.	Field scale; 1-D (vertical into soil profile) on a unit-area basis; crop root zone and deeper vadose zone.
Temporal scale	Long term; continuous; daily time step (some processes are simulated with hourly or less time step).	Long term; continuous; daily time step and not designed to simulate detailed, single-event flood routing.	Storm event and long term; continuous; daily time step.	Long term; continuous; daily time step (water and chemical movement on a sub-hourly time step).
Weather generator	Weather generator; missing input data.	WXGEN weather generator; missing input data; climate customization (e.g. orographic effects, impact of CO_2 levels) and weather forecast.	CLIGEN weather generator or BPCDG (Break Point Climate Data Generator).	CLIGEN weather generator.
Rainfall – excess on overland/water balance/hydrological soil processes	Daily water balance; precipitation, rainfall interception, runoff, ET, soil and plant evaporation, snow melt, subsurface flow (deep	Daily water balance; precipitation; interception and canopy storage; runoff; ET; snow melt; transpiration and sublimation/soil evaporation; subsurface	Daily water balance; precipitation; rainfall interception; runoff; ET; snow evaporation and melt; soil and	Daily water balance; rainfall; infiltration; ET; water redistribution in soil matrix: (i) excess rainfall routed into macropores, (ii) infiltration–

	percolation, lateral and quick return flow), and water table dynamics (no direct linkage to other root-zone soil water processes).	flow (percolation, bypass or crack flow, return flow or base flow from subsurface and groundwater flow).	plant evaporation; root-zone soil water content; percolation and subsurface lateral flow.	excess runoff in absence of macropores, (iii) lateral water into soil surrounding macropores; snow accumulation and melt.
Infiltration and surface runoff	Modified SCS[a] (runoff) curve number (CN); Green and Ampt method.	Developed SCS CN procedure; Green and Ampt; Green–Ampt–Mein–Larson excess rainfall method.	Modified Green–Ampt–Mein–Larson excess rainfall method.	Modified Green and Ampt; infiltration–excess runoff in absence of macropores by coupling a numerical solution of kinematic wave equations with vertically distributed Green and Ampt infiltration.
Peak runoff	Modified Rational method; USDA SCS TR-55 method; snow melt events are considered.	Modified Rational method.	Modified Rational equation similar to EPIC[a] and CREAMS[a] models.	Not simulated.
Water redistribution and subsurface flow	Vertical and horizontal subsurface flows using storage routing and pipe flow equations; groundwater balance includes changes in groundwater volumes, percolation rate, reservoir seepage rate, percolation rate from groundwater storage, and return flow rate.	Kinematic storage model for lateral subsurface flow and empirical relations for groundwater flow; return flow from a shallow, unconfined aquifer.	Percolation from each soil layer; subsurface lateral flow using kinematic storage–discharge model; flow to drainage tiles or ditches (if present).	1–D soil water redistribution based on Richard's equation; macropore flow using Poiseuille's law assuming gravity; lateral water movement using lateral Green and Ampt approach; a pseudo 2–D drainage flow and water table fluctuation.
Potential ET (PET)/ET	Hargreaves and Samani, Penman, Priestley–Taylor, Penman–Monteith, and Baier–Robertson methods.	Hargreaves, Priestley–Taylor, and Penman–Monteith methods.	Penman and Priestley–Taylor methods.	Shuttleworth and Wallace's dual surface version of Penman–Monteith equation.
Soil properties/soil temperature/soil evaporation	Soil physical and hydrological properties; erodibility; daily average soil temperature at soil layers; soil evaporation and plant transpiration based on Ritchie's method.	Soil physical and hydrological properties; erodibility; diurnal and seasonal and variations in soil surface temperature; transpiration and sublimation/soil evaporation based on Ritchie's method.	Soil physical and hydrological properties; roughness; inter-rill and rill erodibility; critical shear stress; soil evaporation and plant transpiration based on Ritchie's method.	Soil physical and hydrological properties; sensible and latent heat transport based on partial mixing and displacement during infiltration and convective-dispersive equation during redistribution; surface energy balance.

Continued

Table 3.1. Continued.

Component/ capability/process/ criterion	APEX	SWAT	WEPP	RZWQM/RZWQM2
Overland sediment/ water erosion and wind erosion	USLE[a]; Onstad–Foster modification of USLE; Modified USLE (MUSLE); two MUSLE variants – MUST and MUSS; MUSLE that allows input coefficients (MUSI); RUSLE[a]; RUSLE2; wind erosion; nutrients/pesticides transported with sediment using modified loading function.	MUSLE; nutrient/pesticide/ bacteria/water quality parameters transported with sediment using modified loading function.	Closed-form solutions to a steady-state form of the hillslope sediment continuity equation; equations for inter-rill detachment, shear stress in rills, deposition and transport, and sheet and overland flow in rill and inter-rill areas.	Model does not have an erosion component; sediment transport is linked with surface flow predictions; chemicals in solution and adsorbed on sediment move off the field with predicted surface discharge and sediment load.
Manure erosion	Direct estimates of organic nutrient and C losses based on erosion equation MUST.	Not simulated.	Not simulated.	Not simulated.
Nutrients/pesticides	N cycles: crop N uptake, denitrification, mineralization, immobilization, nitrification, ammonia volatilization, sediment organic N transport, and NO_3-N losses; P cycles: solution and sediment phases; pesticides from the GLEAMS[a]; enhanced C and N cycling.	Two pools inorganic N, NH_4^+ and NO_3, and three pools organic forms of N; three pools inorganic and three pools organic forms of P; pesticide from the GLEAMS; bacteria, growth/die-off patterns of pathogens.	Not simulated.	C and N transformations within soil profile; N mineralization, nitrification, immobilization, denitrification and volatilization; pesticide transformation and degradation on plant surfaces, residue, soil surface and in soil profile; non-uniform mixing model for chemical transfer to runoff and partial displacement for matrix transport.
Channel/flood-plain routing	A daily time step average flow method for long–term and variable storage coefficient (VSC) flood routing method for a short time interval.	Variable storage routing or Muskingum river routing kinematic wave model.	Water balance including infiltration, ET, percolation, interception, and surface depressional storage in the same manner as overland flow areas.	Not simulated.

Channel/flood-plain sediment	Variation of Bagnold's sediment transport equation for sediment routing using a daily time step average flow or VSC method.	Simplified Bagnold's stream power concept for deposition/ degradation; in-stream nutrient loadings using the QUAL2E[a] model; in-stream pesticides, bacteria and heavy metals.	Detachment, transport and deposition of sediment within permanent channels or ephemeral gullies using a steady-state sediment continuity; flow depth and hydraulic shear stress using CREAMS regression equations.	Not simulated.
Water bodies/ reservoir/ impoundment routing	Amount of flood storage is determined by the storage volume between the principal and emergency spillways to accommodate a variety of structures; inflow is derived from the sub-area plus all other contributing sub-areas.	Ponds and wetlands receive inflow from a fraction of the sub-basin area; depressions/potholes (contribute only to runoff); reservoirs (inflow, outflow, rainfall on surface, evaporation, seepage from bottom and diversions).	Outflow hydrographs and sediment concentration for farm ponds or terraces, culverts, filter fences, straw bales, drop and emergency spillways and perforated risers.	Not simulated.
Water bodies/ reservoir/ impoundment sediment	Sediment and attached nutrient and pesticide are deposited in reservoirs, but soluble materials are considered conservative.	Mass balance model for N, P and pesticides for sediment transport into and out of water bodies, and bacterial die-off process.	Deposition assumes complete mixing and adjusted for stratification, non-homogeneous concentrations, and impoundment shape; a continuity mass balance for sediment outflow concentration.	Not simulated.
Crop growth/land cover/plants	Single crop growth model; about 100 annual and perennial crops; potential growth, water use, N and P uptake, and phenological development; stresses due to water, nutrients, etc.; constraints on biomass accumulation, root growth and yield; plant competition component.	Single plant growth model; annual and perennial crops and land cover; potential growth; removal of root-zone water and nutrients; plant use of N and P, transpiration and biomass/yield production; stresses caused by water, nutrients and temperature.	Plant growth model similar to the EPIC; annual and perennial crops; potential growth, harvest, silage and grazing; a unimodal or bimodal potential curve for rangeland grazing, burning and herbicide application.	A generic or DSSAT4.0[a] plant growth and crop production model (maize, soybean and wheat in RZWQM; 19 crops in RZWQM2); CO_2 assimilation, C allocation, dark respiration, periodic tissue loss, plant mortality, root growth, water and N uptake.

Continued

Table 3.1. Continued.

Component/ capability/process/ criterion	APEX	SWAT	WEPP	RZWQM/RZWQM2
Residue decomposition and management	C- and N-enriched organic matter is split into microbial (or active), slow and passive compartments; residues added to surface or below ground are split into metabolic and structural compartments according to their N and lignin contents.	Decomposition and mineralization in the first soil layer; a daily decay rate constant as a function of C:N ratio and C:P ratio of residues, temperature and soil water content.	Decomposition of flat. above-ground and submerged residues and dead roots for fallow, mono. double, rotation, strip, and mixed cropping practices; cropland and rangeland management (e.g. shredding or cutting, etc.).	Fast and slow pools of surface residue decomposition as a function of residue N, residue moisture and daily air temperature; crop residues can be used for the soil surface after harvest.
Agricultural practices/ evaluation of best management practices (BMPs) and total maximum daily loads (TMDLs)	Planting/beginning of growing season, harvest, biomass removal, grazing, crop rotation, tillage, irrigation (furrow and sprinkler), manure handling, fertilizer and pesticide applications, liming, furrow disking, drainage systems, buffer strips, terraces, waterways, cover crops; can be used for both sediment and nutrient TMDLs.	Planting/beginning of growing season, harvest, grazing, harvest and kill, biomass removal, tillage, fertilization and pesticide applications, crop rotation, cover crops, filter strip, irrigation, tile drainage, point source loadings, runoff from urban areas; can be used for sediment and nutrient TMDLs.	Cropping management, surface cover, irrigation (sprinkler or furrow) with depletion level and fixed-date scheduling, tillage effects on soil detachment and transport, and suitable for sediment TMDLs.	Management that influences the state of the root zone, including tillage, fertilizer, pesticide and manure applications; crop planting; irrigation (furrow, sprinkler, and drip systems); dynamic N; soil surface reconsolidation; decomposition and bio-incorporation of surface residues; ridge-tilled and no-tilled systems.
Economics	A crop budget and accounting subsystem; costs of producing and marketing crops; changes in inputs that respond to management, soil quantity and quality, and climate; alternative policies and programmes.	Not simulated.	Not simulated.	Not simulated.

[a]CREAMS, Chemicals, Runoff, and Erosion from Agricultural Management Systems; DSSAT4.0, Decision Support System for Agrotechnology Transfer 4.0; EPIC, Erosion Productivity Impact Calculator; GLEAMS, Groundwater Loading Effects on Agricultural Management Systems; HRUs, Hydrological Response Units; QUAL2E, Enhanced Stream Water Quality Model; RUSLE, Revised Universal Soil Loss Equation; SCS, USDA Soil Conservation Service; TR-55, SCS procedures for calculating storm runoff volume, peak rate of discharge, hydrographs and storage volumes required for flood-water reservoirs; USLE, Universal Soil Loss Equation.

APEX

The APEX model is a flexible and dynamic tool for simulating management and land-use impacts for single fields (like the EPIC model), as well as for a whole farm or watershed that may be subdivided into many homogeneous sub-areas (such as fields, soil types, different land uses, landscape positions, management, weather or any other desired configuration). The model operates on a daily time step, with some processes being simulated on an hourly or smaller time steps, for performing long-term simulations. In APEX, sub-areas or Hydrologic Landuse Units (HLUs) include the land-phase components of the hydrological cycle, e.g. climate, infiltration, surface runoff, ET, water redistribution, subsurface flow, return flow, erosion, plant growth, erosion, nutrient/pesticide cycling and management. These control the amounts of water, sediment and nutrient/pesticide loadings by linking the sub-areas to each other with respect to the water-routing direction towards a single or multiple watersheds or farm outlets. The major components of the EPIC model are weather simulation, hydrology, erosion, sedimentation, nutrient cycling, pesticide fate, crop growth, soil temperature, tillage, economics and plant environment control. In addition to these EPIC functions, the APEX model has components for routing water, sediment, nutrients and pesticides across complex landscapes and channel systems to the watershed outlet (Table 3.1). Groundwater and reservoir components have been incorporated in addition to the routing algorithms. The APEX model can be configured for novel land-management strategies or BMPs such as impacts of filter strips on sediment and other pollutants, grazing patterns of multiple herds and intensive rotational grazing scenarios, terrace systems, manure management from feed yards and dairies with or without lagoons, land application of manure removed from livestock feedlots or waste storage ponds, etc. The model can be used in assessing the effects of global climate or CO_2 changes, designing environmentally safe and economic landfill sites, designing biomass production systems for energy, and other spin-off applications. A complete description of APEX components can be found in Williams *et al.* (2008b).

SWAT

The SWAT model is a continuous-time, river basin or watershed scale, and quasi-physically based simulation model for predicting the impact of land-management practices on water, sediment and agricultural chemical yields in large complex ungauged watersheds with varying soils, land-use and management conditions over long periods of time. The SWAT model operates on a daily time step, and the model is not designed to simulate single-event-based flood routing. SWAT simulates landscape processes and streamflow with a high level of spatial detail by allowing the river/watershed to be divided into sub-basins or sub-watersheds, which are grouped based on climate, HRUs, ponds, groundwater and the main channels. The HRUs are lumped land areas within the sub-basin comprising unique land cover, soil and management combinations. The HRUs represent percentages of the sub-watershed area and are not identified spatially within a SWAT simulation. A watershed can be also subdivided into only sub-watersheds that are characterized by dominant land use, soil type and management. In SWAT, the land-phase component of the hydrological cycle controls the amount of water, sediment, nutrient/pesticide and bacterial loadings to the main channel in each sub-basin; the second component of the hydrological cycle is the water or routing phase of the hydrological cycle which controls the movement of water, sediments and nutrients/pesticides through the channel network of the watershed to the outlet. Major model components include weather simulation (including climate change), hydrology, soil temperature and properties, plant growth, erosion, nutrients, pesticides, bacteria and pathogens, and land management or BMPs (Table 3.1). A complete description of SWAT components can be found in Neitsch *et al.* (2005).

WEPP

The WEPP model is a process-based, distributed-parameter and continuous-time simulation model that is capable of predicting distributions of net soil loss and deposition for a wide range of time periods and spatial scales. The model can be operated on a daily time step and single storm event basis. The WEPP hillslope version simulates erosion along a single hillslope profile, while the WEPP watershed version is built as an extension of the WEPP hillslope model that can be used to estimate runoff and sediment yield of a small watershed (Flanagan and Nearing, 1995; Ascough *et al.*, 1997). The model is composed of several components (Table 3.1), which take into account climate, surface runoff, infiltration, percolation, evaporation, transpiration, rill and inter-rill erosion, sediment transport and deposition, soil consolidation, residue and canopy effects on soil detachment and infiltration, surface sealing, rill hydraulics, plant growth, residue decomposition, snow melt, frozen soil effects on infiltration and erodibility, climate and tillage effects on soil properties, effects of soil random roughness and contour effects – including the potential overtopping of contour ridges. A watershed in the WEPP is partitioned into hillslopes and a channel network that includes channel segments and impoundments. Water and sediment from one or more hillslopes can feed into the channel network. A hillslope can be further divided into overland flow elements (OFEs), within which soils, vegetation and management conditions are assumed to be homogeneous. Because the WEPP hillslope and watershed routines are used for the overland flow portion of the area, and channels and impoundments, respectively, the WEPP model may not be applied to areas with permanent channels – such as classical gullies and perennial streams. In highly dissected landscapes with several different and distinct slope shapes, several hillslopes need to be simulated, either as separate runs or as a single watershed simulation at the watershed interface. A complete description of WEPP components can be found in Flanagan and Nearing (1995) and Flanagan *et al.* (2001).

RZWQM

The RZWQM model is a comprehensive, one-dimensional (1-D) (i.e. vertical into the soil profile) process-based model that simulates major physical, chemical and biological processes in an agricultural crop growth and production system, as well as the movement of water, nutrients and agrochemicals over, within and below the crop root zone of a unit area of an agricultural cropping system. The model best describes areas where rainfall, soil and crop conditions are uniform, while it may need to run for different conditions in heterogeneous situations to characterize the combination. A complete description of RZWQM components can be found in Ahuja *et al.* (2000). The RZWQM model operates on a daily time step, while hydrological processes and chemical movement in the soil are simulated at sub-hourly time steps (5-60 min). The model can be used as a tool for assessing the environmental impact of alternative management strategies on a field-by-field basis and for predicting management effects on crop production and water quality. The major sub-modules or processes in the RZWQM model (Table 3.1) include: physical processes, e.g. hydrology and macropore flow, soil heat flow, tile drain, etc.; plant growth processes and plant production; soil chemical processes, e.g. soil inorganic environment in support of nutrient processes, chemical transport and pesticide processes; nutrient processes, e.g. C and N transformations within the soil profile; pesticide processes, e.g. transformations and degradation of pesticides on plant surfaces, in plant residues, on the soil surface and in the soil profile; and agricultural management processes, e.g. BMPs influencing the state of the root zone. The RZWQM model does not have an erosion component, which limits its utility in describing runoff- or sediment-transported pollutants.

Model inputs

APEX

The farm or watershed in the APEX model involves several fields or sub-areas or HLUs, and the heterogeneity of a watershed/farm is determined by the number of sub-areas. Each sub-area is linked with each other sub-area, starting from the most distant sub-area towards the watershed outlet. The input data of the APEX model are presented in Table 3.2.

Table 3.2. Input parameters of the APEX (Agricultural Policy Environmental eXtender) model.

Input data file	Input parameters
Data file list	• A list of input file names and descriptions: site, sub-areas or Hydrologic Landuse Units (HLUs), operation schedules, soil, daily weather, monthly weather stations, wind stations, field operations, crop parameter, fertilizer, pesticides, data for TR-55[a] runoff estimation, herds of animals and grazing, equation parameters and coefficients, and output option.
Simulation control file	• Option of model run control: run length; daily, monthly or annual output; weather variables; potential evapotranspiration (PET) method; stochastic SCS[a] curve number (CN) estimator; peak rate estimate; automatic heat unit scheduling; option for variable SCS CN with soil water, or a constant CN for all storms; runoff estimation method; pesticide output in mass and concentration; enrichment ratio method; soluble P runoff estimate equation; N and P plant uptake concentration; manure application; air quality analysis; flood routing; atmospheric CO_2; field capacity/wilting point estimation; pest damage scaling factor; channel capacity; wind erosion adjustment factor; routing threshold; variable storage coefficient (VSC) routing; water erosion; USLE[a] crop management channel factor, etc.
Site	• Watershed location, watershed elevation, peak runoff rate–rainfall energy adjustment factor, CO_2 in the atmosphere, N in irrigation water, manure application rate to supply N and P uptake rate, etc.
Sub-area or HLU	• Sub-area number, soil number, operation schedule, feeding area, manure application area, watershed area, daily weather station, land-use number, CN-CN2 option, outflow release method, distance from outlet to most distant point, etc. • Channel depth, mainstream channel slope, Manning '*n*' coefficient for channel and upland, average upland slope and slope length, channel slope of routing reach, channel length and depth of routing reach, bottom and top width of channel of routing reach, channel Manning '*n*' of routing reach, channel USLE crop management and erodibility factors, etc. • Buffer/flood plain width and length, reservoir-related parameters, irrigation option, water stress factor to trigger automatic irrigation, drainage option, fertilizer and solid manure application data, erosion control practice factor, lagoon information, grazing information, etc.
Operation schedule	• Land-use number; auto irrigation, commercial fertilizer, manure deposition by animals, solid and liquid manure and lime application; time of operation; tillage and crop number; plant population; runoff fraction; time of operation as fraction of growing season, etc.
Soil	• Soil albedo, hydrological group, initial soil water content, minimum and maximum depth to water table, initial water table height, available groundwater storage, maximum groundwater storage, groundwater residence time, return flow/return flow plus deep percolation, maximum number of soil layers, options of soil weathering and soil grouping, minimum thickness of maximum layer, minimum profile thickness, fraction of organic carbon in biomass and passive pools, depth to bottom of layer, moist and dry bulk densities, soil water content at wilting point, soil water content at field capacity, sand and silt contents, coarse fragment content, saturated and lateral hydraulic conductivities, initial soil water storage, etc.

Continued

Table 3.2. Continued.

Input data file	Input parameters
	• Initial organic N, pH, electrical conductivity, cation exchange capacity (CEC), organic C, calcium carbonate content, initial soluble N and P concentrations, crop residue, P sorption ratio, initial organic P, exchangeable K, litter, N and C contents in biomass and humus contents, etc.
Daily weather	• Daily precipitation (point/spatially distributed), solar radiation (optional), minimum and maximum temperatures, relative humidity (optional) and wind speed (optional) for the simulation period.
Monthly weather	• Monthly weather statistics of a single weather station, which also generate weather when no daily weather data are available.
Wind	• Monthly wind weather statistics of a single wind weather station (crucial if wind erosion, dust distribution and air quality from feedlots are simulated): average monthly wind speed and monthly % wind from 16 wind directions.
Tillage	• Tillage equipment, machine efficiency, mixing efficiency, tillage depth, ridge height and interval, operation option, harvest efficiency or pesticide application efficiency, fraction of soil compacted, cost of equipment, cost of repair and operation, etc.
Crop	• Crop number; biomass:energy ratio and decline rate parameter; harvest index; optimal and minimum temperature for plant growth; leaf area index and decline rate parameter; maximum crop height and root depth; fraction of N, P and K in yield; seed cost; prices for yield and forage yield; N, P and K uptake parameters; wind erosion-related parameters; vapour pressure deficit; salinity, water use conversion to biomass; fraction of root weight, etc.
Fertilizer	• Fertilizer name, mineral and organic N and P fractions, mineral K fraction, ammonia N fraction, organic C fraction, salt fraction, and cost.
Pesticide	• Pesticide name, solubility, half-life in soil and foliage, wash-off fraction, organic C absorption coefficient, and cost.
Herd/grazing	• Number of cows in herd, manure number, fraction of day that herd is in feeding area, daily grazing rate per animal unit, daily manure and urine production per animal unit, herd number, buy/sell information, number of animals in herd after buy/sell, etc.
Coefficients/constant	• S (sigmoid)-shaped curve related parameters, miscellaneous parameters and coefficients of equations in the APEX.

[a]SCS, USDA Soil Conservation Service; TR-55, SCS procedures for calculating storm runoff volume, peak rate of discharge, hydrographs and storage volumes required for flood-water reservoirs; USLE, Universal Soil Loss Equation.

Two methods have been developed for constructing the sub-areas: a manual method and the SWAT model interface method. The SWAT model interface method, which utilizes a GIS to develop input data files (using a digital elevation model (DEM) and other GIS data), can facilitate the accuracy for each sub-area configuration in the watershed. The APEX model facilitates the use of soil, crop, equipment, fertilizer and pesticide databases and associated input parameter information. In addition, the model provides the equation parameters and coefficients that are implemented within APEX. A complete description of APEX input parameters was reported by Steglich and Williams (2008).

SWAT

The SWAT model requires input data at the watershed, sub-basin and HRU levels (Table 3.3). The model also allows input data for reservoir, lake water quality and point sources at any point along the channel network. The three most common techniques used to discretize a watershed are included: (i) a grid-cell configuration similar to AGNPS (Young *et al.*, 1987), to ANSWERS (Beasley *et al.*, 1980) and to the WEPP grid version (Foster and Lane, 1987); (ii) a representative hillslope configuration technique similar to APEX (Williams *et al.*, 1998) and the WEPP hillslope version (Lane and Nearing, 1989); and (iii) a sub-watershed configuration similar to the WEPP watershed version (Foster and Lane, 1987), HYMO (a problem-oriented computer language for hydrological modelling; Williams and Hann, 1973) and SWRRB. A complete description of SWAT input parameters can be found in Neitsch *et al.* (2004).

WEPP

The major inputs to the WEPP hillslope model include climate, slope, soil and cropping/management information (Table 3.4). Soil properties may be specified for eight different layers within the maximum profile depth of 1.8 m. The cropping/management input requires land use (e.g. agriculture, range or forest) to be identified by users, and for each land use information is needed about the specific plants present and the management practices used. In addition to the input data required to run WEPP on each hillslope, the WEPP watershed model requires all information from each hillslope, including watershed configuration, channel topography, channel soils, channel management practices, climate data, channel hydraulic characteristics, channel irrigation and impoundments present in the watershed (Table 3.4). A complete description of WEPP model input parameters was reported by Flanagan and Livingston (1995).

RZWQM

As the RZWQM model is complex and primarily physically based, extensive input for model initialization and parameterization is required (Ahuja *et al.*, 2000). At a minimum, the model requires break-point rainfall or irrigation data, meteorological data, a site description including data on soil horizon and soil parameters at each depth increment, dry mass of residue on the surface, pesticide data, crop parameters, management information, tile drain information (if simulated), initial soil conditions at each depth increment, initial management details and initial nutrient inputs (Table 3.5). The RZWQM model also provides: (i) many default values for input parameters (e.g. plant specific growth parameters); (ii) schemes to estimate soil hydraulic properties from minimum data; (iii) some initialization wizards (e.g. organic N pool initialization); (iv) an extensive help system; and (v) databases for soil properties, pesticide, crop growth and management scenarios. The CLIGEN weather generator is included to generate daily solar radiation, wind, humidity and rainfall data where observations are not available. The enhanced Windows-based interface facilitates data input, parameterization and graphical display of simulation results.

Model adaptation or enhancement

To date, the APEX, RZWQM, SWAT and WEPP models have all undergone extensive adaptation or enhancements to provide improved simulation of specific processes or model components. Parallel progression

Table 3.3. Input parameters of the SWAT (Soil and Water Assessment Tool) model.

Domain	Input data file	Input parameters
Watershed level input	Watershed file	• Information related to modelling options, climate inputs, databases and output specifications.
	Watershed configuration	• List of input file names. • Watershed subunits: (i) sub-basins (unlimited number of Hydrological Response Units (HRUs), one per sub-basin required), and one pond (optional); (ii) reach/main channel segments (one per sub-basin); (iii) impoundments on main channel network (optional); and (iv) point sources (optional).
	Basin (land-area modelling)	• Water balance: snowfall temperature, snow-melt base temperature, melt factors, snowpack temperature lag factor, snow water content, option of potential evapotranspiration (PET) methods, soil evaporation and plant uptake compensation factors, leaf area index, initial soil water storage, depth to impervious layer, etc. • Surface runoff: option of rainfall/runoff/routing and daily SCS[a] curve number (CN) calculation, plant evapotranspiration (ET) CN coefficient, option of daily maximum half-hour rainfall value, runoff lag coefficient, peak rate adjustment factor, etc. • Nutrient cycling: rainfall N, N and P rate factors for humus mineralization of active organic nutrients, denitrification exponential rate coefficient and threshold water content, N and P uptake distribution parameters, nitrate and P percolation coefficients, P availability index, residue decomposition coefficient, etc. • Pesticide cycling: pesticide percolation coefficient. • Algae/carbonaceous biochemical oxygen demand/dissolved oxygen: option of sub-basin water quality. • Bacteria: die-off factor and growth factors for persistent and less persistent bacteria in soil solution adsorbed to soil particles, on foliage, in stream and in water bodies; fraction of manure applied to land areas; bacterial soil partitioning and percolation coefficients; etc.
	Basin (reaches modelling)	• Option of channel water routing (variable or Muskingum methods), calibration coefficients, weighting factor, fraction of transmission losses, evaporation and peak rate adjustment factors, linear and exponent parameters for sediment routing, in-stream water quality option, etc.

	Precipitation (observed or generated)	• Daily precipitation (when CN is chosen; up to 18 precipitation files and each file can hold data for up to 300 stations); sub-daily precipitation (when Green and Ampt method is chosen).
	Solar radiation (observed or generated)	• Daily solar radiation; the file can hold data for up to 300 stations.
	Temperature (observed or generated)	• Daily minimum and maximum temperatures (up to 18 files and each file can hold data for up to 150 stations).
	Wind (observed or generated)	• Daily average wind speed (when Penman–Monteith method for PET[a] is chosen); the file can hold data for up to 300 stations.
	Relative humidity (observed or generated)	• Daily relative humidity (when Penman–Monteith and Priestley–Taylor methods for PET are chosen); the file can hold data for up to 300 stations.
	PET[a]	• Daily PET values (optional).
	Weather forecast	• Statistical data needed to generate daily climate data during the forecast period (optional).
	Database files	• Required information and parameters stored in five database files: land cover/plant growth, tillage, pesticide, fertilizer and different types of urban areas.
	Water quality	• General parameters needed for QUAL2E[a] nutrient transformation in the main channel (optional).
Sub-basin level input	Sub-basin	• Area, location, elevation, number of weather (e.g. precipitation, temperature, etc.) records, number of weather forecast regions, etc. • Tributary channel: length, width, alluvium hydraulic conductivity, Manning '*n*' coefficient, etc. • Pond input data, water-use management data, weather variables (e.g. rainfall, temperature, etc.) adjustment options, CO_2 concentration, etc. • Number of HRUs; information about HRU soil, soil chemical, land-use management and groundwater data files, etc.
	Weather generator	• Statistical data needed to generate representative daily climate data.
	Pond/wetland	• Information for impoundments (optional): fraction of sub-basin area draining into ponds, initial volume of water; initial and equilibrium sediment concentration in pond water, hydraulic conductivity of the pond bottom, etc.
	Water use	• Information for consumptive use (optional): average daily water removal from the pond, reach and shallow and deep aquifers.
	Main channel	• Length, width, depth, slope, alluvium effective hydraulic conductivity, erodibility and cover factors, Manning '*n*', width:depth ratio, and base-flow alpha factor for bank storage.
	Stream water quality	• Specific parameters needed for pesticide and QUAL2E nutrient transformations in the main channel (optional).

Continued

Table 3.3. Continued.

Domain	Input data file	Input parameters
HRU level input	HRU	• Topographic characteristics: fraction of sub-basin area contained in HRU, average slope length, slope length for lateral subsurface flow, average slope steepness. • Water flow: lateral flow travel time; fraction of HRU area draining into potholes, flood-plain and riparian areas; depth of impervious layer. • Erosion: sediment concentration in lateral and groundwater flow, soil evaporation and plant uptake compensation factors, organic N and P enrichment ratio. • Land cover: maximum canopy storage, initial residue cover, Manning '*n*' for overland flow, etc. • Depressional storage areas: average daily outflow to main channel from tile flow, maximum volume of water stored in the pothole, equilibrium sediment concentration in pothole, etc.
	HRU soil	• Soil hydrological group, maximum rooting depth, fraction of porosity, potential or maximum crack volume, texture of soil layer, moist bulk density, available water capacity of the soil layer, depth from soil surface to bottom of layer, saturated hydraulic conductivity, organic C content, rock fragment content, moist soil albedo, USLE[a] soil erodibility factor, etc.
	HRU soil chemicals	• Information about initial nutrient (N and P) and pesticide levels of the soil.
	HRU management	• Information for planting, harvest, irrigation applications, nutrient applications, pesticide applications and tillage operations (optional).
	HRU groundwater	• Initial depth of water in shallow and deep aquifers, groundwater delay time, base-flow alpha factor, initial groundwater height, etc.
Reservoir (optional)	Reservoir	• Sub-basin number with which the reservoir is associated, operational month/year, reservoir surface area, initial reservoir volume, initial and equilibrium sediment concentration, hydraulic conductivity of the reservoir bottom, etc.
	Water quality	• Beginning month of mid-year nutrient settling period, N and P settling rates, initial concentration of organic and soluble P in reservoir, chlorophyll a production coefficient, etc.

[a]QUAL2E, Enhanced Stream Water Quality Model; SCS, USDA Soil Conservation Service; USLE, Universal Soil Loss Equation.

Table 3.4. Input parameters of the WEPP (Water Erosion Prediction Project) model.

Model	Input data file	Input parameters
Hillslope	Climate	• Daily precipitation, storm duration, maximum storm intensity, time to maximum storm intensity, maximum and minimum temperatures, solar radiation, wind speed and direction, and dew point temperature. • The CLIGEN weather generator (optional; continuous or single storm climate files, or TR-55[a] design single storm climate file) requires a stations file and a database file. • Break-point precipitation input (optional).
	Slope	• Number of overland flow elements (OFEs), number of slope points on the OFE, length of the OFE, slope orientation, length, steepness at points down the profile and profile aspect, representative width for a hillslope profile.
	Soil	• Number of OFEs, soil albedo, number of soil layers, soil depth, initial soil water content, soil textures, effective hydraulic conductivity (calculated internal to the WEPP), rock content, per cent organic matter, cation exchange capacity (CEC), inter-rill erodibility, rill erodibility and critical shear stress.
	Cropping/management	• A group of data required for each section: (i) information (e.g. number of OFEs, years of simulation, etc.), (ii) plant growth parameters, (iii) operation (e.g. tillage and other implement parameters), (iv) OFE- or channel-specific initial conditions and parameters, (v) surface effects (e.g. tillage sequences, etc.), (vi) irrigation, (vii) contouring parameters, (viii) drainage parameters, and (ix) yearly management information and scenarios.
Watershed	Hillslope information	• Number of hillslopes, simulation period, information about watershed climate, hillslope input climate, hillslope area and particle size distribution. • Runoff duration, time of concentration, alpha value for peak calculation, runoff depth and volume, soil detachment and deposition, sediment concentration for each particle class, fraction of sediment in each particle class, etc.
	Watershed structure/ configuration	• Number and Identification of all hillslopes, channels or impoundments draining from the top or laterally from the left or right. • Restrictions applied to the watershed configuration: (i) hillslopes – fed by: nothing; feed: channels, impoundments; (ii) channels – fed by: channels, impoundments, hillslopes; feed: channels, impoundments, nothing (outlet); and (iii) impoundments – fed by: channels, hillslopes; feed: channels, nothing (outlet).
	Channel slope	• Similar to the hillslope slope input. • Number of channels, aspect, width, number of slope points for the channel, slope steepness at point, and length of the channel.
	Channel soil	• Each channel's soil characteristics (identical to the hillslope soil input file in which the number of channels replaces the number of OFEs).
	Channel climate	• Only one climate file for all channels (identical to the hillslope climate file).

Table 3.4. Continued.

Model	Input data file	Input parameters
	Channel management	• Each channel management practices (identical to the hillslope management file in which the number of channels replaces the number of OFEs).
	Watershed channel file	• Choice of runoff peak calculation method (modified EPIC[a] or CREAMS[a]), channel shape and hydraulic/routing parameters (e.g. rating curve at the outlet, Manning '*n*' coefficient, erodibility factor, critical shear stress, etc.), control structure parameters, etc.
	Impoundment (if present)	• Number of impoundments, input parameters for each outflow structure (e.g. drop spillway, perforated riser, culvert, emergency spillway, user specified stage–discharge relationship, etc.), etc.
	Channel irrigation (if used)	• Irrigation dates and application rates for each channel element (similar to the hillslope irrigation input file in which OFEs would be replaced by channel elements).

[a]CREAMS, Chemicals, Runoff, and Erosion from Agricultural Management Systems; EPIC, Erosion Productivity Impact Calculator; TR-55, USDA SCS (Soil Conservation Service) procedures for calculating storm runoff volume, peak rate of discharge, hydrographs and storage volumes required for flood-water reservoirs.

in enhancement of various GIS linkage and other interface tools has emerged to support the integration of topographic, land-use, soil and other digital data into the models.

APEX

Some examples of APEX modifications or adaptations in user interfaces are presented in Table 3.6. Several non-GIS-enabled interfaces developed by different authors include a DOS-based Universal Text Integration Language (UTIL) interface (Williams *et al.*, 2006), and two Windows-based interfaces, WinAPEX (Magre *et al.*, 2006; Steglich and Williams, 2008) and Interactive APEX (i_APEX). Other adaptations in user interfaces include: the SWAT-APEX interface called SWAPP (Saleh and Gallego, 2007); an enhanced version of SWAPP and the Comprehensive Economic and Environmental Optimization Tool (CEEOT-SWAPP) (Saleh *et al.*, 2008); the CEEOT-Macro Modeling System (MMS) (Osei *et al.*, 2004); a software package that interfaces the SWAT with the APEX in ArcGIS platform (APEX-SWAT); a combination of ArcGIS and the WinAPEX modelling system called the WinAPEX-GIS; and an interface as an extension to the ArcGIS 9.× called the ArcAPEX (Tuppad *et al.*, 2009).

SWAT

Some examples of SWAT modifications are presented in Table 3.7. These include the SWAT-G model (Lenhart *et al.*, 2002, 2005), the Extended SWAT (ESWAT) model (van Griensven and Bauwens, 2003, 2005) and the Soil and Water Integrated Model (SWIM) (Krysanova *et al.*, 1998, 2005). SWAT modifications also include: the SWAT-M (Du *et al.*, 2005); the SWAT-N (Pohlert *et al.*, 2007); the SWAT-variable source area (VSA) (Easton *et al.*, 2008); modification of SWAT's groundwater component (SWATMOD) (Sophocleous *et al.*, 1999); development of a simplistic field-scale vegetative filter strip (VFS) sub-model to enhance SWAT's capability of evaluating the effectiveness of VFS at watershed scale (White and Arnold, 2009); modification of SWAT2005 to provide faster percolation through the soil substrate and recharge of the aquifer in karst watershed (Baffaut and Benson, 2009); and incorporation of new shallow water table-depth algorithms into SWAT2005 (Moriasi *et al.*, 2009). Notable adaptations in the SWAT interfaces include: the ArcView-SWAT (AVSWAT) interface tool (Di Luzio *et al.*, 2004a,b); the most recent version of AVSWAT called AVSWAT-X; the Automated Geospatial

Table 3.5. Input parameters of the RZWQM (Root Zone Water Quality Model).

Processes/data file	Input parameters
Infiltration, water redistribution, and macropore flow	Break-point rainfall: a minimum of two pairs of rainfall amounts and times. Daily meteorology: maximum and minimum air temperature, wind run, solar radiation and relative humidity. Soil crust saturated hydraulic conductivity, soil texture, horizon delineation, bulk density, soil water retention curves or, 1/3 or 1/10 bar soil water contents, initial soil water content. Lateral sorptivity reduction factor, macroporosity, effective soil radius, fraction dead-end macropores, average radius of cylindrical pores, length and width of cracks, depth of cracks.
ET (evapotranspiration)	Albedo of dry and wet soils, albedo at crop maturity, albedo of fresh residue, pan coefficient for pan evaporation, dry mass of surface residue.
Tile drainage	Drain depth and spacing, radius of drains, water table leakage rate and lateral saturated hydraulic conductivity.
Heat transport	Soil textural class, dry volumetric soil heat capacity and initial soil temperature.
Plant growth	Maximum N uptake rate, photosynthate to respire, specific leaf density, plant density, propagule age effect, seed age effect, maximum rooting depth, minimum leaf stomatal resistance, N sufficiency index and luxurious N uptake factor.
Organic matter/N cycling	Fast and slow residue pools, fast humus pool, transition humus pool, stable humus pool, aerobic and anaerobic heterotrophs pools, autotrophs pool, initial urea-N, initial NO_3-N, and initial NH_4-N.
Pesticide processes	Freundlich sorption coefficient and exponent, parameters related to kinetic and irreversibly bound pesticide sorption, acid/base dissociation constants, parameters related to pesticide wash-off from foliage and mulch, pesticide half-life (foliar, residue, soil surface, and soil subsurface), half-life adjustment coefficient for soil depth, half-life adjustment coefficients for soil temperature and water content, and metabolite formation fraction
Chemical transport	Non-uniform mixing factor (based on soil type, surface roughness, and cover conditions), fraction microporosity and diffusion rate.
Agricultural management	Management timing (e.g. fertilizer application date, tillage date), management or application type and quantity (e.g. quantity of fertilizer surface broadcast, chisel plough), and initial surface residue properties (e.g. C:N ratio, dry mass of residue, age of residue).

Watershed Assessment (AGWA) ArcView-based interface tool (Miller *et al.*, 2007) for the SWAT2000 and Kinematic Runoff and Erosion (KINEROS2) model (Woolhiser *et al.*, 1990); and the SWAT interface compatible with ArcGIS version 9.×, called the ArcSWAT (Olivera *et al.*, 2006). A variety of other tools have been developed to support executions of the SWAT simulations, for instance: the Interactive SWAT (i_SWAT) developed by the Center for Agricultural and Rural Development, Iowa (Gassman *et al.*, 2007) to support SWAT simulations using a Windows interface with a Microsoft Access database; the Conservation Reserve Program (CRP) Decision Support System (CRP-DSS) (Rao *et al.*, 2006); SWAT Calibration and Uncertainty Procedures (SWAT-CUP); and a generic interface (iSWAT) program (Abbaspour *et al.*, 2007) that automates parameter selection and aggregation for the iterative SWAT calibration.

WEPP

The complexity and spatio-temporal variability of parameters involved in erosion processes require managing large quantities of data for the WEPP model at the watershed scale.

Table 3.6. APEX (Agricultural Policy Environmental eXtender) modification or adaptation in model interfaces.

Reference/source	Modification/adaptation in model interfaces	Purpose	Description/comments
Osei *et al.* (2004)	CEEOT[a]-Macro Modeling System (CEEOT-MMS)	Macro-level policy assessments	The CEEOT-MMS integrated APEX, Farm Economic Model (FEM), supporting data sets, and an automated interface between the models and databases.
Magre *et al.* (2006); Steglich and Williams (2008)	WinAPEX	Watershed builder and editing tools for APEX version 0604	The interface facilitates users through a series of screens to construct input data for individual sub-areas and assess the impacts of alternative scenarios on APEX outputs.
Saleh and Gallego (2007)	SWAT[a]-APEX (SWAPP) in ArcVIEW 3.x environment	APEX simulation within SWAT	The automated program converts SWAT files to-and-from the APEX format and simulates SWAT and APEX simultaneously. SWAPP can be used to convert SWAT-GIS data layers created by the ArcView SWAT (AVSWAT) interface (Di Luzio *et al.*, 2004a,b) to the formats of APEX and EPIC data files.
Saleh *et al.* (2008)	CEEOT-SWAPP	Estimates of net farm returns and economic indicators	An enhanced version of SWAPP that supports an expanded interface between FEM and APEX and/or SWAT to estimate net farm returns and other economic indicators for different representative farms.
Gassman *et al.* (2009)	i_APEX	Automatic input file builder and execution program to support APEX	i_APEX performs essentially the same functions as the predecessor run_APEX for a user-defined set of APEX simulations, including automatic management of the input data, execution of each APEX run and storage of the selected model outputs.
Gassman *et al.* (2009)	APEX-SWAT interface	Software package that interfaces SWAT with APEX within the ArcGIS environment	The interface provides overall modelling support similar to that in SWAPP and takes advantage of the ArcGIS. It allows stand-alone APEX and SWAT simulations, as well as integrated APEX-SWAT scenarios.
Gassman *et al.* (2009)	WinAPEX-GIS	Combined ArcGIS and WinAPEX modelling system	WinAPEX-GIS has been developed and is being used to build input files and execute a 64-bit version APEX (version 0806, not publically released) for the Bosque River watershed, Texas.
Tuppad *et al.* (2009)	ArcAPEX	ArcGIS 9.x-based user interface	ArcAPEX is designed to automate APEX parameterization using topographic, hydrological, land-use and soil spatial data sets, and APEX's databases (e.g. plant, tillage, fertilizer, pesticide and weather). ArcAPEX provides direct integration with SWAT using the ArcSWAT interface.

[a]CEEOT, Comprehensive Economic and Environmental Optimization Tool; SWAT, Soil and Water Assessment Tool.

Table 3.7. SWAT (Soil and Water Assessment Tool) modification or adaptation in model interfaces.

Reference/source	Modification/ adaptation in model interfaces	Purpose	Description/comments
Krysanova *et al.* (1998, 2005)	SWIM[a]	Combination of the SWAT model's hydrological and MATSALU[a] model's nutrient cycling components	Further SWIM enhancements include enhanced capability for forest systems (Wattenbach *et al.*, 2005), and developed routines for wetlands and riparian zones (Hatterman *et al.*, 2006).
Sophocleous *et al.* (1999)	SWAT and MODFLOW interface	SWAT modification for simulating surface water, groundwater, and stream–aquifer interactions	SWAT and the groundwater model MODFLOW[a] (McDonald and Harbaugh, 1988) with stream–aquifer interaction routines were modified and linked into a comprehensive basin model known as SWATMOD.
Lenhart *et al.* (2002, 2005); Eckhardt and Ulbrich (2003)	SWAT-G	SWAT modification for improved flow in low mountain range catchments	The modifications include enhanced percolation, hydraulic conductivity and interflow functions in SWAT99.2, an improved erosion loss estimation method, and a more detailed accounting of CO_2 effects on leaf area index and stomatal conductance.
van Griensven and Bauwens (2003, 2005)	ESWAT[a]	Modified time step in hydrology and erosion components	The modifications incorporate sub-hourly precipitation inputs and infiltration, runoff and erosion loss estimates. A river routing module on an hourly time step interfaces with a water quality component, and with multi-objective calibration and autocalibration modules.
Di Luzio *et al.* (2004a,b)	AVSWAT	The SWAT-GIS interface in ArcView 3.x GIS (AVSWAT)	The first SWAT-GIS interface was built within the GRASS GIS (Srinivasan and Arnold, 1994). AVSWAT facilitates input parameterization and execution of SWAT2000, which was incorporated within the US EPA[a] Better Assessment Science Integrating point and Nonpoint Sources (BASINS) software package. The AVSWAT-X version facilitates soil data input from the USDA NRCS STATSGO and SSURGO soil databases, automatic sensitivity, calibration and uncertainty analysis for the SWAT2005.
Du *et al.* (2005)	SWAT-M	Application to landscape with tiles/potholes	The modification enhances the capability of simulating landscapes with tiles and potholes.
Olivera *et al.* (2006)	ArcSWAT	The SWAT interface in the ArcGIS 9.x	The interface uses a geo-database approach and a programming structure consistent with Component Object Model (COM) protocol.

Continued

Table 3.7. Continued.

Reference/source	Modification/ adaptation in model interfaces	Purpose	Description/comments
Borah *et al.* (2007)	SWAT and DWSM[a]	Enhanced storm-event hydrological simulation in SWAT	A combination of an extended SCS curve number (CN) procedure for rainfall-excess computations and a robust analytical solution of the kinematic wave equations from the storm-event DWSM model.
Pohlert *et al.* (2007)	SWAT-N	Enhanced nitrogen simulation for SWAT	An extension of SWAT algorithms from a detailed nitrogen turnover model for predicting nitrogen leaching.
Miller *et al.* (2007)	AGWA	Alternative ArcView-based interface tool	The tool supports data input generation for both SWAT2000 and KINEROS2, including options for soil input from the SSURGO, STATSGO or FAO global soil maps.
Easton *et al.* (2008)	The SWAT–VSA	Distribution of overland flow in ways consistent with VSA hydrology	The modification provides a simple means of capturing spatially variant saturation-excess runoff processes from the landscape.
Ouessar *et al.* (2009)	The SWAT–WH (SWAT for Water Harvesting)	Simulation of main hydrological processes in arid environments	The SWAT was modified for water balance assessments in arid watersheds with water-harvesting systems.

[a]DWSM, Dynamic Watershed Simulation Model; EPA, US Environmental Protection Agency; FAO, UN Food and Agriculture Organization; ESWAT, Extended SWAT; GRASS GIS, Geographic Resources Analysis Support System GIS; KINEROS2, Kinematic Runoff and Erosion model; MATSALU, a system of four simulation models for a mesoscale agricultural watershed and the ecosystem of a sea bay developed in Estonia for the Matsalu Bay (Baltic Sea) and the Bay ecosystem; MODFLOW, modular finite-difference groundwater flow model; NRCS, USDA Natural Resources Conservation Service; SCS, USDA Soil Conservation Service; SSURGO, Soil Survey Geographic Database; STATSGO, State Soil Geographic Database, now renamed US General Soil Map (STATSGO2); SWIM, Soil and Water Integrated Model.

Some examples of the adaptation of WEPP in interfaces or as other modifications are presented in Table 3.8. These include: GeoWEPP or GeoWEPP ArcX (Renschler, 2003); the Web-based interfaces for WEPP (Elliot, 2004); the Web-based Erosion Risk Management Tool (ERMiT) and WEPP (Robichaud *et al.*, 2007); GeoWEPP-SWAT (Renschler and Lee, 2005); the Dynamic soil erosion model based on WEPP (DWEPP) (Bulygina *et al.*, 2007); the Geospatial Modelling of Soil Erosion (GEMSE) interface (Baigorria and Romero, 2007); and the GeoWEPP interface in the ArcGIS 9.× environment (Minkowski and Renschler, 2010).

RZWQM

Some examples of RZWQM adaptations or enhancements are presented in Table 3.9. Notable enhancements of the model include: (i) the incorporation of DSSAT's crop growth modules (e.g. the RZWQM-CROPGRO and RZWQM-CERES (Crop Environment Resource Synthesis) hybrid models, and the RZWQM-DSSAT4.0) to provide state-of-the-science plant growth simulation (Ma *et al.*, 2005, 2006, 2008); (ii) the coupling of RZWQM and the Simultaneous Heat and Water (SHAW) model (RZ-SHAW) to simulate surface energy balance and frozen soils (Flerchinger *et al.*, 2000); (iii) the enhancement of the RZWQM water movement module (RZWFLO) to simulate fluctuating water tables in the root zone and in tile drainage conditions (Johnsen *et al.*, 1995); (iv) the extension of the soil profile to 30 m to utilize RZWQM simulation results in a groundwater flow model for evaluating management effects on groundwater contamination at the regional scale; (v) the modification of RZWQM to simulate tile flow under controlled drainage and subsurface lateral flow below the tile (Ma *et al.*, 2007b,c); and (vi) enhancement in expressing a portion of pesticide directly into tile flow via a fraction of macropores (Fox *et al.*, 2004). In addition to improvement in the Windows-based interface for facilitating model parameterization and post-data analysis with experimental data, the RZWQM model has been linked to GIS for watershed modelling, e.g. RZWQM-ArcVIEW 3.× GIS (Wang and Cui, 2004), AgSimGIS-RZWQM in the ArcGIS 8.3 environment (Ascough *et al.*, 2003) and the MARIA (Management of Agricultural Resources through Integrated Assessment)-GIS model and RZWQM in the ArcGIS 9 environment (Ascough *et al.*, 2005). RZWQM simulation results were also used to develop a database for decision support and economic analysis (Heilman *et al.*, 2006).

Other Watershed Models

The major components or capabilities, processes and modelling approaches of the ANSWERS-2000, AnnAGNPS, DHSVM, DWSM, MIKE SHE and SHETRAN models are summarized in Table 3.10.

The ANSWERS-2000 model, which was developed from the original event-based ANSWERS model, is a distributed, physically based, continuous, watershed-scale and upland planning model (Beasley *et al.*, 1980). The model evaluates the effectiveness of agricultural and urban BMPs in terms of reducing sediment and nutrient delivery to streams, as well as assessing the leaching of nitrogen through the root zone. It utilizes components from the WEPP, EPIC and GLEAMS models, and a series of other subroutines to evaluate sediment and nutrient delivery to streams and nitrate leaching. However, the model does not have channel sediment and chemical routing processes. A user interface aids in the selection of parameters, including information from soil surveys, climate and, topographic and land-use maps, as well as providing some default values. ANSWERS-2000 can also utilize data exported from ArcView, ArcInfo or similar GIS packages.

The AnnAGNPS model is a distributed, physically based, watershed-scale and multi-event modification of the AGNPS (Agricultural Nonpoint-Source Pollution) model (Young *et al.*, 1987), with improved technology and additional new features to analyse the impact of non-point source pollutant loadings from agricultural watersheds. It expands the capabilities of the single-event AGNPS model, and other models that simulate additional processes have been integrated into AnnAGNPS within the AGNPS suite of modules. AnnAGNPS

Table 3.8. WEPP (Water Erosion Prediction Project) modification or adaptation in model interfaces.

Reference/ source	Modification/ adaptation in model interfaces	Purpose	Description/comments
Renschler (2003)	GeoWEPP	Geospatial interface for WEPP (GeoWEPP) in ArcView 3.x environment	A GIS-based graphical user interface (GUI) for preparing model inputs, running WEPP hillslope and watershed models, and analysing outputs. Watershed configuration (e.g. channels, representative hillslope) is based on the TOPAZ[a] code. The GeoWEPP ArcX includes scaling theory to derive topographical input parameters based on a digital elevation model (DEM).
Elliot (2004)	WEPP Internet interface	Internet interfaces developed to more easily predict soil erosion for a wide range of climatic and forest/rangeland conditions	A suite of a set of Internet interfaces to run WEPP for many common forest conditions on the USDA Forest Service Internet servers using a Web browser, including: the WEPP-Road interface for predicting road erosion and sediment delivery; Disturbed-WEPP for predicting erosion from hillslopes disturbed by forest operations, prescribed fire or wildfire; and Rock-Clime for generating weather inputs for WEPP-Road and Disturbed-WEPP interfaces or WEPP stand-alone applications.
Renschler and Lee (2005)	GeoWEPP-SWAT[a] model	A single assessment tool that allows for more detailed, spatially explicit assessment of beat management practices (BMPs) over both short- and long-term temporal scales	The GeoWEPP-SWAT model utilizes GIS or precision farming data sets of topography, soils and land use to automatically derive WEPP model input. and uses WEPP model output as point sources into SWAT. The GeoWEPP-SWAT model provides a mechanism for applying WEPP to larger watershed scales.
Baigorria and Romero (2007)	Geospatial Modelling of Soil Erosion (GEMSE) interface	Assessment of erosion hot spots in an Andean watershed	A Windows-based software interface that integrates GIS with the WEPP hillslope model.
Bulygina *et al.* (2007)	Dynamic soil erosion model based on WEPP (DWEPP)	Tool for assessing erosion rates and dynamics based on WEPP sediment source and sink terms	DWEPP provides a dynamic version of WEPP based on kinematic wave routing of the runoff module of the KINEROS2[a] and WEPP sediment source term equations coded with a fully dynamic solution.

Table 3.8.

Reference/ source	Modification/ adaptation in model interfaces	Purpose	Description/comments
Robichaud *et al.* (2007)	ERMiT[a] and WEPP	ERMiT Web- and WEPP-based application to estimate erosion	User inputs are processed by ERMiT to combine rain event variability with spatial and temporal variability of hillslope burn severity and soil properties, which are then used as WEPP inputs. ERMiT produces a distribution of rain event erosion rates with a probability of occurrence, and erosion rate distributions for post-fire hillslopes treated by seeding and with straw mulch and erosion barriers – such as contour-felled logs or straw wattles.
Dun *et al.* (2009)	Modified WEPP	Enhanced capability of simulating forest watershed hydrology and erosion	The algorithms and subroutines of WEPP version 2004.7 are modified for better forest subsurface hydrological processes such as percolation and subsurface lateral flow (i.e. WEPP version 2008.9).
Minkowski and Renschler (2010)	GeoWEPP ArcGIS 9.x	GeoWEPP in the ArcGIS 9.x environment	GeoWEPP uses digital geo-referenced information such as DEM and topographic maps to delineate the watershed.

[a]ERMiT, Web-based Erosion Risk Management Tool; KINEROS2, Kinematic Runoff and Erosion model; SWAT, Soil and Water Assessment Tool; TOPAZ, TOpographic PArameteriZation.

includes: (i) a 1-D channel model to integrate the impact of upland loadings and channel characteristics; (ii) a stream corridor computer model to predict bank erosion and failures, bank mass wasting, bed aggradation and degradation, etc.; (iii) a watershed-scale stream-network water-temperature model; and (iv) a model to quantify the impact of pollutant loadings on spawning and rearing habitats. A number of modules for preparing input data that support the AGNPS databases are included: (i) TOPAZ (TOpographic PArameteriZation) modules (TOPAZ and TOPAGNPS) (Garbrecht and Martz, 1999) to generate cell and stream network information from a watershed DEM and provide the topographic-related information; (ii) a program to determine the topographic-related input parameters and format the TOPAGNPS output; (iii) a weather generator; (iv) a graphical input editor; (v) a visual interface program to view the TOPAGNPS-related GIS data; and (vi) an output processor to analyse the results from AnnAGNPS in tabular or GIS format.

The DHSVM model is a process-based, distributed-parameter, physically based hydrological model for simulating runoff, erosion and sediment transport processes in forested, mountainous watersheds. The model provides a dynamic representation of watershed processes at a spatial scale described by user-defined DEM data, and accounts for the effects of topography and land cover, including roads, by explicitly representing the spatial distribution of stream and road networks, stream and road morphology, soil properties, soil depth, vegetation properties and elevation. The DHSVM model has a mass-wasting component for slope failure prediction and downslope redistribution of material released from slope failures. The mass-wasting component in the model is stochastic in nature

Table 3.9. RZWQM (Root Zone Water Quality Model) modification or adaptation in model interfaces.

Reference/ source	Modification/adaptation in model interfaces	Purpose	Description/comments
Johnsen *et al.* (1995)	Enhanced water movement module (RZWFLO) of the RZWQM	Effects of agricultural management systems on water quality under fluctuating water tables and tile drainage conditions	The RZWFLO and a numerical model, WAFLOWM, that contains an optimizing dynamic gridding scheme were evaluated using the mass-conservative simulation technique for solving the 1-D soil water flow equation under fluctuating water tables in the root zone and tile drainage conditions.
Flerchinger *et al.* (2000)	RZWQM-SHAW (RZ-SHAW) model	RZWQM coupled with the SHAW[a] model to address wintertime processes and soil freezing	Incorporating snow, soil heat and soil freezing routines from the SHAW model into the RZWQM model extends its applicability to simulating winter conditions as well as heat and water transfer through canopy, stubble and residue layers.
Ascough *et al.* (2003)	AgSimGIS coupled with RZWQM in the ArcGIS 8.3 environment	Simulation of strategic planning scenarios across spatially variable agricultural land units	The AgSimGIS couples the GIS framework to the modified (i.e. quasi-distributed parameter) of RZWQM and facilitates a multifunctional system that provides an interface between users and the software, manages spatially referenced data, and interprets geo-referenced spatial data.
Wang and Cui (2004)	RZWQM linked with ArcVIEW 3.x GIS	Enhanced use of RZWQM's lumped structure to account for soil heterogeneity and associated non-uniformity of management practices across the field	RZWQM linked with ArcVIEW 3.x provides a quasi-distributed structure and can be executed cell-by-cell where a cell represents a smaller homogeneous area of the field.
Ascough *et al.* (2005)	MARIA-GIS[a] in ArcGIS 9 environment	Enhanced interface for hydrological modelling across spatially variable agricultural landscapes	The MARIA-GIS offers a spatial framework for integrating the modified RZWQM with interaction between simulated land areas via overland runoff and run-on.
Ma *et al.* (2005)	RZWQM-CROPGRO hybrid model	RZWQM-CROPGRO hybrid model for yield components and phenology simulations within RZWQM	The CROPGRO plant growth model of DSSAT3.5[a] is linked to RZWQM. A FORTRAN subroutine facilitates data and information transfer between RZWQM and CROPGRO models.
Ma *et al.* (2006)	RZWQM-CERES[a] hybrid model	The RZWQM–CERES-Maize hybrid model for maize production	RZWQM-CERES-Maize version 3.5 offers RZWQM users access to a rigorous new plant growth model and provides CERES-Maize users with a tool to address soil and water quality issues under different cropping systems.
Fang *et al.* (2008)	Enhancement of the RZWQM-CERES hybrid model	Enhanced RZWQM-CERES capability for evaluating interactions between N and irrigation management	The modified model evaluates different N application rates at a fixed irrigation level in a double cropping system, as well as auto-irrigation based on soil water depletion and associated N uptake and leaching.
Ma *et al.* (2008)	RZWQM-DSSAT 4.0[a]	Evaluation of simulated effects of N management and soil microbes on soil N balance and crop production	RZWQM-DSSAT4.0 enhances the applicability of RZWQM2 with dynamic and constant soil microbial populations for soil C and N balances.

[a]CERES, Crop Environment Resource Synthesis model; DSSAT, Decision Support System for Agrotechnology Transfer; MARIA-GIS, Management of Agricultural Resources through Integrated Assessment-GIS model; SHAW, Simultaneous Heat and Water model.

Table 3.10. Summary of the AnnAGNPS, ANSWERS-2000, DHSVM, DWSM, MIKE SHE, and SHETRAN models.[a]

Component/ process/criteria/ capability	AnnAGNPS	ANSWERS-2000	DHSVM	DWSM	MIKE SHE	SHETRAN
Model components/ processes	Climate; runoff; infiltration; evapotranspiration (ET); snow melt; subsurface flow; overland and channel sediment; nutrient and pesticide; reservoir routing; irrigation; pollutant sources accounting.	Climate; infiltration; runoff; percolation; ET; overland detachment, sediment; channel routing; N and P transformations; nutrient losses through uptake, runoff and sediment.	Climate; interception; throughfall; infiltration- and saturation-excess runoff; ET; transpiration; snow accumulation and melt; channel/road ditch flow; multi-layer unsaturated flow; subsurface flow; hillslope/forest roads/channel erosion and sediment; mass wasting.	Distributed rainfall, hyetograph for each overland and rainfall excess; surface and subsurface flow; soil erosion; sediment transport; agrochemical mixing and transport; channel erosion and routing; sediment and agrochemical routing through reservoirs; detention basins; alternative ground covers; tile drains.	Climate; interception, ET; overland and channel flow; unsaturated and saturated flow; snow melt; channel/ surface-aquifer exchanges; advective-dispersive solute transport; geochemical processes; crop growth and root zone N; erosion; irrigation; urban runoff.	Climate; interception; ET; evaporation and transpiration; overland/overbank/ channel runoff; snowpack and snow melt, subsurface flow; aquifers; river–subsurface water interactions; erosion and overland/ channel transport; overland/channel/ subsurface multiple, reactive solutes transport.
Watershed representation/ spatial scale	Homogenous land areas (cells); reaches; impoundments.	Square grids (not exceeding 1 ha) with uniform hydrological characteristics (soils, crop management, topography, etc.); channel elements; 1-D simulations.	Grid cells; cell size determines spatial resolution, relatively high spatial resolution (typically 10–150 m); stream channels are linked through grid cells.	Topographic-based natural boundaries or sub-watersheds, including 1-D overland elements (spatially distributed), channel segments and reservoir units.	2-D rectangular/square overland grids; 1-D channels; 1-D unsaturated and 3-D saturated flow layers.	Catchment scale; a single complete river basin, or a group of contiguous basins; 2-D overland grids; 1-D channels; bank elements; 3-D variably saturated flow layers.

Continued

Table 3.10. Continued.

Component/ process/criteria/ capability	AnnAGNPS	ANSWERS-2000	DHSVM	DWSM	MIKE SHE	SHETRAN
Temporal scale	Long-term; continuous; daily or sub-daily time step.	Long-term; continuous; daily and 30 s time step for days with and without rainfall, respectively.	Long-term; storm-event or continuous; hourly or longer time step.	Storm event; variable constant time steps.	Long-term; storm-event and variable time steps depending on numerical stability.	Long-term; continuous; less than 2 h time steps, which are reduced to as short as a few minutes during and immediately after heavy rainfall.
Rainfall – excess on overland/ water balance/ hydrological soil processes	Precipitation; snow melt; irrigation; infiltration; surface and channel runoff; ET; subsurface flow in two-layer soil system (tillage depth and user-defined second layer).	Precipitation; interception; surface retention; infiltration; surface and channel runoff; ET; percolation within the root zone in a single homogeneous soil layer.	Precipitation; two-layer interception and ET; throughfall; soil evaporation; transpiration; two-layer energy balance; overland and channel/road ditch/culvert flow; multi-layer 1-D unsaturated flow; 3-D saturated or subsurface flow; lateral subsurface flow; return flow.	Spatially varying rainfall or rainfall event; interception; infiltration; depression storage; overland and channel runoff; propagation of flood waves; subsurface and reservoir flows.	Precipitation; interception and ET; overland and channel flow; unsaturated flow using a 1-D implicit finite difference form of Richard's equation, 1-D explicit finite-difference gravity drainage approach or a simplified linear water balance approach; saturated flow; channel/ surface–aquifer exchanges.	Precipitation; interception; ET; evaporation and transpiration; overland/overbank/ channel runoff; snowpack development and snow melt, variably saturated subsurface flow and storage; confined, unconfined and perched aquifers; subsurface–channel water interactions.
Infiltration/surface runoff/overland water routing	SCS[b] curve number (CN) procedure similar to SWRRB[b] and EPIC[b] models;	Green and Ampt; Manning's and continuity equations (temporarily	Static or dynamic infiltration-excess and saturation-excess mechanisms; snow	Simple CN procedure or extensive interception, and Smith-Parlange infiltration	A 2-D finite-difference diffusive wave approximation of Saint-Venant equations or a	A 2-D routing based on the diffusive wave approximation of Saint-Venant equations, allowing

	SCS TR-55[b] method for peak flow.	variable and spatially uniform) solved by an explicit numerical scheme.	accumulation and melt using a two-layer energy balance model; routing using an explicit cell-by-cell finite-difference solution of kinematic wave approximation to Saint-Venant equations, or a unit hydrograph approach.	proce dure; routing based on kinematic wave approximation of Saint-Venant equations or shallow water wave using analytical and approximate shock-fitting solutions.	simplified sub-catch-ment-based kinematic routing approach.	backwater effects to be modelled.
Water redistribution and subsurface flow	Lateral subsurface flow using Darcy's equation; tile drain using Hooghoudt's equation and parallel drain approximation.	Unsaturated zone drainage using Darcy's gravity flow; subsurface flow using tile drainage coefficient and groundwater or interflow release fraction; a groundwater component was added.	Vertical unsaturated flow using Darcy's equation; a transient, 3-D saturated subsurface flow; lateral flow; return flow; groundwater recharge and discharge; a cell-by-cell routing using a kinematic or diffusion approximation.	Kinematic storage equation (similar to the SWAT[b]); lateral subsurface flow and tile-drain contribu-tions; groundwater flows are lumped into subsurface flows.	A 3-D implicit finite-difference solution of groundwater flow equation or a simple 2-D linear reservoir approach; the 3-D approach (confined and unconfined conditions) is numerically equivalent to MODFLOW[b].	Variably saturated flow equation (3-D); combinations of confined, unconfined and perched aquifers; transfers between subsurface and river waters; groundwater seepage discharge.
ET and plant/soil evaporation	Penman equation.	Adaptation of the Ritchie's method.	Two-layer model: evaporation from wet vegetation, and transpiration from dry vegetation using Penman–Monteith approach; a soil-physics-based approach for soil evaporation.	Not simulated (ET is assumed to be negligible during a storm event).	Actual ET as a function of potential eva-potranspiration (PET) input data, leaf area index (LAI) and soil moisture content.	Penman or Penman–Monteith equations or as fraction of PET (input data) rate; soil evaporation and transpiration are sink terms for 3-D subsurface flow equation.

Continued

Table 3.10. Continued.

Component/ process/criteria/ capability	AnnAGNPS	ANSWERS-2000	DHSVM	DWSM	MIKE SHE	SHETRAN
Overland sediment/water erosion	Modified RUSLE[b] for sheet and rill erosion; Hydro-geomorphic USLE[b] (HUSLE) model for sediment delivery ratio; deposition; tillage-induced ephemeral gully erosion.	Raindrop detachment using rainfall intensities and USLE factors; overland sediment using unit-width flow and USLE factors; transport and deposition using modified Yalin's equation.	Hillslopes/forest road erosion; sediment routing using a four-point finite-difference solution of 2-D mass conservation equation; slope failure; mass redistribution.	Raindrop erosion using relationship in terms of rainfall intensity squared and a reduction factor; overland discharge; sediment routing based on sediment transport capacity concept combined with mass conservation (continuity) equations.	Erosion and deposition for overland flow (cohesive transport); development of depressions and rills in a catchment.	Erosion by raindrop and leaf drip impacts, and overland flow; deposition; landslide and gully erosion; overbank transport; overland transport using advection-dispersion equation (2-D); riverbed and bank erosion; bed deposition.
Nutrients/ chemicals/ pesticides/ carbon/solutes	Dissolved and adsorbed chemicals; N and P, and organic C; nutrients/ pesticides are tracked using NRCS[b] soil database and crop information; reach routing includes fate and transport of N and P, pesticides and organic C.	Soil N (active organic, stable organic, nitrate and ammonium) and P pools (stable and active mineral, organic and exchangeable P); mineralization; ammonification; nitrification; denitrification; losses through uptake, runoff and sediment.	Not simulated.	Mixing of nutrients and pesticides and transport of dissolved and adsorbed forms in overland planes and channel segments using approximate analytical solutions of spatially and temporally varying continuity equations.	Solving numerically advection-dispersion equation for dissolved conservative solutes in surface, soil and groundwater, respectively; MIKE SHE can be used in combination with MIKE 11 for kinetic water quality processes.	Mobile/immobile advection-dispersion equation (3-D) for surface and subsurface solute transport (adsorption, radioactive decay, deposition, plant uptake, etc.); advection-dispersion equation for solute transport in 1-D channel network (adsorption, radioactive decay, overbank transport, etc.).

Channel routing/ runoff in channel	Manning's equation is numerically solved for hydraulic parameters and TR-55 for peak flow in trapezoidal and compound cross sections.	Manning's and continuity equations (temporally variable and spatially uniform) solved by an explicit, backward difference solution.	Rainfall interception, overland and subsurface flow contributions; routing based on a simple, robust linear storage algorithm, or a Muskingum–Cunge scheme.	Kinematic wave approximation of Saint-Venant equations or shallow water wave using analytical and approximate shock-fitting solutions.	A simple 1-D Muskingum–Cunge routing approach and implicit finite-difference 1-D approximation of Saint-Venant equations (kinematic, diffusive and fully dynamic).	A 1-D routing based on the diffusive wave approximation of Saint-Venant equations, allowing backwater effects to be modelled.
Channel erosion/ sediment	Modified Einstein's equation for sediment transport and Bagnold's equation for the sediment transport capacity of flow.	Not simulated.	Channel/roadside ditch sediment from debris flows or hillslope lateral inflow; discharge using a linear reservoir routing; transport capacity using Bagnold's equation; routing using a four-point finite-difference solution of the 2-D conservation of mass equation.	Stream-bed scouring; routing of eroded soil or sediment based on a physically based sediment transport capacity concept combined with mass conservation (continuity) equations.	MIKE SHE, together with MIKE 11, can be used for channel sediment; transport rates and bed level changes (non-cohesive transport).	Using advection-dispersion equation (transport in 1-D channel network) with terms for deposition and erosion and for infiltration into bed.
Reservoir/ impoundment routing and sediment	Average outflow during runoff event based on permanent pool storage and stage, runoff volume, and coefficients derived from elevation–storage relationships.	Not simulated.	Outflow for channel segment at culvert location is added to surface water for re-infiltration or overland flow; sediment discharges through culvert to the channel/hillslope.	Reservoir routing using the storage-indication or modified PULS method; inflowing sediment into a lake, reservoir or impoundment is assumed to be trapped and not routed through these units.	MIKE SHE, together with MIKE 11, can be used for reservoir, weirs, culverts, regulating structures and other user-defined structures.	No information.

Continued

Table 3.10. Continued.

Component/process/criteria/capability	AnnAGNPS	ANSWERS-2000	DHSVM	DWSM	MIKE SHE	SHETRAN
Evaluation of agricultural practices/best management practices (BMPs) and TMDLs[b]	Effects of BMPs (agricultural practices, ponds, grassed waterways, irrigation, tile drainage, vegetative filter strips, riparian buffers, etc.), and suitability for both sediment and nutrient total maximum daily loads (TMDLs).	Long-term effectiveness of BMPs in controlling runoff, sediment, and nutrient losses from agricultural/smaller watersheds, and may be used for both sediment and nutrient TMDLs.	No information on agricultural management; effects of management on sediment generation and transport, and may be used for sediment TMDLs.	Effect of some BMPs such as detention basins, alternative ground covers and tile drains, and may be used for both sediment and nutrient TMDLs.	Nutrient and pesticide management, decision support system for irrigation management, drainage, and may be used for both sediment and nutrient TMDLs.	Decision support system for agricultural basin, agricultural policy, fertilization, and may be used for both sediment and nutrient TMDLs.

[a]AnnAGNPS, Annualized Agricultural NonPoint Source; ANSWERS-2000, Areal Nonpoint Source Watershed Environment Response Simulation; DHSVM, Distributed Hydrology Soil Vegetation Model; DWSM, Dynamic Watershed Simulation Model; MIKE SHE, a development from the Système Hydrologique Européen (SHE) model; SHETRAN, Système Hydrologique Européen (SHE) TRANsport.
[b]EPIC, Erosion Productivity Impact Calculator; MODFLOW, modular finite-difference groundwater flow model (McDonald and Harbaugh, 1988); NRCS, USDA Natural Resources Conservation Service; RUSLE, Revised Universal Soil Loss Equation; SCS, USDA Soil Conservation Service; SWAT, Soil and Water Assessment Tool; SWRRB, Simulator for Water Resources in Rural Basins; TR-55, SCS procedures for calculating storm runoff volume, peak rate of discharge, hydrographs and storage volumes required for flood-water reservoirs; USLE, Universal Soil Loss Equation.

and results in an event probability of failure. However, the DHSVM model does not incorporate water-quality parameters or nutrient/chemical processes.

Developed at the Illinois State Water Survey, DWSM is a storm-event, distributed and physically based model for simulations of surface and subsurface storm water runoff, propagation of flood waves, soil erosion, and entrainment and transport of sediment and agricultural chemicals in primarily agricultural watersheds during a single or a series of rainfall events. Each of the three major components (i.e. hydrology, soil erosion and sediment and nutrient/pesticide) provides routing schemes based on approximate analytical solutions of physically based equations that preserve the dynamic behaviours of water, sediment and accompanying chemical movements within a watershed. The model has no routines for ET calculation, and rainfall excess and infiltration rates on overland are computed based on the assumption that ET losses are negligible during a storm event. DWSM routing schemes are adequately formulated to analyse extreme single-event storms and evaluate structural BMPs designed to withstand severe actual or designed single-event storms.

MIKE SHE is a deterministic, fully distributed, watershed-scale (ranging from less than one to several thousand square km) and physically based hydrological and water quality model that is capable of simulating water, sediment and water-quality parameters in two-dimensional (2-D) overland grids, 1-D channels, 1-D unsaturated and three-dimensional (3-D) saturated flow layers. MIKE SHE was originally derived from the Système Hydrologique Européen (SHE) model (Abbott *et al.*, 1986a,b), which was developed by a consortium of the Danish Hydraulic Institute (DHI), the British Institute of Hydrology and the French consulting company SOGREAH. The MIKE SHE model has been coupled with the MIKE 11 hydraulic modelling system to incorporate complete dynamic wave formulation of the Saint-Venant equations. However, the multidimensional flow-governing equations with numerical solution schemes may make the model computationally intensive and subject to numerical instabilities for long-term continuous simulations in medium-to-large watersheds. MIKE SHE (version 2008) provides a comprehensive Windows-based GUI and numerical model engines, including a model database, and pre- and post-processing utilities that work within the GUI. The model also has a GIS interface allowing for the preparation and presentation of model input and output in the GIS environment. Because MIKE SHE requires extensive input data for the model parameterization, it is difficult to apply to ungauged watersheds.

SHETRAN is a physically based, distributed, deterministic, integrated surface and subsurface modelling system to simulate water flow, sediment transport and contaminant transport at the catchment scale. The model was developed by integrating: (i) upgraded ET, snow melt and surface water flow components from the SHE model; (ii) a variably saturated subsurface flow component; (iii) a revised sediment transport component of the SHESED model (a physically based, distributed erosion and sediment yield component for the SHE hydrological modelling system; Wicks and Bathurst, 1996) that allows surface erosion, multi-fraction transport on the ground surface and in stream channels, and deterministic mass wasting; and (iv) a component for combined surface/subsurface multiple, reactive solute transport (Ewen, 1995). The SHETRAN model is distributed as either a Windows-based GUI version for the rapid set-up of a river basin using GIS data, or as a standard version that uses text-based files. Like the MIKE SHE model, the large requirement for input data and computational time may restrict use of the SHETRAN model to small, extensively instrumented catchments.

Summary

The compilation of available recent information on model components, processes and modelling methods reveals that most of the ten selected models are comprehensive, pseudo-physically based, distributed parameters (except for the field-scale RZWQM model), and continuous-time (except for the event-based DWSM) in character. The models include a wide spectrum of land-phase components of the hydrological cycle in the watershed soil–vegetation–atmosphere system, as well as the routing phase of the hydrological cycle, which controls the movement of sediment, nutrients and agrochemicals (except WEPP) in soil, across complex landscapes and in channel systems (except for RZWQM, and the no-channel sediment/nutrient routing processes in ANSWERS-2000). The APEX, SWAT, WEPP, AnnAGNPS, ANSWERS-2000, DHSVM, MIKE SHE and SHETRAN watershed models are suitable for the study of both long-term conditions and single-event storms. However, owing to its use of daily time steps, the SWAT model may not simulate single-event storms adequately. Among the watershed-scale models, the APEX, SWAT, WEPP, AnnAGNPS, ANSWERS-2000 and MIKE SHE are particularly useful for assessments of climate variability and hydrological changes and agricultural management practices. Depending on factors such as the objectives of the hydrological watershed modelling, watershed soil–crop–climate characteristics, environmental and management concerns, and the availability of data, the information summarized in Tables 3.1 and 3.10 could prove valuable for selecting the most suitable model for a range of criteria or aspects of modelling. Some of these criteria for selecting a given model include: the climatic and hydrological conditions for model suitability; model structure, capabilities and limitations; required input data; data acquisition and processing tools; desired temporal and spatial scales; expected accuracies and uncertainties; agricultural management or BMP implementation; computer resources and user's skills, etc.

As described for the APEX, SWAT, WEPP and RZWQM models (Tables 3.6–3.9) major directions in model enhancements since the year 2000 have been to: (i) enhance model capabilities by including many physical, chemical and biological processes; (ii) extend the different modelling processes or components from 1-D to 2-D or 3-D; (iii) expand field scale to landscape and watershed scales through the incorporation of GIS functionalities and databases into GIS packages – as well as for evaluating the impact of climate change; (iv) enhance computational techniques using more powerful computing resources, or by upgrading the model using better computer technology (e.g. better modularization; watershed discretization or configuration and parameterization tools; incorporation of uncertainty analysis, multi-objective functions for modelling; more comprehensive calibration–validation processes, etc.); and (v) integrate or combine components or processes of complementary models that facilitate the simultaneous use of these models. Although the development and sophistication of these existing watershed models, together with the advent of and rapid progress in computer and geo-information technology, continue at present and will continue in the years to come, the selection of most appropriate model for an application and for a certain watershed remains quite a challenging task.

Climate change and its influence on soil hydrology are uncertain because of the large uncertainties associated with climate change predictions. Broadly, in the current state of knowledge of the hydrological consequences of climate change, modelling approaches brought by hydrological and atmospheric scientists to the forefront of research include: the development and use of general circulation models (GCMs) to provide future global climate scenarios under the effect of increasing atmospheric CO_2 or other greenhouse gases; downscaling techniques (e.g. nested regional climate models (RCMs) and statistical methods) for downscaling GCM output to scales compatible with hydrological models; and hydrological models to simulate the effects of climate change on hydrological regimes. The application of these climate change scenarios in impact assessment is still limited because of the uncertainty in projections at regional or small scales, and the lack of spatial specificity in these projections. If substantial changes in precipitation patterns take place in the next several decades, hydrological models will be needed to accurately account for water, soil moisture, erosion and productivity at the changed frequency and intensity of precipitation. In addition to the potential for shifts in land use necessary to accommodate a new climate change, these models will need to take into account increases in atmospheric CO_2 or other greenhouse gases and their impacts on complex agricultural systems, and crop response to changed moisture patterns, soil temperatures and length of the growing season.

Acknowledgements

We thank New Mexico State University Agricultural Experiment Station for support. We also thank Dr Deva Borah, Woolpert, Inc., Virginia and Dr Manoj K. Jha, North Carolina Agricultural and Technical State University, North Carolina, for their constructive suggestions.

References

Abbaspour, K.C., Yang, J., Maximov, I., Siber, R., Bogner, K., Mieleitner, J., Zobrist, J. and Srinivasan, R. (2007) Modelling hydrology and water quality in the pre-alpine/alpine Thur watershed using SWAT. *Journal of Hydrology* 333, 413–430.

Abbott, M.B., Bathurst, J.C., Cunge, J.A., O'Connell, P.E. and Rasmussen, J. (1986a) An introduction to the European hydrological system – Système Hydrologique Européen, "SHE", 1: History and philosophy of a physically-based, distributed modeling system. *Journal of Hydrology* 87, 45–59.

Abbott, M.B., Bathurst, J.C., Cunge, J.A., O'Connell, P.E. and Rasmussen, J. (1986b) An introduction to the European hydrological system – Système Hydrologique Européen, "SHE", 2: Structure of a physically-based, distributed modeling system. *Journal of Hydrology* 87, 61–77.

Ahuja, L.R., Rojas, K.W., Hanson, J.D., Shaffer, M.J. and Ma, L. (2000) *The Root Zone Water Quality Model: Modeling Management Effects on Water Quality and Crop Production.* Water Resources Publications, Highlands Ranch, Colorado.

Arnold, J.G. and Fohrer, N. (2005) SWAT2000: Current capabilities and research opportunities in applied watershed modeling. *Hydrological Processes* 19, 563–572.

Arnold, J.G., Williams, J.R. and Maidment, D.R. (1995) Continuous-time water and sediment-routing model for large basins. *Journal of Hydraulic Engineering* 121, 171–183.

Arnold, J.G., Srinivasan, R., Muttiah, R.S. and Williams, J.R. (1998) Large area hydrologic modeling and assessment, Part I: Model development. *Journal of the American Water Resources Association*, 73–89.

Arnold, J.G., Gassman, P.W. and White, M.J. (2010) New developments in the SWAT ecohydrology model. In: Tollner, E.W. and Saleh, A. (eds) *21st Century Watershed Technology: Improving Water Quality and Environment Conference Proceedings*, 21–24 February 2010, Universidad EARTH, Costa Rica. ASABE Publication No. 701P0210cd, American Society of Agricultural and Biological Engineers. St Joseph, Michigan.

Ascough, J.C. II, Baffaut, C., Nearing, M.A. and Liu, B.Y. (1997) The WEPP watershed model: I. Hydrology and erosion. *Transactions of the ASABE* 40, 921–933.

Ascough, J.C. II, Green, T.R., Cipra, J.E., Vandenberg, B.C., Flynn, R.L., Norman, J.B., Ahuja, L.R. and Ma, L. (2003) AgSimGIS: Integrated GIS and Agricultural System Modeling, In: *Proceedings of the 2003 ESRI International Users Conference.* Available at: http://proceedings.esri.com/library/userconf/proc03/p0309.pdf (accessed 2 March 2010).

Ascough, J.C. II, Green, T.R., Ahuja, L.R., Ma, L. and Vandenberg, B.C. (2005) *Spatial Water Quality Modeling Framework Development Using ArcGIS 9.* ASABE Paper No. 052082, American Society of Agricultural and Biological Engineers, St Joseph, Michigan. Available at: http://ddr.nal.usda.gov/bitstream/10113/27325/1/IND44179077.pdf (accessed 2 March 2010).

Baffaut, C. and Benson, V.W. (2009) Modeling flow and pollutant transport in a karst watershed with SWAT. *Transactions of the ASABE* 52, 469–479.

Baigorria, G.A. and Romero, C.C. (2007) Assessment of erosion hotspots in a watershed: integrating the WEPP model and GIS in a case study in the Peruvian Andes. *Environmental Modelling and Software* 22, 1175–1183.

Beasley, D.B., Huggins, L.F. and Monke, E.J. (1980) ANSWERS: a model for watershed planning. *Transactions of the ASABE* 23, 938–944.
Bingner, R.L., Theurer, F.D. and Yuan, Y. (2009) *AnnAGNPS Technical Processes: Documentation Version 5.0.* Available at: http://www.wsi.nrcs.usda.gov/products/w2q/h&h/tools_models/agnps/downloads.html (accessed 2 March 2010).
Borah, D.K. and Bera, M. (2003) Watershed-scale hydrologic and nonpoint-source pollution models: review of mathematical bases. *Transactions of the ASABE* 46, 1553–1566.
Borah, D.K. and Bera, M. (2004) Watershed-scale hydrologic and nonpoint-source pollution models: review of applications. *Transactions of the ASABE* 47, 789–803.
Borah, D.K., Xia, R. and Bera, M. (2002) DWSM – a dynamic watershed simulation model. In: Singh, V.P. and Frevert, D.K. (eds) *Mathematical Models of Small Watershed Hydrology and Applications.* Water Resources Publications, Highlands Ranch, Colorado, pp. 113–166.
Borah, D.K., Bera, M. and Xia, R. (2004) Storm event flow and sediment simulations in agricultural watersheds using DWSM. *Transactions of the ASABE* 47, 1539–1559.
Borah, D.K., Arnold, J.G., Bera, M., Krug, E.C. and Liang, X.-Z. (2007) Storm event and continuous hydrologic modeling for comprehensive and efficient watershed simulations. *Journal of Hydrologic Engineering* 12, 605–616.
Bouraoui, F. and Dillaha, T.A. (1996) ANSWERS-2000: runoff and sediment transport model. *Journal of Environmental Engineering* 122, 493–502.
Bouraoui, F. and Dillaha, T.A. (2000) ANSWERS-2000: non-point-source nutrient planning model. *Journal of Environmental Engineering* 126, 1045–1055.
Brown, L.C. and Barnwell, T.O. Jr (1987) *The Enhanced Water Quality Models QUAL2E and QUAL2E-UNCAS: Documentation and User Manual.* EPA document EPA/600/3–87/007. US EPA (Environmental Protection Agency), Athens, Georgia.
Bulygina, N.S., Nearing, M.A., Stone, J.J. and Nichols, M.H. (2007) DWEPP: a dynamic soil erosion model based on WEPP source terms. *Earth Surface Processes and Landforms* 32, 998–1012.
Carsel, R.F., Mulkey, L.A, Lorber, M.N. and Baskin, L.B. (1985) The pesticide root zone model (PZRM): a procedure for evaluating pesticide leaching threats to groundwater. *Ecological Modeling* 30, 49–69.
Cochrane, T.A. and Flanagan, D.C. (1999) Assessing water erosion in small watersheds using WEPP with GIS and digital elevation models. *Journal of Soil and Water Conservation* 54, 678–685.
Crawford, N.H. and Linsley, R.K. (1966) *Digital Simulation on Hydrology: Stanford Watershed Model IV.* Technical Report No. 39, Department of Civil Engineering, Stanford University, Palo Alto, California.
Di Luzio, M., Arnold, J.G. and Srinivasan, R. (2004a) Integration of SSURGO maps and soil parameters within a geographic information system and nonpoint source pollution model system. *Journal of Soil and Water Conservation* 59, 123–133.
Di Luzio, M., Srinivasan, R. and Arnold, J.G. (2004b) A GIS-coupled hydrological model system for the watershed assessment of agricultural nonpoint and point sources of pollution. *Transactions in GIS* 8, 113–136.
Du, B., Arnold, J.G., Saleh, A. and Jaynes, D.B. (2005) Development and application of SWAT to landscapes with tiles and potholes. *Transactions of the ASABE* 48, 1121–1133.
Dun, S., Wu, J.Q., Elliot, W.J., Robichaud, P.R., Flanagan, D.C., Frankenberger, J.R., Brown, R.E. and Xu, A.C. (2009) Adapting the Water Erosion Prediction Project (WEPP) model for forest applications. *Journal of Hydrology* 366, 46–54.
Easton, Z.M., Fuka, D.R., Walter, M.T., Cowan, D.M., Schneiderman, E.M. and Steenhuis, T.S. (2008) Re-conceptualizing the soil and water assessment tool (SWAT) model to predict runoff from variable source areas. *Journal of Hydrology* 348, 279–291.
Eckhardt, K. and Ulbrich, U. (2003) Potential impacts of climate change on groundwater recharge and streamflow in a central European low mountain range. *Journal of Hydrology* 84, 244–252.
Elliot, W.J. (2004) WEPP Internet interfaces for forest erosion prediction. *Journal of the American Water Resources Association* 40, 299–309.
Ewen, J. (1995) Contaminant transport component of the catchment modelling system SHETRAN. In: Trudgill, S.T. (ed.) *Solute Modelling in Catchment Systems.* Wiley, London, pp. 417–441.
Ewen, J., Parkin, G. and O'Connell, P.E. (2000) SHETRAN: distributed river basin flow and transport modeling system. *Journal of Hydrologic Engineering* 5, 250–258.
Fang, Q., Ma, L., Yu, Q., Malone, R.W., Saseendran, S.A. and Ahuja, L.R. (2008) Modeling nitrogen and water management effects in a wheat-maize double-cropping system. *Journal of Environmental Quality* 37, 2232–2242.

Flanagan, D.C. and Livingston, S.J. (1995) *WEPP User Summary: USDA Water Erosion Prediction Project.* NSERL Report No. 11, USDA ARS National Soil Erosion Research Laboratory, West Lafayette, Indiana.

Flanagan, D.C. and Nearing, M.A. (1995) *USDA Water Erosion Prediction Project Hillslope and Watershed Model Documentation.* NSERL Report No. 10, USDA ARS National Soil Erosion Research Laboratory, West Lafayette, Indiana.

Flanagan, D.C., Ascough, J.C. II, Nearing, M.A. and Laflen, J.M. (2001) The water erosion prediction project (WEPP) model. In: Harmon, R.S. and Doe, W.W. III (eds) *Landscape Erosion and Evolution Modeling.* Kluwer Academic/Plenum Publishers, New York, pp. 145–199.

Flanagan, D.C., Gilley, J.E. and Franti, T.G. (2007) Water erosion prediction project: development history, model capabilities, and future enhancements. *Transactions of the ASABE* 50, 1603–1612.

Flerchinger, G.N., Aiken, R.M., Rojas, K.W. and Ahuja, L.R. (2000) Development of the Root Zone Water Quality Model (RZWQM) for over-winter conditions. *Transactions of the ASABE* 43, 59–68.

Foster, G.R. and Lane, L.J. (1987) *User Requirements: USDA Water Erosion Prediction Project (WEPP).* NSERL Report No. 1, USDA ARS National Soil Erosion Research Laboratory, West Lafayette, Indiana.

Fox, G.A., Malone, R.W., Sabbagh, G.J. and Rojas, K. (2004) Interrelationship of macropores and subsurface drainage for conservative tracer and pesticide transport. *Journal of Environmental Quality* 33, 2281–2289.

Garbrecht, J. and Martz, L.W. (1999) *An Overview of TOPAZ: an Automated Digital Landscape Analysis Tool for Topographic Evaluation, Drainage Identification, Watershed Segmentation, and Subcatchment Parameterization.* Report No. GRL99-1, Grazinglands Research Laboratory, USDA ARS, El Reno, Oklahoma. Available at: http://homepage.usask.ca/~lwm885/topaz/overview.html (accessed 2 March 2010).

Gassman, P.W., Reyes, M.R., Green, C.H. and Arnold, J.G. (2007) The soil and water assessment tool: historical development, applications and future research directions. *Transactions of the ASABE* 50, 1211–1250.

Gassman, P.W., Williams, J.R., Wang, X., Saleh, A., Osei, E., Hauck, L.M., Izaurralde, R.C. and Flowers, J.D. (2009) *The Agricultural Policy Environmental eXtender (APEX) Model: An Emerging Tool for Landscape and Watershed Environmental Analyses.* Technical Report 09-TR 49, Center for Agricultural and Rural Development, Iowa State University, Ames, Iowa.

Hatterman, F.F., Krysanova, V., Habeck, A. and Bronstert, A. (2006) Integrating wetlands and riparian zones in river basin modeling. *Ecological Modelling* 199, 379–392.

Heilman, P., Malone, R.W., Ma, L., Hatfield, J.L., Ahuja, L.R., Ayen, J., Boyle, K. and Kanwar, R.S. (2006) Decision support for nitrogen management in tile-drained agriculture. In: Voinov, A., Jakeman, A.J. and Rizzoli, A.E. (eds) *Proceedings of the iEMSs Third Biennial Meeting:"Summit on Environmental Modelling and Software".* International Environmental Modelling and Software Society, Manno, Switzerland. Available at: www.iemss.org/iemss2006/papers/s2/139_Heilman_1.pdf (accessed 2 March 2010).

Johnsen, K.E., Liu, H.H., Dane, J.H., Ahuja, L.R. and Workman, S.R. (1995) Simulating fluctuating water tables and tile drainage with a modified root zone water quality model and a new model WAFLOWM. *Transactions of the ASABE* 38, 75–83.

Jones, J.W., Hoogenboom, G., Porter, C.H., Boote, K.J., Batchelor, W.D., Hunt, L.A., Wilkens, P.W., Singh, U., Gijsman, A.J. and Ritchie, J.T. (2003) The DSSAT cropping system model. *European Journal of Agronomy* 18, 235–265.

Knisel, W.G. (ed.) (1980) *CREAMS: A Field Scale Model for Chemicals, Runoff, and Erosion from Agricultural Management Systems.* Conservation Research Report No. 26, US Department of Agriculture, Washington, DC.

Krysanova, V., Müller-Wohlfeil, D.-I. and Becker, A. (1998) Development and test of a spatially distributed hydrological/water quality model for mesoscale watersheds. *Ecological Modelling* 106, 261–289.

Krysanova, V., Hatterman, F. and Wechsung, F. (2005) Development of the ecohydrological model SWIM for regional impact studies and vulnerability assessment. *Hydrological Processes* 19, 763–783.

Lane, L.J. and Nearing, M.A. (1989) *USDA Water Erosion Prediction Project: Hillslope Profile Model Documentation.* NSERL Report No. 2, USDA ARS National Soil Erosion Research Laboratory, West Lafayette, Indiana.

Lenhart, T., Eckhardt, K., Fohrer, N. and Frede, H.-G. (2002) Comparison of two different approaches of sensitivity analysis. *Physics and Chemistry of the Earth* 27, 645–654.

Lenhart, T., Van Rompaey, A., Steegen, A., Fohrer, N., Frede, H.-G. and Govers, G. (2005) Considering spatial distribution and deposition of sediment in lumped and semi-distributed models. *Hydrological Processes* 19, 785–794.

Leonard, R.A., Knisel, W.G. and Still, D.A. (1987) GLEAMS: groundwater loading effects of agricultural management systems. *Transactions of the ASABE* 30, 1403–1418.

Ma, L., Hoogenboom, G., Ahuja, L.R., Nielsen, D.C. and Ascough, J.C. II (2005) Evaluation of the RZWQM-CROPGRO hybrid model for soybean production. *Agronomy Journal* 97, 1172–1182.

Ma, L., Hoogenboom, G., Ahuja, L.R., Ascough, J.C. II and Saseendran, S.A. (2006) Evaluation of the RZWQM-CERES-Maize hybrid model for maize production. *Agricultural Systems* 87, 274–295.

Ma, L., Ahuja, L.R. and Malone, R.W. (2007a) Systems modeling for soil and water research and management: current status and needs for the 21st century. *Transactions of the ASABE* 50, 1705–1713.

Ma, L., Malone, R.W., Heilman, P., Ahuja, L.R., Kanwar, R.S., Karlen, D.L., Cambardella, C.A. and Saseendran, S.A. (2007b) RZWQM simulation of long-term crop production, water, and nitrogen balances in northeast Iowa. *Geoderma* 140, 247–259.

Ma, L., Malone, R.W., Heilman, P., Jaynes, D., Ahuja, L.R., Saseendran, S.A., Kanwar, R.S. and Ascough, J.C. II (2007c) RZWQM-simulated management effects of crop rotation, tillage, and controlled drainage on crop production and nitrate-N loss in drain flow. *Geoderma* 140, 297–309.

Ma, L., Malone, R.W., Jaynes, D.B., Thorp, K. and Ahuja, L.R. (2008) Simulated effects of nitrogen management and soil microbes on soil N balance and crop production. *Soil Science Society of America Journal* 72, 1594–1603.

Magre, M., Williams, J.R., Harman, W.L., Gerik, T., Greiner, J., Francis, L., Steglich, E. and Meinardus, A. (2006) *WinAPEX Users Guide Version 1.0.* Blackland Research and Extension Center, Temple, Texas. Available at: http://winapex.brc.tamus.edu/media/17085/winapexv0604_nov2006.pdf (accessed 2 March 2010).

Malone, R.W., Ahuja, L.R., Ma, L., Wauchope, R.D., Ma, Q. and Rojas, K.W. (2004) Application of the Root Zone Water Quality Model (RZWQM) to pesticide fate and transport: an overview. *Pest Management Science* 60, 205–221.

McDonald, M.G. and Harbaugh, A.W. (1988) A modular three-dimensional finite-differences ground-water flow model. In: *US Geological Survey, Techniques of Water-Resources Investigations (TWRI), Book 6 – Modeling Techniques*, Chapter Al, Denver, Colorado. Available at: http://pubs.usgs.gov/twri/twri6a1/ (accessed 10 February 2011).

Miller, S.N., Semmens, D.J., Goodrich, D.C., Hernandez, M., Miller, R.C., Kepner, W.G. and Guertin, D.P. (2007) The automated geospatial watershed assessment tool. *Environmental Modelling and Software* 22, 365–377.

Minkowski, M. and Renschler, C. (2010) *GeoWEPP for ArcGIS 9.x Full Version Manual (for GeoWEPP Version 2.2008 or Earlier).* Department of Geography, The State University of New York at Buffalo. Available at: http://www.geog.buffalo.edu/~rensch/geowepp/arcgeowepp/GeoWEPP%20for%20ArcGIS%209%20Manual.pdf (accessed 2 March 2010).

Moriasi, D.N., Arnold, J.G., Vaarquez-Amiaile, G.G., Engel, B.A. and Rossi, C.G. (2009) Incorporation of a new shallow water table depth algorithm into SWAT2005. *Transactions of the ASABE* 52, 771–784.

Neitsch, S.L., Arnold, J.G., Kiniry, J.R., Williams, J.R. and King, K.W. (2002a) *Soil and Water Assessment Tool Theoretical Documentation, Version 2000*, GSWRL Report 02-01, Grassland, Soil and Water Research Laboratory, USDA ARS/BRC Report 02-05, Blackland Research Center, Texas Agricultural Experiment Station, Temple, Texas/TWRI Report TR-191, Texas Water Resources Institute, College Station, Texas. Available at: http://swatmodel.tamu.edu/media/1290/swat2000theory.pdf (accessed 2 March 2010).

Neitsch, S.L., Arnold, J.G., Kiniry, J.R., Srinivasan, R. and Williams, J.R. (2002b) *Soil and Water Assessment Tool User's Manual, Version 2000*, GSWRL Report 02-02, Grassland, Soil and Water Research Laboratory, USDA ARS/BRC Report 02-06, Blackland Research Center, Texas Agricultural Experiment Station, Temple, Texas/TWRI Report TR-192, Texas Water Resources Institute, College Station, Texas. Available at: http://swatmodel.tamu.edu/media/1294/swatuserman.pdf (accessed 2 March 2010).

Neitsch, S.L., Arnold, J.G., Kiniry, J.R., Srinivasan, R. and Williams, J.R. (2004) *Soil and Water Assessment Tool Input/Output File Documentation, Version 2005.* Grassland, Soil and Water Research Laboratory, USDA ARS/Blackland Research Center, Temple, Texas Agricultural Experiment Station, Texas. Available at: http://swatmodel.tamu.edu/media/1291/SWAT2005io.pdf (accessed 2 March 2010).

Neitsch, S.L., Arnold, J.G., Kiniry, J.R. and Williams, J.R. (2005) *Soil and Water Assessment Tool Theoretical Documentation, Version 2005.* Grassland, Soil and Water Research Laboratory, USDA ARS/Blackland Research Center, Texas Agricultural Experiment Station, Temple, Texas. Available at: http://swatmodel.tamu.edu/media/1292/SWAT2005theory.pdf (accessed 2 March 2010).

Olivera, F., Valenzuela, M., Srinivasan, R., Choi, J., Cho, H., Koka, S. and Agrawal, A. (2006) ArcGIS-SWAT: a geodata model and GIS interface for SWAT. *Journal of the American Water Resources Association* 42, 295–309.

Osei, E., Du, B., Bekele, A., Hauck, L., Saleh, A. and Tanter, A. (2004). *CEEOT-MMS: A Macro Modeling System for Environmental Assessment.* Technical Report TR0409, Texas Institute for Applied Environmental Research. Tarleton State University, Stephenville, Texas.

Ouessar, M., Bruggeman, A., Abdelli, F., Mohtar, R.H., Gabriels, D. and Cornelis, W.M. (2009) Modelling water-harvesting systems in the arid south of Tunisia using SWAT. *Hydrology and Earth System Sciences* 13, 2003–2021.

Pohlert, T., Huisman, J.A., Breuer, L. and Freude, H.-G. (2007) Integration of a detailed biogeochemical model into SWAT for improved nitrogen predictions: model development, sensitivity, and GLUE analysis. *Ecological Modelling* 203, 215–228.

Rao, M., Fan, G., Thomas, J., Cherian, G., Chudiwale, V. and Awawdeh, M. (2006) A web-based GIS decision support system for managing and planning USDA's Conservation Reserve Program (CRP). *Environmental Modeling and Software* 22, 1270–1280.

Refsgaard, J.C. and Storm, B. (1995) MIKE SHE. In: Singh, V.P. (ed.) *Computer Models of Watershed Hydrology.* Water Resources Publications, Highlands Ranch, Colorado, pp. 809–846.

Renschler, C.S. (2003) Designing geo-spatial interfaces to scale process models: the GeoWEPP approach. *Hydrological Processes* 17, 1005–1017.

Renschler, C.S. and Lee, T. (2005) Spatially distributed assessment of short- and long-term impacts of multiple best management practices in agricultural watersheds. *Journal of Soil and Water Conservation* 60, 446–456.

Robichaud, P.R., Elliot, W.J., Pierson, F.B., Hall, D.E. and Moffet, C.A. (2007) Predicting postfire erosion and mitigation effectiveness with a web-based probabilistic erosion model. *Catena* 71, 229–241.

RZWQM Development Team, Hanson, J.D., Ahuja, L.R., Shaffer, M.D., Rojas, K.W., DeCoursey, D.G., Farahani, H. and Johnson, K. (1998) RZWQM: simulating the effects of management on water quality and crop production. *Agricultural Systems* 57, 161–195.

RZWQM Team (1992) *Root Zone Water Quality Model, Version 1.0, Technical Documentation.* Great Plains Systems Research (GPSR) Unit Technical Report No. 2, USDA ARS GSPR, Fort Collins, Colorado.

Saleh, A. and Gallego, O. (2007) Application of SWAT and APEX using the SWAPP (SWAT-APEX) program for the Upper North Bosque River watershed in Texas. *Transactions of the ASABE* 50, 1177–1187.

Saleh, A., Osei, E. and Gallego, O. (2008) Use of CEEOT-SWAPP modeling system for cost-effective targeting and evaluation of environmental pollutants. In: Tollner, E.W. and Saleh, A. (eds) *21st Century Watershed Technology: Improving Water Quality and Environment Conference Proceedings*, 21–24 February 2010, Universidad EARTH, Costa Rica. ASABE Publication No. 701P0210cd, American Society of Agricultural and Biological Engineers. St Joseph, Michigan.

Shaffer, M.J. and Larson, W.E. (1987) *NTRM, A Soil–Crop Simulation Model for Nitrogen, Tillage, and Crop Residue Management.* USDA ARS Conservation Research Report No. 34-1, National Technical Information Service, Springfield, Virginia.

Singh, V.P. (1995) *Computer Models of Watershed Hydrology.* Water Resources Publications, Highlands Ranch, Colorado.

Singh, V.P. and Frevert, D.K. (2002a) *Mathematical Models of Large Watershed Hydrology.* Water Resources Publications, Highlands Ranch, Colorado.

Singh, V.P. and Frevert, D.K. (2002b) *Mathematical Models of Small Watershed Hydrology and Applications.* Water Resource Publications, Highlands Ranch, Colorado.

Sophocleous, M.A., Koelliker, J.K., Govindaraju, R.S., Birdie, T., Ramireddygari, S.R. and Perkins, S.P. (1999) Integrated numerical modeling for basin-wide water management: the case of the Rattlesnake Creek basin in south-central Kansas. *Journal of Hydrology* 214, 179–196.

Srinivasan, R. and Arnold, J.G. (1994) Integration of a basin-scale water quality model with GIS. *Journal of the American Water Resources Association* 30, 453–462.

Steglich, E.M. and Williams, J.R. (2008) *Agricultural Policy/Environmental eXtender Model: User's Manual Version 0604.* BREC Report No. 2008-16, Blackland Research and Extension Center, Temple, Texas.

Stone, J.J., Lopes, V.L. and Lane, L.J. (1990) Watershed erosion prediction project (WEPP) model: Hydrologic and erosion calculations. In: Higgins, R.E., Jones, E.B., Singh, R. and Rechard, P.A. (eds) *Watershed Planning and Analysis in Action.* American Society of Civil Engineers, Reston, Virginia, pp. 184–190.

Tuppad, P., Winchell, M.F., Wang, X., Srinivasan, R. and Williams, J.R. (2009) ARCAPEX: ARCGIS interface for Agricultural Policy Environmental eXtender (APEX) hydrology/water quality model. *International Agricultural Engineering Journal* 18, 59–71.

van Griensven, A. and Bauwens, W. (2003) Multiobjective autocalibration for semidistributed water quality models. *Water Resources Research* 39, 1348–1357.

van Griensven, A. and Bauwens, W. (2005) Application and evaluation of ESWAT on the Dender basin and Wister Lake basin. *Hydrological Processes* 19, 827–838.

Wang, X. and Cui, P. (2004) Linkage of Arcview GIS with the RZWQM. *Journal of Spatial Hydrology* 4, 1–16.

Wattenbach, M., Hatterman, F., Weng, R., Wechsung, F., Krysanova, V. and Badeck, F. (2005) A simplified approach to implement forest eco-hydrological properties in regional hydrological modelling. *Ecological Modelling* 187, 49–50.

White, M.J. and Arnold, J.G. (2009) Development of a simplistic vegetative filter strip model for sediment and nutrient retention at the field scale. *Hydrological Processes* 23, 1602–1616.

Wicks, J.M. and Bathurst, J.C. (1996) SHESED: a physically based, distributed erosion and sediment yield component for the SHE hydrological modelling system. *Journal of Hydrology* 175, 213–238.

Wigmosta, M.S., Vail, L.W. and Lettenmaier, D.P. (1994) A distributed hydrology-vegetation model for complex terrain. *Water Resources Research* 30, 1665–1669.

Wigmosta, M.S., Nijssen, B., Storck, P. and Lettenmaier, D.P. (2002) The Distributed Hydrology Soil Vegetation Model. In: Singh, V.P. and Frevert, D.K. (eds) *Mathematical Models of Small Watershed Hydrology and Applications.* Water Resource Publications, Highlands Ranch, Colorado, pp. 7–42.

Williams, J.R. (1995) The EPIC model. In: Singh, V.P. (ed.) *Computer Models of Watershed Hydrology.* Water Resources Publications, Highlands Ranch, Colorado, pp. 909–1000.

Williams, J.R. and Hann, R.W. (1973) *HYMO: Problem Oriented Computer Language for Hydrologic Modeling: User's Manual.* Technical Publication ARS-S-9, USDA ARS, Washington, DC.

Williams, J.R. and Izaurralde, R.C. (2006) The APEX model. In: Singh, V.P. and Frevert, D.K. (eds) *Watershed Models.* CRC Press, Taylor and Francis Group, Boca Raton, Florida, pp. 437–482.

Williams, J.R., Jones, C.A. and Dyke, P.T. (1984) A modeling approach to determining the relationship between erosion and soil productivity. *Transactions of the ASABE* 27, 129–144.

Williams, J.R., Nicks, A.D. and Arnold, J.G. (1985) Simulator for water resources in rural basins. *Journal of Hydraulic Engineering* 111, 970–986.

Williams, J.R., Arnold, J.G., Srinivasan, R. and Ramanarayanan, T.S. (1998) APEX: a new tool for predicting the effects of climate and CO_2 changes on erosion and water quality. In: Boardman, J. and Favis-Mortlock, D. (eds) *Modeling Soil Erosion by Water.* Springer-Verlag, Berlin, pp. 441–449.

Williams, J.R., Wang, E., Meinardus, A., Harman, W.L., Siemers, M. and Atwood, J.D. (2006) *APEX Users Guide v.2110.* Texas A&M University, Texas Agricultural Extension Service, Blackland Research Center, Temple, Texas.

Williams, J.R., Arnold, J.G., Kiniry, J.R. Gassman, P.W. and Green, C.H. (2008a) History of model development at Temple, Texas. *Hydrological Sciences Journal* 53, 948–960.

Williams, J.R., Izaurralde, R.C. and Steglich, E.M. (2008b) *Agricultural Policy/Environmental eXtender (APEX) model: Theoretical Documentation Version 0604.* BREC Report No. 2008–17, Texas AgriLife Extension Service, Texas A&M System, Blackland Research and Extension Center, Temple, Texas.

Wischmeier, W.H. and Smith, D.D. (1978). *Predicting Rainfall Erosion Losses – A Guide to Conservation Planning.* Agriculture Handbook No. 537, US Department of Agriculture, Washington, DC.

Woolhiser, D.A., Smith, R.E. and Goodrich, D.C. (1990) *KINEROS: A Kinematic Runoff and Erosion Model: Documentation and User Manual.* Publication ARS-77, USDA ARS, Washington, DC.

Young, R.A., Onstad, C.A., Bosch, D.D and Anderson, W.P. (1987) *AGNPS: Agricultural Nonpoint-Source Pollution Model – A Watershed Analytical Tool.* Conservation Research Report No. 35, USDA ARS, Washington, D.C.

4 Modelling Agricultural Management Systems with APEX

Xiuying Wang, Pushpa Tuppad and Jimmy R. Williams*

Introduction

The Agricultural Policy Environmental eXtender (APEX) model (Williams and Izaurralde, 2006) was developed to simulate whole farms and watersheds. It is an outcome of extensive biophysical/environmental/hydrological model development conducted over the past four decades by the US Department of Agriculture Agricultural Research Service (USDA ARS) and Texas A&M System's Texas AgriLife Research (formerly Texas Agricultural Experiment Station), located in Temple, Texas. The model simulates the hydrological, biological, chemical and meteorological processes of complex farming systems involving multiple crops, soil types, field delineations, and structural and agronomic conservation practices across the landscape (Williams and Izaurralde, 2006). The APEX model and its predecessor, the Environmental Policy Integrated Climate (EPIC) model (Williams, 1995), have had a long history of use in the simulation of agricultural and environmental processes, as well as in agricultural technology and government policy (Gassman *et al.*, 2005, 2010). APEX extended the EPIC model's ability by allowing the user to simulate several related sub-areas instead of a single area, while routing water, sediment, nutrients and pesticides from sub-area to sub-area across complex landscapes and channel systems to the watershed outlet. With this capability, APEX allows the assessment of various conservation practices, including terraces, grass waterways, strip cropping, buffer strips, feedlots, animal waste lagoons and water retention structures. The APEX model has a GIS (geographic information system) interface tool, ArcAPEX, which is designed to use the ArcGIS 9.× GIS platform to generate inputs and execute the APEX model (Tuppad *et al.*, 2009).

Since its inception, the APEX model has emerged as one of the key scientific tools used worldwide for assessing and evaluating a variety of environmental and water resource issues at various spatial and temporal scales. The model has been applied to evaluate environmental and effectiveness of filter strips and furrow dyking in a playa lake system (Willis, 2008), sediment ponds and wetlands on atrazine losses (Harman *et al.*, 2004), of conservation tillage on erosion and nutrients (Chung *et al.*, 2002), reforestation, grazing scenarios and reservoirs on soil erosion (Wang, E. *et al.*, 2006), and manure management scenarios (Flowers *et al.*, 1996; Osei *et al.*, 2000) on water quality. Besides water quantity and quality, the model is used to assess the

* Corresponding author: jwilliams@brc.tamus.edu

impacts of different agricultural management systems on crop yield and of soil carbon sequestration potential in relation to climate change (Thomson *et al.*, 2006). The model is being used as a field-scale modelling tool to address the on-site benefits obtained from USDA conservation programme expenditures, and APEX outputs are integrated with the Soil and Water Assessment Tool (SWAT) (Arnold *et al.*, 1998) watershed model to estimate off-site water-quality effects for a national assessment of cropland by the Conservation Effects Assessment Project (CEAP) (Mausbach and Dedrick, 2004; Wang, X. *et al.*, 2006; USDA NRCS, 2007). Applications of the APEX model are further summarized in Table 4.1.

To understand the capabilities of the APEX model, an overview of the environmental processes on an agricultural landscape is helpful. Farmers prepare the soil, add fertilizer and/or organic amendments such as manure, plant seeds or seedlings, cultivate for weed or water-flow control, apply chemicals for pest control, irrigate as needed and then harvest the crop. Weather events occur and affect crop production both positively and negatively, and result in beneficial and adverse impacts of farming activities on the environment. Properties of the soil, such as bulk density, organic matter content and soil microbial populations, change over time and also affect crop yields and environmental outcomes. The atmosphere contains nitrogen, carbon, sulfur and other elements that are deposited in the soil, and then taken up by the crop and/or partially lost with water running off the field. As a result of the interaction between the farmer's production activities, soil properties and weather events, some soil, crop residues and fertilizers and other chemicals are carried off the field by water and wind, either adhered to soil particles or in water-soluble form. Nutrients and pesticides are also lost from the field through leaching below the root zone. Over time, the chemical make-up and physical structure of the soil may change. Organic material may build up in the soil, or it may degrade. All of these processes are accounted for in the APEX model in a very detailed fashion.

The objectives of this chapter are to describe the major components of the APEX model and demonstrate its application using a case study for evaluating agricultural management practices for a 104 km^2 sub-watershed within the Bosque River watershed in central Texas. The Bosque River eventually drains into Lake Waco, which serves as the primary drinking water supply for the greater Waco area and provides water for agricultural production, recreational fishing and swimming. The watershed is facing a suite of water-quality issues resulting from sediment, nutrient and bacterial loadings which are potentially derived from improperly managed cropland and grazing land, and from applied dairy waste and effluent discharge from waste-water treatment plants. The study sub-watershed is predominantly cropland. The APEX model was to be calibrated and validated for monthly streamflow, sediment yield and nutrient losses at a monitoring station within the sub-watershed, and the calibrated model was then used for long-term scenario analysis to evaluate the benefits of alternative agricultural management practices.

Materials and Methods

Overview of the APEX model

The APEX model operates on a continuous basis using a daily time step. APEX simulates the basic biological, chemical, hydrological and meteorological processes of farming systems and their interactions (Fig. 4.1). The major components simulated on an individual sub-area include weather, hydrology, soil erosion, nutrients (nitrogen, phosphorus, carbon), pesticide fate, crop growth, soil temperature, tillage, plant environment control (drainage, irrigation, liming) and economics. These functions are adopted from the EPIC model (Williams, 1995). The routing mechanisms in APEX can route water, sediment, nutrients and pesticides across landscapes through channels, flood plains and reservoirs to the watershed outlet. The APEX groundwater component partitions flow between deep percolation and return flow. APEX also has a grazing component which provides flexibility to simulate a confined or partially confined

Table 4.1. Review of applications of the APEX model.

Reference	Study location/s	Study objectives	Results	Comments
Impact Assessment Group, 2000	• Representative household simulations, Kenya and Mali, Africa • Sondu River Watershed in western Kenya directly draining into Lake Victoria	Assess the runoff and erosion impacts of crop rotations and practices	• The impacts of land allocations between districts among different dairy management scenarios were environmentally neutral for runoff and erosion • Estimated substantial reduction in sediment yield due to conversion of cropland to native grass • Discharge increased by 23% from the traditional smallholder dairy to the current land use • Full adoption technology land use increased the sediment load into Lake Victoria by 6% compared with traditional land use	Baseline situation considered all dairy production with unimproved forage and minimal use of modern technology
Osei *et al.*, 2000	Upper North Bosque River Watershed, Texas	Evaluate alternative manure management scenarios, including relatively complex combinations of farm-level landscapes, cropping systems and/or management practices for addressing water quality/ environmental issues associated with intensive livestock production.	• Both P-based scenarios resulted in a reduction in P losses.	Dairy manure applied based on N rate is the baseline scenario with which all other scenarios were compared
Harman *et al.*, 2004	AquillaWatershed, Hill County, Central Texas	Assess the impacts of agricultural practices (sediment ponds, no-till, filter strips, split application wetlands, banding of the pesticide, conservation tillage) on atrazine losses	Reductions of atrazine losses: • Sediment ponds: 0.09% • Grass filter strips: 0.14%, • Banding at a 25% rate: 0.40% • Wetlands: 0.45% • Conservation, no-till and split applications between autumn and spring: only marginally effective	Major limitation: the absence of long-term in-stream and reservoir atrazine measurements for model validation
Thomson *et al.*, 2006	Huang-Hai Plain, China	Assess impacts of climate change on crop yield and soil carbon sequestration potential of different agricultural management systems	• Projected increases in precipitation had a positive impact on both crop yields and soil organic carbon (SOC) • No-till and double-cropping systems resulted in higher simulated SOC compared with a conventional tillage system	Need more field and experimental data to validate the model estimates

Continued

Table 4.1. Continued.

Reference	Study location/s	Study objectives	Results	Comments
Wang, E. *et al.*, 2006	North-west China	Investigate soil erosion effects of alternative land uses: all grass, all forest, all cropland, partial grazing, tree and grass in two different proportions, and a reservoir	• The reservoir was the most effective practice to control water runoff (approx. 56%) and erosion, followed by reforestation	APEX databases required minor modifications to reflect conditions of the study area in China
Wang *et al.*, 2007	Alto, Texas	Assess the performance of the model in simulating flow, sediment yield and herbicide losses	• APEX simulated streamflow, sediment and herbicide losses reasonably well for all nine watersheds studied	Highlights the applicability of the APEX model, without calibration, for simulating water quantity and quality responses of forested watersheds and silvicultural practices within them
Willis, 2008	Texas High Plains	Quantify the impact of existing (range cotton, sorghum, wheat and maize) as well as on-farm conservation practices (50 m filter strips and furrow dyking) on playa hydroperiod and maximize water storage volume	• As eroded sediment gradually filled the playa, the average number of wet days for playas adjacent to either cotton or wheat decreased by 60% and 25%, respectively • Filter strips reduced the rate at which wet days were lost by 10%, and extended the number of years that the playa maintains some water by 20% • Field dyking was slightly more effective than buffers but prevented water runoff into the playa	The study focused on the ecological impacts of agricultural management
Wang *et al.*, 2008	Two watersheds (W2 & W3) at the USDA Deep Loess Research Station near Treynor, Iowa	Assess the benefits of ridge tillage over conventional tillage	• Ridge tillage reduced surface runoff by 36–39% and sediment yield by 82–86% compared with conventional tillage • Ridge tillage increased maize grain yield by at least 3.8%	Model calibration, validation and scenario analysis
Wang *et al.*, 2009	Shoal creek, Texas	Evaluate the performance of the APEX model and its capability in simulating conservation practices on military landscapes	• Gully plugs and contour soil ripping reduced runoff by 52% and sediment yield by 86%	APEX application to assessing the impacts of erosion control practices implemented in an Army military reservation
Yin *et al.*, 2009	China	Assess the impacts of woodland in combination with grass and terraces, and ditches, on runoff and sediment yield	• Mixed woodland/grass with horizontal terraces reduced runoff by 35% and sediment yield by 84% • Woodland with horizontal-level ditches reduced runoff by 37% and sediment yield by 89%	• Variable-based sensitivity analysis • Automatic calibration • Scenario analysis with fallow as baseline

area of feeding, intensive rotational grazing, cropland grazing after harvest, etc. Table 4.2 summarizes the hydrology and water erosion components, and the options available in APEX. These options allow broader applicability, improved accuracy and convenience by providing methods that are more accurate in certain locations, perform well with limited inputs and/or are commonly used by users. For a complete description of the APEX model see Williams and Izaurralde (2006).

Hydrology

The hydrology component of APEX simulates daily runoff volume, peak runoff rate, subsurface flow, percolation below the soil profile, evapotranspiration and snow melt. Rainfall is intercepted by the plant canopy, which is estimated by considering the above-ground plant material and leaf area index of the plant stand. If snow is present, it may be melted on days when the second soil layer temperature exceeds 0 °C. Snow is melted as a function of the daily maximum air temperature and the snowpack temperature. Runoff can be calculated directly using a modification of the USDA NRCS (US Department of Agriculture Natural Resources Conservation Service) curve number (CN, also known as the runoff curve number) method (Mockus, 1969; USDA NRCS, 2004) or indirectly with the Green and Ampt infiltration equation (Green and Ampt, 1911). The CN technique is often used in APEX applications because the required inputs for this method are generally available for most parts of the USA. In the CN method, the retention parameter, *S*, is related to CN. It varies among watersheds because of differences in soils, land use, management and slope, and it varies with time within a

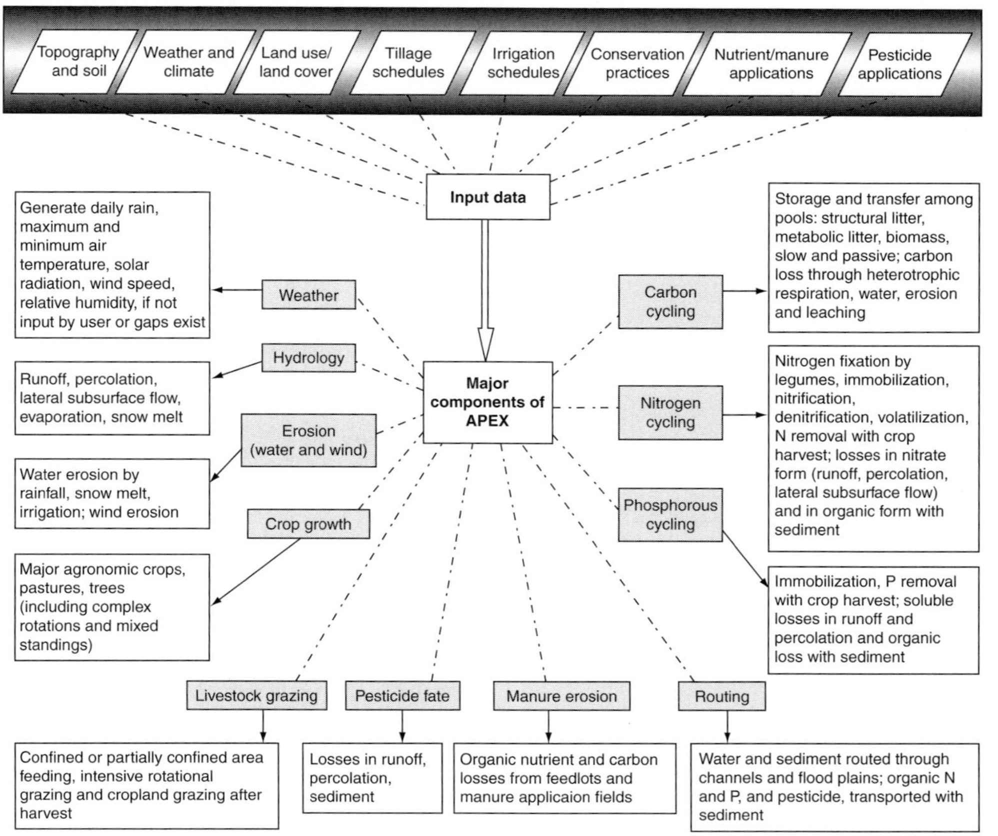

Fig. 4.1. Major processes simulated in the APEX model.

Table 4.2. APEX model hydrology and water erosion components.

Component	Method/principle	Reference
Hydrology	Curve number (CN)	Mockus, 1969; USDA NRCS, 2004
Surface runoff	1. Variable daily CN, non-linear CN/SW (soil water) with SW depth weighting	Williams, 1995
	2. Variable daily CN, linear CN/SW with no SW depth weighting	
	3. Non-varying CN used for all storms	
	4. Variable daily CN, SMI (soil moisture index)[a]	Kannan *et al.*, 2008; Wang *et al.*, 2009
	5. Green and Ampt	Green and Ampt, 1911
Peak runoff rate	1. Modified Rational formula	Williams, 1995
	2. SCS TR-55 peak rate estimate	USDA SCS, 1986
Subsurface flow	Storage routing and pipe flow equations	
Potential evapotranspiration	1. Penman–Monteith	Monteith, 1965
	2. Penman	Penman, 1948
	3. Priestley–Taylor	Priestley and Taylor, 1972
	4. Hargreaves[a]	Hargreaves and Samani, 1985
	5. Baier–Robertson	Baier and Robertson, 1965
Water erosion	1. Universal Soil Loss Equation (USLE)	Wischmeier and Smith, 1978
	2. Onstad–Foster modification of USLE (AOF)	Onstad and Foster, 1975
	3. Revised USLE (RUSLE)	Renard *et al.*, 1997
	4. Modified USLE (MUSLE)	Williams, 1975
	5. MUSLE variation 1 (MUST)[a]	Williams, 1995
	6. MUSLE variation 2 (MUSS)	Williams, 1995
	7. MUSLE with input coefficients (MUSI)	
	8. RUSLE2	Foster, 2005

[a]Options used in this modelling study.

watershed because of changes in soil water content. A sound continuous soil moisture accounting procedure is necessary in models using the CN method. Four options are provided in APEX for calculating the retention parameter S (Table 4.2). Option 4 – variable daily CN soil moisture index – is often used in APEX applications (Wang *et al.*, 2009) because it performs well over a wide range in soil properties. Option 4 was used in this modelling study too, where the calculation of S accounts for plant evapotranspiration. APEX computes S daily using the following equation:

$$S = S_{prev} + PET \times \exp(-CNIC \times \frac{S_{prev}}{S_{max}}) - P_{prev} + Q_{prev} + Q_{return} + Q_{drainage} + SSF + PRK \quad (4.1)$$

where S is the retention parameter for a given day (mm), S_{prev} is the retention parameter on the previous day (mm), PET is the potential evapotranspiration for the day (mm d^{-1}), $CNIC$ is the weighting coefficient used to calculate the retention coefficient for daily CN calculations depending on plant evapotranspiration, S_{max} is the maximum value the retention parameter can achieve (mm), which is associated with CN_1 for moisture condition 1 (dry), P_{prev} is the rainfall reaching the ground surface after plant interception on the previous day (mm), Q_{prev} is the runoff on the previous day (mm), Q_{return} is quick return flow on the previous day (mm), $Q_{drainage}$ is drainage flow on the previous day (mm), SSF is the lateral subsurface flow on the previous day (mm) and PRK is percolation on the previous day (mm).

The peak runoff rate can be estimated using the modified Rational formula

(Williams, 1995) or the SCS TR-55 method (USDA SCS, 1986). A stochastic element is included in the Rational equation to allow realistic simulation of peak runoff rates, given only daily rainfall and monthly rainfall intensity.

The subsurface flow component computes vertical and horizontal subsurface flow simultaneously using storage routing and pipe flow equations. The storage routing technique allows flow from a soil layer when soil water content exceeds field capacity. Water drains from the layer as a function of layer storage and saturated conductivity until the storage returns to field capacity. Vertical and horizontal flows are partitioned as a function of the vertical flow travel time and the horizontal travel time. The travel time is a function of soil water storage and saturated flow rate. Horizontal flow is partitioned between quick return flow and subsurface flow based on the ratio of upland slope length to reach channel length. As the ratio approaches one (very small hillslope watersheds) all of the subsurface flow remains below ground and enters the adjacent sub-area soil water storage. Conversely, as the ratio approaches zero all of the subsurface flow resurfaces as quick return flow.

Pipe flow is common in forested watersheds and must be simulated to properly account for the rapid vertical and horizontal flow. Flow through pipes created by decayed roots, animals, etc. is not included in the storage routing. Vertical and horizontal pipe flows are partitioned based on the inflow rate, the vertical pipe flow rate and the horizontal pipe flow rate. Horizontal pipe flow is added to quick return flow.

The vertical percolation flows to groundwater storage, which is further subject to partitioning into return flow and deep percolation that is assumed to be lost from the system. Return flow is added to channel flow from the sub-area. The APEX groundwater component partitions flow between deep percolation and return flow using the groundwater storage residence time and a partitioning coefficient. Return flow stops when storage is below a threshold storage.

The Penman and Penman–Monteith methods (Penman, 1948; Monteith, 1965) require all five weather parameters, including precipitation, air temperature, solar radiation, wind speed and relative humidity, as input. If wind speed, relative humidity and solar radiation data are not available, the Hargreaves (Hargreaves and Samani, 1985) or Priestley–Taylor (Priestley and Taylor, 1972) methods provide options that give realistic results in most cases. The Baier–Robertson (Baier and Robertson, 1965) method developed in Canada performs well in cold climates. Evaporation from soils and plants is computed separately, as described by Ritchie (1972). Potential soil water evaporation is estimated as a function of potential evaporation and leaf area index. Actual soil water evaporation is estimated by using exponential functions of soil depth and water content. Plant water evaporation is simulated as a linear function of potential evaporation and leaf area index.

Soil erosion

The APEX water-induced erosion component simulates erosion caused by runoff from rainfall and irrigation. Eight options are available to the user for calculating water erosion (Table 4.2). The general formula is based on the Universal Soil Loss Equation (USLE) (Wischmeier and Smith, 1978) with different energy factor estimations. The USLE depends strictly upon rainfall as an indicator of erosive energy, whereas the modified USLE (MUSLE) and its variations use runoff variables (runoff volume and peak runoff rate) to simulate erosion and sediment yield. The Onstad–Foster equation contains a combination of the USLE and MUSLE energy factors. For example, the MUST (Williams, 1995) (used in this modelling study) equation is shown below:

$$\begin{aligned} Y &= X \times EK \times CVF \times PEC \\ &\quad \times SL \times ROKF \\ X &= 2.5 \times (Q \times q_p)^{0.5} \end{aligned} \tag{4.2}$$

where Y is the sediment yield (t ha^{-1}) on a given day, EK is the soil erodibility factor, CVF is the crop management 'C' factor, PEC is the erosion control practice factor, SL is the slope length and steepness factor, and $ROKF$ is the coarse fragment factor, Q is the runoff volume (mm), and q_p is the peak runoff rate (mm h^{-1}).

Wind erosion is calculated with the Wind Erosion Continuous Simulator (WECS) (Potter *et al.*, 1998), which requires the daily distribution of wind speed. The approach estimates potential wind erosion for a smooth bare soil by integrating the erosion equation through a day using the wind speed distribution. The potential erosion is then adjusted according to soil properties, surface roughness, vegetative cover, and distance across the field in the wind direction (Potter *et al.*, 1998).

Crop growth

The APEX plant growth and plant competition capabilities provide a very flexible basis for simulating crop rotations and other cropping/vegetation systems, such as cover crops, double cropping, plant and weed competition, pastures and tree growth. A single model is used in APEX for simulating all the crops considered (about 100), with a unique set of values characterizing each crop. These crop parameters are packaged in the model's databases.

Phenological development of the crop is based on daily heat unit accumulation. The daily gains of plant biomass are proportional to the daily photosynthetically active radiation intercepted by the plant canopy. Daily growth may be affected by atmospheric CO_2 concentration and physiological stresses caused by water, temperature, N, aeration in the root zone, soil strength and aluminium content.

Annual crops grow from planting date to harvest date or until the accumulated heat units equal the potential heat units for the crop. Perennial crops maintain their root systems throughout the year, although they may become dormant after frost. They start growing when the average daily air temperature exceeds their base temperature. The model is also capable of simulating mixed plant stands (up to ten crops can grow in the same space and time). Plant competition was built into the APEX crop growth model using algorithms contained in the Agricultural Land Management Alternatives with Numerical Assessment Criteria (ALMANAC) model (Kiniry *et al.*, 1992). Plants compete for light, water and nutrients.

Tillage

The tillage component of the model has the functions of mixing nutrients and crop residues within the tillage depth, converting standing residue to flat residue, and simulating change in bulk density, ridge height and surface roughness. The tillage mixing equation is:

$$X(1) = (1 - EF) \times X_0(1) + EF \times SMX_0 \times Z / TLD \tag{4.3}$$

where X is the amount of the material in layer 1 after mixing (kg ha^{-1}), X_0 is the amount of the material before mixing (kg ha^{-1}), EF is the mixing efficiency of the tillage operation, TLD is the tillage depth (m), SMX_0 is the sum of the material in the TLD before mixing (kg ha^{-1}), and Z is the depth to the bottom of the plough layer (m). Converting standing residue to flat residue is accomplished with the equation:

$$STD = STD_0 \times \exp(-56.9 \times TLD \times EF) \tag{4.4}$$

where STD_0 and STD are the standing residue weights before and after tillage (t ha^{-1}).

Carbon cycling routine

The APEX model incorporates carbon and nitrogen algorithms similar to those of the Century model (Parton *et al.*, 1987, 1993, 1994; Vitousek *et al.*, 1994) to distribute C and N across soil layers into several pools: metabolic litter, structural litter, active, slow and passive humus (Izaurralde *et al.*, 2006, 2007). Soil carbon sequestration is estimated as a function of climatic conditions, soil properties and management practices. The following modifications were performed when implementing the Century model carbon cycling into APEX: (i) movement of organic materials from surface litter to subsurface layers is estimated by the leaching equations currently in APEX; (ii) temperature and water controls affecting transformation rates are calculated with equations currently in APEX; (iii) the surface litter fraction in APEX has a slow compartment but no passive compartment; and (iv) the lignin concentration is modelled as a sigmoidal function of plant age.

Nitrogen cycling

The N cycle simulated in APEX includes atmospheric N inputs, fertilizer and manure N applications, crop N uptake, mineralization, immobilization, nitrification, denitrification, ammonia volatilization, organic N transport on sediment and nitrate-nitrogen (NO_3-N) losses in leaching, surface runoff, lateral subsurface flow and tile flow. Denitrification is a function of temperature and water content (Williams, 1995), with the requirement of anaerobic conditions and a carbon source. Nitrification, the conversion of ammonia N to NO_3-N, is estimated based on the first-order kinetic rate equation of Reddy *et al.* (1979). Atmospheric emissions of N gases from the soil profile simulated in APEX include N_2 and nitrous oxide (N_2O) and ammonia volatilization. Volatilization is estimated simultaneously with nitrification. The organic N loss is estimated using a modified loading function (Williams and Hann, 1978) that considers sediment yield, organic N loss in the soil surface and an enrichment ratio. The soluble N loss is estimated by considering the change in concentration (Williams, 1995). The concentration in a soil layer decreases exponentially as a function of flow volume.

Phosphorus cycling

Soluble P runoff loss is estimated as a function of the concentration of labile P in the topsoil layer, runoff volume and a linear adsorption isotherm. Sediment transport of P is estimated with a modified loading function originally developed by McElroy *et al.* (1976). The P mineralization and immobilization routines in APEX were developed by Jones *et al.* (1984). The mineralization model is a modification of the Production of Arid Pastures Limited by Rainfall and Nitrogen (PAPRAN) mineralization model (Seligman and van Keulen, 1981). Mineralization from the fresh organic P pool is estimated as the product of the mineralization rate and the fresh organic P content. Mineralization of organic P associated with humus is estimated for each soil layer as a function of soil water content, temperature and bulk density. Mineral P is transferred among three pools: labile, active mineral and stable mineral. Fertilizer P is labile (available for plant use) at application but may be quickly transferred to the active mineral pool.

Routing component

For APEX simulation, a watershed can be subdivided into multiple sub-areas, each assumed to be homogeneous in terms of soil, slope, land use, management and weather. Each sub-area is associated with a channel for routing. APEX sub-areas are functionally equivalent to sub-basins in the SWAT model with respect to defining watershed connectivity. For larger watersheds or complex landscape, GIS tools are available for delineating sub-area boundaries based upon a digital elevation model (DEM). For example, the ArcAPEX tool (Tuppad *et al.*, 2009) can be used to define the sub-area boundaries either by using a DEM or by importing user-defined sub-area boundaries and streams. The DEM-based sub-area delineation implements the single-flow direction algorithm used in ESRI (Environmental Systems Research Institute) software (Jenson and Domingue, 1988) to generate the required flow direction and flow accumulation raster data sets used in watershed delineation. The user may also manually define the sub-areas using the procedure described in Williams *et al.* (2006). A downstream sub-area is identified if the distance from the sub-area outlet to the most distant point of the sub-area is greater than the routing reach length.

APEX has two options for routing water through channels and flood plains: a daily time step average flow method and a short time interval complete flood routing method. The complete flood routing approach simulates dynamic streamflow, whereas the daily time step method can only estimate daily water yield. Sediment is routed through the channel and the flood plain separately. The sediment routing equation is a variation of Bagnold's sediment transport equation (Bagnold, 1977), which estimates the transport concentration capacity as a function of velocity. Organic forms of N and P, and adsorbed pesticide, are transported by sediment and are routed using an enrichment ratio approach. The enrichment ratio is estimated as the ratio of the mean sediment particle size distribution of the outflow divided by that of the inflow. Organic N and P mineralization in the channels is not considered because, in general, the travel time is short. Mineral nutrient and soluble pesticide losses occur only if flow is lost within the reach. The pesticide routing approach is the same as that described for nutrients.

Description of study watershed and model inputs

The study watershed is a sub-watershed within the Bosque River watershed in central Texas. Tonk Creek (TC) and Wasp Creek (WC) drain this sub-watershed and have a combined drainage area of 104 km^2 (Fig. 4.2). Watershed elevation ranges from 174 to 294 m. Long-term annual precipitation averages 871 mm. The major soil series include Crawford (fine clayey), Denton (fine silty), Slidell (fine clayey), Bolar (gravelly loamy clay) and Aledo (gravelly loamy clay).

The inputs used in this study are listed in Table 4.3. The sub-watershed was divided into 102 sub-areas (Fig. 4.2) based on the 10 m DEM using the GIS-based APEX interface – ArcAPEX (Tuppad *et al.*, 2009). The watershed is predominantly cropland with maize (*Zea mays* L.) and winter wheat (*Triticum aestivum* L.) being the major crops. A 3-year conventionally tilled 'maize–maize–winter wheat' rotation was simulated on all croplands. Management scheduling data for cropland, rangeland and pasture land, including

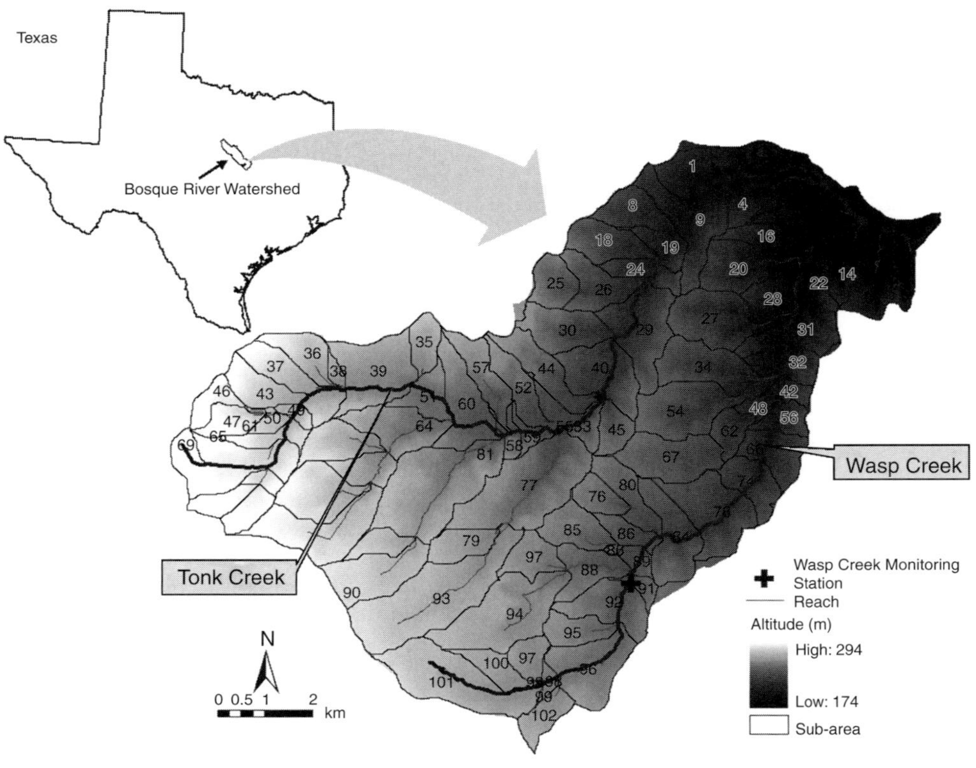

Fig. 4.2. Sub-area delineation and stream network for the study sub-watershed of the Bosque River Watershed in central Texas. The location of the sub-area within the whole watershed is indicated on the outline map for Texas (top).

Table 4.3. Input used in APEX model set up to define topographic, soil, land-use and weather parameters for the study watershed (a sub-area of the Bosque Watershed in Texas).

Input	Resolution	Source[a]
DEM[b]	10 m	National Elevation Data Set, USGS
Land use/land cover data	1:24,000	USDA NRCS
Soils	1:24,000	SSURGO
Rainfall	Daily	Rain gauge at the monitoring site on Tonk Creek, maintained by TIAER
Temperature (minimum and maximum)	Daily	Cooperative weather station
Solar radiation, relative humidity and wind data	Monthly statistics (used to generate daily data)	Weather station at McGregor, Texas from APEX weather database

[a]SSURGO, Soil Survey Geographic; TIAER, Texas Institute of Applied Environmental Research; USDA NRCS, US Department of Agriculture National Resources Conservation Service; USGS, US Geological Survey.
[b]DEM, digital elevation model.

the type and dates of tillage, and the type, dates and rates of fertilizer application, were obtained from local 'soil and water conservation district' personnel. Maize was planted with a planter at the end of February and harvested in early August. Wheat was planted in October and harvested in June. The tillage system consisted of one field cultivation operation before planting for seedbed preparation and disc and sweep chisel tillage to incorporate applied fertilizer. Fertilizers were broadcast at the rates of 108 kg N ha^{-1} and 39 kg P ha^{-1} before planting the maize and at the rates of 73 kg N ha^{-1} and 34 kg P ha^{-1} 2 weeks before planting the wheat. Both rangeland and pasture land were simulated as grazed. Fertilizers were applied to pasture land at the rates of 59 kg N ha^{-1} and 46 kg P ha^{-1}, while rangeland was not fertilized. These represent typical management operations of producers in the area.

APEX calibration and validation

The measured monthly flow, sediment and nutrient data at the WC monitoring station, available from October 1995 to December 1999, were used for calibration of APEX and data from January 2000 to March 2003 were used for model validation. The model was set up to run from 1993. Model options used in this study were the NRCS curve number method for runoff estimation, the variable daily CN soil moisture index method for daily CN estimation, the modified Rational equation for peak flow, the Hargreaves method for potential evapotranspiration and a variation of the modified USLE – the MUST equation (Williams, 1995) – for erosion/sedimentation estimation (see Table 4.2).

This study builds on the previous study by Tuppad *et al.* (2009), which was to demonstrate the application of ArcAPEX, a GIS-based APEX user interface that integrates enhanced GIS capabilities and algorithms with APEX databases, input and output management. In the previous study, although the model performed well for the calibration period, it did not replicate the validation period well. Tuppad *et al.* (2009) pointed out the need for further investigation to improve the modelling results. In this study, we collected the crop grain yield data from the USDA National Agricultural Statistics Service (NASS) and readjusted the plant populations based on crop grain yield comparisons. Fine-tuning in CN values and timing of tillage were revisited and modified. The model parameters adjusted are listed in Table 4.4.

Statistical measures based on monthly values, including mean, standard deviation (SD), R^2, Nash–Sutcliffe efficiency (NSE) (Nash and Sutcliffe, 1970), per cent error or per cent bias (PBIAS), and ratio of the root mean square error to the standard deviation of observed data (RSR) were used to evaluate the model performance based on criteria suggested by Moriasi *et al.* (2007).

Table 4.4. Calibration parameters for the APEX model, their range and the actual calibrated values.

Parameter	Description	Range (Williams *et al.*, 2006)	Calibration value
Curve number	SCS[a] Runoff Curve number	±10%	Reduced by −8%
Parm42	SCS curve number index coefficient (regulates effect of potential evapotranspiration in driving the SCS curve number retention parameter)	0.5–1.5	1.2
Parm46	RUSLE[a] 'C' factor coefficient in exponential residue function in residue factor	0.5–1.5	1.0
Parm18	Sediment routing exponent	1–1.5	1.0
Parm19	Sediment routing coefficient (sets potential sediment concentration when flow velocity is 1.0 m s^{-1})	0.01–0.05	0.01
Parm29	Biological mixing efficiency (simulates mixing in topsoil by earthworms, etc.)	0.1–0.5	0.3
Parm31	Maximum depth for biological mixing (m)	0.1–0.3	0.3
Parm14	Nitrate leaching ratio (ratio of nitrate concentration in surface runoff to that in the percolate)	0.1–1	0.2
Parm35	Denitrification soil water threshold (fraction of field capacity soil water storage to trigger denitrification)	0.9–1.1	0.9
Parm72	Volatilization/nitrification partitioning coefficient	0.05–0.5	0.4
Parm8	Soluble P runoff coefficient (ratio of P concentration in sediment to that in water)	10–25	25
Parm59	P upward movement by evaporation coefficient	1–20	3

[a]RUSLE, Revised Universal Soil Loss Equation; SCS, USDA Soil Conservation Service.

Scenario analysis

Differences in weather, soil, topographic and land-management characteristics of watersheds contribute to different watershed runoff, sediment yield and crop productivity. Therefore, statistical comparisons among treatments (e.g. different tillage systems) are hardly isolated by using only field data, which inevitably involves factor interaction. The calibrated model was run for a 30-year period (1977 to 2006) to establish the baseline condition against which the conservation practice scenarios, including no-till cropping, furrow dyking and contour farming, were evaluated for sediment yield, total nitrogen, total phosphorus and crop grain yield. Details of these practices can be found in USDA NRCS (2007); they were simulated individually, and all inputs except the parameters used to represent a practice were held constant.

No-till was represented in the APEX model by excluding all tillage operations, replacing row crop planters for maize and drills for winter wheat with no-till planters and no-till drills. Furrow dyking was simulated by building furrow dykes during the planting of maize and removing them after harvest. The simulated furrow dykes were spaced 1 m apart and offset at 150 mm in height. Contouring was represented by the USLE conservation support practice factor (PEC) and CN. A PEC value of 1.0 in the baseline condition was altered to 0.6 or 0.5 depending on the average upland slope of the sub-area (Schwab *et al.*, 1995; Arabi *et al.*, 2008). The CN was reduced by three from the baseline condition (Arabi *et al.*, 2008).

Results and Discussion

Flow, sediment and nutrients

The simulated mean and standard deviation of the monthly streamflow at the WC monitoring location compared closely to the measured values for both the calibration and validation periods (Table 4.5). The model overestimated sediment during the calibration period by 20% whereas it underestimated sediment by 23% during the validation period. Predicted average monthly total nitrogen (TN) and total phosphorus (TP) values were close to the observed values during both calibration and validation periods, with PBIAS within 10% of expected TP for the validation period. NSE values ranged from 0.59 to 0.83, R^2 values from 0.60 to 0.87 and RSR values from 0.41 to 0.63, considering all the constituents evaluated for both the calibration and validation periods (Table 4.5). Based on the statistical criteria (established based on values of NSE, RSR and PBIAS) for establishing satisfactory water quality model performance as proposed by Moriasi *et al.* (2007), model performance is satisfactory for monthly flow, sediment and nutrient losses for both the calibration and validation periods.

Simulated monthly streamflow matched well in trend and quantity with observed values (Fig. 4.3). Simulated monthly sediment yield followed, in general, the pattern of observed sediment yield (Fig. 4.4). There were no big runoff events in both 1995 and 1996, but this period was followed by the relatively high flow events of February 1997 (Fig. 4.3). The observed sediment yield in February 1997 is about 60% of the total sediment yield in 1997. The model captured the sediment yield response to the wet condition followed by a period of dry condition; however, it overestimated the sediment yield in February 1997 by 32%. Total N loss was overestimated in March 1997 and March 2000, although APEX underestimated the total N loss in March 2001 (Fig. 4.5). The model realistically captured the evolution of total P losses during the 90-month period of measurement (Fig. 4.6). The annual sediment yield (Fig. 4.7) was over-predicted in 1997, mostly as a result of the over-prediction from February to June. The simulated annual total N loss was 15.8 ± 10.2 kg ha^{-1} compared with the measured value of 15.4 ± 8.4 kg ha^{-1} based on available annual values (1996–2002). The simulated annual total P losses ranged from 0.09 to 1.62 kg ha^{-1} with an annual mean of 0.65 kg ha^{-1}; measured annual values ranged from 0.08 to 1.34 kg ha^{-1}, with an average annual total P loss of 0.66 kg ha^{-1} from 1996 to 2002. APEX captured the annual flow, sediment and nutrient losses reasonably well (Fig. 4.7).

Table 4.5. Measured versus APEX-simulated streamflow, sediment, total nitrogen and total phosphorus yields at Wasp Creek monitoring station in the Bosque River study sub-watershed in Texas for the calibration period (October1995 to December 1999) and validation period (January 2000 to March 2003).

Parameter	Calibration/ Validation	Measured Mean	Measured SD	Simulated Mean	Simulated SD	PBIAS[a] (%)	NSE[a]	R^2	RSR[a]
Flow (m^3 s^{-1})	Calibration	0.04	0.05	0.03	0.06	−10.2	0.81	0.87	0.44
	Validation	0.03	0.03	0.04	0.03	15.4	0.64	0.70	0.59
Sediment (Mg	Calibration	0.05	0.16	0.06	0.15	20.0	0.82	0.82	0.42
ha^{-1} month^{-1})	Validation	0.05	0.08	0.04	0.06	−23.2	0.59	0.60	0.63
Total nitrogen	Calibration	1.31	1.56	1.33	2.06	1.3	0.65	0.82	0.58
(kg ha^{-1} month^{-1})	Validation	1.15	1.15	1.25	1.14	8.7	0.59	0.63	0.63
Total phosphorus	Calibration	0.06	0.13	0.06	0.14	8.1	0.83	0.84	0.41
(kg ha^{-1} month^{-1})	Validation	0.05	0.07	0.04	0.05	−18.3	0.64	0.66	0.59

[a]NSE, Nash–Sutcliffe efficiency; PBIAS, per cent bias; RSR, ratio of root mean square error to standard deviation of observed data.

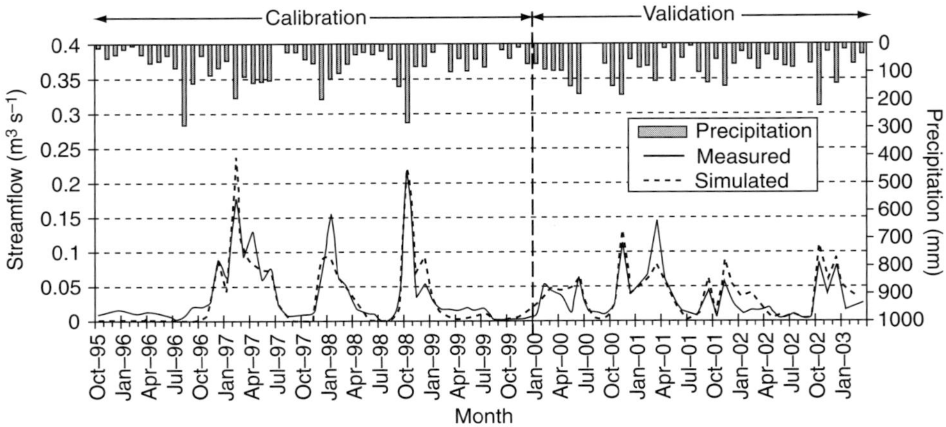

Fig. 4.3. Monthly precipitation and measured versus simulated streamflow at the Wasp Creek monitoring station within the Bosque River Watershed in central Texas.

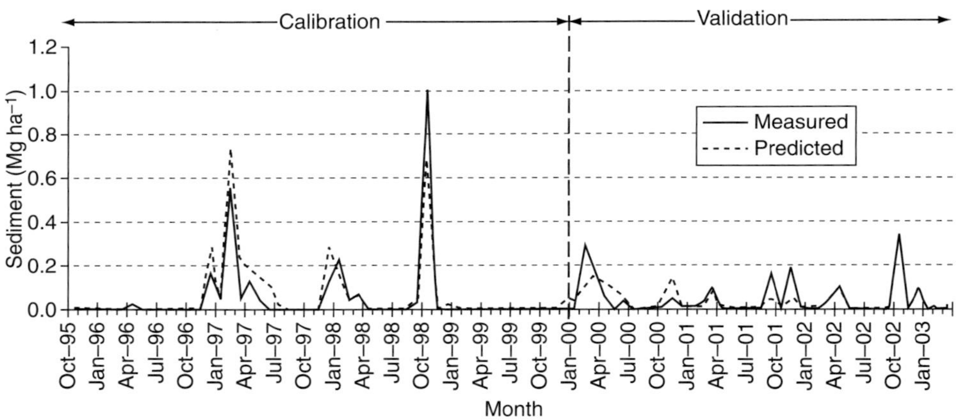

Fig. 4. 4. Measured and simulated monthly sediment yield at the Wasp Creek monitoring station within the Bosque River Watershed in central Texas.

Crop productivity

APEX-simulated crop yields for the two major grain crops (maize and winter wheat) were compared with those from NASS. Maize yields in the maize–winter wheat rotation during 1995–2003 averaged 4.6 ± 1.6 Mg ha^{-1} (n = 9), while APEX simulated yields of 5.0 ± 1.0 Mg ha^{-1}. Observed winter wheat yields averaged 2.5 ± 0.4 Mg ha^{-1} (n = 9) while simulated yields averaged 2.6 ± 0.25 Mg ha^{-1}. The relative errors between simulated and observed yields are within 10%.

Scenario analysis

No-till, furrow dyke and contour systems reduced sediment by 44%, 23% and 54%, respectively (Table 4.6), at the watershed outlet. Chichester and Richardson (1992) reported a 90% reduction in sediment with no-till compared with a chisel-till system on watersheds near Riesel, Texas. Dickey *et al.* (1984) reported 44–90% reduction in sediment with no-till. The no-till farming system generally retains crop residues on the soil surface, which protects the surface against

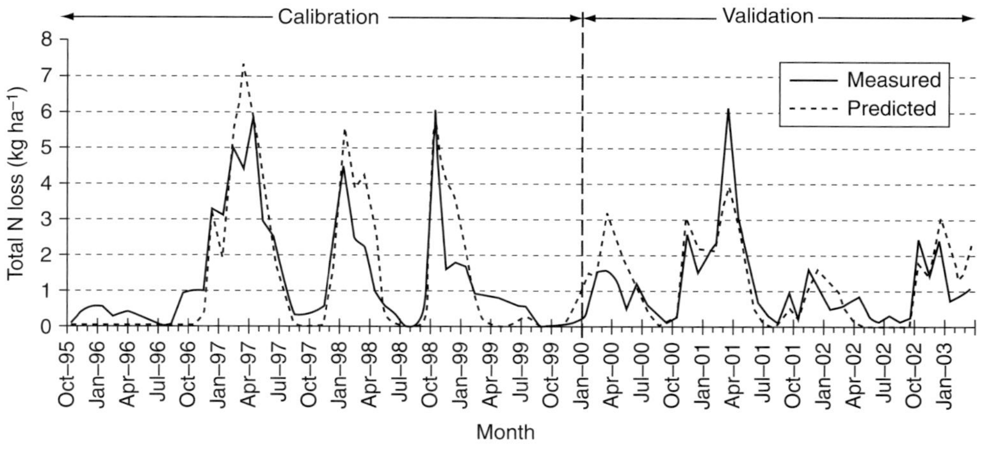

Fig. 4.5. Measured and simulated monthly total nitrogen loss at the Wasp Creek monitoring station within the Bosque River Watershed in central Texas.

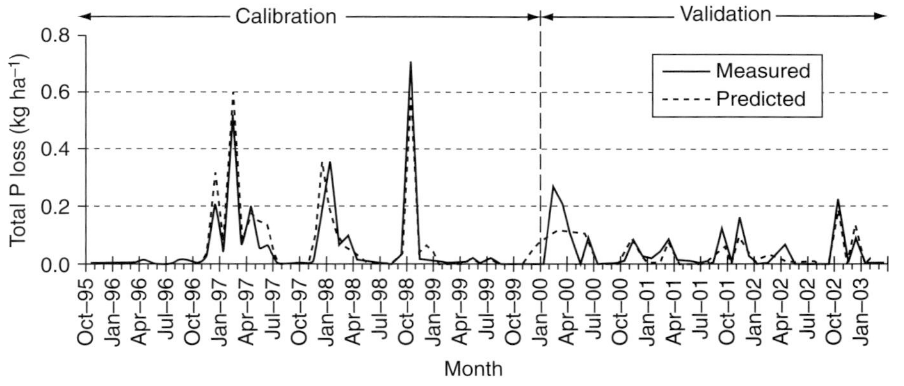

Fig. 4.6. Measured and simulated monthly total phosphorus loss at the Wasp Creek monitoring station within the Bosque River Watershed in central Texas.

raindrop impact. Furrow dyke and contouring farming increases infiltration capacity and reduces water loss. By permitting higher rates of infiltration, erosion is reduced (Hackwell *et al.*, 1991; Truman and Nuti, 2009). In most cases, contour farming can reduce soil loss by as much as 50% (USDA NRCS, 2001). The simulated sediment yield results in the present study agree with the findings in the literature.

Total N losses were reduced by 1%, 2% and 4% in no-till, furrow dyke and contour farming, respectively (Table 4.6). No-till resulted in a reduction of 31% total P loss. The model predicted that furrow dyking reduced total P loss by 25% and that contour farming reduced total P loss by 41% at the watershed outlet. These alternative tillage practices are effective in reducing erosion and sediment production; therefore, less sediment and related pollutants may be transported. However, increased infiltration may increase the transportation potential for soluble pollutants. Greater reductions were predicted for total P than for total N as P is mostly associated with the sediment phase. The no-till practice, as simulated by APEX,

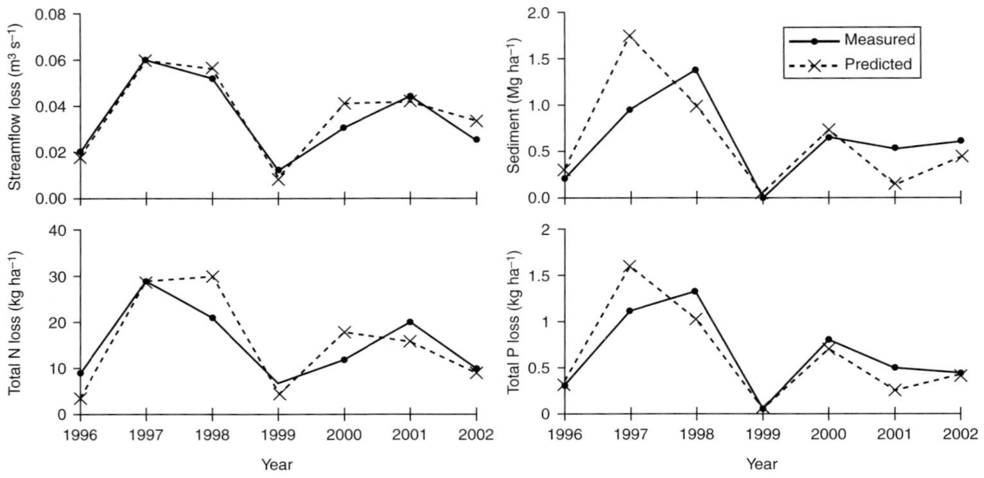

Fig. 4.7. Measured and simulated annual flow, sediment and nutrient losses at the Wasp Creek monitoring station within the Bosque River Watershed in central Texas.

Table 4.6. Long-term annual average simulated sediment, total nitrogen and total phosphorus losses from no-till, furrow dyking and contour farming scenarios compared with conventional tillage in the baseline scenario at the Bosque River study sub-watershed outlet in Texas, from 1977 to 2006.

	Baseline	No-till (benefit)	Furrow dyke (benefit)	Contour (benefit)
Sediment yield (Mg ha^{-1})	0.63	0.35 (44.4%)	0.48 (23.3%)	0.29 (53.6%)
Total nitrogen loss (kg ha^{-1})	7.87	7.80 (0.9%)	7.74 (1.6%)	7.57 (3.9%)
Total phosphorus loss (kg ha^{-1})	0.57	0.40 (30.3%)	0.43 (24.6%)	0.33 (41.4%)
Maize grain yield (Mg ha^{-1})	5.23	5.20 (–0.6%)	5.27 (0.8%)	5.25 (0.4%)
Wheat grain yield (Mg ha^{-1})	2.62	2.59 (–1.1%)	2.64 (0.8%)[a]	2.57 (–1.9%)

[a]Wheat yield when maize, in rotation, had furrow dykes.

slightly decreased crop grain yields compared with conventional tillage (Table 4.6). Furrow dyking had a positive effect on crop yields, and contour farming slightly increased maize yield but reduced wheat yield as predicted by APEX. Yield response to the tillage system used has inconsistent reports in the literature. Logsdon *et al.* (1999) reported that maize yield was significantly reduced and soybean yield was significantly increased under no-till management compared with the chisel tillage system. Maize yield losses ranging from 0.9% to 4.3% with no-till compared with chisel-till were reported in Vyn *et al.* (2002). Positive yield responses to furrow dyking have been reported in Nuti *et al.* (2009); however, it has been reported that furrow dyking could lead to reduced yields under higher rainfall because of nutrient leaching (McFarland *et al.*, 1991; Wiyo *et al.*, 2000).

Summary

The APEX model was tested using field-measured data from an agriculture-dominated sub-watershed (including Tonk Creek and Wasp Creek) of the Bosque River Watershed in central Texas. APEX was calibrated and validated for monthly streamflow, sediment yield, total N and total P losses using the data from a monitoring station at Wasp Creek. The R^2

values ranged from 0.82 to 0.87 during the calibration period (October 1995 to December 1999) and from 0.60 to 0.70 during the validation period (January 2000 to March 2003). The NSE values ranged from 0.59 to 0.83, values of RSR from 0.41 to 0.63 and values of PBIAS were within ±25% during both the calibration and validation periods. The results indicate that the performance of the model is satisfactory for monthly flow, sediment and nutrient losses based on the statistical criteria proposed by Moriasi *et al.* (2007). The percentage errors between the simulated and reported maize and wheat grain yields from NASS were 9% and 3%, respectively. Scenario analysis (1977–2006) indicated substantial reduction in sediment – by 44%, 23% and 54% due to no-till, furrow dyking and contour farming, respectively. Total P losses were reduced by 31%, 25% and 41% due to no-till, furrow dyking and contour farming, respectively. However, the benefit on total N loss was not significant, and there were only very slight changes in crop yields between these cropping systems compared with the conventional system in the baseline scenario. Results from scenario analysis are in line with findings in the literature. The study shows that the APEX model is capable of predicting streamflow, sediment yield, nutrient losses and crop yields under different agricultural management systems.

The APEX model has proved to be a useful tool for simulating different management scenarios at field, farm or watershed scales, as reviewed in this study. The model is being expanded to include a hail occurrence/damage component, a feedlot dust component, and an optional method based on the nearest-neighbour concept (Nemes *et al.*, 2006) for estimating hydraulic conductivity, field capacity, and wilting point as a function of soil texture and organic C. More comprehensive approaches to estimate CO_2, O_2 and N_2O fluxes in the soil–plant–atmosphere system are currently under development; these would make the current denitrification routine more mechanistic. An improvement to the APEX grazing component is being worked on to include preferential grazing and weight gain and loss. As the use of the APEX model expands to accommodate an ever-increasing range of environmental problems and conditions, the model needs to include an improved subsurface tile drainage routine similar to the approach used in the DRAINMOD model (Skaggs, 1999, 2007), incorporate a kinematic wave methodology for a water routing scheme as described in Borah *et al.* (1980, 2007), and incorporate a routine to simulate bacterial die-off and transport similar to the methodology in SWAT. The bacterial routine will provide capabilities necessary for source tracking of bacteria and for assessing the impacts of management practices on the fate and transport of bacteria.

References

Arabi, M., Frankenberger, J.R., Engel, B.A. and Arnold, J.G. (2008) Representation of agricultural conservation practices with SWAT. *Hydrological Process* 22, 3042–3055.

Arnold, J.G., Srinivasan, R., Muttiah, R.S. and Williams, J.R. (1998) Large area hydrologic modeling and assessment Part I: Model development. *Journal of the American Water Resources Association* 34, 73–89.

Bagnold, R.A. (1977) Bed-load transport by natural rivers. *Water Resources Research* 13, 303–312.

Baier, W. and Robertson, G.W. (1965) Estimation of latent evaporation from simple weather observations. *Canadian Journal of Plant Science* 45, 276–284.

Borah, D.K., Prasad, S.N. and Alonso, C.V. (1980) Kinematic wave routing incorporating shock fitting. *Water Resource Research* 16, 529–541.

Borah, D.K., Arnold, J.G., Bera, M., Krug, E.C. and Liang, X. (2007) Storm event and continuous hydrologic modeling for comprehensive and efficient watershed simulations. *Journal of Hydrologic Engineering* 12, 605–616.

Chichester, F.W. and Richardson, C.W. (1992) Sediment and nutrient loss from clay soils as affected by tillage. *Journal of Environmental Quality* 21, 587–590.

Chung, S.W., Gassman, P.W., Gu, R. and Kanwar, R.S. (2002) Evaluation of EPIC for assessing tile flow and nitrogen losses for alternative agricultural management systems. *Transactions of the ASABE* 45, 1135–1146.

Dickey, E.C., Shelton, D.P., Jasa, P.J. and Peterson, T.R. (1984) Tillage, residue, and erosion on moderately sloping soils. *Transactions of the ASABE* 27, 1093–1099.

Flowers, J.D., Williams, J.R. and Hauck, L.M. (1996) *Livestock and the Environment: A National Pilot Project. NPP Integrated Modeling System: Calibration of the APEX Model for Dairy Waste Application Fields in Erath County, Texas.* TIAER Report No. PR96-07. Stephenville, Texas: Tarleton State University, Texas Institute for Applied Environmental Research. Available at: http://tiaer.tarleton.edu/pdf/PR9607.pdf (accessed 6 July 2011).

Foster, G.P. (2005) *Science Documentation. Revised Universal Soil Loss Equation, Version 2 (RUSLE 2).* National Sedimentation Laboratory, US Department of Agriculture, Agricultural Research Service, Oxford, Massachusetts.

Gassman, P.W., Williams, J.R., Benson, V.W., Izaurralde, R.C., Hauck, L., Jones, C.A., Atwood, J.D., Kiniry, J. and Flowers, J.D. (2005) *Historical Development and Applications of the EPIC and APEX Models.* Working Paper 05-WP 397. Center for Agricultural and Rural Development, Iowa: Iowa State University, Ames. Iowa. Available at: http://www.card.iastate.edu/publications/synopsis.aspx?id=763 (accessed 28 September 2007).

Gassman, P.W., Williams, J.R., Wang, X., Saleh, A., Osei, E., Hauck, L., Izaurralde, R.C. and Flowers, J.D. (2010) The Agricultural Policy Environmental extender (APEX) model: An emerging tool for landscape and watershed environmental analyses. *Transactions of the ASABE* 53, 711–740.

Green, W.H. and Ampt, G.A. (1911) Studies on soil physics: 1. Flow of air and water through soils. *Journal of Agricultural Science* 4, 1–24.

Hackwell, S.G., Rochester, E.W., Yoo, K.H., Burt, E.C. and Monroe, G.E. (1991) Impact of reservoir tillage on water intake and soil erosion. *Transactions of the ASABE* 34, 436–442.

Hargreaves, G.H. and Samani, Z.A. (1985) Reference crop evapotranspiration from temperature. *Applied Engineering in Agriculture* 1, 96–99.

Harman, W.L., Wang, E. and Williams, J.R. (2004) Reducing atrazine losses: water quality implications of alternative runoff control practices. *Journal of Environmental Quality* 33, 7–12.

Impact Assessment Group (2000) Impact Methods to Predict and Assess Contributions of Technology (IMPACT), Final Report, 2000. Impact Assessment Group, Texas Agricultural Experiment Station, Texas A&M University System in Collaboration with Kenyan Minister of Agriculture, Kenya Agricultural Research Institute, International Livestock Research Institute, Institute of Rural Economy, Institute of Sahel. Research under USAID Grant No. PCE-G-00-97-00051-00. Available at: http://cnrit.tamu.edu/IMPACT (accessed 11 February 2011).

Izaurralde, R.C., Williams, J.R., McGill, W.B., Rosenberg, N.J. and Quiroga Jakas, M.C. (2006) Simulating soil C dynamics with EPIC: model description and testing against long-term data. *Ecological Modelling* 192, 362–384.

Izaurralde, R.C., Williams, J.R., Post, W.M., Thomson, A.M., McGill, W.B., Owens, L.B. and Lal, R. (2007) Long-term modeling of soil C erosion and sequestration at the small watershed scale. *Climatic Change* 80, 73–90.

Jenson, S.K. and Domingue, J.O. (1988) Extracting topographic structure from digital elevation data for geographic information system analysis. *Photogrammetric Engineering and Remote Sensing* 54, 1593–1600.

Jones, C.A., Cole, C.V., Sharpley, A.N. and Williams, J.R. (1984) A simplified soil and plant phosphorus model: I. Documentation. *Soil Science Society of America Journal* 48, 800–805.

Kannan, N., Santhi, C., Williams, J.R. and Arnold, J.G. (2008) Development of a continuous soil moisture accounting procedure for curve number methodology and its behaviour with different evapotranspiration methods. *Hydrological Processes* 22, 2114–2121.

Kiniry, J.R., Williams, J.R., Gassman, P.W. and Debaeke, P. (1992) A general, process-oriented model for two competing plant species. *Transactions of the ASABE* 35, 801–810.

Logsdon, S.D., Kaspar, T.C. and Cambardella, C.A. (1999) Depth-incremental soil properties under no-till or chisel management. *Soil Science Society of America Journal* 63, 197–200.

Mausbach, J.M. and Dedrick, A.R. (2004) The length we go: measuring environmental benefits of conservation practices in the CEAP. *Journal of Soil and Water Conservation* 59(5), 96A.

McElroy, A.D., Chiu, S.Y., Nebgen, J.W., Aleti, A. and Bennett, F.W. (1976) *Loading Functions for Assessment of Water Pollution from Nonpoint Sources.* EPA 600/2-76-151. US Environmental Protection Agency, Washington, DC.

McFarland, M.L., Hons, F.M. and Saladino, V.A. (1991) Effects of furrow diking and tillage on corn grain yield and nitrogen accumulation. *Agronomy Journal* 83, 382–386.

Mockus, V. (1969) Hydrologic soil-cover complexes. In: *SCS National Engineering Handbook, Section 4, Hydrology*. US Department of Agriculture, Soil Conservation Service, Washington, DC, pp. 10.1–10.24.

Monteith, J.L. (1965) Evaporation and environment. In: *Proceedings of the 19th Symposium of the Society for Experimental Biology*. Cambridge University Press, New York, pp. 205–233.

Moriasi, D.N., Arnold, J.G., Van Liew, M.W., Binger, R.L., Harmel, R.D. and Veith, T. (2007) Model evaluation guidelines for systematic quantification of accuracy in watershed simulations. *Transactions of the ASABE* 50, 885–900.

Nash, J.E. and Sutcliffe, J.V. (1970) River flow forecasting through conceptual models. Part 1: A discussion of principles. *Journal of Hydrology* 10, 282–290.

Nemes, A., Rawls, W.J. and Pachepsky, Ya.A. (2006) Use of a non-parametric nearest-neighbor technique to estimate soil water retention. *Soil Science Society of America Journal* 70, 327–336.

Nuti, R.C., Lamb, M.C., Sorensen, R.B. and Truman, C.C. (2009) Agronomic and economic response to furrow diking tillage in irrigated and non-irrigated cotton (*Gossypium hirsutum* L.). *Agricultural Water Management* 96, 1078–1084.

Onstad, C.A. and Foster, G.R. (1975) Erosion modeling on a watershed. *Transactions of the ASABE* 18, 288–292.

Osei, E., Gassman, P.W., Jones, R.D., Pratt, S.J., Hauck, L.M., Beran, L.J., Rosenthal, W.D. and Williams, J.R. (2000) Economic and environmental impacts of alternative practices on dairy farms in an agricultural watershed. *Journal of Soil and Water Conservation* 55, 466–472.

Parton, W.J., Schimel, D.S., Cole, C.V. and Ojima, D.S. (1987) Analysis of factors controlling soil organic matter levels in Great Plains grasslands. *Soil Science Society of America Journal* 51, 1173–1179.

Parton, W.J., Scurlock, J.M.O., Ojima, D.S., Gilmanov, T.G., Scholes, R.J., Schimel, D.S., Kirchner, J., Menaut, J.-C., Seastedt, T., Garcia Moya, E., Kamnalrut, A. and Kinyamario, J.I. (1993) Observations and modelling of biomass and soil organic matter dynamics for the grassland biome worldwide. *Global Biogeochemical Cycles* 7, 785–809.

Parton, W.J., Ojima, D.S., Cole, C.V. and Schimel, D.S. (1994) A general model for soil organic matter dynamics: sensitivity to litter chemistry, texture and management. In: *Quantitative Modeling of Soil Forming Processes*. Soil Science Society of America Special Publication No. 39, Madison, Wisconsin, pp. 147–167.

Penman, H.L. (1948) Natural evaporation from open water, bare soil, and grass. In: *Proceedings of the Royal Society of London, A* 193, 120–146.

Potter, K.N., Williams, J.R., Larney, F.J. and Bullock, M.S. (1998) Evaluation of EPIC's wind erosion submodel using data from southern Alberta. *Canadian Journal of Soil Science* 78, 485–492.

Priestley, C.H.B. and Taylor, R.J. (1972) On the assessment of surface heat flux and evaporation using large-scale parameters. *Monthly Weather Review* 100, 81–92.

Reddy, K.R., Khaleel, R., Overcash, M.R. and Westerman, P.W. (1979) A nonpoint-source model for land areas receiving animal wastes: II. Ammonia volatilization. *Transactions of the ASABE* 22, 1398–1404.

Renard, K.G., Foster, G.R., Weesies, G.A., McCool, D.K. and Yoder, D.C. (1997) *Predicting Soil Erosion by Water: A Guide to Conservation Planning with the Revised Universal Soil Loss Equation (RUSLE)*. Agriculture Handbook No. 703, US Department of Agriculture, Agricultural Research Service, Washington, DC. Available at: http://www.techtransfer.osmre.gov/NTTMainSite/Library/hbmanual/rusle/ah_703.pdf (accessed 11 February 2011).

Ritchie, J.T. (1972) A model for predicting evaporation from a row crop with incomplete cover. *Water Resources Research* 8, 1204–1213.

Schwab, G.O., Fangmeier, D.D. and Elliot, W.J. (1995) *Soil and Water Management Systems*, 4th edn. John Wiley, New York, pp. 108–111.

Seligman, N.G. and van Keulen, H. (1981) PAPRAN: a simulation model of annual pasture production limited by rainfall and nitrogen. In: Frissel, M.J. and van Veen, J.A. (eds) *Proceedings, Workshop: Simulation of Nitrogen Behaviour of Soil–Plant Systems*. Centre for Agricultural Publishing and Documentation, Wageningen, The Netherlands, pp. 192–221.

Skaggs, R.W. (1999) Drainage simulation models. In: Skaggs, R.W. and van Schilfgaarde, J. (eds) *Agricultural Drainage*. Agronomy Monograph 38, ASA (American Society of Agronomy)/CSSA/SSSA, Madison, Wisconsin, pp. 469–500.

Skaggs, R.W. (2007) Criteria for calculating drain spacing and depth. *Transactions of the ASABE* 50, 1657–1662.

Thomson, A.M., Izaurralde, R.C., Rosenberg, N.J. and He, X. (2006) Climate change impacts on agriculture and soil carbon sequestration potential in the Huang-Hai Plain of China. *Agriculture, Ecosystems and Environment* 114, 195–209.

Truman, C.C. and Nuti, R.C. (2009) Improved water capture and erosion reduction through furrow diking. *Agricultural Water Management* 96, 1071–1077.

Tuppad, P., Winchell, M., Wang, X., Srinivasan, R. and Williams, J.R. (2009) ArcAPEX: ArcGIS interface for Agricultural Policy Environmental eXtender (APEX) hydrology/water quality model. *International Agricultural Engineering Journal* 18, 59–71.

USDA NRCS (US Department of Agriculture, Natural Resources Conservation Service) (2001) Conservation Practice Standard: Contour Farming, (Acre), Code 330. Available at: http://efotg.sc.egov.usda.gov// references/public/AL/al330.pdf (accessed 11 February 2011).

USDA NRCS (2004) Chapter 10 – Estimation of direct runoff from storm rainfall. In: *NRCS National Engineering Handbook (NEH), Part 630 – Hydrology*, USDA NRCS, Washington, DC, pp. 10.1–10.22. Available at: http://policy.nrcs.usda.gov/viewerFS.aspx?hid=21422 (accessed 11 February 2011).

USDA NRCS (2007) Conservation Effects Assessment Project (CEAP). Washington, DC. Available at: http:// www.nrcs.usda.gov/Technical/nri/ceap/ (accessed 11 February 2011)

USDA SCS (USDA Soil Conservation Service) (1986) *Urban Hydrology for Small Watersheds*. Technical Release 55 (TR-55), Washington, DC. Available at: http://www.hydrocad.net/tr-55.htm (accessed 11 February 2011).

Vitousek, P.M., Turner, D.R., Parton, W.J. and Sanford, R.L. (1994) Litter decomposition on the Mauna Loa environmental matrix, Hawaii: patterns, mechanisms, and models. *Ecology* 75, 418–429.

Vyn, T., West, T. and Steinhardt, G. (2002) *Use No-till in Delayed Planting of Both Corn and Soybean*. Corny News Network, Purdue University Department of Agronomy, West Lafayette, Indiana. Published at the Chat 'n Chew Cafe. Available at: http://www.agry.purdue.edu/ext/corn/news/articles.02/Delayed_ Plant_Notill-0515.html (accessed 11 February 2011).

Wang, E., Chang, X., Williams, J.R. and Cheng, X. (2006) Predicting soil erosion for alternative land uses. *Journal of Environmental Quality* 35, 459–467.

Wang, X., Potter, S.R., Williams, J.R., Atwood, J.D. and T. Pitts (2006) Sensitivity analysis of APEX for national assessment. *Transactions of the ASABE* 49, 679–688.

Wang, X., Saleh, A., McBroom, M.W., Williams, J.R., and Yin, L. (2007) Test of APEX for nine forested watersheds in east Texas. *Journal of Environmental Quality* 36, 983–995.

Wang, X., Gassman, P.W., Williams, J.R., Potter, S. and Kemanian, A.R. (2008) Modeling the impacts of soil management practices on runoff, sediment yield, maize productivity, and soil organic carbon using APEX. *Soil and Tillage Research* 101, 78–88.

Wang, X., Hoffman, D.W., Wolfe, J.E., Williams, J.R. and Fox, W.E. (2009) Modeling the effectiveness of conservation practices at Shoal Creek watershed, TX, using APEX. *Transactions of the ASABE* 52, 1181–1192.

Williams, J.R. (1975) *Sediment Yield Prediction with Universal Equation Using Runoff Energy Factor*. ARS-S-40. US Department of Agriculture, Agricultural Research Service, Washington, DC.

Williams, J.R. (1995) The EPIC Model. In: Singh, V.P. (ed.) *Computer Models of Watershed Hydrology*. Water Resources Publications, Highlands Ranch, Colorado, pp. 909–1000.

Williams, J.R. and Hann, R.W. (1978) *Optimal Operation of Large Agricultural Watersheds with Water Quality Constraints*. Technical Report No. 96, Texas Water Resources Institute, Texas A&M University, College Station, Texas.

Williams, J.R. and Izaurralde. R.C. (2006) The APEX model. In: Singh, V.P. and Frevert, D.K. (eds) *Watershed Models*. CRC Press, Taylor and Francis Group, Boca Raton, Florida, pp. 437–482.

Williams, J.R., Wang, E., Meinardus, A., Harman, W.L., Siemers, M. and Atwood, J.D. (2006) *APEX Users Guide. V.2110*. Blackland Research Center, Texas Agricultural Experiment Station, Texas Agricultural Extension Service, Texas A&M University, Temple, Texas.

Willis, D. B. (2008) *Cost-effectiveness of On-farm Conservation Practices to Protect Playa Lake Hydroperiod in the Texas High Plains*. Selected paper presented at the Southern Agricultural Economics Association Annual Meetings, 3–6 February 2008, Dallas, Texas. Available at: http://ageconsearch.umn.edu/ bitstream/6753/2/sp08wi11.pdf (accessed 11 February 2011).

Wischmeier, W.H. and Smith, D.D. (1978) *Predicting Rainfall Erosion Losses, a Guide to Conservation Planning*. Agriculture Handbook No. 537, US Department of Agriculture, Washington, DC.

Wiyo, K.A., Kasomekera, Z.M. and Feyen, J. (2000) Effect of tied-ridging on soil water status of a maize crop under Malawi conditions. *Agricultural Water Management* 45, 101–125.

Yin, L., Wang, X., Pan, J. and Gassman, P.W. (2009) Evaluation of APEX for daily runoff and sediment yield from three plots in the Upland Huaihe River watershed, China. *Transactions of the ASABE* 52, 1833–1845.

5 Application of the WEPP Model to Hillslopes and Watersheds in the USA

Dennis C. Flanagan*

Introduction

Soil erosion continues to be the largest threat to maintaining sustainable agricultural production. Detachment and removal of soil and sediment by the erosive forces of wind and water degrade on-site soil quality, as well as off-site water quality and air quality. Erosion removes the upper layers of soils, which are often those that have the greatest organic matter content and nutrient- and water-holding capacities. Cultivated and eroded soils often have reduced aggregate stability and a propensity to seal at the surface, thus reducing water and air intake into the plant root zone, and increasing runoff and associated erosion. Runoff water, as well as sediment particles transported in that water, usually contains agricultural chemicals – nutrients (nitrogen (N), phosphorus (P) and others) and pesticides, which can have an impact on downstream water bodies and drinking water supplies. Problems such as silting in of rivers and harbours, eutrophication in the Great Lakes and hypoxia in the Gulf of Mexico are largely related to surface and subsurface runoff, sediment losses and chemical losses from high-input and high-productivity agricultural systems. Sediment remains the largest pollutant of water by volume, and results in billions of dollars in dredging costs each year to maintain drainage and navigability in rivers and other water bodies.

Soil erosion caused by water (raindrop impact, overland water flow, channel erosion, gully erosion) is globally the largest erosion problem on agricultural lands, though in some climatic regions erosion due to wind detachment and transport is the greater concern. Managing agricultural lands to control and minimize soil erosion is an important mission, which is carried out by landowners, farmers, and federal, state and local conservation agency personnel.

In order to determine the effect of different land management practices on potential soil loss, some type of predictive technology or modelling is usually employed. Historically, in the USA and, subsequently, throughout much of the world, the empirical Universal Soil Loss Equation (USLE) (Wischmeier and Smith, 1978) or revisions to it (the Revised Universal Soil Loss Equation – RUSLE; Renard *et al.*, 1997) have been used to estimate long-term average annual soil loss resulting from sheet and rill erosion by water. However, with increasing concerns related to runoff and off-site sediment losses, new process-based technology has been developed by the US Department of Agriculture (USDA) to provide additional

* Corresponding author: Dennis.Flanagan@ars.usda.gov

information on land management impacts not only on soil erosion, but also on runoff from hillslope profiles as well as small watersheds (Flanagan *et al.*, 2007). The Water Erosion Prediction Project (WEPP) was initiated in 1985, and currently provides modelling capabilities to assess spatial and temporal soil loss, the impacts of various types of land management and the implementation of soil conservation practices on runoff, erosion and sediment delivery from slope profiles as small as a few square metres, up to small watersheds of several hundred hectares in size (Flanagan and Nearing, 1995; Flanagan *et al.*, 2001). This chapter will provide information on the background to the WEPP model, the physical processes simulated, the available user interfaces and applications of the model for a variety of hillslope and watershed situations. Some novel uses of WEPP will also be presented and discussed.

Materials and Methods

WEPP model description

WEPP is a physically based soil erosion prediction model, written in Fortran computer language, and implemented on personal computers within a variety of interface programs, including a stand-alone Windows application, a geographic information system (GIS)-linked extension, and a variety of Web-based interfaces accessible via the Internet and a Web-browsing program (e.g. Internet Explorer, Firefox, Safari, Chrome, etc.).

The model mathematically simulates the important physical processes that result in surface runoff, soil erosion, and sediment transport and delivery. Major components of WEPP include a stochastic weather generator (CLIGEN; Nicks *et al.*, 1995) and components addressing surface hydrology, winter hydrology, water balance and percolation, subsurface hydrology, soil, plant growth, residue management and decomposition, overland flow hydraulics, hillslope erosion, irrigation, watershed channel hydrology and erosion and surface impoundments (Flanagan and Nearing, 1995).

Prediction of infiltration and surface runoff is of critical importance, as subsequent erosion by flow, sediment transport and delivery depend upon accurate estimation of storm runoff and peak runoff rate. WEPP uses a Green–Ampt Mein–Larson (GAML) approach (Mein and Larson, 1973), adjusted for unsteady rainfall with multiple times to ponding (Chu, 1978), to determine the cumulative infiltration during a rainstorm event. Water from rainfall that ponds on the soil surface when rainfall rate exceeds infiltration rate is termed 'rainfall excess'. Water stored in surface depressions is subtracted from rainfall excess, then the total rainfall excess during a storm event with multiple time-accounting periods is summed to give the total runoff volume. Peak runoff rate is subsequently computed using either a solution to the kinematic wave equation by the method of characteristics (Stone *et al.*, 1992), or an approximation of the kinematic wave solution (Stone *et al.*, 1995).

The hillslope erosion component of WEPP uses a steady-state sediment continuity equation of the form:

$$\frac{dG}{dx} = D_f + D_i \qquad (5.1)$$

where G is sediment load (kg s^{-1} m^{-1}), x is distance downslope (m), D_f is rill erosion rate (kg s^{-1} m^{-2}) and D_i is inter-rill sediment delivery to the rill (kg s^{-1} m^{-2}) (Foster *et al.*, 1995). The total runoff depth and the peak runoff rate determined in the hydrology component of WEPP are used to determine the effective duration of storm runoff, and erosion is assumed to occur for a storm during this time period at the characteristic (peak) runoff rate. Flow shear stress for computing rill detachment and sediment transport capacity is computed using the peak runoff rate and rill flow hydraulics. Inter-rill sediment delivery to the rill is a function of effective rainfall intensity during the period of rainfall excess, inter-rill runoff rate, adjusted inter-rill erodibility (including slope and cover effects), and a sediment delivery ratio that depends upon the surface roughness and inter-rill sediment particle characteristics. Rill erosion rate is positive and in a detachment mode when flow shear stress exceeds critical shear stress and sediment load is less than the sediment transport capacity. For these conditions, an excess flow shear stress equation is used in which rill detachment rate is a function of the difference between the flow shear stress exerted on the soil and the critical shear stress, an adjusted rill erodibility parameter and the ratio of the sediment load to the sediment transport capacity. For other conditions in which the sediment load is greater than the transport capacity, negative (–) soil loss, or deposition, will be computed using a different equation, where D_f is a function of the flow discharge rate, an equivalent particle fall velocity, and the difference between the sediment transport capacity and the sediment load. Complete details of the WEPP model hillslope erosion computations can be seen in Foster *et al.* (1995) and Flanagan *et al.* (2001). Minimal input files needed to conduct a hillslope simulation are for climate, soil, slope, cropping/management and project run configuration (I/O specifications). Additional data files for irrigation (sprinkler or furrow, fixed date or depletion scheduling) may also be needed if water is applied to the land area simulated in that way.

When applied to a small watershed, WEPP simulates hydrology, soil loss and sediment yield for the hillslope profiles, channels and impoundments that make up the catchment. All of the hillslope model routines relating to water balance, crop growth, infiltration, etc. are also applied to each of the individual channels. In a typical simulation, all of the hillslopes are simulated first, and the runoff and sediment loss information gained from each storm event through a simulation is stored in an external 'pass' file. These pass files are then read by the watershed-runoff and sediment-routing routines as inputs of water and sediment to the top or sides of a channel, as determined by the channel structure file.

A steady-state sediment continuity equation is also used for the channel erosion computations:

$$\frac{dG}{dx} = D_F + D_L \qquad (5.2)$$

where G is the sediment load in the channel flow (kg s^{-1} m^{-1}), x is the channel segment distance downslope (m), D_F is the detachment or deposition of sediment by flow in the channel (kg s^{-1} m^{-2}), and D_L is the lateral sediment inflow from adjacent hillslopes (kg s^{-1} m^{-2}). Sediment can enter the top of a channel from a contributing hillslope, another channel or an impoundment. Detachment by flow will be predicted to occur when the flow sediment transport capacity exceeds the sediment load, and the flow shear stress acting on the channel bed exceeds the critical shear stress. The channel erosion routines can account for erosion downwards through an erodible layer, typically a loose-tilled soil layer for ephemeral gullies, until a non-erodible layer is encountered. At that point, the channel will widen until the shear stress acting on the channel walls is equal to the reduced-flow shear stress of the shallower flow, and detachment will stop. The new channel geometry will be used until either a larger runoff event occurs to widen the channel, or a tillage operation occurs that is assumed to fill the area back to its original configuration and depth of erodible soil. Complete details on the WEPP channel hydrology and erosion routines can be seen in Ascough *et al.* (1995, 1997).

The impoundment component of WEPP allows for simulation of structures such as drop spillways, perforated risers, culverts, emergency spillways or open channels, rock-fill check dams, filter fences or straw-bale check dams, or user-specified structures with input stage-discharge relationships. Impoundment simulation in WEPP utilizes storm runoff and sediment input from an upslope hillslope or channel, conducts a hydraulic simulation of water ponded in the structure and any outflow, a simulation of sedimentation and selective particle class deposition, and then output from the impoundment of water flow and sediment load and composition. Sediment filling of the available impoundment volume is also tracked and reported. Lindley *et al.* (1995) provide complete details of the WEPP surface impoundment routine computations.

Input files necessary to conduct WEPP watershed simulations include a climate file, and files on slope, soil and cropping/management for each hillslope and channel. Additionally, channel-specific input parameters must be described, and if there are impoundments present, then an impoundment input parameter file must also be provided. A watershed structure file describes how each of the hillslopes, channels and impoundments are connected to each other. If the hillslopes will be run separately from the channel/impoundment routing, hillslope pass files containing detailed information on all storm runoff and sediment losses during the simulation must be created for use during watershed simulation. Details on all model input file formats can be found in Flanagan and Livingston (1995), with additional information available at the WEPP website of the National Soil Erosion Laboratory (NSERL) of the USDA Agricultural Research Service (ARS) (http://www.ars.usda.gov/Research/docs.htm?docid=10621).

WEPP model interfaces and applicability

The WEPP Fortran science model runs from flat ASCII input files, and produces a set of flat ASCII output files, dependent upon those desired and specified by the user. This type of input/output (I/O) formatting allows a wide variety of user interface programs to be developed, which can be designed and tailored to the specific application of the model.

The main software typically used for running the WEPP model is a stand-alone Windows application graphical user interface (GUI) written in the C++ programming language (Flanagan *et al.*, 1998). This GUI allows the creation, editing and simulation of both hillslope profile WEPP projects and small watershed projects. Additionally, it has a project-set functionality that allows a user to easily create sets of hillslope profile projects that may be used to compare different tillage systems, crop rotations, conservation practices, soil effects, etc.

Plate 1 shows a screen capture of a hillslope project in the Windows GUI. The slope profile is graphically depicted in either a 2-D or 3-D side view. The profile has three layers (soil, slope, cropping/management) that correspond to three of the four main input files to the model. Each of these layers is 'hot', and when double clicked on with the mouse cursor will open up the editing screens for each. Additionally, at the top of the screen is an icon of the sun with a cloud partially over it, which represents the climate input to the model. When this icon is clicked on, it will bring up the climate input editor screen.

Model simulations can be conducted for any number of years, and typically 50–100-year periods are recommended to obtain good long-term average annual soil loss estimates, as well as values for return period and risk analysis. A sample 100-yr simulation was run and the results are displayed in Plate 1, both in a text box at the upper right of the screen and by image shading in the centre (slope) profile layer. For this example, average annual precipitation was 1111 mm yr^{-1}. When seen in colour, rates of soil loss are shown in shades of red in this layer, while rates of sediment deposition are shown in shades of green. The mouse cursor can also be hovered over any location in the centre profile layer and the predicted erosion rate at that point will be displayed. For example, in Plate 1 the cursor is at a location representing 51.8 m downslope, which has a predicted average annual erosion rate of 11 kg m^{-2}.

Additional model outputs, both text and graphical, can be accessed by clicking on the buttons at the bottom of the screen entitled 'Soil Loss Graph', 'Graphical Output', 'Return Periods' and 'Text Output'. The Soil Loss Graph button will display a line graph of the slope profile, with the predicted detachment and deposition rates as another line plotted away from the profile line. The Return Period button will display a table screen containing the values for daily precipitation, runoff volume, peak runoff rate and sediment leaving the profile by their return periods in years, and there is also the capability to graph these values (Plate 2). Pressing the Graphical Output button will bring up a screen allowing line, bar, scatter and area plots of over 100 model variables through a simulation period. For example, in Plate 3, graphs showing above-ground live biomass, daily runoff and daily sediment loss versus time from a 5-year simulation period are depicted. There are a number of text outputs available from WEPP model simulations, including a main output file that can provide storm-by-storm, monthly, annual or average annual results for predicted runoff, soil loss and sediment yield. Additionally, detailed outputs for plant growth and residue status, soil parameters, water balance, winter hydrology and crop yields can also be generated as desired.

Small watershed simulations up to a recommended maximum area of about 260 ha can also be conducted utilizing another option within the WEPP Windows interface. The user can import an image file (JPEG, TIFF or BMP format), and then scale it to known distances in the image. Editing tools in the graphics window allow the addition of channels, hillslopes and impoundments, and changing their locations and sizes. All information on slope, soil, climate, cropping/management, channel parameters and impoundment parameters must be either selected from existing files in the databases or entered by the user in this interface. Plate 4 shows a screen capture of an example agricultural watershed consisting of three hillslopes (H1 – continuous lucerne (alfalfa) management, H2 and H3 – a maize (corn) and soybean crop rotation with an autumn (Fall) mouldboard ploughing management system), and one channel (C1 – grass waterway). Tabular results from a 10-year simulation are shown in the pop-up window at the bottom left of the image, as well as in the component details table on the right side of the screen. Also, colour shading of the hillslope areas in the graphics window shows relative amounts of soil loss (in shades of red when viewed on a colour monitor) and sediment deposition (in shades of green when viewed on a colour monitor), allowing the user to assess the regions of greatest soil loss and rerun the simulation with alternative cropping/management inputs for the various hillslopes.

While relatively simple watersheds consisting of a few elements can be easily simulated with the WEPP Windows interface, when moving to larger and more complicated catchments the work necessary to accurately create and parameterize the simulation area becomes much more difficult. Because of this, other interfaces that utilize geospatial information (particularly topographic digital elevation data) have been created for larger WEPP watershed model applications. One of these is called GeoWEPP (Geospatial interface for the Water Erosion Prediction Project) (Renschler *et al.*, 2002; Renschler, 2003). GeoWEPP is an ArcView 3.× or ArcGIS 9.× extension that allows users to import their own spatial topographic information (from a DEM – digital elevation model), spatial soil data (if available) and spatial land-use data (if available). The software was originally developed at the NSERL, but now is cooperatively maintained and updated at the State University of New York in Buffalo (http://www.geog.buffalo.edu/rensch/geowepp/).

Plate 5 shows the main GeoWEPP interface screen running under ArcGIS 9.×, with a small watershed near West Lafayette, Indiana delineated. In this example, the DEM being used is from the US Geological Survey (USGS), which has nationwide coverage at a resolution of 30 metres. The interface uses the TOPAZ (TOpographic PArameteriZation) topographic delineation software developed by Garbrecht and Martz (1995) first to delineate the channel network, then to delineate the watershed boundary, sub-catchments and flow paths once the user has selected an outlet point on a channel cell. When satisfied with the watershed configuration, the user then proceeds to set up WEPP model simulations, usually for representative hillslopes (sub-catchment areas) along with channel routing, as well as for all flow paths within the watershed for spatial soil loss. A flow path is a route that water takes within a sub-catchment, beginning at a cell having no inflow to it (local high point) down to and terminating at a channel cell. When run in the hillslope mode only, simulation time is relatively rapid, because runoff and erosion for only a few representative profiles and channel segments are needed; the results provided are predicted runoff and sediment delivery from each sub-catchment, channel and the catchment as a whole. Relative amounts of sediment loss from each hillslope region are displayed in a new map layer within the GIS. When run in the flow-path mode, simulation time is much longer, because the WEPP model is run for every individual flow path, and there may be hundreds or thousands of these, depending upon the size of the watershed. If spatial land-use and soil layers are available, the interface can simulate the variability in runoff and soil loss down the flow paths as these differences occur. WEPP simulates soil detachment or deposition at a minimum of 100 points down a slope profile (or flow path), and predicted values from the flow paths are then translated (in some cases merged where there is convergent flow) and used to create a spatial soil loss map (Plate 6). Cochrane and Flanagan (1999) provide more details.

Following a simulation, a user can examine the sediment loss map layer (from the hillslope method run) to see which area is estimated to contribute the most sediment, and also examine the spatial soil loss map (from the flow-path method run) to identify 'hot spots' where erosion is maximum. Subsequent model runs can then be performed in which management for some/all sub-catchments and/or channels can be modified to attempt to reduce predicted runoff and soil loss. One current limitation of GeoWEPP software is that it lacks the capability for a user to enter man-made features such as terrace berms or to include impoundments within a watershed.

While GeoWEPP is a very helpful tool for conducting detailed watershed analyses with WEPP, it does have some disadvantages. One major disadvantage is the need for the ESRI (Environmental Systems Research Institute) ArcGIS system and licence, which can cost in excess of US$1500. Additionally, a user must be familiar with ArcGIS as well as with the necessary tools and procedures to obtain, create and manipulate the various geospatial data layers, in particular the topographic DEMs, background images, soil layer and land-use layer. These tasks and GIS knowledge can be daunting for a field agency staff member, who is only interested in running a quick and easy erosion model simulation using commonly available data.

Because of this, a web-based WEPP GIS interface was created that allows a user to easily locate the area of interest concerned on a US map, delineate channel networks using the nationwide USGS DEM, use the closest climate data station, specify soils and management for a small watershed, run a simulation and then view soil loss and sediment yield results (Flanagan *et al.*, 2004). This system is accessible at http://milford.nserl.purdue.edu. For its basic Web GIS, this interface uses the open-source MapServer environment, available from the University of Minnesota (http://mapserver.gis.umn.edu). The same TOPAZ software (Garbrecht and Martz, 1995) described earlier is used for the channel, watershed and hillslope (sub-basin) delineation. The user first selects the state, zooms in to the county, then further zooms in to the specific area of interest (e.g. farm fields or an experimental site); the screen display data are obtained from Microsoft Research Maps (formerly TerraServer) website (http://msrmaps.com). Custom software called TopazPrep, coded in C++ and an open-source scripting language called PHP, is used to extract a region of the DEM to process. Data stored on the NSERL server and used in this application are the National Elevation Data set from USGS with 30 m coverage for the entire USA clipped by state, the National Land Cover Data set from USGS, STATSGO (US General Soil Map), soil data from NRCS (USDA Natural Resources Conservation Service), and climate station data for about 2600 weather stations from the CLIGEN weather generator model database. Once a watershed has been identified, and the user has specified desired inputs and outputs, a second custom program called Prepwepp, also written in C++ and PHP, is used to generate WEPP model inputs from the DEM information, land use, soils and TOPAZ watershed delineation results. The CLIGEN weather generator is also called by Prepwepp to create a climate input for the location. The WEPP model is then run for the representative hillslopes and channels, as well as for all flow paths, identically to the procedures used in the GeoWEPP program. After execution of the WEPP model, the Prepwepp program scans the output files and produces GIS map layers for hillslope sediment loss and spatial soil loss (by cell), and also produces additional tabular output summarizing the results.

Plate 7 shows results from an example WEPP application of this web-based WEPP GIS interface to a small watershed in southeast Illinois. In this 10-year simulation, climate information for Fairfield, Illinois was used, along with a Proctor silt loam soil and maize–soybean autumn mulch tillage land management. Results from the flow-path WEPP model simulations can also be displayed by clicking on the 'Soil Loss – Flowpaths' layer box. In addition to the graphical outputs, a text window is also available and can be seen at the bottom left of this image, which details the runoff, soil loss and sediment yield from each sub-catchment and channel.

The WEPP Web-GIS interface was developed as a prototype of a system that could be enhanced and tailored to specific watershed applications or locations. For example, the US Forest Service Rocky Mountain Research Station was recently awarded a grant from the US Army Corps of Engineers to develop a modified version of a WEPP Web-GIS to simulate runoff and sediment losses from forested regions bordering the Great Lakes (W.J. Elliot, Moscow, Idaho, 2009, personal communication). Also, NRCS could use the NSERL prototype as a template, where desired climate, soils, channel and land-management/cropping databases specific for each state could be populated and maintained by each state NRCS office. Other Web-based interfaces also available at the same website allow for simple WEPP hillslope cropland simulations, as well as buffer strip and strip cropping model applications. The Forest Service has also developed a number of targeted Web-based interfaces specific for application to forest roads, timber harvest regions and areas disturbed by wildfire burns (Elliot, 2004).

WEPP hillslope applications for conservation planning

A major original targeted scale of application of WEPP was for hillslope profiles, similar to the application of USLE to profiles in a field. A hillslope profile typically begins at the top of a watershed boundary, and ends at a channel, fence row or other field boundary. An example hillslope profile in a small watershed catchment is shown in Fig. 5.1, with some of the processes occurring and being simulated by the WEPP model illustrated.

WEPP is a distributed parameter model, meaning that it can simulate spatial heterogeneities in soils and land-use management down a hillslope profile. Figure 5.2 illustrates a uniform slope profile that has three different soil types as well as a cropped region in the top portion of the profile and a grass buffer at the bottom. In the model, this would be simulated using four Overland Flow Elements (OFEs), on which a complete water

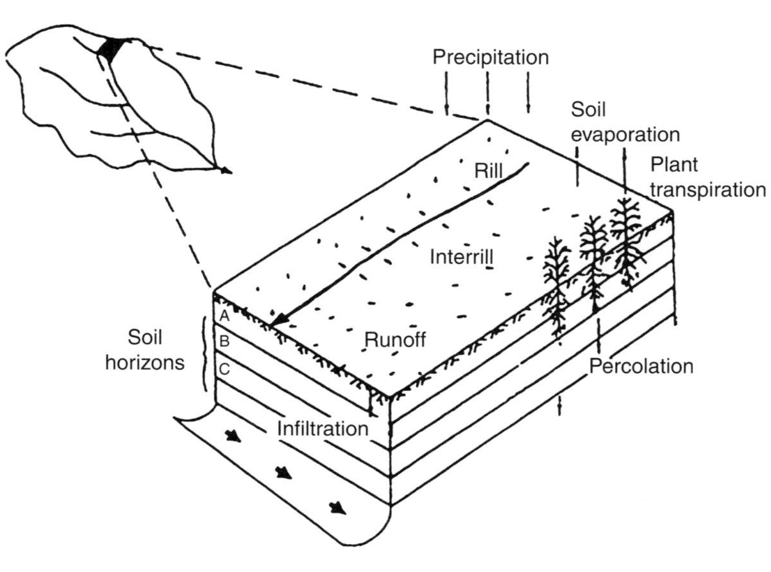

Fig. 5.1. An example hillslope profile, with its location in a small watershed shown. Some of the important processes simulated in the WEPP model water balance are identified here (from Savabi and Williams, 1995).

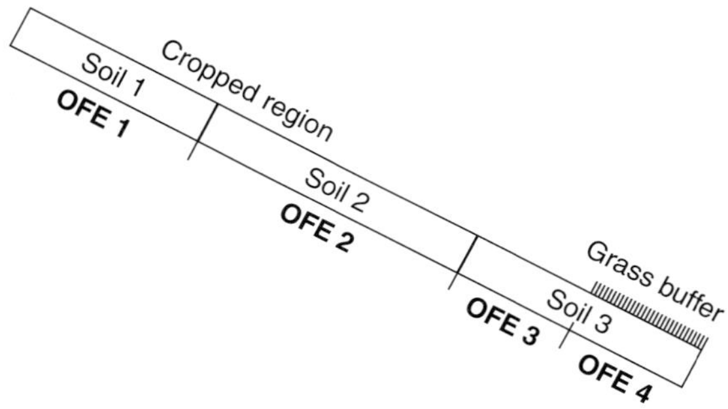

Fig. 5.2. A uniform slope profile, composed of two different cropping/management regions (a cropped region and a grass buffer strip) and three different soils. In a WEPP model hillslope simulation, this profile would be composed of four overland flow elements (OFEs), as shown here.

balance, plant growth, infiltration and erodibility parameter updating, and other factors, would be tracked. Soil detachment or sediment deposition is computed by WEPP at 100 points on each OFE for each storm event, and these spatial values are summed through time, and then used to calculate monthly, annual and average annual soil loss at these points and over the entire profile area. The model can simulate up to ten different OFEs, allowing for the effects of multiple soils on a landscape, or agricultural practices such as strip cropping or buffer strips. The model databases include the NRCS STATSGO soils data parameterized for WEPP, approximately 2600 climate stations and a number of basic cropping/management scenarios.

Plate 8 is a photograph showing a typical agricultural hillslope in Indiana that can be simulated with the WEPP model. On the left is a tilled agricultural field, bordered by forest and a fence row. At the bottom of this tilled area, there is a narrow grass vegetation area as well as a grassed waterway that removes water and sediment from the field. A possible location for a representative hillslope profile is illustrated in the image as the solid black line down the hill to the channel. If a conservation agency staff member were in this field doing planning, that member could manually make slope measurements on that profile using a tape measure and clinometer or, alternatively, do a detailed survey using optical or GPS survey equipment, or could estimate profile slopes from USGS topographic maps.

Results and Discussion

Management systems for erosion control on a hillslope: WEPP model evaluations

WEPP model simulations are needed to approximate the hill located in north-eastern Indiana shown in Plate 8. We assume that the total profile length shown is 32 m, with the grass buffer strip being 2 m long. Measurements in the field indicate that average slope of this profile is 6%, and, while there is some variation in the topography, it can be reasonably approximated as a uniform slope profile. There are two soils in this field: a Glynwood silt loam that extends from the top field boundary down 18 m, and a Blount silt loam on the remainder, and default parameters are available for both in the WEPP database for Indiana. The tolerable soil loss value (T-value) for these soils is 6.7 Mg ha^{-1} (USDA SCS, 1982). The normal tillage system is an autumn or Fall chisel ploughing and a secondary spring tillage after both the maize and soybean crops that are grown in rotation. The closest weather station is Waterloo, Indiana, about 2 km away.

WEPP model v2010.1 simulations conducted for the default slope profile without any buffer strip at all found that predicted soil loss and sediment yield from this profile were 22 Mg ha^{-1}, which exceeded the T-value by a factor of three. Adding the 2 m grass buffer strip that is currently present in the field reduced the off-site sediment loss to

12.9 Mg ha^{-1}, but on the eroding portion of the slope above the grass strip average annual soil loss was still 19 Mg ha^{-1} and maximum soil loss was 35 Mg ha^{-1}. Other alternative management systems and crop rotations need to be explored to see if the predicted soil loss for this profile can be reduced to the T-value or less. A number of different management systems in WEPP were next run for the profile, to see what their impact would be, both with and without the grass buffer. The results are summarized in Table 5.1.

In these example simulations, one can see that both the no-till and continuous lucerne systems reduced predicted soil loss below the T-value of 6.7 Mg ha^{-1}. If the T-value is considered for sediment loss instead of for soil erosion on the eroding portion of the slope, then, in addition, the long cropping rotation (maize–soybean, wheat, lucerne, 4 years) with the grass buffer strip would also be an acceptable management system. The length of the grass buffer strip could also be increased. Results from simulations with the existing crop management and buffer lengths of 4, 6 and 8 m are also shown in Table 5.1. Only a buffer length of 8 m reduced predicted average annual sediment loss below 6.7 Mg ha^{-1} but, even then, the soil loss on the eroding portion of the slope is large, at 16.2 Mg ha^{-1}. Other possible configurations that could be tried might include strip cropping, where the existing crop rotation is alternated with one or more strips of lucerne down the hill, for example.

WEPP application on a small watershed

WEPP model watershed simulations are at a more complicated level than those for the hillslopes that were examined in the previous section. As shown in Plate 9A, watersheds can be composed of hillslopes (H1–H5), channels (C1, C2) and impoundments (I1). When structuring a small-field watershed by hand within the WEPP Windows interface, it is necessary to create rectangular hillslope regions

Table 5.1. WEPP model 100-year simulation results for example profile in Indiana.

Cropping/Management System	Predicted average annual runoff (mm)	Predicted average annual soil loss (Mg ha^{-1})	Predicted average annual sediment loss (Mg ha^{-1})
Existing autumn chisel-ploughed maize–soybean rotation	150	22.4	22.4
Existing management with 2 m grass buffer strip	138	19.2	12.9
Autumn chisel-ploughed maize–soybean, wheat, lucerne (4 years)	152	7.6	7.6
Autumn chisel-ploughed maize–soybean, wheat, lucerne (4 years) with 2 m grass buffer	140	6.9	5.1
No-till maize–soybean	171	2.2	2.2
No-till maize–soybean with 2 m grass buffer	156	2.0	1.9
Continuous lucerne, replant every 5 years	154	3.6	3.6
Continuous lucerne with 2 m grass buffer	142	3.4	2.8
Existing management with 4 m grass buffer strip	133	18.3	9.2
Existing management. with 6 m grass buffer strip	131	17.3	7.2
Existing management with 8 m grass buffer strip	130	16.2	5.7

and straight-line channel sections (Plate 9B) to approximate actual field polygons and curving lines (Plate 9A).

A steeply sloping and highly dissected portion of a field is shown in the photograph in Fig. 5.3, with six hillslopes (H1–H6) and three channel sections (C1–C3) identified for structuring a WEPP model simulation. Additionally, hillslopes H4 and H6 have two regions of different management on them – perennial native cover at the top, and cultivated soil at the bottom – so these two hills would need to be simulated as two OFEs. This relatively simple watershed was constructed within the WEPP Windows interface (Plate 10), and a 10 year simulation run using a Des Moines, Iowa climate, a Duncanon soil and continuous maize with autumn mouldboard-ploughing management. For the native vegetation, a continuous grass management data file was used.

This small watershed had an area of 0.25 ha, predicted average annual runoff of 258 m^3 and an average annual sediment yield of 14.0 Mg ha^{-1}. If a no-till maize system is implemented in place of the existing heavy tillage, a subsequent simulation found that average annual runoff decreased to 211 m^3 and average annual sediment yield to 1.8 Mg ha^{-1}. Additional alternatives that could be explored here would be to modify channel parameters to simulate a grassed waterway in place of the bare soil channels (C1–C3). However, this alternative only slightly decreased watershed sediment yields to 13.8 Mg ha^{-1} under the conventional tillage and 1.7 Mg ha^{-1} under the no-till maize system.

WEPP application on a large watershed

WEPP can also be applied to larger watersheds, with a recommended maximum size of about 260 ha. However, in some cases, such as in rangeland areas, the maximum size may be two or three times larger. The main limitation currently in WEPP is with channel erosion processes – the model was designed to simulate mainly Hortonian overland flow and channel routing of that runoff, rather than to simulate base flow in perennial channels or erosion processes there. More recent model enhancements have improved the ability of

Fig. 5.3. A photograph showing a small watershed composed of six hillslopes (H1–H6) and three channel sections (C1–C3). Hillslopes 4 and 6 would be further subdivided into two overland flow elements with permanent vegetation at the top above cultivated agricultural land.

the model to simulate forest hydrology processes and subsurface lateral flow due to a low permeability subsurface layer of rock or other material (Dun *et al.*, 2009), thereby making better predictions of runoff in forest streams.

With either GeoWEPP or the Web-based WEPP GIS interface, larger watersheds can be simulated using either the hillslope or flow-path methods, or both. If the main interest is in spatial soil erosion estimates using the flow-path method, then the accuracy of the entire catchment runoff and sediment yield predictions is of lower importance, and areas much larger than 260 ha may be simulated. However, if there is a need for watershed runoff and sediment yields from the hillslope method, the user needs to be aware that those predictions will be likely to become less reliable as the watershed area becomes larger and channel processes are more dominant.

In an example exercise, a forested region in Jefferson County, Colorado that had experienced extensive burning from wildfires was simulated using the GeoWEPP interface. A watershed of 209 ha in size, draining into the south-west corner of Cheesman Lake, was delineated, and a land-use layer of no-burn, low-severity burn, moderate-severity burn and high-severity burn that had been provided by the Forest Service was used to set and parameterize the soil and management inputs. From the TOPAZ analysis of the DEM, 50 hillslopes, 21 channels and 586 flow paths were identified. A 10-year WEPP model simulation was run, and the results are shown in Plate 11. The average annual precipitation volume in the catchment was 902551 m^3 and average annual discharge at the outlet was 115949 m^3. Average annual sediment delivery was 1348 Mg, and on a per unit area was 6.4 Mg ha^{-1}. The sediment delivery ratio for the watershed was 0.433, indicating that more than half of the sediment detached from the hillslopes was predicted to deposit in the channels. The hillslopes with the highest predicted sediment losses were largely on the southern and western portions of the catchment (Plate 11), and, if the fire there had been more recent, those areas might have been identified for the greatest remediation efforts. The Forest Service regularly uses the WEPP model with its online interfaces, as well as with GeoWEPP, to identify critical areas for remediation from wildfire burns, as well as from other forest disturbance activities.

Novel uses of WEPP technology

In addition to WEPP model applications for conservation and remediation planning on agricultural and forested hillslope profiles and small watersheds, there have been several other novel uses of the erosion prediction technology. Two of the major ones are described here.

Rocky Flats Nuclear Weapons Plant site

The WEPP model was applied by the Actinide Migration Evaluation (AME) group, who worked on determining how actinide elements (uranium, plutonium, americium) moved in the environment, and how contaminated landscapes could be remediated at the 24 km^2 Rocky Flats Nuclear Weapons Plant site in Colorado, following its closure in 1989 and its assignment to the US Environmental Protection Agency (EPA) Superfund list (Clark *et al.*, 2006). The major product of the plant was the plutonium 'pit', commonly known as the trigger for a nuclear bomb. The soils on portions of the landscape had been heavily contaminated with uranium, plutonium and americium, and initial estimates by the Department of Energy (DOE) were that the clean-up of the site would take 70 years and cost more than US$37 billion. However, research studies determined that plutonium and americium formed insoluble oxides that adhered to organic matter and soil particles, and then were detached, transported and redeposited through the processes of wind and water erosion. To estimate the water sediment and contaminant transport, WEPP was applied to the hillslope areas and linked with the HEC-6T model from the US Army (MBH Software, 2002). On-site measurements of contaminants in the soil were used along with local climate information and various land-management scenarios to determine the most appropriate remediation practices to implement at reasonable cost. In addition to removal of the most highly contaminated soils to a depth of 1 m, and replacement with clean soil, a site-wide storm-water pollution prevention plan was developed and implemented, and consisted of a variety of practices, including use of straw bales, hydro-mulching, riprap channel linings and some constructed wetlands. The final cost of the site clean-up was only US$7 billion and was completed in just 10 years, removing a US$600 million annual liability from the DOE's budget (Clark *et al.*, 2006).

Iowa Daily Erosion Project

Iowa is one of the largest agricultural states in the USA, and because of its humid climate and rolling topography can experience significant runoff and soil erosion. Researchers at Iowa State University, the University of Iowa and the ARS obtained grant funding to develop a Web-based erosion prediction technology to provide near real-time estimates across the state of daily precipitation, runoff and soil erosion on a spatial basis by township (Cruse *et al.*, 2006). In this project, NEXRAD (NEXt generation RADar) radar precipitation data from the previous day are downloaded from the National Weather Service and processed to provide a break-point precipitation input climate file for WEPP (appended to a continuous string of these weather inputs) for each township. Information from the NRCS National Resources Inventory (NRI) sample points is used to provide soil and cropping management inputs for the model (USDA NRCS, 2010). All existing cropping management for NRI points within a township are run, and average runoff and soil loss values determined. Results are then posted to the project's public website (Plate 12), and can be easily accessed and used as desired for the entire state or for a specific township (users can view runoff or soil loss for a specific date, time period, etc.).

Summary

WEPP is a powerful tool allowing the simulation of a variety of land-management practices, at almost any location within the USA, utilizing large pre-built databases for climate, soils, management and topography, and estimation of runoff, soil loss and sediment yield from hillslope profiles and (usually) small watersheds. As was demonstrated in this chapter, very common uses of the model are to evaluate a range of alternative crop types, rotations, tillage systems or other conservation-management strategies, such as buffer strips or grassed waterways on agricultural areas. Similar types of evaluations can be made in other land-management systems, for example impacts of rates of timber harvest, road traffic, fire intensity or mulching levels on forested sites.

The model has been extensively tested and evaluated by scientists both within the USA and in a number of foreign countries, and it has been applied to a number of land-management issues. WEPP model development continues in a number of important areas. During the past several years, a major effort has been to combine water erosion components from WEPP with wind erosion components from the Wind Erosion Prediction System (WEPS) to create a single process-based field-scale tool for predicting both water and wind erosion in the same simulation. Prototypes of the Wind and Water Erosion Model (WWEM) are in various stages of testing and validation, for possible eventual use by field office personnel in NRCS.

Several additional areas of planned development for WEPP include addition of the ability to predict tillage translocation erosion, improvement of the frost and thaw adjustments to the soil erodibility parameters, and improved prediction of ephemeral gully erosion in watershed simulations. University collaborators, working together with WEPP project staff, are also developing improvements for the model's channel-flow routing and erosion predictions so that it can be applied with more confidence in larger-size catchments. Addition of the ability to simulate nutrient and pesticide losses in surface runoff and associated with sediment is also a possible area of enhancement to the WEPP model under evaluation.

Maintenance, user support and WEPP model enhancement work continues at the USDA Agricultural Research Service's National Soil Erosion Research Laboratory in West Lafayette, Indiana.

References

Ascough, J.C. II, Baffaut, C., Nearing, M.A. and Flanagan, D.C. (1995) Chapter 13. Watershed model channel hydrology and erosion processes. In: Flanagan, D.C. and Nearing, M.A. (eds) *USDA Water Erosion Prediction Project: Hillslope Profile and Watershed Model Documentation.* NSERL Report No. 10, National Soil Erosion Research Laboratory, USDA Agricultural Research Service, West Lafayette, Indiana.

Ascough, J.C. II, Baffaut, C., Nearing, M.A. and Liu, B.Y. (1997) The WEPP watershed model: I. Hydrology and erosion. *Transactions of the ASAE* 40, 921–933.

Chu, S.T. (1978) Infiltration during an unsteady rain. *Water Resources Research* 14, 461–466.

Clark, D.L., Janecky, D.R. and Lane, L.J. (2006) Science-based cleanup of Rocky Flats. *Physics Today* 59(9), 34–40.

Cochrane, T.A. and Flanagan, D.C. (1999) Assessing water erosion in small watersheds using WEPP with GIS and digital elevation models. *Journal of Soil and Water Conservation* 54, 678–685.

Cruse, R.M., Flanagan, D.C., Frankenberger, J.R., Gelder, B.K., Herzmann, D., James, D., Krajewski, W., Kraszewski, M., Laflen, J.M. and Todey, D. (2006) Daily estimates of rainfall, water runoff, and soil erosion in Iowa. *Journal of Soil and Water Conservation* 61, 191–199.

Dun, S., Wu, J.Q., Elliot, W.J., Robichaud, P.R., Flanagan, D.C., Frankenberger, J.R., Brown, R.E. and Xu, A.C. (2009) Adapting the Water Erosion Prediction Project (WEPP) model for forest applications. *Journal of Hydrology* 366, 46–54.

Elliot, W.J. (2004) WEPP Internet interfaces for forest erosion prediction. *Journal of the American Water Resources Association* 40, 299–309.

Flanagan, D.C. and Livingston, S.J. (eds) (1995) *Water Erosion Prediction Project (WEPP) Version 95.7: User Summary.* NSERL Report No. 11, National Soil Erosion Research Laboratory, USDA Agricultural Research Service, West Lafayette, Indiana.

Flanagan, D.C. and Nearing, M.A. (eds) (1995) *USDA Water Erosion Prediction Project Hillslope Profile and Watershed Model Documentation.* NSERL Report No. 10, National Soil Erosion Research Laboratory, USDA Agricultural Research Service, West Lafayette, Indiana.

Flanagan, D.C., Fu, H., Frankenberger, J.R., Livingston, S.J. and Meyer, C.R. (1998) A Windows interface for the WEPP erosion model. ASAE Paper No. 98-2135, American Society of Agricultural Engineers, St Joseph, Michigan.

Flanagan, D.C., Ascough J.C. II, Nearing, M.A. and Laflen, J.M. (2001) Chapter 7. The Water Erosion Prediction Project (WEPP) model. In: Harmon, R.S. and Doe, W.W. III (eds) *Landscape Erosion and Evolution Modeling.* Kluwer Academic/Plenum Publishers, New York, pp. 145–199.

Flanagan, D.C., Frankenberger, J.R. and Engel, B.A. (2004) Web-based GIS application of the WEPP model. ASAE Paper No. 042024, American Society of Agricultural Engineers, St Joseph, Michigan.

Flanagan, D.C., Gilley, J.E. and Franti, T.G. (2007) Water Erosion Prediction Project (WEPP): development history, model capabilities, and future enhancements. *Transactions of the ASABE* 50, 1603–1612.

Foster, G.R., Flanagan, D.C., Nearing, M.A., Lane, L.J., Risse, L.M. and Finkner, S.C. (1995) Chapter 11. Hillslope erosion component. In: Flanagan, D.C. and Nearing, M.A. (eds) *USDA Water Erosion Prediction Project: Hillslope Profile and Watershed Model Documentation.* NSERL Report No. 10, National Soil Erosion Research Laboratory, USDA Agricultural Research Service, West Lafayette, Indiana.

Garbrecht, J. and Martz, L.W. (1995) *TOPAZ: An Automated Digital Landscape Analysis Tool for Topographic Evaluation, Drainage Identification, Watershed Segmentation and Subcatchment Parameterization: Overview.* ARS-NAWQL 95-1, USDA Agricultural Research Service, Durant, Oklahoma.

Lindley, M.R., Barfield, B.J. and Wilson, B.N. (1995) Chapter 14. Surface impoundment element model description. In: Flanagan, D.C. and Nearing, M.A. (eds) *USDA Water Erosion Prediction Project: Hillslope Profile and Watershed Model Documentation.* NSERL Report No. 10, National Soil Erosion Research Laboratory, USDA Agricultural Research Service, West Lafayette, Indiana.

MBH Software. (2002) *Sedimentation in Stream Networks (HEC-6T): User Manual.* Mobile Boundary Hydraulics, Clinton, Mississippi. Available at: http://www.mbh2o.com/docs/HEC6T/H6TUM5h_SecondPrinting.pdf (accessed February 2010).

Mein, R.G. and Larson, C.L. (1973) Modeling infiltration during a steady rain. *Water Resources Research* 9, 384–394.

Nicks, A.D., Lane, L.J. and Gander, G.A. (1995) Chapter 2. Weather generator. In: Flanagan, D.C. and Nearing, M.A. (eds) *USDA Water Erosion Prediction Project: Hillslope Profile and Watershed Model*

Documentation. NSERL Report No. 10, National Soil Erosion Research Laboratory, USDA Agricultural Research Service, West Lafayette, Indiana.

Renard, K.G., Foster, G.R., Weesies, G.A., McCool, D.K. and Yoder, D.C. (1997) *Predicting Soil Erosion by Water: A Guide to Conservation Planning with the Revised Universal Soil Loss Equation (RUSLE).* Agriculture Handbook No. 703, US Department of Agriculture, Washington, DC.

Renschler, C.S. (2003) Designing geo-spatial interfaces to scale process models: the GeoWEPP approach. *Hydrological Processes* 17, 1005–1017.

Renschler, C.S., Flanagan, D.C., Engel, B.A. and Frankenberger, J.R. (2002) *GeoWEPP: The Geospatial Interface to the Water Erosion Prediction Project.* ASAE Paper No. 022171, American Society of Agricultural Engineers, St Joseph, Michigan.

Savabi, M.R. and Williams, J.R. (1995) Chapter 5. Water balance and percolation. In: Flanagan, D.C. and Nearing, M.A. (eds) *USDA Water Erosion Prediction Project: Hillslope Profile and Watershed Model Documentation.* NSERL Report No. 10, National Soil Erosion Research Laboratory, USDA Agricultural Research Service, West Lafayette, Indiana.

Stone, J.J., Lane, L.J. and Shirley, E.D. (1992) Infiltration and runoff simulation on a plane. *Transactions of the ASAE* 35, 161–170.

Stone, J.J., Lane, L.J., Shirley, E.D. and Hernandez, M. (1995) Chapter 4. Hillslope surface hydrology. In: Flanagan, D.C. and Nearing, M.A. (eds) *USDA Water Erosion Prediction Project: Hillslope Profile and Watershed Model Documentation.* NSERL Report No. 10, National Soil Erosion Research Laboratory, USDA Agricultural Research Service, West Lafayette, Indiana.

USDA NRCS (USDA Natural Resources Conservation Service) (2010) National Resources Inventory. Washington, DC. Available at: http://www.nrcs.usda.gov/technical/NRI/ (accessed March 2010).

USDA SCS (USDA Soil Conservation Service) (1982) *Soil Survey of DeKalb County Indiana.* Washington, DC.

Wischmeier, W.H. and Smith, D.D. (1978) *Predicting Rainfall Erosion Losses: A Guide to Conservation Planning.* Agriculture Handbook No. 537, US Department of Agriculture, Washington, DC.

6 Application of the WEPP Model to Some Austrian Watersheds

Andreas Klik,* Khaled Hardan and Hans-Peter Nachtnebel

Introduction

Soil erosion is a major factor causing land degradation and deterioration of environmental quality. Inappropriate land use and poor land management are often the triggers for high amounts of runoff and soil loss. Worldwide, over 80% of the world's agricultural lands are estimated to be affected to some degree by erosion (Larson *et al.*, 1983; El-Swaify, 1994). The GLASOD (Global Assessment of Soil Degradation) survey (Oldeman *et al.*, 1990) has indicated that more than 10^9 ha of the land surface of the earth are currently experiencing serious soil degradation as a result of water erosion, and, if it is assumed that the mean rate of soil loss from these areas is about 50 t ha^{-1} $year^{-1}$, the total annual soil loss would be of the order of 50 × 10^9 tons a year. Pimental *et al.* (1995) have provided a similar estimate of the current global rate of annual soil loss at 75 × 10^9 tons a year. In Europe, erosion is the major threat to the soil resource. A report of the Council of Europe, using revised GLASOD data (van Lynden, 1995), estimated that 12 million ha of land in Europe (including part of the former Soviet Union), or approximately 10% of the area considered, is strongly or extremely degraded by water erosion (Jones *et al.*, 2004). European Union (EU) estimates (EEA, 2000) have indicated that as a result of climate change the water erosion risk is expected to increase by the year 2050 by about 80% of EU agricultural areas. This increase will mainly take place in the areas where soil erosion is currently severe (EEA, 1999).

To estimate the current soil erosion risk of an area and to develop soil conservation strategies for sustainable soil and land use, computer simulation models are useful tools. For planning purposes, it is important that these models reliably predict the occurrence, intensity and impacts of erosive events. Under- or over-assessment of these events will result in the wrong design of soil protection measures and/or in extreme damages which would have been avoidable with correct estimations.

Since the mid-1970s, several soil erosion models have been developed, mainly to assess the effects of different management practices and crop rotations on runoff and soil loss from agricultural fields. CREAMS (Chemicals, Runoff, and Erosion from Agricultural Management Systems; Knisel, 1980), ANSWERS (Areal Nonpoint Source Watershed Environment Response Simulation; Beasley *et al.*, 1980; Beasley and Huggins, 1991), AGNPS (Agricultural NonPoint Source; Young *et al.*, 1987), GLEAMS (Groundwater Loading Effects on Agricultural Management Systems; Leonard *et al.*, 1987),

* Corresponding author: andreas.klik@boku.ac.at

EPIC (Erosion Productivity Impact Calculator; Sharpley and Williams, 1990; Williams *et al.*, 1990), OPUS (an integrated model for transport of non-point-source pollutants; Ferreira and Smith, 1992) and RUSLE (Revised Universal Soil Loss Equation; Renard *et al.*, 1997) are probably the most important. In most of these models, soil erosion is calculated based on the concept formulated in the Universal Soil Loss Equation (USLE; Wischmeier and Smith, 1965, 1978). In the last decade, many physically based erosion models were developed which can describe the physical mechanisms of infiltration and runoff, and can simulate the individual components of the entire erosion process by solving the corresponding equations; therefore, it is argued that they have a wider range of applicability. Such models are also generally better in assessing both spatial and temporal variability of the natural erosion process. These kinds of simulation models are represented by WEPP (Water Erosion Prediction Project model; Lane and Nearing, 1989; Flanagan and Nearing, 1995), EUROSEM (European Soil Erosion Model, Morgan *et al.*, 1992, 1993), EROSION-2D/3D (Schmidt, 1991; Schmidt *et al.*, 1997) and LISEM (Limburg Soil Erosion Model; De Roo, 1994). However, all models differ greatly in terms of their complexity, their inputs and requirements, the processes they represent and the manner in which these are represented, the scale of their intended use and the types of output they provide (Merritt *et al.*, 2003).

The objectives of this study were to evaluate the performance of the hillslope and watershed version of the WEPP model by comparing measured with simulated amounts of runoff and soil loss/sediment yield. The predictive capability was investigated for two spatial scales: for natural runoff plots with an area of 60 m^2 and for a large (0.66 km^2) agricultural watershed, both located in the eastern part of Austria.

Materials and Methods

Overview of the model

Introduction

The Water Erosion Prediction Project (WEPP; Lane and Nearing, 1989) was initiated in the 1980s to develop a new-generation prediction technology to serve as an improved tool for soil and water conservation planning and assessment. It is a continuous, process-based, soil erosion prediction model based on the fundamentals of infiltration theory, hydrology, soil physics, plant science, hydraulics and erosion mechanics. It consists of nine major components: climate generation; winter process; irrigation; hydrology; soils; plant growth; residue composition; hydraulics of overland flow; and erosion and deposition. The model can be applied both to single hillslopes of a maximum length of about 100 m and to small watersheds of up to about 2.5 km^2 (Foster and Lane, 1987). In the latter case, the watershed is divided into representative hillslope elements and channels (the 'open-book method').

For this study, the WEPP hillslope and watershed version 2008.907 was used. The model structure is similar to that of other physically based erosion models working on the same scheme:

1. Precipitation is described by rainfall depth, duration and peak intensity. The inter-rill erosion process is driven by the effective rainfall intensity as well as by the inter-rill runoff rate, which are calculated for each runoff event. In general terms, these can be considered as erosivity, which characterizes the ability of raindrops to detach and remove soil particles from the bare soil surface. The rainfall impact is reduced by plant cover or canopy cover.
2. Soil erodibility, which describes the resistance of the soil against the erosive force, is usually derived from surface and soil characteristics.
3. The infiltration approach used is based on the Hortonian overland flow concept (Horton, 1933). This means that the amount of water available at the soil surface during each time step of simulation is the positive difference between the actual rainfall intensity and infiltration rates.
4. The ability of surface runoff to detach soil particles from the soil surface is calculated from the inter-rill runoff rate and two additional soil erodibility factors.
5. The total amount of sediment which can be moved by overland flow along a hillslope

per unit time is limited by its available transport capacity.

6. The amount of deposited sediment is calculated from the positive difference between the sum of sediment input from upslope, the sediment eroded by raindrop and flow impact, and the available transport capacity of overland flow in the actual hillslope segment.

Description of the processes

INFILTRATION. For simulation of the infiltration rate (i) for unsteady rainfall, the Green and Ampt (1911) approach is used as presented by Chu (1978):

$$i = K_e \cdot \left(1 + \frac{(n - \theta_0) \cdot \psi}{I}\right) \tag{6.1}$$

where i is the actual infiltration rate (m s^{-1}), K_e the effective hydraulic conductivity of the wetted zone (m s^{-1}), n the effective porosity (m^3 m^{-3}), θ_0 the initial saturation (m^3 m^{-3}), ψ the average capillary tension or matric potential of the wetting front (m) and I the cumulative infiltration depth (m).

This equation describes the approach of the actual infiltration rate i to the hydraulic conductivity K_e when I approximates infinity. The main assumptions of this approach are the piston-like entry of the water into the soil and a sharply defined wetting front which separates the fully saturated and unsaturated zones. The driving parameters of the Green and Ampt model are the matric potential ψ of the wetting front, the soil moisture deficit ($n - \theta_0$) and the effective hydraulic conductivity K_e. The wetting front term is calculated from the soil type, the soil water content and the soil bulk density using a pedo-transfer function modified from the one developed by Rawls and Brakensiek (1983). The moisture deficit is determined in a similar manner from empirical functions which were developed during extensive WEPP rainfall simulation studies (Elliott *et al.*, 1989). The effective hydraulic conductivity is calculated from sand and clay contents and the cation exchange capacity of the topsoil (USDA ARS, 1994; Flanagan and Livingston, 1995).

RUNOFF ROUTING. Dynamic infiltration-hydrograph models for overland flow consist of an infiltration function that computes the infiltration rate as it varies with time from an unsteady rainfall input and a routing function that transforms rainfall excess into flow depths on a flow surface. The choice of the infiltration function is somewhat arbitrary, but the routing function is generally some form of the Saint-Venant shallow-water equations. One such form, the kinematic wave model, has been shown to be a valid approximation for most overland flow cases (Woolhiser and Liggett, 1967).

The WEPP model uses two methods of computing the peak discharge: a semi-analytical solution of the kinematic wave model (Stone *et al.*, 1992) and an approximation of the kinematic wave model. The first method is used when WEPP is run in a single-event mode, while the second is used when WEPP is run in a continuous simulation mode.

The kinematic equations for flow on a plane are the continuity equation:

$$\frac{\partial h}{\partial t} + \frac{\partial q}{\partial x} = v \tag{6.2}$$

and a depth–discharge relationship:

$$q = \alpha \cdot h^m \tag{6.3}$$

where h is the depth of flow (m), q the discharge per unit width of the plane (m^3 m^{-1}–s^{-1}), t the time (s), v the runoff or rainfall excess rate (m s^{-1}), α the Chézy depth–discharge coefficient (m$^{1/2}$ s^{-1}), m the depth–discharge exponent and x the distance from top of plane (m).

Both of these equations are solved analytically by the method of characteristics (Eagleson, 1970) and rewritten as differential equations on characteristic curves on the xth plane:

$$\frac{dh}{dt} = v(t) \tag{6.4}$$

and

$$\frac{dx}{dt} = \alpha \cdot m \cdot h(t)^{m-1} \tag{6.5}$$

These equations are solved together with the infiltration calculations by using a Runge–Kutta iteration scheme. The recession limb of the hydrograph is calculated until the routed runoff volume equals 95% of the total infiltration excess volume, or the discharge rate equals 10% of the peak discharge rate. The approximate method used for calculation of runoff volume, peak runoff rate and runoff duration is based on empirical relationships among these parameters developed from kinematic wave simulations.

EROSION AND DEPOSITION. The movement of the sediment along a hillslope is described on the basis of the steady-state sediment continuity equation (Foster and Meyer, 1972):

$$\frac{dG}{dx} = D_f + D_i \quad (6.6)$$

where x represents the distance downslope (m), G is the sediment load (kg s^{-1} m^{-1}), D_f is the net rill detachment (kg s^{-1} $^{-2}$) and D_i is the inter-rill erosion rate (kg s^{-1} m^{-2}). Inter-rill erosion rate is considered to be independent of distance, which means that it occurs at a constant rate down the slope. Rill erosion D_f is positive for detachment and negative for deposition.

The inter-rill detachment D_i is calculated by:

$$D_i = K_{iadj} \cdot l \cdot q \cdot SDR_{RR} \cdot F_{nozzle} \cdot (r_s/w) \quad (6.7)$$

where K_{iadj} is the adjusted inter-rill soil erodibility (kg s m^{-4}), I is the effective rainfall intensity (m s^{-1}), q is the inter-rill runoff rate (m s^{-1}), SDR_{RR} is the sediment delivery ratio as a function of the random roughness, the row side slope and the inter-rill sediment particle size distribution, F_{nozzle} is an adjustment factor to account for sprinkler irrigation nozzle energy variation (–), and r_s and w are the rill spacing and width (m) (Foster *et al.*, 1995).

Erosion processes in rills are determined by the equation:

$$D_c = K_r \cdot (\tau_f - \tau_c) \quad (6.8)$$

where D_c is the detachment capacity of rill flow (kg s^{-1} m^{-2}), K_r is the rill erodibility of the soil (s m^{-1}), τ_f is the flow (hydraulic) shear stress acting on soil particles (Pa) and τ_c is the critical shear stress to initiate particle detachment (Pa). Rill detachment is zero if the shear stress is less than the critical shear stress of the soil.

The inter-rill erosion rate is always greater than or equal to zero and is added to the rill erosion rate. A rill spacing of 1 m is assumed if no rills are specified by the user. Whether detachment or deposition occurs in a rill segment is decided by the sign of the rill erosion rate D_f.

In the case of erosion:

$$D_f = D_c \cdot \left(1 - \frac{G}{T_c}\right) \text{ (if } G < T_c) \quad (6.9)$$

In the case of deposition:

$$D_f = \frac{\beta \cdot v_f}{q} \cdot (T_c - G) \text{ (if } G \geq T_c) \quad (6.10)$$

where D_f is net detachment or deposition (kg s^{-1} m^{-2}), and T_c is the transport capacity (kg s^{-1} m^{-1}), β is the raindrop-induced turbulence coefficient (assigned a value of 0.5 for rain-impacted rill flows), v_f is the effective fall velocity for sediment particles (m s^{-1}) calculated by Stoke's Law, and q is the discharge per unit width (m^2 s^{-1}).

The WEPP model uses a simplified form of the Yalin (1963) transport-capacity equation developed by Finkner *et al.* (1989):

$$T_c = k_t \cdot \tau_f^{3/2} \quad (6.11)$$

where k_t is the transport coefficient (m$^{1/2}$ s^2 kg$^{-1/2}$).

Description of the study sites

Soil erosion plots

Between 1994 and 1997, nine erosion plots were installed at three sites in eastern Austria. The first site, Mistelbach, is situated 50 km north-east of Vienna. The region is one of the warmest but also one of the driest parts of Austria. The second research site is situated at Pixendorf, about 50 km west of Vienna, located on a north-facing hillslope. The third site, Pyhra, is 80 km west of Vienna in a landscape characterized by gentle to fairly steep slopes (Fig. 6.1). Average annual precipitation of the sites ranges between 645 and 944 mm, and average air temperature between 9.4 and 10.2 °C (Table 6.1). Soils are classified as Typic

Fig. 6.1. Location of the investigated sites in Austria.

Table 6.1. Main characteristics of the investigated sites in Austria.

Parameter	Mistelbach	Pixendorf	Pyhra	Petzenkirchen
Location	lat 48° 34' N long 16° 34' E	lat 48° 17' N long 15° 58' N	lat 48° 09' N long 15° 41' E	lat 48° 09' N long 15° 09' E
Elevation (m asl)	260	225	300	256–325
Average annual rainfall (mm)	645	685	944	801
Average annual temperature (°C)	9.7	10.2	9.4	9.3
Area (m^2)	45	60	60	6619
Average slope (%)	13.2	5.0	15.2	8
Soil texture	Silt loam	Silt loam	Loam	Loam, silt loam, silt clay loam
Sand (%)	12.6	30.6	36.9	7–40
Silt (%)	69.5	63.1	41.3	48–75
Clay (%)	17.9	6.3	21.8	12–33
Organic matter content (%)	2.0	2.1	2.1	2.2–5.3
Cation exchange capacity ($cmol_c\ kg^{-1}$)	12	22	15	11–22
Rock fragments (%)	0	0	0	1

Argiudoll (Mistelbach and Pyhra) and Entic Hapludoll (Pixendorf), with mixed mineralogy and a mesic temperature regime. Soil textures are silt loam and loam (USDA SCS, 1992) with slopes between 5 and 15%. Details of the crop rotations at the sites during the experiment are compiled in Table 6.2.

The erosion plots were 3 or 4 m wide and 15 m long, and bordered by 20 cm high stainless steel metal sheets. At the lower end of the plot, surface runoff and soil loss were collected in a trough and then diverted by a 100 mm PVC pipe to an Automated Erosion Wheel (AEW; Klik *et al.*, 2004). The design of this AEW is similar to a tipping bucket with a quadratic shape (550 × 550 mm) and a horizontal axle. The AEW consists of four equal sections. Each section contains approximately 5 l, resulting in a resolution of each tip of 0.08 mm for 60 m^2 plots. A magnetic sensor system was used for continuous runoff measurement. As soon as the AEW was turning,

Table 6.2. Crop rotations at Mistelbach, Pixendorf and Pyhra, 1994–2003.

Year	Mistelbach	Pyhra	Pixendorf
1994	Maize	Maize	–
1995	Winter wheat	Winter wheat	–
1996	Sugarbeet	Maize	–
1997	Summer barley	Winter wheat	Maize
1998	Sunflower	Maize	Winter wheat
1999	Winter wheat	Winter wheat	Maize
2000	Maize	Maize	Winter wheat
2001	Winter wheat	Winter wheat	Sugarbeet
2002	Maize	Maize	Oats
2003	Winter wheat	Winter wheat	Maize

one magnet mounted on the wheel passed the fixed magnetic sensor on the frame. This produced a signal to a data acquisition system. In addition to the electronic data logging system, tippings can also be counted mechanically with a magnetic cumulative counter, or simply by the rewinds of a string on a spool mounted on the axle.

The soil–water suspension (runoff) was divided by an adapted multi-tube divisor that took 3.3% of the runoff as a sample; this was collected in a 60 l collection tank. After each erosive storm, the collection tank was emptied and the runoff sample was brought to the laboratory, weighed and dried until constancy of mass was achieved to determine sediment concentration. Based on the continuously measured runoff data from the data logging system and the sediment concentration in the sample the amount of soil loss from the plots was calculated for each erosive event.

Erosion plots were operated only during the growing season (April to October/November). Runoff and soil-loss measurements for the period 1994–2003 (Mistelbach and Pyhra) and 1997–2003 (Pixendorf) were used for this study. Overall, 26 plot years were available for the analyses, with a total of 82 runoff and 72 erosion events. During the growing seasons of 2002 and 2003, soil water contents were measured at Mistelbach using FDR (Frequency Domain Reflectometry) sensors (®Enviroscan system). Two access tubes were inserted outside the erosion plots into the soil approximately 10 m above the upper end of the plots. In each access tube, ten sensors were placed at 10 cm increments to a soil depth of 100 cm. Soil moisture was measured continuously at 15 min time intervals and data were stored on a data acquisition system (Hofmann, 2005). Daily soil moisture readings at 7 a.m. were used to calculate the water profile content (0–100 cm) by weighted summing up of the readings from the different measuring depths.

The WEPP model requires climate input data on a daily basis, including precipitation (amount, duration, intensity), temperature, solar radiation and relative humidity. For the three sites, precipitation and air temperature were measured in 5 min intervals mainly during the growing season from April to October. Missing climatic data for the winter periods as well as solar radiation and humidity values that were not measured at the sites were obtained from nearby meteorological stations run by the Hydrological Office of Lower Austria. For the WEPP climate file, precipitation was input as break-point data. Daily rainfall was described by cumulative time and rainfall depth.

The necessary soil parameters were analysed in laboratory and field investigations. Soil texture was determined using a combined wet sieving and pipette method (ÖNORM L-1068; ÖNORM, 1988), organic carbon content was analysed by a combination of dry combustion with a CN-Analyzer (Fa. Elementar, vario MAX) and the Scheibler method (ÖNORM L-1084; ÖNORM, 1999). The determination of cation exchange capacity (CEC) was carried out using the unbuffered salt extraction method (Rhoades, 1982). The baseline effective hydraulic conductivity, inter-rill erodibility, rill erodibility, critical flow hydraulic shear stress values and albedo were estimated as described in the WEPP user manual (Flanagan and Livingston, 1995).

Management input files were built for the three sites from the management records obtained from the experimental fields. The different crop and tillage parameters were selected from the model database and user manual (Flanagan and Livingston, 1995) and adapted to Austrian farming conditions whenever necessary. Slope data were obtained from topographic survey of the area.

The watershed study

The fourth study site, at Petzenkirchen, is under agriculture and located about 100 km west of Vienna, Austria (Fig. 6.1). It represents the typical environmental conditions of arable and forest lands in the pre-Alpine mountains where the Molasse sediment is considered the main component of the geological bedrock material. The watershed is about 1.5 km long and 0.6 km wide and covers an area of 66 ha. The elevation ranges from 256 m to 325 m asl and average slope is about 8%.

For runoff measurement, the watershed was equipped with an H-flume combined with a continuously operating ultrasonic sensor (Fa. Endress and Hauser) and a data-logging system. Runoff samples were taken automatically during a runoff event, controlled by discharge. Additionally, soil–water suspension samples were taken manually at weekly intervals. Sediment was filtered and dried until constancy of mass. Sediment concentration and total sediment load were determined for each event. Runoff and sediment yield data from October 2001 to December 2006 are included in this study.

Mean annual air temperature for the study area is 9.3 °C. Mean annual precipitation was 801 mm, with the lowest monthly values recorded during January–February and the highest monthly values during July–August. The majority of the precipitation occurred between the beginning of March and the middle of September, and comprised approximately 70% of the yearly total. Mean annual discharge is 21 s^{-1}. According to the Hydrological Atlas of Austria, the actual evapotranspiration (ET) of this region is around 630 mm (Dobesch, 2003). A weather station is located about 1 km away from the watershed that measures rainfall, temperature, solar radiation and wind speed with high temporal discretization.

Soils in the catchment area are moderately deep, varying between 1 and 2 m depth. According to the FAO (UN Food and Agriculture Organization) classification systems five different soils can be distinguished in the watershed, including Cambisol, calcic Cambisol, Planosol, gleyic Cambisol and Gleysol (FAO, 1974). Gleyic Cambisol and Planosol are considered to be the major soils in the watershed. Soil texture, rock fragment content and organic matter content were determined in the laboratory using the above-described methods and are shown in Table 6.1. Albedo and CEC were estimated from the WEPP user summary (Flanagan and Livingston, 1995) and from a pedo-transfer function (Breeuwsma *et al.*, 1986). WEPP erodibility values (K_i, K_r and τ_c) and effective hydraulic conductivity were computed using Eqns 6.1, 6.7 and 6.8. A restricting layer at a deeper soil depth was included in the soil file. The water discharge at base flow in the watershed was simulated by setting a restrictive layer with low permeability. Daily meteorological and hydrological data in the catchment have been monitored by the Austrian Federal Agency for Water Management, Institute for Land and Water Management Research (ILWMR) in Petzenkirchen.

According to the annual land-use survey 92% of the area was classified as arable land, while 8% was classified as grass and forest land, including a marginal portion (<0.5%) of paved and unpaved roads. However, maize, winter wheat and summer barley crops occupied the major part of the watershed, with the crop rotation patterns shown in Table 6.3 (Strauss, 2004). All information about agricultural practices in the catchment including tillage, crop type, planting and harvest dates were verified by field inspection twice a year by ILWMR.

The generation of the digital elevation model (DEM) for the site – which had a grid size of 10 m – was done by an automatic interpretation of aerial orthophotos (1:15,000). Based on the DEM, the watershed was divided into eight hillslopes and three channels (Fig. 6.2). The hillslopes had areas between 2.8 and 12.3 ha with slopes from 7.5 to 12.0% (Table 6.4). They were connected through three channels with lengths ranging from 452 to 847 m. For each sub-watershed, a representative hillslope was selected and then divided into different overland flow elements (OFE) corresponding to the existing soil/vegetation conditions (Hardan, 2010).

Sensitivity analysis and model validation

To calibrate the model, sensitivity analyses were performed for the erosion plot simulations. Sensitivity analysis was done for assessing the

Table 6.3. Crop rotation patterns in the Petzenkirchen watershed (2001–2006).

Crop rotation (CR)	2001	2002	2003	2004	2005	2006
CR 1	Maize	Summer barley	Maize	Winter wheat	Maize	Winter wheat
CR 2	Maize	Winter wheat	Maize	Winter wheat	Summer barley	Maize
CR 3	Winter wheat	Maize	Winter wheat	Maize	Winter barley	Maize
CR 4	Winter wheat	Maize	Summer barley	Maize	Summer barley	Maize
CR 5	Winter barley	Winter wheat	Summer barley	Maize	Winter wheat	Summer barley
CR 6	Maize	Winter wheat	Maize	Winter wheat	Summer barley	Maize
CR 7	Maize	Winter wheat	Maize	Summer barley	Maize	Winter wheat
CR 8	Winter wheat	Winter barley	Maize	Summer barley	Winter wheat	Maize
CR 9	Summer barley	Winter wheat	Maize	Winter wheat	Maize	Summer barley

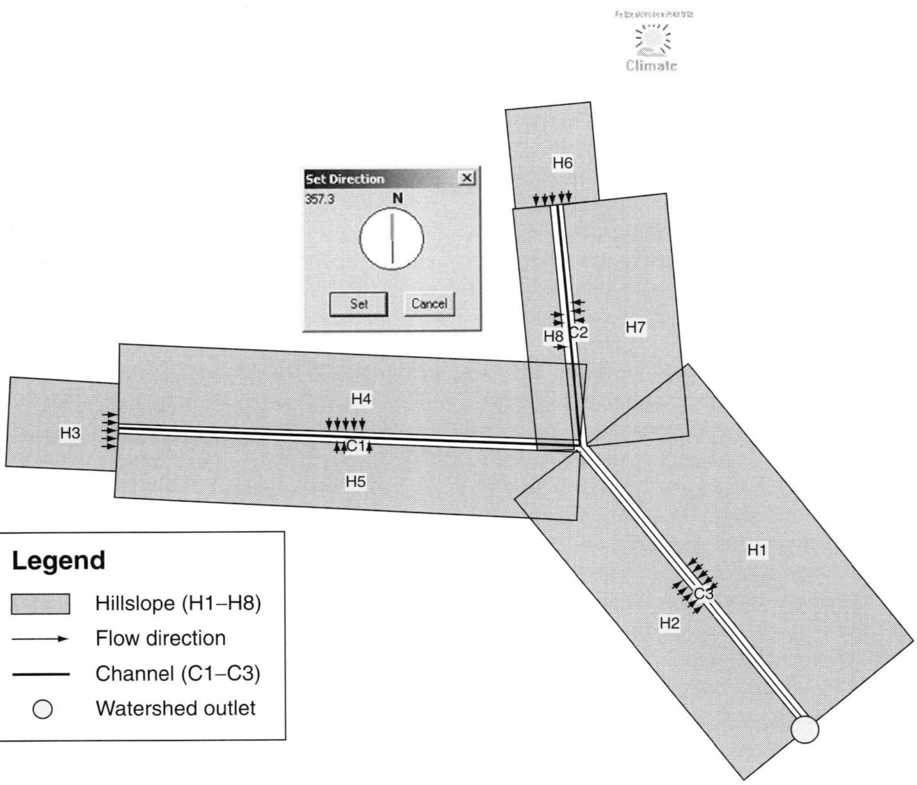

Fig. 6.2. Layout of the Petzenkirchen watershed, showing hillslopes and channels.

prediction errors of complex modelling systems. Usually, prediction errors produced by complex modelling systems are influenced by all input and structural errors. Assuming that structural errors are negligible due to the intensive validation of the mathematical functions in the models, the errors may give hints of input errors, which may result mainly from estimation errors resulting from inadequate representation of the spatial variability of the input parameters (Warrick and Nielsen, 1980; Buchter *et al.*, 1991). The sensitivity analysis used in this study was done by changing the value of one input parameter within an acceptable range and observing the

Table 6.4. Characteristics of the hillslopes and channels of the Petzenkirchen watershed.

Hillslope/channel	Length (m)	Width (m)	Area (ha)	Average slope (%)
Hillslopes				
H1	235.6	659.4	15.60	11.3
H2	145.3	659.4	9.58	9.9
H3	200.6	165	3.31	8.3
H4	145	848.3	12.30	12.0
H5	124.1	846.1	10.50	10.5
H6	182.4	154.0	2.81	11.8
H7	188	455.0	8.55	12.6
H8	70.6	455.0	3.21	7.5
Total hillslopes			65.86	
Channels				
C1	846.9	1.5	0.13	2.7
C2	452.1	1.5	0.07	4.2
C3	659.4	2.0	0.13	2.6
Total channels			0.33	
Overall			66.19	

runoff and sediment yield output. With all other parameters held constant, this one-dimensional analysis assumes that the input parameters are independent of each other.

The sensitivity measure used in this study was proposed by Nearing *et al.* (1990) and is based on the deterministic sensitivity concept formulated by McCuen (1973). The dimensionless sensitivity ratio S is formulated as:

$$S = \frac{(O_2 - O_1)/O_{12}}{(I_2 - I_1)/I_{12}} \quad (6.12)$$

where I_1 and I_2 are the least and greatest values of input used, I_{12} is the average of I_1 and I_2, O_1 and O_2 are the outputs for the two input values, and O_{12} is the average of the two outputs. The parameter S is a function of the chosen input range for non-linear response (Fig. 6.3). S less than zero indicates that an increase in input corresponds to a decrease in the output and $S > 1$ indicates that input and output are positively correlated. The sensitivity analyses were performed for the erosion plot data for the first eight erosive events.

First, the effective hydraulic conductivity was adjusted to obtain a good agreement between measured and observed runoff values. When the hydrological part of the simulation was correct, then the erodibility parameters were adjusted if necessary. The input parameters that showed negligible variation were not calibrated and were taken as model default values. Once the model was calibrated, it was run with the calibrated parameters, and the runoff and sediment yield were predicted for the validation period.

The simulated results were evaluated by visual inspection of the graphs that plotted the range of observed and simulated values for all events. The root mean square error (*RMSE*; Thomann, 1982) and two indices for evaluating model efficiency/performance: the Wilmott index (*d*; Wilmott, 1981) and the Nash–Sutcliffe model efficiency (*NSE*; Nash and Sutcliffe, 1970) were computed as criteria for goodness of fit.

The *RMSE* is defined as:

$$RMSE = \left(\frac{\sum_{i=1}^{n} (O_i - P_i)^2}{n} \right)^{1/2} \quad (6.13)$$

where O_i and P_i are the observed and predicted values for the ith pair, and n is the total number of paired values. The smaller the RMSE, the closer simulated values are to observed values.

Willmott's index of agreement, d, reflects the degree to which the observed value is

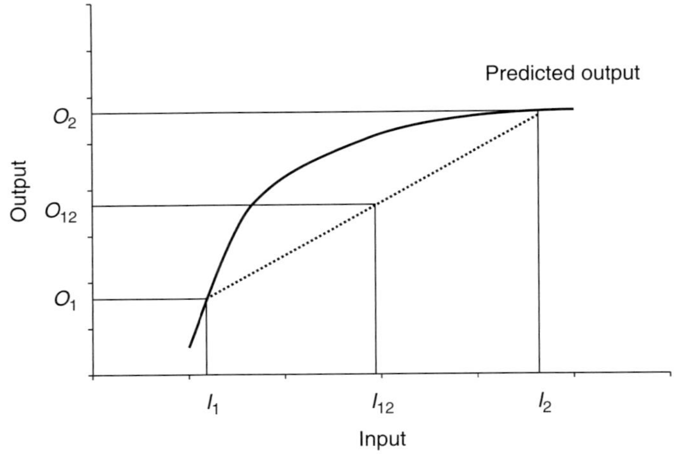

Fig. 6.3. Graphical representation of the sensitivity ratio *S* used in model sensitivity analysis.

accurately estimated by the simulated variate. It varies between 0.0 and 1.0, where a computed value of 1.0 indicates perfect agreement between observed and predicted observations and a value of 0.0 indicates complete disagreement:

$$d = 1 - \frac{\sum_{i=1}^{n}(P_i - O_i)^2}{\sum_{i-1}^{n}(|P_i - \bar{O}| + |O_i - \bar{O}|)^2} \quad (6.14)$$

where $\bar{O}$ is the mean of observed values.

The Nash–Sutcliffe efficiency, *NSE*, was calculated as follows:

$$NSE = 1 - \frac{\sum_{i=1}^{n}(O_i - P_i)^2}{\sum_{i=1}^{n}(O_i - \bar{O})^2} \quad (6.15)$$

The model efficiency can range from $-\infty$ to 1.0, and the closer the value is to 1.0, the better are the individual predictions. A value of one indicates a perfect fit, while a value of zero indicates that the model results are no better than the mean measured value. A negative *NSE* results when there is a greater difference between measured and simulated values than between measured values and the mean of measured values (Warner *et al.*, 1997). However, a shortcoming of the Nash–Sutcliffe statistic is that it does not perform well in periods with low flow. It works well when the coefficient of variation for the data is large (Pandey *et al.*, 2008). Also, a time lag between measured and simulated data produces low *NSE* values.

The WEPP model was only calibrated for the erosion plot simulations. For the Petzenkirchen watershed, inputs for effective hydraulic conductivity, inter-rill and rill erodibility, as well as critical shear stress, were determined using the recommended equations in the user manual. The simulation runs were then performed without calibration.

Results

Measurement results

Measurements from the erosion plots

During the investigation period, the variability of rainfall was high at the sites equipped with erosion plots. At Mistelbach, the annual rainfall ranged between 390 mm during 2003 and 794 mm during 1996; at Pixendorf it ranged between 461 mm during 2003 and 926 mm during 2002; and at Pyhra it ranged between 736 during 2003 and 1161 mm during 2002. As most of the erosive events occur in the eastern part of Austria between the beginning of May and the end of October (Strauss *et al.*, 1995; Klik and Truman, 2003), the amount of rainfall and its kinetic energy during this time period is the driving force for the erosion process. In Table 6.5, annual precipitation from May to October and corresponding rainfall erosivity factors, *R*, for that period are shown. The *R* factors were calculated using equations described by Renard *et al.* (1997) and Brown and Foster (1987). At Mistelbach, the highest erosivities were calculated for 1994 and 1995, and at Pixendorf and Pyhra they were for 2000 and 2002 (Table 6.5). All of these years had rainfall amounts below the average value, which indicates the occurrence of a few high-intensity storms. This is especially true for 1994, when at Mistelbach three heavy rainstorm events occurred in May, June and July, which produced 51.7, 115.5 and 53.1 mm of rain with maximum 30 min intensities of 65.4, 79.6 and 47.6 mm h^{-1}, respectively. These three storms were responsible for 34% of the annual rainfall.

At Mistelbach, Pixendorf and Pyhra, 26, 30 and 26 runoff-producing rainstorms were observed, respectively. Runoff was measured in 50% (Mistelbach) to 80% (Pyhra) of the years. With increasing rainfall, the frequency of runoff and total runoff amounts increased. Average annual soil losses ranged between 6.88 Mg ha^{-1} (Pixendorf) and 8.86 Mg ha^{-1} (Mistelbach), with high variability within the years. Soil loss reached extreme values when erosive storms occurred in periods when there was still a seedbed condition, or when the plants had not yet developed sufficient soil cover. When vegetation and soil cover were well developed and/or rainfall events with low intensity occurred, even if they produced large rainfall depths, soil losses less than 0.1 Mg ha^{-1} were measured.

At Mistelbach, the highest soil losses were observed in 1994, 1998 and 2002. In all of these years row or tuber crops were planted (maize, sugarbeet or sunflower; Table 6.2) and the erosion resulted from events starting from April until mid-June. After that time, rainfall produced only runoff, with nearly no soil loss. The year 2002 was extraordinary, with high rainfall amounts in June (77 mm at Mistelbach, 145 mm at Pyhra and 199 mm at Pixendorf) and at the beginning of August (between 145 mm at Mistelbach and 334 mm at Pyhra). However, only the events in June led to erosion, while in August, despite large rainfall amounts (up to 266 mm within 7 days at Pyhra), the kinetic energy was too small to initiate the erosion process. Figure 6.4 shows the relationship between runoff and soil loss values over all three of the erosion plots.

The data show that soil erosion is an extreme event process. Most of the soil loss/sediment yield is produced by only a few high-intensity rainfall events. This confirms the necessity of continuous simulation models which are able to reliably assess the impacts of storm events on a daily or event basis.

Measurements from the watershed

Average rainfall in the investigated watershed was higher than at the sites with the runoff plots. Yearly precipitation during the study period ranged from 591 to 1119 mm, with an average of 801 mm (Table 6.6). An average base flow of 2 L d^{-1} (corresponding to a runoff depth of 0.3 mm) was measured daily. Only in 8% of all measurements did runoff exceed 1 mm (66 L d^{-1}), and in less than 1% did it exceed a daily value of 5 mm. No correlation between annual surface runoff and sediment yield could be found. Overall, the sediment yield data were very small, with annual values between 0.16 and 26.4 Mg (Table 6.6). For the whole 66.19 ha watershed the average sediment yield of 7.8 Mg resulted in net erosion of only 0.12 Mg ha^{-1}.

Sensitivity analyses

The sensitivity analysis of WEPP parameters showed that the effective hydraulic conductivity the baseline inter-rill and rill erodibility, and the critical hydraulic shear stress were most sensitive to surface runoff and/or soil loss. The tested range of these parameters and their average sensitivity ratios are presented for the data from the three erosion plot sites (Table 6.7).

Besides these investigated parameters, WEPP predictions are also highly sensitive to rainfall intensity and duration, soil surface roughness and slope steepness (Schroeder, 2000). The sensitive soil input parameters were calibrated and adjusted as described in the model calibration section above. The calibrated values of these parameters are presented in

Table 6.5. Summary of annual (P) and seasonal rainfall ($P_{May–Oct}$), rainfall erosivity factor (*R*) and runoff and soil loss amounts from the three natural runoff (erosion) plots investigated.

	Mistelbach					Pixendorf					Pyhra				
Year	P (mm)	$P_{May–Oct}$ (mm)	$R_{May–Oct}$ (kJ m^{-2} mm h^{-1})	Runoff (mm)	Soil loss (t ha^{-1})	P (mm)	$P_{May–Oct}$ (mm)	$R_{May–Oct}$ (kJ m^{-2} mm h^{-1})	Runoff (mm)	Soil loss (t ha^{-1})	P (mm)	$P_{May–Oct}$ (mm)	$R_{May–Oct}$ (kJ m^{-2} mm h^{-1})	Runoff (mm)	Soil loss (t ha^{-1})
1994	642	468	417	82.0	53.50	–	–		–	–	835	459	176	18.0	67.50
1995	601	378	105	0.0	0.0	–	–		–	–	893	525	39	0.0	0.0
1996	794	566	74	11.1	3.25	–	–		–	–	1121	847	70	111.1	0.15
1997	637	427	52	2.7	0.0	759	500	72	57.7	3.19	987	607	105	76.4	0.65
1998	668	541	92	41.1	19.77	704	558	82	1.7	1.42	773	477	57	1.1	1.62
1999	649	417	85	0.0	0.0	792	533	93	29.4	20.49	1159	774	157	23.2	0.06
2000	706	636	59	0.0	0.0	542	344	130	10.3	17.49	848	542	214	32.1	12.61
2001	648	449	58	0.0	0.0	610	387	32	0.0	0.0	934	590	67	0.0	0.0
2002	718	495	71	23.5	12.10	926	649	237	17.8	1.41	1161	857	375	108.0	3.67
2003	390	214	17	0.0.	0.0	461	304	32	24.8	4.10	736	469	83	3.6	0.03
Average	645	432	103	16.0	8.86	685	467	97	21.7	6.88	944	615	134	37.4	8.63
SD	99	95	107	25.5	16.20	147	116	65	19.4	7.18	149	147	97	42.2	19.97

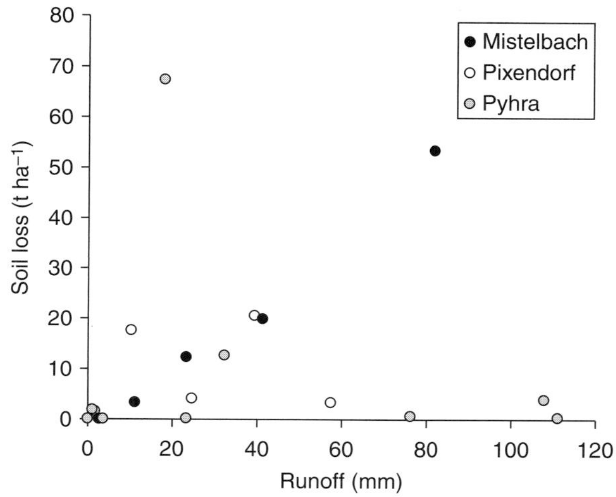

Fig. 6.4. Relationship between runoff and soil loss values for the three erosion plots investigated.

Table 6.6. Annual rainfall (P), runoff and sediment yield amounts for the Petzenkirchen watershed.

Year	P (mm)	Runoff (mm)	Sediment yield (t)
2002	1119	271.3	7.08
2003	591	158.8	3.97
2004	774	93.4	26.40
2005	790	152.3	1.19
2006	736	164.2	0.16
Average	801	168.0	7.76
SD	174	57.6	9.63

Table 6.8. In this study, the effective hydraulic conductivity and the baseline inter-rill erodibility were increased for Mistelbach and Pyhra compared with the data described in the WEPP user manual, while the chosen hydraulic conductivity data for Mistelbach agreed well with the measured data at this site (Salvador Munoz, 2003).

Surface runoff

As the erosion plots were installed only during the growing season, simulated runoff events during this period were included in this study. During the whole simulation period, between 26 and 30 events occurred at each site, resulting in overall 83 runoff events (Table 6.9). Note that for both runoff and erosion events, not all events that were observed were simulated, and not all events that were simulated were observed. For example, for erosion, 72 events were observed and 40 simulated; 32 of these simulated events occurred when events were also observed, but eight events were simulated without observed erosion. For surface runoff, 47 of the 59 simulated runoff events occurred in reality and 12 events were only simulated – without being observed. Table 6.10 gives the statistics for the analysis of observed and simulated surface runoff values. Results for the calibration, the validation and for the whole simulation period are displayed in Table 6.10. It can be seen that during the calibration period (the first eight erosive events) much higher runoff occurred than during the following storms. During all three periods, the runoff values that were predicted by simulation were found to be close to the measured (observed) values, with similar standard deviations. Linear regression analysis between measurements and simulations showed that WEPP was able to estimate surface runoff well (correlation coefficients between 0.72 and 0.83). At Mistelbach and Pixendorf, runoff was under-predicted on average by 11 and 18%, respectively, while at Pyhra it was over-predicted by 5% (Table 6.10). Besides the mean values, the standard deviations between observations and simulations exhibited good agreement. This similarity reveals that the range and distribution of the predicted event runoff volumes resembled

Table 6.7. Sensitivity ratio (S) for effective hydraulic conductivity (K_e), baseline inter-rill (K_i) and rill erodibility (K_r), and critical flow hydraulic shear (τ_c) in the WEPP for the soil erosion sites investigated.

Parameter	Range of test values	Sensitivity ratio S for soil loss			Sensitivity ratio S for runoff		
		Mistelbach	Pixendorf	Pyhra	Mistelbach	Pixendorf	Pyhra
K_e (mm h^{-1})	1–20	−0.5028	−0.9337	−0.3702	−0.4650	−0.4054	−0.7272
K_i (kg s m^{-4})	450,000–12,000,000	0.2141	0.7406	0.5024	–	–	–
K_r (kg s m^{-4})	0.01–0.1	0.3973	0.3456	0.4228	–	–	–
τ_c (Pa)	0.5–3.5	−0.4714	−0.5132	−0.4040	–	–	–

Table 6.8. Calibrated soil input parameters for the WEPP model for the erosion plot sites investigated.

Parameter	Mistelbach	Pixendorf	Pyhra
Effective hydraulic conductivity K_e (mm h^{-1})	9.0	3.5	16.5
Baseline inter-rill soil erodibility K_i (kg s m^{-4})	15,201,519	3,880,000	12,794,400
Baseline rill soil erodibility K_r (s m^{-1})	0.0318	0.007	0.0049
Critical flow hydraulic shear τ_c (Pa)	3.5	3.06	3.68

those of the measured values. The average magnitude of error for the different sites was within 11 mm per event and was about 7 mm when taking into account all data. Model efficiencies (NSE) for the three sites were more than 0.64 and Willmott's index *d* was higher than 0.92. This demonstrates that WEPP performed well at predicting runoff amounts; this has also been determined in other studies (Soto and Diaz-Fierros, 1998; Raclot and Albergel, 2006; Centeri *et al.*, 2009).

Figure 6.5 presents a plot of measured and WEPP-estimated values of surface runoff from the erosion plots. From the overall 83 runoff events, only 59 were simulated (predicted). Between nine and 14 observed runoff events were not predicted by the simulation, which means that 30–52% of the runoff events were simulated as 'non-events'. At Mistelbach, Pixendorf and Pyhra, 50% of all events were smaller than 0.5, 1.0 and 2.0 mm, respectively (Table 6.9). Therefore, the predicted non-events had no significant influence on the runoff because the missed events were very small. In contrast, maximum runoff values were very well predicted in Mistelbach. They were underestimated in Pixendorf and were slightly overestimated in Pyhra.

During the simulation period of 6 years in the Petzenkirchen watershed, average daily runoff (base flow) amounted to 0.5 mm, which was mainly produced by lateral flow. Considering average runoff from the watershed WEPP results showed good agreement with the measurements (Table 6.10). In Fig. 6.6, all daily runoff data from the watershed are displayed. More than 99% of the runoff data were smaller than 5 mm, and only a few events exceeded this value. Overall, without calibration, WEPP overestimated runoff from the watershed by 26%. A Willmott's index *d* of 0.73 points out that WEPP simulates daily runoff satisfactorily. Owing to the large amounts of data that are similar to or smaller than the mean value (73%) the Nash–Sutcliffe model efficiency produced a negative value.

Average yearly runoff volumes from the runoff plots ranged between 16 mm (Mistelbach) and 37 mm (Pyhra). Maximum yearly runoff reached 58 mm in Pixendorf, 82 mm in Mistelbach and 111 mm in Pyhra (Table 6.5). Yearly runoff volumes were predicted well (Fig. 6.7; Table 6.10). Considering data from all sites, the model under-predicted annual runoff by approximately 10%.

Table 6.9. Number and size of observed and simulated runoff events at the three erosion plot sites investigated as well as number of non-events.

	Mistelbach		Pixendorf		Pyhra	
Parameter	Observed	Simulated	Observed	Simulated	Observed	Simulated
Number of events	27	20	30	23	26	16
Maximum runoff (mm)	55.0	55.3	52.5	43.7	75.0	86.1
Median runoff (mm)	0.5	0.2	1.0	0.03	2.0	0.02
Number of events observed and simulated	13		21		13	
Number of events observed but not simulated	14		9		13	
Number of events simulated but not observed	7		2		3	
Maximum runoff (mm)	7.2		5.9		13.2	
Median runoff (mm)	1.8		1.5		2.1	

The average magnitude of error was 13 mm. Taking into account the large runoff variations between years, these results were quite acceptable. Runoff from the watershed yielded a yearly mean of 144 mm and a predicted mean of 141 mm, with an average error of 107 mm (Table 6.10). WEPP estimations of annual runoff volumes were reliable with a regression line near unity (Table 6.10).

Soil loss

For the erosion plots, WEPP simulated soil loss values reasonably and most of the estimations were within the range of the observed values (Fig. 6.8). As the calibration period included the major erosive events during the whole simulation period, the calibrated WEPP inputs were responsible for a very good agreement between observations and predictions. This can be seen in the resemblance of measured and simulated average soil losses and standard deviations in Table 6.11). Linear regression showed that soil loss was underestimated by 41% in Pyhra and by 22% in Mistelbach). Soil loss from very small runoff events was apparently over-predicted (positive intercepts). For Pixendorf, the regression line was nearly unity, and for all sites the intercept was small.

The phenomenon of soil erosion models overestimating small events and underestimating large events was explained by Nearing (1998), and has been shown in many other studies (Ghidey *et al.*, 1995; Zhang *et al.*, 1996; Tiwari *et al.*, 2000; Shen *et al.*, 2009). Therefore, uncalibrated use of these models is not advisable (Klik *et al.*, 1998). Nevertheless, when the input parameters were calibrated, WEPP was able to simulate small as well as large soil losses with good accuracy (Fig. 6.8). This is supported by studies from Bhuyan *et al.* (2002) and Stolpe (2005). The model efficiencies (*NSE*) for the three soil erosion sites ranged between 0.71 and 0.73 and were similar to the values for runoff. Willmott's index was higher than 0.91, which underlines the good model performance when calibrated. The correct estimation of large erosion events is very important because the major part of soil erosion and sediment yield is produced by only a few erosive events with high rainfall intensity (Klik, 2003). These few extreme events were then responsible for costly on-site and off-site damages.

During the simulation period, combined for the three erosion plot sites, 72 erosion events were observed and 36 events were simulated (Table 6.12), but only 32 of these events were predicted for days when runoff was measured. At all the sites, these large events were very well covered by simulation, which is shown by the agreement of the maxima between observations

Table 6.10. Summary statistics of observed and WEPP-simulated runoff values for events for all of the investigated sites.

Parameter	Mistelbach		Pixendorf		Pyhra		All plot sites		Petzenkirchen	
	Observed	Simulated	Observed	Simulated	Observed	Simulated	Observed	Simulated	Observed	Simulated
Single events										
Calibration period										
Average (mm)	10.1	9.5	6.4	6.4	22.7	19.1	13.1	11.7	–	–
SD (mm)	16.8	17.5	16.3	13.5	14.4	25.5	24.4	26.5	–	–
Validation period										
Average (mm)	2.6	3.1	2.6	2.6	7.5	9.1	4.0	4.6	–	–
SD (mm)	5.1	4.8	3.3	4.4	10.4	17.5	7.0	10.5	–	–
Simulation period										
Average (mm)	4.5	4.7	3.4	3.5	11.9	12.0	6.3	6.4	0.5	0.5
SD (mm)	10.0	10.1	8.5	7.8	17.3	21.1	12.8	14.2	0.7	1.2
RMSE (mm)	4.9		3.6		10.9		6.9		0.9	
Model efficiency, *NSE*[a]	0.78		0.85		0.64		0.71		–0.83	
Willmott's *d*	0.94		0.96		0.92		0.93		0.73	
Regression results										
Intercept	0.69		0.59		0.61		0.27		–0.13	
Slope	0.89		0.82		1.05		0.97		1.26	
Correlation coefficient	0.7852		0.8295		0.7204		0.7632		0.4854	
Annual values										
Average (mm)	16.0	16.7	21.7	19.0	37.4	41.4	25.6	25.9	144.5	140.8
SD (mm)	25.5	22.8	19.4	18.9	42.2	43.1	33.7	32.7	74.4	85.8
RMSE (mm)	9.7		6.9		18.6		13.0		106.8	
Model efficiency, *NSE*	0.86		0.94		0.81		0.85		0.66	
Willmott's *d*	0.96		0.97		0.95		0.96		0.92	
Regression results										
Intercept	3.50		2.60		3.96		2.99		2.84	
Slope	0.83		0.86		0.91		0.90		0.99	
Correlation coefficient	0.8594		0.8906		0.8214		0.8535		0.7436	

[a]*NSE*, Nash–Sutcliffe efficiency.

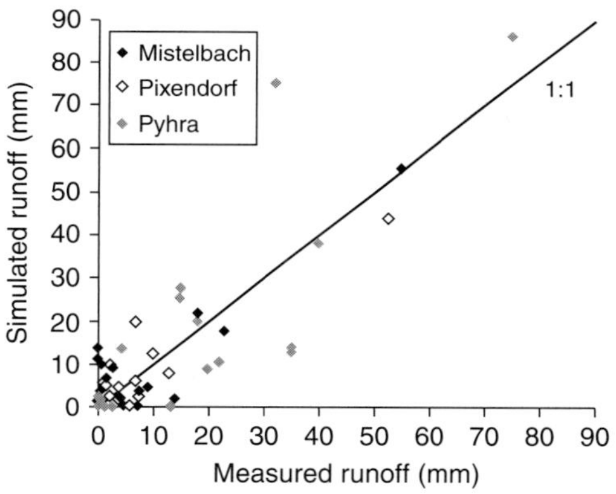

Fig. 6.5. WEPP predicted versus observed event runoff from the three erosion plots investigated.

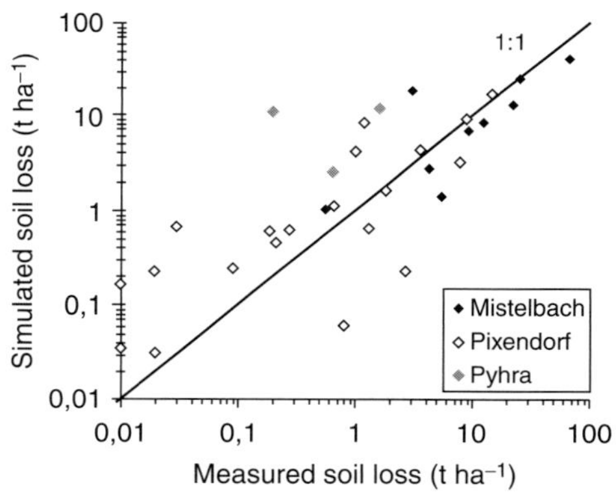

Fig. 6.8. WEPP predicted versus observed soil loss from the three erosion plots investigated.

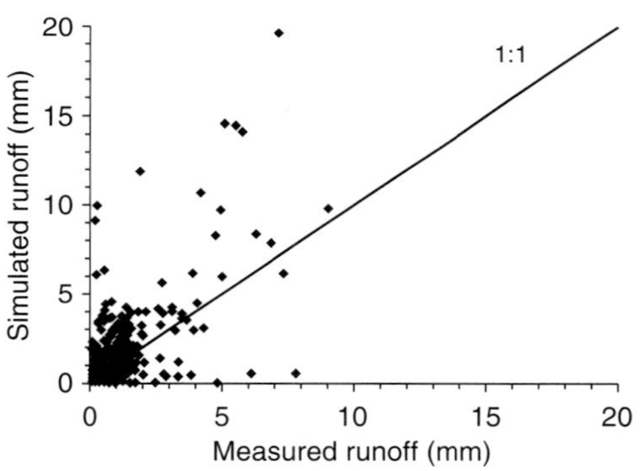

Fig. 6.6. WEPP predicted versus observed event runoff from the Petzenkirchen watershed.

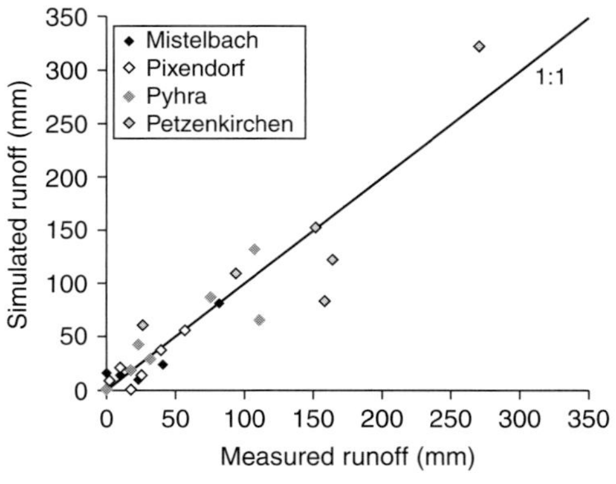

Fig. 6.7. WEPP predicted versus observed annual runoff amounts from all of the erosion plots investigated and from the Petzenkirchen watershed.

and measurements. The median values of between 0.05 and 0.06 t ha^{-1} indicate that most of the measured (observed) soil erosion events were very small. Therefore, the error in simulating or not simulating these very small events does not greatly influence the total amount of sediment yield. As for the runoff events, some of the soil loss events (between eight and 18) were not predicted, i.e. they were simulated as 'non-events'.

For the Petzenkirchen watershed, 30 runoff events causing sediment yield could be used for this study. The simulated values were over-predicted for small as well as for large events (Fig. 6.9). On average, the simulations exceeded the measurements by 105% (Table 6.11). This can be partly attributed to the surface runoff, which was also overestimated (by 26%, Table 6.10). A decrease in runoff will result in reduced soil loss, but presumably the reduction in runoff will not fully generate the necessary decrease in sediment yield to match observed values. A calibration of the sensitive input parameters will certainly improve the simulation results. Jetten *et al.* (1999) stated that calibration is imperative for small- and medium-scale watersheds where the influence of the spatial variability on the runoff and erosion process strongly influences the simulation. Calibration should be performed not only on sediment yield data observed at the watershed outlet, but also on runoff and erosion data obtained within the watershed. In addition, the a priori knowledge

Table 6.11. Summary statistics of observed and WEPP-simulated soil loss (Mistelbach, Pixendorf and Pyhra) and sediment yield values (Petzenkirchen) for events.

	Mistelbach		Pixendorf		Pyhra		All plot sites		Petzenkirchen[a]	
Parameter	Observed	Simulated	Observed	Simulated	Observed	Simulated	Observed	Simulated	Observed	Simulated
Single events										
Calibration period										
Average (t ha^{-1})	6.20	4.44	0.36	1.54	8.54	5.43	4.90	3.74	–	–
SD (t ha^{-1})	9.68	8.13	0.45	2.71	22.29	13.41	22.29	13.41	–	–
Validation period										
Average (t ha^{-1})	1.21	1.36	1.50	1.35	0.82	1.17	1.21	1.30	–	–
SD (t ha^{-1})	2.93	3.84	3.35	3.51	1.84	3.29	2.86	3.57	–	–
Simulation period										
Average (t ha^{-1} or tons)	2.46	2.13	1.20	1.39	2.97	2.31	2.13	1.91	3.58	5.53
SD (t ha^{-1} or tons)	5.88	5.42	2.94	3.33	12.30	7.71	7.61	5.61	6.27	11.88
RMSE (t ha^{-1} or tons)	3.21		1.59		5.77		3.76		16.30	
Model efficiency, *NSE*[b]	0.71		0.73		0.73		0.71		−6.01	
Willmott's *d*	0.92		0.94		0.91		0.91		0.56	
Regression results										
Intercept	0.21		0.18		0.68		0.53		2.37	
Slope	0.78		0.99		0.59		0.65		2.05	
Correlation coefficient	0.7185		0.7745		0.8489		0.7743		0.4701	
Annual values										
Average (t ha^{-1} or tons)	8.86	7.68	6.88	7.56	8.63	7.69	8.52	7.65	7.76	15.83
SD (t ha^{-1} or tons)	16.20	13.47	7.18	7.01	19.97	12.52	16.42	11.72	9.63	17.61
RMSE (t ha^{-1} or tons)	5.76		4.55		9.30		6.95		25.80	
Model efficiency, *NSE*	0.87		0.94		0.80		0.82		−0.44	
Willmott's d	0.89		0.90		0.92		0.94		0.83	
Regression results										
Intercept	0.72		2.69		2.08		1.97		1.82	
Slope	0.79		0.73		0.59		0.67		1.81	
Correlation coefficient	0.8916		0.6878		0.9475		0.8722		0.9738	

[a]Sediment yield values in tons.
[b]*NSE*, Nash–Sutcliffe efficiency.

Table 6.12. Number of observed and WEPP-simulated soil loss events at the three erosion plot sites investigated, as well as number of non-events.

	Mistelbach		Pixendorf		Pyhra	
Parameter	Observed	Simulated	Observed	Simulated	Observed	Simulated
Number of events observed and simulated	26	10	28	23	18	7
Maximum soil loss (t ha^{-1})	25.60	24.53	14.80	17.16	67.50	40.83
Median soil loss (t ha^{-1})	0.05	0.00	0.06	0.05	0.05	0.00
Number of events observed and simulated	8		20		4	
Number of events observed but not simulated	18		8		14	
Number of events simulated but not observed	2		3		3	
Maximum soil loss (t ha^{-1})	1.51		0.30		8.85	
Median soil loss (t ha^{-1})	0.30		0.12		1.16	

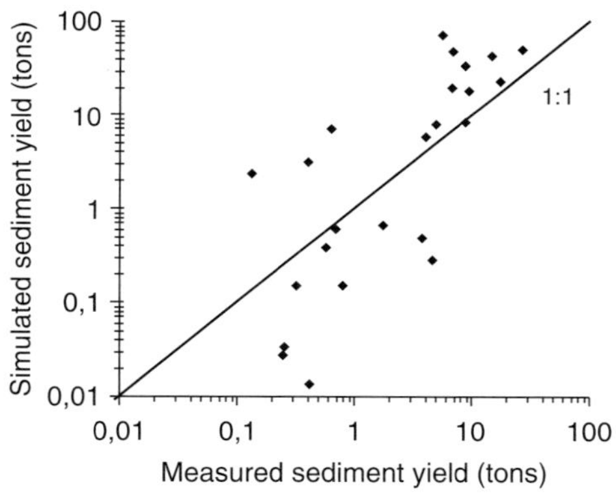

Fig. 6.9. WEPP predicted versus observed sediment yield from the Petzenkirchen watershed.

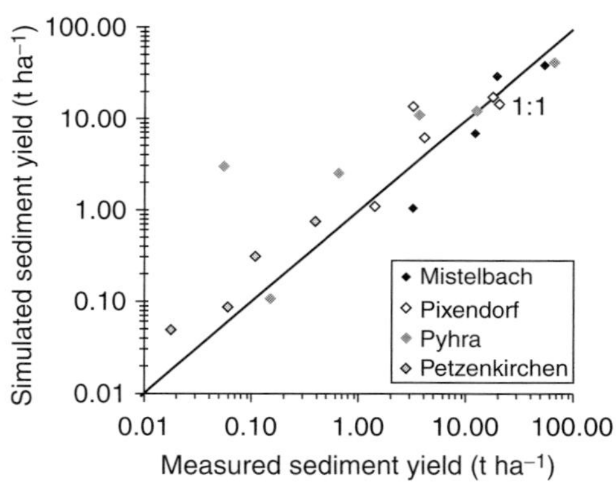

Fig. 6.10. WEPP predicted versus observed annual sediment yield values from all four of the investigated sites.[1]

of the terrain derived from field observations of runoff and erosion patterns, an analysis of the DEM, and aerial photographs or remote sensing may greatly improve the accuracy and predictive capability of assessment of spatial patterns of runoff, erosion and deposition (Desmet and Govers, 1997; Takken *et al.*, 2001; Jetten *et al.*, 2003).

Measured average annual soil yields were 8.9 Mg ha^{-1} in Mistelbach, 6.9 Mg ha^{-1} in Pixendorf, and 8.6 Mg ha^{-1} in Pyhra (Table 6.5). In the Petzenkirchen watershed a mean sediment yield of 7.8 Mg (Table 6.6) or a net soil loss of 0.12 Mg ha^{-1} was observed. For all the soil erosion plots, WEPP calculated about 33% lower soil yields than observed values, while for the watershed it over-predicted sediment yield by 81% (Table 6.11). For the watershed, it needs to be noted that the overall values were very small and reached a maximum of only 26.4 Mg or 0.40 Mg ha^{-1}, respectively. Overall, WEPP performed better in calculating annual values of erosion than it did for single storm events, because for all of the erosion plot sites together the slope of the regression line increased from 0.65 to 0.67, while for the watershed it decreased from 2.05 to 1.81. In Fig. 6.10, measured and simulated annual soil loss/sediment yields are plotted for all four of the investigated sites.

Profile water contents

Considering the investigation period (1994–2003), the year 2002 was the wettest in Mistelbach, with extreme rainfall events in August (715 mm), while 2003 was the driest year (390 mm), with low precipitation amounts throughout the whole year. This difference in rainfall amounts and pattern greatly influenced the temporal distribution and variability of profile water contents in both years. Figures 6.11 and 6.12 display the temporal distribution of the measured as well as the WEPP-simulated profile water contents for the two periods investigated (16 May to 18 September 2002, and 3 April to 21 July 2003). In both 2002 and 2003, WEPP overestimated soil water contents in spring and underestimated them in summer. The decrease in simulated soil profile water contents was steeper than from the measured values. This suggests that WEPP possibly estimated higher evapotranspiration rates than those actually occurring in the field. The standard error (*RMSE*) was calculated as 30 mm in 2002 and 37 mm in 2003. Changes of the relevant parameters in the plant input file could reduce this error. Soto and Diaz-Fierros (1998) and Pieri *et al.* (2007) found better agreement between measured and simulated soil water contents. Nevertheless, the overall trend of profile water content and its variability could be simulated with acceptable accuracy.

Summary

The Water Erosion Prediction Project (WEPP) model was used in this study to predict runoff and soil loss on plot and small watershed scales. The essential problem is how and with what accuracy hydrological and erosion processes can be simulated by this physically based model.

As a first step, the basic model concepts and fundamental equations were discussed. In a second step, a one-dimensional sensitivity analysis was performed. This showed the high sensitivity of infiltration-related parameters in runoff and soil loss assessments. This behaviour is of particular importance because of the close relationship between these two parameters. Other crucial parameters included rainfall intensity and duration, soil surface roughness and slope steepness. The third step of this study was the application of the WEPP

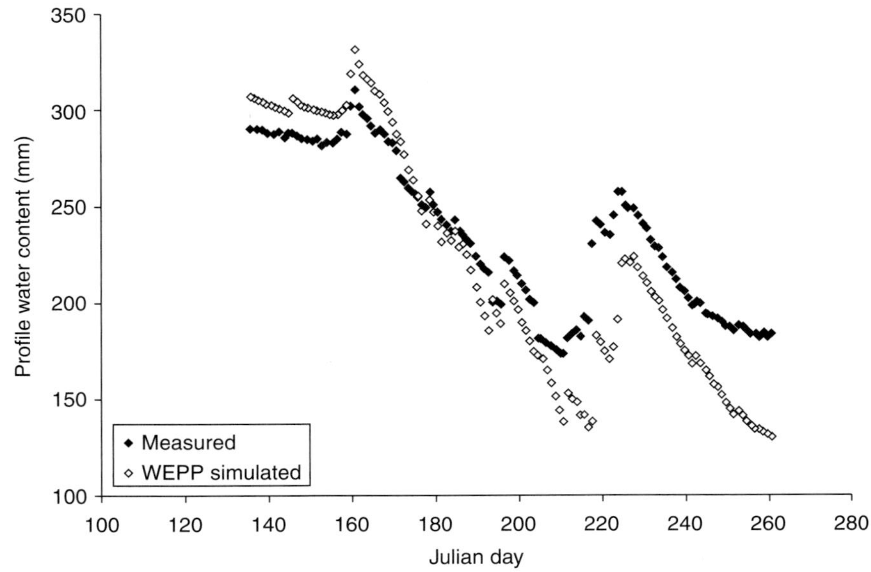

Fig. 6.11. Temporal distribution of soil profile water content (0–100 cm) under maize from 16 May to 18 September 2002 at the Mistelbach erosion plot.

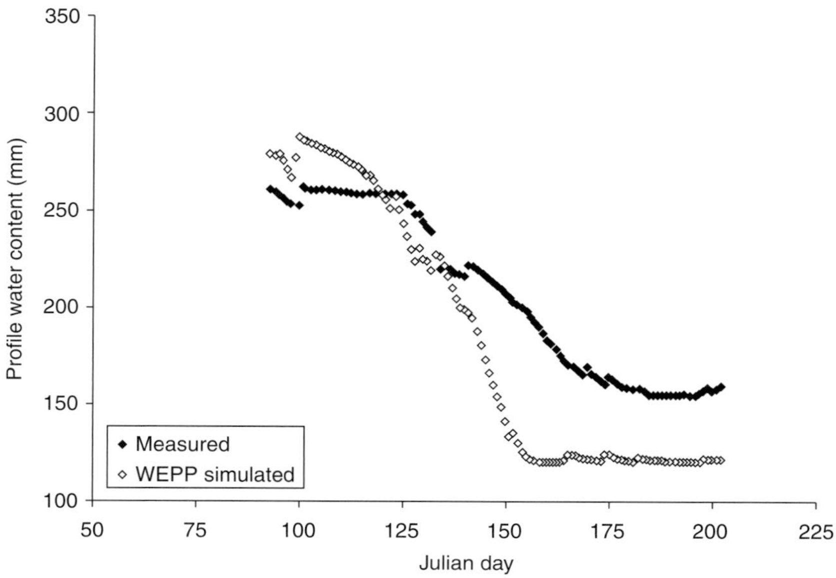

Fig. 6.12. Temporal distribution of soil profile water content (0–100 cm) under winter wheat from 3 April to 21 July 2003 at the Mistelbach erosion plot.

model to different spatial scales. The calibrated model was used to simulate runoff and soil loss from natural runoff plots operated since 1994 and 1997 at three sites in the eastern part of Austria. With the uncalibrated model, runoff and sediment yield from a large 66 ha agricultural watershed were calculated and compared with actual 5-year measurements.

Judged by the Nash–Sutcliffe efficiency parameter and the Willmott's index of agreement the runoff predictions were more accurate when the model was calibrated. Without calibration, the model had the tendency to under-predict the runoff volume from plots, but showed good agreement at the watershed scale. WEPP underestimated soil losses from plots without the calibration of soil erodibility input data. With calibration, simulated values correlated well with observed data. This was underlined by the high values of the Nash–Sutcliffe parameter and Willmott's index. The accuracy and reliability of prediction were shown to improve from an event to an annual to an average annual basis owing to the high variability of soil loss rates. The model performed well in assessing both event totals and average annual values.

For the watershed with very low sediment yield data, WEPP simulated values were well correlated to measurements. For all sites, the simulation model calculated mean values of runoff and soil loss rates as well as standard deviations similar to observations. This similarity reveals that the model predicts not only the amount but also the variability between events.

A comparison between two growing seasons of simulated and measured soil profile water contents indicated that WEPP overestimated soil water contents in the spring and underestimated them in the summer. This can be attributed to the too high evapotranspiration rates calculated by the model.

Uncalibrated use of soil erosion models is not advisable because the spatial variability of runoff and erosion processes within watersheds can strongly influence the simulation results. WEPP derives its strengths from being a process-based model and has the capability to assess spatial and temporal distribution of runoff, soil loss and deposition. These explicit estimations are a sound basis on which to design appropriate soil conservation measures and to most effectively control runoff and sediment yield. Calibration should not only be performed on sediment yield data observed at the watershed outlet, but also on runoff and erosion data obtained within the watershed. In addition, a priori knowledge of the terrain may greatly improve the accuracy and predictive capability of assessing the spatial patterns of runoff, erosion and deposition.

Future work will concentrate on the application and verification of the WEPP model for arid and semi-arid regions, where most of the necessary input data are not easily available and need to be measured on the ground or derived from airborne sources. In WEPP as in most of the existing soil erosion models, only rill and inter-rill erosion processes, including ephemeral gullies, are represented. In many areas, especially in regions where soil erosion is the biggest threat to the soil, the incision and formation of gullies strongly contributes to soil loss and sediment yield production. The incorporation of this process would increase the wider applicability of WEPP.

Acknowledgements

The data from the Petzenkirchen watershed have been collected and provided for this study by the Institute for Land and Water Management Research (ILWMR) in Petzenkirchen, Austria. This study would not have been carried out without the support of Peter Strauss.

Note

[1] Sediment yield is the amount of sediment leaving the plot or watershed. In the case of the plots these are more or less the same as soil loss, but in the case of the watershed soil loss can be much higher than sediment yield because of deposition within the watershed.

References

Beasley, D.B. and Huggins, L.F. (1991) *ANSWERS (Areal Nonpoint Source Watershed Environment Response Simulation) User's Manual.* Publication No. 5, Agricultural Engineering Department, University of Georgia, Coastal Plain Experiment Station, Tifton, Georgia.

Beasley, D.B., Huggins, L.F. and Monke, E.J. (1980) ANSWERS, a model for watershed planning. *Transactions of the ASABE* 23, 938–944.

Bhuyan, S.J., Kalita, P.K., Janssen, K.A. and Barnes, P.L. (2002) Soil loss predictions with three erosion simulation models. *Environmental Modelling and Software* 17, 137–146.

Breeuwsma, A., Woesten, J.H.M., Vleeshouwer, J.J., van Slobbe, A.M. and Bouma, J. (1986) Derivation of land qualities to assess environmental problems from soil surveys. *Soil Science Society of America Journal* 50, 186–190.

Brown, L.C. and Foster, G.R. (1987) Storm erosivity using idealized intensity distributions. *Transactions of the ASABE* 30, 379–386.

Buchter, B., Aina, P.O., Azari, A.S. and Nielsen, D.R. (1991) Spatial variability along transects. *Soil Technology* 4, 297–314.

Centeri, C., Barta, K., Gergely, J., Szalai, Z. and Biro, Z. (2009) Comparison of EUROSEM, WEPP, and MEDRUSH model calculations with measured runoff and soil-loss data from rainfall simulations in Hungary. *Journal of Plant Nutrition and Soil Science* 72, 789–797.

Chu, S.T. (1978) Infiltration during an unsteady rain. *Water Resources Research* 14, 461–466.

De Roo, A.P.J., Vesseling, C.G., Cremers, N.H.D.T., Offermans, R.J.E., Ritsema, C.J. and Oostindie, K. (1994) LISEM: a new physically-based hydrological and soil erosion model in a GIS environment. In: Olive, L.J., Loughran, R.J. and Kesbys, J.A. (eds) *Variability in Stream Erosion and Sediment Transport, Proceedings of the Canberra Symposium*, December 1994. IAHS Publication No. 224, pp. 439–448.

Desmet, P.J.J. and Govers, G, (1997) Two-dimensional modelling of the within-field variation in rill and gully geometry and location related to topography. *Catena* 29, 283–306.

Dobesch, H. (2003) Mittlere jährliche potentielle Evapotranspiration. In: Bundesministerium für Land- und Forstwirtschaft, Umwelt und Wasserwirtschaft (ed.) *Hydrologischer Atlas Oesterreichs*, 1. Lieferung, Kartentafel 3.2. Bundesministerium für Land- und Forstwirtschaft, Umwelt und Wasserwirtschaft, Vienna.

Eagleson, P.S. (1970) *Dynamic Hydrology.* McGraw-Hill, New York.

EEA (European Environment Agency) (1999) *Environment in the European Union at the Turn of the Century*. EEA, Copenhagen, Denmark.

EEA (2000) *Down to Earth: Soil Degradation and Sustainable Development in Europe – a Challenge for the 21st Century*. Environmental Issue Report No.16, EEA, Copenhagen, Denmark.

Elliott, W.J., Liebenow, A.M., Laflen, J.M. and Kohl, K.D. (1989) *A Compendium of Soil Erodibility Data from WEPP Cropland Soil Field Erodibility Experiments 1987 and 1988*. NSERL Report No. 3, The Ohio State University and USDA Agricultural Research Service (USDA ARS) National Soil Erosion Research Laboratory, West Lafayette, Indiana.

El-Swaify, S.A. (1994) State of the art for assessing soil and water conservation needs and technologies. In: Napier, T.L., Camboni, S.M. and El-Swaify, S.A. (eds) *Adopting Conservation on the Farm*. Soil and Water Conservation Society, Ankeny, Iowa, pp. 13–27.

FAO (UN Food and Agriculture Organization) (1974) *Soil Map of the World*, Vol. 1. Legend, Paris.

Ferreira, V.A. and Smith, R.E. (1992) *OPUS: An Integrated Simulation Model for Transport of Nonpoint Source Pollutants at the Field Scale. Volume II, User Manual*. Publication No. ARS-98, USDA ARS, Washington, DC.

Finkner, S.C., Nearing, M.A., Foster, G.R. and Gilley, J.E. (1989) Calibrating a simplified equation for modelling sediment transport capacity. *Transactions of the ASAE* 32, 1545–1550.

Flanagan, D.C. and Livingston, S.L. (1995) *USDA-Water Erosion Prediction Project. WEPP User Summary*. NSERL Report No. 11. USDA ARS National Soil Erosion Research Laboratory, West Lafayette, Indiana.

Flanagan, D.C. and Nearing, M.A. (eds) (1995) *USDA Water Erosion Prediction Project (WEPP) Version 95.7, Hillslope Profile and Watershed Model Documentation*. NSERL Report No. 10, USDA ARS National Soil Erosion Research Laboratory, West Lafayette, Indiana.

Foster, G.R. and Lane, L.J. (1987) *USDA-Water Erosion Prediction Project (WEPP), User Requirements*. NSERL Report No. 1, USDA ARS National Soil Erosion Research Laboratory, West Lafayette, Indiana.

Foster, G.R. and Meyer, L.D. (1972) A closed-form soil erosion equation for upland areas. In: Shen, H.W. (ed.) *Sedimentation*. Colorado State University, Fort Collins, Colorado, pp. 12.1–12.19.

Foster, G.R., Flanagan, D.C., Nearing, M.A., Lane, L.J., Risse, L.M. and Finkner, S.C. (1995) Chapter 11. Hillslope erosion component. In: Flanagan, D.C. and Nearing M.A. (eds) *USDA Water Erosion Prediction Project (WEPP) Version 95.7, Hillslope Profile and Watershed Model Documentation*. NSERL Report No. 10, USDA ARS National Soil Erosion Research Laboratory, West Lafayette, Indiana.

Ghidey, F., Alberts, E.E. and Kramer, L.A. (1995) *Comparison of Runoff and Soil Loss Predicitions from the WEPP Hillslope Model to Measured Values for Eight Cropping and Management Treatments*. Paper No. 95-2383, American Society of Agricultural Engineers, St Joseph, Michigan.

Green, W.H. and Ampt, G.A. (1911) Studies on soil physics. I. Flow of air and water through soils. *Journal of Agricultural Science* 4, 1–24.

Hardan, K. (2010) Water harvesting as a tool for sustainable water Management in a semi-arid region. Dissertation. Institut für wasserwirtschaft, Hydrologie und Konstruktiven Wasserbau. Universität für Bodenkultur Wien, Austria.

Hofmann, J. (2005) Auswirkungen unterschiedlicher Bodenbearbeitungssysteme auf die Bodengesundheit. Dissertation. Institut für Hydraulik und landeskulturelle Wasserwirtschaft, Universität für Bodenkultur Wien, Austria.

Horton, R.E. (1933) The role of infiltration in the hydrological cycle. *Transactions of the American Geophysical Union* 14, 446–460.

Jetten, V., De Roo, A. and Favis-Mortlock, D. (1999) Evaluation of field-scale and catchment-scale soil erosion models. *Catena* 37, 521–541.

Jetten, V., Govers, G. and Hessel, R. (2003) Erosion models: quality of spatial predictions. *Hydrological Processes* 17, 887–900.

Jones, R.J.A., LeBissonnais, Y., Bazzoffi, P., Sanchez Diaz, J., Düwel, O., Loj, G., Øygarden, L., Prasuhn, V., Rydell, B., Strauss, P., Berényi Üveges, J., Vandekerckhove, L. and Yordanov, Y. (2004) Nature and Extent of Soil Erosion in Europe, Working Group 2, Final Report, Technical Working Group on Erosion. European Commission Joint Research Centre, Brussels, Belgium.

Klik, A. (2003) Einfluss unterschiedlicher Bodenbearbeitung auf Oberflaechenabfluss, Bodenabtrag sowie auf Naehrstoff- und Pestizidaustraege. *Oesterreichische Wasserwirtschaft* 55(5–6), 89–96.

Klik, A. and Truman, C.C. (2003) What is a typical heavy rainstorm? In: Gabriels, D. and Cornelis, W. (eds) *25 Years of Assessment of Erosion, Proceedings of the International Symposium*, 22–26 September 2003, Ghent, Belgium, pp. 93–98.

Klik, A., Zartl, A.S., Hebel, B. and Schmidt, J. (1998) *Comparing RUSLE, EROSION 2D/3D, and WEPP Soil Loss Calculations with Four Years of Observed Data*. Presentation at the 1998 International Annual

Meeting Sponsored by the American Society of Agricultural Engineers. Paper No. 982055, ASAE, St Joseph, Michigan.

Klik, A., Sokol, W. and Steindl, F. (2004) Automated erosion wheel: a new measuring device for field erosion plots. *Journal of Soil and Water Conservation* 59, 116–121.

Knisel, W.G. (ed.) (1980) *CREAMS: A Field-Scale Model for Chemicals, Runoff, and Erosion from Agricultural Management Systems*. Conservation Research Report No. 26, US Department of Agriculture, Washington, DC.

Lane L.J. and Nearing, M.A. (eds) (1989) *USDA-Water Erosion Prediction Project (WEPP): Hillslope Profile Model Documentation*. NSERL Report No. 2, USDA ARS National Soil Erosion Research Laboratory, West Lafayette, Indiana.

Larson, W.E., Pierce, F.J. and Dowdy, R.H. (1983) The threat of soil erosion to long-term crop production. *Science* 219, 458–465.

Leonard, R.A., Knisel, W.G. and Still, D.A. (1987) GLEAMS: Groundwater Loading Effects of Agricultural Management Systems. *Transactions of the ASAE* 30, 1403–1418.

McCuen, R.H. (1973) The role of sensitivity analysis in hydraulic modelling. *Journal of Hydrology* 18, 37–53.

Merritt, W.S., Letcher, R.A. and Jakeman, A.J. (2003) A review of erosion and sediment transport models. *Environmental Modelling and Software* 18, 761–799.

Morgan, R.C.P., Quinton, J.N. and Rickson, R.J. (1992) *The European Soil Erosion Model (EUROSEM): Documentation and Manual, Version 1.0*. Silsoe College, Cranfield University, Silsoe, UK.

Morgan, R.C.P., Quinton, J.N. and Rickson, R.J. (1993) *EUROSEM, a User Guide, Version 2*. Silsoe College, Cranfield University, Silsoe, UK.

Nash, J.E. and Sutcliffe, J.E. (1970) River flow forecasting through conceptual model. Part I. A discussion of principles. *Journal of Hydrology* 10, 282–290.

Nearing, M.A. (1998) Why soil erosion models over-predict small soil losses and under-predict large soil losses. *Catena* 32, 15–22.

Nearing, M.A., Deer-Ascough, L. and Laflen, J.M. (1990) Sensitivity of the WEPP hillslope version soil erosion model. *Transactions of the ASAE* 33, 839–849.

Oldeman, L.R., Hakkeling, R.T.A. and Sombroek, W.G. (1990) World Map of the Status of Human-Induced Soil Degradation: An Explanatory Note. Working Paper 90/7, ISRIC (International Soil Reference and Information Centre) – World Soil Information, Wageningen, The Netherlands.

ÖNORM (Österreichisches Normungsinstitut) (1988) ÖNORM L-1068: Physikalische Bodenuntersuchungen – Bestimmung der Korngroessenverteilung des mineralischen Feinbodens. Vienna.

ÖNORM (1999) ÖNORM L-1084: Chemische Bodenuntersuchung – Bestimmung von Carbonat. Vienna.

Pandey, A., Chowdary, V.M., Mal, B.C. and Billib, M. (2008) Runoff and sediment yield modelling from a small agricultural watershed in India using the WEPP model. *Journal of Hydrology* 348, 305–319.

Pieri, L., Bittelli, M., Wu, J.Q., Dun, S., Flanagan, D.C., Rossi Pisa, P., Ventura, F. and Salvatorelli, F. (2007) Using the Water Erosion Prediction Project (WEPP) model to simulate field-observed runoff and erosion in the Apennines mountain range, Italy. *Journal of Hydrology* 336, 84–97.

Pimental, D., Harvey, C., Resosudarmo, P., Sinclair, K., Kurz, D., McNair, M., Crist, S., Shpritz, L., Fitton, L., Saffouri, R. and Blair, R. (1995) Environmental and economic costs of soil erosion and conservation benefits. *Science* 267, 1117–1123.

Raclot, D. and Albergel, J. (2006) Runoff and water erosion modeling using WEPP on a Mediterranean cultivated catchment. *Physics and Chemistry of the Earth* 31, 1038–1947.

Rawls, W.J., and Brakensiek, D.L. (1983) A procedure to predict Green and Ampt infiltration parameters. In: *Proceedings of the American Society of Agricultural Engineers Conference on Advances in Infiltration*. American Society of Agricultural Engineers, St Joseph, Michigan, pp. 102–112.

Renard, K.G., Foster, G.R., Weesies, G.A., McCool, D.K. and Yoder, D.C, (1997) *Predicting Soil Erosion by Water: A Guide to Conservation Planning With the Revised Universal Soil Loss Equation (RUSLE)*. Agriculture Handbook No. 703, US Department of Agriculture, Washington, DC.

Rhoades, J.D. (1982) Cation exchange capacity. In: Klute, A., Miller, R.H. and Keeney, D.R. (eds) *Methods of Soil Analysis. Part 2. Chemical and Microbiological Properties*. Agronomy Monograph No. 9 (Part 2), 2nd edn. American Society of Agronomy, Madison, Wisconsin, pp. 149–157.

Salvador Munoz, J. (2003) Spatial and temporal distribution of soil properties with an agricultural used watershed. Master's Thesis, Institute of Hydraulics and Rural Water Management. University of Natural Resources and Applied Life Sciences Vienna, Austria.

Schmidt, J. (1991) A mathematical model to simulate rainfall erosion. In: Bork, H.-R., De Ploey, J. and Schick, A.P. (eds) *Erosion, Transport and Deposition Processes – Theories and Models. Catena*, Supplement 19, pp. 101–109.

Schmidt, J., Michael, A., Schmidt, W. and von Werner, M. (1997) EROSION 2D/3 – ein Computermodell zur Simulation der Bodenerosion durch Wasser. Sächsisches Landesamt für Umwelt und Geologie, Sächsische Landesanstalt für Landwirtschaft, Dresden.

Schroeder, A. (2000) WEPP, EUROSEM, E-2D: results of applications at the plot scale. In: Schmidt, J. (ed.) *Soil Erosion – Application of Physically Based Models*. Springer Verlag, Berlin-Heidelberg, pp. 199–250.

Sharpley, A.N. and Williams, J.R. (eds) (1990) *EPIC – Erosion/Productivity Impact Calculator. 1. Model Documentation*. USDA Technical Bulletin No. 1768, Washington, DC.

Shen, Z.Y., Gong, Y.W., Li, Y.H., Hong, Q., Xu, L. and Liu, R.M. (2009) A comparison of WEPP and SWAT for modelling soil erosion in the Zhangjiachong Watershed in the Three Gorges Reservoir Area. *Agricultural Water Management* 96, 1435–1442.

Soto, B. and Diaz-Fierros, F. (1998) Runoff and soil erosion from areas of burnt scrub: comparison of experimental results with those predicted by WEPP model. *Catena* 31, 257–270.

Stolpe, N.B. (2005) A comparison of the RUSLE, EPIC and WEPP erosion models as calibrated to climate and soil of south-central Chile. *Acta Agriculturae Scandinavica Section B – Soil and Plant* 55, 2–8.

Stone, J.J., Lane, L.J. and Shirley, E.D. (1992) Infiltration and runoff simulation on a plane. *Transactions of the ASAE* 35, 161–170.

Strauss, P. (2004) *Floodrisk – Teilprojekt Bodenerosion (Schwebstoffquelle)*. Endbericht, Institutes für Kulturtechnik und Bodenwasserhaushalt, Petzenkirchen, Austria.

Strauss, P., Auerswald, K., Blum, W.E.H. and Klaghofer, E. (1995) Erosivitaet von Niederschlaegen. Ein Vergleich Oesterreich-Bayern. *Zeitschrift fuer Kulturtechnik und Landentwicklung* 36, 304–309.

Takken, I., Govers, G., Jetten, V., Nachtergaele, J., Steegen, A. and Poesen, J. (2001) The effect of tillage on runoff and erosion patterns. *Soil and Tillage Research* 61, 55–60.

Thomann, R.V. (1982) Verification of water quality models. *Journal of the Environmental Engineering Division* 108, 923–940.

Tiwari, A.K., Risse, L.M. and Nearing, M.A. (2000) Evaluation of WEPP and its comparison with USLE and RUSLE. *Transactions of the ASAE* 43, 1129–1135.

USDA ARS (US Department of Agriculture Agricultural Research Service) (1994) *WEPP-Erosion Prediction Model. Version 94.3 User Summary*. NSERL Report No. 8, USDA ARS National Soil Erosion Research Laboratory, West Lafayette, Indiana.

USDA SCS (US Department of Agriculture, Soil Conservation Service) (1992) *Keys to Soil Taxonomy*. SMSS (USDA Soil Management Support Services) Technical Monograph No. 19, 5th edn, Pocahontas Press, Blacksburg, Virginia.

van Lynden, G.W.J. (1998) *European Soil Resources: Current Status of Soil Degradation in Europe: Causes, Impacts and Needs for Action*. Nature and Environment No. 71, Council of Europe, Strasbourg, France.

Warner, G.S., Stake, J.D., Guillard, K. and Neafsey, J. (1997) Evaluation of EPIC for a shallow New England soil. II. Soil nitrate. *Transactions of the ASAE* 40, 585–593.

Warrick, A.W. and Nielsen, D.R. (1980) Spatial variability of soil physical properties in the field. In: Hillel, D. (ed.) *Applications of Soil Physics*. Academic Press, New York, pp. 319–344.

Williams, J.R., Dyke, P.T., Fuchs, W.W., Benson, V.W., Rice, O.W. and Taylor, E.D. (1990) *EPIC – Erosion/Productivity Impact Calculator: 2. User Manual*. USDA Technical Bulletin No. 1768, Washington, DC.

Willmott, C.J. (1981) On the validation of models. *Physical Geography* 2, 184–194.

Wischmeier, W.H. and Smith, D.D. (1965) *Predicting Rainfall Erosion Losses from Cropland East of the Rocky Mountains – Guide for Selection of Practices for Soil and Water Conservation*. Agriculture Handbook No. 282, US Department of Agriculture, Washington, DC.

Wischmeier, W.H. and Smith, D.D. (1978) *Predicting Rainfall Erosion Losses – Guide to Conservation Planning*. Agriculture Handbook No. 537, US Department of Agriculture, Washington, DC.

Woolhiser, D.A. and Liggett, J.A. (1967) Unsteady, one-dimensional flow over a plane – the rising hydrograph. *Water Resources Research* 3, 753–771.

Yalin, M.S. (1963) An expression for bed-load transportation. *Journal of the Hydraulics Division, Proceedings of the American Society of Civil Engineers* 89, 221–250.

Young, R.A., Onstad, C.A., Bosch, D.D. and Anderson, W.P. (1987) *AGNPS – Agricultural Nonpoint Source Pollution Model, a Watershed Analysis Tool*. USDA Agricultural Research Service, Washington, DC.

Zhang, X.C., Nearing, M.A., Risse, L.M. and McGregor, K.C. (1996) Evaluation of WEPP runoff and soil loss predictions using natural runoff plot data. *Transactions of the ASAE* 39, 855–863.

7 Application of the SWAT Model for Ecohydrological Modelling in Germany

Martin Volk,* Nicola Fohrer, Britta Schmalz and Antje Ullrich

Introduction

The Soil and Water Assessment Tool (SWAT) model (Arnold *et al.*, 1998; Arnold and Fohrer, 2005) has proved to be an effective tool for assessing water resource and non-point-source pollution problems for a wide range of scales and environmental conditions across the globe (Gassman *et al.*, 2007). Many of the applications have been driven by the needs of various government agencies, particularly in the USA and the European Union (EU), that require direct assessments of anthropogenic, climate change and other influences on a wide range of water resources, or exploratory assessments of model capabilities for potential future applications. SWAT has also been used extensively in Europe, including in projects supported by different European Commission (EC) agencies.

Several models, including SWAT, were used to quantify the impacts of climate change for five different watersheds in Europe within the Climate Hydrochemistry and Economics of Surface-water Systems (CHESS) project, which was sponsored by the EC Environment and Climate Research Programme (Boorman, 2003). A suite of nine models, also including SWAT, was tested in 17 different European watersheds as part of the EUROHARP project, which was sponsored by the EC Energy, Environment and Sustainable Development (EESD) Programme (Kronvang *et al.*, 2009). The goal of the research was to assess the ability of the models to estimate non-point-source nitrogen and phosphorus losses to both freshwater streams and coastal waters. The EUROHARP project was presented in nine scientific papers included in a themed issue of the *Journal of Environmental Monitoring* in 2009 (11(3), pp. 503–609), which was another main achievement of the project; the editorial introduction to this (Kronvang *et al.*, 2009) gives an outline of the nine papers presented. The wide use of SWAT in Europe is documented by the European international SWAT conferences, which are reported for the 1st conference (held in Giessen, Germany in 2005) in a special issue of *Hydrological Processes* (19(3), with an introductory paper by Arnold and Fohrer, 2005). The proceedings for the succeeding conferences are also published: the 2nd, held in Bari, Italy, in 2003 (Srinivasan *et al.*, 2003), the 3rd, held in Zurich, Switzerland, in 2005 (Srinivasan *et al.*, 2005), the 4th, held in Delft, The Netherlands, in 2007 (UNESCO-IHE, 2007), the 5th, held in Boulder, Colorado, in 2009 (Srinivasan, 2009) and the 6th, held in Seoul, Korea, in 2010 (Kim and Srinivasan, 2010). In addition, a SWAT model developer's workshop was held in 2008 in Potsdam, Germany, and the proceedings were published in a special

* Corresponding author: Martin.volk@ufz.de

issue of *Hydrological Sciences Journal* (53(5)), with 12 papers that include an introductory review (Krysanova and Arnold, 2008).

For the special requirements of catchments in Europe, SWAT has been modified, supplemented or has formed the basis for new model developments. Krysanova *et al.* (1998) published the Soil and Water Integrated Model (SWIM), which is based on the hydrological components of SWAT and the nutrient modules of the model MATSALU (Krysanova *et al.*, 1989). MATSALU was named after the study area Matsalu Bay in Estonia. SWIM (Krysanova *et al.*, 1998) is often used for simulating the impact of climate change on water and nutrient fluxes as well as within the context of the implementation of the European Water Framework Directive (WFD; EC, 2000). For instance, Krysanova *et al.* (2005) reported the impacts of 12 different climate scenarios on the hydrological balance and crop yields for a 30,000 km^2 watershed in the state of Brandenburg in Germany using the SWIM model. Further uncertainty analysis of climate change was performed by Krysanova *et al.* (2007) for the 100,000 km^2 Elbe River basin in eastern Germany, based on an interface between a downscaled General Circulation Model (GCM) scenario and SWIM.

The modification of surface and river processes, especially a reduction in the time step of the rainfall/runoff module to a user-defined fraction of an hour, and the development of an hourly river routing and water quality module, led to the publication of the model ESWAT by van Griensven and Bauwens (2005). SWAT-G (Eckhardt *et al.*, 2002) was compiled for the application in low mountain range areas with high proportions of interflow. Lenhart *et al.* (2005) added an improved sediment concept to SWAT-G, and an extensive sensitivity analysis for SWAT-G was presented by Lenhart *et al.* (2002). Autocalibration of SWAT has been carried out by Eckhardt *et al.* (2005) and van Griensven and Bauwens (2003), and the model has also been integrated into an interdisciplinary modelling tool – as shown in Weber *et al.* (2001) and Fohrer *et al.* (2001) – to study the effects of land-use change.

Pohlert *et al.* (2007a) extended SWAT with algorithms from a detailed nitrogen turnover model to enhance model performance with regard to the prediction of nitrogen leaching. The new model, which is further referred to as SWAT-N, includes algorithms for decomposition, growth of nitrifying bacteria, nitrification, nitrificatory as well as denitrificatory N emissions, N uptake by plants and N transport due to water fluxes. The model was tested with a lysimeter data set of a long-term fertilization experiment including crop rotation conducted in eastern Germany. Pohlert *et al.* (2007b) evaluated the performance of the SWAT-N model for discharge and nitrate predictions at the mesoscale Dill catchment in Germany for a 5-year period. They concluded that the model efficiency of SWAT-N was sufficient for the assessment of scenarios for daily discharge predictions. SWAT-N can be employed without further calibration for nitrate load simulations on both a weekly and monthly basis with an acceptable degree of accuracy. However, the efficiency of the model for daily nitrate load was insufficient, which can be attributed to both data uncertainty (i.e. point-source effluents and actual farming practice) and structural errors.

A review within the SWAT literature database (SWAT, 2010) shows that at the time of writing (September 2010) 56 published studies are related to applications of SWAT in Germany. Several of these studies use SWAT for simulating the impact of land-use and land-management changes on hydrology, water quality and nutrient fluxes (Weber *et al.*, 2001; Haverkamp *et al.*, 2005), mostly in the context of the WFD (Fohrer *et al.*, 2005; Hörmann *et al.*, 2005; Volk *et al.*, 2007, 2008, 2009; Huang *et al.*, 2009). Recently, SWAT has been also applied in combination with hydrodynamic models to assess the impact of global change on aquatic habitats (Kiesel *et al.*, 2009).

This chapter gives an overview of the application of SWAT for ecohydrological modelling in Germany. Because of its importance related to the application of SWAT in Europe and Germany, the next section gives an overview on the WFD programme, followed by the introduction of the SWAT model. After an overview of studies on the simulation of land management and tillage on hydrology and river water quality, the following section describes the results and challenges of model-based scenario simulations for river basin management. The two studies that are presented were carried out within the context of the WFD: one is a 3500 km^2 agricultural river basin known as the Upper Ems River Basin; the other is a 50 km^2 lowland catchment in northern Germany (Kielstau catchment). The following section includes supporting studies in which the work on sensitivity analyses of management parameters in SWAT and on the impact of water-quality monitoring strategies and load-estimation methods on model calibration and evaluation is presented.

The European Water Framework Directive

At the European level, the WFD (2000/60/EC; EC, 2000) adopted in 2000 is the first European directive to explicitly recognize the importance of the interdependency between aquatic ecosystems and their socio-economic values; it advocates a more integrated river basin approach to water policy. Investments and water resource allocations in river basin management plans are guided by their cost-effectiveness and cost recovery (Brouwer *et al.*, 2008).

The implementation of the WFD poses significant new challenges to water managers, planning authorities, researchers and stakeholders (Hirschfeld *et al.*, 2005; Jessel and Jacobs, 2005). This is accomplished through the following: (i) a focus on river basins and surface water bodies as reference units (instead of administrative units); (ii) the consideration of natural scientific as well as socio-economic aspects; and (iii) an emphasis on stakeholder interaction. Van der Helm (2003) states that 'the entire procedure and water policy have to be submitted to active involvement of all interested parties', or, in other words, large-scale stakeholder interaction. Member states are required to promote active involvement of all interested parties in the implementation of the WFD (Article 14), which states that all waters in the EU should reach a good status by 2015 (EC, 2000). 'Good status' is a general term meaning the status achieved by a surface water body when both the ecological status and its chemical status are at least good, or for groundwater when both its quantitative status and chemical status are at least good. In Germany, the 'good status' corresponds to water quality class II according to Germany's Working Group of the Federal States on Water Problems Issues (LAWA, 1998). The LAWA water-quality classification corresponds to the classification used by the WFD. The limit values for selected nutrients are listed in Table 7.1.

To achieve this goal, the member states need to set up river basin districts, each one with a management plan that includes a programme of measures that will achieve good status in the most cost-effective manner. This involves an evaluation of different policy measures with respect to their effects on nutrient reducing procedures, habitat suitability and economic consequences, upon which policy makers can base their decisions. Decision support systems are supposed to support the implementation of the Programme of Measures (PoM) according to the WFD, with special focus on impact analysis, including nutrient reduction (contribution to the environmental objectives) and economic analysis (costs of the measure) (Volk *et al.*, 2010). In general, the methodological approach for decision support structures in the implementation of the PoM is in six phases (see Fig. 7.1). The successful implementation of the WFD requires appropriate mathematical models and other tools to manage the different phases of the planning procedure and to support decision making at various steps of the implementation process (Rekolainen *et al.*, 2003).

Table 7.1. Limit values for nitrogen and phosphorus, water quality class II (LAWA, 1998).

Nutrient	Total N	NO_3-N	NO_2-N	NH_4-N	Total P
Maximum concentration (mm l^{-1})	≤3.0	≤2.5	≤0.1	≤0.3	≤0.15

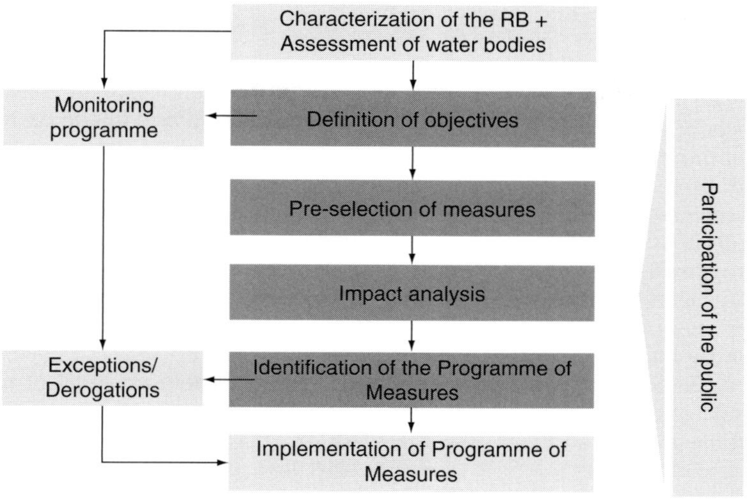

Fig. 7.1. Structure of the Programme of Measures (PoM) with steps in the planning process (RB = river basin) (Rode *et al.*, 2008).

Materials and Methods: the SWAT Model

The SWAT model is considered to be one of the most suitable models for predicting the long-term impacts of land management measures on water, sediment and agricultural chemical yield (nutrient loss) in large complex watersheds with varying soils, land-use and management conditions (Arnold and Fohrer, 2005; Behera and Panda, 2006; Gassman *et al.*, 2007). It has gained international acceptance as a robust interdisciplinary watershed modelling tool (Gassman *et al.*, 2007). SWAT is a partly physically based, partly conceptual, continuous-time river basin model with spatial distributed parameters operating on a daily time step. It is not designed to simulate detailed, single-event flood routing (Neitsch *et al.*, 2002).

The SWAT model integrates all relevant ecohydrological processes, including water flow, nutrient transport and turnover, vegetation growth and land-use and water management at the sub-basin scale. Consequently, the watershed is subdivided into sub-basins based on the number of tributaries. The size and number of sub-basins is variable, depending on the stream network and the size of the entire watershed. Sub-basins are further disaggregated into classes of Hydrological Response Units (HRUs), whereby each unique combination of the underlying geographical maps of soils, land use, slope, etc., form one class. HRUs are the spatial units where flows, sediment, nutrients, bacteria and pesticides are calculated, which are then aggregated and summed for each sub-basin. Water and constituent loadings exported from each HRU are then routed to the corresponding sub-watershed outlet. The HRUs in SWAT are spatially implicit; i.e. their exact position in the landscape is unknown, and thus a single HRU can represent multiple disaggregated areas within a given sub-basin that consist of homogeneous soil, land-use and management characteristics (Di Luzio *et al.*, 2005; Neitsch *et al.*, 2005).

The water balance for each HRU is represented by four storages that include snow, soil profile, shallow aquifer and deep aquifer. The soil profile can be subdivided into as many as ten soil layers. Soil water processes include evaporation, surface runoff, infiltration, plant uptake, lateral flow and percolation to lower layers (Arnold and Allen, 1996; Neitsch *et al.*, 2005). The surface runoff from daily rainfall is estimated with a modification of the SCS curve number method from the United States Department of Agriculture Soil Conservation Service (USDA SCS) (Arnold and Allen, 1996; Neitsch *et al.*, 2002).

Nitrogen movement and transformation are simulated as a function of the nitrogen cycle (Jha *et al.*, 2004; Neitsch *et al.*, 2004). The SWAT model monitors five different pools of nitrogen in the soils: two inorganic (ammonium (NH_4^+) and nitrate (NO_3^-)) and three organic (fresh organic nitrogen (associated with crop residue and microbial biomass) and active and stable organic nitrogen (associated with the soil humus)). Nitrogen is added to the soil by fertilizer, manure or residue application, fixation by bacteria and precipitation (Neitsch *et al.*, 2005). Nitrogen losses occur by plant uptake and by surface runoff in the solution and the eroded sediment (Jha *et al.*, 2004; Neitsch *et al.*, 2005).

The SWAT crop growth sub-model (Neitsch et al., 2004) is based on the Environmental Policy Impact Climate (EPIC) crop growth module (Williams *et al.*, 1989). EPIC originally stood for 'Erosion Productivity Impact Calculator', but the name was later changed (Williams *et al.*, 1996, 2008; Gassman *et al.*, 2005). EPIC is a comprehensive field-scale model that was originally developed to simulate the impact of soil erosion on crop productivity and has now evolved into a comprehensive agricultural management, field-scale and non-point-source loading model (Williams, 1990; Williams *et al.*, 1996, 2008; Gassman *et al.*, 2005). The SWAT crop growth sub-model features a generic crop growth routine that supports the simulation of dozens of crops, grasses and trees in either monocultures or complex crop rotations, including rotations with winter cover crops.

The conservation and management practice algorithms of SWAT are also based in part on approaches used in EPIC, and are defined by specific management operations (e.g. the beginning and end of the growing season, timing of tillage operations, as well as the timing and amount of application of fertilizer, pesticide and irrigation). These management operations can be simulated in any HRU, as appropriate. The operations are, in turn, defined by specific management parameters (e.g. tillage depth, biological soil mixing efficiency, etc.) (Neitsch *et al.*, 2002).

Results and Discussion

Overview: simulation of effects of land management and tillage on hydrology and river water quality

Owing to the insufficient water quality of many European streams, environmental programmes such as the WFD were implemented to obtain acceptable ecological and chemical water quality for groundwater and surface water bodies (EC, 2000; Rekolainen *et al.*, 2003). The main nutrient inputs to EU waters come from non-point-source pollution, mainly from intensive agricultural activities (Behrendt *et al.*, 1999). Therefore, alternative land-management practices are increasingly used to reduce non-point-source pollution. Arabi *et al.* (2008) developed and evaluated a method for the representation of several agricultural conservation practices ('Best Management Practices'; BMPs) with SWAT. These include seven practices that are installed in upland areas: contour farming, strip cropping, parallel terraces, cover crops, residue management, field borders and filter strips. In contrast, grassed waterways, lined waterways and grade stabilization structures are practices that are implemented within small channels. The representation procedure entails identifying hydrological and water-quality processes that are affected by practice implementation, selecting SWAT parameters that represent the affected processes, performing a sensitivity analysis to ascertain the sensitivity of model outputs to selected parameters, adjusting the selected parameters based on the function of conservation practices, and verifying the reasonableness of the SWAT results. The methods developed in the study by Arabi *et al.* (2008) can be applied with other watershed models that employ similar underlying equations to represent hydrological and water-quality processes. White and Arnold (2009) improved the algorithms being developed for SWAT to better represent BMPs (filter strips and buffers).

Another option for reducing non-point-source pollution is the reduction of soil tillage intensity, which positively affects numerous soil properties, such as aggregate stability, macroporosity and saturated hydraulic conductivity, and consequently increases infiltration rates and reduces surface runoff, nutrient loss and soil erosion (Jones *et al.*, 1969; Pitkänen and Nuutinen, 1998; Schmidt *et al.*, 2001; Kirsch *et al.*, 2002; Pandey *et al.*, 2005; Tripathi *et al.*, 2005). Alternative land-management practices may include reduced tillage such as conservation tillage, e.g. without deep ploughing, field preparation just before planting or no-tillage (direct drilling) (Sullivan, 2003; KBDS, 2011). In Germany, the implementation of alternative tillage systems is increasingly supported by agro-environmental programmes. In the German State of Saxony, for instance, conservation tillage and mulch seeding on arable land have increased from < 1% to about 27% during 1994 to 2004 with support from the Saxonian Program for Environmental Agriculture (LfL, 2011). A number of field studies have illustrated the positive effects of conservation tillage and no-tillage practices on water and material fluxes at the field local level (e.g. Sloot *et al.*, 1994; King *et al.*, 1996; Schmidt *et al.*, 2001), but these effects need to be assessed at the watershed level to guide river basin management programmes, as WFD has claimed (Kirsch *et al.*, 2002; Chaplot *et al.*, 2004; Pandey *et al.*, 2005; Behera and Panda, 2006; Bracmort *et al.*, 2006). Therefore, watershed models are useful tools and have been used for decades to evaluate non-point-source pollution and the short- and long-term impacts of alternative management practices.

Gassman *et al.* (2007) point out that a key strength of SWAT is a flexible framework that allows the simulation of a wide variety of conservation practices and other BMPs, such as fertilizer and manure application rate and timing, cover crops (perennial grasses), filter strips, conservation tillage, irrigation management, flood-prevention structures, grassed waterways and wetlands. The majority of conservation practices can be simulated in SWAT with straightforward parameter changes. Hence, many studies have used SWAT to evaluate the effects of land-use scenarios and BMPs on water quality and soil loss (Saleh *et al.*, 2000; Fohrer *et al.*, 2001, 2005; Santhi *et al.*, 2001, 2003; Kirsch *et al.*, 2002; Vache *et al.*, 2002; Chaplot *et al.*, 2004; Pandey *et al.*, 2005; Tripathi *et al.*, 2005; Arabi *et al.*, 2006, 2007; Behera and Panda, 2006, Bracmort *et al.*, 2006; Rode *et al.*, 2008; Volk *et al.*, 2008).

Working on the watershed scale – for which the SWAT model is developed – means that the required input data are often aggregated in terms of temporal scale (e.g. daily climate data). In contrast, land-management parameters, i.e. tillage, fertilization, crop rotation, etc., can be included in high resolution and detail owing to the modular structure of the model and its historical development based on the EPIC model (Benson *et al.*, 1988; Neitsch *et al.*, 2002; Arnold and Fohrer, 2005; Gassman *et al.*, 2007). Furthermore, modelling evaluations of conservation management effects at the watershed scale are limited by the lack of management operation data.

Land-use and land-management scenarios to improve hydrology and water quality

In the following two sections two case studies are presented that developed and analysed land-use and land-management scenarios to improve hydrology and water quality by using SWAT. First, a SWAT application in the Upper Ems River Basin in north-western Germany is presented. The basin is a predominantly flat landscape with widespread permeable sandy soils. The Ems basin is situated in one of the most intensive agricultural regions in Europe. Arable land covers approximately 77.2% of the area (the average in Germany is 50%; BMELV, 2008), which has led to a dramatic loss of landscape diversity. The proportions of the other land-use types are 9.9% for forest, 8.9% for urban areas, 3.9% for pasture and 0.1% for other areas. Intensive livestock production has contributed to severe environmental problems, as evidenced by the nitrogen value of water quality class II being exceeded by a factor three to four for some Ems River gauges.

The second case study applied SWAT in the Kielstau catchment, which has a size of approximately 50 km^2 and is located in the North German lowlands. Land use is dominated by arable land and pasture. The arable land area occupies over 56%, and pasture over 26% of the catchment area. The dominant soils of the Kielstau catchment are Stagnic Luvisols and Haplic Luvisols. In the catchment, diffuse source pollution of nutrients comes mainly from fertilizer application or from animal husbandry in the vicinity of the river, as well as from urban areas. The combination of these diffuse sources and point sources – such as six waste-water treatment plants – influences the in-stream water quality considerably (Schmalz *et al.*, 2007).

The objective of these two studies was to determine land-use scenarios with the corresponding land- (or water-) management practices to achieve the limit value for nitrogen and phosphorus concentrations of water quality class II (see Table 7.1; LAWA, 1998).

Case study of Upper Ems River Basin

Volk *et al.* (2008) present the results of the FLUMAGIS project, in which they developed a spatial decision support system (SDSS) to support the implementation of the WFD. FLUMAGIS is an acronym for 'Interdisziplinäre Entwicklung von Methoden und Werkzeugen für das FLUsseinzugsgebietsMAnagement mit Geo-InformationsSystemen' (Interdisciplinary development of methods and tools for the planning process and measurement control for river basin management with geoinformation systems) (FLUMAGIS, 2009). The rivers in the intensively cropped Upper Ems River basin showed total nitrogen concentrations in excess of 5–10 mg l^{-1}, i.e. two to three times higher than the defined environmental target for the WFD, with some values exceeding the target value by up to four times. An objective of the project was to find a land-use and land-management scenario that would reduce the total nitrogen concentration to meet the WFD requirements for good ecological and chemical status (Volk *et al.*, 2009).

Within the FLUMAGIS project, Volk *et al.* (2009) developed consecutive land-use and management scenarios on the basis of policy instruments such as the support of agro-environmental measures by Common Agricultural Policy and regional landscape development programmes. The model simulations were done by using SWAT. Table 7.2 summarizes the agro-environmental measures that were considered in the model. These scenarios developed successively in the direction of a target scenario that would finally nearly match the water-quality objective of the WFD. The areas for the necessary land-use changes were mainly chosen by catchment characteristics, including permeability of the soils and groundwater table, and by the degree of human impairment, including river channel regulation and nutrient leaching.

In order to achieve the good ecological status for nitrogen at the Upper Ems River, the nitrogen concentration has to be reduced by 50% of the mean annual average. This would require substantial and expensive changes of land-use and land-management intensity as well as changes of the river morphology. Figure 7.2 shows the current conditions and the target scenario (Scenario 8) that best matches the requirements of water quality class II for nitrogen. In addition to river channel changes, this scenario includes the reduction of conventional arable land from 77% to 33% conventional arable land + 13% conservation farming (46% arable land in total), and increases of pasture from 4% to 15%, afforestation from 10% to 21%, and protected wetlands from 0% to 9%. However, this is not realistic from an economic point of view, because these drastic cuts would be so strong for farmers that most of them would have to abandon their operations. Agro-economic calculations have shown, for instance, that the above-mentioned changes in the flood plains alone would cost between 500 and 800 euros per hectare per year (31.6 million euros a year) (Volk *et al.*, 2008) depending on regional soil qualities and management intensities. This measure is expected to result in intense conflicts with affected farmers (Volk *et al.*, 2008).

The results of SWAT scenario calculations showed that drastic measures, which are unrealistic from a socio-economic point of view, would be needed to achieve the water-quality targets in the basin. The example showed additionally that achieving the WFD targets is only possible with a consideration of regional landscape and land-use distinctions. A related problem yet to be addressed is the

Table 7.2. Implementation of agro-environmental measures for the Upper Ems River, Germany in the SWAT model.

Scenario	Measure	Implementation in the model
1	Increase of pasture/decrease of arable land	Modification of the land-use file (Shape-files). Allocation of the land-use type pasture on former arable land.
2	Extensification of pasture	Reduction of livestock density to 1.4 livestock units per hectare by modification of the management scenarios. Reduction of the amount of fertilizers.
3	Afforestation of arable land	Modification of the land-use file (Shape-files). Allocation of the land-use type forest on former arable land.
4/5	Implementation of conservation tillage practices	No ploughing, only cultivator and harrow (reduction of tillage depth), reduction of mixing efficiency.
	Extensification of arable land	Reduction of the amount of applied mineral fertilizers in accordance with 0.7 livestock units.
	Modification of crop rotation schemes	Implementation of complex management scenarios over several years with short fallow periods.
6	Oxbow reconnection (improvement of river morphology)	Increase of the river length at the HRU[a] level. Modification of Manning's roughness coefficient '*n*' (CH_N(2)) for main channel flow.
	Riparian buffer strips	Modification of the SWAT parameter FILTERW.
7	Abandonment of the flood-plain use	Conversion of arable land and pasture in riparian zones to areas without management. Change of flood-plain pasture into wetland (according to soil and groundwater conditions).
8	Additional increase of pasture/ decrease of arable land	Conversion of arable land to pasture (randomly chosen).

[a]HRU, Hydrological Response Unit.

general lack of measured water-quality data with which to calibrate and validate water-quality models such as SWAT. This adds considerable uncertainty to already complicated and nearly unpredictable situations. Thus, improved strategies for water-quality monitoring and data accessibility must be established (see the section on 'Supporting Studies').

Case study of Kielstau River Basin

A large number of measurement campaigns and analyses related to agricultural pollutants have been conducted in the lowland catchment of the Kielstau River in northern Germany. Schmalz and Fohrer (2010) give a conclusion of the main studies and their results in this investigation area. Since 2005, the Department of Hydrology and Water Resources Management of the Ecology Centre of Kiel University has investigated the water balance and water quality in detail. At the catchment outlet, water levels, flow velocity and discharge, as well as water quality and suspended sediment, were measured. In addition, water-quality and sediment measurements were made at the tributary Moorau (i) in the catchment (e.g. at the outlet), (ii) at the Kielstau main stream along longitudinal river transects, (iii) at the Moorau along longitudinal river transects, and (iv) at drainage systems. Further activities concern river morphology, bed load and erosion measurements and mapping. A drained riparian wetland used as grassland was chosen to investigate the close interactions between groundwater, drainage ditches and the stream.

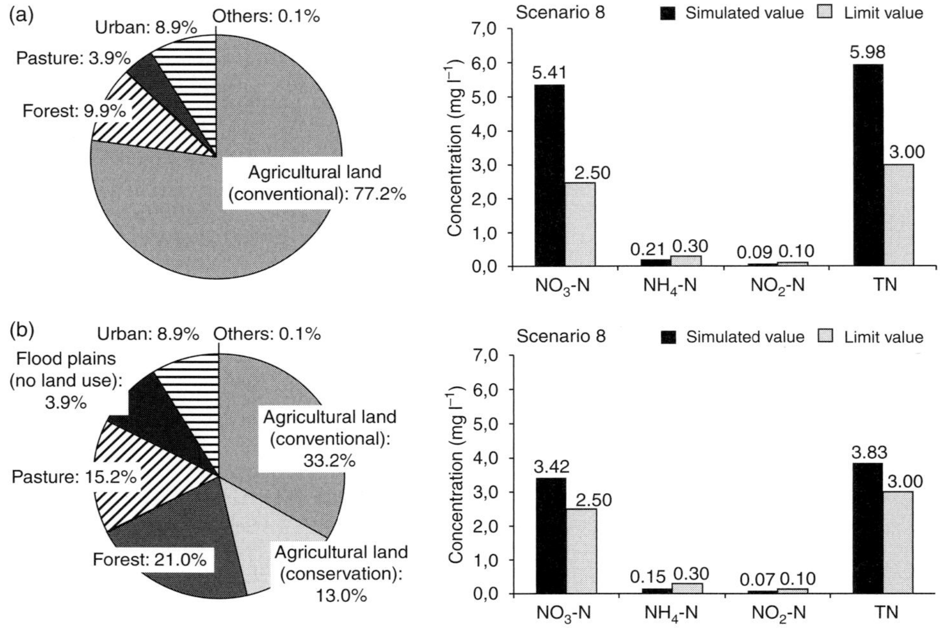

Fig. 7.2. Current land-use distribution and nitrogen concentrations (A, Scenario 0) and (B, Scenario 8) with the land-use distribution and simulated nitrogen concentrations to achieve good ecological status for the Upper Ems River in Germany. Total nitrogen (TN) values include 0.26 mg l^{-1} organic nitrogen for (a) and 0.2 mg l^{-1} organic nitrogen for (b).

Most of the 50 km^2 area is used as arable land (56%) mainly with winter wheat, oilseed rape and maize as well as pasture (26%). The water quality at the catchment outlet showed median total nitrogen (TN) concentrations of 5.50 mg l^{-1} for 2006–2008 on a daily resolution with a range from 1.90 to 15.60 mg l^{-1}, and median nitrate-N concentrations of 4.49 mg l^{-1} for 2006–2008 on a daily time step with a range from 1.35 to 10.05 mg l^{-1}. The objectives of the studies were (i) to find a land-use and land-management scenario that would reduce the nutrient concentration to meet the WFD requirements for good ecological and chemical status and (ii) to investigate the impact of a prospective increasing use of bioenergy plants in the catchment. The model simulations were carried out using SWAT.

In order to reduce nutrient concentrations and to achieve water-quality improvements in the Kielstau catchment, Lam *et al.* (2011) developed consecutive land-use and management scenarios considering BMPs. Measures including extensive land-use management, grazing management practice, field buffer strips and nutrient management plans were developed to investigate their impacts on pollutant discharge from diffuse sources (Table 7.3). Five scenarios were simulated, and discharge, sediment and nutrient loads for 2006–2008 were compared with the baseline, which is characterized by the actual land use and management.

The model results showed that the implementation of BMPs was very effective in reducing NO_3-N and total N load (Fig. 7.3). However, the impacts of these BMPs were not significant on the reduction of sediment and total P load owing to the specific lowland characteristics of the area, e.g. flat topography and low surface runoff. Changes in sediment loads due to BMPs are so low in absolute number that they cannot be evaluated with the given model accuracy for the sediment simulations. The same is true for the reduction of sediment-bound phosphorus.

Table 7.3. Description of best management practices (BMPs) simulated by the SWAT model for the Kielstau catchment, Germany (Lam *et al.*, 2010b).

Measure	Implementation in the model	Scenario
Extensive land-use management	Combination of different crop rotations and tillage	ELUM
Nutrient management plan	Decrease of nutrient application (both fertilizer and manure) in agricultural land	NMP
Grazing management practice	Reduction of livestock density from 2 LU[a] ha^{-1} to 1.1 LU ha^{-1} and no fertilizer application on pasture land	GZM
Field buffer strip	Application of 10 m edge-of-field buffer strips on arable and pasture land along main channel segments	FBS
Combination scenarios	Combination of the most efficient scenarios	CBN

[a]LU, livestock units.

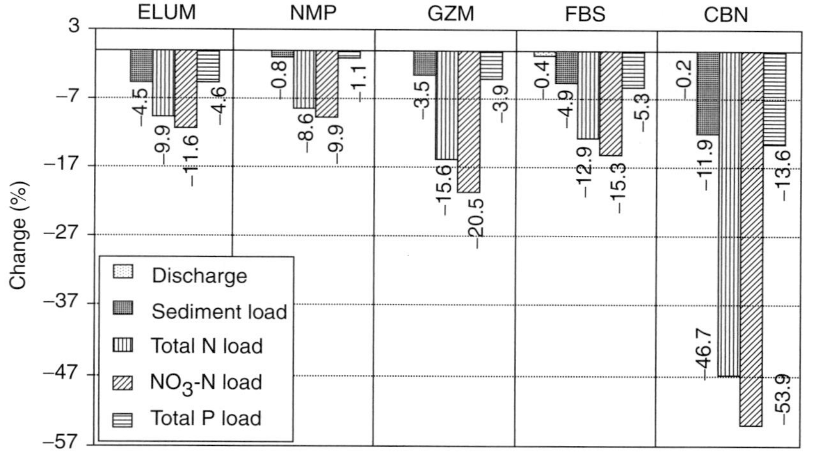

Fig. 7.3. Average annual reduction in sediment and nutrient load at the outlet of the Kielstau catchment, Germany by implementing four best management practices (BMPs) (after Lam *et al.*, 2010b). ELUM, extensive land-use management; NMP, nutrient management plan; GZM, grazing management practice; FBS, field buffer strip; CBN, combination scenarios.

The results also indicated that the combined scenarios provided the most load reduction in the average annual load for NO_3-N and total N at the outlet of the catchment: 53.9% and 46.7%, respectively. Moreover, the concentrations of NO_3-N and total N decreased significantly and reached nearly the LAWA (1998) class II (moderately polluted) at the outlet of the watershed (Fig. 7.4). Thus, these improvements nearly met the WFD target value for water quality by implementing simultaneous BMPs. The combined scenarios provided the highest load reduction of nutrients, but also the highest corresponding costs per year (Fig. 7.4). The trade-off between cost and effectiveness of BMPs will be helpful for policy makers and stakeholders in identifying suitable BMPs for improving water quality in the Kielstau catchment (Lam *et al.*, 2011).

In a further simulation, the potential environmental effects of an increased production of energy crops on the catchment were also evaluated. The two most important energy crops in the catchment are maize silage and oilseed rape. In the Kielstau catchment, shares of 13.3% of the total area (15.7% of the agricultural area) for oilseed rape and 10.3% of the total area (12.1% of the agricultural area) for maize silage have been captured (own land-use mapping

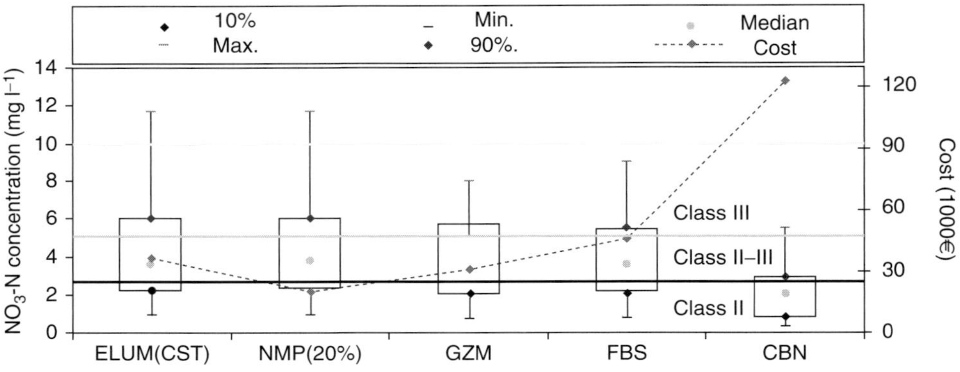

Fig. 7.4. Concentration of NO_3-N (mg l^{-1}) at the Kielstau catchment (Germany) outlet under different best management practices (BMPs) and their positioning within LAWA (1998) water quality classes. The solid lines indicate the respective water quality classes II (moderately polluted) and II–III (critically polluted) and III (heavily contaminated). The dotted line represents the cost of BMPs (after Lam *et al.*, 2010b). ELUM(CST), an extensive land-use management which is a combination of conservation tillage and a winter rye/rye rotation; NMP, nutrient management plan with nutrient application rates (both mineral fertilizer and manure) deceased to 20%; GZM, grazing management practice; FBS, field buffer strip; CBN, combination scenarios.

summer in 2008; Golon, 2009). Different land-use scenarios simulated increased as well as decreased extents of cultivation of maize silage and oilseed rape. The scenarios represented the following changes to the actual land use for 2008: Scenario 1, one-third of the cropland is maize silage; Scenario 2, one-third in each case of the cropland is maize silage and oilseed rape, and Scenario 3, there are no energy crops, with only about 6.5 % of the cropland as maize for fodder production (Table 7.4).

These three land-use scenarios and their effects were evaluated according to the resulting river NO_3-N loads and concentrations from the respective model runs. The land-use scenarios resulted in relative changes of the output variables from −6.4 to +5.4% for mean annual NO_3-N loads, and from −4.2 to +4.6% for mean annual NO_3-N concentrations. The scenarios that implemented an increased extent of cultivation for the two energy crops resulted in higher mean annual NO_3-N loads and concentrations. Simulated lower shares of these crops than those observed in the catchment in 2008 resulted in reduced values for both variables (Golon, 2009; Table 7.5, Fig. 7.5).

Supporting Studies

Efficient and accurate ecohydrological simulations require knowledge about the sensitivity of the model, adaptation options of the model to specific basin characteristics and the impact of water-quality monitoring strategies on model calibration and evaluation. This section presents some examples of such supporting studies in Germany that help to improve the accuracy and efficiency of SWAT model simulations.

Sensitivity analysis

With regard to evaluations of the effects of BMPs at the watershed scale, knowledge is needed of the sensitivity of models like SWAT to management parameters and practice to improve the efficiency of model parameterization and the quality of model calibration. Furthermore, potential simulation uncertainties based on ranges of realistic parameter values and on influences of scale need to be understood because simulated effects

Table 7.4. Land-use scenarios and actual land use (base scenario) compared as share of arable land (%) for the Kielstau catchment, Germany (Golon, 2009).

Crops	Maize silage	Oilseed rape	Winter wheat	Winter barley
Actual land use 2008	16	21	34	24
Scenario 1	31	21	25	18
Scenario 2	31	31	19	14
Scenario 3	7	0	55	33

Table 7.5. Average daily NO_3-N loads (kg day^{-1}) and average daily NO_3-N concentrations (mg l^{-1}) of the base scenario (actual land use 2008) and Scenarios 1–3 for the SWAT modelled catchment outlet, and relative changes (%) compared with the base scenario for the Kielstau catchment, Germany (Golon, 2009). Scenario 1, 1/3 maize silage; Scenario 2, 1/3 each maize silage and oilseed rape; Scenario 3, no energy crops.

Parameter	Base scenario	Scenario 1	Scenario 2	Scenario 3
Average daily NO_3-N loads	209.6	215.5	220.9	196.2
Change in NO_3-N load (%) from base scenario	–	2.8	5.4	–6.4
Average daily NO_3-N concentrations	5.8	6.0	6.1	5.6
Change in NO_3-N load (%) from base scenario	–	2.6	4.6	–4.2

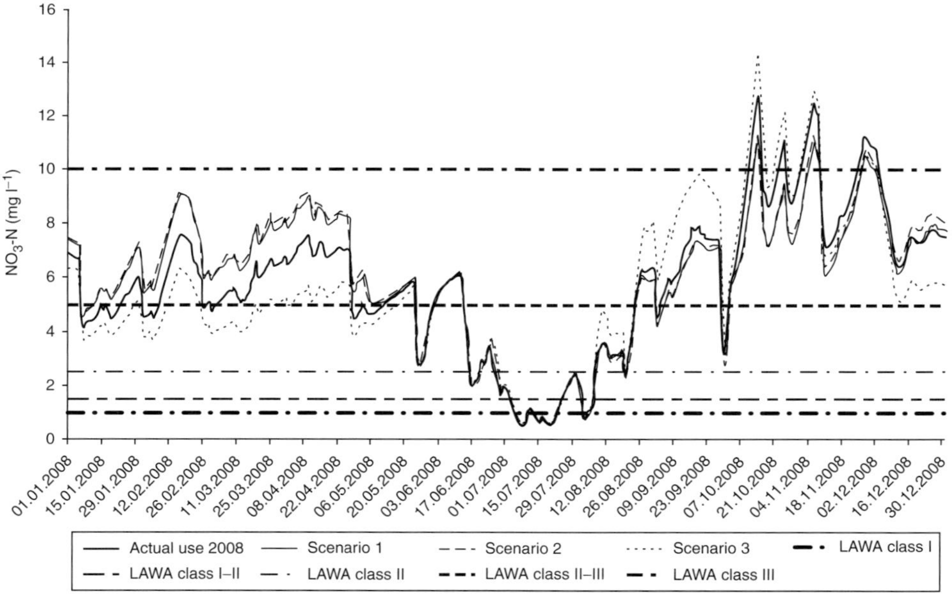

Fig. 7.5. NO_3-N concentrations (mg l^{-1}) model output for base scenario (actual land use 2008) and Scenarios 1–3 for the Kielstau catchment, Germany. Category thresholds for river chemical burden of NO_3-N concentrations according to LAWA (1998) (Golon, 2009). Scenario 1, 1/3 maize silage; Scenario 2, 1/3 each maize silage and oilseed rape; Scenario 3, no energy crops.

often drive financial and political decisions (Onatski and Williams, 2003).

Ullrich and Volk (2009) presented a study in which the first step consists of a sensitivity analysis for conservation management parameters in SWAT, specifically tillage depth, mechanical soil mixing efficiency, biological soil mixing efficiency, USDA SCS curve number (CN), Manning's roughness coefficient '*n*' for overland flow, the USLE (Universal Soil Loss Equation) support practice factor (P), and filter strip width. With this analysis, they aimed to improve model parameterization and calibration efficiency. In contrast to less sensitive parameters, such as tillage depth and mixing efficiency, they parameterized sensitive parameters such as CN in detail. Further analysis consisted of varying management practices, including conventional tillage, conservation tillage and no-tillage for different crops, i.e. spring barley, winter barley and sugarbeet, and varying operation dates. Results showed that the model is very sensitive to applied crop rotations, crop growth period, soil cover characteristics of applied crops and, in some cases, even to small variations in management practices.

Enhancing SWAT – incorporating tile drains and depressions

Hydrological models need to be adapted to specific hydrological characteristics of the catchment in which they are applied. In lowland regions of northern Germany and in many other parts of the world, tile drains and depressions are prominent features of the landscape, though these are often neglected in hydrological modelling on the catchment scale. Kiesel *et al.* (2010) showed how these lowland features can be implemented into SWAT. For obtaining the necessary input data, results from a GIS (geographic information system) method were used to derive the location of artificial (tile) drainage areas. Another GIS method was developed to evaluate the spatial distribution and characteristics of landscape depressions. In the study catchment of Kiesel *et al.* (2010), 31% of the watershed area is artificially drained, which heavily influences groundwater processes. Landscape depressions are also common over the 50 km^2 study area and have considerable retention potential with an estimated surface area of 582 ha. The work evaluated the extent to which these two processes affect model performance. Accordingly, three hypotheses were formulated and tested through a stepwise incorporation of drainage and depression processes into an auto-calibrated default set-up: (i) integration of artificial drainage alone; (ii) integration of depressions alone; and (iii) integration of both processes combined.

The results show a strong improvement of model performance when artificial drainage is included, while the depression set-up only induces a slight improvement. The incorporation of the two landscape characteristics combined led to an overall enhancement of model performance and the strongest improvement of the coefficient of determination (R^2), root mean square error (RMSE) and Nash–Sutcliffe efficiency (NSE) of all set-ups. In particular, summer rainfall events with high intensity, winter flows and the hydrograph's recession limbs are depicted more realistically.

Combined simulation of point- and diffuse-source pollution

Combined assessments of the potential environmental impacts of point- and diffuse-source pollution at regional scales are necessary to achieve the sustainable development of natural resources such as land and water. Hence, Lam *et al.* (2010) aimed with their study at an evaluation of the long-term impact of point- and diffuse-source pollution on nitrate load in the lowland catchment of the Kielstau River using SWAT. In addition, their second objective was to determine the contribution of point and diffuse sources to nitrate load in the entire catchment.

The water quality of the study catchment is influenced not only by the predominant agricultural land use in the catchment of cropland and pasture, but also by six municipal waste-water treatment plants. Diffuse entries as well as punctuated entries from the waste-water treatment plants were implemented in the model set-up. The model was first calibrated and then validated in a daily time step. The values of the NSE for the simulations of flow and nitrate load ranged from 0.68 to 0.75 for the calibration period and from 0.76 to 0.78 for the validation period. These statistical results revealed that the SWAT model performed satisfactorily in simulating daily flow and nitrate load in a lowland catchment in northern Germany. The results showed that diffuse sources are the main contributor to nitrate load in the entire catchment, accounting for about 95% of the total nitrate load, while only 5% results from point sources. The model results also indicated that agriculture is the dominant contributor of diffuse sources, and that the percentage of agricultural land area is highly positively correlated with nitrate load at the different sub-basins. The area covered by forest was found to be negatively correlated with nitrate load.

Analysing the impact of water-quality monitoring strategies and load-estimation methods on model calibration and evaluation

Model-based predictions of the impact of land management practices on nutrient loading require measured nutrient-flux data for model calibration and evaluation. Consequently, uncertainties in the monitoring data resulting from sample collection and load-estimation methods influence the calibration and, thus, the parameter settings that affect the modelling results. To investigate this influence, Ullrich and Volk (2010) compared three different time-based sampling strategies and four different load-estimation methods for model calibration. For their study, they used SWAT on the intensively managed loess-dominated Parthe watershed (315 km^2) in central Germany.

The results showed that NO_3–N load estimations differ considerably depending on sampling strategy, load-estimation method and period of interest. Within the study period, the annual NO_3–N load-estimation values for the daily composite data set had the lowest values, ranging between 9.8% and 15.7% of maximum deviations from the mean of all applied methods. In contrast, annual estimation results for the sub-monthly and the monthly data set varied in greater ranges – between 24.9% and 67.7%.

To show differences among the sampling strategies, Ullrich and Volk (2010) calculated the percentage deviation of mean load estimations of sub-monthly and monthly data sets as related to the mean estimation value of the composite data set. For NO_3–N, the maximum deviation was 64.5% for the sub-monthly data set in the year 2000. They used average monthly NO_3–N loads of the daily composite data set to calibrate the model to achieve satisfactory simulation results (NSE of 0.52). Using the same parameter settings with sub-monthly and monthly data sets, the NSE dropped to 0.42 and 0.31, respectively. Considering the different results from the monitoring strategy and the load-estimation method, they recommended both the implementation of optimized monitoring programmes and the use of multiple load-estimation methods to improve water-quality characterization and provide appropriate model calibration and evaluation data.

Summary

This chapter introduces the structure and application of the Soil and Water Assessment Tool (SWAT) model. Besides a general overview, it focuses on examples for the application of SWAT in Germany. The model has been used to evaluate the impact of land-use and land-management options on hydrology and water quality at different scales. This has been done especially to support the achievement of the environmental targets of the WFD. Results and challenges of model-based scenario simulations for river-basin management are shown in the examples of both a 3500 km^2 agricultural river basin (the Upper Ems River Basin) and a 50 km^2 lowland catchment in northern Germany (the Kielstau catchment). The chapter includes also an overview of supporting studies in which work on sensitivity analyses of management parameters in SWAT and the impact of water-quality monitoring strategies and load-estimation methods on model calibration and evaluation is presented. The findings and developments from these studies can be useful for other SWAT users worldwide to improve the interpretation and quality of their model simulations.

The examples presented from Germany have shown that SWAT is a suitable tool for simulating the impact of land use and land management on hydrology and water quality in medium-to-large river basins in various landscapes. This is important because efficient tools are needed to support the implementation of spatial, environmental and river-basin planning and management programmes to find land-use and management configurations that could help to mitigate water pollution and sustain soil conservation. This becomes even more crucial with regard to increased biofuel crop production. This increase is expected to substantially affect land-use patterns in Europe and thus also control the implementation of land-use-related measures to improve water quality.

The studies showed that especially area-related measures, such as changed tillage operations, fertilizer applications, etc., can be described reasonably; however, more sensitivity analysis is required to answer the question: how detailed do we have to be to parameterize management operations in SWAT for large-area applications? In addition, measures based on linear structures (such as riparian zones) or spatially explicit measures are not represented satisfactorily and need to be improved. Nevertheless, such modelling experiments help us to understand river system behaviour better, especially to identify areas of the highest diffuse pollution. Knowing these sources and hot-spot areas, it is easier to identify useful measures for reducing actual nutrient loads in the river network and for achieving the 'good ecological status' as determined by the WFD. A dynamic catchment model taking into account water and nutrient processes as a function of vegetation, land use and human impacts, and driven by climate conditions, can provide a very functional tool for creating a river-basin management plan taking into account possible changes with which the basin could be confronted in the future.

In general, the lack of long time series of water-quality data with daily time steps and higher spatial resolution has limited our capacity to evaluate the simulations – which represents a general problem, and results in uncertainty. In addition, the existing monitoring programmes for water quality in Europe are not yet suitable for delivering a sound database for the simulations (Jarvie *et al.*, 1997; EEB/WWF, 2005; Allan *et al.*, 2006). The main reasons for that are: (i) the high costs of the required procedures, which result in sparse water sampling (every 2–5 weeks), and (ii) the sometimes insufficient cooperation between the relevant authorities, NGOs (non-governmental organizations) and research institutes.

What we learned from the scenario simulations is that, taking economic aspects into account, it will be almost impossible to achieve the environmental objectives of the WFD in our intensively used agricultural study area up to the year 2015. The results suggest that the achievement of the WFD environmental targets is only possible with a consideration of regional landscape and land-use distinctions (different natural conditions, intensively used areas, areas with decreasing land-use intensity, etc.), which would be more realistic. A 'balanced' approach could be also taken into account in which we could ask if it is possible to balance out areas of pollution with areas with no or only less pollution. However, the success of land-use-related measures to improve water quality will also depend on the future increase of biofuel crop production, which generates specialized land-management patterns to maximum biomass production. This could lead to conflicts between water protection and energy needs.

River-basin management requires a spatially distributed representation of basin hydrology and nutrient transport processes. To accomplish this, recent work deals with the enhancement of SWAT to simulate water flow across discretized landscape units (Arnold *et al.*, 2010). According to the results of the current studies, the SWAT model structure more closely reflects the complex controls on infiltration, runoff generation, run-on and subsurface flow without requiring large computational resources or detailed parameterization.

References

Allan, I.J., Brana, B., Greenwood, R., Mills, G.A., Roig, B. and Gonzalez, C. (2006) A "toolbox" for biological and chemical monitoring requirements for the European Union's Water Framework Directive. *Talanta* 69, 302–322.

Arabi, M., Govindaraju, R.S., Hantush, M.M. and Engel, B.A. (2006) Role of watershed subdivision on modelling the effectiveness of Best Management Practices with SWAT. *Journal of the American Water Resources Association* 42, 513–528.

Arabi, M., Govindaraju, R.S. and Hantush, M.M. (2007) A probabilistic approach for analysis of uncertainty in the evaluation of watershed management practices. *Journal of Hydrology* 333, 459–471.

Arabi, M., Frankenberger, J.R., Engel, B.A. and Arnold, J.R. (2008) Representation of agricultural conservation practices with SWAT. *Hydrological Processes* 22, 3042–3055.

Arnold, J.G. and Allen, P. (1996) Estimating hydrologic budgets for three Illinois watersheds. *Journal of Hydrology* 176, 57–77.

Arnold, J.G. and Fohrer, N. (2005) SWAT2000: current capabilities and research opportunities in applied watershed modelling. *Hydrological Processes* 19, 563–572.

Arnold, J.G., Srinivasan, R., Muttiah, R.S. and Williams, J.R. (1998) Large area hydrologic modeling and assessment. 1. Model development. *Journal of the American Water Resources Association* 34, 73–89.

Arnold, J.G., Allen, P.M., Volk, M., Williams, J.R. and Bosch, D.D. (2010) Assessment of different representations of spatial variability on SWAT model performance. *Transactions of the ASABE* 53, 1433–1443.

Behera, S. and Panda, R.K. (2006) Evaluation of management alternatives for an agricultural watershed in a sub-humid subtropical region using a physical process based model. *Agriculture, Ecosystems and Environment* 113, 62–72.

Behrendt, H., Huber, P., Opitz, D., Scholz, G. and Uebe, R. (1999) *Nährstoffbilanzierung der Flussgebiete Deutschlands.* UBA-Bericht, Leibniz-Institut für Gewässerökologie und Binnenfischerei im Forschungsverbund Berlin e.V., Berlin.

Benson, V.W., Bogusch H.C. Jr and Williams, J.R. (1988) Evaluating alternative soil conservation and crop tillage practices with EPIC. In: Unger, P.W., Sneed, T.V., Jordan, W.R. and Jensen, R. (eds) *Proceedings of the International Conference on Dryland Agriculture – A Global Perspective*, Texas Agricultural Experiment Station (TAES), Amarillo/Bushland, Texas, pp. 91–94.

BMELV (Bundesministerium für Ernährung, Landwirtschaft und Verbraucherschutz) (2008) *Land- und Forstwirtschaft in Deutschland: Daten und Fakten.* Available at: http://www.bmelv-statistik.de/fileadmin/sites/100_W_daten/Land-_und_Forstwirtschaft_Daten_und_Fakten.pdf (accessed 3 March 2011).

Boorman, D.B. (2003) Climate, Hydrochemistry and Economics of Surface-water Systems (CHESS): adding a European dimension to the catchment modelling experience developed under LOIS. *The Science of the Total Environment* 314/316, 411–437.

Bracmort, K.S., Arabi, M., Frankenberger, J.R., Engel, B.A. and Arnold, J.G. (2006) Modelling long-term water quality impact of structural BMPs. *Transactions of the ASABE* 49, 367–374.

Brouwer, R., Hofkes, M. and Linderhof, V. (2008) General equilibrium modelling of the direct and indirect economic impacts of water quality improvements in the Netherlands at national and river basin scale. *Ecological Economics* 66, 127–140.

Chaplot, V., Saleh, A., Jaynes, D.B. and Arnold, J.G. (2004) Predicting water, sediment and NO_3–N loads under scenarios of land-use and management practices in a flat watershed. *Water, Air and Soil Pollution* 154, 271–293.

Di Luzio, M., Arnold, J.G. and Srinivasan, R. (2005) Effect of GIS data quality on small watershed stream flow and sediment simulation. *Hydrological Processes* 19, 629–650.

EC (European Commission) (2000) European Parliament and the Council of the European Union. Directive 2000/60/EC establishing a framework for the Community action in the field of water policy. *Official Journal of the European Communities* L327, 1–72.

Eckhardt, K., Haverkamp, S., Fohrer, N. and Frede, H.-G. (2002) SWAT-G, a version of SWAT99 modified for application to low mountain range catchments. *Physics and Chemistry of the Earth* 27, 641–644.

Eckhardt, K., Fohrer, N. and Frede, H.-G. (2005) Automatic model calibration. *Hydrological Processes* 19, 651–658.

EEB/WWF (European Environmental Bureau and WWF European Policy Office) (2005) EU Water Policy: Making the Water Framework Directive. The quality of national transposition and implementation of the Water Framework Directive at the end of 2004. EEB, Brussels. Available at: http://www.eeb.org (accessed March 2005).

FLUMAGIS (2009) Interdisciplinary development of methods and tools for the planning process and measurement control for river basin management with geoinformation systems. Available at: http://www.flumagis.de/english/e_index.htm (accessed 19 January 2009).

Fohrer, N., Haverkamp, S., Eckhardt, K. and Frede, H.-G. (2001) Hydrologic response to land use changes on the catchment scale. *Physics and Chemistry of the Earth* 26, 577–582.

Fohrer, N., Haverkamp, S. and Frede, H.-G. (2005) Assessment of the effects of land use patterns on hydrologic landscape functions: development of sustainable land use concepts for low mountain range areas. *Hydrological Processes* 19, 659–672.

Gassman, P.W., Williams, J.R., Benson, V.R., Izaurralde, R.C., Hauck, L.M., Jones, C.A., Atwood, J.D., Kiniry, J.R. and Flowers, J.D. (2005) *Historical Development and Applications of the EPIC and APEX Models.* Working paper 05-WP 397. Center for Agricultural and Rural Development, Iowa State University, Ames, Iowa. Available at: http://www.card.iastate.edu/publications/DBS/PDFFiles/05wp397.pdf (accessed 18 February 2011).

Gassman, P.W., Reyes, M.R., Green, C.H. and Arnold, J.G. (2007) The soil and water assessment tool: historical development, applications, and future research directions. *Transactions of the ASABE* 50, 1211–150.

Golon, J. (2009) Environmental effects of varied energy crop cultivation scenarios on a lowland catchment in Northern Germany – a SWAT approach. MSc thesis, Ecology Centre, Kiel University, Germany. Available at: http://www.hydrology.uni-kiel.de/lehre/abschlussarbeiten/2009_golon_msc.pdf (accessed 17 February 2011).

Haverkamp, S., Fohrer, N. and Frede, H.-G. (2005) Assessment of the effect of land use patterns on hydrologic landscape functions: a comprehensive GIS-based tool to minimize model uncertainty resulting from spatial aggregation. *Hydrological Processes* 19, 715–727.

Hirschfeld, J., Dehnhardt, A. and Dietrich, J. (2005) Socioeconomic analysis within an interdisciplinary spatial decision support system for an integrated management of the Werra River Basin. *Limnologica* 35, 234–244.

Hörmann, G., Horn, A. and Fohrer, N. (2005) The evaluation of land-use options in mesoscale catchments. Prospects and limitations of eco-hydrological models. *Ecological Modelling* 187, 3–14.

Huang, S., Hesse, C., Krysanova, V. and Hattermann, F. (2009) From meso- to macro-scale dynamic water quality modelling for the assessment of land use change scenarios. *Ecological Modelling* 220, 2543–2558.

Jarvie, H.P., Neal, C. and Tappin, A.D. (1997). European land-based pollutant loads to the North Sea: an analysis of the Paris Commission data and review of monitoring strategies. *The Science of the Total Environment* 194/195, 39–58.

Jessel, B. and Jacobs, J. (2005) Land use scenario development and stakeholder involvement as tools for watershed management within the Havel River Basin. *Limnologica* 35, 220–233.

Jha, M., Gassman, P.W., Secchi, S., Gu, R. and Arnold, J.G. (2004) Effect of watershed subdivision on SWAT flow, sediment, and nutrient predictions. *Journal of the American Water Resources Association* 40, 811–825.

Jones, J.N., Moody, J.E. and Lillard, J.H. (1969) Effects of tillage, no-tillage, and mulch on soil water and plant growth. *Agronomy Journal* 94, 299–304.

KBDS (Konservierende Bodenbearbeitung/Direktsaat in Sachsen) (2011) Konservierende Bodenbearbeitung. Available at: http://kbd-sachsen.de/1d2e3d699b0916924e7ddf6ef706e2df/konservierende-bodenbearbeitung/konservierende-bodenbearbeitung.html (accessed 17 February 2011).

Kiesel, J., Hering, D., Schmalz, B. and Fohrer, N. (2009) A transdisciplinary approach for modelling macroinvertebrate habitats in lowland streams. IAHS Publication No. 328, 24–33.

Kiesel, J.D., Fohrer, N., Schmalz, B. and White, M.J. (2010) Incorporating landscape depressions and tile drainages of a northern German lowland catchment into a semi-distributed model. *Hydrological Processes* 24, 1472–1486.

Kim, N.W. and Srinivasan, R. (eds) (2010) *2010 [6th] International SWAT Conference, Proceedings*, 4–6 August 2010, Seoul, Korea. Korea Institute of Construction Technology, Goyang-Si Gyeonggi-Do, Korea [and others]. Available at: http://swatmodel.tamu.edu/media/33774/swat2010-proceedings.pdf (accessed 17 February 2011).

King, K.W., Richardson, C.W. and Williams, J.R. (1996) Simulation of sediment and nitrate loss on a vertisol with conservation tillage practices. *Transactions of the ASAE* 39, 2139–2145.

Kirsch, K., Kirsch, A. and Arnold, J.G. (2002) Predicting sediment and phosphorus loads in the Rock River Basin using SWAT. *Transactions of the ASAE* 45, 1757–1769.

Kronvang, B., Borgvang, S.A. and Barkved, L.J. (2009) Towards European harmonised procedures for quantification of nutrient losses from diffuse sources – the EUROHARP project. *Journal of Environmental Monitoring* 11, 503–505.

Krysanova, V. and Arnold, J.G. (2008) Advances in ecohydrological modelling with SWAT – a review. *Hydrological Sciences Journal* 53, 939–947.

Krysanova, V., Meiner, A., Roosaare, J. and Vasilyev, A. (1989) Simulation modelling of the coastal waters pollution from agricultural watershed. *Ecological Modelling* 49, 7–29.

Krysanova, V., Müller-Wohlfeil, D.I. and Becker, A. (1998) Development and test of a spatially distributed hydrological/water quality model for mesoscale watersheds. *Ecological Modelling* 106, 261–289.

Krysanova, V., Hattermann, F. and Wechsung, F. (2005) Development of the ecohydrological model SWIM for regional impact studies and vulnerability assessment. *Hydrological Processes* 19, 763–783.

Krysanova, V., Hattermann, F. and Wechsung, F. (2007) Implications of complexity and uncertainty for integrated modelling and impact assessment in river basins. *Environmental Modelling and Software* 22, 701–709.

Lam, Q.D., Schmalz, B. and Fohrer, N. (2010) Modelling point and diffuse source pollution of nitrate in a rural lowland catchment using the SWAT model. *Agricultural Water Management* 97, 317–325.

Lam, Q.D., Schmalz, B. and Fohrer, N. (2011) Water quality modelling of Best Management Practices in a North German lowland catchment. *Environmental Monitoring and Assessment* (in press).

LAWA (Länderarbeitsgemeinschaft Wasser) (1998) *Beurteilung der Wasserbeschaffenheit von Fließgewässern in der Bundesrepublik Deutschland – Chemische Gewässergüteklassifikation.* Kulturbuchverlag, Berlin.

Lenhart, T., Eckhardt, K., Fohrer, N. and Frede, H.-G. (2002) Comparison of two different approaches of sensitivity analysis. *Physics and Chemistry of the Earth* 27, 645–654.

Lenhart, T., Van Rompaey, A., Steegen, A., Fohrer, N., Frede, H.-G. and Govers, G. (2005) Considering spatial distribution and deposition of sediment in lumped and semi-distributed models. *Hydrological Processes* 19, 785–794.

LfL (Landesanstalt für Landwirtschaft) (2011) Landwirtschaft in Sachsen. Available at: http://www.landwirtschaft.sachsen.de/Landwirtschaft/index.html (accessed 17 February 2011).

Neitsch, S.L., Arnold, J.G., Kiniry, J.R., Williams, J.R. and King, K.W. (2002) *Soil and Water Assessment Tool Theoretical Documentation, Version 2000.* GSWRL Report 02-01, Grassland, Soil and Water Research Laboratory, USDA ARS (Agricultural Research Service)/BRC Report 2-05,,Blackland Research and Extension Center, Texas Agricultural Experiment Station, Temple, Texas/TWRI Report TR-191, Texas Water Resources Institute, College Station, Texas.. Available at: http://swatmodel.tamu.edu/media/1290/swat2000theory.pdf (accessed 17 February 2011).

Neitsch, S.L., Arnold, J.G., Kiniry, J.R., Srinivasan, R. and Williams, J.R. (2004) *Soil and Water Assessment Tool Input/Output File Documentation. Version 2005.* USDA ARS Grassland, Soil and Water Research Laboratory/Blackland Research Center, Texas Agricultural Experiment Station, Temple, Texas. Available at: http://swatmodel.tamu.edu/media/1291/SWAT2005io.pdf (accessed 17 February 2011).

Neitsch, S.L., Arnold, J.G., Kiniry, J.R. and Williams, J.R. (2005) *Soil and Water Assessment Tool Theoretical Documentation, Version 2005.* USDA ARS Grassland, Soil and Water Research Laboratory/Blackland Research Center, Texas Agricultural Experiment Station, Temple, Texas. Available at: http://swatmodel.tamu.edu/media/1292/SWAT2005theory.pdf (accessed 17 February 2011).

Onatski, A. and Williams, N. (2003) Modelling model uncertainty. *NBER Working Paper* 9566, National Bureau of Economic Research, Cambridge, Massachusetts. Available at: http://ideas.repec.org/p/nbr/nberwo/9566.html (accessed 17 February 2011).

Pandey, V.K., Panda, S.N. and Sudhakar, S. (2005) Modelling of an agricultural watershed using remote sensing and a geographic information system. *Biosystems Engineering* 90, 331–347.

Pitkänen, J. and Nuutinen, V. (1998) Earthworm contributions to infiltration and surface runoff after 15 years of different soil management. *Applied Soil Ecology* 9, 411–415.

Pohlert, T., Huisman, J.A., Breuer, L. and Frede, H.-G. (2007a) Integration of detailed biogeochemical model into SWAT for improved nitrogen predictions – Model development, sensitivity, and GLUE analysis. *Ecological Modelling* 203, 215–228.

Pohlert, T., Breuer, L., Huisman, J.A. and Frede, H.-G. (2007b) Assessing the model performance of an integrated hydrological and biogeochemical model for discharge and nitrate load predictions. *Hydrology and Earth System Sciences* 11, 997–1011.

Rekolainen, S., Kämäri, J. and Hiltunen, M. (2003) A conceptual framework for identifying the need and role of models in the implementation of the Water Framework Directive. *International Journal of River Basin Management* 1, 347–352.

Rode, M., Klauer, B., Petry, D., Volk, M., Wenk, G. and Wagenschein, D. (2008) Integrated nutrient transport modelling with respect to the implementation of the European WFD: the Weiße Elster case study, Germany. *Water SA* 34, 490–496.

Saleh, A., Arnold, J.G., Gassman, P.W., Hauck, L.M., Rosenthal, W.D., Williams, J.R. and McFarland, A.M.S. (2000) Application of SWAT for the Upper North Bosque River watershed. *Transactions of the ASAE* 43, 1077–1087.

Santhi, C., Arnold, J.G., Williams, J.R., Dugas, W.A., Srinivasan, R. and Hauck, L.M. (2001) Validation of the SWAT model on a large river basin with point and nonpoint sources. *Journal of the American Water Resources Association* 37, 1169–1188.

Santhi, C., Srinivasan, R., Arnold, J.G. and Williams, J.R. (2003) A modelling approach to evaluate the impacts of water quality management plans implemented in the Big Cypress Creek Watershed. In: *Second Conference on Watershed Management to Meet Emerging TMDL Environmental Regulations*, Albuquerque, New Mexico, pp. 384–394.

Schmalz, B. and Fohrer, N. (2010) Ecohydrological research in the German lowland catchment Kielstau. IAHS Publication No. 336, 115–120.

Schmalz, B., Tavares, F. and Fohrer, N. (2007) Assessment of nutrient entry pathways and dominating hydrological processes in lowland catchments. *Advances in Geosciences* 11, 107–112.

Schmidt, W., Zimmerling, B., Nitzsche, O. and Krück, St. (2001) Conservation tillage – a new strategy in flood control. In: Marsalek, J., Watt, E., Zeman, E. and Sieker, H. (eds) *Advances in Urban Stormwater and Agricultural Runoff Source Controls. Proceedings of the NATO Advanced Research Workshop on Source Control Measures for Stormwater Runoff*, 8–12 November 2000, St Marienthal-Ostritz, Germany. NATO Science Series 74, IV. Earth and Environmental Sciences – Vol. 6. Kluwer Academic Publishers, Dordrecht, The Netherlands, pp. 287–292.

Sloot, P.H.M., Jeldres, D.I. and Morimoto, K. (1994) Calibration of the erosion-productivity model EPIC for three soil tillage systems in the Secano Interior of Chile. *Presentation at the 8th Conference of the International Soil Conservation Organisation* (ISCO), New Delhi.

Srinivasan, R. (ed.) (2007) *2007 4th International SWAT Conference, Proceedings*, 4–6 July 2007, Delft, the Netherlands. United Nations Educational, Scientific and Cultural Organization – Institute for Water Education, Delft. Available at: http://www.brc.tamus.edu/swat/4thswatconf/docs/4thConfProceedings.pdf (accessed 17 February 2011).

Srinivasan, R., Jacobs, H. and Jensen, R. (eds) (2003) *SWAT2003: 2nd International SWAT Conference, Proceedings*, 1–4 July 2003, Bari, Italy. TWRI Technical Report No. 266, Texas Water Resources Institute, Texas A&M University, College Station, Texas. Available at: http://www.brc.tamus.edu/swat/2ndswatconf/2ndswatconfproceeding.pdf (accessed 4 February 2007).

Srinivasan, R., Jacobs, H., Day, D. and Abbaspour, K. (eds) (2005) *2005 3rd International SWAT Conference, Proceedings*, 11–15 July 2005, Zurich, Switzerland. EAWAG (Swiss Federal Institute for Environmental Science and Technology)/SWAT Soil and Water Assessment Tool) Team, Texas A&M University, College Station, Texas. Available at: www.brc.tamus.edu/swat/3rdswatconf/SWAT%20Book%203rd%20Conference.pdf (accessed 14 February 2007).

Sullivan, P. (2003) Conservation tillage. ATTRA Publication No. CT105, Appropriate Technology Transfer for Rural Areas, National Center for Appropriate Technology, Fayetteville, Arkansas. Available at: http://attra.ncat.org/attra-pub/PDF/consertill.pdf (accessed 19 January 2009).

SWAT (2010) SWAT Literature Database for Peer-Reviewed Journal Articles. Available at: https://www.card.iastate.edu/swat_articles/ (accessed 25 September 2010).

Tripathi, M.P., Panda, R.K. and Raghuwanshi, N.S. (2005) Development of effective management plan for critical subwatersheds using SWAT model. *Hydrological Processes* 19, 809–826.

Twigg, K., Swyden, C. and Srinivasan, R. (eds) (2009) *2009 5th International SWAT Conference, Proceedings*, 5–7 August 2009, Boulder, Colorado. TWRI Technical Report No. 356, Texas Water Resources Institute, Texas A&M University, College Station, Texas. Available at: http://twri.tamu.edu/reports/2009/tr356.pdf (accessed 17 February 2011).

Ullrich, A. and Volk, M. (2009) Application of the Soil and Water Assessment Tool (SWAT) to predict the impact of alternative management practices on water quality and quantity. *Agricultural Water Management* 96, 1207–1217.

Ullrich, A. and Volk, M. (2010) Influence of different nitrate-N monitoring strategies on load estimation as a base for model calibration and evaluation. *Environmental Monitoring and Assessment* 171, 513–527.

Vache, K.B., Eilers, J.M. and Santelmann, M.V. (2002) Water quality modeling of alternative agricultural scenarios in the U.S. Corn Belt. *Journal of the American Water Resources Association* 38, 773–787.

van der Helm, R. (2003) Challenging future studies to enhance EU's participatory river basin management. *Physics and Chemistry of the Earth* 28, 563–570.

van Griensven, A. and Bauwens, W. (2003) Multiobjective autocalibration for semidistributed water quality models. *Water Resources Research* 39, 1348–1356.

van Griensven, A. and Bauwens, W. (2005) Evaluation of the application of ESWAT for integrated river water quality modeling. *Hydrological Processes* 19, 827–838.

Volk, M., Hirschfeld, J., Schmidt, G., Bohn, C., Dehnhardt, A., Liersch, S. and Lymburner, L. (2007) A SDSS-based ecological-economic modelling approach for integrated river basin management on different scale levels – the Project FLUMAGIS. *Water Resources Management* 21, 2049–2061.

Volk, M., Hirschfeld, J., Dehnhardt, A., Schmidt, G., Bohn, C., Liersch, S. and Gassman, P.W. (2008) Integrated ecological–economic modelling of water pollution abatement options in the Upper Ems River Basin. *Ecological Economics* 66, 66–76.

Volk, M., Liersch, S. and Schmidt, G. (2009) Towards the implementation of the European Water Framework Directive? Lessons learned from water quality simulations in an agricultural watershed. *Land Use Policy* 26, 580–588.

Volk, M., Lautenbach, S., van Delden, H., Newham, L.T.H. and Seppelt, R. (2010) How can we make progress with decision support systems in landscape and river basin management? Lessons learned from a comparative analysis of four different decision support systems. *Environmental Management* 46, 834–849.

Weber, A., Fohrer, N. and Möller, D. (2001) Long-term land use changes in a mesoscale watershed due to socio-economic factors – effects on landscape structures and functions. *Ecological Modelling* 140, 125–140.

White, M.J. and Arnold, J.G. (2009) Development of a simplistic vegetative filter strip model for sediment and nutrient retention at the field scale. *Hydrological Processes* 23, 1602–1616.

Williams, J.R. (1990) The erosion productivity impact calculator (EPIC) model: a case history. *Philosophical Transactions of the Royal Society London B* 329, 421–428.

Williams, J.R., Jones, C.A., Kiniry, J.R. and Spanel, D.A. (1989).The EPIC crop growth model. *Transactions of the ASAE* 32, 497-511.

Williams, J.R., Nearing, M., Nicks, A., Skidmore, E., Valentin, C., King, K. and Savabi, R. (1996) Using soil erosion models for global change studies. *Journal of Soil and Water Conservation* 51, 381–385.

Williams, J.R., Arnold, J.G., Kiniry, J.R., Gassman, P.W. and Green, C.H. (2008) History of model development at Temple, Texas. *Hydrological Sciences Journal* 53, 948–960.

8 Spatially Distributed Hydrological Modelling in the Illinois River Drainage Area, Arkansas, Using SWAT

Dharmendra Saraswat* and Naresh Pai

Introduction

Spatially distributed watershed models are increasingly being used for understanding and managing the effect of natural and human activities on sediment and nutrient loadings in recent years (Li *et al.*, 2010; Zhang *et al.*, 2010). In the last four decades, advancement in computer-based hydrological modelling reflects the incorporation of the needs and concepts for a distributed hydrological model proposed by Freeze and Harlan (1969). The development of the first distributed hydrological model, SHE (Système Hydrologique Européen) (Abbott *et al.*, 1986), has been followed by that of a large number of distributed models for use across point, field and watershed scales (Srivastava *et al.*, 2007). The availability of numerous watershed models has also necessitated the development of model usage guidelines. Borah and Bera (2003) reviewed flow-governing equations and solution methods used in 11 watershed-scale models with a view to selecting the most appropriate model for an application. This preliminary investigation was followed up with a detailed investigation of two long-term continuous simulation models, the Soil and Watershed Assessment Tool (SWAT) and the Hydrological Simulation Program-Fortran (HSPF), which are part of the US Environmental Protection Agency's (US EPA's) Better Assessment Science Integrating Point and Nonpoint Sources (BASINS) modelling system (Borah and Bera, 2004; USEPA, 2007). The investigation resulted in the identification of the strengths and improvements required for enhancing the model's predictive power (Borah and Bera, 2004). Out of the two long-term continuous simulation models studied by Borah and Bera, SWAT has consistently undergone review and enhancement of its capabilities. The latest version of SWAT, SWAT2009, was released in January 2010, with several enhancements (Arnold *et al.*, 2010). The global adoption of the SWAT model can be gauged by the publication of more than 680 peer-reviewed journal articles and hundreds of articles published in conference proceedings (Gassman *et al.*, 2010).

To gain confidence in the reliability of models for predicting streamflow and yields of its constituents, sensitivity analysis, calibration and validation have been considered essential (Grayson and Bloschl, 2000; White and Chaubey, 2005). A majority of watershed models rely on a combination of statistical and/or graphical techniques to evaluate objective functions for model calibration (Moriasi *et al.*, 2007). In several studies reported before the

* Corresponding author: dsaraswat@uaex.edu

new millennium, watershed models were calibrated and validated at one site, i.e. the drainage outlet of a watershed (Refsgaard, 2000; Qi and Grunwald, 2005). This approach has been considered problematic for extrapolating watershed response to all other locations within the watershed because of the uniqueness of model parameters for individual watersheds (Beven, 2000). Therefore, spatially distributed calibration and validation have been used to account for hydrological patterns in the sub-watersheds (Santhi *et al.*, 2001; Van Liew and Garbrecht, 2003; Qi and Grunwald, 2005; White and Chaubey, 2005; Cao *et al.*, 2006; Engeland *et al.*, 2006; Bekele and Nicklow, 2007; Zhang *et al.*, 2008; Li *et al.*, 2010).

Variation in climatic conditions, soils, watershed geology, topography, land use and in the nested/independent availability of monitoring gauges, among other influences, introduces additional confounding factors that require verification of distributed hydrological patterns both at the watershed outlet and within the watershed. Our objectives for this study were to calibrate and validate hydrological processes on a long-term continuous basis at a total of seven monitoring gauges, including the drainage outlet of the Illinois River Drainage Area in Arkansas (IRDAA). The study watershed is also of interstate concern for the EPA owing to a lawsuit filed by the State of Oklahoma citing human health risks resulting from the application of poultry litter as a fertilizer source on pasture lands in the watershed.

Materials and Methods

Overview of the SWAT model

The SWAT model is a long-term, continuous watershed-scale, quasi-physical model developed by the US Department of Agriculture Agricultural Research Service (USDA ARS) (Arnold *et al.*, 1998; Qi and Grunwald, 2005). The model has been used worldwide to assess the impact of management practices – including climate and vegetative changes, reservoir management, groundwater withdrawals and water transfer – on water, sediment and agricultural chemical transport in river basins. The SWAT model is an outgrowth of a number of models developed by the USDA ARS. The current SWAT model is considered to be a direct descendant of the Simulator for Water Resources in River Basins (SWRRB) model (Gassman *et al.*, 2007). Other models that have enhanced the SWRRB model's capability to deal with water-quality problems on a watershed scale include the Chemicals, Runoff, and Erosion from Agricultural Management Systems (CREAMS) (Knisel, 1980), the Groundwater Loading Effects on Agricultural Management Systems (GLEAMS) (Leonard *et al.*, 1987) and the Environment Policy Integrated Climate (EPIC) model (Williams *et al.*, 1984). The SWAT model has been in the public domain since its inception in the 1990s, and has undergone expansion in its capabilities through inputs from researchers from all around the world. The documentation for some of the major enhancements in the previous versions of the model (SWAT94.2, 96.2, 98.1, 99.2, 2000 and 2005) is available at the SWAT website maintained by the Spatial Sciences Laboratory at Texas A&M University (SWAT, 2010). The model's handling of spatially and temporally distributed input data has also been simplified over the years with the development of input interfaces using Windows (Visual Basic), Geographic Resources Analysis Support System-Geographic Information System (GRASS-GIS), ArcView and ArcGIS software.

The model permits the simulation of different physical processes through eight major components: climate, hydrology, nutrient cycling/pesticide fate, erosion/sedimentation, land cover/plant growth, management practices, channel processes/routing and pond/reservoir routing (Neitsch *et al.*, 2005). The spatial heterogeneity of a study area is represented by division of the model into a number of sub-watersheds (also referred to as sub-basins). This enables differences in evapotranspiration (ET) for various crops, soils etc. to be accounted for (Arnold *et al.*, 1998). Each sub-watershed is further subdivided into a series of non-spatially explicit Hydrological Response Units (HRUs), which represent unique combinations of land cover, soil type and slope within a sub-watershed. The model simulates soil water content, surface runoff, nutrient cycles, sediment yield, crop growth and management practices for each HRU, and then aggregates them for the sub-watershed using a weighted average (Abbaspour *et al.*, 2007). Flow and water-quality variables are routed by HRUs and sub-watersheds to the watershed outlet.

The hydrological routines in the model are used for simulating snowfall and snow melt, vadose zone processes (i.e. infiltration, evaporation, plant uptake, lateral flows and percolation) and groundwater movement (Zhang *et al.*, 2010). The model simulates the hydrological cycle using the water-balance equation. Precipitation is simulated in the form of both rainfall and snowfall. Snow routing is accomplished through three different approaches, the degree day (DD), DD plus the elevation band (Fontaine *et al.*, 2002) and the energy balance-based SNOW17 models (Zhang *et al.*, 2008). The amount of surface runoff is estimated from daily rainfall data using the modified USDA Soil Conservation Service (SCS) curve number (CN) method (Mishra and Singh, 2003). The Green–Ampt method (Green and Ampt, 1911) is an alternative option for estimating surface runoff and infiltration by using sub-daily weather data. Three methods of estimating potential evapotranspiration (PET) are available in the model: the Priestley–Taylor (Priestley and Taylor, 1972), Hargreaves (Hargreaves and Samani, 1985) and Penman–Monteith (Allen *et al.*, 1998) methods. A soil database is used to obtain information on soil type, texture, depth and hydrological classification. Channel routing is simulated using either the variable-storage method or the Muskingum method; both methods are variations of the kinematic wave model (Chow *et al.*, 1988). The model estimates infiltration by subtracting runoff from precipitation. Infiltration moves into the soil profile where it is routed through the soil layers. A storage routing flow coefficient is used to predict flow through each of the ten soil layers, with flow occurring when a layer exceeds field capacity. When water percolates past the bottom layer, it enters the shallow aquifer zone (Arnold *et al.*, 1993). Channel transmission loss and pond/reservoir seepage replenish the shallow aquifer while it interacts directly with the stream. Flow to the deep aquifer system is effectively lost and cannot return to the stream. The irrigation algorithm developed for SWAT allows irrigation water to be transferred from any reach or reservoir to any other in the watershed. Based on surface runoff calculated using either the SCS runoff equation or the Green–Ampt method, excess surface runoff not lost to other functions makes its way to the channels, where it is routed downstream.

The model is also capable of simulating sediment and chemical yields processes. Only the hydrological component of the model is described in this study. Therefore, users are encouraged to refer to Nietsch *et al.* (2005) for gaining theoretical understanding of the modelling constituents apart from the hydrology component in the SWAT model.

The Illinois River Drainage Area in Arkansas (IRDAA)

This section discusses the approach taken to conducting spatially distributed hydrology modelling using a case study of the IRDAA, a part of the Illinois River Watershed (IRW) in north-west Arkansas, using the latest version of the SWAT model (SWAT2009). The watershed has a drainage area of approximately 1960 km^2 and covers portions of Benton (39.5%), Crawford (0.2%) and Washington (60.3%) counties (Fig. 8.1). The watershed is urbanizing at an accelerated pace, as can be seen by an increase in population of 47.1% and 26.9% during 2000–2009 in Benton and Washington counties, respectively (US Census Bureau, 2010). The watershed is home to a number of animal enterprises, including poultry, swine and cattle (beef and dairy). Historically, manure/litter generated from confined animal operations has been applied to pasture as fertilizer. The leading suspected causes of water-quality impairment have been identified as siltation/turbidity, pathogens, total phosphorus (TP), and nitrate-nitrogen (NO_3-N), to which are attributed negative impacts on designated usages, such as aquatic life and primary contact (ADEQ, 2008).

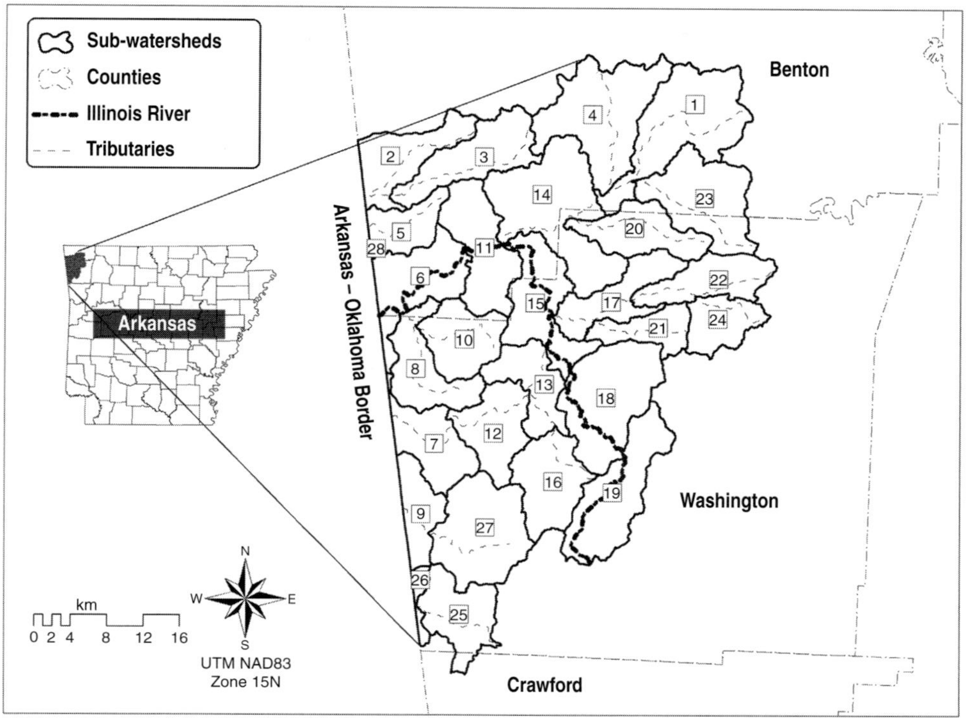

Fig. 8.1. Location of the Illinois River Drainage Area in Arkansas. Also shown are the sub-watershed locations delineated by the SWAT model.

Model set-up

Sub-watersheds and HRU delineation

The set-up of the SWAT model was initiated by partitioning the IRDAA into 28 sub-watersheds that matched the 12-digit Hydrologic Unit Code (HUC) boundaries defined by the US Geological Survey (USGS). Use of a predefined sub-watershed boundary for delineation was desirable because it ensured repeatability of the model set-up and also allowed output to be obtained at the outlet of each 12-digit HUC sub-watershed. The sub-watershed delineation was followed by the creation of multiple HRUs within each sub-watershed using a threshold of 10%, 5% and 0% for soil, land cover and slope, respectively (Gitau *et al.*, 2007). A total of 876 HRUs were delineated that had combinations of soil, land cover and slope greater than the selected percentages within the sub-watersheds.

Input data

Elevation data

A digital elevation model (DEM) at 10 m spatial resolution was used to describe topographic conditions (i.e. minimum and maximum elevation, slope, slope length, etc.) for each of the sub-watersheds in the SWAT model. HRUs in the SWAT model can be defined using an average slope per sub-watershed or multiple slope categories. To better represent the diversity of slope in the study area, multiple slope classes were used as follows: 0–1% (16.3%), 1–3% (11.8%), 3–8% (38.9%) and greater than 8% (33.0%).

Land use

Variation in sub-watershed pollutant contribution is primarily a function of the type and distribution of its temporal land uses

(Miller *et al.*, 2002). All previous versions of the SWAT model treated land cover as constant within HRUs throughout the model run period. However, in the newer version of the SWAT model (SWAT2009), land-use distribution for each HRU can be updated during the model run (Arnold *et al.*, 2010). Land-use change during the model run has the potential to represent watershed responses, such as runoff and pollutant loads, and to respond dynamically to these changes. The land-use change over the modelling period was represented by extracting information from multiple years of remotely sensed land-use land-cover (LULC) imagery data. The LULC imagery data for the years 1992, 1993, 1999, 2001, 2004 and 2006 from two different sources were used. Classification inconsistencies between multiple years of imagery data were resolved before inputting the data into the model. The 1992 LULC data were used as base year data for setting up HRUs in the SWAT model. A note of caution is that HRU area fractions within a sub-watershed should sum to one in order to be consistent with their definition in the latest version of the SWAT model (Arnold *et al.*, 2010).

Soils

Soil characteristics of the study area were obtained from the Soil Survey Geographic (SSURGO) soil database for each county. Soil series comprising the largest relative land area by per cent within the watershed were Enders (25.2%), Captina (15.5%), Nixa (15.5%) and Clarksville (11.3%). These soil series were classified into hydrological group C, except for Clarksville, which was in hydrological group B. Soils classified into hydrological groups B and C have moderate and low infiltration rates, respectively (USDA SCS, 1972). The SWAT model uses soil hydrological group and land-use type to determine the default CN for each HRU (Neitsch *et al.*, 2005)

Weather

Historical daily precipitation for 13 years (1981–1993) was obtained from the National Climatic Data Center (NCDC) for three national weather stations (NWS) located in close proximity to the IRDAA. The NEXt generation RADar (NEXRAD) data were used to obtain precipitation values during 1994 to 2008. A long historical set of climate data was useful in adequate parameterization of the weather database of the model. All NEXRAD grid points falling within a sub-watershed were aggregated; an average value was calculated; and assigned to pseudo weather stations located at the centroid of each sub-watershed using a tool developed by Zhang and Srinivasan (2010). ET was calculated within the model using the Penman–Monteith (Monteith, 1965) method, while channel routing was simulated using the variable-storage method.

Ponds

Using the National Hydrography Dataset (http://nhd.usgs.gov) water-body layer, 9019 ponds were identified. As the SWAT model allows only one pond per sub-watershed, pond-related information was aggregated on a sub-watershed basis. The average minimum depth, drainage area and storage volume of ponds was determined as per the recommendation of Deal *et al.* (1997).

Management data

Because a majority of the study area (99.8%) falls in Benton and Washington counties, management data collection pertained to these two counties only. Management data on cattle grazing and manure deposition, poultry litter application, pig manure application and urban lawn management were determined based on consultation with county agents, subject matter specialists and published literature.

Sensitivity analysis, calibration and validation

Measured flow data from a total of seven gauges were used for model calibration and validation. Out of the seven gauges, five were used for model calibration. Four calibration gauges belonged to the USGS: Illinois River near Siloam Springs (USGS 07195430); Baron Fork at Dutch Mills (USGS 07196900); Flint

Creek at Springtown (USGS 07195800); and Osage Creek (Elm Springs-USGS 07195000). One gauge station belonged to the Arkansas Water Resources Center (AWRC) at Ballard Creek. Flink Creek at Springtown was not used for validation as a result of the unavailability of data during the validation period. A total of six gauges were utilized for model validation: four were common to calibration and validation, and of the two other gauges one was operated by the USGS at Illinois River at Savoy (USGS 07194800) and another was operated by the AWRC at Moore's Creek. The spatially distributed nature of monitoring gauges can be seen on Fig. 8.2.

Sensitivity analysis

Sensitivity analysis was performed using a combination of Latin-Hypercube (LH) and one-factor-at-a-time (OAT) sampling methods embedded in the SWAT2009 model. The model parameters that were given the top-five ranking for flow were: ALPHA_BF, SURLAG, RCHRG_DP, ESCO and CN2 (Table 8.1). Readers may refer to the SWAT manual (Neitsch *et al.*, 2005) for further description of these parameters.

Calibration and validation

A review of 115 SWAT hydrological calibration and/or validation results reported in the literature indicates that, despite the use of some type of graphical and/or statistical model evaluation criteria, no commonly accepted guidance has been established for modellers (Gassman *et al.*, 2007; Moriasi *et al.*, 2007). Moriasi *et al.* (2007) also pointed out that the model evaluation criteria used in the past failed to account for measurement uncertainty resulting from relative lack of data. The present study utilized a combination of model evaluation criteria proposed by Santhi *et al.* (2001) and Moriasi *et al.* (2007) for calibrating and validating model performance.

The model was run from 1992 to 2008. The years from 1992 to 1995 were included as 'warm-up' for the model to generate reasonable initial values. It has been reported that a warm-up period for SWAT is required to initialize and then approach reasonable starting values for model variables. Past studies have used 5-year (Mamillapalli, 1998), 3-year (White and Chaubey, 2005) and 2-year (Tolson and Shoemaker, 2007) warm-up periods to ensure that model parameters are

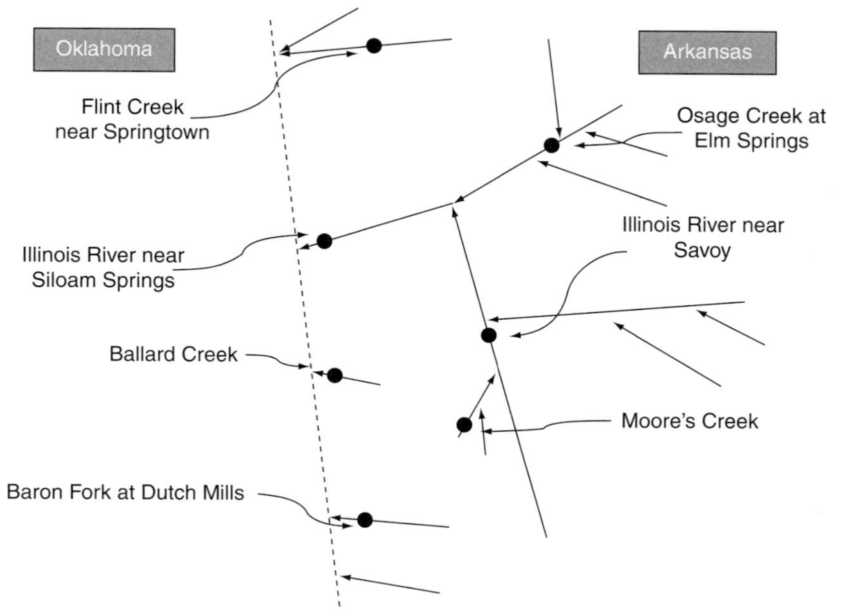

Fig. 8.2. Line map of monitoring gauge locations in the Illinois River Drainage study area in Arkansas (figure not to scale).

Table 8.1. Top ranked SWAT parameters for flow.[a]

Parameter	Parameter location file	Definition
ALPHA_BF	.gw	Base flow alpha factor
SURLAG	.bsn	Surface runoff lag coefficient
RCHRG_DP	.gw	Deep aquifer percolation fraction
ESCO	.hru	Soil evaporation compensation factor
CN2	.mgt	Initial SCS[b] runoff curve number (CN) for moisture condition II (= CN2)

[a]See SWAT Manual for further details of SWAT parameters (Neitsch *et al.*, 2005).
[b]SCS, USDA Soil Conservation Service.

stabilized and are representative of conditions in the watershed. In this study, model simulations were initialized in 1992, thus providing a 4-year warm-up period. The next 10 years (1996–2005) of simulations were used to systematically match simulated and observed data by making suitable parameter adjustments. The last 3 years (2006–2008) were used for validating the model by comparing it with a data set that was not used during calibration.

During calibration and validation, measured data were compared with specific model outputs. Total flow was calibrated first until measured, and simulated data met a threshold relative error (RE) value of 15% on an annual time step (Santhi *et al.*, 2001). The RE was measured as follows:

$$RE(\%) = \frac{O - P}{O} \times 100 \qquad (8.1)$$

where O is the measured and P is the predicted data.

The measured and predicted total flow was separated into base flow and surface runoff using the digital automatic filter developed by Arnold and Allen (1999). Evaluation guidelines for hydrological simulations prescribed by Moriasi *et al.* (2007) pertain to monthly total flow only. However, the model was calibrated for base flow and surface flow as well to ensure that these constituents of the model were properly simulated. In general, Moriasi *et al.* (2007) suggested that base flow should be calibrated to within 20%, while no specific guidelines are provided for surface flow.

The SWAT model was further evaluated on monthly time steps by calculating the coefficient of determination (R^2), Nash-Sutcliffe efficiency (Nash and Sutcliffe, 1970; NSE), RMSE-observations standard deviation ratio (Moriasi *et al.*, 2007; RSR), and Per cent bias (Moriasi *et al.*, 2007; PBIAS) using evaluation guidelines provided by Moriasi *et al.* (2007) (Table 8.2).

A more stringent test of model performance was assessed during the validation phase (January 2006–December 2008) as no parameter adjustments are allowed during this phase. The model's predictive capabilities were put to a rigorous test because the second driest (2007) and wettest (2008) periods of the entire study period were observed during the validation phase.

Results and Discussion

Annual calibration

The goal of annual calibration was to adjust the input parameters such that the SWAT output was within a predetermined range of measured data. Model calibrations were started at the most upstream headwater sub-watershed gauge (Osage Creek at Elm Springs) followed by the most downstream gauge (Siloam Springs). Other hydrologically disconnected calibration sites (Flint Creek, Ballard Creek and Baron Fork) were calibrated simultaneously. Hydrologically connected gauges (Moores Creek and Savoy) were used only for validation because the error had already been minimized on the water body during the calibration period. An exception was made for Osage Creek gauge, which was used for calibration and validation, owing to point-source impacts (Ekka *et al.*, 2006).

Table 8.2. General performance rating used for evaluating monthly SWAT model results for the Illinois River Drainage Area in Arkansas (based on performance rating table in Moriasi *et al.*, 2007).

Rating	NSE[a]	RSR[a]	PBIAS (%)[a]		
			Streamflow	Sediment	N, P
Very good	$0.75 \le NSE \le 1.00$	$0.00 \le RSR \le 0.50$	$PBIAS \le \pm 10$	$PBIAS \le \pm 15$	$PBIAS \le \pm 25$
Good	$0.65 \le NSE \le 0.75$	$0.50 \le RSR \le 0.60$	$\pm 10 \le PBIAS \le \pm 15$	$\pm 15 \le PBIAS \le \pm 30$	$\pm 25 \le PBIAS \le \pm 40$
Satisfactory	$0.50 \le NSE \le 0.65$	$0.60 \le RSR \le 0.70$	$\pm 10 \le PBIAS \le \pm 25$	$\pm 30 \le PBIAS \le \pm 55$	$\pm 40 \le PBIAS \le \pm 70$
Unsatisfactory	$NSE \le 0.50$	$RSR \le 0.70$	$PBIAS \ge \pm 25$	$PBIAS \ge \pm 55$	$PBIAS \ge \pm 70$

[a]NSE, Nash–Sutcliffe efficiency; PBIAS, per cent bias; RSR, root mean square error (RMSE)-observations standard deviation ratio.

Parameters identified during sensitivity analysis and from previous published SWAT studies at multi-sites were used to minimize the relative error statistics during calibration. Streamflow calibration involved total flow adjustments followed by surface and base flow. Once total flow was within 15% of the observed annual average, the model parameters were adjusted so as to reproduce proper surface runoff and base-flow volumes. The model was able to simulate for flow (<11.4% RE) satisfactorily within the thresholds recommended by Santhi *et al.* (2001) (Table 8.3). Overall, model output for flow on an annual time step followed comparable ranges to those reported in the literature for watersheds located in northwest Arkansas (White and Chaubey, 2005).

Monthly calibration and validation

The results of monthly calibration/validation at all seven monitoring gauges are summarized in Table 8.4. For monthly total flow predictions during the calibration period, R^2 ranged from 0.42 (Ballard Creek) to 0.85 (Baron Fork), NSE ranged from 0.51 (Flint Creek) to 0.78 (Ballard Creek), PBIAS ranged from 2.11% (Ballard Creek) to 7.85% (Baron Fork), and RSR ranged from 0.53 (Baron Fork) to 0.72 (Flint Creek). Based on the performance evaluation guidelines suggested by Moriasi *et al.* (2007), the simulated total flow values at all five gauges during the calibration period can be described as 'satisfactory' to 'very good', with the exception of the RSR value at Flint Creek. Also, these results are comparable to or better than those obtained in multi-site SWAT calibration studies by White and Chaubey (2005) and Qi and Grunwald (2005). During the validation period, the relationships between measured and simulated data for total flow for the different statistics were: R^2, 0.71 to 0.85; NSE, 0.50 to 0.85; PBIAS, −11.40 to 22.7; and RSR, 0.12 to 0.70. Thus, the model's performance during the validation period for total flow was similar to that during the calibration period, thereby adding confidence in the model's predictive ability. 'Satisfactory' to 'very good' performance suggests that, despite spatial variation across sub-watersheds, the model was able to capture total flow response adequately in the IRDAA.

A comparison of the observed and simulated monthly total flow at Osage

Table 8.3. Annual flow calibration (1996–2005) results for use of the SWAT model in the Illinois River Drainage Area in Arkansas.

		Average (mm)		Statistics
Gauge	Output	Measured	Simulated	RE (%)[a]
Flint Creek	Total flow	359	354	1.4
	Surface flow	186	165	11.4
	Base flow	174	179	−2.9
Siloam Springs	Total flow	303	306	−1.0
	Surface flow	146	147	−0.7
	Base flow	157	156	0.8
Baron Fork	Total flow	354	350	1.2
	Surface flow	252	240	4.7
	Base flow	102	103	−0.9
Ballard Creek[b]	Total flow	539	547	−1.7
	Surface flow	105	99	5.9
	Base flow	433	422	2.6
Osage Creek (Elm Springs)	Total flow	1048	1089	3.7
	Surface flow	463	469	1.3
	Base flow	572	620	7.9

[a]RE, relative error.
[b]Note: Ballard Creek was only calibrated for 2003–2005 owing to lack of data.

Table 8.4. Monthly flow calibrations and validation results.

		Calibration statistics (1996–2005)[a]				Validation statistics (2006–2008)			
Gauge	Output	R^2	NSE	PBIAS	RSR	R^2	NSE	PBIAS	RSR
Flint Creek	Total flow	0.73	0.51	3.96	0.72				
	Base flow	0.45	–	6.40	1.56		Calibration only		
	Surface flow	0.76	0.55	1.71	0.67				
Siloam Springs	Total flow	0.73	0.53	2.71	0.68	0.76	0.56	8.99	0.12
	Base flow	0.65	0.25	−3.42	0.86	0.68	0.46	7.67	0.87
	Surface flow	0.67	0.37	9.25	0.79	0.69	0.47	3.49	0.78
Baron Fork	Total flow	0.85	0.71	7.85	0.53	0.71	0.50	6.98	0.70
	Base flow	0.76	0.55	−16.42	0.67	0.31	0.38	19.17	0.95
	Surface flow	0.85	0.71	17.06	0.53	0.74	0.69	−0.19	0.68
Ballard Creek	Total flow	0.42	0.78	2.11	0.63	0.81	0.66	−11.40	0.68
	Base flow	0.43	–	3.14	4.11	0.44	–	−38.39	7.0
	Surface flow	0.51	0.88	−2.12	0.5	0.80	0.63	26.11	0.62
Osage Creek	Total flow	0.78	0.57	3.86	0.65	0.85	0.63	22.7	0.60
	Base flow	0.70	0.09	5.04	0.95	0.73	0.30	37.9	0.83
	Surface flow	0.78	0.57	2.27	0.65	0.82	0.57	7.82	0.65
Savoy	Total flow					0.71	0.85	1.92	0.53
	Base flow		Validation only			0.19	0.51	−14.31	0.89
	Surface flow					0.75	0.87	7.87	0.49
Moores	Total flow					0.75	0.56	−1.72	0.65
	Base flow		Validation only			0.31	0.07	8.91	0.94
	Surface flow					0.80	0.61	0.21	0.61

[a]NSE, Nash–Sutcliffe efficiency; PBIAS, per cent bias; R^2, coefficient of determination; RSP, root mean square error (RMSE)-observations standard deviation ratio.

Creek (Elm Springs), Ballard Creek and the watershed outlet (Illinois River near Siloam Springs) is presented in Figs 8.3–8.5. The simulated monthly total flows at Ballard Creek did not very closely follow the observed values, whereas, for the other two gauges, the simulated flows were close to the observed values. Therefore, except at Ballard Creek, the model simulations for base and surface flow for the rest of the gauges could be considered satisfactory (Table 8.4). It is clear from other statistics at Ballard Creek that the model simulations failed to capture base flow at this gauge. The lower R^2, NSE and RSR statistics for monthly base flow calibration could be attributed to the inability of SWAT to correctly account for the timing of subsurface flow and recharge rates.

Because Ballard Creek is a headwater sub-watershed, it was presumed in the model that the monitoring gauge receives input only from a single sub-watershed. However, a recent study has suggested that the Ballard Creek sub-watershed is located in a high karst sensitivity zone (TNC, 2007). Karst hydrology is typically characterized by springs, sinkholes and the loss of streams that provide direct connection between the surface and groundwater (Baffaut and Benson, 2009). Thus, modelling errors in this sub-watershed could also be attributed, in part, to the flow that it receives from outside the sub-watershed, which is not accounted for in the model owing to lack of any observed data. The simulated total flow at the most downstream of the gauges (the watershed outlet at Illinois River near Siloam Springs; Fig. 8.5) largely followed temporal patterns but failed to capture certain large peaks in the spring (e.g. for April 2000, February 2001). The timing of most peaks was consistent with observed data. The results show (Table 8.4) that the model performed satisfactorily or better for total flow, and for its constituents – base flow and surface flow – at the watershed outlet.

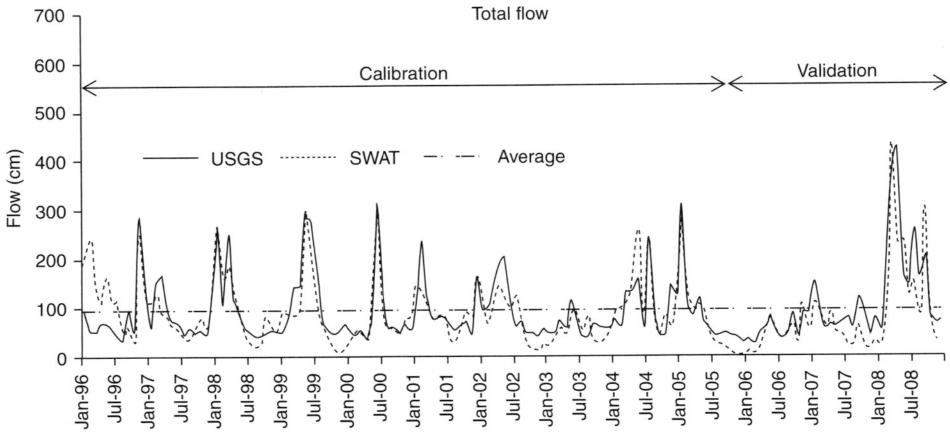

Fig. 8.3. Observed and SWAT-simulated monthly total flow at Osage Creek (Elm Springs), Illinois River Drainage Area, Arkansas.

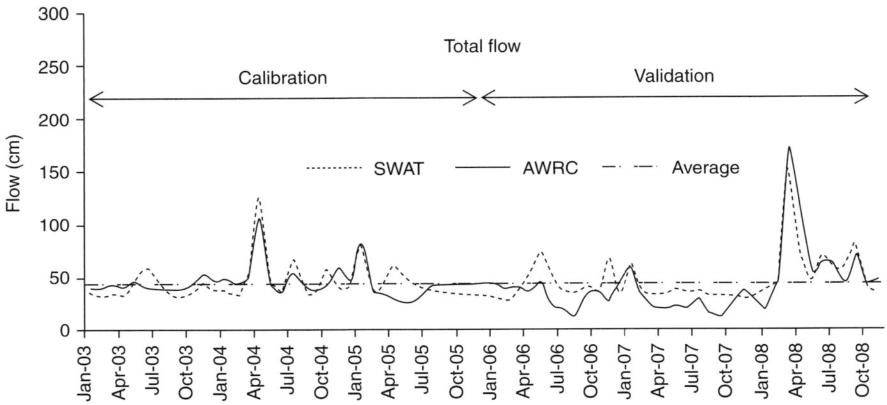

Fig. 8.4. Observed and SWAT-simulated monthly total flow at Ballard Creek, Illinois River Drainage Area, Arkansas.

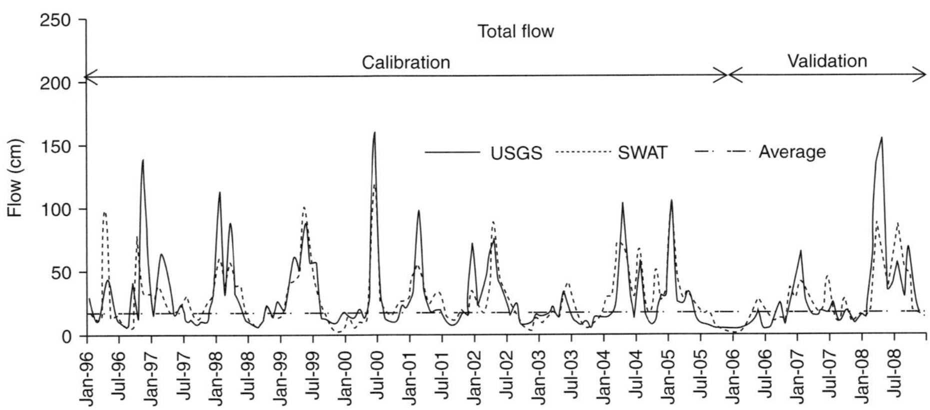

Fig. 8.5. Observed and SWAT-simulated monthly total flow at Illinois River near Siloam Springs, Arkansas.

Summary

Spatially distributed hydrological modelling is becoming possible with the increasing availability of hydrological data, and also with continued improvements in the simulation capabilities of long-term, continuous watershed-scale, quasi-physical models. In this chapter, we used the SWAT model to simulate hydrology in the Illinois River Drainage Area in Arkansas (IRDAA). The results demonstrate that spatially distributed model simulations helped in understanding hydrological processes across sub-watersheds within the IRDAA. The model calibration and validation were carried out for total flow and its constituents: surface flow and base flow. The calibration procedure for hydrology involved a fairly rigorous, 10-year (January 1996–December 2005) statistical and graphical (not presented) comparison of measured and simulated total, base and surface flows. Subsequently, the model was validated using 3 years (January 2006–December 2008) of measured data at the same gauges.

Overall, the model was able to capture the hydrology of the IRDAA satisfactorily, which is evident from the PBIAS values of less than 20%. However, the model was unable to properly simulate patterns of base flows except at the watershed outlet, which is likely to be due to its inability to accurately simulate the timing of subsurface flows and recharge rates.

This study showed that multi-site calibration and validation provide a means to gain more confidence in the model's predictive abilities at the outlet and at other spatially distributed locations within the watershed. We found that hydrological patterns at the outlet may differ from one or more internal locations. In such conditions, calibrating only at the outlet may limit our understanding of the model's performance. Qi and Grunwald (2005) reported that the reliability of hydrological simulations is enhanced by using multiple calibration and validation stations. Our study supports their finding and recommends the calibration and/or validation of model simulations at multiple monitoring sites using the available observed data. We found that the power of a distributed model can be fully exercised if calibrations are conducted at locations where sub-watershed characteristics deviate from general trends in the watershed. One of the future uses of this model is also to simulate for pollutant loads in the IRDAA, and a properly calibrated hydrological model is expected to enhance the success of subsequent sediment and nutrient simulations owing to a lower propagation of errors related to the simulation of hydrological processes by the model.

The case study included in this chapter suggests that the SWAT model is a useful tool for conducting spatially varied hydrological evaluation of large watersheds. Although the strength of the model lies in its capability to simulate both upland and transport processes in one simulation framework, realistically, each of these processes is a simplified understanding of the reality and thus eligible for improvement. Gassman *et al.* (2007) have provided an excellent discussion on improving certain routines in the model to enhance flow and transport across the landscape for simulating connected landscape units within each sub-watershed. Arnold *et al.* (2010) have discussed the latest improvements for enhancing overland water and pollutant simulation capabilities. They have reported that, to predict the impact of landscape position on management practices and environmental response, routines have been developed to evaluate the impact of riparian buffers, potholes, flood plains and variable runoff source areas. New input file functionality has been developed in the SWAT2009 model to allow users to schedule management information on a Julian day and calendar year basis, thereby making these operations independent of cropping rotations or heat unit scheduling. Comprehensive information on the improvements made in the SWAT2009 model is in the offing from the model developers. These model improvements point towards a future where revised modelling routines may likely allow the simulation of watershed responses in a realistic manner.

Acknowledgements

Special thanks are due to the Arkansas Natural Resource Commission (ANRC) for funding this project through the Environmental Protection Agency's (EPA's) 319 Grant Program. We would like to thank state and federal agencies for providing the much needed data for this work. Inputs from several of the University of Arkansas subject matter specialists and extension agents have contributed immensely in fine tuning of the management data. We hereby acknowledge their cooperation and thank all of them individually for their invaluable comments/ suggestions. We would also like to thank the Environmental Task Force (ETF) of the University of Arkansas for organizing several fruitful discussions on critical comparisons of SWAT model output with monitoring efforts carried out in the IRDAA. Last but not least, our thanks are due to SWAT interface/model developers, namely Dr Raghavan Srinivasan, Dr Jeff Arnold and his team at the USDA ARS, Temple, Texas, for promptly addressing our modelling needs. Thanks are also due to two anonymous reviewers whose comments helped in improving this manuscript.

References

Abbaspour, K.C., Yang, J., Maximov, I., Siber, R., Bogner, K., Mieleitner, J., Zobrist, J. and Srinivasan, R. (2007) Modeling hydrology and water quality in the pre-alpine/alpine Thur watershed using SWAT. *Journal of Hydrology* 333, 413–430.

Abbott, M.B., Bathurst, J.C., Cunge, J.A., O'Connell, P.E. and Rasmussen, J. (1986) An introduction to the European Hydrologic Système Hydrologique Européen, SHE 1: History and philosophy of a physically-based distributed modeling system. *Journal of Hydrology* 87, 45–59.

ADEQ (Arkansas Department of Environmental Quality) (2008) List of impaired waterbodies (303(d) list). Available at: http://www.adeq.state.ar.us/water/branch_planning/pdfs/303d_list_2008.pdf (accessed 10 June 2010).

Allen, R.G., Pereira, L.S., Raes, D. and Smith, M. (1998) *Crop Evapotranspiration – Guidelines for Computing Crop Water Requirements.* FAO Irrigation and Drainage Paper No. 56. UN Food and Agriculture Organization (FAO), Rome.

Arnold, J.G. and Allen, P.M. (1999) Automated methods for estimating baseflow and groundwater recharge from stream flow records. *Journal of the American Water Resources Association* 35, 411–424.

Arnold, J.G., Allen, P.M. and Bernhardt, G. (1993) A comprehensive surface-groundwater flow model. *Journal of Hydrology* 142, 47–69.

Arnold, J.G., Srinivasan, R., Muttiah, R.S. and Allen, P.M. (1998) Large-area hydrologic modeling and assessment: Part I. Model development. *Journal of the American Water Resources Association* 34, 73–89.

Arnold, J.G., Gassman, P.W. and White, M.J. (2010) New developments in the SWAT ecohydrology model. In: *Conference Proceedings, 21st Century Watershed Technology: Improving Water Quality and Environment*, 21–24 February 2010. ASABE Publication No. 701P0210cd, American Society of Agricultural and Biological Engineers, St Joseph, Michigan.

Baffaut, C. and Benson, V.W. (2009) Modeling flow and pollutant transport in a karst watershed with SWAT. *Transactions of the ASABE* 52, 469–479.

Bekele, G.E. and Nicklow, W.J. (2007) Multi-objective automatic calibration of SWAT using NSGA-II. *Journal of Hydrology* 34, 165–176.

Beven, K.J. (2000) *Rainfall-Runoff Modeling: the Primer.* John Wiley and Sons, New York.

Borah, D.K. and Bera, M. (2003) Watershed-scale hydrologic and nonpoint-source pollution models: review of mathematical bases. *Transactions of the ASABE* 46, 1553–1566.

Borah, D.K. and Bera, M. (2004) Watershed-scale hydrologic and nonpoint-source pollution models: review of applications. *Transactions of the ASABE* 47, 789–803.

Cao, W., Bowden, W.B., Davie, T. and Fenemor, A. (2006) Multi-variable and multi-site calibration and validation of SWAT in a large mountainous catchment with high spatial variability. *Hydrological Processes* 20, 1085–1099.

Chow, V.T., Maidment, D.R. and Mays, L.W. (1988) *Applied Hydrology.* McGraw-Hill, New York.

Deal, C., Edwards, J., Pellmann, N., Tuttle, R. and Woodward, D. (1997) *Ponds – Planning, Design, Construction.* Agriculture Handbook No. 590. USDA Natural Resource Conservation Service, Washington, DC.

Ekka, S.A., Hagard, B.E., Matlock, M.D. and Chaubey, I. (2006) Dissolved phosphorus concentrations and sediment interactions in effluent-dominated Ozark streams. *Ecological Engineering* 26, 375–391.
Engeland, K., Braud, I., Gottschalk, L. and Leblois, E. (2006) Multi-objective regional modelling. *Journal of Hydrology* 327, 339–351.
Fontaine, T.A., Cruickshank, T.S., Arnold, J.G. and Hotchkiss, R.H. (2002) Development of a snowfall–snowmelt routine for mountainous terrain for Soil and Water Assessment Tool (SWAT). *Journal of Hydrology* 262, 209–223.
Freeze, R.A. and Harlan, R.L. (1969) Blueprint for a physically-based, digitally-simulated hydrologic response model. *Journal of Hydrology* 9, 237–258.
Gassman, P.W., Reyes, M.R., Green, C.H. and Arnold, J.G. (2007) The soil and water assessment tool: historical development, applications, and future research directions. *Transactions of the ASABE* 50, 1211–1250.
Gassman, P.W., Williams, J.R., Wang, X., Saleh, A., Osei, E., Hauck, L.M., Izaurralde, R.C. and Flowers, J.D. (2010) The Agricultural Policy/Environmental eXtender (APEX) Model: an emerging tool for landscape and watershed environmental analysis. *Transactions of the ASABE* 53, 711–740.
Gitau, M.W., Srivastava, R.K. and Chaubey, I. (2007) Watershed response modeling in Arkansas priority watersheds: experience with SWAT autocalibration. ASABE Paper No. 072171, American Society of Agricultural and Biological Engineers, St Joseph, Michigan.
Grayson, R. and Bloschl, G. (2000) Spatial modeling of catchment dynamics. In: Grayson, R. and Bloschl, G. (eds) *Spatial Patterns in Catchment Hydrology: Observations and Modeling.* Cambridge University Press, Cambridge, UK, pp. 51–81.
Green, W.H. and Ampt, G.A. (1911) Studies on soil physics, 1. The flow of air and water through soils. *Journal of Agricultural Sciences* 4, 11–24.
Hargreaves, G. and Samani, Z.A. (1985) Reference crop-evapotranspiration from temperature. *Applied Engineering in Agriculture* 1, 96–99.
Knisel, W.G. (ed.) (1980) *CREAMS, a Field Scale Model for Chemicals, Runoff, and Erosion from Agricultural Management Systems.* Conservation Research Report No. 26. US Department of Agriculture, Washington, DC.
Leonard, R.A., Knisel, W.G. and Still, D.A. (1987) GLEAMS: Groundwater Loading Effects on Agricultural Management Systems. *Transactions of the ASAE* 30, 1403–1428.
Li, X., Weller, D.E. and Jordan, T.E. (2010) Watershed model calibration using multi-objective optimization and multi-site averaging. *Journal of Hydrology* 380, 277–288.
Mamillapalli, S. (1998) Effect of spatial variability on river basin stream flow modeling. PhD thesis, Purdue University, West Lafayette, Indiana.
Miller, S.N., Kepner, W.G., Mehaffey, M.H., Hernandez, M., Miller, R.C., Goodrich, D.C., Devonald, K., Heggem, D.T. and Miller, W.P. (2002) Integrating landscape assessment and hydrologic modeling for land cover change analysis. *Journal of the American Water Resources Association* 38, 915–929.
Mishra, S.K. and Singh, V.P. (2003) *Soil Conservation Service Curve Number (SCS-CN) Methodology.* Kluwer Academic Publishers, Dordrecht, The Netherlands.
Monteith, J.L. (1965) Evaporation and the environment. In: *The State and Movement of Water in Living Organisms,* XIXth Symposium, Society for Experimental Biology, Swansea. Cambridge University Press, Cambridge, UK, pp. 205–234.
Moriasi, D.N., Arnold, J.G., Van Liew, M.W., Bingner, R.L., Harmel, R.D. and Veith, T.L. (2007) Model evaluation guidelines for systematic quantification of accuracy in watershed simulations. *Transactions of the ASABE* 50, 885–900.
Nash, J.E. and Sutcliffe, J.V. (1970) River flow forecasting through conceptual models part 1- A discussion of principles. *Journal of Hydrology* 10, 282–290.
Neitsch, S.L., Arnold, J.G., Kiniry, J.R. and Williams, J.R. (2005) *Soil and Water Assessment Tool Theoretical Documentation, Version 2005.* Grassland, Soil and Water Research Laboratory, USDA Agricultural Research Service/Blackland Research Center, Texas Agricultural Experiment Station, Temple, Texas. Available at: http://swatmodel.tamu.edu/media/1292/SWAT2005theory.pdf (accessed 17 February 2011).
Priestley, C.H.B. and Taylor, R.J. (1972) On the assessment of surface heat flux and evaporation using large-scale parameters. *Monthly Weather Review* 100, 81–92.
Qi, C. and Grunwald, S. (2005) GIS-based hydrologic modeling in the Sandusky watershed using SWAT. *Transactions of the ASAE* 48, 169–180.

Refsgaard, J.C. (2000) Towards a formal approach to calibration and validation of models using spatial data. In Grayson, R. and Bloschl, G. (eds) *Spatial Patterns in Catchment Hydrology: Observations and Modeling.* Cambridge University Press, Cambridge, UK, pp. 329–367.

Santhi, C., Arnold, J.G., Williams, J.R., Hauck, L.M. and Dugas, W.A. (2001) Applications of a watershed model to evaluate management effects on point and nonpoint source pollution. *Transactions of the ASAE* 44, 1559–1570.

Srivastava, P., Migliaccio, K.W. and Simunek, J. (2007) Landscape models for simulating water quality at point, field, and watershed scales. *Transactions of the ASABE* 50, 1683–1693.

SWAT (Soil and Watershed Assessment Tool) (2010) Official SWAT web site. Available at: http://swatmodel.tamu.edu/ (accessed 9 October 2010).

TNC (The Nature Conservancy) (2007) Karst area sensitivity map for Northwest Arkansas: Washington County. Available at: http://www.nwarpc.org/pdf/GIS-Imagery/KASM_WASHINGTON_CO.pdf (accessed 15 August 2010).

Tolson, B.A. and Shoemaker, C.A. (2007) Cannonsville Reservoir Watershed SWAT2000 model development, calibration and validation. *Journal of Hydrology* 337, 68–86.

US Census Bureau (2010) Available at: http://www.census.gov/ (accessed 17 February 2011).

USDA SCS (1972) Hydrology Section 4, Chapter 4-10. In: *National Engineering Handbook.* U.S. Department of Agriculture, Soil Conservation Service, Washington, DC.

US EPA (2007). BASINS 4.0–Fact Sheet. US Environmental Protection Agency, Washington, DC. Available at: http://www.epa.gov/waterscience/basins/fs-basins4.html (accessed 15 August 2010).

Van Liew, M. and Garbrecht, J.D. (2003) Hydrologic simulation of the Little Washita River experimental watershed using SWAT. *Journal of the American Water Resources Association* 39, 413–426.

White, K.L. and Chaubey, I. (2005) Sensitivity analysis, calibration, and validations for a multisite and multivariable SWAT model. *Journal of the American Water Resources Association* 41, 1077–1089.

Williams, J.R., Jones, C.A. and Dyke, P.T. (1984) A modeling approach to determining the relationship between erosion and soil productivity. *Transactions of the ASAE* 27, 129–144.

Zhang, X. and Srinivasan, R. (2010) GIS-based spatial precipitation estimation using next generation radar and raingauge data. *Environmental Modelling and Software* 25, 1781–1788.

Zhang, X., Van Liew, M. and Srinivasan, R. (2008) Multi-site calibration of the SWAT model for hydrologic modeling. *Transactions of the ASABE* 51, 2039–2049.

Zhang, X., Srinivasan, R. and Van Liew, M. (2010) On the use of multi-algorithm, genetically adaptive multi-objective method for multi-site calibration of the SWAT model. *Hydrological Processes* 24, 955–969.

9 Hydrological Modelling: A Case Study of the Kosi Himalayan Basin Using SWAT

A.K. Gosain,* Sandhya Rao and A. Mani

Introduction

A watershed, catchment or basin is an area with a common natural drainage course. In other words, it is an area over which hydrological processes are integrated. The behaviour of each process is controlled by its own characteristics as well as by its interaction with other processes active in the watershed. The predominant hydrological processes include rainfall, interception, evapotranspiration, snow melt, surface runoff, infiltration, percolation and subsurface flow. During the last four decades researchers have been actively involved in formulating mathematical models to represent the various hydrological processes prevalent in a watershed, and there is now a plethora of mathematical models whose details are available in literature. These models vary from empirical models for the evaluation of a single flood event to the more recent comprehensive distributed models.

A detailed description of the various categories of models is provided here, with the objective of familiarizing the reader with the advantages and disadvantages of various models. It may be emphasized that, although information on local water availability as well as its variability in time is essential for proper watershed planning and management, measurement of these quantities in terms of flows is not financially viable at such scales. In this context, hydrological simulation modelling is a very effective tool which can allow assessment of these quantities at the watershed scale. One such model, namely, the Soil and Water Assessment Tool (SWAT), is described briefly and its performance in comparison with other similar models evaluated.

A case study of the Kosi basin in the eastern Himalayas using the SWAT model is then presented. The study evaluates the impact of climate change on streamflow in the Upper Kosi basin by using a Regional Climate Model (RCM) coupled with SWAT. The potential impacts of climate change on water yield and other hydrological budget components are quantified by driving SWAT with current and future climate scenarios.

'Sustainable' water resources management is going to be the key question in view of both ever-increasing population and the possible impacts of climate change. It is unwise to think of adaptation and coping strategies for climate change as something new, or devoid of basic water-management practices. However, currently, there are no management options that are uniquely suited for adaptation to climate change that would be potentially different from those already employed for coping with

* Corresponding author: gosain@civil.iitd.ac.in

contemporary climate variability. The only substantive difference is whether one adopts a more conventional and incremental 'no regrets' approach, or a more anticipatory and 'precautionary' approach. Integrated Water Resources Management (IWRM) is now the encompassing paradigm for adaptation to contemporary climate variability, and it is the prerequisite for coping with the still uncertain consequences of global warming, the climate changes associated with it and their repercussions on the water cycle. A framework of IWRM that can be readily adopted is described.

Materials and Methods

Overview of hydrological models

Mathematical models in hydrology were developed during the second half of the 19th century mainly to estimate the maximum flow for designing drainage systems, flood flow at outlets, etc.

One of the earliest and simplest hydrological models is the rational formula (Mulvaney, 1851) used for small catchments. This formula quantitatively expresses flood flow rates in relation to rainfall and watershed area. The method is based on the concept of the 'time of concentration', which is the time required for water to flow from the most remote point of the area to the outlet. It is assumed that when the duration of the storm equals the time of concentration all parts of the watershed are contributing to the discharge at the outlet and that this results in maximum flow corresponding to the intensity of the rainfall. Later, the unit hydrograph concept was developed on the basis of the principle of superposition, and introduced by Sherman (1932). In the 1950s, a systems approach was used for the analysis of complex dynamic systems. The response function was obtained from the analysis of input and output data and represented by mathematical expressions. The response function carried no physical significance in regard to the system.

Studies on non-linear systems were also carried out based on volterra integrals, orthogonal polynomials (Amorocho and Orlob, 1961) and piecewise linearizations (Todini and Wallis, 1977). Advanced statistical techniques were also applied for expressing the unit hydrograph. Subsequent developments of these techniques were satisfactory from a mathematical point of view, but lost their connection with the real hydrological system. Although these techniques helped the further development of the unit hydrograph, they failed to incorporate many other subsystems active in the rainfall-runoff process.

During the 1960s, continuous hydrological simulation was introduced through conceptual models. These models are continuous volume-accounting models based on water balance. They have proved to be useful for studying catchment response over time to a wide variety of weather sequences. An early application of this type of model was described by Pereira *et al.* (1962), and a large number of these models subsequently came to light, such as that of Dawdy and O'Donnell (1965), the Stanford Watershed Model IV (Crawford and Linsley, 1966), the Stream flow Synthesis and Reservoir Regulation (SSARR) model (Rockwood and Nelson, 1966), the Sacramento model (Burnash *et al.*, 1973), etc. The basic functioning of these models is controlled by parameters that represent the processes of the drainage system and are estimated by an optimization procedure.

In many cases, when parameter estimates are made based on the objective functions, unrealistic values may be obtained, owing to errors in measurement of data and in the description of the various processes of the model. In addition, the data observability conditions could not always be guaranteed (Sorooshian and Gupta, 1983). This lack of any real 'physical' connection between model parameters and reality prompted Freeze and Harlan (1969) to propose a mathematical model based on distributed physical knowledge of the phenomena, and describing surface flow, flow in the unsaturated zone and flow under the water table by means of differential equations.

The next logical step in modelling was to obtain more reliable estimates of runoff from ungauged catchments by linking parameter values to catchment characteristics. The parameters which have physical significance can be measured from field experiments.

1

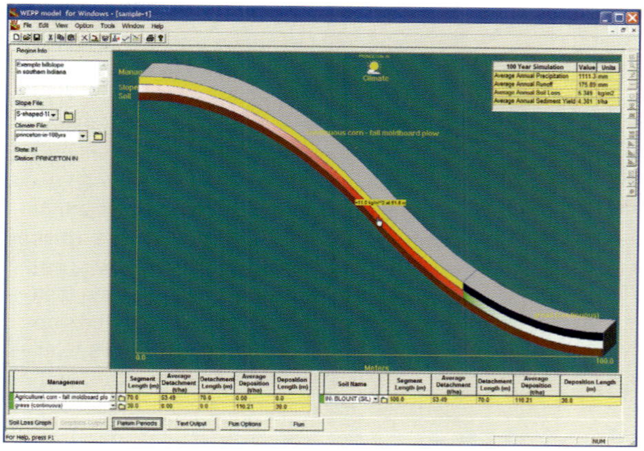

Plate 1. WEPP (Water Erosion Prediction Project) model Windows interface, showing 100-year simulation results for an S-shaped slope profile near Princeton in southern Indiana. The management layer at the top of the three-part profile (management, slope and soil) contains two sections: continuous maize (corn) using an autumn mouldboard plough tillage system at the top, and a continuous grass buffer strip at the bottom of the slope. Results are shown both in the text box (upper right) and by image shading in the centre (slope) profile layer, where rates of soil loss are shown in shades of red and rates of sediment deposition in shades of green. The mouse cursor can also be hovered over any location in the centre profile layer and the predicted erosion rate at that point displayed, as shown for 51.8 m downslope, with a predicted average annual erosion rate of 11 kg m^{-2}.

2

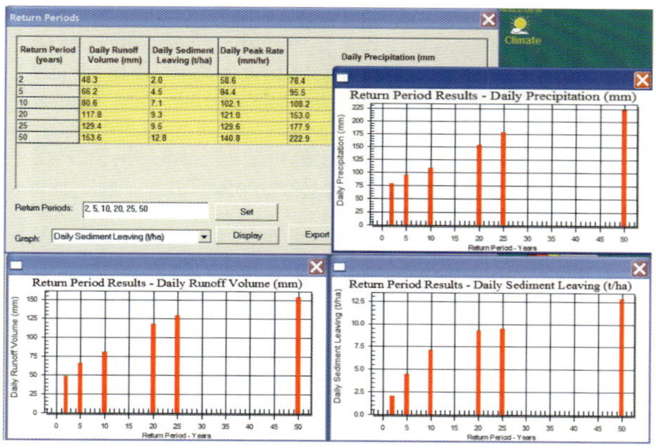

Plate 2. WEPP (Water Erosion Prediction Project) model Windows interface Return Period output screen, showing results from a 100-year simulation period for an S-shaped slope profile near Princeton in southern Indiana, USA, with graphs of daily precipitation, runoff volume (as depth in mm) and sediment loss by return period.

3

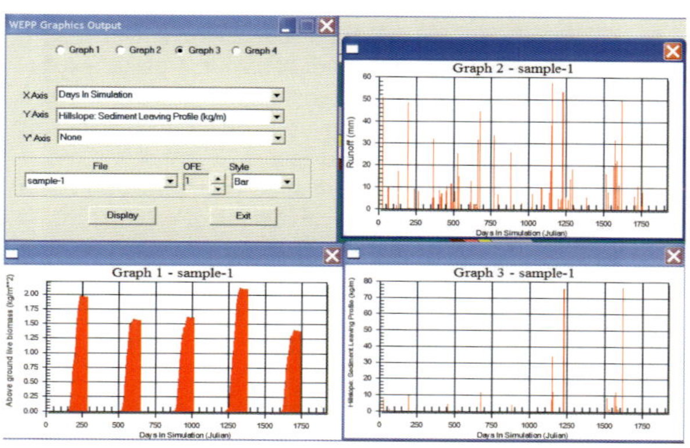

Plate 3. WEPP (Water Erosion Prediction Project) model Windows interface Graphical Output screen, showing results from a 5-year simulation for an S-shaped slope profile near Princeton in southern Indiana, with graphs for above-ground live biomass (Graph 1), runoff depth (Graph 2) and sediment loss (Graph 3) versus time.

4

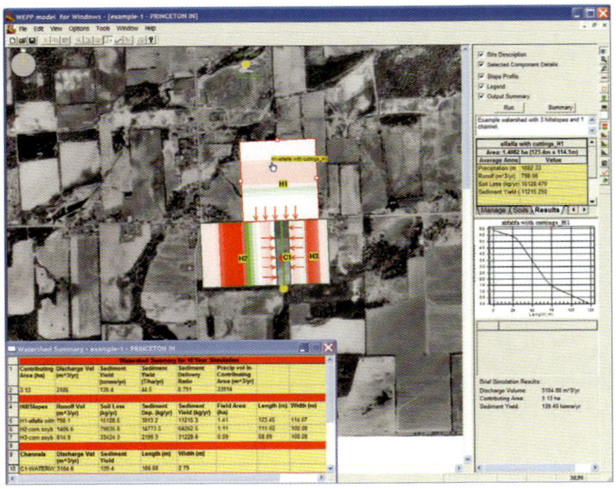

Plate 4. WEPP (Water Erosion Prediction Project) model Windows interface with results from a small agricultural watershed simulation displayed. Three hillslopes (H1, lucerne (alfalfa); H2 and H3, maize (corn)–soybean) contribute runoff and sediment to a single channel (C1, a grass waterway) under an autumn mouldboard ploughing management system. Tabular results from a 10-year simulation are shown in the window at bottom left and component details in the table on the right. Colour shading of the hillslope areas in the graphics window shows relative amounts of soil loss in shades of red and sediment deposition in shades of green.

5

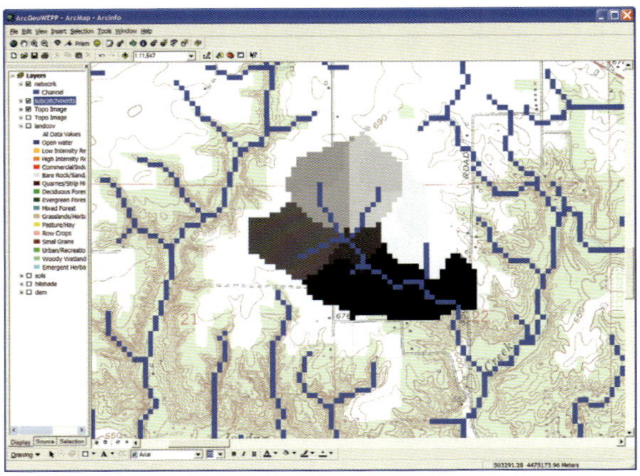

Plate 5. GeoWEPP (Geo-spatial interface for Water Erosion Prediction Project) interface screen, showing a small watershed near West Lafayette, Indiana, delineated utilizing U.S. Geological Survey 30 m DEM (digital elevation model) topographic data. Channel networks, as well as watershed sub-catchments, were determined using the TOPAZ (TOpographic PArameteriZation) delineation program.

6

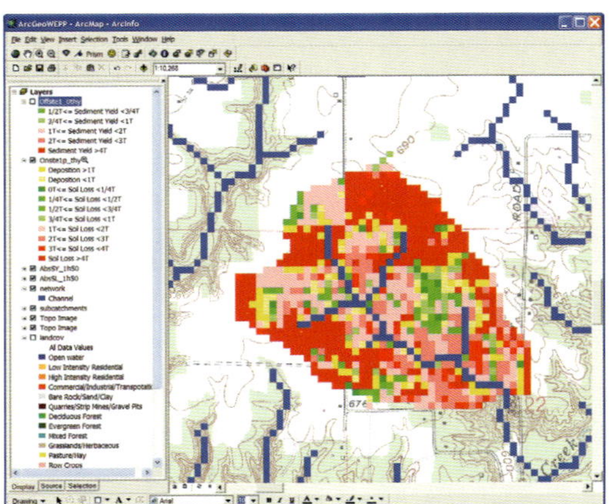

Plate 6. Spatial soil erosion results predicted by the WEPP (Water Erosion Prediction Project) model and displayed in the GeoWEPP ArcGIS interface. In colour, erosion results are scaled by a user-selected Tolerable soil loss value (T-value), and soil loss values exceeding T are displayed in shades of red, those below T in shades of green, and depositional regions in shades of yellow.

7

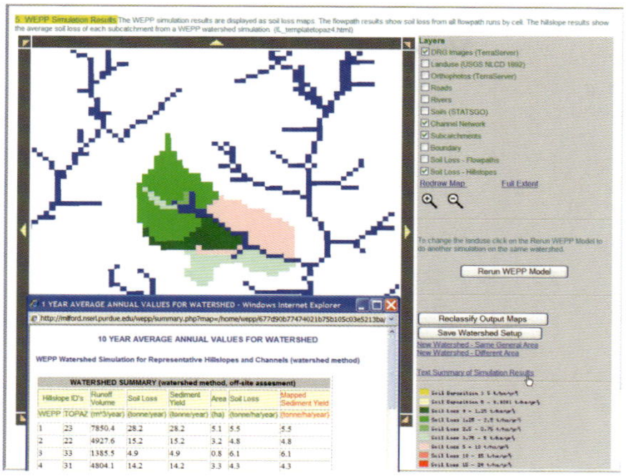

Plate 7. Spatial hillslope sediment loss results predicted by the WEPP (Water Erosion Prediction Project) model for a small watershed in Illinois, and displayed in the web GIS interface described in the text. Climate information for Fairfield, Illinois was used, with a Proctor silt loam soil and maize–soybean autumn mulch tillage land management. When viewed in colour, results are scaled by a user-selected tolerable soil loss value (T-value), and soil loss values exceeding T are displayed in shades of red, those below T are in shades of green, and depositional regions are shown in shades of yellow.

8

Plate 8. A photograph showing a typical agricultural field where the WEPP (Water Erosion Prediction Project) model could be applied. This landscape from north-eastern Indiana has a tilled area bordered by forest and a fence row, and a narrow grass vegetation (buffer) area and grassed waterway that removes water and sediment from the field at the bottom. It could be simulated as a WEPP hillslope with a representative profile (possibly along the solid black line drawn from centre left to centre bottom).

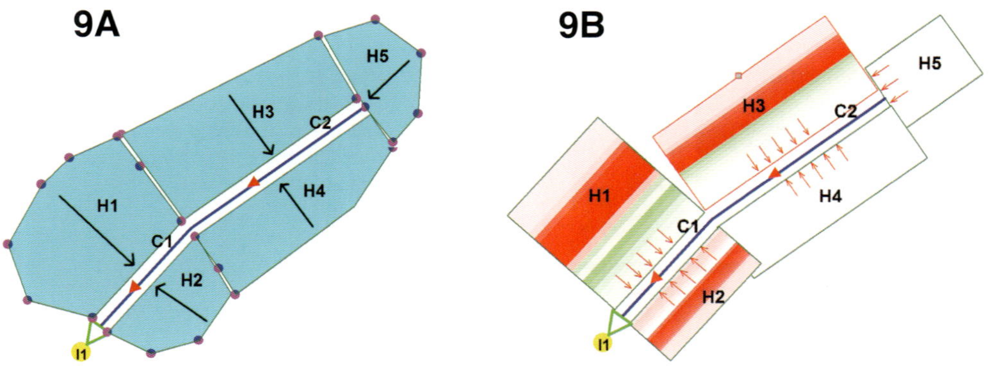

Plate 9. A small watershed, showing actual sub-catchment areas as polygons (A) with flow directions for the five hillslopes (H1–H5), two channels (C1, C2) and one impoundment (I1). The actual WEPP (Water Erosion Prediction Project) model simulates these areas as rectangles (B), and spatial soil loss values can be displayed down the profiles. In this simulation, H1 and H3 were similar to the profile in Fig. 5.2 with heavy tillage cropped regions and grass strips; H2 had heavy tillage, and H4 and H5 had no-till cropping. Shading in (B) indicates degree of soil erosion.

10

Plate 10. Results from a 10-year WEPP (Water Erosion Prediction Project) model simulation, with continuous maize (corn) with autumn mouldboard-ploughed tillage on hillslopes H1, H2, H3 and H5, and two overland flow elements (OFEs) of grass and maize (corn) strips on hillslopes H4 and H6.

11

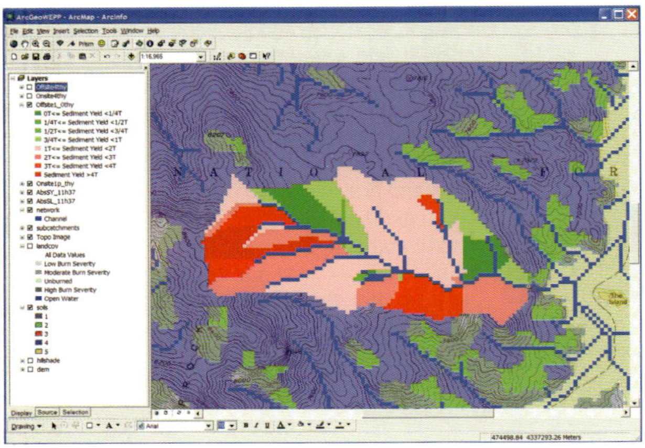

Plate 11. Application of WEPP (Water Erosion Prediction Project) and GeoWEPP (Geo-spatial interface for WEPP) to a burned forested region in Jefferson County, Colorado draining into the south-west portion of Cheesman Lake. Results from a 10-year WEPP simulation shown here indicate the highest sediment losses from the southern and western sub-catchments.

12

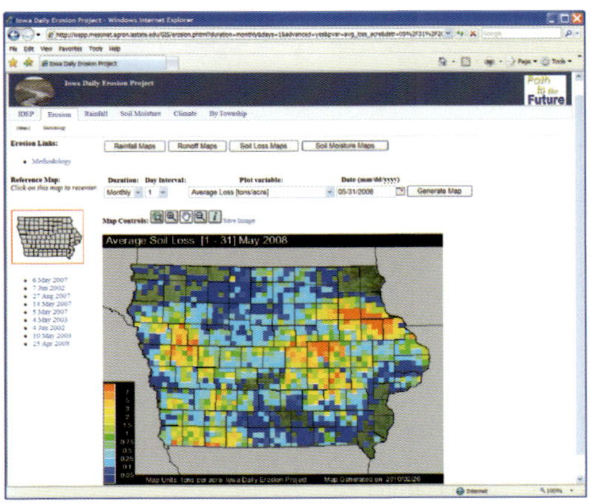

Plate 12. Results from the Iowa Daily Erosion Project website, showing predicted average soil loss by township in Iowa for the month of May 2008 (see http://wepp.mesonet.agron.iastate.edu).

13

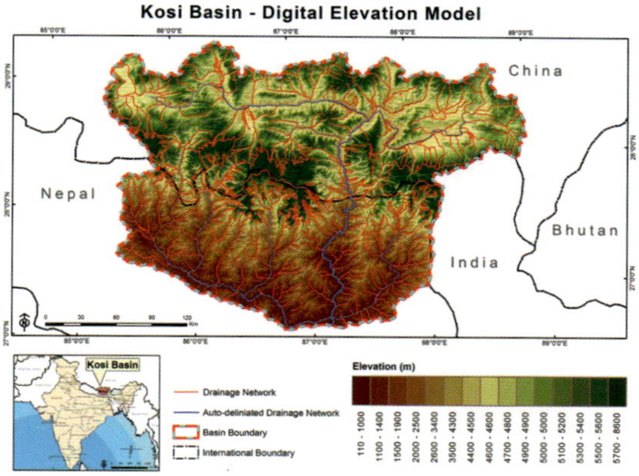

Plate 13. Digital elevation model (DEM) of the Kosi River basin showing the automatically delineated watershed (basin boundary) at Chatara-Kothu (in the south-east, see Plate 14) and drainage network generated from the DEM. The drainage network (shown in blue) was generated using a threshold value of 100,000 ha; the actual drainage is shown in red.

14

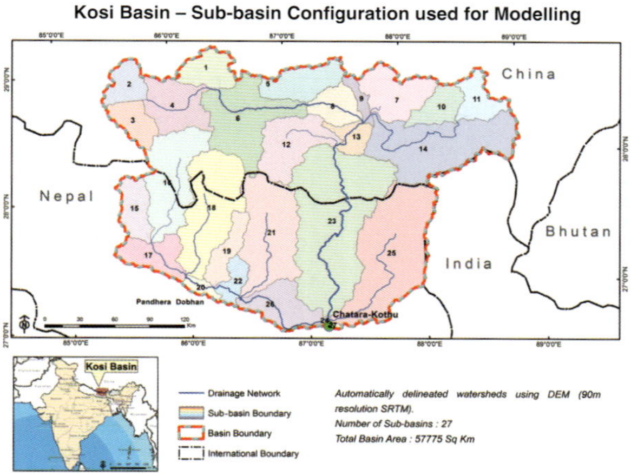

Plate 14. Subdivision of the Kosi River basin into 27 sub-areas delineated in the basis of the drainage network generated from the digital elevation map (DEM) using a threshold of 100,000 ha.

15A

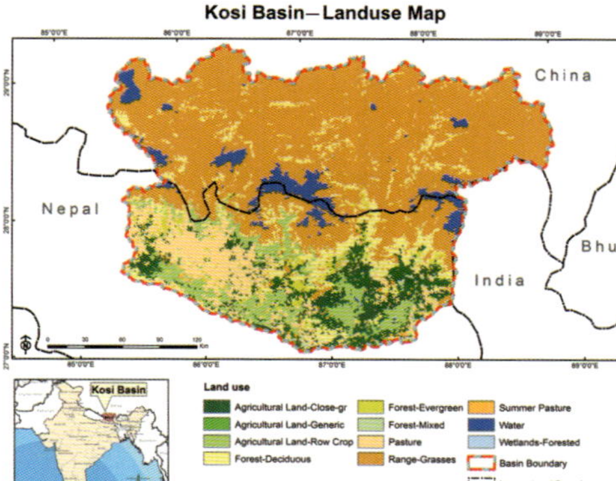

15B

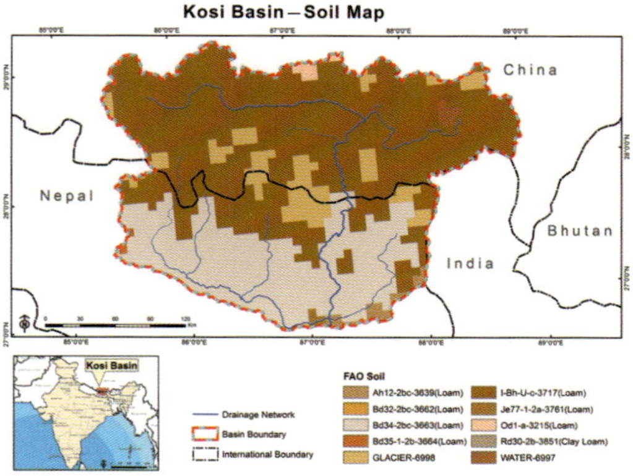

Plate 15. Land use (A) and soil type (B) geographical information system (GIS) layers of the Kosi River basin.

16

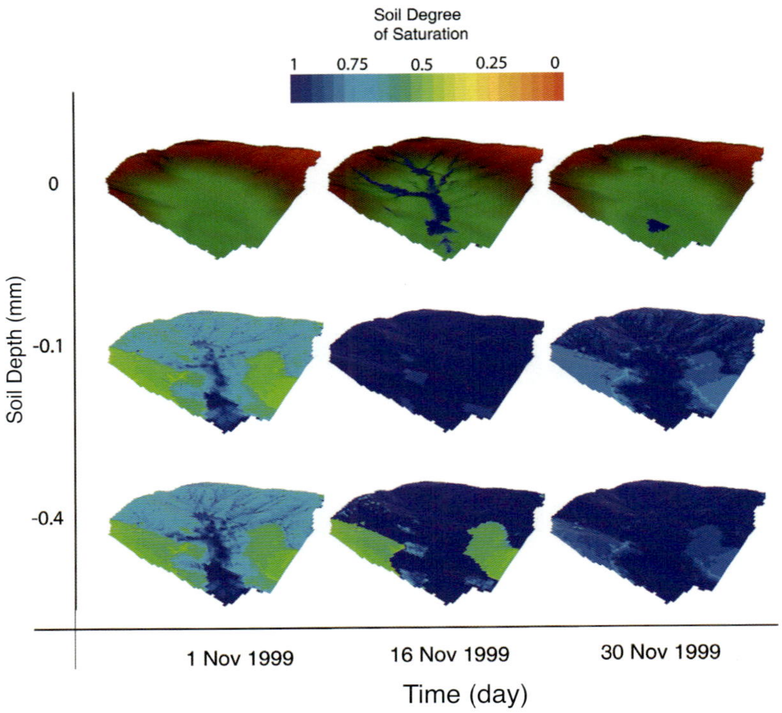

17

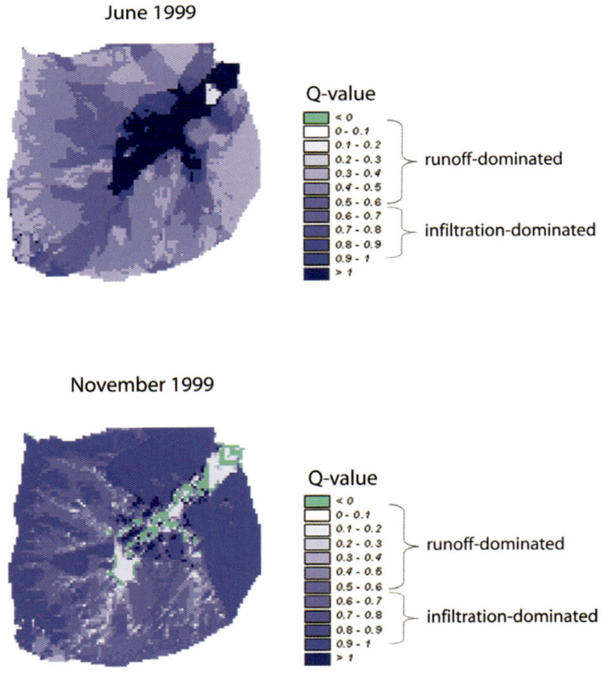

Plate 16. Soil water contents (degree of saturation) at the Spescia catchment, Serra de' Conti, Central Italy, on different dates in November 1999 and at different soil depths.

Plate 17. Analysis of runoff-generating mechanisms showing areas of Hortonian and Dunnian flow (top for the June 1999 runoff event and bottom for the November 1999 event). The *Q* value is the normalized rate of water exchange between the surface and the subsurface.

18

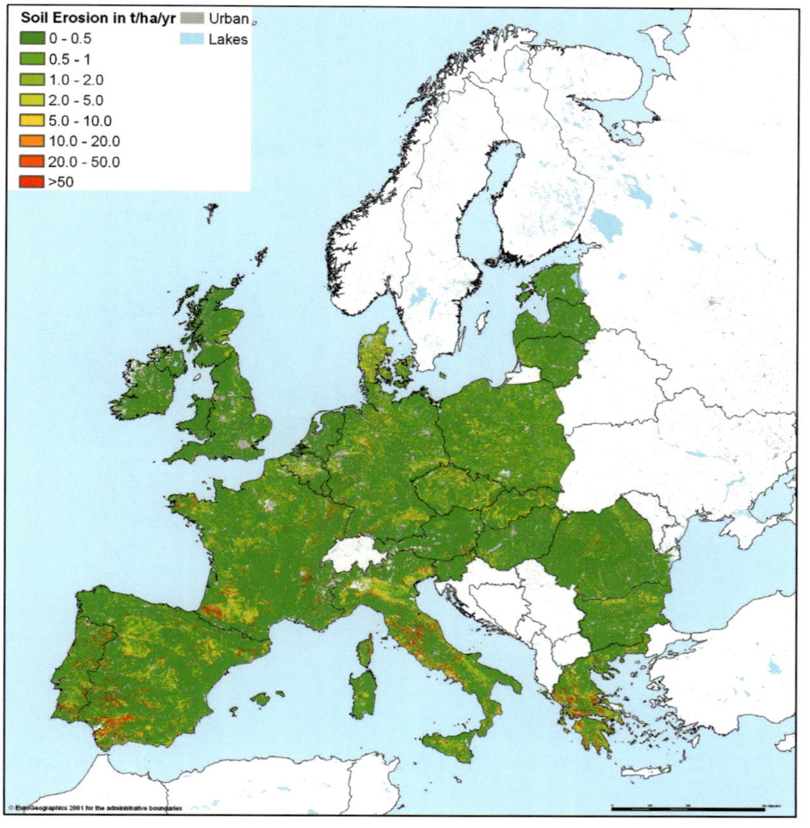

19A

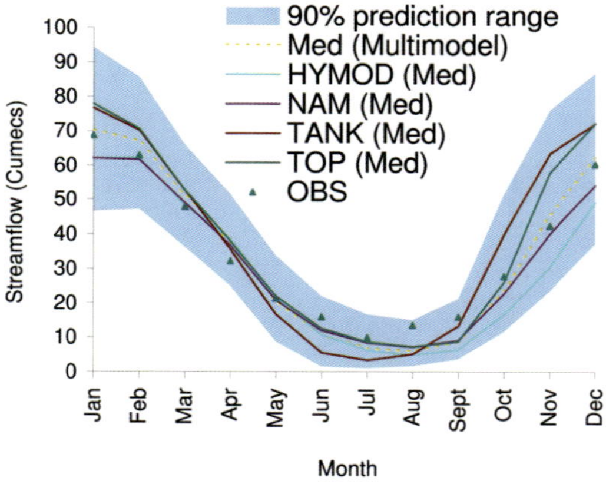

19B

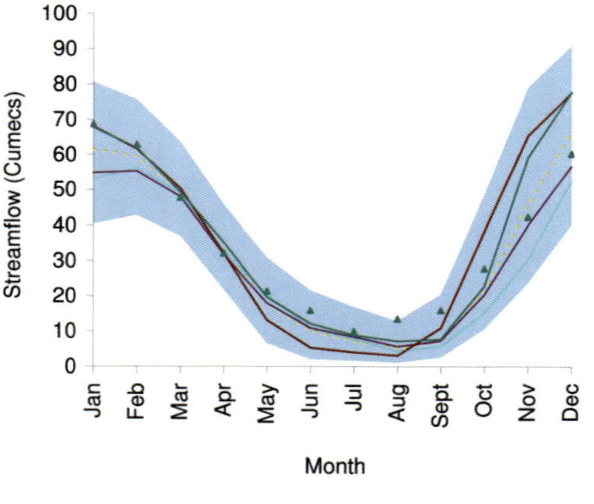

Plate 18. Pan-European Soil Erosion Risk Assessment in Europe (PESERA) (Adapted from The PESERA Map, Version 1 October 2003, PESERA Project, European Commission Joint Research Centre (JRC), Ispra, Italy).

Plate 19. The prediction interval (upper 95% and lower 5%) for streamflow simulated by a number of behavioural parameter sets of four hydrological models forced with climate inputs from the Hadley Centre Regional Climate model HadCM3 and the Intergovernmental Panel on Climate Change (IPCC) A2 emissions scenario for two periods, namely (A) 1970–1990 and (B) 2020–2029, is shown in the shaded region. The median prediction from each individual hydrological model is shown as lines for the Boyne River basin in Ireland (cumec, $m^3\ s^{-1}$). See Chapter 16 for further details.

20

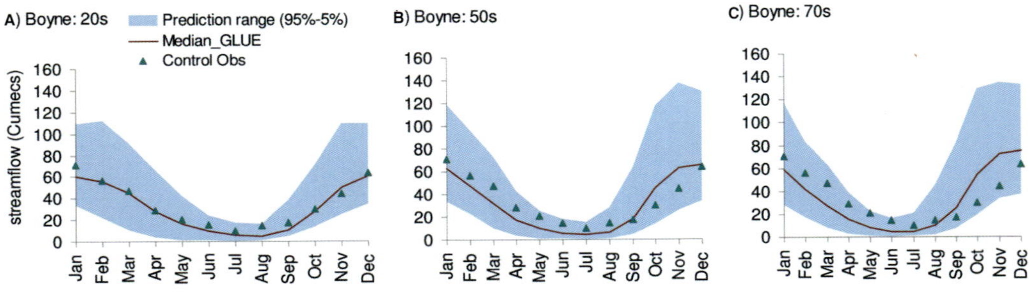

21

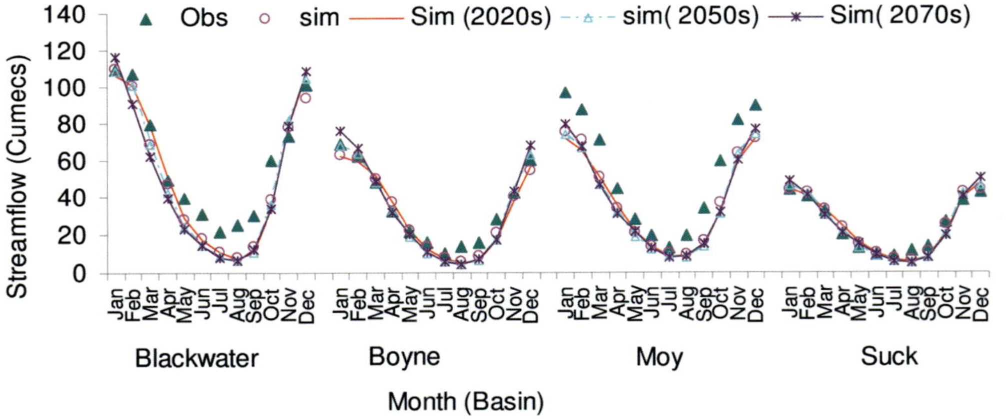

Plate 20. The total uncertainty envelope for streamflow predictions derived from six climate scenarios (three global climate models with the A2 and B2 emissions scenarios from the Intergovernmental Panel on Climate Change) and four hydrological models for the Boyne catchment in Ireland for (A) the 2020s, (B) the 2050s and (C) the 2070s. Control observations are for the 1970–1990 period and the median prediction used GLUE (Generalized Likelihood Uncertainty Estimation) (cumec, $m^3\ s^{-1}$). See Chapter 16 for further details.

Plate 21. Median predictions of streamflow derived from four hydrological models forced with downscaled output from three global climate models and two emissions scenarios (A2 and B2 emissions scenarios from the Intergovernmental Panel on Climate Change) for the Blackwater, Boyne, Moy and Suck catchments in Ireland. Control observations are for the 1970–1990 period and the median prediction used GLUE (Generalized Likelihood Uncertainty Estimation) (cumec, $m^3\ s^{-1}$). See Chapter 16 for further details.

However, this development faced difficulty, mainly on account of catchment heterogeneity. Runoff-generating processes vary spatially in a pattern determined by many physical, vegetative and topographic features. This phenomenon of spatial heterogeneity has been taken care of by distributed models, and the era of the distributed model is ongoing.

On the basis of the historical development of the hydrological models, modelling approaches may be classified as black-box models, conceptual models and deterministic models – although this categorization cannot be rigidly followed because there is considerable overlap between the various classes of models.

Black-box models

Black-box models describe, mathematically, the relation between input (precipitation) and output (runoff) by establishing a statistical correspondence between the two, but without describing the physical process by which they are related. These models are often successful within the range of data available/collected and analysed from a region. The reason is that the mathematical structure carries with it an implicit representation of the underlying physical system. Beyond the range of analysed data, the prediction depends only on mathematical techniques, as the physical significance is lost. The recent Artificial Neural Networks (ANN) models belong to this category (Anmala *et al.*, 2000; Uvo *et al.*, 2000; Hsu *et al.*, 2002, Sudheer *et al.*, 2002; Riad *et al.*, 2004).

Deterministic models

These models are based on complex physical theory. They are necessarily distributed because of the non-linear partial differential equations used to describe hydrological processes. It has been noted that analytical solutions are generally not available for solving these equations. Hence, resort must be made to the adoption of partial differential equations, including the finite difference method (Freeze, 1971), finite element methods (Beven, 1977; Ross *et al.*, 1979) and integral finite difference and boundary integral methods, which are difficult to implement and are time-consuming. Simplifications have been made to these methods and kinematic wave theory has been used as an alternative. The models offer the ability to simulate complete runoff and the effect of catchment changes, which is particularly important in case of resource management. A noteworthy aspect of deterministic models is that they offer an internal view of the process, which enables improved understanding of the hydrological system.

One of the most well-known distributed models of this category is the Système Hydrologique Européen (European Hydrological System), SHE (Abbott *et al.*, 1986). SHE is an advanced physically based, distributed modelling system developed collaboratively by the Institute of Hydrology, the Danish Hydraulic Institute and SOGREAH (the French acronym for 'Grenoble Hydraulics Studies and Applications Company'), France. The model achieves the spatial distribution of catchment parameters in the horizontal direction through an orthogonal grid network, and in the vertical direction by a column of horizontal layers at each grid square. Each process of the hydrological cycle (snow melt, canopy interception, evapotranspiration, overland and channel flow, unsaturated and saturated subsurface flow) is modelled either by finite difference representations of the theoretical partial differential equations of mass, momentum and energy conservation, or by empirical equations derived from independent experimental research. Interception is modelled by a modified Rutter model (Rutter *et al.*, 1971), which is essentially an accounting procedure for canopy storage. Evapotranspiration is estimated by the Penman–Monteith equation (Monteith, 1965). Unsaturated subsurface flow is modelled by the one-dimensional Richards equation using an implicit finite difference solution. Overland and channel flow is evaluated by simplifications of the Saint-Venant equations and saturated zone flow by the two-dimensional Boussinesq equation.

Conceptual models

These models serve as a trade-off between the deterministic and black-box approaches. Conceptual models are formulated by a number of conceptual elements, each of which is a simplified representation of one process element of the system being modelled. Each element of the model is generally described by a non-linear reservoir with an equation for outflow:

$$S = K \times Q^n \qquad (9.1)$$

where S is storage, Q is outflow and K and n are constants. The basic advantage of this non-linear form of modelling is that it reflects the true nature of hydrological systems, which cannot be adequately described using a linear model.

The functioning of the model is controlled by the parameters of the different processes. Hence, assigning proper values to these parameters is absolutely essential for obtaining accurate results for the specific area being modelled. Based on the representation of the parameters, conceptual models are further classified as lumped models and distributed models. Lumped models include spatially averaged watershed characteristics, whereas distributed models incorporate spatial variability.

LUMPED CONCEPTUAL MODELS. A lumped model is one in which the spatial variations of watershed characteristics are generally ignored. Precipitation is considered to be spatially uniform throughout the watershed. Average values of watershed characteristics, i.e. vegetation, soils, geology or topography, are utilized. Hence, the results produced by these models display the average watershed conditions. The basis of lumped models is the continuity equation, that is, the water balance equation. These models attempt to describe three basic processes within any watershed, namely:

- loss of water from storage to the atmosphere through evaporation or by lateral flow across the watershed topographic boundaries;
- storage of water in soil, vegetation, aquifers and streams; and
- routing of water over the surface or through the soil and aquifers, from within the basin to the outfall.

A large number of lumped models are available and described in the literature.

The SSARR model (Rockwood, 1958) was developed for the planning, design and operation of water control projects in the Colombia River basin in the Pacific North-west USA. Later on, Anderson (1967) modified the original version of the model. The model is a general-purpose, continuous-simulation model in which runoff is determined as a per cent of precipitation, which, in turn, is based on soil moisture status indexed by SMI (the soil moisture index) and rainfall intensity. The different components of the flow are determined and routed through successive increments of reservoir-type storage.

The Stanford Watershed Model (Linsley and Crawford, 1960) is a general-purpose model which simulates daily flow from precipitation using infiltration, unit hydrograph and recession functions. It has been further improved by including soil moisture budgeting, evapotranspiration and flow-routing techniques (Crawford and Linsley, 1966). The model has undergone several modifications based on the conceptualization of various researchers. Some of these models – which are either improvements of the Stanford Watershed Model or have used similar concepts – include the Hydrocomp Simulation Program (HSP), the Kentucky Watershed Model (KWM), the Kentucky self-calibrating system (OPSET), the Texas Watershed Model (TWM) and the National Weather Service River Forecast Model (NWSRFS).

A parallel development of the model by Sugawara (1961), which is popularly known as Tank model, simulated water movement in the system using simple linear reservoirs arranged in series and in parallel. The model is defined by different storage tanks of interception, soil moisture and groundwater storages at different depths in the profile. A model was also developed by Boughton (1966) for estimating water yield from catchments in dry regions. The United States Department of Agriculture (USDA) Hydrograph Laboratory model has been developed mainly to acquire knowledge on the interactions between agricultural activities and the hydrology of small rural watersheds (Holtan *et al.*, 1975). This model utilizes many parameters, most of which are measurable. The most sensitive parameters of these are soil depth, root depth, evapotranspiration rates, rainfall distribution and intensity, and storage routing coefficients.

The (UK) Institute of Hydrology (IUH) lumped model (Nash and Sutcliffe, 1970) was designed to produce hourly estimates of streamflow from hourly catchment precipitation and hourly potential evaporation derived from meteorological data using the Penman formula. The Simple Lumped Reservoir Parametric (SLURP) model was developed to provide an alternative to complex hydrological models for Canadian basins (Kite, 1978). Evaporation is computed using Morton's (1983) complementary relationship areal evapotranspiration model, and the infiltration process was represented by the Philip formula (Philip, 1954).

Some limitations of lumped models (Franchini and Pacciani, 1991) are:

- Average values of the watershed characteristics are utilized to represent the various processes of the hydrological cycle. Spatial heterogeneities are not well reproduced by average parameters. By taking the average value of a certain parameter, the process is (implicitly) averaged. Because of the non-linearity and threshold values, this can lead to significant error which, in turn, affects simulation accuracy.
- When a model is calibrated based on the available historical records, any bias existing in the data is transferred to the set of optimized parameter values. This feature restricts the applicability of the model to other catchments where a different bias may be present in the data.
- Normally, the model parameters are optimized for some rainfall-runoff events over a given watershed and the optimized values, at best, represent the watershed only for the events used in the optimization. As soon as the set of rainfall-runoff events changes, the optimum parameter values also change.
- Most lumped models have some degree of interdependence between the parameters. Thus, the parameter values attained through the optimization are not necessarily the best estimate of physical values, but simply a set of numbers that give best fit to the data within the constraints imposed.

DISTRIBUTED CONCEPTUAL MODELS. Distributed models take the spatial variability of watershed properties into account. The underlying principle in these models is to discretize the watershed into a number of zones that are hydrologically similar. Discretization can be attained either through the Representative Elementary Area (REA), Hydrological Response Unit (HRU) or Grouped Response Unit (GRU) concept. The REA is equivalent to the representative elementary volume concept (Freeze and Cherry, 1979). The size of the element within a watershed is defined in such a manner that within-element statistics can be considered insignificant for modelling purposes. An alternative method for describing the spatial variability is by means of the HRU. The HRU is considered to be homogeneous, with a distinct hydrological response. The distinction can be made on the basis of vegetative cover, soil type, slope and aspect. The grouping can also be on the basis of zones of uniform meteorology or on the basis of grid cells – which is convenient for integrating with map coordinates and remotely sensed data. Runoff generation processes such as snow melt, infiltration and surface runoff are modelled separately for each unit. A separate set of parameter values has to be specified for each unit. The computed yield is then routed through one unit to another to obtain the total catchment yield. A significant aspect is that geographical position within a watershed is preserved. Distributed models are well suited for:

1. evaluating the effects of land-use change within a watershed;
2. evaluating the effects of spatially variable inputs;
3. simulating water quality and sediment yield on a watershed basis; and
4. simulating the hydrological response of ungauged catchments where no data are available for calibration.

The focus of the various models may differ with respect to the initial intent with which the model was developed. There are some models that consider only infiltration and surface-flow processes. Huggins and Monke (1968) used a grid system to delineate watershed elements in a distributed parameter model. They applied this concept to two areas in Indiana with a grid size of 7.5 m × 7.5 m. The slope direction for each element was used to route the runoff from one element to two adjoining elements. Computed runoff from each element was then integrated using a finite difference form of the continuity equation relating moisture supply, storage and outflow. The interception process was evaluated using the Horton interception equation (Horton, 1919) and infiltration by the Holtan equation (Holtan, 1961). Soil moisture was updated after considering the balance between infiltrated and the drained moisture. This work led to the development of a very comprehensive watershed model: ANSWERS (Areal Nonpoint Source Watershed Environment Response Simulation).

The ANSWERS model is physically based and simulates runoff and sediment transport at the watershed level (Beasley *et al.*, 1980). The watershed is divided into a set of square grids with a cell size of 1 to 4 ha. Runoff is the difference between rainfall and the sum of interception, retention and infiltration. Surface flow is routed from cell to cell using the continuity equation, together with Manning's uniform velocity equation. The unique feature of the ANSWERS model is that it continuously simulates infiltration as flow moves downstream, while considering transmission losses. This feature is effective on flat slopes, where runoff is slow and overland flow distances are extremely long.

A linearized distributed model was developed by Bravo *et al.* (1970) to estimate catchment runoff. In this model, the catchment is partitioned into sub-areas of simple shapes, and surface runoff from each sub-area is determined by solving one-dimensional equations of flow. The partial source area concept of Engman and Rogowski (1974) requires an intimate knowledge of the subsurface characteristics – which control infiltration and exfiltration rates. In this approach, the sub-area of the watershed expands in time and space depending on the storm characteristics and distribution of infiltration capacity. The kinematic wave equation is used for routing the flow from the sub-area.

The Finite Element Storm Hydrograph Model (FESHJVT) was presented by Ross *et al.* (1979) for determining the hydrological impact of land-use change in a watershed. It utilizes a one-dimensional finite-element scheme to simulate overland flow and channel flow. Galerkin's residual method is used to solve the kinematic equations of one-dimensional transient flow. The catchment is divided into channel and slope elements in a similar way to that of the Institute of Hydrology Distributed Model (IHDM) (Beven *et al.*, 1987), except that the hillslope planes are not rectangular. Infiltration losses are estimated using the Holtan (1961) equation with parameters varying between HRUs based on categories of soil type and land use.

A simple runoff simulation model was developed by Borah (1989) for small watersheds, which had simpler equations and fewer parameters. The parameters used in the simulation are the USDA Soil Conservation Service (SCS) runoff curve number (CN) and Manning's roughness coefficient. The watershed is divided into a number of representative overland and channel-flow elements to account for the non-uniformities in topographic, soil and land-use characteristics. The characteristics are considered to be uniform within each of these elements. The model simulates the runoff by using the SCS runoff curve number procedure (USDA SCS, 1972). The water-routing scheme is based on the kinematic wave approximation of the Saint-Venant or shallow-water equations governing unsteady free surface flow. Besides the topographic and rainfall data, the model requires only the runoff curve number and Manning's roughness coefficient.

The Kinematic Runoff and Erosion (KINEROS) model is an event-based, distributed rainfall-runoff model developed specifically for semi-arid regions (Woolhiser *et al.*, 1990). It is a Hortonian model used for simulating hillslope infiltration. It performs unsteady routing of overland and channel flow using kinematic routing which is based on simplifications of the Saint-Venant equations of shallow-water flow (Linsley *et al.*, 1982). Like most kinematic wave models, KINEROS uses a watershed characterization as a dendritic network of overland flow planes, channels and ponds.

The following are the major problems that have been involved in using distributed models:

- The large quantity of input data required often renders them inefficient for everyday operational hydrology.
- There is often insufficient information available about the physical characteristics of the basin to evaluate the parameters of physically based models at the required scale (Loague and Freeze, 1985).
- There is insufficient understanding of the processes of runoff generation at the catchment scale to build truly physically based models.
- Some studies have demonstrated that simple models are as successful as complex models (Pilgrim and McDermott, 1982; Loague and Freeze, 1985).

SEMI-DISTRIBUTED MODELS. In order to overcome the difficulties being faced with distributed models, researchers started developing semi-distributed models as a compromise between lumped models and fully distributed models (Williams *et al.*, 1985; Arnold *et al.*, 1993). In these models, algorithms are simple but physically based, and spatial heterogeneity is represented by means of observable physical characteristics of the basin, such as land use, soils and topography, etc.

Spatial variability in hydrological processes, particularly those that give rise to rapid runoff during and immediately following rain, has been taken into account in the model of Beven and Kirkby (1979). Then came an era when issues of the water quality of point and non-point sources were addressed through hydrological models. The Chemicals, Runoff and Erosion from Agricultural Management Systems (CREAMS) model is one such model that simulated agricultural contributions to water pollution (Knisel, 1980). It is specifically designed as an agricultural field-scale model and, as a result, has limited routing capabilities. The model contains three major components: hydrology, water quality and sedimentation. The main processes included in the hydrology component are surface runoff, percolation and evapotranspiration. Runoff volume is predicted using the SCS curve number technique. Ritchie's evapotranspiration model is applied to estimate evapotranspiration (Ritchie, 1972). The percolation component uses a storage routing technique to predict flow through the root zone. The hydrological component of the Simulator for Water Resources in Rural Basins (SWRRB) (Arnold *et al.*, 1990) and Erosion Productivity Impact Calculator (EPIC) (Izaurralde *et al.*, 2006) models have been derived from CREAMS.

It has been reported that the semi-distributed approach is better than the lumped approach (Kite and Kouwen, 1992). The major advantage of the semi-distributed approach is that relating the parameter values to land cover characteristics provides a method of investigating the impact of land-use changes, and allows the model to be more easily transferred to other basins. Arnold *et al.* (1993) developed a comprehensive surface water and groundwater flow model, which was incorporated into the later version of their model: the Soil and Water Assessment Tool (SWAT). The main objective of this model is to predict the impact of management changes on total water supplies. The model simulates four control volumes, namely surface, soil profile or root zone, shallow aquifer and deep aquifer.

Several limitations were faced when considering the application of many of the hydrological models that had been developed to agricultural areas. At the watershed scale, these limitations included inappropriate scale, inability to perform continuous-time simulation, inadequate maximum number of sub-areas and the inability to characterize the watershed or river basin in enough spatial detail. The SWAT model (Arnold *et al.*, 1998) was developed to provide continuous-time simulations with a high level of spatial detail by allowing the further division of a watershed into hundreds or thousands of sub-watersheds. The land area in a sub-basin is then further divided into HRUs, which are portions of sub-basin that possess unique land use, management and soil attributes. The SWAT model operates on a daily or hourly time step and is designed to evaluate management effects on water quality, sediment and agricultural chemical yield in large, ungauged basins, as it requires minimal calibration.

The SWAT model

The development of SWAT is the result of a commendable effort by the USDA Agricultural Research Service (ARS) over more than 30 years. Gassman *et al.* (2007) provides a comprehensive review of the historical development of this model, its salient features and an account of its performance. The following sections provide a few excerpts from the paper.

Figure 9.1 depicts the historical evolution of the SWAT model and its linkages with other models, such as the CREAMS model (Knisel, 1980), the Groundwater Loading Effects on Agricultural Management Systems (GLEAMS) model (Leonard *et al.*, 1987), and the Environmental Policy Impact Climate (EPIC) model (Izaurralde *et al.*, 2006), which was originally known as the Erosion Productivity Impact Calculator (Williams, 1990). The current SWAT model is an improvement of the SWRRB model (Arnold and Williams, 1987), which was designed to simulate management impacts on water and sediment movement for ungauged rural basins.

The SWAT model is based on a command structure for routing runoff and chemicals through a watershed. These commands allow the user to route flows through streams and reservoirs, combine flows and input measured data (e.g. weather) and point-source loading. The major components of SWAT include hydrology, weather, sedimentation, soil temperature, crop growth, nutrients, pesticides and agricultural management. The minimum weather inputs required by SWAT for simulation of the hydrological process are maximum and minimum daily temperature and precipitation. Sediment yield is estimated by the Modified Universal Soil Loss Equation (MUSLE; Williams, 1975). Daily average soil temperature is simulated using the maximum and minimum annual air temperatures, surface temperature and damping depth.

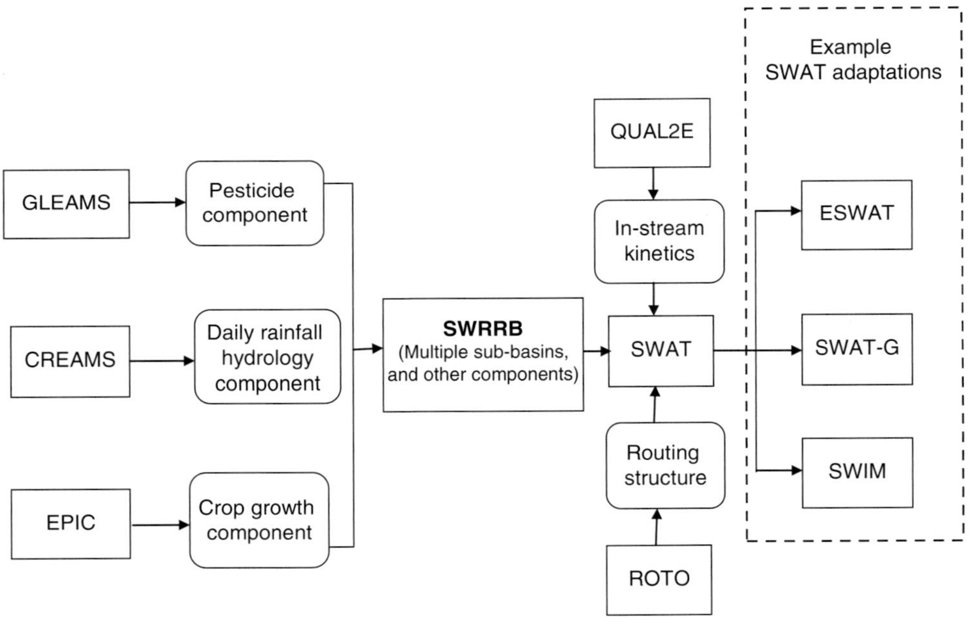

Fig. 9.1. Schematic of SWAT (Soil and Water Assessment Tool) developmental history, including selected SWAT adaptations (redrawn from Gassman *et al.*, 2007). CREAMS, Chemicals, Runoff and Erosion from Agricultural Management Systems; EPIC, Environmental Policy Impact Climate (originally the Erosion Productivity Impact Calculator); ESWAT, Extended SWAT; GLEAMS, Groundwater Loading Effects on Agricultural Management Systems; QUAL2E, Enhanced Stream Water Quality Model; ROTO, Routing Outputs to Outlet; SWAT-G, SWAT modification for improved flow in low mountain range catchments; SWIM, the Soil and Water Integrated Model; SWRRB, Simulator for Water Resources in Rural Basins.

There has been another important development in the SWAT family in the form of the APEX (Agricultural Policy Environmental eXtender) model, which is a field-scale model (Williams *et al.*, 1995) designed to simulate edge-of-field nutrient concentration, runoff volume and nutrient loadings from specific field-management practices. Similar to the HRUs in SWAT, the fields within APEX are portions of a subbasin that possess unique land use, management and soil attributes. There have been considerable improvements in the APEX model in terms of its capability to simulate a wide range of management practices, cropping systems and other land uses over whole farms or small watersheds (Gassman *et al.*, 2010).

The APEX model has also been integrated with SWAT through the ArcGIS interface. In the new version of SWAT, provision has been made for the two models to interact with each other. Interactions between fields simulated by APEX are made possible (e.g. one field can represent a filter strip for an up-gradient field). APEX also simulates weather, hydrology, soil temperature, erosion/sedimentation, nutrient cycling, tillage, field management practices, crop management and growth, pesticide and nutrient fate, and transport, as well as the costs and returns of the various management practices. APEX is applicable to a wide range of soils, climates and cropping systems.

Additional advantages of using APEX at the field level when compared with SWAT are:

1. In contrast to the SWAT HRUs, the field units within APEX have a spatial relationship and can be routed in a specified order.
2. The simulation of filter strips in SWAT is currently done through adjustment of coefficients that are based on the academic literature, whereas in APEX 'filter strips' are simulated based on physically based functions.
3. APEX is a micro-level model capable of simulating detailed field conditions, such as management practices related to farm animal production, the economic impacts of BMPs (best management practices) and wind erosion. Such functions are currently not available in SWAT.

There have been many studies that have demonstrated the combined application of APEX and SWAT by tackling the space with detailed information using APEX and integrating with the bigger system using SWAT. Saleh *et al.* (2000), Osei *et al.* (2000) and Gassman *et al.* (2001) have reported on such studies that have used the capabilities of the combined SWAT and APEX models to simulate environmental baselines and BMPs at the field level with APEX, and integrated these results from APEX with the remaining land uses within a watershed using SWAT. Such requirements are going to grow in future on account of increasing interventions being made on space either because of increasing demands or on account of adaptation strategies to cope with the impacts of climate change.

Use of SWAT and comparison with other models

A very elaborate account of the applications of SWAT across the globe and also its comparison with some of the other popular hydrological models has been given by Gassman *et al.* (2007). Gosain *et al.* (2005) made use of SWAT for estimating the return flows from the introduction of an irrigation project in a sub-basin of the Krishna River in South India. Assessment of return flow is a very challenging task and, very often, the estimates used are erroneous and lead to incorrect strategies. Gosain *et al.* (2006) also used SWAT for evaluating the impact of climate change on 12 Indian river systems using data from the HadRM2 (Hadley Centre Regional Model 2) regional climate model data.

Comparisons of SWAT with the Dynamic Watershed Simulation Model (DWSM) (Borah *et al.*, 2004) and the Hydrologic Simulation Program – Fortran (HSPF) model (Bicknell *et al.*, 1997) have been made by Borah and Bera (2003, 2004). They concluded that SWAT scores over other models for continuous simulations in predominantly agricultural watersheds. Shepherd *et al.* (1999) evaluated 14 models for estimating phosphorus loss from a lowland watershed in the UK and found that SWAT was the most suitable. Streamflow predictions of SWAT and HSPF have been compared by Van Liew *et al.* (2003)

on eight nested agricultural watersheds. They found SWAT to be more consistent than HSPF in estimating streamflow for different climatic conditions. Saleh and Du (2004) found the average daily flow, sediment loads and nutrient loads simulated by SWAT to be closer than those simulated by HSPF to values measured at five sites for the upper North Bosque River watershed in Texas. Singh *et al.* (2005) studied the Iroquois River watershed in eastern Illinois and western Indiana, and found the SWAT flow predictions to be better than the corresponding HSPF estimates. Nasr *et al.* (2007) found simulation of the hydrology of Belgium's Jeker River basin by both SWAT and the MIKE SHE model (Refsgaard and Storm, 1995) to be acceptable and comparable. However, they observed that MIKE SHE predicted the overall variation of river flow slightly better. Srinivasan *et al.* (2005) compared the performance of the Soil Moisture Distribution and Routing (SMDR) model (Cornell University, 2003) for the FD36 experimental watershed in east central Pennsylvania with that of SWAT and found that SWAT estimated flow more accurately than SMDR on a seasonal basis.

Results and Discussion

Case study of Kosi: a Himalayan basin

The case study evaluated the impact of climate change on streamflow in the upper Kosi basin by using an RCM coupled with the SWAT model. The potential impacts of climate change on water yield and other hydrological budget components were quantified by driving SWAT with current and future climates (Gosain *et al.*, 2010).

The Kosi river system

Kosi is a transboundary river between Nepal and India and is one of the largest tributaries of the Ganga (Ganges). The river along with its tributaries, drains a total area of 69,300 km^2 up to its confluence with the Ganga in India (29,400 km^2 in Tibet, 30,700 km^2 in Nepal and 9200 km^2 in India). The river basin is surrounded by the ridges separating it from the Brahmaputra in the north, the Gandaki in the west, the Mahananda in the east and by the Ganga in the south. One major tributary of Kosi is the Arun, which has the major part of its journey in Tibet. Seven major tributaries join together to form the Saptakoshi River, which is popularly known as the Khoshi (Kosi in Sanskrit). The Kosi River carries heavy silt during the monsoon season (Wikipedia, 2011).

Use of GIS as preprocessor in SWAT

The realization existed all along that the lumped models that have already been described are not in a position to represent the spatial variability inherent in hydrological systems. However, this spatial variability was at most tackled by subdividing the system into subsystems and subsequently treating the subsystem in a lumped manner. A major difference was made to this option with the advent of the geographic information system/s (GIS). Presently, all distributed and semi-distributed models use GIS to handle spatial variability. Use ranges from the construction of terrain to the derivation of drainage parameters to the dissemination of model results to end users. Consequently, the effective use of a hydrological model requires adequate understanding of related GIS concepts used in the models. The following sections describe the use of GIS as a preprocessor in the SWAT model.

DIGITAL REPRESENTATION OF TOPOGRAPHY. A digital elevation model (DEM) is an ordered array of numbers that represents the spatial distribution of elevations above some datum, either sampled at discrete points, or the average elevation over a specified segment of landscape. A digital terrain model (DTM) is an ordered array of numbers that represent the spatial distribution of terrain attributes such as slope, soil type, land use, etc.; elevation is one of these attributes. In that sense, a DEM is a subset of a DTM. However, there are many ways of structuring the network of elevation data, namely, triangular (or triangulated) irregular network (TIN), square grid network (raster) or contour-based network.

It is the raster form that has become very popular in hydrology, mainly due to the ease with which computer algorithms are implemented using this structure.

WATERSHED DELINEATION. Defining the watershed and the drainage network is one of the basic requirements of applying hydrological models. There are many algorithms that have been proposed for automatic delineation, but it is the D8 approach that has become most popular (O'Callaghan and Mark, 1984; Jenson and Dominique, 1988). In this approach, the four grid data matrices that are required include Elevation, Flow Direction, Flow Accumulation and Stream Link grids.

To create an accurate representation of Flow Direction, and therefore of accumulated flow, it is essential to use a data set that is free of sinks. Therefore, the DEM is processed to remove all sinks to obtain a depression-less DEM. Flow Direction calculates the direction of flow out of each cell into one of its eight neighbours (Fig. 9.2). It is encoded to correspond to the orientation of one of the eight cells that surround the cell 'X', as shown in Fig. 9.2.

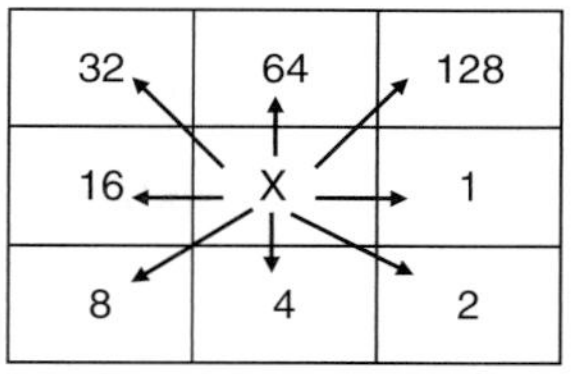

Fig. 9.2. Representation of flow direction in a watershed/drainage network using the D8 approach (algorithm) (see text for full description).

The cell elevation is compared with that of its eight neighbours, and the Flow Direction is assigned with a look-up table defining the most likely direction. Next, the Flow Accumulation for each cell is calculated by accumulating the weight for all cells that flow into each downslope cell. The result of the stream accumulation is used to create a stream network by applying a threshold value to subset cells with a high accumulated flow. Each cell is assigned a value equal to the number of cells that flow to it. Cells having a flow accumulation of zero correspond to ridges or hilltops.

The pattern formed by values of the Flow Accumulation grid larger than a certain threshold will form a fully connected drainage network. Determination of threshold value is subjective and the channel network so derived depends only on the DEM and not on hydrological information. Links are the sections of a stream channel connecting two successive junctions, a junction and the outlet, or a junction and the drainage divide. The result of the Stream Link can be used as the source grid of the watershed to create drainage basins that correspond to the branches of a stream network.

Finally, automatic delineation of watersheds is done by using the Elevation grid as input; the source grid can be a grid of target outflow points or the output from the Stream Link procedure. Watersheds of different sizes can be delineated by giving different threshold values while building the stream links. The watershed boundary of the Kosi river determined using the automatic delineation procedure is illustrated in Plate 13. Also presented in Plate 13 is the DEM of the basin. This has been used to generate the drainage network (shown in blue) that corresponds to a threshold value of 100,000 hectares. The actual drainage is shown in red.

The next step after the generation of the drainage network is to subdivide the automatically delineated watershed into sub-areas. This subdivision is dictated by the stream links established through the generation of the drainage network, and is determined by the value of the threshold used. The smaller the value, the denser is the drainage network and, consequently, the larger is the number of sub-areas. Plate 14 presents the 27 sub-areas delineated on the basis of the drainage network created using a 100,000 ha threshold.

LAND USE/LAND COVER AND SOIL TYPE LAYERS. Other essential input data that are required for hydrological modelling are land use/land cover information. Currently, a usual source of this information is satellite-interpreted land use. Plate 15A shows the broad land-use categories for the Kosi watershed area. The information is transformed into a digital layer (raster) to be used in the model.

The SWAT model also requires the soil profile and its associated characteristics, which the model uses to maintain the soil status on a continuous timescale. This is a very important input, which influences the simulation process drastically. Invariably such large-scale soil information is usually missing, but can always be improved upon through additional observations. In the case of the Kosi basin, the NBSS&LUP (Indian National Bureau of Soil Survey and Land Use Planning) soil map has been used to build a digital soil map of the watershed. The digitized soil layer is shown in Plate 15B. Through the soil profile, the model incorporates the variation in movement and allocation of water corresponding to the changing characteristics of the soil, which makes it possible to enhance the simulation capability of the model.

HYDROLOGICAL RESPONSE UNITS. The next step of the preprocessing using GIS is to obtain the HRUs, the basic modelling units under each sub-basin. The HRUs are obtained by the GIS overlay process using the land use and soil layers within each sub-basin. The number of HRUs in a sub-basin is dependent on the number of unique combinations of land use and soil type. Each HRU participates in generating the response to the natural inputs, and the state of its moisture condition is kept track of continuously in time.

ADDITIONAL FUNCTIONALITY OF GIS. GIS is also used to incorporate many other natural or man-made features which are required to be introduced into the natural terrain of the watershed for simulation of their behaviour. These features can be in the form of water bodies (tanks, ponds, check dams, trenches, etc.). Under some situations, man-made features that are proposed to be incorporated may also be introduced.

Yet another segment of the preprocessing is to identify the location of the precipitation and/or weather stations available in and around the area to be modelled. This is required to assign the individual weather data measurement stations to the nearby sub-basins to incorporate the spatial variability of the weather inputs.

Climate change impact assessment on water resources

Simulated climate outputs from the HadRM3 (Hadley Centre Regional Model 3) regional climate model for the present (1961–1990) and for a future period (2071–2100) for two different socio-economic scenarios from the IPCC (Intergovernmental Panel on Climate Change) SRES (Special Report on Emissions Scenarios) scenarios. Both of these greenhouse gas emission scenarios are characterized by regionally focused development, but one has priority for economic issues (the A2 scenario) and the other for environmental issues (the B2 scenario).

In order to fulfil the objective of assessing the impact of climate change on the water resources of the Kosi basin, weather data are required. For this purpose, data generated in transient experiments by the UK Hadley Centre for Climate Prediction at a resolution of 0.44° × 0.44° latitude by longitude grid points were obtained from the Indian Institute of Tropical Meteorology (IITM) at Pune. The daily weather data on maximum and minimum temperature, rainfall, solar radiation, wind speed and relative humidity at all the grid locations were processed. The RCM grid was superimposed on the sub-basins to derive the weighted means of the inputs for each of the sub-basins. The centroid of each sub-basin is then taken as the location for the weather station to be used in the SWAT model.

The SWAT model has been run using the HadRM3 (PRECIS) baseline (1961–1990) and the A2 (2071–2100) and B2 (2071–2100) IPCC SRES scenarios (30 years). Although the model generates very detailed outputs at the spatial and temporal scales, in the present analyses only some of the components that are considered important have been selected and reported. Table 9.1 depicts the average annual values for the selected water balance components for the control baseline (1961–1990) and A2 and B2 (2071–2100) scenarios using HadRM3 data. The outputs for these two scenarios depict the possible impacts on various water balance components, such as runoff, soil moisture, actual evapotranspiration, etc., with respect to the baseline scenario.

Table 9.1. Average annual water balance components (mm) simulated for the Upper Kosi River basin using SWAT and HadRM3-simulated IPCC SRES scenarios.[a]

	IPCC SRES scenario		
Water balance components (mm)	Baseline	A2	B2
Precipitation	1966.70	2153.60	2012.40
Snowfall	1225.40	945.24	988.80
Snowmelt	259.42	456.84	451.84
Groundwater recharge	161.59	289.14	238.80
Water yield	786.50	1338.88	1170.94
Surface runoff	582.04	965.24	860.22
Actual evapotranspiration	143.90	205.30	187.80
Potential evapotranspiration	333.60	452.10	402.10

[a]HadRM3, Hadley Centre Regional Model; IPCC SRES, Intergovernmental Panel on Climate Change Special Report on Emissions Scenarios.

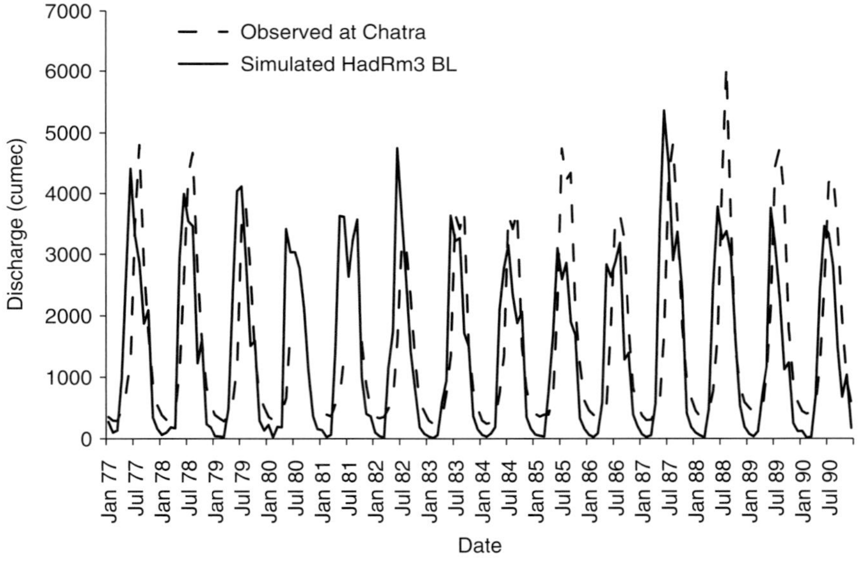

Fig. 9.3. Observed versus simulated discharge for the Kosi River basin at Chatra, Nepal, using SWAT and Hadley Centre Regional Model HadRM3-simulated baseline (BL) scenario from IPCC SRES (Intergovernmental Panel on Climate Change Special Report on Emissions Scenarios) (cumec, $m^3 s^{-1}$).

Validation of the SWAT model

Model validation is an important part of the simulation procedure. In the present study, it has been performed by using the baseline weather data available from the HadRM3 simulation (in the absence of actual observed weather data for the basin). The simulated discharge corresponding to the simulated precipitation was compared with the actual observed flow data at Chatra for the period 1977–1990, and is presented in Fig. 9.3. In fact, the ideal validation requires the availability of actual observed precipitation and other weather parameters, which were not available. However, even when using the simulated baseline weather data to generate the simulated flow series, the validation is reasonably good. It may also be mentioned that, although the simulated flow series is plotted with the observed series, it is not expected to have the same chronology, and the validation should be confined to comparing the extent of fluctuation of high and low flows.

New Requirements of Hydrological Models

Water managers' routine activities include water allocation among multiple and often competing uses, minimization of risk, and adaptation to changing circumstances such as variability in water storage levels and water demand due to seasonal effects and/or population growth. A wide range of adaptation techniques have been applied over many decades: capacity expansion (e.g. building new reservoirs), changing the operating rules for existing water supply systems, managing water demand and changing institutional practices are some of these. Within this context, historical climate and hydrological records provide the basis for the determination of reliable water yields and the assessment of flood and drought risk. Underpinning these investigations is the assumption that the statistical properties (e.g. averages and standard deviation) of the climatic and hydrological variables remain constant over time. The prospect of climate change means that the key climate and hydrological variables will change, as will water demand. Climate-induced effects may be non-linear, and carry the potential for surprises beyond those already incorporated into water supply system designs and existing water management strategies (Appleton, 2004).

Adaptation to the impact of climate change on water resources

Identifying the practical implications of climate change for the water sector and seeking to cope with the long-term effects of climate change need to be accomplished through the IWRM approach.

Most of the hydrological systems around the world have been disturbed through man-made actions such as deforestation, the construction of dams, land-use changes, exploitation of marginal lands, etc. The implementation of IWRM in such systems – already subject to intervention – has already become very complex. To add to the problem, the impacts of climate change also need to be considered and have complicated the process further. The DPSIR approach (Schulze, 2005) considers various attributes that need to be considered to incorporate climate change issues into enhanced hydrological models. The method categorizes all the factors, natural or anthropogenic, that will be affected by climate change, into:

- Drivers (D), e.g. changes in inter- and intra-seasonal climate variability, or
- Pressures (P), i.e. causes of hydrological changes, including irreversible regional-scale climate change (e.g. changes in precipitation amounts, sequences, intensities), or
- States (S) of the hydrology, such as changes in quantity and quality (physical, chemical and biological) of streamflow and its seasonal distribution, as well as the states of wetlands, constructed dams or groundwater recharge levels, or
- Impacts (I), i.e. the positive or negative environmental, social and economic consequences, such as the degradation of terrestrial and aquatic ecosystems, the hydrological amplification of extreme climate events or the increased need for reliable water supplies, or
- Responses (R), i.e. the reactions of societies to the D, P, S or I components of the DPSIR approach, which can be via statements/resolutions from international forums, national responses, and/or by putting into practice new/existing concepts.

The recommended IWRM approach is presented in Fig. 9.4. It involves a comprehensive approach to modelling hydrological processes, because perturbations in the drivers of these processes (e.g. changed precipitation, temperature and net radiation patterns, as well as enhanced CO_2 feedbacks

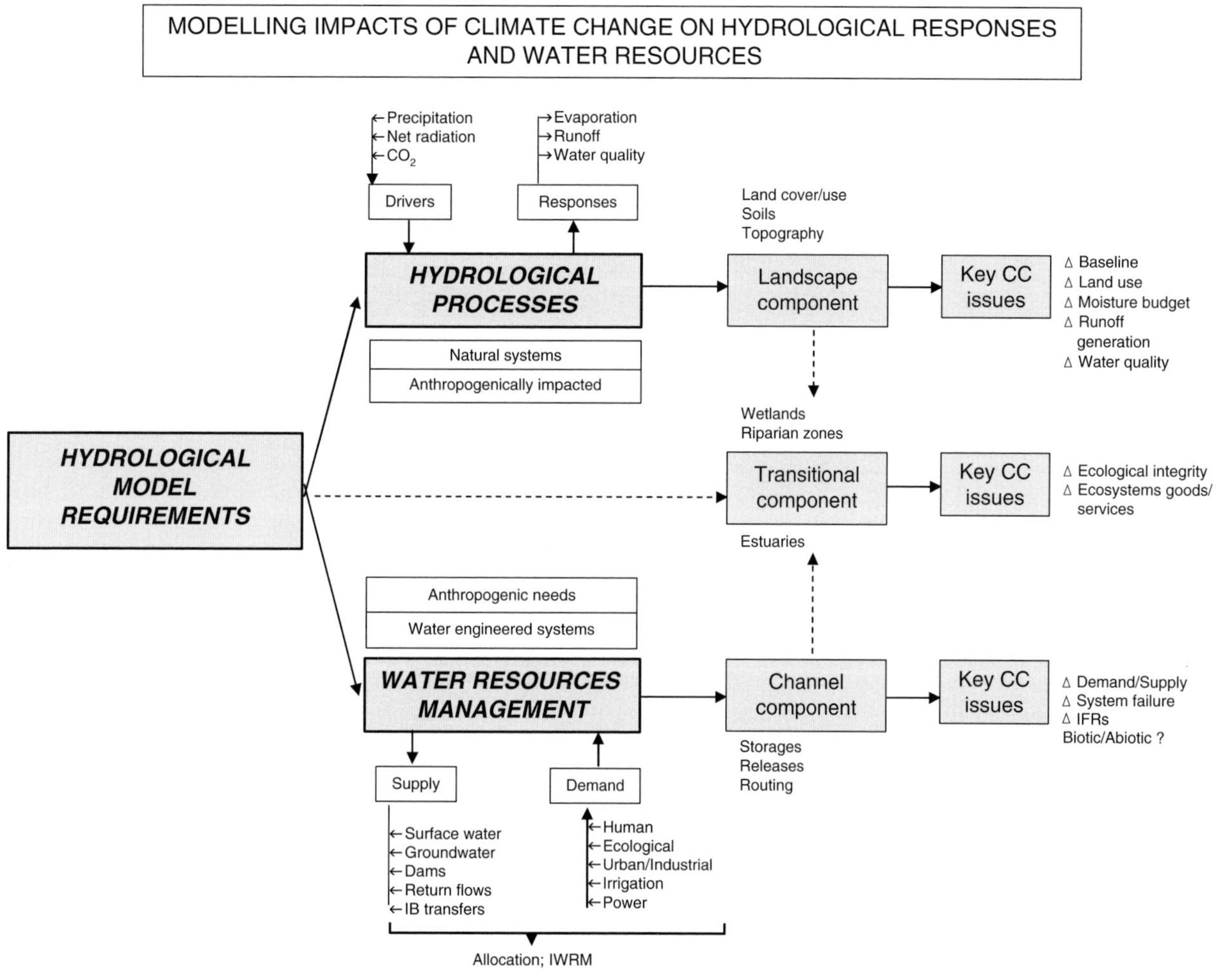

Fig. 9.4. Requirements from hydrological model under conditions of climate change (CC). IB, inter-basin; IFRs, in-stream flow requirements; IWRM, Integrated Water Resources Management. Adapted from Schulze, 2005.

on transpiration) will result in entirely new hydrological regimes without even considering any man-made interventions. Man-made interventions, in the form of land-use changes and soil and water conservation structures, have further implications for the hydrological regime, but can be handled by the present models.

The second component of IWRM revolves around the water resources practitioner/manager whose responsibility is to balance the supply of water (be it from rivers, groundwater, impoundments, return flows or water transfers) with demand for water (from a basic human and ecological needs perspective and in relation to the requirements for the urban/industry sectors, for power and for irrigation), so that allocations can be made in a sustainable manner. The manager will use the channel component of the catchment through controls of storage, releases and routing of water. Climate change will impose the additional challenges that would generally revolve around the engineering issues of changes in supply/demand and limits to the design of hydraulic structures in regard to system failure, as well as around the environmental consequences of changes in hydrological regimes, including in-stream flow requirements (IFRs) and other abiotic/biotic effects downstream. There is also the transitional component of the hydrological system, which represents the changes happening in wetlands, riparian zones and estuaries. Another important aspect is the ecosystem and its interaction with other subsystems. All these requirements need to be incorporated into the hydrological models.

As far as the SWAT model is concerned, some of these needs for tools and information to help water and land managers to assess and manage the impacts of climate variability and change have already been incorporated into the model. Users can conduct watershed-based studies of the potential implications of climate variability and change for water and land resources. Specifically, SWAT provides flexible capabilities for creating climate-change scenarios, allowing users to quickly assess a wide range of 'what if' questions about how weather and climate could affect their systems. The existing capabilities of SWAT for assessing the effects of land-use change and management practices have been enhanced to assess the coupled effects of climate and land-use changes. However, many other requirements – as described above – are yet to be incorporated.

Summary

Out of the available hydrological models, the semi-distributed models are the ones that are useful from various viewpoints and may be adopted for a variety of applications. Selection of a specific model has always been a difficult task. Besides the capability of the model, the capability of the model user is equally important.

The wide range of SWAT applications that have been described here underscores the fact that the model is a very flexible and robust tool that can be used to simulate a variety of watershed problems. The process of configuring SWAT for a given watershed has also been greatly facilitated by the development of GIS-based interfaces, which provide a straightforward means of translating digital land-use, topographic and soil data into model inputs. In addition, enhancement of the model to cater for climate variability and change has also improved its usability. The ability of SWAT to replicate hydrological and/or pollutant loads at a variety of spatial scales on an annual or monthly basis has been confirmed in numerous studies. As it is a public-domain model, some users have addressed weaknesses in SWAT through modifications of its components to support more accurate simulation of specific processes or regions, or by interfacing it with other models. Both of these trends are expected to continue. The SWAT model will continue to evolve in response to the needs of the ever-increasing worldwide user community and to provide improved simulation accuracy of key processes. A major challenge of the ongoing evolution of the model will be meeting the desire for additional spatial complexity while maintaining ease of use.

References

Abbott, M.B., Bathurst, J.C., Cunge, J.A., O'Connell and Rasmussen, J. (1986) An introduction to the European Hydrological System – Système Hydrologique Européen, "SHE", 2: Structure of a physically based, distributed modeling system, *Journal of Hydrology* 87, 61–77.

Amorocho, J. and Orlob, G.T. (1961) *Non-linear Analysis of Hydrologic Systems.* Contribution 40, Water Resources Centre, University of California, Berkeley, California.

Anderson, J.A. (1967) *Computer Application to System Analysis, Lower Mellong River.* US Army Engineer Division, North Pacific, Portland, Oregon.

Anmala, J., Zhang, B. and Govindaraju, R.S. (2000) Comparison of ANNs and empirical approaches for predicting watershed runoff. *Journal of Water Resources Planning and Management, American Society of Civil Engineers* 126, 156–166.

Appleton, B. (ed.) (2004) *Climate Changes the Water Rules: How Water Managers Can Cope with Today's Climate Variability and Tomorrow's Climate Change.* Dialogue on Water and Climate, Delft, The Netherlands.

Arnold, J.G. and Williams, J.R. (1987) Validation of SWRRB: simulator for water resources in rural basins *Journal of Water Resources Planning and Management, American Society of Civil Engineers* 113, 243–256.

Arnold, J.G., Williams, J.R., Griggs, R.H. and Sammons, N.B. (1990) *SWRRB: A Basin Scale Simulation Model for Soil and Water Resources Management.* Texas A&M University Press, College Station, Texas.

Arnold, J.G., Allen, P.M. and Bernhardt, G. (1993) A comprehensive surface-ground water flow model. *Journal of Hydrology* 142, 47–69.

Arnold, J.G., Srinivasan, R., Muttiah, R.S. and Williams, J.R. (1998) Large area hydrologic modeling and assessment Part I: Model development. *Journal of the American Water Resources Association* 34, 73–89.

Beasley, D.B., Huggins, L.F. and Monke, E.J. (1980) ANSWERS: a model for watershed planning. *Transactions of the ASAE* 23, 938–944.

Beven, K.J. (1977) Hillslope hydrographs by the finite element method. *Earth Surface Processes* 2, 13–28.

Beven, K.J. and Kirkby, M.J. (1979) A physically based variable contributing area model of basin hydrology. *Hydrological Sciences Bulletin* 24, 43–69.

Beven, K.J. Calver, A. and Morris, E.M. (1987) *The Institute of Hydrology Distributed Model.* Hydrology, Report No. 98, Institute of Wallingford, UK.

Bicknell, B.R., Imhoff, J.C., Donigian, A.S. and Johanson, R.C. (1997) *Hydrological Simulation Program - FORTRAN (HSPF): User's Manual for Release 11.* Publication No. EPA-600/R-97/080. US Environmental Protection Agency, Athens, Georgia.

Borah, D.K. (1989) Runoff simulation model for small watersheds. *Transactions of the ASAE* 32, 887–886.

Borah, D.K. and Bera, M. (2003) Watershed-scale hydrologic and nonpoint-source pollution models: review of mathematical bases. *Transactions of the ASAE* 46, 1553–1566.

Borah, D.K. and Bera, M. (2004) Watershed-scale hydrologic and nonpoint-source pollution models: review of applications. *Transactions of the ASAE* 47, 789–803.

Borah, D.K., Bera, M. and Xia, R. (2004) Storm event flow and sediment simulations in agricultural watersheds using DWSM. *Transactions of the ASAE* 47, 1539–1559.

Boughton, M.E. (1966) A mathematical model for relating runoff to rainfall with daily data. *Transactions of the Institution of Engineers, Australia, Civil Engineering, CE* 8, 83–97.

Bravo, S.C.A., Harley, B.M., Perkins, F.E. and Eagleson, P.S. (1970) *A Linear Distributed Model of Catchment Runoff.* Hydrodynamics Laboratory Report No. 123, MIT Department of Civil Engineering, Cambridge, Massachusetts.

Burnash, R.J.C., Ferral, R.L. and McGuire, R.A. (1973) *A Generalized Streamflow Simulation System: Conceptual Model for Digital Computers.* Technical Report, Joint Federal and State River Forecast Center, US National Weather Service/California Department of Water Resources, Sacramento, California.

Cornell University (2003) *SMDR: The Soil Moisture Distribution and Routing Model. Documentation Version 2.0, January 2003.* Soil and Water Laboratory, Biological and Environmental Engineering Department, Cornell University, Ithaca, New York. Available at: http://soilandwater.bee.cornell.edu/research/VSA/papers/SMDR-doc.pdf (accessed 21 February 2011).

Crawford, N.H. and Linsley, R.K. (1966) *Digital Simulation in Hydrology, Stanford Watershed Model IV.* Technical Report No. 39, Department of Civil Engineering, Stanford University, Stanford, California.

Dawdy, D.R. and O'Donnell, T. (1965) Mathematical models of catchment behaviour. *Journal of the Hydraulics Division: Proceedings of the American Society of Civil Engineers* 91, 123–137.

Engman, E.T. and Rogowski, A.S. (1974) A partial area model for storm flow synthesis. *Water Resources Research* 10, 464–472.

Franchini, M. and Pacciani, M. (1991) Comparative analysis of several conceptual rainfall-runoff models, *Journal of Hydrology* 122, 161–219.

Freeze, R.A. (1971) Three dimensional transient, saturated-unsaturated flow in a ground water basin. *Water Resources Research* 7, 347–366.

Freeze, R.A. and Cherry, J.A. (1979) *Ground Water.* Prentice Hall, Eaglewood Cliffs, New Jersey.

Freeze, R.A. and Harlan, R.L. (1969) Blueprint for a physically based digitally simulated hydrologic response model. *Journal of Hydrology* 9, 237–258.

Gassman, P.W., Abraham, J., Saleh, A., Keplinger, K. and Williams, J.R. (2001) Simulation of nutrient losses from chicken litter applications in East Central Texas with APEX. Presented at: *American Society of Agricultural Engineers Annual Meeting, Sacramento, California.*

Gassman, P.W., Reyes, M.R., Green, C.H. and Arnold, J.G. (2007) The Soil and Water Assessment Tool: historical development, applications, and future research directions: Invited Review Series. *Transactions of the ASABE* 50, 1211–1250.

Gassman, P.W., Williams, J.R., Wang, X., Saleh, A., Osei, E., Hauck, L., Izaurralde, C. and Flowers, J. (2010) The Agricultural Policy Environmental eXtender (APEX) model: an emerging tool for landscape and watershed environmental analyses. *Transactions of the ASABE* 53, 711–740.

Gosain, A.K., Rao, S., Srinivasan, R. and Gopal Reddy, N. (2005) Return-flow assessment for irrigation command in the Palleru River basin using SWAT model. *Hydrological Processes* 19, 673–682.

Gosain, A.K., Rao, S. and Basuray, D. (2006) Climate change impact assessment on hydrology of Indian river basins. *Current Science* 90, 346–353.

Gosain, A.K., Shrestha, A.B. and Rao, S. (2010) *Modelling Climate Change Impact on the Hydrology of the Eastern Himalayas: Climate Change Impact and Vulnerability in the Eastern Himalayas – Technical Report 4.* International ICIMOD (Centre for Integrated Mountain Development), Kathmandu, Nepal.

Holtan, H.N. (1961) *A Concept of Infiltration Estimates in Watershed Engineering.* Report 41-51, USDA Agricultural Research Service, Washington, DC.

Holtan, H.N., Stiltner, G.J., Hensen, W.H. and Lopez, N.C. (1975) *USDAHL-74: Revised model of Watershed Hydrology*, Technical Bulletin No. 1518, USDA Agricultural Research Service, Washington, DC.

Horton, R.E. (1919) Rainfall interception. *Monthly Weather Review* 147, 603–623.

Hsu, K., Gupta, H.V., Gao, X., Sorooshian, S. and Imam, B. (2002) Self-Organizing Linear Output Map (SOLO): an artificial neural network suitable for hydrologic modelling and analysis. *Journal of Hydrology* 38, 1–17.

Huggins, L.F. and Monke, E.J. (1968) A mathematical model for simulating the hydrologic response of a watershed. *Water Resources Research* 4, 529–539.

Izaurralde, R.C., Williams, J.R., McGill, W.B., Rosenberg, N.J. and Quiroga Jakas, M.C. (2006) Simulating soil C dynamics with EPIC: model description and testing against long-term data. *Ecological Modelling* 192, 362–384.

Jenson, S.K. and Dominique, J.O. (1988) Extracting topographic structure from digital elevation data for geographic information system analysis. *Photogrammetric Engineering and Remote Sensing* 54, 1593–1600.

Kite, G.W. (1978) Development of a hydrologic model for a Canadian watershed. *Canadian Journal of Civil Engineering* 5, 126–134.

Kite, G.W. and Kouwen, N. (1992) Watershed modeling using land classifications. *Water Resources Research* 28, 3193–3200.

Knisel, W.G. (ed.) (1980) *CREAMS, a Field Scale Model for Chemicals, Runoff, and Erosion from Agricultural Management Systems.* Conservation Research Report No. 26, US Department of Agriculture, Washington, DC.

Leonard, R.A., Knisel, W.G. and Still, D.A. (1987) GLEAMS: groundwater loading effects of agricultural management systems *Transactions of the ASAE* 30, 1403–1418.

Linsley, R.K. and Crawford, N.H. (1960) Computation of synthetic storm flow record on a digital computer, *International Association of Scientific Hydrology Publication* 51, 526–538.

Linsley, R.K., Kohler, M.A. and Paulhus, J.L.H. (1982) *Hydrology for Engineers.* McGraw Hill, New York.

Loague, K.M. and Freeze, R.A. (1985) A comparison of rainfall-runoff modelling techniques on small upland catchments, *Water Resources Research* 21, 229–248.
Monteith, J.L. (1965) Evaporation and environment. In: Fogg, G.E. (ed.) *The State and Movement of Water in Living Organisms*. Academic Press, New York, pp. 205–234.
Morton, F.I. (1983) Operational estimates of areal evapotranspiration and their significance to the science and practice of hydrology. *Journal of Hydrology* 66, 1–76.
Mulvaney, T.J. (1851) On the use of self-registering rain and flood gauges in making observations of the relations of rainfall and flood discharges in a given catchment. *Transactions of the Institution of Civil Engineers of Ireland* 4, 18–33.
Nash, J.E. and Sutcliffe, J.V. (1970) River flow forecasting through conceptual models, Part 1 – A discussion of principles, *Journal of Hydrology* 10, 282–290.
Nasr, A., Bruen, M., Jordan, P., Moles, R., Kiely, G. and Byrne, P. (2007) A comparison of SWAT, HSPF, and SHETRAN/GOPC for modeling phosphorus export from three catchments in Ireland. *Water Research* 41, 65–1073.
O'Callaghan, J.F. and Mark, D.M. (1984) The extraction of drainage networks from digital elevation data. *Computer Vision, Graphics, and Image Processing* 28, 323–344.
Osei, E., Gassman, P. and Saleh, A. (2000) *Livestock and the Environment: A National Pilot Project: CEEOT-LP Modeling for Upper Maquoketa River Watershed, Iowa*. Technical Report Texas No. PR0003/Texas Institute for Applied Environmental Research Report, Tarleton State University, Stephenville, Texas.
Pereira, H.C., McCulloch, J.S.G., Dagg, M., Hosegood, P.H. and Pratt, M.A.C. (1962) A short term method for catchment basin studies, *East African Agricultural and Forestry Journal* 27(Special Issue), 4–7.
Philip, J.R. (1954) An infiltration equation with physical significance, *Soil Science* 77, 153–157.
Pilgrim, D.H. and McDermott, G.E. (1982) Design floods for small rural catchments in eastern New South Wales, *Transactions of the Institution of Engineers, Australia, Civil Engineering, CE* 24, 226–234.
Refsgaard, J.C. and Storm, B. (1995) MIKE SHE. In: Singh, V.J. (ed.) *Computer Models in Watershed Hydrology*. Water Resources Publications, Highland Ranch, Colorado, pp. 809–846.
Riad, S., Mania, J., Bouchaou, L. and Najjar, Y. (2004) Predicting catchment flow in a semi-arid region via an artificial neural network technique. *Hydrological Processes* 18, 2387–2393.
Ritchie, J.T. (1972) Model for predicting evaporation from a row crop with incomplete cover. *Water Resources Research* 8, 1204–1213.
Rockwood, D.M. (1958) Columbia basin streamflow routing by computer, *Journal of the Waterways and Harbors Division, American Society of Civil Engineers* 84, Paper No. 1874.
Rockwood, D.M. and Nelson, M.L. (1966) Computer application to streamflow synthesis and reservoir regulation. In: *Proceedings, IV International Conference on Irrigation and Drainage*.
Ross, B.B., Contractor, D.N. and Shanholz, V.O. (1979) A finite element model of overland and channel flow for assessing the hydrological impact of land-use change, *Journal of Hydrology* 41, 11–30.
Rutter, A.J., Kershaw, K.A., Morton, A.J. and Robins, P.C. (1971) A predictive model of rainfall interception in forests. 1 Derivation of the model from observations in a plantation of Corsican pine. *Agricultural Meteorology* 9, 367–384.
Saleh, A. and Du, B. (2004) Evaluation of SWAT and HSPF within BASINS program for the upper North Bosque River watershed in central Texas. *Transactions of the ASAE* 47, 1039–1049.
Saleh, A., Arnold, J.G., Gassman, P., Hauck, L., Rosenthal, W.D., Williams, J.R. and McFarland, A. (2000) Application of SWAT model for Upper North Bosque River Watershed. *Transactions of the ASAE* 43, 1077–1087.
Schulze, R.E. (ed.) (2005) *Climate Change and Water Resources in Southern Africa, Studies on Scenarios, Impacts, Vulnerabilities and Adaptation*. WRC Report No. 1430/1/05, Water Research Commission, Pretoria, RSA.
Shepherd, B., Harper, D. and Millington, A. (1999) Modelling catchment-scale nutrient transport to watercourses in the U.K. *Hydrobiologia* 395/396, 227–237.
Sherman, L.K. (1932) Streamflow from rainfall by the unit graph method. *Engineering News Record* No. 108, 501–505.
Singh, J., Knapp, H.V., Arnold, J.G. and Demissie, M. (2005) Hydrological modeling of the Iroquois River watershed using HSPF and SWAT. *Journal of the American Water Resources Association* 41, 343–360.
Sorooshian, S. and Gupta, V.K. (1983) Automatic calibration of conceptual rainfall runoff models: the question of parameter observability and uniqueness, *Water Resources Research* 19, 260–268.

Srinivasan, M.S., Gerald-Marchant, P., Veith, T.L., Gburek, W.J. and Steenhuis, T.S. (2005) Watershed-scale modeling of critical source areas of runoff generation and phosphorus transport. *Journal of the American Water Resources Association* 41, 361–375.

Sudheer, K.P., Gosain, A.K. and Ramasastri, K.S. (2002) A data driven algorithm for constructing ANN based rainfall-runoff models. *Hydrological Processes* 16, 1325–1330.

Sugawara, M. (1961) An analysis of runoff structure about several Japanese rivers, *Japanese Journal of Geophysics* 2(4), 1–76.

Todini, E. and Wallis, J.R. (1977) Using CLS for daily or longer period rainfall-runoff modelling. In: Ciriani, T.A., Maione, U. and Wallis, J.R. (eds) *Mathematical Models for Surface Water Hydrology*. John Wiley, Chichester, UK, pp. 149–168.

USDA SCS (1972) *National Engineering Handbook, Hydrology Section 4*. US Department of Agriculture Soil Conservation Service, Washington, DC, Chapters 4–10.

Uvo, C.B., Tölle, U. and Berndtsson, R. (2000) Forecasting discharge in Amazonia using artificial neural networks. *International Journal of Climatology* 20, 1495–1507.

Van Liew, M.W., Arnold, J.G. and Garbrecht, J.D. (2003) Hydrologic simulation on agricultural watersheds: choosing between two models. *Transactions of the ASAE* 46, 1539–1551.

Wikipedia (2011) Koshi River. Available at: http://en.wikipedia.org/wiki/Kosi_River (accessed 21 February 2011).

Williams, J.R. (1975) Sediment routing for agricultural watersheds. *Water Resources Bulletin* 11, 965–974.

Williams, J.R. (1990) The erosion productivity impact calculator (EPIC) model: a case history. *Philosophical Transactions of the Royal Society of London B* 329, 421–428.

Williams, J.R., Nicks, A.D. and Arnold, J.G. (1985) Simulation for water resources in rural basins, *Journal of Hydraulic Engineering* 111, 970–986.

Williams, J.R., Jones, C.A., Gassman, P.W. and Hauck, L.M. (1995) Simulation of animal waste management with APEX. In: McFarland, J. (ed.) *Innovations and New Horizons in Livestock and Poultry Manure Management*. Texas A&M University, Texas Agricultural Extension Service, Austin, Texas, pp. 22–26.

Woolhiser, D.A., Smith, R.E. and Goodrich, D.C. (1990) *A Kinematic Runoff and Erosion Manual. Documentation and User Manual*. ARS Report 77. USDA Agricultural Research Service, Washington, DC.

10 Deep Percolation from Surface Irrigation: Measurement and Modelling Using the RZWQM

Carlos G. Ochoa,* Alexander G. Fernald and Steven J. Guldan

Introduction

Deep percolation from irrigation plays a key role in groundwater supply by replenishing shallow aquifers at local and regional scales. Schmidt and Sherman (1987) stated that deep percolation from irrigation is a significant source of groundwater recharge beneath large irrigated areas in California. Willis and Black (1996) also found that irrigation on highly permeable soils is associated with excessive deep percolation and shallow water table formations in the Macquarie Valley in Australia. Fernald and Guldan (2006) concluded that infiltration from surface irrigation can be a significant source of shallow aquifer recharge in an agricultural valley of northern New Mexico.

A simple approach for determining deep percolation below the root zone is the use of the water balance method (Sammis *et al.*, 1982; Jaber *et al.*, 2006; Ochoa *et al.*, 2007). In this method, water applied (irrigation and rainfall) is measured, evapotranspiration and the change in soil water storage are either calculated or estimated, and the deep percolation is the only unknown variable (Ben-Asher and Ayars, 1990). When reliable field observations of different water balance components are available, this method provides a good characterization of the different water inflow–outflow relationships occurring at the field scale. However, the difficulties associated with conducting extensive field data collection and extrapolating point-scale results to larger spatial scales have encouraged researchers to rely on simulation tools. An increase in the use of computer models for simulating the transport of water and agricultural chemicals in soils and groundwater has occurred over the last two decades. One numerical model that is available for simulating vadose-zone hydrological processes is the Root Zone Water Quality Model (RZWQM). This model, developed by the USDA ARS (US Department of Agriculture Agricultural Research Service), is capable of integrating important physical, biological and chemical processes occurring in the vadose zone with diverse crop management practices (Ahuja *et al.*, 2000).

The RZWQM has been used to simulate various hydrological processes, including soil water content (Starks *et al.*, 2003), runoff (Ma *et al.*, 1998), macropore flow (Cameira *et al.*, 2000; Kozak *et al.*, 2007), evapotranspiration (Ma *et al.*, 1999; Alves and Cameira, 2002) and deep percolation (Ochoa *et al.*, 2007). The model has also been used in numerous studies where water and nutrient transport processes are combined. Examples are subsurface drainage and pesticide transport (Fox *et al.*, 2004), saturated

* Corresponding author: carochoa@nmsu.edu

hydraulic conductivity and pesticide transport (Ellerbroek *et al.*, 1998), macropore flow and herbicide leaching (Malone *et al.*, 2003), and soil water content and pesticide transport (Azevedo *et al.*, 2000). In addition, the effects of management practice on water and nutrient transport have been evaluated with the RZWQM. For example, studies have included the effects of tillage on water and nitrate-nitrogen movement (Kumar *et al.*, 1999), crop rotation and tillage effects on drain flow (Ma *et al.*, 2007), tillage effects on carbon and nitrogen content (Karlen *et al.*, 1998), and the effects of nitrogen management on water quality (Azevedo *et al.*, 1997). More than 200 publications, mostly peer reviewed, that are related to the use of RZWQM for simulating various physical, chemical and biological processes are listed on the USDA ARS web site (USDA ARS, 2011).

Irrigation water that percolates below the root zone in highly permeable soils may rapidly reach the shallow aquifer and generate a transient rise in the water table. This type of setting is commonly found in irrigated valleys along the Rio Grande in northern New Mexico, where surface irrigation water often exceeds plant consumptive use, becoming deep percolation below the root zone, which in the end results in a temporary rise in the water table (Ochoa *et al.*, 2009).

Improving our abilities to characterize surface water and groundwater interactions helps to enhance our understanding of the mechanisms involved in water transport through the vadose zone and in shallow aquifer recharge in flood-plain irrigated valleys. Over the last three decades, diverse studies have focused on researching the linkage between surface water and groundwater, and the mechanisms of aquifer recharge (Sophocleous, 2001; De Vries and Simmers, 2002; Healy and Cook, 2002; Scanlon *et al.*, 2003; Rimon *et al.*, 2007). However, most of these studies rely on observations made in the upper portion of the root zone or at the lower boundary or the water table of the vadose zone, and do not directly show the connection between surface water and groundwater.

We present results from a 2-year study (2005–2006) undertaken to characterize the relationships between surface irrigation, deep percolation and a shallow aquifer for a lucerne/grass field in an irrigated valley in northern New Mexico, USA. The scarcity of field-measured data used in physically based models is still a major limitation for obtaining accurate estimates of hydraulic variables (Mallants *et al.*, 1998). In this study, field measurements of different water budget components and the RZWQM were used to characterize water movement through and drainage below the upper 1 metre soil zone. In addition, shallow groundwater level measurements were used to characterize deep percolation effects on the water table. The objectives of the study were to: (i) measure deep percolation following different amounts of irrigation, (ii) evaluate the performance of the model RZWQM in simulating deep percolation, and (iii) characterize the shallow aquifer response to deep percolation from irrigation.

Methods

Site description

Our study was conducted in an experimental field, at an altitude of 1733 m above sea level, at New Mexico State University's Alcalde Sustainable Agriculture Science Center (Alcalde Science Center), 8 km north of Española, New Mexico. The Alcalde Science Center is located in the agricultural corridor between the Alcalde main irrigation canal and the Rio Grande in the northern part of the Española basin (Fig. 10.1). Within this corridor, traditional irrigation systems are used to divert water from the Rio Grande into a main irrigation canal, from which it is then distributed in the valley, primarily for surface (border and furrow) irrigation of crop fields. Most crops grown in the valley are lucerne, pasture grass and apples. Common soils in the irrigated corridor include Fruitland sandy loam, Werlog clay loam and an Abiquiu-Peralta complex (Soil Survey Staff, USDA NRCS, 2008). The average annual precipitation for the study site is 251 mm. The average maximum annual temperature is 20.1 °C, and the average minimum annual temperature is 1.1 °C. Normally, the maximum temperature occurs during the month of July and the minimum temperature

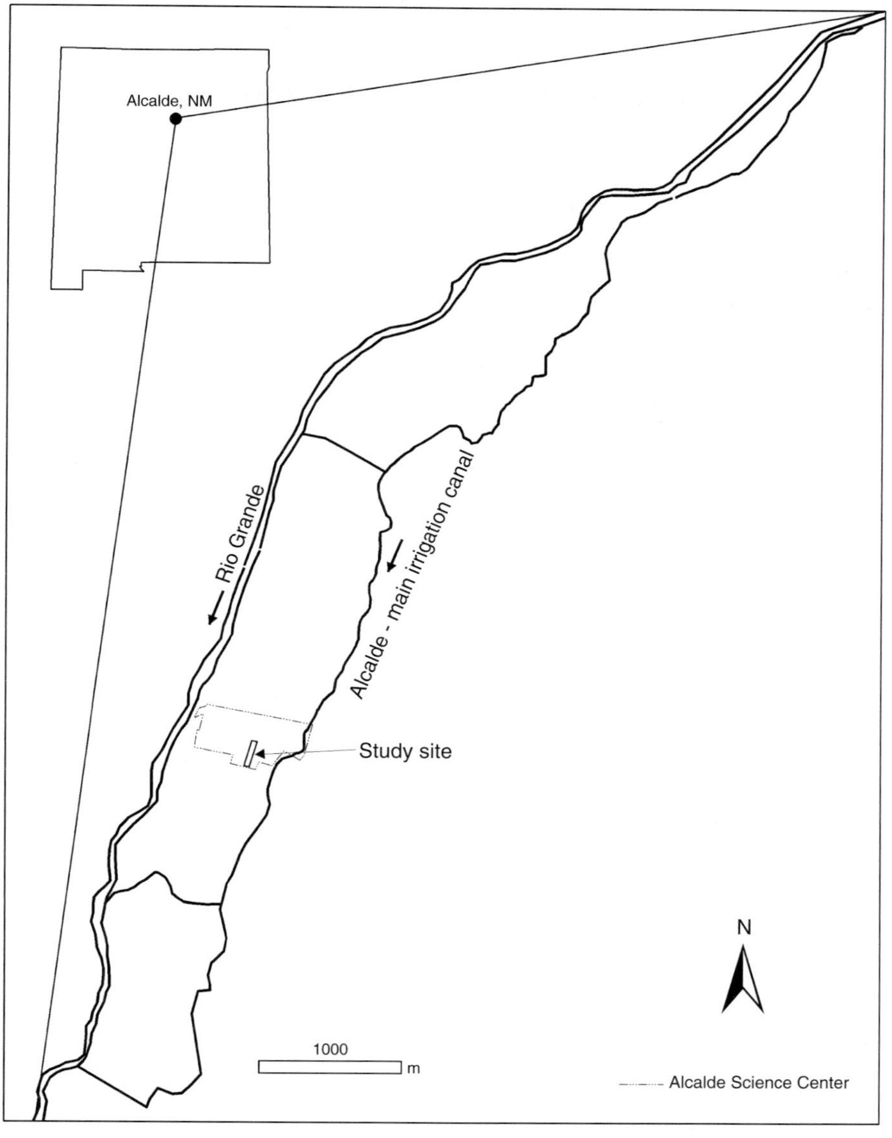

Fig. 10.1. Field study site in the agricultural corridor between the Alcalde main irrigation canal and the Rio Grande in New Mexico.

occurs during the month of January (WRCC, 2006). The Alcalde Science Center overlies a shallow unconfined aquifer with depth to water table ranging from 1.5 to 10 m depending on proximity to the river and measured at the lowest level before the irrigation season, which runs from March to November. Regional groundwater flow is mostly influenced by the Rio Grande and by Rio de Truchas and Cañada de Las Entrañas, which are important tributaries coming from the Sangre de Cristo Range on the east side of the basin, which drain in the vicinity of the Truchas Peaks (Daniel B. Stephen and Associates, 2003). The shallow groundwater flow at the study site is influenced by the Alcalde main irrigation canal and by deep percolation from irrigation contributions (Ochoa *et al.*, 2007; Fernald *et al.*, 2010). During winter when no water is flowing in the main canal, the flow paths follow the river flow (north to south); during the irrigation season, when there is water flowing in the ditch, the flow paths orient more towards the river (Fernald and Guldan, 2006).

Experimental design

The field experiment was conducted in a lucerne field located in the mid-section of the Alcalde Science Center. The field has an area of 0.7 ha and it is subdivided into three 12 m by 190 m strips. Four test pits (1 m by 1m by 1.25 m depth approximately) for soil characterization and sensor installation were excavated in the middle strip of the field. Prior to the experiment, a weather station was installed in the north-east corner of the lucerne field. Also, three experimental wells were installed in the north-east corner (Well 1), in the north-west corner (Well 2), in the midfield (Well 3), and on the western edge (Well 4) of the field (Fig. 10.2). An old stand of lucerne (*Medicago sativa* L.), planted in 1998, and several species of intermixed grasses dominate the crop-field landscape. In this experimental field, depth to water table ranges from 3.3 to 5.0 m throughout the year. The soil type in the field is Fruitland sandy loam, classified as coarse loamy, mixed, superactive, calcareous, mesic Typic Torriorthents (Soil Survey Staff, USDA NRCS, 2008).

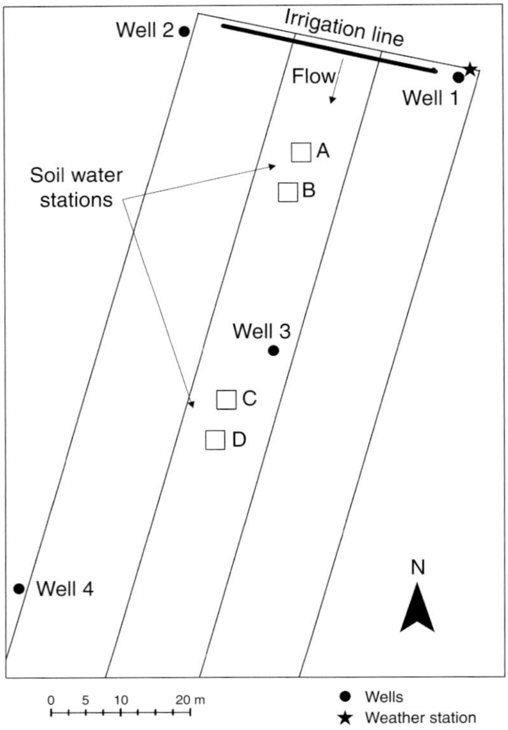

Fig. 10.2. Layout of the study site at the Alcalde Science Center, New Mexico.

Field data collection

Field measurements for characterizing weather conditions, soil physical properties, soil water content, irrigation and shallow groundwater level fluctuations were taken during the 2-year lifespan of this experiment. Sensors in the weather station were programmed to collect hourly measurements of air and soil temperature, wind speed and direction, incoming short-wave (solar plus sky) radiation, relative humidity and rainfall. Rainfall data were obtained from the NOAA (National Oceanic and Atmospheric Administration) gauge at the Alcalde Science Center. Soil samples were collected at 0.125, 0.375, 0.625, 0.875 and 1.125 m depth at the four test pits excavated in the lucerne field for the four soil water content stations A–D (see Fig. 10.2); these samples were used to determine soil bulk density and soil texture. Soil bulk density was calculated using the method after Blake and Hartge (1986) and soil texture was determined using an LS230 (Beckman Coulter, Inc.; Fullerton, California) laser diffraction particle size analyser. Soil samples for determining soil texture were air dried for at least 48 h and passed through a 2 mm sieve. A fractionator was used to partition each soil sample into two subsamples. Soil particle size was determined with the laser diffraction particle size analyser using about 0.3 g of soil from each subsample. Soil texture results obtained from each subsample were averaged by soil depth.

Soil water content data were collected using time domain reflectometry (TDR) systems (Campbell Scientific Inc.; Logan, Utah). The four soil water content stations

(A, B, C and D), each with a nest of five TDR sensors, were installed in the four test pits that had been excavated (see Fig. 10.2). The five TDR sensors were installed at an angle to cover the upper 1.25 m soil profile in each pit (Fig. 10.3).

Measurements of soil volumetric water content (θ) were collected at 3 min intervals during irrigation and throughout 24 h after the end of irrigation, and at hourly intervals thereafter. The use of TDR technology to measure soil volumetric water content is based on the unique electrical properties of water. For instance, the dielectric constant of water (about 80) is much greater than the dielectric constants of air (1) and of the remaining soil solid components (from 2 to 7). Thus, a reliable estimate of soil water content can be obtained by measuring the dielectric constant of the soil (Topp, 1993). The following third-degree polynomial equation is used to convert the dielectric constant (*Ka*) into soil volumetric water content (Topp *et al.*, 1980):

$$\theta = (5.03 \times 10^{-2}) + (2.92 \times 10^{-2} Ka) - (5.5 \times 10^{-4} Ka^2) + (4.3 \times 10^{-6} Ka^3) \quad (10.1)$$

A wide range of soils have shown a close correlation between θ and *Ka* when using Topp's equation (Dalton, 1992). However, calibration may be required for some specific soils (Teixeira *et al.*, 2003). We used the gravimetric water content method to calibrate the TDR sensor data from each soil water station.

We used a correction factor (f_c) to calibrate TDR data by soil depth at each soil water station:

(See Equation 10.2. at the bottom of the page.)

The correction factor was determined by subtracting the gravimetric water content from TDR water content data measured before and at the end of (after) irrigation.

A total of nine surface (border) irrigation events were applied during the 2 years

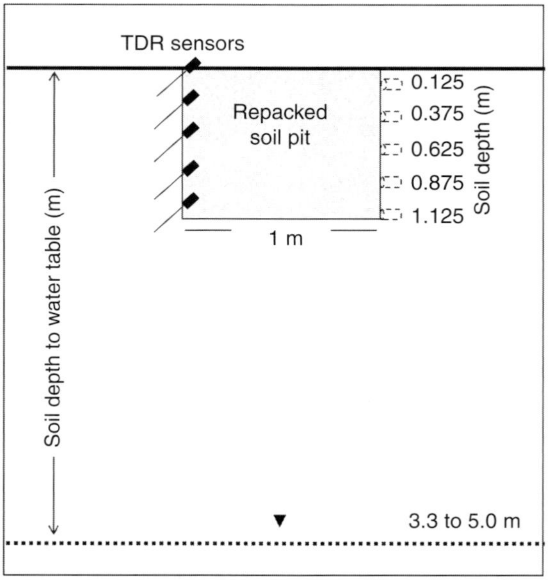

Fig. 10.3. Schematic representation of TDR sensor installation and soil sample collection, illustrating depth to water table at the study site.

$$f_c = \frac{[(\theta_{before}\_TDR) - (\theta_{before}\_Gravimetric)] + [(\theta_{after}\_TDR) - (\theta_{after}\_Gravimetric)]}{2} \quad (10.2)$$

of the experiment. A propeller flow meter (McCrometer, Inc.; Hemet, California) was used to measure the total volume of water applied during each irrigation event. The four experimental wells (50 mm diameter) installed in the lucerne field were located at a variable distance from the water source. Wells 1, 2, 3 and 4 were installed at 2, 3.5, 40 and 85 m away from the irrigation source, respectively. The four wells were equipped with pressure transducers, attached to data-loggers, to monitor shallow groundwater level fluctuations during irrigation and throughout the year. Shallow groundwater-level data were collected at hourly intervals.

Water balance method for determining deep percolation

A daily water balance method (DWBM) was used for determining deep percolation (*DP*). Lucerne roots can grow below 2 m (Kohl and Kolar, 1976; Abdul-Jabbar *et al.*, 1982; Dudley *et al.*, 1994), but most of the root mass and water uptake is in the top 1 m of the soil. In a study conducted by Abdul-Jabbar *et al.* (1982), most of the lucerne root mass was found in the upper 0.45 m. A study conducted in central Idaho showed that up to 80% of the total water withdrawn by lucerne roots in the upper 2.3 m of the soil came from the top 1 m (Kohl and Kolar, 1976). Visual observations during sensor installation in this experiment also showed that all of the grass roots and most of the lucerne root mass were present in the upper 0.6 m of the soil profile. This was corroborated during pit excavation (to 1.8 m depth) for sensor retrieval at the end of the experiment. Thus, based on the literature review and visual observations, we hypothesized that most of the water passing below the upper 1 m of the soil can be considered deep percolation water which, under sufficient, non-stressed irrigation conditions, will not be used by the lucerne and grass plants.

We calculated *DP* below the upper 1 m soil depth using the following water balance equation:

$$DP = SWC_i + IRR + R - SWC_{fc} - RO - ET \tag{10.3}$$

where *DP* is the deep percolation (mm day^{-1}), SWC_i is the soil initial water content (mm), *IRR* is the irrigation depth (mm), *R* is the rainfall (mm), SWC_{fc} is the soil water content at field capacity (mm), *RO* is the field runoff (mm) and *ET* is the evapotranspiration (mm).

The *DP* was determined as the total amount of water passing below the upper 1 m of soil depth after each irrigation event. Measurements of soil volumetric water contents, collected at each soil water measuring station, were used to determine initial (SWC_i) and field capacity (SWC_{fc}) soil water contents in the upper 1 m soil profile. SWC_i was calculated based on θ data collected before the onset of irrigation. SWC_{fc} was obtained based on average θ collected 24 h after the end of irrigation. Evapotranspiration (*ET*) was calculated by the FAO-56 Penman–Monteith equation (Allen *et al.*, 1998) using daily averaged data collected from our weather station and with rainfall data from the NOAA gauge.

(See Equation 10.4 at the bottom of the page.) where ET_0 is the reference evapotranspiration (mm day^{-1}), R_n is the net radiation at the crop surface (MJ m^{-2} day^{-1}), *G* is the the soil heat flux density (MJ m^{-2} day^{-1}), *T* is the mean daily air temperature at 2 m height (°C), u_2 is the wind speed at 2 m height (m s^{-1}), e_s is the saturation vapour pressure (kPa), e_a is the actual vapour pressure (kPa), $e_s - e_a$ is the saturation vapour pressure deficit (kPa), Δ is the slope of the vapour pressure curve (kPa $°C^{-1}$) and γ is the psychrometric constant (kPa $°C^{-1}$).

Elevated roads and raised berms that surround the lucerne field prevented irrigation water from running off the field; thus, runoff was considered negligible. No rainfall was observed during any of the irrigation events.

$$ET_0 = \frac{0.408 \times \Delta \times (R_n - G) + \gamma \times \frac{900}{T + 273} \times u_2 \times (e_s - e_a)}{\Delta + \gamma \times (1 + 0.34 \times u_2)} \tag{10.4}$$

Root zone water quality model (RZWQM)

Model overview

The RZWQM is an integrated physical, biological and chemical process model developed by a team of USDA ARS scientists. The one-dimensional RZWQM was created to simulate physical, chemical and biological processes in agricultural crop production systems. The model can simulate plant growth, water flow and solute transport in the vadose zone (Ahuja *et al.*, 2000). It consists of six main processes, namely physical, chemical, nutrient, pesticide, plant growth and crop management. The RZWQM's physical processes comprise a large number of hydrological processes: water infiltration, solute transport, water and solute movement through the soil matrix and macropores, soil heat flow, fluctuating water table, tile drain, soil evaporation and crop transpiration (Ahuja *et al.*, 2000).

Using RZWQM for simulating deep percolation

For this study, we focused on the water infiltration, soil water redistribution and evapotranspiration processes of RZWQM. As described by Ahuja *et al.* (2000) infiltration rates are calculated for homogeneous or for layered soil profiles subdivided into 1 cm increments. The RZWQM uses a modified equation after Green and Ampt (1911), in which the infiltration rate is divided by a viscous resistance correction factor of 2. This equation is represented as:

$$\frac{V}{2} = \overline{K}_S \times \frac{\tau_c + H_0 + Z_{wf}}{Z_{wf}} \tag{10.5}$$

where V is the infiltration rate (cm h^{-1}), 2 is the viscous resistance correction factor, $\overline{K}_s$ is the average effective hydraulic conductivity in the wetting zone (cm h^{-1}), τ_c is the suction head at the wetting front (cm), H_0 is the surface ponding depth (cm) and Z_{wf} is the wetting front depth (cm).

When infiltration rate is greater than rainfall, the model uses the rainfall rate. When the rainfall rate is greater than the infiltration rate, excess rainfall is considered runoff. For calculating water redistribution in between irrigation or rainfall events, RZWQM uses the Richards equation (Richards, 1931):

$$\frac{\partial \theta}{\partial t} = \frac{\partial}{\partial z} \times \left[K(\theta) \times \left(\frac{\partial \Psi}{\partial z} + 1 \right) \right] \tag{10.6}$$

where θ is the water content, t is time, z is the elevation above a vertical datum, K is the unsaturated hydraulic conductivity and ψ is the pressure head.

The RZWQM initial condition for the soil water pressure head is known to be a function of soil depth, so the initial condition for time is set to zero and the initial condition for soil depth is greater than or equal to zero. The surface boundary condition is considered to be an evaporative flux until the surface pressure head falls below –20,000 cm, then the condition changes to a constant head. The boundary condition at the bottom of the soil profile can be set as pressure head, constant flux or a unit gradient (Ahuja *et al.*, 2000). Potential evapotranspiration (PET) is calculated using the Farahani (1994) extended version of the Shuttleworth and Wallace (1985) model. This extended model describes the evapotranspiration process under no-till or minimum till practices that leave a portion of the crop residue on the soil surface. It explicitly defines a partially covered soil and predicts evaporation from the bare soil fraction of the substrate, the residue fraction of the substrate and transpiration from the canopy.

Model parameters

The RZWQM can be initialized with values for texture and bulk density as the minimum soil properties input data (Abrahamson *et al.*, 2005). As part of the initial parameterization, the model requires the description of the crop selection, break-point rainfall, daily meteorology, crop residue, soil properties, soil water content data and water applied (Cameira *et al.*, 1998). We used RZWQM2 v.1.80.2009 for simulating deep percolation below the upper 1 m root zone following flood irrigation. RZWQM provides a user-friendly Windows interface for entering input parameters. Figure 10.4a and b show the precipitation and weather data entry

(a)

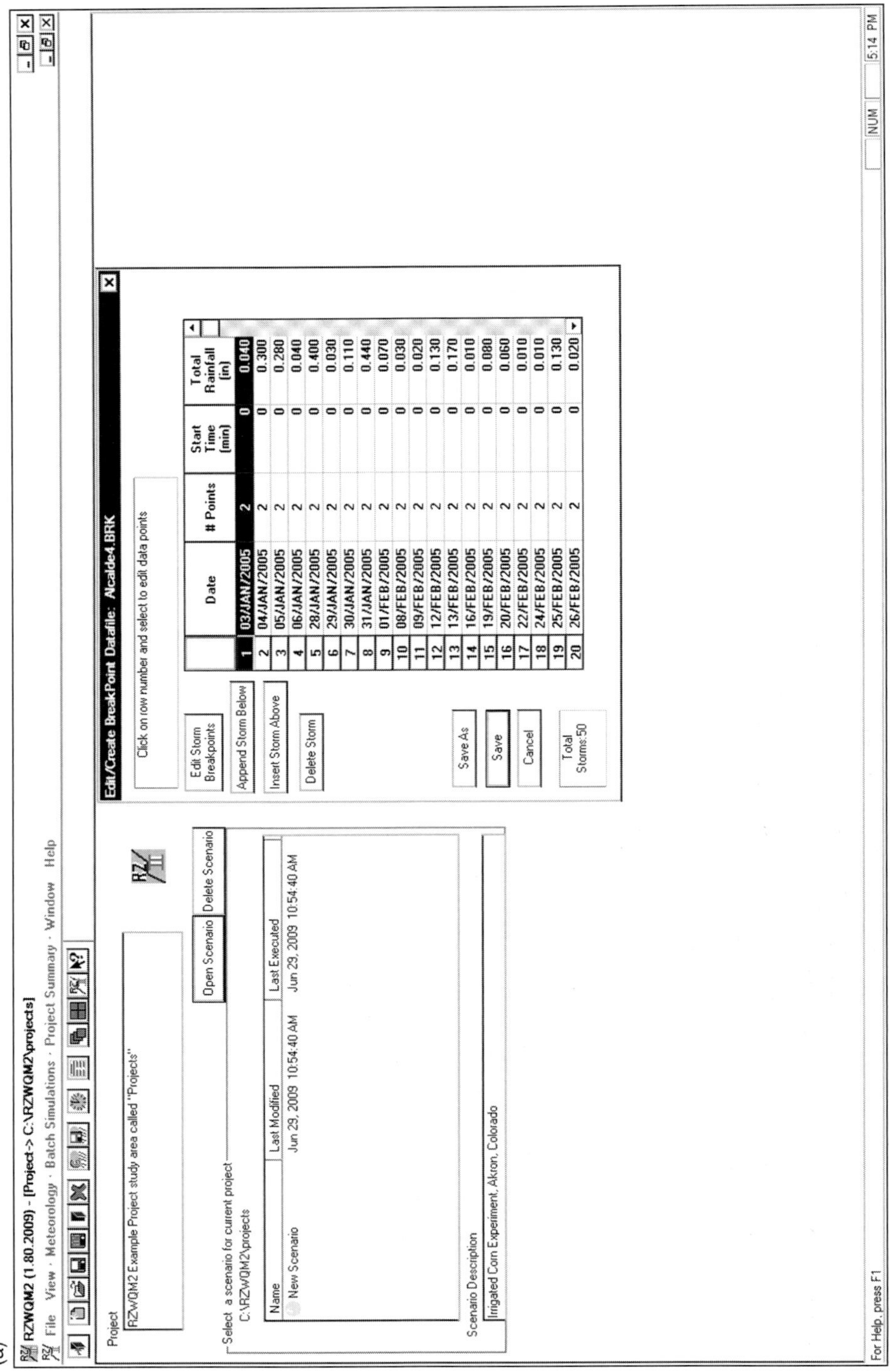

Fig. 10.4. (a) Example of rainfall data entry window in the RZWQM.

(b)

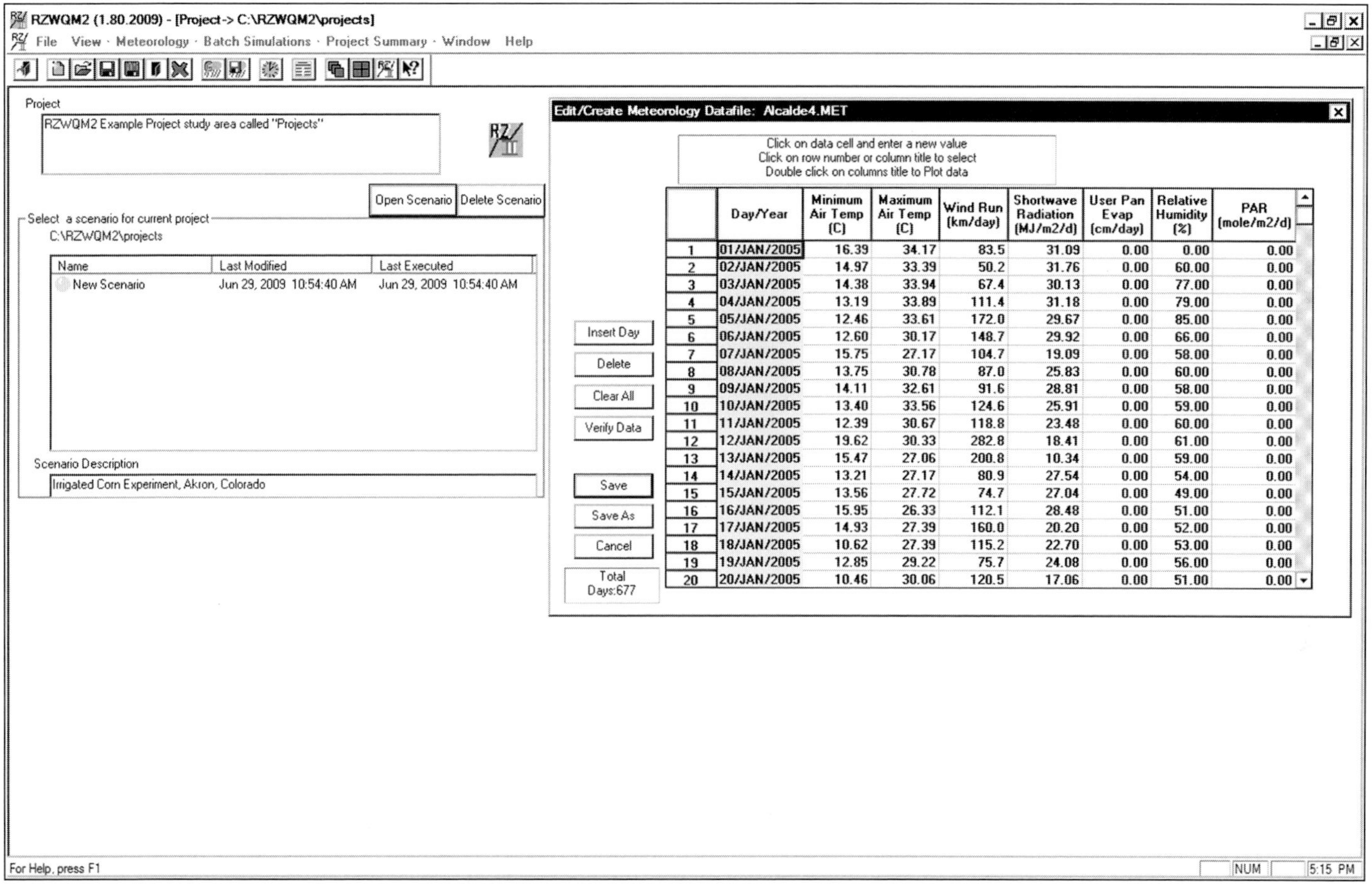

Fig. 10.4. (b) Example of weather data entry window in the RZWQM.

(c)

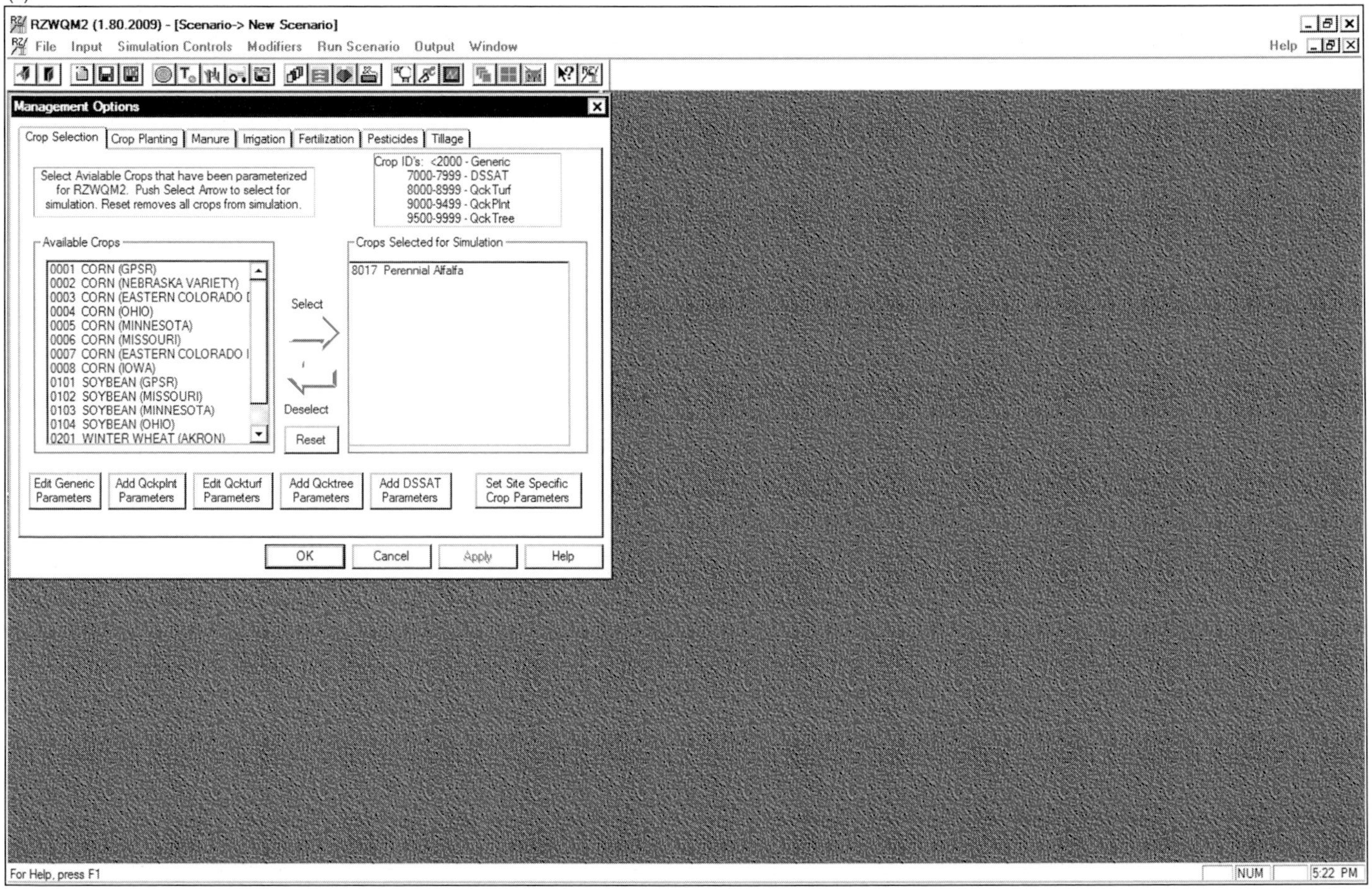

Fig. 10.4. (c) Example of the crop selection data entry window in the management options tab in the RZWQM.

windows, respectively, used in this project. Figure 10.4c also shows the RZWQM window used for entering management options, including crop and irrigation data.

Field data collected at the lucerne field were used to parameterize the model. For crop selection, we used lucerne (8017 Perennial Alfalfa) and entered the following values for model parameterization: height of lucerne at cutting 40 cm, crop height after cut 5 cm, the earliest day of the year when lucerne would come out of dormancy 16 March 2005, and height of lucerne on the first day of the year 10 cm. For break-point rainfall data (the model requires rainfall in inches), we used data from the NOAA gauge. For daily meteorology data, we converted hourly data collected to daily values. Field-based weather parameters entered were: minimum and maximum air temperature (°C), wind run (km d^{-1}), short-wave radiation (MJ m^2 m^{-1}) and relative humidity (%). The period of record used for both rainfall and meteorology data went from 1 January 2005 through to 31 December 2006.

For calculating evapotranspiration values, the model requires crop residue data. We collected crop residue samples after four lucerne harvestings. Samples were collected using a 1 m by 1 m square frame in five randomly selected areas near the soil water content sensors. Crop residue samples were weighed and the average of all sample weights over the four collection dates was used as the model input for crop residue (ton ha^{-1}).

For soil profile characterization in the model, we divided the upper 1 m soil profile into four soil horizon depths of 0.25 m each. Data collected for soil physical properties data were averaged by depth across the four pits and entered as model input parameters for soil bulk density, sand, clay and silt content (Table 10.1).

For soil water content, we used TDR data for entering the soil initial water conditions and water content at field capacity. Input data for soil initial water content at the different horizon depths were obtained by averaging TDR data collected at the different soil water content stations on 1 January 2005. Input parameters for soil water content at field capacity for each 0.25 m soil depth were

Table 10.1. Soil physical properties used in RZWQM.

Horizon depth (m)	Bulk density (Mg m^{-3})	Sand (%)	Silt (%)	Clay (%)
0–0.25	1.40	60.5	35.6	3.9
0.25–0.50	1.44	61.5	34.5	4.0
0.50–0.75	1.34	68.8	28.4	2.8
0.75–0.99	1.40	70.8	26.6	2.6

obtained by averaging TDR data collected at 24 hours after the end of each irrigation event. It is noteworthy that ponding lasting between 2 and 4 h was present following all irrigation events. However, the TDR data showed that a semi-steady state in soil water content was reached for all sensors in less than 16 h following the end of all irrigations. For water applied, we used irrigation depth calculated after each irrigation event applied to the lucerne field. The total amount of water applied over the nine irrigation events in the two (2005–2006) irrigation seasons was 2139 mm.

MACROPORE FLOW. Macropores represent a relatively small proportion of the total soil volume, yet they play an important role in chemical and water transport through the soil matrix. They can be defined as the pore fraction between soil aggregates, which can include biopores, wormholes and decayed root channels, as well as shrinkage cracks in the soil. The RZWQM is capable of simulating hydrological processes with and without macropore flow data. Macropore content and structure can be greatly influenced by soil management practices and the interaction between the biota and the soil (Lal and Shukla, 2004), and play an important role in soil water transport in crop fields (Cameira *et al.*, 2000). The extensive root system of crops like lucerne can affect soil physical properties by increasing macropore content and water transport rates through the soil profile (Rasse *et al.*, 2000). High infiltration rates in a 5-year-old lucerne field due to a well-developed macropore system, primarily resulting from the decomposition of a long and dense rooting system, were reported by Meek *et al.* (1990). At our study field, a 7-year-old stand of lucerne generates a strong likelihood of a well-developed macropore system. Long roots and root channels were observed during the pit excavations, but no measurements of macropores were made. Therefore, we decided to run RZWQM with and without the macropore flow option. Because macropore data were unavailable, model default values for total macroporosity were used. Total macroporosity is defined as the volume of macropores divided by the soil bulk volume, and typically ranges from 0 to 0.1% of the soil bulk volume. We used the model's suggested value of 0.5% for when no data are available.

Model sensitivity analysis

A series of sensitivity analyses were conducted to determine which parameters affected RZWQM performance the most for calculating deep percolation. Individual variations of the input parameters, including total irrigation, initial soil water content, soil bulk density, macroporosity and saturated hydraulic conductivity, were conducted over an expected range of values to calculate *DP* output values. An equation after Walker *et al.* (2000) was used to calculate relative sensitivity index (*RSI*) for the selected parameters and to determine the influence of individual parameters on estimates of average deep percolation for the simulation period. The *RSI* equation used is as follows:

$$RSI = \frac{(y_2 - y_1)}{(x_2 - x_1)} \times \frac{\bar{x}}{\bar{y}} \quad (10.7)$$

where *RSI* is the relative sensitivity index, x_1 and x_2 are the input parameter minimum (x_1) and maximum (x_2) values tested, $\bar{x}$ is the average value of x_1 and x_2, y_1 and y_2 are the model output values corresponding to x_1 and x_2, and $\bar{y}$ is the average value of y_1 and y_2.

The *RSI* allows the determination of the sensitivity of RZWQM to changes in selected parameters. The greater the sensitivity index, the greater the impact of the evaluated input parameters on the estimation of *DP*. If the *RSI* value is positive, this indicates a positive relationship between the input parameter and the *DP* output value. An inverse relationship between the input parameter and the output value occurs if the *RSI* is negative (Walker *et al.*, 2000). In addition, we compared field-measurement-based deep percolation (DWBM *DP*) results with those results simulated by the RZWQM. A literature search revealed that simulated results after calibration to within 10% to 20% of field observations can be considered an acceptable degree of model error (Farahani *et al.*, 1999). Field observation errors are normally greater than 10%; therefore, simulated errors cannot be less than that (Hanson *et al.*, 1999). A variable degree of error can be obtained when comparing field observations with RZWQM-simulated results, depending upon whether the model is calibrated or not. An error degree ranging from –18% to 88% was reported by Hanson *et al.* (1999) before calibration of the RZWQM for different crop production indicators. After calibration, simulations were within 5% to 20% of measured values. Abrahamson *et al.* (2005) reported calibrated simulations of tile drainage and leached nitrate to be within 15% of field values. Farahani *et al.* (1999) reported that, after model calibration, seasonal estimates of seepage below the root zone were within 3% of measured results. The use of the macropore option in the RZWQM can improve the prediction of different hydrological variables (Malone *et al.*, 2004). As reported by Ghidey *et al.* (1999), when using the macropore option in the model, soil water predictions were within 15% to 20% of measured values.

The goal of this investigation was not to calibrate the RZWQM, but rather to test its performance with the minimum input data and to compare the uncalibrated model results with the field observations. An arbitrary target of 25% difference between the deep percolation results calculated by the DWBM and the non-calibrated RZWQM was established. The percentage degree of error (% *D*) was calculated using an equation after Oliver and Smettem (2005):

$$\%D = \frac{(Meas_{DP} - Model_{DP})}{(Meas_{DP})} \times 100 \quad (10.8)$$

where %*D* is the degree of error (%), $Meas_{DP}$ is the field-measurement-based deep percolation calculated by the DWBM (mm) and $Model_{DP}$ is the RZWQM-predicted deep percolation (mm).

Results and Discussion

Field observations of weather, soil physical properties, soil water content, irrigation applied and water table response to irrigation were evaluated at the lucerne experimental field during the irrigation seasons of 2005 and 2006.

Weather conditions

In general, weather data collected in the study area were similar for the two years of the experiment. Precipitation data collected from the NOAA station between 1 January 2005 and 31 December 2006 totalled 637.3 mm. Total precipitation in 2005 was 324.9 mm, and in 2006 it was 312.4 mm. In 2005, the lowest precipitation of 0.3 mm was observed during November and December, and the highest precipitation of 74.9 mm was observed in August. In 2006, no precipitation was observed in February and the highest precipitation of 99.6 mm was observed in August (Fig. 10.5). Hourly data collected by the on-site weather station showed a maximum air temperature of 37.1°C on 19 July 2007 at 4.00 p.m., and a minimum temperature of −19.62°C recorded on 8 December 2005 at 5.00 a.m. Soil temperature measured next to our weather station at the 5 to 15 cm depth showed a maximum value of 29.91 °C on 8 June 2005 and a minimum value of −1.58 °C on 17 December 2005.

Soil properties

Soil physical properties obtained for soil samples collected from different test excavated pits were different across field locations, but were also different across soil depths (Table 10.2). The highest soil bulk density value of 1.63 Mg m^{-3} was observed at 0.375 m depth in the station B pit. The lowest soil bulk density value of 1.24 Mg m^{-3} was observed at 0.125 m depth in the station C pit. Average soil bulk density across locations and soil depths was 1.38 Mg m^{-3}. In general, the across-location average sand content increased and silt and clay contents decreased with increasing soil depth. The sand (85.2%) content of soil was much higher than the silt (13.1%) and clay (1.7%) contents at the 1.125 m soil depth in test pit D (see Table 10.2).

Deep percolation with the DWBM

Field observations of different parameters used for calculating deep percolation below the upper 1 m root zone varied throughout the 2-year experiment (Table 10.3). Time of irrigation (t) ranged from 6.0 to 12 h.

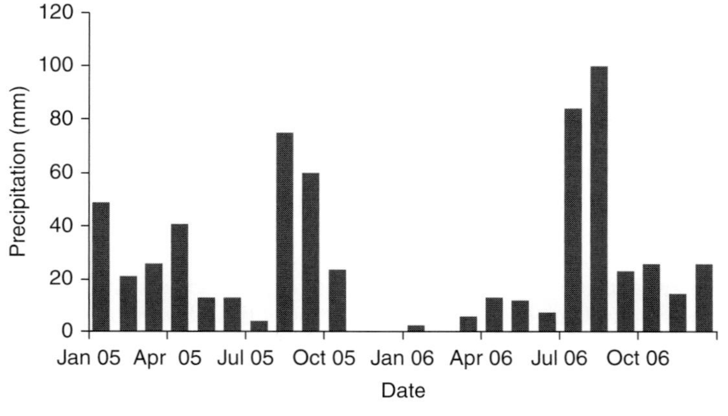

Fig. 10.5. Total monthly rainfall at the study site during 2005–2006.

Table 10.2. Soil physical properties in the experimental field.

Test pit	Soil depth (m)	Bulk density (Mg m^{-3})	Sand (%)	Clay (%)	Silt (%)
A	0.125	1.48	58.3	37.7	4.0
B	0.125	1.43	56.5	39.2	4.4
C	0.125	1.24	61.6	34.1	4.3
D	0.125	1.44	65.7	31.3	3.0
Mean ± SD		1.40 ± 0.11	60.5 ± 4.05	35.6 ± 3.55	3.9 ± 0.63
A	0.375	1.44	71.0	26.4	2.6
B	0.375	1.63	61.4	35.3	3.3
C	0.375	1.44	49.9	43.6	6.6
D	0.375	1.27	64.0	32.7	3.4
Mean ± SD		1.44 ± 0.15	61.6 ± 8.79	34.5 ± 7.12	4.0 ± 1.77
A	0.625	1.41	66.8	30.2	3.0
B	0.625	1.33	71.6	25.9	2.5
C	0.625	1.34	69.8	27.5	2.8
D	0.625	1.27	67.4	29.9	2.8
Mean ± SD		1.34 ± 0.06	68.9 ± 2.22	28.4 ± 2.03	2.8 ± 0.20
A	0.875	1.5	65.3	31.7	3.0
B	0.875	1.47	78.5	19.5	2.0
C	0.875	1.34	66.1	30.9	3.1
D	0.875	1.3	73.1	24.4	2.5
Mean ± SD		1.40 ± 0.10	70.7 ± 6.25	26.6 ± 5.75	2.6 ± 0.50
A	1.125	1.24	56.9	38.6	4.6
B	1.125	1.29	77.3	20.8	2.0
C	1.125	1.31	57.2	39.2	3.6
D	1.125	1.31	85.2	13.1	1.7
Mean ± SD		1.29 ± 0.03	69.1 ± 14.35	27.9 ± 13.07	3.0 ± 1.34

Initial conditions of soil water content (SWC_i) in the upper 1 m zone varied across irrigations, but a similar trend across years was observed (Table 10.3). The driest soil conditions before irrigation were at the beginning of the irrigation season and a maximum value of about 190 mm was reached towards the end of the irrigation season. Soil water content ranged from 124 to 193 mm in year 2005 and from 120 to 190 mm in year 2006. The lowest soil initial water content values of 124 and 120 mm were observed during the first irrigations in 2005 (19 May) and 2006 (21 April). Irrigation depth varied across irrigations and ranged from 125 to 390 mm. Soil water content at field capacity (SWC_{fc}) remained relatively constant with an average value of 278 mm. Calculated evapotranspiration (*ET*) ranged from 4 to 9 mm. A high correlation (r = 0.94) between the amount of water applied (*IRR*) and deep percolation (DWBM *DP*) was observed. The highest deep percolation value of 235.3 mm was observed after the highest irrigation application of 389.5 mm on 7 June 2006. The lowest deep percolation value of 24.9 mm corresponded to the lowest irrigation application of 124.5 mm on 21 Sep 2006. Initial soil water content showed an effect on deep percolation during the irrigations in 2005, as previously reported by Ochoa *et al.* (2007), this trend was not shown during irrigation applications in 2006. Evapotranspiration did not show a significant effect on deep percolation (Table 10.3).

Deep percolation with the RZWQM

Deep percolation below the upper 1 m root zone was simulated using the RZWQM. A sensitivity analysis of selected parameters was performed and simulations of deep percolation with and without macropores were compared with results obtained by the DWBM. The sensitivity analysis results showed that total irrigation during the 2-year experiment was the most influential parameter when simulating deep percolation with the RZWQM (Table 10.4). Greater total irrigation effects on deep percolation have also been reported by Ellerbroek *et al.* (1998). Average soil bulk density, saturated hydraulic conductivity, total macropores and soil initial water content showed a minimal effect on estimates of RZWQM deep percolation (Table 10.4). A minimal effect of saturated hydraulic conductivity on the estimation of *DP* was also reported by Oliver and Smettem (2005) and Ellerbroek *et al.* (1998).

In general, the RZWQM results for deep percolation (*DP*) with macropores (104 ± 74.0) were in closer agreement with the DWBM *DP* (112 ± 67.4) values than with the uncalibrated RZWQM results without macropores (75 ± 52.7). The results showed that, during six out of a total of nine irrigations applied throughout the experiment, deep percolation results estimated with the RZWQM with macropores were within the set target of a 25% degree of error. Only during two irrigations (19 May and 15 June 2005) were the RZWQM without macropores estimates of deep percolation within 25% of the DWBM results (Table 10.5). The RZWQM without macropores consistently under-predicted deep percolation by the DWBM. Deep percolation simulations with the RZWQM with the macropores option

Table 10.3. Deep percolation after irrigation by the daily water balance method (DWBM).

Date	Time (h)[a]	SWC_i (mm)[a]	*IRR* (mm)[a]	SWC_{fc} (mm)[a]	*ET* (mm)[a]	DWBM *DP* (mm)[a]
19 May 2005	7.0	124	216	280	8	52
15 Jun 2005	7.9	140	246	278	9	100
6 Jul 2005	6.9	193	219	280	8	124
27 Jul 2005	8.7	171	298	283	4	182
1 Sep 2005	7.3	187	175	278	8	77
21 Apr 2006	11.0	120	317	279	8	150
7 Jun 2006	12.0	130	390	276	8	235
2 Aug 2006	6.0	190	154	275	4	65
21 Sep 2006	6.5	180	125	273	7	25

[a]*DP*, deep percolation; *ET*, evapotranspiration; *IRR*, depth of water applied; SWC_i, initial soil water content; SWC_{fc}, soil water content at field capacity; Time, time of irrigation.

Table 10.4. Sensitivity analysis of selected parameters used in the RZWQM for calculating deep percolation.

Parameter	Base value	Input values[a]			Output values[a]			Relative sensitivity index (RSI)
		x_1	x_2	$\bar{x}$	y_1	y_2	$\bar{y}$	
Total irrigation depth (mm)	2049	0	2738.9	1369.5	3.3	1244.2	623.7	0.99
Soil bulk density (Mg m^{-3})	1.395	1.2	1.6	1.4	1251.4	1219.8	1235.6	−0.09
Saturated hydraulic conductivity (mm h^{-1})	36.75	10	100	55	1123.5	1137.4	1130.4	0.01
Total macropores (m^3 m^{-3})	0.0005	0	0.01	0.0005	1049.3	1196.0	1122.7	0.07
Soil initial water content (m^3 m^{-3})	0.17	0.05	0.28	0.17	1182.0	1226.7	1204.3	0.03

[a]In the equation used to calculate *RSI* (Eqn 10.7), x_1 is the input parameter minimum, x_2 is the input parameter maximum (x_2) values tested, $\bar{x}$ is the average value of x_1 and x_2, y_1 and y_2 are the model output values corresponding to x_1 and x_2, and $\bar{y}$ is the average value of y_1 and y_2.

over-predicted *DP* for the first two irrigation events in 2005 (19 May and 15 June) and for the second irrigation in 2006 (7 June).

Water level response

Water level response to surface irrigation input was determined based on water level fluctuations recorded in the wells installed in the lucerne field. In order to avoid interference from irrigation water applied to other crop fields near the experimental field, these other crop fields were not irrigated for at least 4 days before and at least 3 days after irrigation of the lucerne field. Given the relatively small spatial and temporal resolution of the experiment, regional groundwater flow was considered negligible. Well number 3 was discarded owing to technical difficulties that included malfunctioning of the data-logger and pressure transducer during most of the life of the experiment; thus, data from Wells 1, 2 and 4 are presented here. Groundwater level fluctuations were observed in all three wells in response to deep percolation from different irrigation depths (Table 10.6). In general, it was noted that higher irrigation

Table 10.5. Daily water balance method (DWBM) calculated and RZWQM-simulated deep percolation (*DP*) with degree of error (*D*).

		RZWQM macropores		RZWQM no macropores	
Irrigation date	DWBM *DP* (mm)	*DP* (mm)	%*D*	*DP* (mm)	%*D*
19 May 2005	52	68	–29.3	45	13.6
15 Jun 2005	100	114	–14.4	93	6.2
6 Jul 2005	124	95	23.5	79	35.8
27 Jul 2005	182	170	6.4	124	32.0
1 Sep 2005	77	55	28.7	41	47.0
21 Apr 2006	150	148	1.1	108	27.9
7 Jun 2006	235	244	–3.7	165	30.0
2 Aug 2006	65	31	52.8	20	69.6
21 Sep 2006	25	12	53.0	2	90.7
Mean	112	104	–	75	–
SD	67.4	74.0	–	52.7	–

Table 10.6. Groundwater level response to different irrigation depths observed in three monitoring wells.

		Well 1		Well 2		Well 4	
Date	Irrigation depth (mm)	Peak water level (m)	Time to peak (h)	Peak water level (m)	Time to peak (h)	Peak water level (m)	Time to peak (h)
19 May 2005	216	0.14	15	0.16	18	0.13	25
15 June 2005	250	0.35	9	0.21	11	0.09	16
6 July 2005	219	0.36	11	0.29	12	NA[a]	NA
27 July 2005	298	0.38	10	0.30	13	0.22	16
1 Sep 2005	198	0.24	8	0.17	11	0.13	13
24 April 2006	199	0.16	19	0.17	19	0.09	24
7 June 2006	390	0.31	13	0.34	13	0.21	17
2 Aug 2006	154	0.04	18	0.03	19	0.06	19
21 Sep 2006	125	0.01	24	0.01	24	0.01	24
Mean	227.7	29.4	10.6	22.6	13	14.3	17.5
SD	78.8	0.1	5.3	0.1	4.6	0.1	4.5

[a]NA, data not available.

depths yielded greater rises in groundwater level. A rise in water level of 380 mm was observed during a 298 mm irrigation event on 27 July 2005. Minimal or muted response was observed during water applications with less than 160 mm depth (Table 10.6).

During irrigation events, the onset of rise in water level varied in the different monitoring wells, as illustrated during the irrigation event of 7 June 2006 (Fig. 10.6). Water levels in all three wells started rising towards the end of irrigation, observed first in Wells 1 and 2, followed by Well 4. The delay in time to respond observed in Well 4 is attributed to the greater distance existing between the water source and the well, which may affect the intake opportunity time.

Data on the water level fluctuations in Well 2 throughout the irrigation seasons of 2005 and 2006 are presented in Fig. 10.7. A rise in water level in response to deep percolation inputs was observed during most of the irrigations applied. Other water table spikes shown in Fig. 10.7 are associated with irrigations applied in crop fields adjacent to the experimental field. Precipitation was not considered as an important factor in any of the water table spikes observed. This is based on the smallest water table rise of 0.01 m that was in response to a 125 mm irrigation (see Table 10.6), which is higher by far than the highest precipitation record of 50.8 mm observed on 12 August 2005 (data not shown). Water table seasonal fluctuations were observed during both years of the experiment. A peak water level rise of about 1 m was observed towards the middle of the irrigation season, which started during late March in both years. Also, the highest water level rises (see Table 10.6) were observed during those irrigation events when the seasonal water table was relatively high. The seasonal fluctuation of the water table is mainly attributed to deep percolation from irrigation and canal seepage, which together represented 33% of the total water diverted into the Alcalde main irrigation ditch (Fernald *et al.*, 2010).

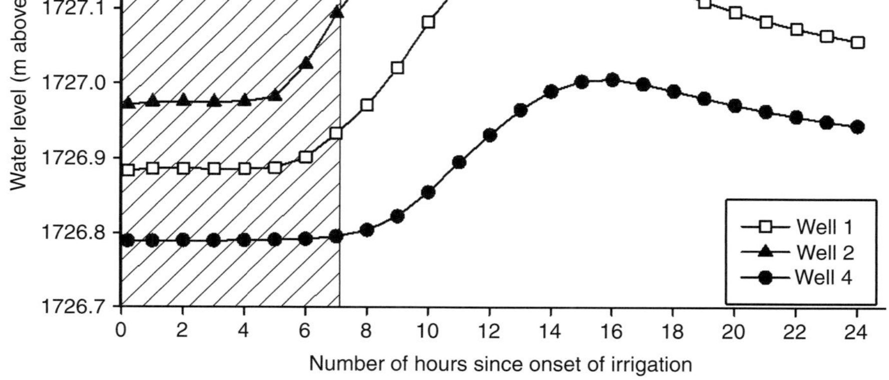

Fig. 10.6. Water level fluctuations in the three experimental wells at the study site in response to the irrigation event of 7 June 2006.

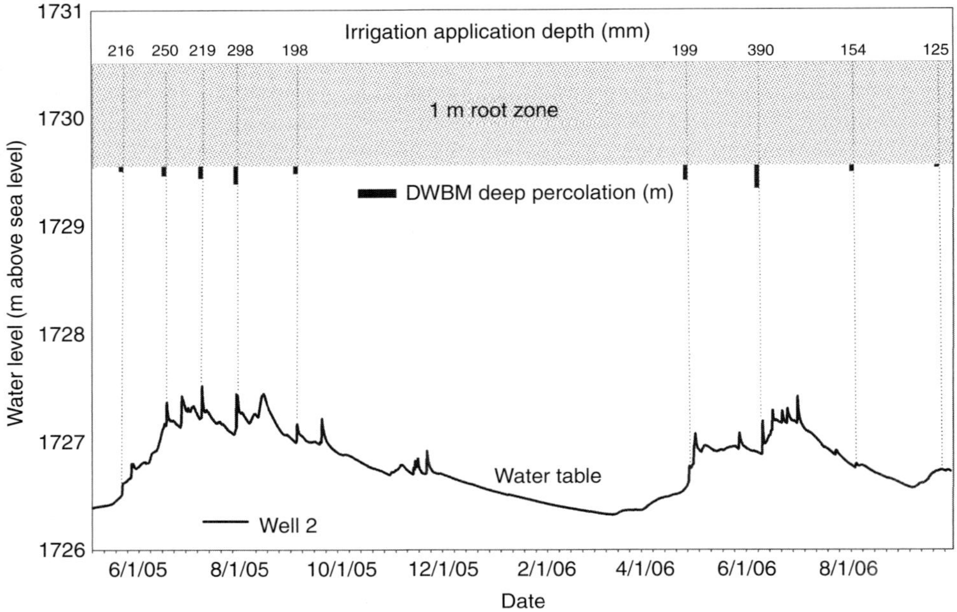

Fig. 10.7. Seasonal water level fluctuations in one experimental well (Well 2) at the study site showing water level response to deep percolation from irrigation inputs (DWBM, Daily Water Balance Method).

Summary

Deep percolation below the root zone and water level response to percolation inputs were evaluated in a lucerne/grass field following different applications of flood irrigation. A daily water balance method (DWBM) was used to calculate deep percolation based on field observations. The Root Zone Water Quality Model (RZWQM) was run with and without macropores to simulate deep percolation in the field. Continuous measurements of shallow groundwater fluctuations were used to characterize water level response to deep percolation. Calculated by the DWBM, deep percolation ranged from 20% to 61% of the amount of water applied in different irrigation events. Similar to reports by Willis *et al.* (1997) and Jaber *et al.* (2006), the more water applied, the more deep percolation was observed. Also, antecedent soil water content seemed to affect deep percolation, but it was not as influential as irrigation water applied. The RZWQM with the macropores option adequately simulated deep percolation, which ranged from 9% to 63% of irrigation water applied. RZWQM-predicted results were in reasonable agreement with those obtained by the DWBM. During six out of a total of nine irrigations, simulated deep percolation by the RZWQM with macropores was within the set target of 25% from measured results.

The results of this study showed that surface (border) irrigation water applied to a lucerne/grass field with sandy loam soil can be rapidly transported downwards through the shallow vadose zone and generate a transient rise in water table. Water level response was found to be highly correlated with the amount of irrigation water applied. The rise in water level in response to deep percolation was variable, ranging from 0.01 m to 0.38 m. It was noted that, during the onset and towards the end of the irrigation season, the local water table was low, contributing to the observed lower water level rises in response to deep percolation. Muted water level response was observed during irrigations of less than 160 mm. Intake opportunity time affected water level response to irrigation, which was more rapid in wells closer to the irrigation source than in wells further from the irrigation source. During this study, a rapid water table rise was observed in response to deep percolation. A greater time of response and lower rises in groundwater level were found by Ochoa *et al.* (2009) using non-vegetated infiltration ponds in the same Fruitland sandy loam soil in fields nearby the study lucerne field. The rapid response of shallow groundwater to deep percolation observed in this study was attributed in part to a macropore flow system, likely to have been created by the ageing lucerne, similar to the findings of Meek *et al.* (1990).

Assumptions were made about the field-averaged soil and water conditions. Some of the limitations noted during this study include the low number and location of field measurement stations, which may not completely reflect soil physical properties and soil water dynamics at the entire field scale. None the less, this study improves our understanding of downward water movement through the shallow vadose zone, and the groundwater response to irrigation in a dominant soil of a flood-plain irrigated valley in northern New Mexico. Based on the reasonable agreement found between calculated and simulated deep percolations without model calibration, we are confident that the RZWQM can be used to simulate soil water dynamics in other crop, soil and water application conditions.

In the near future, we will use the RZWQM to test soil water infiltration and redistribution in an apple orchard in which surface (border) and sprinkler irrigation were alternated. Also, we will test how the one-dimensional RZWQM performs under spatially variable soil texture and variable rates of surface (furrow) irrigation in two oat/grass fields. Although these studies focus mainly on water infiltration and redistribution, it may be of interest to evaluate how well the RZWQM can simulate other physical and chemical processes of importance in these experimental fields. For instance, proper simulation of solute transport can be of great importance given the highly permeable alluvium soils and shallow aquifer prevalent at the study site. Given that RZWQM is a one-dimensional model that simulates several physical and chemical processes in the vadose zone, we believe that if coupled with groundwater models it could help to enhance understanding of surface water–groundwater interactions.

Acknowledgements

The authors gratefully acknowledge the technical assistance of Val Archuleta, David Archuleta, David Salazar, Ciara Cusack and Yeliz Cevik. This material is based upon work supported by the New Mexico Agricultural Experiment Station and by the Cooperative State Research, Education and Extension Service, US Department of Agriculture under Agreement Nos 2005-34461-15661 and 2005–45049–03209.

References

Abdul-Jabbar, A.S., Sammis, T.W. and Lugg, D.G. (1982) Effect of moisture level on the root pattern of alfalfa. *Irrigation Science* 3, 197–207.

Abrahamson, D.A., Radcliffe, D.E., Steiner, J.L., Cabrera, M.L., Hanson, J.D., Rojas, K.W., Schomberg, H.H., Fisher, D.S., Schwartz, L. and Hoogenboom, G. (2005) Calibration of the root zone water quality model for simulating tile drainage and leached nitrate in the Georgia Piedmont. *Agronomy Journal* 97, 1584–1602.

Ahuja, L.R., Rojas, K.W., Hanson, J.D., Shaffer, M.J. and Ma, L. (2000) *Root Zone Water Quality Model: Modeling Management Effects on Water Quality and Crop Production.* Water Resources Publications, Highland Ranch, Colorado.

Allen, R.G., Pereira, L.S., Raes, D. and Smith, M. (1998) *Crop Evapotranspiration: Guidelines for Computing Crop Water Requirements*. Irrigation and Drainage Paper No. 56. UN Food and Agriculture Organization, Rome.

Alves, I. and Cameira, M.R. (2002) Evapotranspiration estimation performance of Root Zone Water Quality Model: evaluation and improvement. *Agriculture Water Management* 57, 61–73.

Azevedo, A.S., Singh, P., Kanwar, R.S. and Ahuja, L.R. (1997) Simulating nitrogen management effects of subsurface drainage water quality. *Agricultural Systems* 55, 481–501.

Azevedo, A.S., Kanwar, R.S. and Pereira, L.S. (2000) Atrazine transport in irrigated heavy- and coarse-textured soils, Part II: Simulations with the root zone water quality model. *Journal of Agricultural Engineering Research* 76, 341–354.

Ben-Asher, J. and Ayars, J.E. (1990) Deep seepage under nonuniform sprinkler irrigation. II: Field data. *Journal of Irrigation and Drainage Engineering* 116, 363–373.

Blake, G.R. and Hartge, K.H. (1986) Bulk density. In: Klute, A. (ed.) *Methods of Soil Analysis, Part I*. Monograph No. 9. ASA/SSSA (American Society of Agronomy/Soil Science Society of America), Madison, Wisconsin.

Cameira, M.R., Sousa, P.L., Farahani, H.J., Ahuja, L.R. and Pereira, L.S. (1998) Evaluation of the RZWQM for the simulation of water and nitrate movement in level-basin, fertigated maize. *Journal of Agricultural Engineering Research* 69, 331–341.

Cameira, M.R., Ahuja, L., Fernando, R.M. and Pereira, L.S. (2000) Evaluating field measured soil hydraulic properties in water transport simulations using the RZWQM. *Journal of Hydrology* 236, 78–90.

Dalton, F.N. (1992) Development of Time Domain Reflectometry for measuring soil water content and bulk soil electrical conductivity. In: Topp, G.C., Reynolds, W.D. and Green, R.E. (eds) *Advances in Measurement of Soil Physical Properties: Bringing Theory Into Practice*. SSSA Special Publication 30, Soil Science Society of America, Madison, Wisconsin, pp. 143–167.

Daniel B. Stephen and Associates (2003) *Jemez y Sangre Regional Water Plan*. Daniel B. Stephen and Associates, Albuquerque, New Mexico.

De Vries, J.J. and Simmers, I. (2002) Groundwater recharge: an overview of processes and challenges. *Hydrogeology Journal* 10, 5–17.

Dudley, L.M., Hanks, R.J., MacAdam, J.W., Mace, R.W. and Low, A.P. (1994) *Use of Saline Waste Water from Electrical Power Plants for Irrigation: 1993 Report. Part 1: Soil, Water and Crop Yields*. Utah Agricultural Experiment Station Research Report 150, Utah State University, Logan, Utah.

Ellerbroek, D.A., Durnford, D.S. and Loftis, J.C. (1998) Modeling pesticide transport in an irrigated field with variable water application and hydraulic conductivity. *Journal of Environmental Quality* 27, 495–504.

Farahani. H.J. (1994) Evaporation and transpiration processes: an update. In: *Root Zone Water Quality Model. Version 2.2*. Technical Report No. 2. Great Plains Systems Research, US Department of Agriculture Agricultural Research Service, Fort Collins, Colorado.

Farahani, H.J., Buchleuiter, G.W., Ahuja, L.R., Peterson, G.A. and Sherrod, L.A. (1999) Seasonal evaluation of the root zone water quality model in Colorado. *Agronomy Journal* 91, 212–219.

Fernald, A.G. and Guldan, S.J. (2006) Surface water–groundwater interactions between irrigation ditches, alluvial aquifers, and streams. *Reviews in Fisheries Science* 14, 79–89.

Fernald, A.G., Cevik, S.Y., Ochoa, C.G., Tidwell, V.C., King, J.P. and Guldan, S.J. (2010) River hydrograph retransmission functions of irrigated valley surface water–groundwater interactions. *Journal of Irrigation and Drainage Engineering* 136, 823–825.

Fox, G.A., Malone, R., Sabbagh, G.J. and Rojas, K. (2004) Interrelationship of macropores and subsurface drainage for conservative tracer and pesticide transport. *Journal of Environmental Quality* 33, 2281–2289.

Ghidey, F., Eugene, E. and Kitchen, N.R. (1999) Evaluation of the Root Zone Water Quality Model using field-measured data from the Missouri MSEA. *Agronomy Journal* 91, 183–192.

Green, W.H. and Ampt, G.A. (1911) Studies on soil physics, 1: the flow of air and water through soils. *Journal of Agricultural Science* 4, 1–24.

Hanson, J.D., Rojas, K.W. and Shaffer, M.J. (1999) Calibrating the Root Zone Water Quality Model. *Agronomy Journal* 91, 171–177.

Healy, R.W. and Cook, P.G. (2002) Using groundwater levels to estimate recharge. *Hydrogeology Journal* 10, 91–109.

Jaber, F.H., Shukla, S. and Srivastava, S. (2006) Recharge, upflux and water table response for shallow water table conditions in southwest Florida. *Hydrological Processes* 20, 1895–1907.

Karlen, D.L, Kumar, A., Kanwar, R.S., Cambardella, T.S. and Colvin, T.S. (1998) Tillage system effects on 15-year carbon-based and simulated N budgets in a tillage-drained Iowa field. *Soil and Tillage Research* 48, 155–165.

Kohl, R.A. and Kolar, J.J. (1976) Soil water uptake by alfalfa. *Agronomy Journal* 68, 536–538.

Kozak, J.A., Aiken, R.M., Flerchinger, G.N., Nielsen, D.C., Ma, L. and Ahuja, L. (2007) Comparison of modeling approaches to quantify residue architecture effects on soil temperature and water. *Soil and Tillage Research* 95, 84–96.

Kumar, A., Kanwar, R.S., Singh, P. and Ahuja, L.R. (1999) Evaluation of the root zone water quality model for predicting water and NO_3-N movement in an Iowa soil. *Soil and Tillage Research* 50, 223–236.

Lal, R. and Shukla, M.K. (2004) *Principles of Soil Physics*. M. Dekker, New York.

Ma, L., Malone, R.W., Heilman, P., Karlen, D.L., Kanwar, R.S., Cambardella, C.A., Saseendran, S.A. and Ahuja, L.R. (2007) RZWQM simulation of long-term crop production, water, and nitrogen balances in Northeast Iowa. *Geoderma* 140, 247–259.

Ma, Q.L., Wauchope, R.D., Hook, J.E., Johnson, A.W., Truman, C.C., Dowler C.C., Gascho, G.J., Davis, J.G., Summer, H.R. and Chandler, L.D. (1998) Influence of tractor wheel tracks and crust/seals on runoff: observations and simulations with the RZWQM. *Agricultural Systems* 57, 77–100.

Ma, Q.L., Hook, J.E. and Wauchope, R.D. (1999) Evapotranspiration predictions: a comparison among GLEAMS, Opus, and RZWQM in a humid and thermic climate. *Agricultural Systems* 59, 41–55.

Mallants, D., Tseng, PH., Vanclooster, M. and Feyen, J. (1998) Predicted drainage for a sandy loam soil: sensitivity to hydraulic property description. *Journal of Hydrology* 206, 136–148.

Malone, R.W., Logdson, S., Shipitalo, M.J., Weatherington-Rice, J., Ahuja, L. and Ma, L. (2003) Tillage effect on macroporosity and herbicide transport in percolate. *Geoderma* 116, 191–215.

Malone, R.W., Ahuja, L.R., Ma, L., Wauchope, R.D., Ma, Q. and Rojas, K.W. (2004) Application of the root zone water quality model (RZWQM) to pesticide fate and transport: an overview. *Pest Management Science* 60, 205–221.

Mankin, K.R. and Koelliker, J.K. (2000) A hydrologic balance approach to saline seep remediation design. *Applied Engineering in Agriculture* 16, 129–133.

Meek, B.D.D., Rolph, W.R., Rechel, D., Carter, E.R. and Carter, L.M. (1990) Infiltration rate as affected by an alfalfa and no-till cotton cropping system. *Soil Science Society of America Journal* 54, 505–508.

Ochoa, C.G., Fernald, A.G., Guldan, S.J. and Shukla, M.K. (2007) Deep percolation and its effects on shallow groundwater level rise following flood irrigation. *Transactions of the ASABE* 50, 73–81.

Ochoa, C.G., Fernald, A.G., Guldan, S.J. and Shukla, M.K. (2009) Water movement through a shallow vadose zone: a field irrigation experiment. *Vadose Zone Journal* 8, 414–425.

Oliver, Y.M. and Smettem, K.R.J. (2005) Predicting water balance in a sandy soil: model sensitivity to the variability of measured saturated and near saturated hydraulic properties. *Australian Journal of Soil Research* 43, 87–96.

Rasse, D.P., Smucker, A.J.M. and Santos, D. (2000) Alfalfa root and shoot mulching effects on soil hydraulic properties and aggregation. *Soil Science Society of America Journal* 64, 725–731.

Richards, L.A. (1931) Capillary conduction of liquids through porous mediums. *Physics* 1, 318–333.

Rimon, Y., Dahan, O., Nativ, R. and Geyer, S. (2007) Water percolation through the deep vadose zone and groundwater recharge: preliminary results based on a new vadose zone monitoring system. *Water Resources Research* 43, 1–12.

Sammis, T.W., Evans, D.D. and Warrick, A.W. (1982) Comparison of methods to estimate deep percolation rates. *Journal of the American Water Resources Association* 18, 465–470.

Scanlon, B.R., Dutton, A. and Sophocleous, M.A. (2003) *Groundwater Recharge in Texas*. The University of Texas at Austin, Bureau of Economic Geology, submitted to Texas Water Development Board.

Schmidt, K.D. and Sherman, I. (1987) Effect of irrigation on groundwater quality in California. *Journal of Irrigation and Drainage Engineering* 113, 16–29.

Shuttleworth, W.J. and Wallace, J.S. (1985) Evaporation from sparse crops – an energy combination theory. *Quarterly Journal of the Royal Meteorological Society* 111, 839–855.

Soil Survey Staff, USDA NRCS (2008) *Official Soil Series Descriptions*. Natural Resources Conservation Service, US Department of Agriculture. Available at http://ortho.ftw.nrcs.usda.gov/cgi-bin/osd/osdname.cgi. (accessed 16 March 2010).

Sophocleous, M.A. (2001) Interactions between groundwater and surface water: the state of the science. *Hydrogeology Journal* 10, 52–67.

Starks, P., Heathman, G.C., Ahuja, L.R. and Ma, L. (2003) Use of limited soil property data and modeling to estimate root zone soil water content. *Journal of Hydrology* 272, 131–147.

Teixeira, W.G., Schrot, G.J., Marques, D. and Huwe. B. (2003) Sampling and TDR probe insertion in the determination of the volumetric soil water content. *Revista Brasileira de Ciencia do Solo* 27, 575–582.

Topp, G.C. (1993) Soil water content. In: Carter, M.R. (ed.) *Soil Sampling and Methods of Analysis.* Lewis Publishers, Ottawa, Canada.

Topp, G.C., Davis, J.L. and Annan, P. (1980) Electromagnetic determination of soil water content: measurements in coaxial transmission lines. *Water Resources Research* 16, 574–582.

USDA ARS (US Department of Agriculture Agricultural Research Service) (2011) List of Published Papers, Technical Reports, and Dissertations Where RZWQM/RZWQM2 Was Used as of Feb 2011. Available at: http://www.ars.usda.gov/SP2UserFiles/Place/54021500/RZWQM/rzwqmpub.pdf (accessed 22 February 2011).

Walker, S.E., Mitchell, J.K., Hirschi, M.C. and Johnsen, K.E. (2000) Sensitivity analysis of the root zone water quality model. *American Society of Agricultural Engineers* 43, 841–846.

Willis, T.M. and Black, A.S. (1996) Irrigation increases groundwater recharge in the Macquarie Valley. *Australian Journal of Soil Research* 34, 837–847.

Willis, T.M., Black, A.S. and Meyer, W.S. (1997) Estimates of deep percolation beneath cotton in the Macquarie Valley. *Irrigation Science* 17, 141–150.

WRCC (Western Regional Climate Center) (2006) Alcalde, New Mexico (290245): Period of Record Monthly Climate Summary, Period of Record 04/01/1953-12/31/2005. Available at: http://www.wrcc.dri.edu/cgi-bin/cliMAIN.pl?nmalca (accessed 16 March 2010).

11 CRITERIA-3D: A Mechanistic Model for Surface and Subsurface Hydrology for Small Catchments

Marco Bittelli,* Alberto Pistocchi, Fausto Tomei, Pier Paolo Roggero, Roberto Orsini, Marco Toderi, Gabriele Antolini and Markus Flury

Introduction

Hydrological models have become important tools for research as well as decision support. The demand for modelling and forecasting of the availability of water resources is driving an increasingly systematic research, in particular because of a widespread access to significant computer power and experimental data. According to Brutsaert (2005), hydrological modelling can be classified into physical, conceptual or systems modelling, and in each of these categories, different modelling approaches can be used depending on the spatial scale of the modelled area, the purpose of the modelling and parameter availability. Some authors favour the use of physically based models, which are commonly based on conservation equations from fluid mechanics, while others argue that the number of model parameters needed and the heterogeneity of hydraulic properties in watersheds makes the application of physically based models impractical.

One of the advantages of implementing physically based models is that physical constraints can be used to reduce the range of model parameters (Loague and Van der Kwaak, 2004). Beven (1993) pointed out that, in the case of distributed models, the same output can be obtained by various combinations of state parameters (such as soil moisture), often making the model ill-defined. The possibility of reducing the range of model parameters and using conservation equations allows for more precise hydrological modelling. Indeed, physically based models provide a consistent way of estimating soil moisture distribution, runoff generation patterns and streamflow (Van der Kwaak, 1999; Van der Kwaak and Loague, 2001). In particular, when experimental data are limited, physically based models allow the exploration of the behaviour of ungauged catchments and the simulation of variables such as surface water depth and soil moisture, for which measurements are rarely available. Furthermore, physically based models allow evaluation of the relative importance of different streamflow-generating mechanisms (Van der Kwaak, 1999).

Physically based models are based on numerical solution of the partial differential equations governing water flow. Although in recent years numerical schemes have become more and more efficient and computing power has increased, physically based models are computationally expensive and their application to large-scale catchments is often not possible on personal computers. Moreover, as pointed out by Vogel and Ippisch (2008), an increase of the modelling scale is typically accompanied by an increase of the spatial

* Corresponding author: marco.bittelli@unibo.it

discretization scale for the numerical solution of the Richards equation. Due to the underlying assumption of local equilibrium between water content and water potential, there is an upper limit of spatial discretization above which the numerical solution becomes biased, therefore requiring a refined computational grid (Vogel and Ippisch, 2008). A refined computational grid applied over a large spatial domain will result in large computational matrices, requiring long computation times and difficulties in the parameterization of hydraulic properties.

Physically based models are therefore most useful for application in small-scale catchments, and can serve as benchmark models for simpler conceptual models that are applied over larger scales. Indeed, it is important to provide methods to calibrate and test conceptual models by comparing their outputs with the results obtained with a physically based model. Overall, physically based models provide useful tools to simulate hydrological systems, to be used when a realistic and consistent behaviour of the system needs to be represented. It is expected that the application of physically based models at increasingly larger scales will increase in coming years as a result of both increase in computing power and the acquisition of distributed data.

Here, we describe an application of a physically based, three-dimensional catchment-scale model. The model is primarily intended to simulate complex hydrological patterns in contexts where simple one-dimensional or two-dimensional schemes are not suitable, and where phenomena related to surface runoff coupled with subsurface flow are of interest. This is especially frequent in agriculture-dominated landscapes. We used a fully integrated, three-dimensional hydrological model that includes both saturated and unsaturated subsurface flow, as well as overland flow. We applied this model to experimental data from a laboratory core infiltration, a field-scale flow and a catchment-scale experiment.

Materials and Methods

Model description

We used the model CRITERIA-3D (Bittelli *et al.*, 2010), which is based on the soil water balance model CRITERIA (Marletto and Zinoni, 1998; Marletto *et al.*, 2005). A flow chart describing the main model components is shown in Fig. 11.1. The model can simulate saturated water flow, unsaturated water flow, surface runoff, soil evaporation, snow accumulation and snow melt, plant water uptake, plant transpiration and topography-dependent solar radiation. It requires spatial information provided by a digital elevation model (DEM), a soil map with parameters for hydraulic properties and a land-use map with Manning's parameters and crop properties. The required weather input data are hourly data for precipitation, temperature, relative humidity, wind velocity and solar radiation, as well as crop parameters. The model solves a general conservation equation:

$$\frac{\partial(W\theta)}{\partial t} = -div(u) + q \tag{11.1}$$

where u is the water flux, W is the total volume, θ is the volumetric water content, t is time and q is the water source or sink. Water flux within the soil is described by the Richards equation:

$$W\frac{d\theta}{dH}\frac{\partial H}{\partial t} = div[K(h)\cdot grad(H)] + q \tag{11.2}$$

where $K(h)$ is the hydraulic conductivity and H is the total hydraulic head. Surface water flow is described by:

$$\frac{\partial(W\theta)}{\partial t} = S\frac{\partial h_s}{\partial t} = S\frac{\partial H}{\partial t} \tag{11.3}$$

where h_s is the surface hydraulic depth and S is the planimetric surface area. By using the Manning equation, and through substitution into Eqn 11.1, the surface flow is described by the parabolic approximation of the Saint-Venant equation:

$$S\frac{\partial H}{\partial t} = div\left[K^*(H,h_s)\cdot grad(H)\right] + q \tag{11.4}$$

where K^* is a conveyance function depending on H and h_s. The Richards equation and the Saint-Venant equation can be written in the unified form:

$$C\frac{\partial H}{\partial t} = div\left[K'\cdot grad(H)\right] + q \tag{11.5}$$

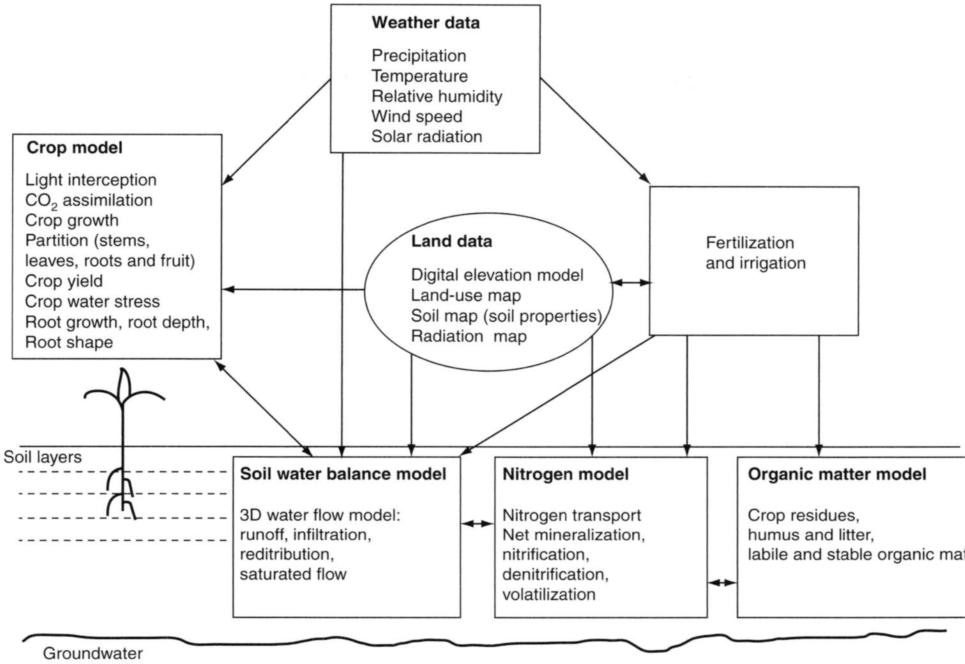

Fig. 11.1. Flow chart with schematic of the main components of the CRITERIA model. Details of the three-dimensional numerical schemes for water flow are described in Bittelli *et al.* (2010).

where *C* is either the surface or a subsurface volume and *K*" is either a conveyance function or the hydraulic conductivity, depending on whether the flow is occurring on the surface or within the soil matrix. A detailed description of the theory used for coupling the surface and the subsurface components is presented in Bittelli *et al.* (2010).

The structure of the model allows coupling with geographic information systems (GIS) and raster data sets, and a Windows Interface for model pre- and post-processing is available. The spatially resolved results (soil moisture, hydraulic heads, flow directions, evapotranspiration, surface runoff) are produced in Windows Access database format. The three-dimensional hydrological component of the model is available as a dynamic link library (DLL), allowing embedding in other computational packages such as comprehensive ecological or agricultural models. A flow chart of the model execution is shown in Bittelli *et al.* (2010).

Model application

The model was originally presented by Bittelli *et al.* (2010) and tested using distributed and integrated response data from a small catchment (4 ha) in Troy, Idaho. Here, we extend the model application to three types of experiments:

1. a laboratory-scale infiltration and redistribution experiment, to test infiltration and redistribution;
2. a field-scale experiment, to test field-scale application of the Richards equation, including evaporation and plant water uptake;
3. a catchment-scale discharge experiment, to test saturated, unsaturated and surface flow components.

The first two experiments were designed to test the accuracy of the Richards equation in one dimension. The third experiment was selected to analyse the model's performance over a larger catchment with different cultivated crops, and to compare it with the Troy watershed of Bittelli *et al.* (2010).

Laboratory-scale infiltration and redistribution experiment

Experiment description

We used experimental data from a one-dimensional infiltration–redistribution experiment done by Ferguson (1959). Ferguson used a Salkum silty clay loam (fine, mixed, mesic ultic Haploxeralf). The soil had a clay content of 30% by weight, but showed no observable shrinking and swelling. Soil material was dried and passed through a 0.5 mm mesh size sieve. The soil was then packed into a rectangular tray (39 cm long, 2.5 cm high, 16 cm wide) (Fig. 11.2). The tray was made of acrylic plastic, except for the top, which was open, and one end of 2.5 cm × 16 cm cross-section which was made of a fine screen covered on the outside with several layers of gauze. During packing, 7.6 cm high side boards were mounted on to the tray and soil was sprinkled through a 1 mm sieve into the tray. The tray was then dropped several times from a small height to achieve a reproducible bulk density. Water was applied to the tray with a tension infiltrometer which was connected to the end of the soil tray. During the infiltration experiments, water was applied to the soil under a constant tension of −0.315 m H_2O.

Water content in the soil profile during infiltration was measured by gamma-ray attenuation. The gamma rays were produced with a 20 millicurie Cs^{137} source and quantified with a NaI crystal detector as described in Ferguson (1959). The position of the gamma-ray beam was fixed, but the soil column was mounted on a movable wagon guided by a track. The wagon could be moved perpendicularly to the gamma-ray beam, thus allowing the measurement of water contents at different positions in the soil column (Ferguson, 1959). We complemented Ferguson's measurement by determining the soil water retention curve of the same soil, which was available from a soil archive at Washington State University, Pullman, Washington. We measured the water retention curve with a pressure plate apparatus at water potential values > −0.3 MPa, and with a vapour pressure method (WP4, Decagon Inc., Pullman, Washington) at water potentials between −0.3 and −1.5 MPa. Ten data points were collected between 0 and −0.3 MPa, and 6 data points were collected between −0.3 and −1.5 MPa. We obtained the parameters for the

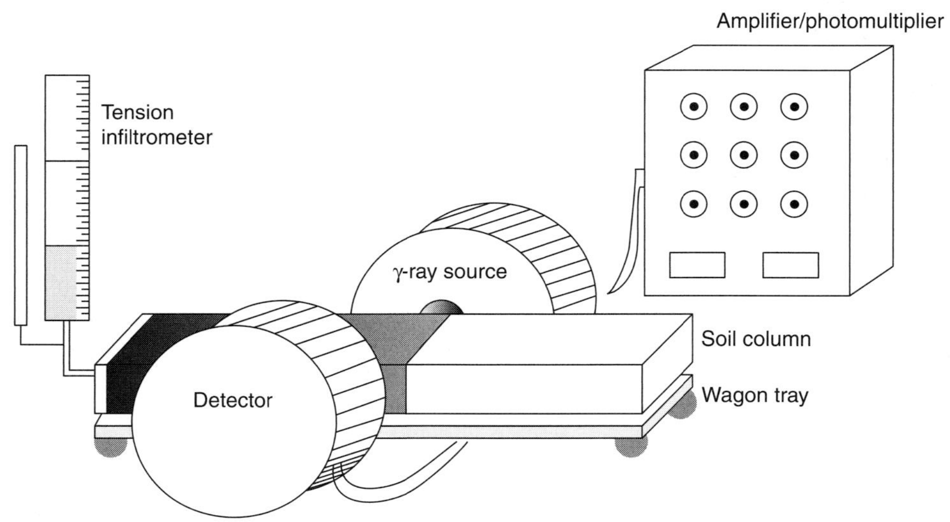

Fig. 11.2. Schematic of experimental set-up for measurement of soil water content during infiltration into a horizontal soil slab (after Ferguson, 1959).

soil moisture characteristic by fitting the van Genuchten (1980) equation to the measured data by a non-linear least-squares procedure implemented by the authors (Marquardt, 1963). Unsaturated conductivity was then calculated by using the Mualem equation (Mualem, 1976). We further measured the saturated conductivity using the constant head method (Klute and Dirksen, 1986).

Numerical simulation

For the numerical simulations, we discretized the horizontal column into 30 elements of Δx = 1 cm length. The total time of the numerical experiment was 18.5 h, equal to the time of the infiltration–redistribution laboratory experiment, with a time step of 1 h. The initial and boundary conditions were $h(x,t = 0) = h_e$ and $\underline{h}(x = 0,t) = h_e$, and $h(x,t) = h_L$ for $x = 0.39$ m. The water potential (h) at the left boundary, $x = 0$, was set to a constant value (h_e) corresponding to the tension applied by the tension infiltrometer. The right boundary, $x = 39$ cm, was set at a fixed potential h_L, corresponding to the measured initial soil water potential ($h_L = -3$ MPa).

Field-scale soil water content experiment

Experimental description

A field monitoring experiment was conducted at the San Pietro Capofiume experimental station (44° 39" N, 11° 37" E) (Hydro-Meteo-Climate Service of Emilia-Romagna Environmental Protection Agency, Bologna, Italy). Weather data were collected with a meteorological station situated 40 m from the soil monitoring station. A soil pit was excavated and soil samples were collected at seven horizons, and standard physical and chemical analyses were performed. Time domain reflectometry (TDR) water content sensors were installed at depths 10, 25, 45, 70, 100, 135 and 180 cm. Table 11.1 shows the soil properties at the site. TDR measurements were made at 1 h intervals with 30 cm three-rod probes. The dielectric permittivity was converted to water content using the Topp *et al.* (1980) equation. The site was also equipped with piezometers to measure groundwater levels. The experimental area was covered by natural perennial grasses. Details of the experimental set-up and the soil properties of the different soil horizons are provided in Tomei *et al.* (2007).

Numerical simulation

For the numerical simulations, we divided the soil profile into 20 layers of 0.10 m. Input variables included weather data, soil data, groundwater level data and crop data. We used published data for the grass crop coefficients and root depths, which are necessary for the evapotranspiration and the plant water uptake modules (Driessen and Konijn, 1992). Table 11.2 shows the crop coefficients for natural perennial grasses at the site. Details of the parameters presented in Table 11.2 can be found in Dottori *et al.* (2007). The maximum root depth was set at −0.65 m, while the beginning of the root system was set at 0.02 m below the surface. The shape of the root system was assumed to be cardioid with a coefficient of deformation equal to 1.9 (Driessen and Konijn, 1992). The initial conditions were set according to the experimental measurements

Table 11.1. Soil properties at the San Pietro Capofiume experimental station, Italy.

Depth (cm)	Sand (%)	Silt (%)	Clay (%)	Bulk density (g cm^{-3})	K_s (cm day^{-1})[a]	θ_s (m^3 m^{-3})[a]
0–15	62	24	14	1.45	60	0.429
15–35	61	23	16	1.55	44	0.4
35–50	67	18	15	1.55	44	0.42
50–80	74	18	8	1.5	44	0.43
80–120	7	56	37	1.2	4	0.43
120–165	9	73	18	1.4	6.5	0.46
165–200	62	27	11	1.4	44	0.36

[a]K_s, saturated hydraulic conductivity; θ_s, saturated volumetric water content.

performed at the site with the TDR sensors. In particular, the initial soil water content and the corresponding water potential (derived from knowledge of the hydraulic properties) were set equal to those measured by the TDR sensors. Atmospheric boundary conditions were set as a prescribed flux depending upon weather conditions at the upper boundary. The lower boundary condition was a fixed potential determined by the groundwater table. The depth of the water table was measured with the piezometers. Two different water table levels were set during the simulation, for two periods, from the beginning of the simulation until the day of the year (DOY) 132, and from DOY 132 to 184; these corresponded to experimental measurements. The simulations were carried out for the time period from July 2004 to February 2005.

Table 11.2. Crop parameters for the natural perennial grasses at the San Pietro Capofiume site in Italy.

Parameter	Value
Plant cycle (days)	365
Min. temperature for growth	2
Degree days for LAI[a] growth (°C)	1400
Degree days for LAI decrease (°C)	200
LAI_min (m^2 m^{-2})	1
LAI_max (m^2 m^{-2})	4
Max. crop coefficient (–)	1
Initial root depth (m)	0.05
Max. root depth (m)	1.2
Root shape (–)	1[b]
Root_deformation parameter (%)	1.7
Root_growth cycle function (–)	1[c]
Max. value for water stress (–)	0.9
Min. value for water stress (–)	0.7
Day degrees for max. water stress (°C)	800
Transpiration compensation factor (–)	0.2

[a]LAI, leaf area index.
[b]Cylindrical.
[c]Logistic.

Catchment-scale experiment

Experimental description

The experiment was performed in the Spescia catchment (Fig. 11.3), located in the Marche region of Central Italy (43° 33" N; 13° 04" E). The dominant crops were durum wheat (*Triticum durum*) and sunflower (*Helianthus annuus*) in rotation, which resulted in bare soil during the autumn. Wheat was sown in November and harvested in July. The wheat harvest was immediately followed by ploughing and the soil was left bare until sunflower seeding in April. Sunflower was harvested in August–September. The average cumulative annual precipitation at the site is 1000 mm. The catchment has a total surface area of 80.8 ha, with total arable land of 70.3 ha and an average slope of 7% (Roggero and Toderi, 2002). An ISCO Area Velocity Flow Meter (Teledyne Isco, Inc., Lincoln, Nebraska) and an automatic sampling unit coupled with a data-logger were used to collect continuous streamflow data; the position of the flow meter is shown in Fig. 11.3. Streamflow was monitored from

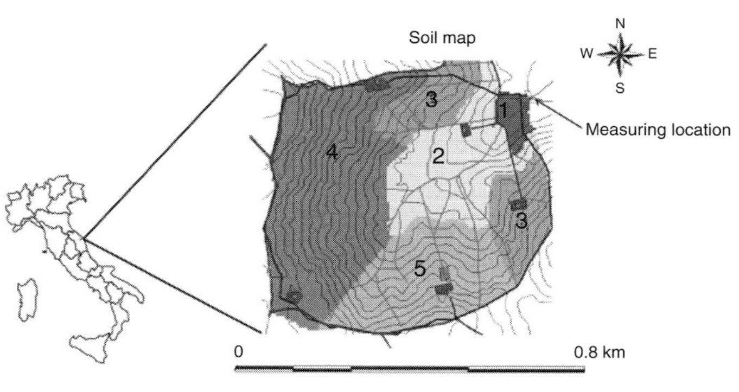

Fig. 11.3. Map of the Spescia catchment, Serra de' Conti, Central Italy, showing topography (5 m equidistance contours), hydrography and soil mapping units (numbers on the map) (from Corti *et al.*, 2006).

November 1998 to December 1999. An automatic weather station (Campbell Scientific Ltd., Logan, Utah) was installed near the basin to record hourly temperature, humidity and rainfall, while wind and radiation data were interpolated based on data collected by adjacent weather stations. The spatial distribution of crops was used to compute spatially averaged parameters for the values of the transpiration and rooting depth.

Streamflow in this catchment is ephemeral. We selected three time periods for our model testing: (i) a typical autumn rainfall sequence, (ii) a typical summer rainfall sequence, and (iii) a short, high-intensity rainfall in summer. As only information on basic soil properties (texture) was available at the site, soil hydraulic properties were obtained by using pedo-transfer functions from the software Rosetta (Schaap *et al.*, 2001), and by using the van Genuchten and Mualem model (Mualem, 1976; van Genuchten, 1980). Manning's surface roughness and surface pond depth were assumed to be homogeneous parameters over the catchment. A calibrated value of 0.03 m was used for the ponding depth and a value of 0.4 for the Manning roughness coefficient.

Numerical simulation

The simulation was performed from 1 March 1998 to December 1999. The initial conditions were set as uniform soil water content (at 0.8 degrees of saturation) on 1 March 1998. The spring and summer 1998 were used for model spin-up, which is the process of adjusting a model to its initial conditions. The comparison of the experimental data with model simulations started in November 1998. Owing to lack of experimental information, we assumed that the catchment had a uniform soil profile of uniform depth. As detailed by Bittelli *et al.* (2010), when simulating a catchment, we assumed that water always exits the catchment at an outlet. In this situation, a constant head at the outlet sufficiently low to ensure drainage from the catchment was used. The simulations were done using a structured grid, with a horizontal spatial resolution of 10 m, and a vertical resolution of 5 cm (close to the surface) and 20 cm (in the deeper layers). The model simulations were compared with the experimental data using the Nash–Sutcliffe (*NS*) coefficient (Nash and Sutcliffe, 1970):

$$NS = 1 - \frac{\sum_{t}^{T}\left(q_o^t - q_m^t\right)^2}{\sum_{t}^{T}\left(q_o^t - q_{ave}^t\right)^2} \tag{11.6}$$

where q_o is observed streamflow, q_m is modelled streamflow, q_{ave} is the average observed streamflow, the superscript t is time and T is the time period of the simulation. When $NS = 1$ there is a perfect match between modelled and observed data, while when $NS = 0$ the model predictions are as accurate as the mean of the observed data, whereas for $NS < 0$ the observed mean is a better predictor than the model.

Results and Discussion

Laboratory-scale infiltration and redistribution

To test the effectiveness of the algorithms in simulating the infiltration and redistribution processes, we compared the results of the model simulation with the data obtained from the infiltration–redistribution experiment. Figure 11.4 depicts the experimental and simulated values of column soil water content, i.e. water content plotted as a function of distance from the surface of infiltration for different times after the beginning of the experiment. The different profiles show the evolution of water content at a given position in the column. The simulation provided good results for the quantification of the infiltration and redistribution rate. As expected, the soil water content increased rapidly as water moved into the column, and then continued to increase at a decreasing rate. The dynamics of the water content variation with time at a distance of $x = 3.5$ cm shows that the water content increased at a high rate over a short time. Specifically, the water content increased from 0.3 m^3 m^{-3} to 0.50 m^3 m^{-3} in a time interval of 1.66 min, and the water content increased progressively at further distance from the infiltration surface as time progressed, as shown in Fig. 11.4. The rapid advance of the wetting

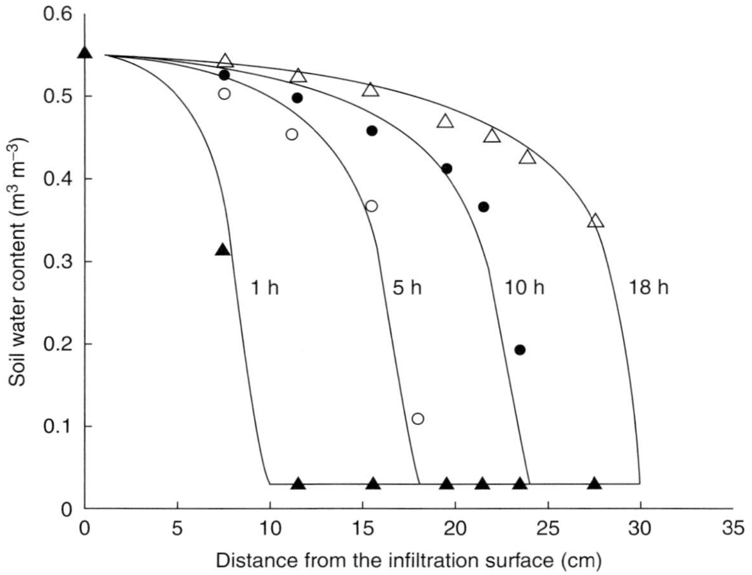

Fig. 11.4. Simulated wetting fronts (from the CRITERIA-3D model) compared with laboratory-scale experimental data (from Ferguson, 1959) at four different times. Lines are simulated data while the symbols are experimental data.

front is the result of the high water potential gradient between the dry soil and the wetting zone. A correct simulation of this process is important because infiltration into dry soil is usually a difficult process to simulate. The simulated values described the experimental data well, with low mass balance errors in the order of 10^{-6} kg m^{-3}.

Field-scale soil water content experiment

Figure 11.5 shows the experimental and simulated soil water contents in the field, as well as precipitation and average air temperature. The experimental soil water content data were well described by the model. The profile at 0.1 m depth displayed more pronounced wetting and drying cycles, due to infiltration and redistribution of water from rainfall and to evaporation and transpiration processes. With increasing soil depth, wetting and drying cycles became less pronounced. The maximum relative error (calculated by computing the difference between the measured soil water content and the simulated water content divided by the measured content and multiplying by 100) was 14.6% for the 0.1 m, 14.6% for the 0.25 m, 10.5% for the 0.7 m, 9.1% for the 1 m, 2.5% for the 1.35 m and 1.8% for the 1.8 m depths. The Nash–Sutcliffe (*NS*) parameter was 0.45 for the first two profiles, 0.51 for the 0.7 m, 0.52 for the 1 m, 0.62 for the 1.35 m and 0.66 for the 1.8 m depths. The higher number of relevant processes (such as infiltration, runoff and soil evaporation) involved close to the surface, made the simulation slightly less precise in the surface layers (between 0 and 0.7 m) than in the deeper layers (between 0.7 and 1.8 m), with relative errors decreasing with increasing soil depths. However, overall, the relative errors were small, so that a good quantification of the magnitude and dynamics of the soil water content in the different soil layers was provided.

Catchment-scale experiment

Integrated response

The streamflow simulated by the model was generally accurate in terms of response time

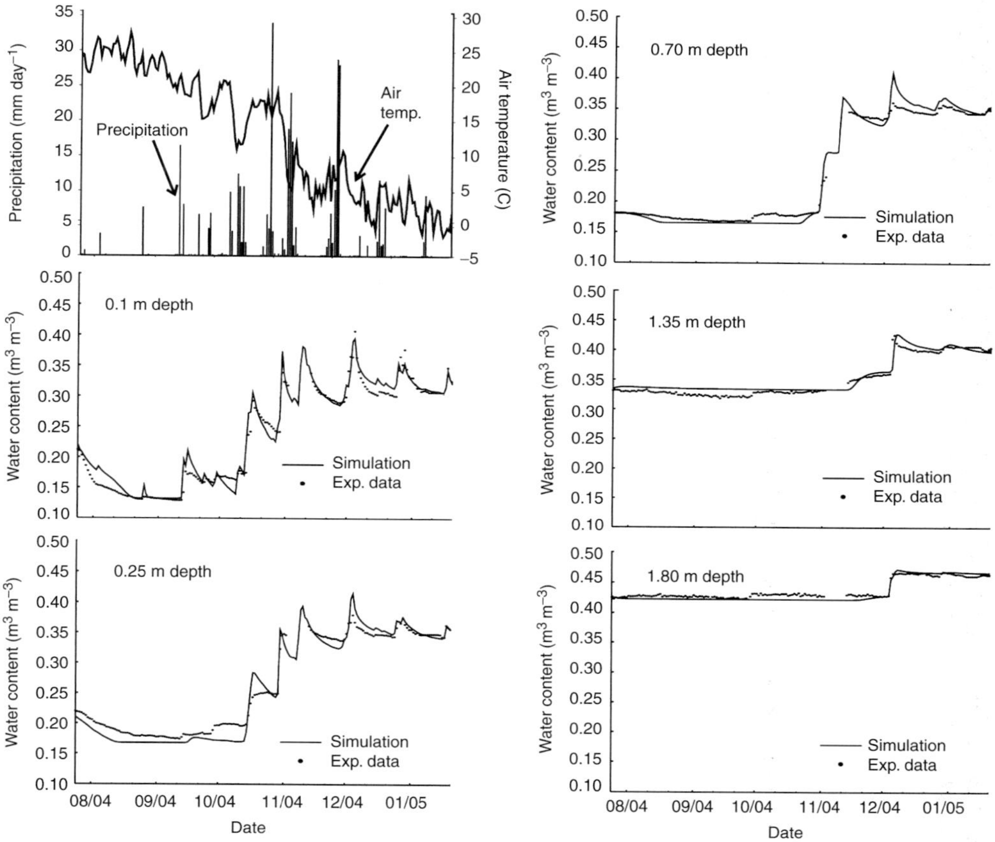

Fig. 11.5. Experimental (Exp.) and CRITERIA-3D-simulated soil water contents at different depths at the San Pietro Capofiume experimental station, Italy. Precipitation and average air temperature are represented in the top-left graph. Dots are field-scale experimental data and lines are simulated soil water content data.

and runoff volumes (Fig. 11.6), with an *NS* parameter of 0.55, showing the ability of the model to describe the hydrological processes in the catchment. An interesting behaviour was found for the event of 16 June 1999; in this case, using hourly rainfall for the simulation did not reproduce the intense runoff peak found on that date. The results obtained using an hourly simulation are displayed in the graph, and show that the simulations underestimated the peak discharge. A better result could be obtained by having a more refined time resolution for the same event, i.e. by simulating the runoff and the peak rainfall with a time step smaller than one hour. This suggestion could not be independently validated, however, as the records of precipitation had hourly steps, while for intense rainfall events – over a few minutes, for example – rainfall intensity may be as high as ten times the corresponding hourly rainfall intensity. Several studies have reported that storm velocity and rainfall intensity had significant effects on the streamflow hydrograph's shape and timing (Hewlett *et al.*, 1977; Foroud *et al.*, 1984; Singh, 1997).

For the purpose of model testing, the same (hourly) rainfall of 33 mm was assumed to occur over 6 minutes, and the model time step was reduced to one minute. This produced a substantial improvement in the simulation of observed runoff (Fig. 11.6, centre bottom graph); indeed, for the event of 16 June 1999,

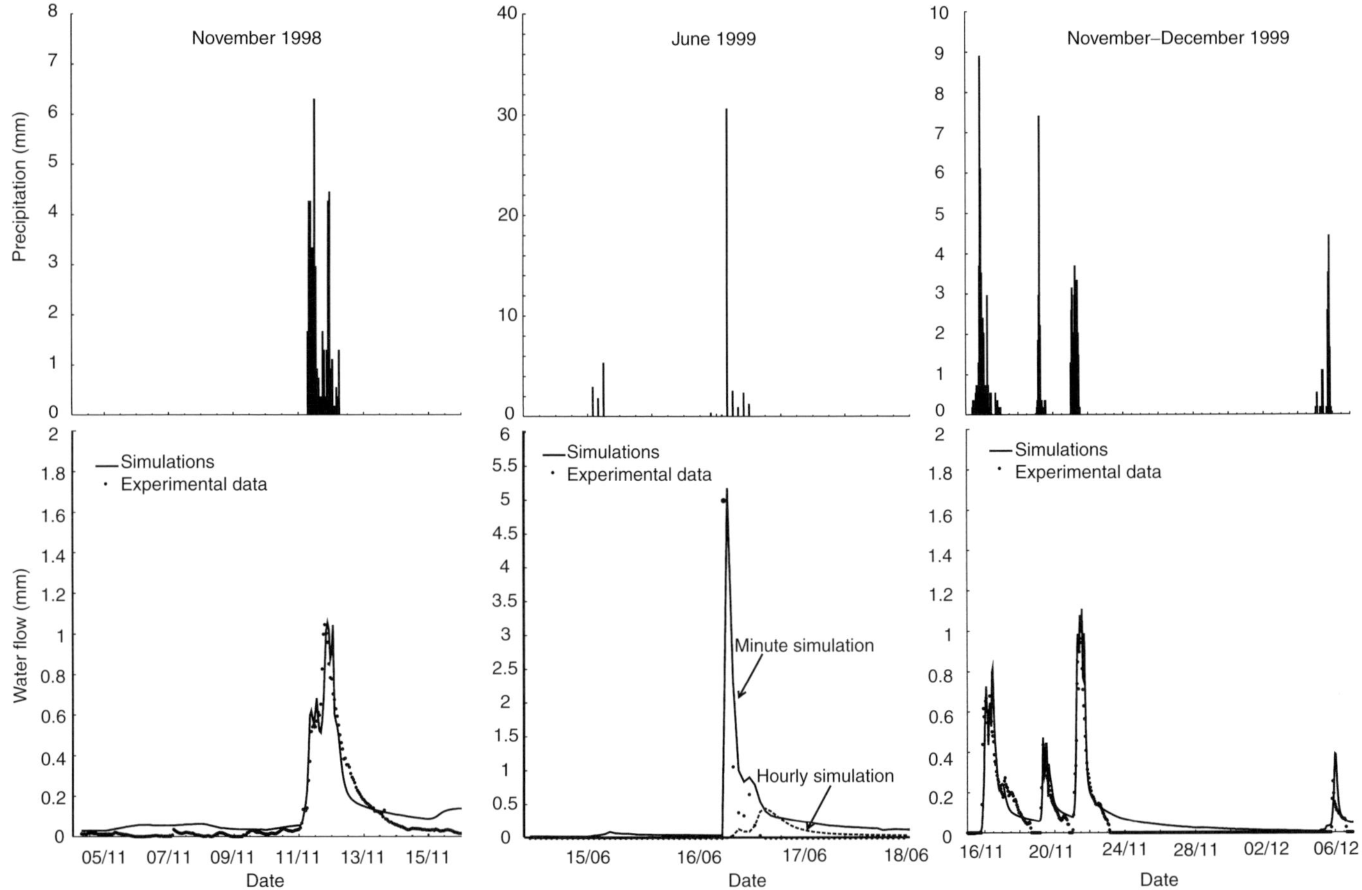

Fig. 11.6. Predicted and observed hourly discharge (water flow) at the Spescia catchment, Serra de' Conti, Central Italy, for three different time periods. The bottom centre graph shows the results of simulations using both an hourly time step (as in the other two graphs) and a minute time step.

using hourly precipitation data, the Nash–Sutcliffe parameter was poor (*NS* = −1.28), but its value was considerably improved (*NS* = 0.28) by using minute-based precipitation data and adjustment of the simulation time step as described above. Note how the minute-based simulation could describe the peak discharge of 5 mm water flux, while the hourly-based simulation did not. These results suggested that such an intense rainfall event can be better simulated by improving rainfall data resolution rather than by tuning model parameters, as can be the case for other hydrological model applications (Koren *et al.*, 1999). For the other two dates on which simulations were run, the simulated streamflow was compared with observed discharge and the Nash–Sutcliffe parameter computed as a statistical index of model performance; for November 1998 *NS* was 0.55, and for November–December 1999 *NS* was 0.35, indicating a good agreement between measured and simulated data.

Distributed response

Plate 16 shows a three-dimensional representation of soil water content. We selected 1, 16 and 30 November 1999 to show representative changes in soil water contents caused by rainfall in November 1999 (Fig. 11.6). The soil depth of 0 m represented surface water, which was divided into moving (runoff) and stationary (ponded) water, as described in detail by Bittelli *et al.* (2010). The soil moisture map on the left-hand side of Plate 16 shows that, on 1 November, there was no ponded or runoff water on the soil and that the soil profile was at different degrees of saturation at different depths. On 16 November, after the first rainfall events, runoff was generated and the map showed water on the soil surface following the patterns determined by the topographic features of the catchment. At 0.1 m depth the soil was close to saturation, because of the infiltration and redistribution, while at the 0.4 m depth the soil had a lower water content. Note the areas in yellow–green corresponding to different soil types, as shown in Fig. 11.3. On 30 November, the surface layer had no runoff water, and only a dark area on the lower left-hand side indicated a ponded area. At deeper depths, the soil was close to saturation in the middle of the catchment, but was at a lower degree of saturation in the side lobes, because of the different soil properties in different regions of the catchment (see Fig. 11.3) and because of the topography of the catchment, which promoted soil water redistribution. As no experimental data on distributed soil water content were available at the site, we could not test the model output in terms of water contents; however, the spatial patterns seem realistic and corresponded to visual observations of the catchment at different periods of the year.

Runoff-generating mechanisms

The development of streamflow is commonly described by two mechanism of runoff: infiltration excess and saturation excess. The infiltration excess mechanism, originally proposed by Horton, and called Hortonian, usually occurs in arid and semi-arid regions owing to high rainfall intensities on soils with low infiltration rates. The saturation excess mechanism, also called Dunnian, constitutes the main mechanism of runoff generation in humid regions which display a high groundwater table and/or fluctuating perched water tables. Runoff is generated when the soil water content is near saturation, due to excess of soil water in the soil surface layer.

A physically based simulation where fluxes are computed based on water potential gradients in each of the three-dimensional vectors allows for an interesting analysis of these mechanisms. We computed the ratio $Q = q^e/q^b$, which is a normalized rate of water exchange between the surface and the subsurface, as originally proposed by Van der Kwaak and Loague (2001). The term q^e is the exchange rate and is negative (or equal to zero) for infiltration and positive for exfiltration, while q^b is the rainfall rate; therefore, the exchange rate is normalized over the rainfall rate. The ratio of q^e to the rainfall rate, q^b, is 1.0 when all the rainfall infiltrates. If $Q = q^e/q^b = 0$, there is no infiltration (all the rainwater becomes runoff); if $q^e/q^b < 0$, exfiltration dominates; if $q^e/q^b > 1$, there is infiltration; and if $0 < q^e/q^b < 1$, there is an intermediate situation where both infiltration and runoff occur at the same time, with runoff being dominant at values closer to zero.

Plate 17 shows the results of water exchange calculations for the two events of November and June 1999. These results show (i) that there was a wide variability of the value of *Q* both in space and between events, and (ii) that different streamflow-generating mechanisms did occur simultaneously across the catchment. For the June event, where soil was initially dry, the catchment showed an exchange rate between 0 and 0.7, which generated infiltration and runoff. In the flat part of the catchment, there was accumulation of water with subsequent dominating infiltration, as shown by the *Q* value of the exchange rate of >1. For the November event, the upstream part of the catchment generally showed an exchange rate between 0 and 0.5, indicating that runoff was dominant. The headwaters and flatter areas, such as the ones close to the catchment outlet, showed predominant infiltration, while the drainage channel showed predominant exfiltration. Therefore, in the channel part of the catchment, the dominating mechanism was saturation excess (Dunnian), which was also confirmed by the soil water content simulation, which showed that the soil was saturated in that part of the catchment. We were not able to define a dominant runoff-generating mechanism, as both Dunnian and Hortonian mechanisms occurred simultaneously depending on the topographical features of the catchment and on the season. The simultaneous occurrence of both Dunnian and Hortonian runoff response has also been reported previously by Van der Kwaak and Loague (2001).

Summary

CRITERIA-3D is a numerical, three-dimensional, physically based model that solves flow equations of surface and subsurface flow applicable for a first-order watershed. The model contains basic physical equations for surface, unsaturated subsurface and saturated subsurface flow. We applied the model to observations obtained from experiments conducted at three different scales: (i) soil water contents in a one-dimensional soil slab obtained from a laboratory-controlled infiltration experiment, (ii) soil water contents in a field soil obtained from a field experiment characterized by one-dimensional flow, and (iii) streamflow data for a hilly catchment of 80.8 ha. The results of the three experimental tests, from the laboratory to the field to the catchment scale, showed the ability of the model to correctly represent water flow and hydrological processes. The model proved capable of representing the water balance in all its components, and the comparison between the experimental and the simulated data was good. Analysis of runoff-generating mechanisms on the catchment scale showed that it was not possible to define a dominant runoff-generating mechanism, but rather that the two main mechanisms (infiltration and saturation excess) occurred simultaneously, depending on topography, soil properties and meteorological conditions.

References

Beven, K. (1993) Prophecy, reality and uncertainty in distributed hydrological modeling. *Advances in Water Resources* 16, 41–51.

Bittelli, M., Tomei, F., Pistocchi, A., Flury, M., Boll, J., Brooks, E.S. and Antolini, G. (2010) Development and testing of a physically based, three-dimensional model of surface and subsurface hydrology. *Advances in Water Resources* 33, 106–122.

Brutsaert, W. (2005) *Hydrology: An Introduction.* Cambridge University Press, Cambridge, UK.

Corti, G., Agnelli, A., Cuniglio, R., Cocco, S. and Orsini, R. (2006) Studio pedologico di dettaglio di due bacini della collina interna marchigiana. In: Esposito, S. and Epifani, C. (eds) *Climagri – Cambiamenti Climatici e Agricoltura. Risultati Conclusivi.* CRA-UCEA, Rome, pp. 129–141.

Dottori, F., Marletto, V., Van Soetendael, M., Tomei, F. and Antolini, G. (2007) *Criteria: Manuale Tecnico.* Hydro-Meteo-Climate Service of Emilia-Romagna Environmental Protection Agency, Bologna, Italy.

Driessen, P.M. and Konijn, N.T. (1992) *Land-use Systems Analysis.* Wageningen Agricultural University, Wageningen, The Netherlands.

Ferguson, A.H. (1959) Movement of soil water as inferred from moisture content measurements by gamma ray adsorption. PhD thesis, Washington State University, Pullman, Washington.

Foroud, N., Broughton, R.S. and Austin, G.L. (1984) The effects of a moving rainstorm on direct runoff. *Water Resources Research* 20, 87–91.

Hewlett, J.D, Fortson, J.C. and Cunningham, G.B. (1977) The effect of rainfall intensity on storm flow and volume. *Water Resources Research* 13, 259–266.

Klute, A. and Dirksen, C. (1986) Hydraulic conductivity and diffusivity: laboratory methods. In: Klute, A. (ed.) *Methods of Soil Analysis. Part 1. Physical and Mineralogical Methods.* American Society of Agronomy, Madison, Wisconsin, pp. 687–734.

Koren, V.I., Finnerty, B.D., Schaake, J.C., Smith, M.B., Seo, D.J. and Duan, Q.Y. (1999) Scale dependencies of hydrologic models to spatial variability of precipitation. *Journal of Hydrology* 217, 285–302.

Loague, K. and Van der Kwaak, J. (2004) Physics-based hydrologic response simulation: platinum bridge, 1958 Edsel, or useful tool. *Hydrological Processes* 18, 2949–2956.

Marletto, V. and Zinoni, F. (1998) The CRITERIA Project: integration of satellite, radar, and traditional agroclimatic data in a GIS-supported water balance modeling environment, In: Dalezios, N.R. (ed.) *Proceedings of the COST 77, 79, 711 International Symposium on Applied Agrometeorology and Agroclimatology*, 24–26 April 1996, University of Thessaly, Volos, Greece. *Special Publication Number 30*, Volos, Greece, pp.173–178.

Marletto, V., Zinoni, F., Criscuolo, L., Fontana, G., Marchesi, S. and Morgillo, A. (2005) Evaluation of down-scaled DEMETER multi-model ensemble seasonal hindcasts in a northern Italy location by means of a model of wheat growth and soil water balance. *Tellus* 57, 488–497.

Marquardt, D.W. (1963) An algorithm for least-squares estimation of non-linear parameters. *Journal of the Society of Industrial and Applied Mathematics* 11, 431–441.

Mualem, Y. (1976) A new model for predicting the hydraulic conductivity of unsaturated porous media. *Water Resources Research* 12, 513–522.

Nash, J.E. and Sutcliffe, J.V. (1970) River flow forecasting through conceptual models – Part I: A discussion of principles. *Journal of Hydrology* 10, 282–90.

Roggero, P.P. and Toderi, M. (2002) *Le Misure Agroambientali: Applicazioni nelle Marche e Analisi di un Caso di Studio Sull'inquinamento da Nitrati di Origine Agricola.* Quaderni 5B, ASSAM (Agenzia Servizi Settore Agroalimentare delle Marche), Ancona, Italy.

Schaap, M.G., Leij, F.J. and van Genuchten, M.T. (2001) Rosetta: a computer program for estimating soil hydraulic parameters with hierarchical pedotransfer functions. *Journal of Hydrology* 251, 163–176.

Singh, V.P. (1997) Effect of spatial and temporal variability in rainfall and watershed characteristics on stream flow hydrograph. *Hydrological Processes* 11, 1649–1669.

Tomei, F., Antolini, G., Bittelli, M., Marletto, V., Pasquali, A. and Van Soetendael, M. (2007) Validazione del modello di bilancio idrico CRITERIA. *Italian Journal of Agrometeorology* 1, 66–67.

Topp, G.C., Annan, J.L. and Davis, A.P. (1980) Electromagnetic determination of soil water content: measurements in coaxial transmission lines. *Water Resources Research* 16, 574–582.

Van der Kwaak, J. (1999) Numerical simulation of flow and chemical transport in integrated surface–subsurface hydrologic systems. PhD thesis, University of Waterloo, Waterloo, Ontario.

Van der Kwaak, J. and Loague, K. (2001) Hydrologic-response simulations for the R-5 catchment with comprehensive physics-based model. *Water Resources Research* 37, 999–1013.

van Genuchten, M.T. (1980) A closed-form equation for predicting the hydraulic conductivity of unsaturated soils. *Soil Science Society of America Journal* 44, 892–898.

Vogel, H.J. and Ippisch, O. (2008) Estimation of a critical spatial discretization limit for solving Richards' equation at large scales. *Vadose Zone Journal* 7, 112–114.

12 Effects of Artificial Drainage on Water Regime and Solute Transport at Different Spatial Scales

Bernd Lennartz,* Manon Janssen and Bärbel Tiemeyer

Introduction

Subsurface drainage is a common agricultural practice to improve aeration and trafficability of soils with shallow groundwater or seasonally perched water tables (Eggelsmann, 1981; Skaggs *et al.*, 1994). These measures significantly alter the soil moisture regime at plot and at the field scale as well as the discharge generation at the catchment scale (Skaggs *et al.*, 1994; Robinson and Rycroft, 1999; Blann *et al.*, 2009). The physical changes resulting from the installation of drainage pipes are basically twofold: (i) the level and the dynamics of the groundwater table are no longer controlled by the soil properties only, but also by the drain depth and the drain spacing, and (ii) the installation of the drainpipes by systematically excavating and refilling trenches generates discontinuities in the spatial distribution of the soil hydraulic properties across the artificially drained site (Kamra *et al.*, 1999). In tile-drained systems, the combination of the shortened residence time of water and substances in the soil – sometimes exacerbated by non-equilibrium flow and transport – and of the direct connection to surface water bodies frequently causes aggravation of the problem of nutrient and pesticide losses from agricultural soils. Artificial drainage is especially widespread in northern Europe, northern America, China and India. In northern Europe, it has reached proportions of 66% (the UK), 87% (the Netherlands) and 97% (Finland) of the arable land (Feick *et al.*, 2005; De la Cueva, 2006). This spatial extent underlines the necessity to thoroughly study water flow and solute transport processes at tile-drained field sites and catchments.

Contrary accounts are available in the literature on the impact of artificial drainage on catchment hydrology, especially on the amount and timing of flood peaks. In summary, some researchers claim that artificial drainage decreases surface runoff rates and flood peaks by reducing the soil moisture and thus enhancing the soil storage capacity (e.g. Burke, 1967; Baden and Eggelsmann, 1970; Iritz *et al.*, 1994; Turtola and Paajanen, 1995), while the others argue that drainage pipes will shorten the pathways to natural streams, especially if drainage channels are built, well maintained (e.g. Robinson, 1990) and act synchronously (Irwin and Whiteley, 1983). At the field scale, evidence for both arguments can be found (Robinson, 1990; Skaggs *et al.*, 1994; Holden *et al.*, 2004). The first argument tends to be true for areas with steep slopes, where surface runoff is reduced by drainage, or under wet conditions, i.e. especially for peat soils.

* Corresponding author: bernd.lennartz@uni-rostock.de

The second argument applies to lowland areas with insignificant differences in elevation. In these cases, channel maintenance by dredging and removal of riparian plants may further shorten the reaction time of the catchment (Robinson, 1990; Dunn and Mackay, 1996). However, the impact of artificial drainage strongly depends on the individual site with its often unique topographic, drainage system and soil characteristics (Dunn and Mackay, 1996; Holden *et al.*, 2004).

Artificial drainage may also increase and accelerate the loss of nutrients, especially that of nitrate and phosphorus, and pesticides from fields. Thus, artificial drainage contributes to the diffuse (or non-point source) pollution of surface water bodies, which is frequently still at an unsatisfactory high level (Kladivko *et al.*, 1999). There is a general agreement that tile drainage increases nitrate losses (Skaggs *et al.*, 1994) for several reasons: nitrate is rarely conveyed by overland flow, but by leaching, and tile drainage shortens the residence time of the leached water in the soil and often creates a direct connection to the surface water resources (Skaggs *et al.*, 1994; Turtola and Paajanen, 1995). Furthermore, the lowering of the groundwater level favours aerobic conditions in the soil and therefore enhances mineralization and nitrification of soil organic matter and fertilizers, while denitrification is inhibited (Gilliam *et al.*, 1999; Elmi *et al.*, 2000). Some factors reported to increase the nitrate losses via tile drainage are high precipitation and high discharge rates (Cambardella *et al.*, 1999; Jaynes *et al.*, 1999; Tomer *et al.*, 2003), narrow drainage spacing (Kladivko *et al.*, 1999, 2004), large drainage depth (Skaggs and Chescheir, 2003), sandy soils (Vinten *et al.*, 1994) and – in several cases – high fertilizer application levels (Jaynes *et al.*, 2001b; Kladivko *et al.*, 2004). Depending on the conditions before installation of the drainage system, tile drainage may reduce overland flow and erosion (Skaggs *et al.*, 1994; Turtola and Paajanen, 1995) and, thus, the export of sorbing substances such as phosphorus and pesticides (Spaling, 1994). However, considerable losses of these substances via tile drainage may occur if preferential flow occurs (Stamm *et al.*, 1998; Laubel *et al.*, 1999; Hodgkinson *et al.*, 2002). Therefore, flow processes and transport mechanisms need to be known to understand the environmental impact of tile drainage under specific conditions.

As tile drainage systems are installed to improve aeration conditions, they are frequently found at sites with loamy and clayey soils. These soils often show a distinct disposition for structure formation and aggregation. As a consequence, a bimodal pore system and preferential flow have frequently been observed at tile-drained fields under both natural and simulated rainfall conditions (Kladivko *et al.*, 1999; Kung *et al.*, 2000b; Köhne and Gerke, 2005; Klaus and Zehe, 2010). A distinction between different kinds of flow and transport patterns in top and lower soil horizons is difficult from tile drainage measurements alone, because the tile drains generate an integrative signal encompassing vertical as well as lateral heterogeneities (Richard and Steenhuis, 1988).

Difficulties in the experimental and theoretical quantification of solute transport at the field scale often originate from soil heterogeneity. Even small horizontal variations in the physical and chemical properties of soils may result in non-uniform flow fields with widely different solute-flux velocities (Kung, 1990; De Rooij and Stagnitti, 2002). One approach to tackling the problem of the spatial variability of solute transport is the concept of a stochastic stream tube model in which the soil is considered as a bundle of individual non-interacting soil columns (Toride and Leij, 1996). The solute transport characteristics of the individual columns, i.e. the local-scale soil properties, may be determined experimentally. The field-scale solute transport process is then represented by integrating the results across the local-scale measurements (Jensen and Refsgaard, 1991; Van Wesenbeeck and Kachanoski, 1991; Sassner *et al.*, 1994). To test the capabilities of assembled local-scale solute transport characteristics, the predicted field situation may be compared with an independently measured integrative signal originating from tile drainage studies, which, under certain conditions, integrates the soil water flux and solute transport at the field scale (Richard and Steenhuis, 1988). This is, however, a simplified consideration, as solute transport at tile-drained field sites consists of two processes: (i) the vertical flow through the unsaturated soil zone, and (ii) the radial flow within the groundwater towards the tile lines. The latter may add significantly to the overall spreading of a solute pulse which is captured in the solute signal from the tile drain but not in the signal assembled from individual (vertical) column tests.

Another approach to improving the understanding of transport processes and to deriving field-scale parameters is the use of tracer experiments. Numerous tracer tests have been conducted on tile-drained fields to study the field-scale transport of solutes (Kung *et al.*, 2000a,b; Jaynes *et al.*, 2001a), particularly pesticides (Czapar *et al.*, 1994; Elliott *et al.*, 2000; Kladivko *et al.*, 2001; Zehe and Flühler, 2001). Especially if soils are structured, a conservative tracer will often be detected in the tile drainage water after only a few millimetres of rainfall (<10mm; e.g. Villholth *et al.*, 1998). Likewise, Brown and van Beinum (2009) could impressively demonstrate that both the peak concentrations of pesticides in the tile effluent and the seasonal load depend on the clay content. In such situations, only a small fraction of the pore system is hydraulically active (Bronswijk *et al.*, 1995; Oygarden *et al.*, 1997), and non-equilibrium conditions between a fast-transporting macropore system and the matrix flow domain prevail (Gerke and Köhne, 2004). Macropore flow occurs when rates of lateral equilibration of water pressure and/or solute concentrations with the matrix are slow compared with the vertical flow rates in the macropores (Jarvis, 2007).

The overall objectives of this study were to assess the impact of artificial drainage on water and solute fluxes at two contrasting tile-drained fields, and to detect scale dependencies of these fluxes. For this, we compared the upscaling of the transport of a conservative tracer from the column to the field scale (Bokhorst experimental site) with the scale transition of nitrate export from the field to the small catchment scale (Dummerstorf experimental site). Both the matrix and the macropore transport processes were considered. For the first site, we applied a stochastic stream tube model using the results from column studies to upscale and compare with the observations of the field scale. For the second site, the translation or transformation of nutrient concentration patterns was investigated with a hierarchical monitoring set-up in combination with a mixing model. Some of the experimental and modelling results have been presented elsewhere, while the synthesis of the various investigations yielded new results and insights into the water flow and solute transport processes in artificially drained landscapes.

Materials and Methods

Experimental sites

The experimental sites are tile drained and located in northern Germany within Pleistocene lowland areas, but differ in their climatic and soil characteristics. Both tile-drainage systems discharge mainly during a 6-month winter period (November to April), owing to the precipitation surplus caused by low temperatures and low evapotranspiration rates. During the growing period in summer, generally only small and infrequent flow events take place following heavy rainfall events. On average, only 2 to 19% of the annual discharge at the different measurement stations occurred during the summer period (Wichtmann, 1994; Tiemeyer *et al.*, 2006). The water table dropped to well below drainage depth and smaller ditches frequently became dry.

Bokhorst experimental site

The first experimental field site, Bokhorst, is situated in north-western Germany in the vicinity of the city of Kiel, federal state of Schleswig-Holstein (Fig. 12.1). Long-term mean annual temperature and rainfall at Bokhorst are 8.3 °C and 825mm, respectively. The site has an average slope of 1%. Tile drains were placed almost perpendicular to the contour lines in 1985. Tile lines were installed at a depth of 1m and with a spacing of 11 to 14.5m. Four tile lines were chosen out of a system of six for monitoring purposes. Assuming that the area of influence of a tile line extends to a distance of one half of drain spacing, the monitored area was estimated to be 0.5ha. The area contributing to the discharge in the north-western part of the field was determined by groundwater wells.

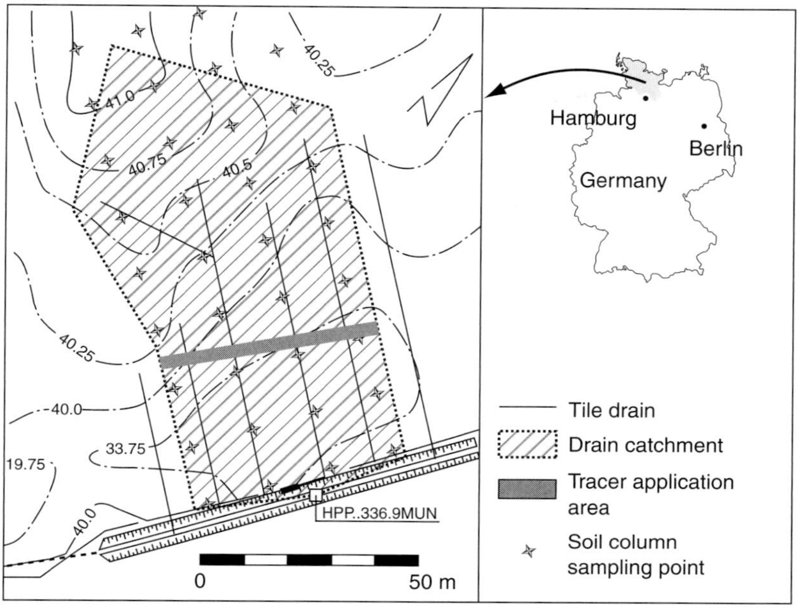

Fig. 12.1. Experimental layout at the Bokhorst site in northern Germany.

The experimental plot was under a 3-year crop rotation with winter rape (harvested in 1992 and again in 1995), winter wheat (1993) and winter barley (1994). Conventional tillage was performed with a mouldboard plough.

The soil was classified as a poorly drained Dystric Gleysol. The experimental area has been identified as heterogeneous in terms of soil type. The clay content decreased with soil depth from 19% at 0–30 cm depth to 10% at 70–110 cm depth, while bulk density increased from 1.46 g cm^{-3} at 0–20 cm depth to 1.95 g cm^{-3} at 70–110 cm depth (Table 12.1). Soil structure changed accordingly from subangular (0–30 cm) and angular (30–95 cm) to coherent (95–160 cm). The high clay content of the soil surface layer induced cracks and fissures during dry periods.

Dummerstorf experimental site

The second experimental field site, Dummerstorf, is located in north-eastern Germany in the vicinity of the city of Rostock, federal state Mecklenburg-Vorpommern (Fig. 12.2). It is part of the small rural catchment of the brook Zarnow, a tributary of the river Warnow, which discharges into the Baltic Sea. Long-term mean annual precipitation and temperature for the experimental site are 665 mm and 8.2 °C, respectively. With elevations ranging from 30 to 50 m above sea level, the catchment is characterized by generally gentle slopes.

Sampling station I is located at the collector drain outlet of a tile drainage plot of 4.2 ha. According to tile drainage maps, the fan-shaped drainage network of plastic tiles is characterized by a drain depth of 1.1 m and drain distances of 12 to 14 m. The crop rotation of the tile drainage plot was maize (harvested in 2003), winter wheat (2004), winter rape (2005) and, again, maize with a mixture of red clover and perennial ryegrass as catch crop (2006). Conventional tillage was applied here, too. The soil at the drainage plot was characterized as a Gleyic Luvisol. Compared with Bokhorst, the topsoil has lower clay content (Table 12.1). Cracks and fissures were not observed, but numerous earthworm burrows and root channels in the subsoil formed preferential flow paths and this was confirmed by a dye tracer experiment (Köhne, J.M. *et al.*, 2006).

The collector drain discharges into a ditch (sampling station II) with a 179 ha catchment

Table 12.1. Soil properties at the two field sites of Bokhorst (Dystric Gleysol) and Dummerstorf (Gleyic Luvisol) in northern Germany.

Depth	Sand content (%)	Clay content (%)	Bulk density (g cm^{-3})	Soil structure
Bokhorst				
0–30 cm	50	19	1.46[a]	Crumbed, subangular
30–70 cm	55	10	1.73[b]	Angular-prismatic
70–110 cm	55	10	1.95	Coherent
Dummerstorf				
0–30 cm	58	12	1.61	Crumbed
30–70 cm	50	12	1.65	Subangular
70–110 cm	65	5	1.66	Subangular

[a]0–20 cm depth
[b]40–70 cm depth

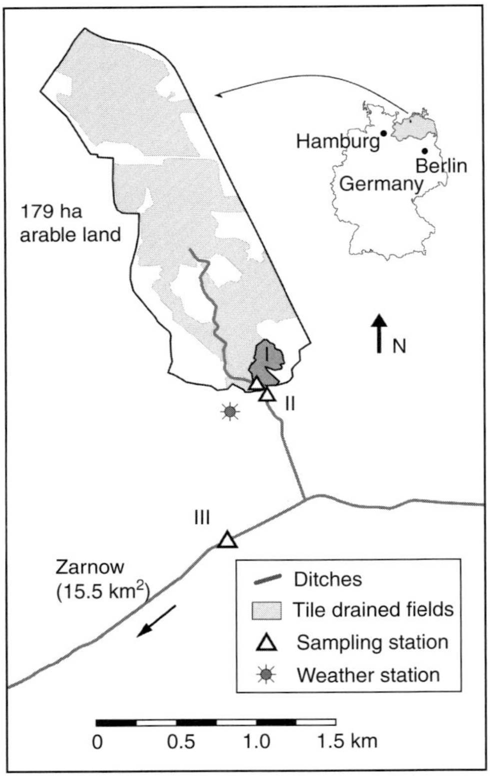

Fig. 12.2. Experimental layout at the Dummerstorf site in northern Germany. Numbers I, II and III refer to sampling stations.

(around 80% tile drainage) used for intensive conventional crop production. Predominant soil units are Luvisols, Gleysols and Cambisols (mainly sandy loams). Predominant crops are winter wheat, winter rape, maize, sugarbeet and winter barley. Similar to the sampling station I, the drainage depth is 1.1 m, and the drainage spacing varies between 8 and 22 m.

The third sampling station (III) is located at the brook Zarnow. The catchment of the brook (15.5 km^2) is extensively artificially drained with tile and ditch drainage, but detailed tile drainage maps were not available. Typically for Mecklenburg-Vorpommern, the Zarnow catchment is dominated by agriculture and sparsely populated: from digital orthophotos, a land-use distribution of 48% arable land, 28% grassland and 14% forests could be derived. Both grassland and forests can frequently be found on organic soils (Histosols), often artificially drained and degraded. Point sources do not exist, as domestic waste water is transferred to Rostock for treatment.

Field experiments

Bokhorst experimental site

Bromide transport field experiments at Bokhorst were performed in 1991/2, 1993/4 and 1994/5. The tracer (dissolved KBr) was applied once every year in late autumn and was monitored in the drainage discharge over a 5-month period. Applied mass was comparable between the tests (16.75, 16.75 and 19.3 kg), but the application area differed: strip-wise application in an area of 167 m^2 across tile drains (Fig. 12.1) during the first year versus uniform application over the entire field during the second and third years.

The four separate tile lines were routed to one monitoring station consisting of a Venturi flume and an automatic sampler. Water table head in the flume was registered on a half-hourly basis. Drain water samples were taken hourly at high flux rates during the first week after tracer application, down to daily at the end of the monitoring period. The chosen sampling scheme and high-frequency sampling (mixed samples instead of grab samples) at high flux rates ensured an exact determination of bromide loads (Tiemeyer *et al.*, 2010). All flux and bromide concentration values from the Bokhorst site presented herein are daily averages. Climate data were retrieved from the meteorological stations of Neumünster (at 10 km distance from the experimental site) and Plön (at 20 km distance).

Dummerstorf experimental site

The discharge and solute concentrations presented in this study were recorded at Dummerstorf in the discharge years 2003/4, 2004/5 and 2005/6. Sampling station I at the tile-drain outlet consisted of a climate station, a Venturi flume for discharge measurement and, like all sampling stations, an automatic sampler for water-quality sampling. At the climate station, temperature, humidity, wind speed and precipitation were recorded. Sampling station II at the ditch was equipped with an ultrasonic water-level and flow-measurement device. Sampling station III at the brook Zarnow was equipped with a pressure sensor. Frequent discharge gauging with an inductive flowmeter especially well adapted to low flow velocities was conducted to develop rating curves for both sampling stations II and III. Discharge and meteorological data were available every 15 min. Water samples were taken with the automatic samplers as daily composite samples resulting from a 3 h sampling interval. During low flow periods, the sampling density was reduced to one to two composite samples a week. All water samples were frozen at −20°C until the analysis for nitrate (NO_3^-) by ion chromatography.

Laboratory column experiment

For column-scale leaching experiments at the Bokhorst site, 35 undisturbed small-sized soil columns (5.7 cm in diameter and 10 cm long) were taken in a 15 × 15 m grid covering the entire experimental field of 0.5 ha (Fig. 12.1). Samples were collected during late autumn 1993 from a soil depth of 5 to 15 cm. The set-up for leaching tests was adapted from Rambow and Lennartz (1993). The apparatus was designed to induce unsaturated water-flow conditions by applying suction at the lower end of the sample (−3 kPa), whereas the soil surface was under atmospheric pressure. A fixed irrigation rate was imposed. Before the application of the bromide tracer, samples were adjusted to steady-state water-flow conditions for at least 48 h. Afterwards, a short bromide (KBr) pulse of 1 ml was applied in two equal parts using a 0.5 ml high performance liquid chromatography (HPLC) syringe. Bromide concentration was 960 mg l^{-1}, which resulted in a bromide mass of 0.960 mg for each column. Automatic sampling of leachate at 4 h intervals corresponding to around 1/25 pore volume was initiated simultaneously at the start of tracer application. Effluent volumes were used to determine the flow rate of individual columns.

Data analysis

Linear regression analysis

The time series of discharge rate and bromide concentration at the Bokhorst experimental site were evaluated by regression analysis in order to identify time periods in which preferential transport contributed considerably to overall solute transport. Data sets were prepared by computing flow rate (Eqn 12.1) and concentration (Eqn 12.2) differences (Lennartz *et al.*, 1999):

$$\Delta Q_j = Q(t_{j+1}) - Q(t_j) \tag{12.1}$$

$$\Delta c_j = c(t_{j+1}) - c(t_j) \tag{12.2}$$

where Q denotes the discharge (mm d^{-1}), c is the solute concentration (mg l^{-1}), and t_j is the

jth time interval (d). Values computed according to Eqns 12.1 and 12.2 were considered for further analysis when the limits $\Delta Q_i > 2.4$ or $\Delta Q_i < -2.4$ mm d^{-1} were reached; these characterized dominant flow peaks only.

Based on the common observation that preferential transport is more relevant and more visible at the early stages of discharge courses, seasonal data sets were separated into successive groups for statistical analysis. The ΔQ_j and Δc_j values of the first 2 months of each discharge period and those of the remaining part of the discharge season were treated separately. As the second part of each season incorporated uncertainties arising from the decreasing solute mass in the system, we expected the relation between flow rates and solute concentrations to be weak in these cases.

Solute loads

From the measured discharge Q (mm d^{-1}) and solute concentration c (mg l^{-1}) at each scale, the daily solute load L (kg ha^{-1} d^{-1}) for the Dummerstorf experimental site was calculated as follows (Tiemeyer *et al.*, 2006):

$$L = \frac{c \times Q}{100} \qquad (12.3)$$

To approximate the 'true' losses as accurately as possible, daily measured nitrate-nitrogen loads of each sampling station and each year were correlated with the daily discharge. The established relationship was then applied to calculate loads for days when only the discharge was measured owing to e.g. frozen pump tubes or a lower sampling frequency. For details on the calculation of losses, readers are referred to Tiemeyer *et al.* (2010).

As a result of crop evapotranspiration and resulting low groundwater levels during summer (>2.0 m below ground surface), and lower solute concentrations, the summer half-years did not significantly contribute to the annual flow and nutrient losses during the study period, as has already been observed in other tile drainage studies (Cambardella *et al.*, 1999; Tan *et al.*, 2002). For this reason, losses are reported for the winter half-years (November to April) only.

Mixing model for nitrate losses

To analyse the flow components and associated nutrient concentrations at different spatial scales at the Dummerstorf experimental site, an automatic hydrograph separation method ('recursive digital filter') was applied to the tile-drain discharge in combination with a simple two-component mixing model (Tiemeyer *et al.*, 2008). The recursive digital filter can distinguish a fast-flow component (interpreted as a high-frequency signal) from the base flow (the low-frequency signal). The filter equation for the fast component, Q_{fast}, at the time step i is given as (Nathan and McMahon, 1990):

$$Q_{fast\_I,i} = \beta Q_{fast\_I,i-1} + \frac{1+\beta}{2}(Q_{total\_I,i} - Q_{total\_I,i-1}) \qquad (12.4)$$

The empirical filter parameter β (–) determines the fractions of the two flow components. The index 'I' (or 'II', 'III') represents the number of the sampling stations. The base flow $Q_{base\_I,i}$ is calculated as the difference between $Q_{total\_I,i}$ and $Q_{fast\_I,i}$.

Hydrograph separation was applied to hourly tile-drain discharge data, and the results were then aggregated to daily values of the base-flow and the fast-flow components to be employed in the mixing equation (Eqn 12.5). The fast-flow component at the larger scales is calculated by multiplying the measured tile-drain discharge Q_{total_I} (given in mm d^{-1}) with the fraction of tile-drained area in the respective catchment (e.g. 75%), while the remainder of the total discharge is assumed to be base flow. Constant NO_3-N concentrations were assumed at the collector drain outlet for both the fast-flow component ($c\text{-}NO_3N_{fast\_I}$, given in mg l^{-1}) and the base-flow component ($c\text{-}NO_3N_{base\_I}$, in mg l^{-1}). The overall concentration $c\text{-}NO_3N_{total\_I}$ (mg l^{-1}) was then calculated as:

$$c\text{-}NO_3N_{total} = c\text{-}NO_3N_{fast}\frac{Q_{fast}}{Q_{total}} + c\text{-}NO_3N_{base}\frac{Q_{base}}{Q_{total}} \qquad (12.5)$$

The fast-flow components in the ditch and in the brook were assumed to consist of tile drainage; the fast-flow concentrations at these scales were thus set equal to $c\text{-}NO_3N_{total\_I}$. The tile-drain discharge component concentrations and base-flow concentrations of the ditch ($c\text{-}NO_3N_{base\_II}$) and the brook ($c\text{-}NO_3N_{base\_III}$) were assumed to be at constant values throughout each winter season, but were allowed to vary between the years to account for different management practices and climatic conditions. As the resulting set of equations for the mixing model could not be solved explicitly, a Monte-Carlo approach was chosen. A total of 25,000 model runs with randomly derived parameter sets were carried out. The Nash–Sutcliffe coefficient (*NSC*; Nash and Sutcliffe, 1970) was used for the evaluation of the modelled NO_3-N concentrations. Based on daily data, a single *NSC* for each measurement station and year as well as the sum of these *NSC*s (ΣNSC) was calculated, and those runs with a ΣNSC of > 2.70 and no single $NSC < 0$ were accepted.

Upscaling of column data to field scale

Our objective was to test the suitability of spatially distributed point measurements and subsequent averaging as a tool for predicting field-scale solute movement. The comparison of the column breakthrough curves with those from the tile drains requires some simplifying assumptions, such as the homogeneity of the soil in the vertical direction. Additionally, we assumed that, once tracer molecules reach the drain depth, they instantaneously enter the tile lines. This simplified assumption neglects the importance of solute spreading resulting from lateral transport processes in the saturated soil zone, as has been shown to be important in other studies (Utermann *et al.*, 1990). However, this assumption may be justified for cases in which preferential flow in the saturated and unsaturated zone overwrites the (expected) pathways resulting from equipotential lines established radially around the tile lines. It has been shown for the Bokhorst site that a tracer that has been applied strip-wise parallel to the tile line takes a more or less direct way to the drain pipe and does not occur later in effluent than a compound that has been applied rectangularly over all tile lines of the field site (Köhne, S. *et al.*, 2006; Gerke *et al.*, 2007).

To upscale the bromide breakthrough curve (BTC) from the column to the field scale, we applied a stochastic stream tube model in which the soil is considered as a bundle of individual non-interacting soil columns. The averaged flux concentration (f^f) from the column leaching tests, measured as a response signal to a small solute pulse, may be regarded as a travel-time probability density function (pdf) of the form (Jury, 1982):

$$f^f(z,I)=\frac{1}{\sigma I\sqrt{2\pi}}\exp\left(-\frac{[\ln(\frac{l}{zI})-\mu]^2}{2\sigma^2}\right) \quad (12.6)$$

where μ denotes the mean and σ the standard deviation of the log-normal distribution, I is the cumulative drainage, z is the soil depth, and l is the distance in flow direction from the soil surface to the depth at which solute concentrations are measured (calibration depth). The pdf of the travel time is assumed to be a unique function of cumulative drainage I.

Results from experimental investigations have demonstrated that the random space functions of the solute transport parameters are often bimodal, as are the assembled field-scale chemographs (Roth *et al.*, 1991; Sassner *et al.*, 1994; Caron *et al.*, 1996). Double-peak distributions reflect the contribution of at least two transport mechanisms to solute movement, one yielding an early unexpected solute breakthrough (preferential) and one forming the expected 'classical' breakthrough apex (matrix flow). Bimodal concentration and frequency distributions may be represented by the weighted sum of two travel-time pdfs (Eqn 12.7) (Jury and Scotter, 1994):

$$f^f(z,I)=\alpha f^1(z,I)+(1-\alpha)f^2(z,I) \quad (12.7)$$

where α denotes the weighting factor, which may assist in assessing the significance of the preferential flow mechanism for the overall flow and transport process. The superscript f in Eqns 12.6 and 12.7

reflects flux-type concentration distributions. A detailed discussion of the theoretical background of travel-time pdfs can be found in Jury and Roth (1990, pp. 63–84).

The individual BTCs from column displacement tests were assembled by computing arithmetic average concentrations. Concentrations were normalized with the surface density of the applied bromide (ρ_A = 0.038 mg cm^{-2}). Equation 12.7 was fitted to the normalized values by means of the NONLIN procedure of SAS (SAS Institute, Inc., 2003) in order to obtain estimates for the weighting factor α, and the means (μ_i) and variances (σ^2_i) of the combined pdfs. The parameter values obtained from the column BTCs were used to predict bromide concentration evolution at a soil depth of 1 m (depth of tile drains) by adjusting the l/z term of Eqn 12.6 in order to compare the results based on small soil samples with the field drain data. Thereby, the transport properties of the soil are considered as vertically homogeneous. Because of the large experimental effort, no soil horizon-wise investigation of transport properties was possible.

Results and Discussion

Bokhorst experimental site

Hydrology at field scale

The discharge hydrographs recorded at the Bokhorst site showed a rapid response to rainfall events during all the study years, indicating the proper functioning of the drainage system (Fig. 12.3a). The quick transformation of rainfall into tile-drain discharge is emphasized by the double-sum curve of precipitation and discharge (Fig. 12.4a). Accordingly, calculations showed that 75% (1991/2), 87% (1993/4) and 68% (1994/5) of the rainfall were transported out of the experimental area through the tile lines. Maximum flow rates of up to 17 mm d^{-1} occurred in general during early winter (December) and during early spring (March). The largest total winter rainfall of 462 mm was registered for the 1994/5 season, but variations in the rainfall/discharge relationship were negligible across the years. Surface runoff was not observed during the study period.

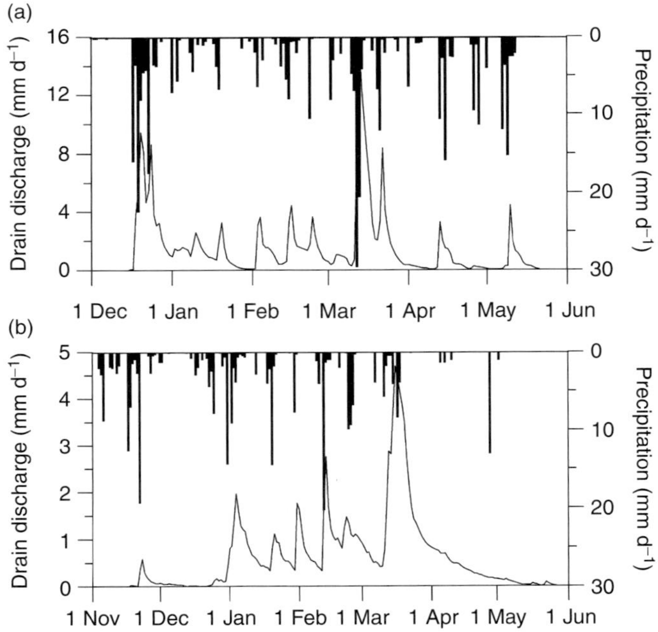

Fig. 12.3. Daily rainfall and drain discharge in 1991/2 at the Bokhorst site (a) and in 2004/5 at the Dummerstorf site (b).

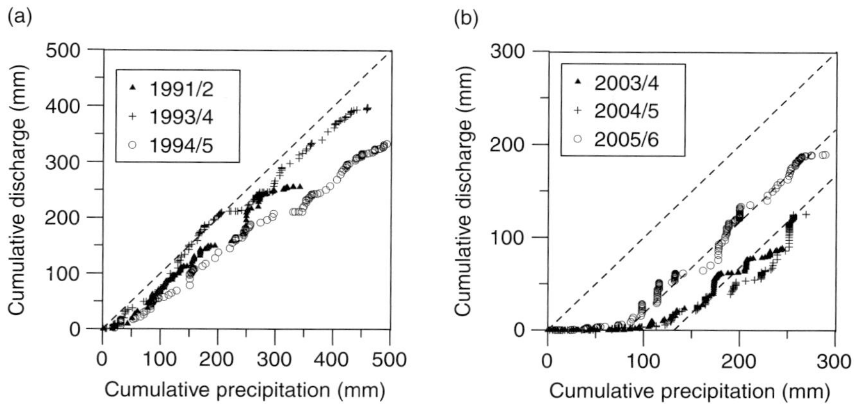

Fig. 12.4. Double sum curves of precipitation and discharge for the Bokhorst site (a) and the Dummerstorf site (b). (1 November to 30 April of each year, except for Bokhorst 1991/2, 1 December to 30 April).

Bromide breakthrough at field scale

Small variations in the rainfall/discharge relationship between the three years investigated (1991/2 to 1994/5) indicated the comparability of the tracer experiments with respect to the general hydrological regime at the experimental site. Bromide concentrations as a function of cumulative drain discharge for all three study years are shown in Fig. 12.5a. All three BTCs showed a similar general shape: a steep peak shortly after application was followed by a long tail that lasted until the end of the monitoring period. The tailed branch of each concentration curve was characterized by several smaller, gradually decreasing peaks in response to later rainfall events. In contrast to the first two years, bromide concentrations displayed a significant rising trend for a second time during the later part of the 1994/5 monitoring period.

Despite differences in the surface area of bromide application, the height and occurrence of the concentration peaks of the bromide BTCs of the first and second years were similar, except that the BTC of the first year did have a few more secondary peaks. We assume that the large bromide concentration of the application solution in the first year was diluted in the saturated zone of the soil profile and also in the drain pipes, because the sampled water contained contributions from four tile lines encompassing the whole area. As a consequence, bromide occurred in the drain outflow at the same levels of concentration as during the subsequent years.

The first bromide peak of the 1994/5 season appeared at a smaller total outflow volumes compared with the other years, and the concentration maximum was lower. Furthermore, a second peak occurred at larger total outflow volumes, suggesting matrix flow as a second dominant mechanism contributing to bromide loss. In summary, in all three years of the investigation, significant preferential solute transport yielded an early solute breakthrough, which was characterized by a short time delay between the first rainfall event following application and the first significant rise in bromide concentrations in drain outflow.

The patterns of bromide loss over time during a season varied considerably between years. In the first two seasons, bromide losses were large at the beginning of the leaching period. In 1993/4 and 1991/2, 50% of the total losses occurred with only 14% and 23% of the total drain discharge of the winter season, respectively. The corresponding cumulative drain discharge fraction for 50% of the seasonal bromide losses during 1994/5 was 51%, indicating relatively uniform leaching of bromide with time. Nevertheless, rapid transport took place during the last study year too, with the onset of bromide loss corresponding to very

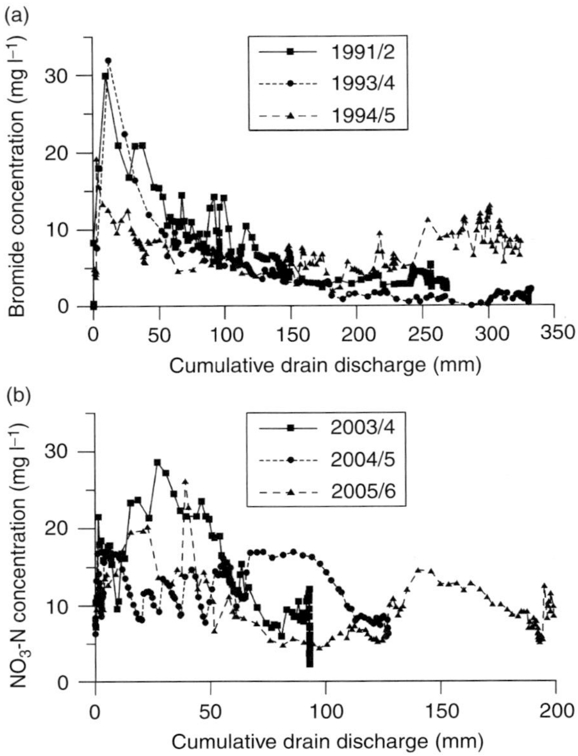

Fig. 12.5. Evolution of (a) the bromide concentrations in drain outflow versus the cumulative drain discharge at the Bokhorst site, and (b) the nitrate-nitrogen concentrations in the drain outflow versus the cumulative drain discharge at the Dummerstorf site.

low drain discharge. If classical matrix flow mechanisms only governed bromide transport, the first appearance of bromide in the drainage water would have occurred after a specific time (drainage) lag. The obvious dominance of matrix flow processes on bromide leaching during the last monitored season was possibly related to the rainfall amount and intensity during the first few weeks following application. Both maximum drain discharge intensity and cumulative outflow volume during the 25 days after the first rainfall were small in 1994/5 compared with the other two years. During the first 25 days of the 1994/5 season, only 51% and 32% of the respective cumulative drain discharge of 1991/2 and 1993/4 were measured. The moderate flow rates during the early stages of the 1994/5 test probably favoured solute movement through the soil matrix rather than along preferential pathways.

Simple linear regression analyses between flow rates and bromide concentrations of the dominant flow peaks were performed. The positive relationship between flow rate and concentration found for the first two months of the first season (1991/2) confirmed event-driven, preferential solute transport. Similar results have been reported by Kladivko *et al.* (1991). The negative slope of the fitted curve for the remainder of the season pointed to a change in the flow process from preferential to matrix flow. In contrast, bromide concentrations decreased with an increase in flow rates during both early and later stages of the last tracer experiment. The regression curves suggested the participation of the soil matrix in the flow processes. These findings were in line with the cumulative bromide loss curve. In conclusion, a small bromide fraction of applied bromide of the 1994/5 season was transported preferentially, whereas the major part was routed through the soil matrix.

The preferential flow observed at the Bokhorst site can partly be explained by disturbances of the soil profile by tile-drain installation. In general, homogeneous, isotropic soil properties are assumed for the generation of flow pathways, and the infiltrated water is expected to percolate vertically towards the groundwater surface (Nieber and Feddes, 1999), as shown in Fig. 12.6a. Under field conditions, however, layering of the soil profile, macropores and anisotropy are likely to alter the water movement. After moving vertically through the topsoil, the water has often been observed to subsequently move horizontally along the interface between the ploughed topsoil and a compacted plough pan (Petersen *et al.*, 1997; Forrer *et al.*, 1999). This process may be facilitated by an anisotropic hydraulic conductivity, which tends to be larger in the horizontal direction in non-structured, layered soils (Dörner and Horn, 2009; Soracco *et al.*, 2010). The installation of tile drains disturbed the developed soil profile (Fig. 12.6b), and the refilled trench probably exhibited a larger porosity and larger hydraulic conductivity than the undisturbed soil, enabling a preferential movement of water and solutes to the drain.

Bromide breakthrough at column scale

That the field-scale bromide BTCs (Fig. 12.5a) demonstrated pronounced preferential transport more or less independently of the rainfall patterns raised the question of whether it would be possible to assess such behaviour or soil properties from point measurements resulting from soil sampling and subsequent laboratory tests. Only in the case of tile-drained field sites with costly instrumentation and long-term investigations is, it possible to derive integrated water and solute signals. In most cases, the site conditions, expense and the time requirements do not allow the acquisition of field-scale data. In these cases, column-scale experiments yielding field-scale parameters would be of interest.

The assembled (or 'large-scale') bromide BTC obtained by computing arithmetic averages from the individual column BTCs is shown in Fig. 12.7 along with envelopes of ± one standard deviation (σ). Analogous to the shape of the individual BTCs, the assembled curve exhibited two distinct peaks, pointing to the contribution of at least two flow domains to the bromide transport behaviour. The first bromide peak at 0.1 to 0.2 pore volumes was due to preferential flow, whereas the second peak at 0.6 to 0.7 pore volumes was due to matrix flow. The latter is considered as the classical advective dispersive transport regime in this study, although the flow situation was presumably not fully equilibrated with respect to solute movement owing to immobile regions (Lennartz and Kamra, 1998). Regions of stagnant water seemed to influence solute movement, resulting in peak apices of pulse-applied tracers occurring significantly before one pore volume of effluent is exchanged.

The value of σ indicated that the uncertainty or range of concentration values reached its maximum shortly after the initiation of the experiment and remained at

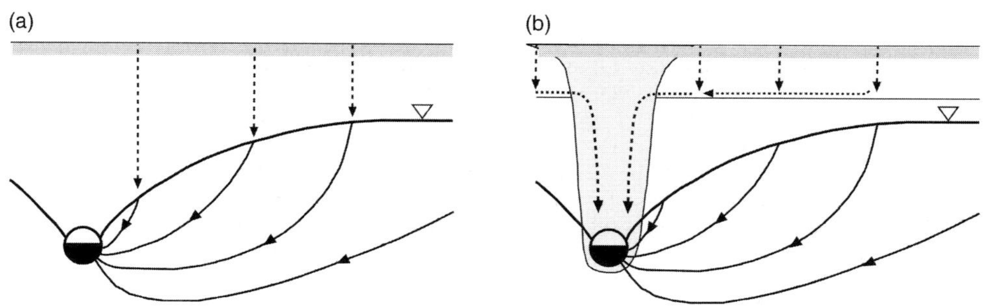

Fig. 12.6. Flow pathways in a drained field site: (a) homogeneous soil, (b) disturbed soil.

a high level during the preferential solute (bromide) peak. The minimum value was reached during the matrix solute peak. The variance during the matrix solute peak was small compared with values reported elsewhere (Sassner *et al.*, 1994).

Scale effects on solute transport

The concentration development from column and field experiments was compared on the basis of the cumulative discharge (Fig. 12.8). Although preferential transport mechanisms contributed to solute losses in both approaches, they were more effective in the field. The time (outflow volume) at which the preferential peak occurred was comparable for the predicted (column) and measured (tile drain) values, whereas the discharge corresponding to the matrix peaks – 220 mm in the column test and at 300 mm in field trial – differed by 80 mm. The strongest difference between the field and laboratory experiments was found for the partitioning between the preferential and the matrix flow domain. An evaluation of the column data revealed that 24% of the recovered bromide originated from the first preferential peak, whereas the analysis of the assembled tile-drain BTCs indicated that 90% of the leached bromide was routed through preferential flow paths.

The poor correlation between the two concentration curves may be attributed to the different flow conditions for each method. In the column transport experiments, unsaturated steady-state flow conditions were established. In contrast, the field flux regime was governed by transient flow conditions depending on the natural rainfall distribution. A varying soil water content might have favoured

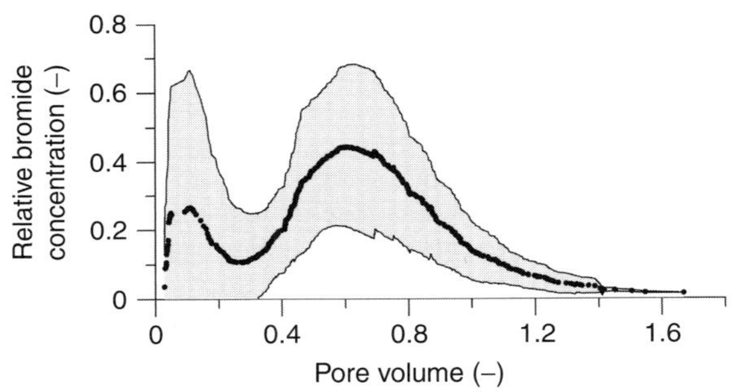

Fig. 12.7. Measured bromide concentrations in column outflow. Dots, mean; embracing lines, mean ± one standard deviation; (–) indicates dimensionless variable.

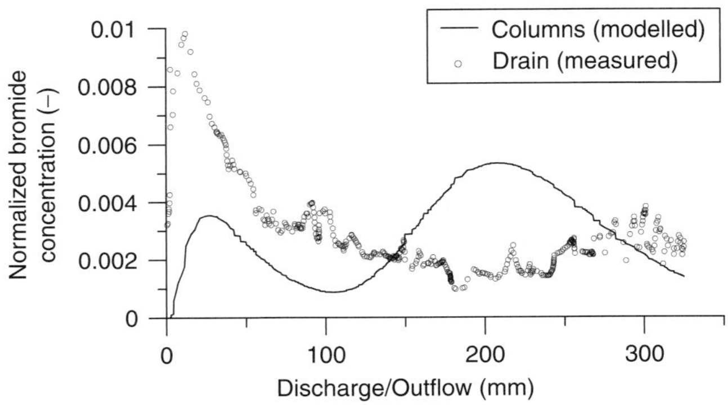

Fig. 12.8. Predicted bromide concentration from multiple column experiments and measured bromide concentration curve in tile drain at the Bokhorst site (average over the three study years). (–) indicates dimensionless variable.

preferential solute transport at the field site. It has been observed in column studies that a conservative tracer as well as pesticides may arrive earlier in the effluent under transient than under steady-state conditions with respect to cumulative drainage (Meyer-Windel *et al.*, 1999), while in other investigations the solute pulse under transient flow conditions lagged behind the one that was moving under steady-state flow (Russo *et al.*, 1989; Porro and Wierenga, 1993). During winter, the groundwater level remained between −80 cm (drain) and −60 cm (between drains) below the soil surface with variations upon rainfall events (water level rise to −60 cm above drains and −30 cm between drains; Wichtmann, 1994). Assuming a hydrostatic equilibrium with the groundwater table, an average matrix potential of −30 cm (≈−3 kPa) was assumed for the sampling depth of the soil columns, which corresponds roughly to field capacity water content at −6.3 kPa. The volumetric soil water content of the topsoil at the Bokhorst site varied between 33.1 and 32.2% at −2.5 and −6.3 kPa, respectively (Wichtmann, 1994). Accordingly, the pressure in the column test was adjusted to a lower value of −3 kPa. Possible differences in the soil water content of the steady-state column experiment and the time- and depth-averaged field soil water content are therefore believed to be of minor importance for the observed differences between the two BTCs (Fig. 12.8).

Soil heterogeneity in the vertical direction has to be taken into consideration as a second reason for the poor correlation between the two BTCs. The change in soil structure from subangular and crumbed peds in the topsoil to angular prismatic interspersed by some worm burrows in the subsoil presumably caused a concentration of the flow paths. Whereas there were some preferential flow regions in the topsoil as determined from the column approach, the structure of the subsoil clearly favoured preferential flow. Thus, we can assume that preferential transport played a much more dominant role in those soil horizons that were not examined in the column experiment. Under these circumstances, the simplifying assumption of a homogeneous soil profile led to incorrect results. The homogeneous transport of water and solutes in the topsoil and the insertion of preferential flow paths in the transition zone between the topsoil and subsoil (often the plough pan) have been documented by dye tracing (e.g. Petersen *et al.*, 1997; Forrer *et al.*, 1999). Considering the extreme predominance of the preferential flow domain as determined from the tile-drain measurements, it is questionable whether other methods, such as soil coring or extraction of soil solution by suction samplers, would have been more suitable for the determination of field solute transport.

From the laboratory and field data presented and their combination using a simple stochastic stream tube model, we conclude that column tests are in general suitable for identifying the processes that are operating at field scale. From the column experiments alone, however, it was not possible to quantify the extent of preferential flow that caused early solute breakthrough in the field trials.

Dummerstorf experimental site

Hydrology

Similar to the Bokhorst experimental site, the tile-drain discharge in general responded quickly to rainfall events at the Dummerstorf site (Fig. 12.3b). While variations in the rainfall/discharge relationship were negligible between the years at Bokhorst, the hydrological conditions differed strongly during the study period in Dummerstorf. Calculations showed that, during the winter season, only 36% (2003/4), 47% (2004/5) and 65% (2005/6) of the rainfall was transported out of the experimental area through the tile lines. This raised the question of the proper functioning of the tile drainage system, but double–sum curves of precipitation and discharge confirmed that there was indeed a period where rainfall was directly transformed into tile-drain discharge and that the system was in good working order (Fig. 12.4b). The difference from the Bokhorst site (Fig. 12.4a) was that the water table obviously needed longer to rise to the drain depth, especially after a record-breaking dry and warm summer such as that of 2003. The maximum flow rates of up to 7 mm d^{-1} were also much lower than at the

Bokhorst site. In addition to the hydrological conditions, the average winter temperatures of 4.0 °C (2003/4), 3.7 °C (2004/5) and 1.7 °C (2005/6) differed strongly between the years. Low temperatures in winter often involve frozen soils and thus increased overland flow, while microbiological activity is reduced.

In 2003/4, with a total winter precipitation of 244 mm, the winter discharge on all scales – $Q_I = 89$ mm, $Q_{II} = 86$ mm, $Q_{III} = 43$ mm – was governed by one major flow event in February (Tiemeyer *et al.*, 2006). In 2004/5, the winter precipitation was slightly higher, at 269 mm, but still below the long-term winter average of 283 mm, and it involved an unusual amount of snow. Furthermore, the summer was wetter in 2004 than in 2003. These conditions resulted in several flow events with discharge sums of $Q_I = 126$ mm, $Q_{II} = 147$ mm and $Q_{III} = 86$ mm. The largest flow event – in March – was caused by snow melt. In the winter of 2005/6, the precipitation (289 mm) was not much higher than during the preceding years and still much lower than the Bokhorst maximum, but none the less caused higher discharge sums ($Q_I = 190$ mm, $Q_{II} = 185$ mm and $Q_{III} = 112$ mm). This last year of the study period exhibited an extensive snow cover and surface runoff when rainfall coincided with snow melt on still frozen ground in March 2006. The significantly lower discharge in the Zarnow catchment compared with the smaller scales during the whole study period might be explained by the potential discrepancy between surface water and groundwater catchments and the presence of non-discharging depressions within the catchment.

Overall, the discharge showed a synchronous behaviour of the three scales: collector drain, ditch and brook (Fig. 12.9). Storm events early in the hydrological year which could not trigger tile drainage did not result in higher flow rates on the larger scales either. At later stages, the onset of tile-drain discharge coincided with the occurrence of the first smaller discharge peaks on the scales of the ditch and the brook, pointing to the importance of the tile drainage for flow generation on the catchment scale. This was further stressed by the close temporal agreement – on the basis of daily data – between the flow rates of the collector drain, the ditch and the brook. Van der Velde *et al.* (2010) found that upscaling the tile-drain

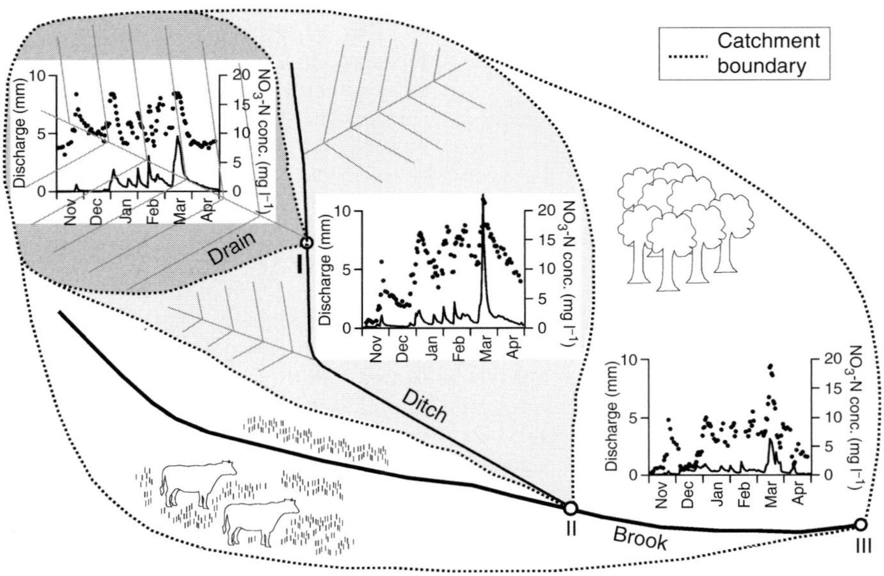

Fig. 12.9. Propagation of the discharge (lines) and nitrate-nitrogen concentration (symbols) signal from a drained plot (4.2 ha), over a ditch (180 ha), to a brook (15.5 km^2) at the Dummerstorf site (2004/5).

discharge contributions to the sub-catchment and catchment scale with a linear flow route-mixing model gave a good prediction of the catchment discharge.

Solute concentrations

Overall, the NO_3-N concentrations were high at all three sampling stations and exceeded the standard for 'good water quality' (2.5 mg l^{-1}) according to LAWA (LAWA, 1998) by far. The drinking water guideline of 11.3 mg l^{-1} (WHO, 2006) was also frequently exceeded (Fig. 12.10). As shown in Fig. 12.9 for the discharge season of 2004/5, the NO_3-N concentrations on all scales depicted a remarkably concurrent behaviour, with high concentrations at high flow rates. This parallelism points to the importance of tile-drain discharge concentrations for the overall water quality on larger scales. However, a scale effect due to dilution, e.g. by less polluted groundwater, land-use diversity, aquatic plant growth and in-stream processes such as benthic denitrification, was observed that caused lower nitrate concentrations at larger scales (Fig. 12.10). A similar scale effect and similar concentrations were observed by Doležal and Kvítek (2004) for a tile-drain outlet and a small stream, and Billy *et al.* (2011) found high NO_3-N concentrations in the drainage water of nested catchments until the 3rd stream order, which were then diluted at the larger scales as a result of diverse land use. However, in 2004/5 the average NO_3-N concentration in the ditch was slightly higher than at the tile-drain outlet, which could be attributed to the management practices at the tile-drained plot in relation to those of the ditch catchment.

The pattern with increasing NO_3-N concentrations at increasing flow rates throughout the whole winter season is rather unusual, but was also observed at the Bokhorst site during 1991/2 (Wichtmann, 1994). Frequently, NO_3-N concentrations and other surface applied solutes have been found to increase with increasing flow rates at the beginning of the discharge season, but then to be diluted by the following flow events until fertilization in the spring (Magesan *et al.*, 1995; Göbel, 1997). Furthermore, long-term measurements at different scales indicated that no consistent relationship could be found between NO_3-N concentrations and stream and tile-drain discharge (Tomer *et al.*, 2003). Although Van Herpe and Troch (2000) observed increasing NO_3-N concentrations with discharge, they found a low R^2, which they attributed to a hysteresis in the NO_3-N concentration–discharge relationship and to varying flow regimes

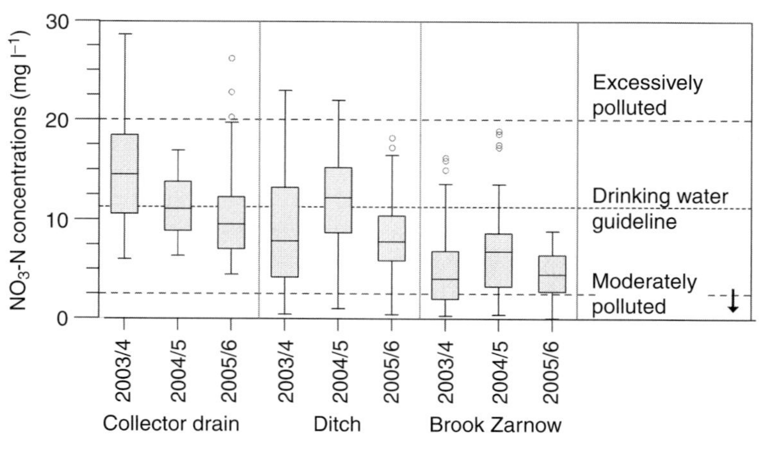

Fig. 12.10. Box plot diagrams of the nitrate-nitrogen concentrations at the collector drain, ditch and brook during the winter half-years of 2003/4, 2004/5 and 2005/6 (Tiemeyer *et al.*, 2006). The lower and upper broken lines indicate the water quality classes II ('moderately polluted', the general target for surface waters in Germany) and IV ('excessively polluted', the worst water quality class; LAWA, 1998), and the middle dashed line indicates the WHO (2006) drinking water guideline (modified and extended from Tiemeyer *et al.*, 2006).

between winter and summer. In contrast to all other events observed in this study, a dilution pattern characterized the main discharge peak in 2005/6, leading to lower NO_3-N concentrations at all scales (Fig. 12.10). At the collector drain, some macropore flow may have diluted the NO_3-N concentrations, which was supported by the occurrence of relatively high phosphorus and potassium concentrations only during this event (Tiemeyer *et al.*, 2009). At the higher scales, frozen-soil induced surface runoff bearing low NO_3-N concentrations may have contributed to the lower concentrations.

The increasing NO_3-N concentrations at increasing flow rates over the entire discharge season can probably be explained by a combination of different factors. First, as indicated by the gap of 30 to 90 kg ha^{-1} between the field balance and the measured loss rates, nitrogen accumulation in the soil can be supposed to be a source of high nitrate concentrations at several successive leaching events. For a detailed calculation of the NO_3-N balance, the reader is referred to Tiemeyer *et al.* (2006). Secondly, stable isotope analysis of the main event in 2003/4 also suggested that the nitrate in the tile-drain discharge mainly originated from fertilization (Deutsch *et al.*, 2006). If fertilizers at the soil *surface* were the only source of NO_3-N in the drain discharge, probably a pattern like that described by Göbel (1997) and Magesan *et al.* (1995), or even a tracer-like pattern as shown in Fig. 12.5a, would have developed. However, the NO_3-N concentrations showed a distinctively different pattern from the bromide concentrations. While the latter were dominated by preferential flow and were relatively insensitive to the actual weather conditions, the NO_3-N concentrations were clearly controlled by the rainfall pattern (Fig. 12.5b). In contrast to the pulse-applied tracer, a decreasing trend was seen only during spring when plant uptake started. Thus, there has to be a fairly constant pool of NO_3-N available for leaching.

To facilitate these high NO_3-N concentrations later in the discharge season, especially in the light of the utilization of manure as a nitrogen source, we assumed that mineralization and nitrification may have taken place even during the winter periods, although we measured neither the mineralization nor the soil temperature. This interpretation was supported not only by comparison of the transport patterns of bromide and NO_3-N (Fig. 12.5), but also by the results of a stable isotope analysis both for the main events during 2003/4 and for two tile drains in a neighbouring catchment. Low $\delta^{18}O$ in the drainage water nitrate indicated that most of the nitrate originated from the nitrification process, and that nitrification took place for most of the year (Deutsch *et al.*, 2005, 2006). Generally, mineralization is thought to slow down and finally cease at low temperatures, which may have contributed to the lower concentrations during the cold winter of 2005/6 (Fig. 12.10). However, there is evidence in the literature that some nitrification might still be possible (Tan *et al.*, 2002) and that the effects of winter temperatures on nitrate leaching have not been profoundly researched (Korsaeth *et al.*, 2003). Similarly, Cambardella *et al.* (1999) assumed that non-growing-season mineralization was greater than expected and resulted in large NO_3-N losses. Recent results even show that mineralization of soil organic N over the winter can make a substantial contribution to the mineral nitrogen pool available for leaching (Zhao *et al.*, 2010).

It can be assumed that the nitrate susceptible for leaching was available and generated in the upper soil layers, which were only rarely reached by the water table so that no denitrification occurred. This nitrate was then displaced to the tile drains by larger precipitation events. In fact, the analysis of stable isotopes did not give any evidence for denitrification in this tile drain; denitrification was only observed in the water of a neighbouring tile for a period of 6 months during summer and autumn (Deutsch *et al.*, 2006).

Owing to the high percentage of tile drainage, processes were similar at the ditch scale apart from the fact that the management and fertilization as well as the timing and amount of tile-drain discharge were inhomogeneous and that some groundwater inflow was present. Stable isotope analysis supported the view that the main source of nitrate was tile-drain discharge, at both the ditch and the brook scale (Deutsch *et al.*, 2006).

At the brook scale, however, the influence of grassland, undrained forests, in-stream processes and possible groundwater inflow has to be considered, and this seemed to change the overall concentration level, but not the patterns observed. Nevertheless, these parallel patterns at all scales indicated an extensive NO_3-N enrichment in the soils of the Zarnow catchment.

Flow rate–solute load relationship and nutrient loss rates

To assess the environmental impact of NO_3-N and other nutrients on downstream water bodies (such as, in this case, the Warnow river and the Baltic Sea), the losses are even more important than the concentrations. The measured losses are shown in Table 12.2.

Figure 12.11 shows the results of the flow rate–solute load regression model for NO_3-N. The two isolines with a slope of 1.0 represent those loads that would theoretically occur at constant concentrations corresponding to the drinking water guideline (11.3 mg l^{-1}; WHO, 2006) and to the upper limit of 'good water quality' (2.5 mg l^{-1}; LAWA, 1998). As lines fitted to the actual data (eyeball best fit) clearly show a slope (b) > 1.0, an increase in the flow rate would lead to a disproportionately high increase in NO_3-N loads on all scales. Moreover, it can be observed that at low flow rates the discharge and the NO_3-N loads were less strongly correlated than at higher flow rates. This was probably because the concentrations exhibit a hysteretic pattern with higher values in the receding limb (compare Van Herpe and Troch, 2000). As observed for other tile-drained catchments (Tomer *et al.*, 2003), high NO_3-N loss rates therefore coincided on all scales with high flow rates on a daily as well as on an annual basis; in that study, the NO_3-N load was also smallest from the largest catchment, but the differences were not as large as they were in our case.

Table 12.2. Nitrate-nitrogen losses (kg ha^{-1}) from the Dummerstorf site in northern Germany (November to April, 2003–2006).

Sampling station	2003/4	2004/5	2005/6
Collector drain	15.7	12.2	3.4
Ditch	16.9	21.8	7.2
Brook Zarnow	20.2	16.5	6.2

Solute loads – mixing model

To assess the importance of different flow pathways, solute loads both with a fast-flow component and with base flow were modelled using a combined hydrograph separation–mixing model. Figure 12.12 shows the cumulated measured NO_3-N loads for the discharge seasons of 2003/4 to 2005/6 and the cumulated loads modelled with the mixing model (Tiemeyer *et al.*, 2008).

At the collector drain, the total NO_3-N losses in the winter half-years (November to April) of the three discharge periods were 15.7, 16.9 and 20.2 kg ha^{-1}, respectively. The simulated total NO_3-N losses were only slightly lower. The fast-flow component clearly dominated the NO_3-N losses from the tile drainage plot: it was responsible for 87, 79 and 80% of total NO_3-N losses during 2003/4, 2004/5 and 2005/6, respectively. These 82% (on an average) of the total losses correspond to 72% of the total discharge. The question now was how could this 'fast' component be interpreted? At Dummerstorf, the interpretation of the fast-flow component as preferential flow was highly unlikely. First, the soil structure did not favour the formation of cracks and fissures as in Bokhorst; this was supported by dye tracer studies showing a uniform dye tracer distribution in the topsoil and only small occurrences of dye in the subsoil. Secondly, the concentration patterns – high phosphorus and potassium concentrations concurring with diluted nitrate and chloride concentrations, pointing to preferential flow – were only found during a single flow event within 3 years (Tiemeyer *et al.*, 2009). Thirdly, preferential flow generally only accounted for a low per cent of the total drain discharge, even when it was a major transport mechanism (Everts and Kanwar, 1990; Laubel *et al.*, 1999). Heppell and Chapman (2006) found a larger fraction of event water by mixing analysis based on the electrical conductivity, but for a clay soil. Thus, the tile drainage signal was interpreted as a mixture of two

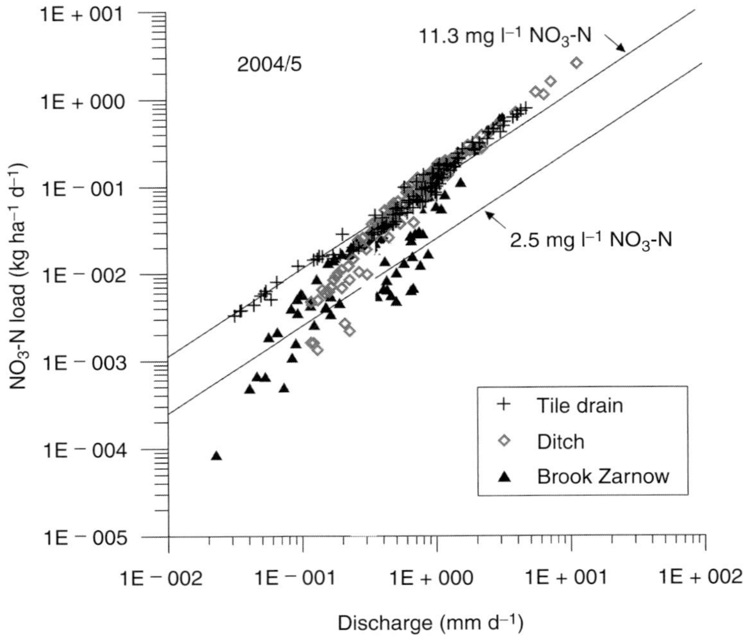

Fig. 12.11. Flow rate–solute load relationship of nitrate-nitrogen at the tile drain, the ditch and Brook Zarnow in 2004/5. The lines represent loads equivalent to the WHO (2006) drinking water guideline (top) and the upper limit of 'good water quality' according to LAWA (1998) (bottom) (modified from Tiemeyer *et al.*, 2006).

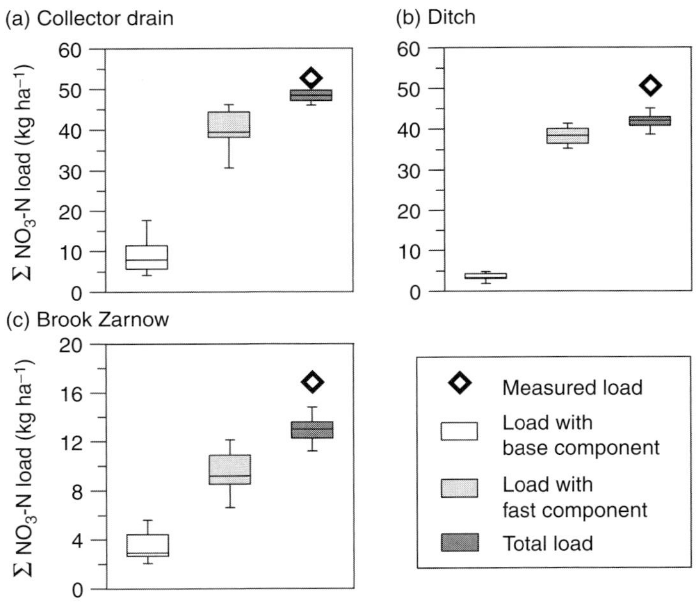

Fig. 12.12. Measured nitrate-nitrogen loads and loads modelled with a combined hydrograph separation mixing model in the (a) collector drain, (b) ditch and (c) Brook Zarnow. Sum of three winter half-years (November to April; discharge seasons 2003/4, 2004/5 and 2005/6).

components from different soil layers. One component was bearing nitrate from the topsoil, which was highly enriched with NO_3-N originating from fertilizers owing to long-term nitrogen balance surpluses (Tiemeyer *et al.*, 2006). This component may still be considered to be 'fast' as both the discharge and the corresponding solute signal characterized by high NO_3-N concentrations reacted very quickly to precipitation. Furthermore, it may have included some preferential flow, as presented for the Bokhorst experimental site; however, the exact quantification of transport processes still needs to be accomplished by tracer tests in this study area. The second, slow component was the tile drainage base flow, which bore lower nitrate concentrations because it originated from deeper soil layers, where denitrification is favoured (Hesterberg *et al.*, 2006). This interpretation was also supported by the results from a study analysing stable isotopes at one discharge event (Kahle *et al.*, 2007). Moravec *et al.* (2010) reported that shallow soil pathways primarily generated winter streamflow, while summer streamflow was dominated by deep soil pathways.

At the ditch scale, the measured NO_3-N losses agreed well with modelled total NO_3-N loads. Underestimated loads during 2004/5 were caused by an underestimation of the NO_3-N concentrations during the main flow event. The fast-flow component governed NO_3-N losses with an average proportion of 92% of total NO_3-N losses. As overland flow could be neglected because of generally gentle slopes, the fast-flow component was assumed to originate from tile-drain discharge, which covered nearly the entire ditch catchment. In contrast, the base-flow component, depicting the groundwater, was of minor importance for nitrate losses.

In the brook catchment, the measured and modelled total NO_3-N losses also agreed well. The fast-flow component, and thus the tile drainage, was responsible for 62 to 88% of total NO_3-N losses. At the most, 25% of the catchment is tile drained. This means that, on average, 73% of the solute losses originated from a maximum of 25% of the catchment area. This proportion was considerably higher than the value given by Behrendt and Bachor (1998), who reported that 47% of the NO_3-N losses from the whole Warnow catchment were exported via tile drains. David *et al.* (1997) found a tile drainage contribution of 68 to 100% to the total nitrate load in a larger catchment in Illinois, but at that site 75% of the area was tile drained. Our results clearly suggest that a reduction of diffuse nutrient pollution in lowland catchments can only be achieved when the nitrate losses in tile-drained areas are reduced, even if tile drains only cover a relatively small part of the catchment.

Summary and Conclusions

At the Bokhorst site, we attempted to transfer transport processes of a conservative tracer from the column to the field scale. The field-scale solute transport characteristics of the 0.5 ha tile-drained Bokhorst field site were determined by two methods. Thirty-five undisturbed small soil columns were taken in a regular grid with a 15 m spacing. Bromide breakthrough curves (BTCs) obtained from classical displacement tests under unsaturated steady-state-flow conditions were assembled by arithmetic averaging and subsequent modelling using a stochastic stream tube approach to obtain a field-scale chemograph. The 'real' field situation was captured in three bromide field tests under natural climate conditions by monitoring the transfer of a surface-applied solute pulse at the tile-drain outlet.

The large-scale BTC as derived from the column displacement test had two distinct concentration apices reflecting the contribution of at least two transport domains (preferential and matrix) on bromide movement through the unsaturated soil. The variance of the mean concentrations as indicated by the standard deviation was large during the preferential peak and decreased during the expected bromide breakthrough. The BTCs from the drain discharge measurements had a pronounced preferential peak but no significant second concentration rise except in one year. Solute losses were large during the preferential peak for the field test only. In the column test, the predominant solute fraction was leached under matrix flow conditions.

The parameter values obtained from fitting a bimodal log-normal probability density function to the mean concentration curve as derived from the column tests were used to predict the bromide concentration development at the drain depth. The discharge volume at which the preferential peak occurred was comparable between predicted (column) and measured (tile drain) values, whereas the discharge corresponding to the matrix apices differed by around 80 mm. The most pronounced deviation between the two approaches was found for the partitioning between the preferential and the matrix flow domains. The model evaluation of the column data showed that 24% of the recovered bromide originated from the fast-transporting flux region, whereas the analysis of the tile-drain BTCs indicated that 90% of the eluted bromide was routed through preferential flow paths. The method of local measurements (column test) and subsequent averaging and upscaling as a tool for characterizing field-scale solute movement seems to be adequate to identify the principal mechanisms involved, even when small samples are used. Nevertheless, the approach seems to be inadequate for estimating precisely the significance of each process in the overall transport situation.

At the Dummerstorf site, a hierarchical sampling approach was adapted in order to analyse the scale transition of a non-conservative solute (NO_3-N) from a drained plot (4.2 ha) over a ditch (180 ha) to a brook catchment (15.5 km^2). The discharge and the NO_3-N concentrations showed a similar and parallel pattern at all three scales. At all the scales, NO_3-N concentrations frequently exceeded the WHO (2006) drinking water guideline and often fell within the worst water quality class in Germany. NO_3-N concentrations nearly always increased at increasing flow rates, indicating the importance of artificial drainage for the larger catchment's hydrology and hydrochemistry. Accordingly, high losses always occurred at high flow rates, and a certain increase of the flow rate evoked a disproportionately high increase of the NO_3-N losses. Generally, the loss rates of NO_3-N were lower at larger scales and, at 3.4 to 21.8 kg ha^{-1}, low to average, although they were relatively high in view of the comparably low flow rates measured. To quantify the role of the different flow components and flow paths for the NO_3-N concentrations at the three scales, an automatic hydrograph separation method was combined with a stochastic two-component mixing model. At the collector drain, the solute signal was interpreted as a mixture of a fast component with high NO_3-N concentrations originating from the enriched topsoil and a slow base-flow component. The fast component was responsible for 63–91% of the total simulated NO_3-N losses at the collector drain. A clear scale effect due to mixing, land-use diversity and in-stream processes causing lower NO_3-N concentrations at larger scales was observed. Tile drainage itself delivered 89–95% of the total losses from the ditch catchment. In the brook catchment, tile-drained sites corresponding to up to 25% of the catchment area were responsible for 54–85% of the losses, which exceeds previous estimates for north-eastern Germany.

Our results showed that tile-drained fields are the key to the reduction of diffuse pollution of lowland catchments. The catchment discharge behaviour and the solute transport processes – which might be dominated by preferential flow as at Bokhorst or by matrix flow as at Dummerstorf – are crucial for the nutrient losses and must therefore be the starting point for any efforts to reduce losses, e.g. by active management of the water table or wetland construction. Future work should concentrate on a refined management of tile-drained field sites, combining reduced fertilizer and agrochemical input with active water management. For example, simply blocking the tile lines in winter and reopening them in spring might lead to non-equilibrium conditions and flushing of nutrients at very high concentrations upon that reopening. Such management approaches should be accompanied by an experimental programme to better understand the pathways and transformations of water and chemicals. Ideally, tracers such as the conservative KBr or the stable isotopes of N and O could be used to improve our understanding of managed artificially drained soils.

References

Baden, W. and Eggelsmann, R. (1970) Hydrological budget of high bogs in the Atlantic region. *Proceedings of the 3rd International Peat Congress 1968*, Quebec. Department of Energy, Mines and Resources, Ottawa, Ontario, pp. 260–311.

Behrendt, H. and Bachor, A. (1998) Point and diffuse load of nutrients to the Baltic Sea by river basins of North East Germany (Mecklenburg-Vorpommern). *Water Science and Technology* 38, 147–155.

Billy, C., Birgand, F., Sebilo, M., Billen, G., Tournebize, J. and Kao, C. (2011) Nitrate dynamics in artificially drained nested watersheds. *Physics and Chemistry of the Earth* (in press, doi:10.1016/j.pce.2008.09.007).

Blann, K.L., Anderson, J.L., Sands, G.R. and Vondracek, B. (2009) Effects of agricultural drainage on aquatic ecosystems: a review. *Critical Reviews in Environmental Science and Technology* 39, 909–1001.

Bronswijk, J.J.B., Hamminga, W. and Oostindie, K. (1995) Rapid nutrient leaching to groundwater and surface water in clay soil areas. *European Journal of Agronomy* 4, 431–439.

Brown, C.D. and van Beinum, W. (2009) Pesticide transport via sub-surface drains in Europe. *Environmental Pollution* 157, 3314–3324.

Burke, W. (1967) Principles of drainage with special reference to peat. *Irish Forestry* 24, 1–7.

Cambardella, C.A., Moorman, T.B., Jaynes, D.B., Hatfield, J.L., Parkin, T.B., Simpkins, W.W. and Karlen, D.L. (1999) Water quality in Walnut Creek watershed: nitrate-nitrogen in soils, subsurface drainage water, and shallow groundwater. *Journal of Environmental Quality* 28, 25–34.

Caron, J., Banton, O., Angers, D.A. and Villeneuve, J.P. (1996) Preferential bromide transport through a clay loam under alfalfa and corn. *Geoderma* 69, 175–191.

Czapar, G.F., Kanwar, R.S. and Fawcett, R.S. (1994) Herbicide and tracer movement to field drainage tiles under simulated rainfall conditions. *Soil and Tillage Research* 30, 19–32.

David, M.B., Gentry, L.E., Kovacic, D.A. and Smith, K.M. (1997) Nitrogen balance in and export from an agricultural watershed. *Journal of Environmental Quality* 26, 1038–1048.

De la Cueva, P. (2006) Identification of agricultural areas in Europe subject to different types of field drainage. MSc thesis, Cranfield University, Silsoe, UK.

De Rooij, G.H. and Stagnitti, F (2002) Spatial and temporal distribution of solute leaching in heterogeneous soils: analysis and application to multisampler lysimeter data. *Journal of Contaminant Hydrology* 54, 329–346.

Deutsch, B., Liskow, I., Kahle, P. and Voss, M. (2005) Variations in the delta N-15 and delta O-18 values of nitrate in drainage water of two fertilized fields in Mecklenburg-Vorpommern (Germany). *Aquatic Sciences* 67, 156–165.

Deutsch, B., Kahle, P. and Voss, M. (2006) Assessing the source of nitrate pollution in water using stable N and O isotopes. *Agronomy for Sustainable Development* 26, 263–267.

Doležal, F. and Kvítek, T. (2004) The role of recharge zones, discharge zones, springs and tile drainage systems in peneplains of Central European highlands with regard to water quality generation processes. *Physics and Chemistry of the Earth* 29, 775–785.

Dörner, J. and Horn, R. (2009) Direction-dependent behaviour of hydraulic and mechanical properties in structured soils under conventional and conservation tillage. *Soil and Tillage Research* 102, 225–232.

Dunn, S.M. and Mackay, R. (1996) Modelling the hydrological impact of open ditch drainage. *Journal of Hydrology* 179, 37–66.

Eggelsmann, R. (1981) *Dränanleitung für den Landbau, Ingenieurbau und Landschaftsbau.* 2nd edn, Verlag Paul Parey, Hamburg/Berlin.

Elliott, J.A., Cessna, A.J., Nicholaichuk, W. and Tollefson, L.C. (2000) Leaching rates and preferential flow of selected herbicides through tilled and untilled soil. *Journal of Environmental Quality* 29, 1650–1656.

Elmi, A.A., Madramootoo, C. and Hamel, C. (2000) Influence of water table and nitrogen management on residual soil NO_3^- and denitrification rate under corn production in sandy loam in Quebec. *Agriculture, Ecosystems and Environment* 79, 187–197.

Everts, C.J. and Kanwar, R.S. (1990) Estimating preferential flow to a subsurface drain with tracers. *Transactions of the ASAE* 33, 451–457.

Feick, S., Siebert, S. and Döll, P. (2005) *A Digital Global Map of Artificially Drained Agricultural Areas.* Frankfurt Hydrology Paper 04, Institute of Physical Geography, Frankfurt University, Frankfurt am Main, Germany.

Forrer, I., Kasteel, R., Flury, M. and Flühler, H. (1999) Longitudinal and lateral dispersion in an unsaturated field soil. *Water Resources Research* 35, 3049–3060.

Gerke, H.H. and Köhne, J.M. (2004) Dual-permeability modeling of preferential bromide leaching from a tile-drained glacial till agricultural field. *Journal of Hydrology* 289, 239–257.

Gerke, H.H., Dusek, J., Vogel, T. and Köhne, J.M. (2007) Two-dimensional dual-permeability analyses of a bromide tracer experiment on a tile-drained field. *Vadose Zone Journal* 6, 651–667.

Gilliam, J.W., Baker, J.L. and Reddy, K.R. (1999) Water quality effects of drainage in humid regions. In: Skaggs, R.W. and Schilfgaarde, J. (eds) Agricultural Drainage. *Agronomy* 38, 801–830.

Göbel, B. (1997) Messung und Modellierung des flächenhaften Wasser- und Stofftransports aus landwirtschaftlich genutzten Flächen auf zwei Maßstabsebenen unter besonderer Berücksichtigung der Bereitstellung bodenkundlicher Daten für die Modellierung. PhD thesis, University of Kiel, Germany.

Heppell, C.M. and Chapman, A.S. (2006) Analysis of a two-component hydrograph separation model to predict herbicide runoff in drained soils. *Agricultural Water Management* 79, 177–207.

Hesterberg, D., de Vos, B. and Raats, P.A.C. (2006) Chemistry of subsurface drain discharge from an agricultural polder soil. *Agricultural Water Management* 86, 220–228.

Hodgkinson, R.A., Chambers, B.J., Withers, P.J.A. and Cross, R. (2002) Phosphorus losses to surface waters following organic manure applications to a drained clay soil. *Agricultural Water Management* 57, 155–173.

Holden, J., Chapman, P.J. and Labadz, J.C. (2004) Artificial drainage of peatlands: hydrological and hydrochemical process and wetland restoration. *Progress in Physical Geography* 28, 95–123.

Iritz, L., Johansson, B. and Lundin, L. (1994) Impacts of forest drainage on floods. *Hydrological Sciences – Journal des Sciences Hydrologiques* 39, 637–661.

Irwin, R.W. and Whiteley, H.R. (1983) Effects of land drainage on stream flow. *Canadian Water Resources Journal* 8, 88–103.

Jarvis, N (2007) A review of non-equilibrium water flow and solute transport in soil macropores: principles, controlling factors and consequences for water quality. *European Journal of Soil Science* 58, 523–546.

Jaynes, D.B., Hatfield, J.L. and Meek, D.W. (1999) Water quality in Walnut Creek watershed: herbicides and nitrate in surface waters. *Journal of Environmental Quality* 28, 45–49.

Jaynes, D.B., Ahmed, S.I., Kung, K.-J.S. and Kanwar, R.S. (2001a) Temporal dynamics of preferential flow to a subsurface drain. *Soil Science Society of America Journal* 65, 1368–1376.

Jaynes, D.B., Colvin, T.S., Karlen, D.L., Cambardella, C.A. and Meek, D.W. (2001b) Nitrate loss in subsurface drains as affected by nitrogen fertilizer rate. *Journal of Environmental Quality* 30, 1305–1314.

Jensen, K.H. and Refsgaard, J.C. (1991) Spatial variability of physical parameters and processes in two field soils, Part III: Solute transport at field scale. *Nordic Hydrology* 22, 327–340.

Jury, W.A. (1982) Simulation of solute transport using a transfer function model. *Water Resources Research* 18, 363–368.

Jury, W.A. and Roth, K. (1990) *Transfer Functions and Solute Movement Through Soil: Theory and Applications*. Birkhäuser, Basel/Boston/Berlin, pp. 63–84.

Jury, W.A. and Scotter, D.R. (1994) A unified approach to stochastic-convective transport problems. *Soil Science Society of America Journal* 58, 1327–1336.

Kahle, P., Tiemeyer, B., Deutsch, B. and Lennartz, B. (2007) Untersuchungen zum Stickstoffaustrag über Dränung in einem nordostdeutschen Tieflandeinzugsgebiet. *WasserWirtschaft* 2007 No. 6, 25–29.

Kamra, S.K., Michaelsen, J., Wichtmann, W. and Widmoser, P. (1999) Preferential solute movement along the interface of soil horizons. *Water Science and Technology* 40, 61–68.

Kladivko, E.J., Van Scoyoc, G.E., Monke, E.J., Oates, K.M. and Pask, W. (1991) Pesticide and nutrient movement into subsurface tile drains on a silt loam soil in Indiana. *Journal of Environmental Quality* 20, 264–270.

Kladivko, E.J., Grochulska, J., Turco, R.F., Van Scoyoc, G.E. and Eigel, J.D. (1999) Pesticide and nitrate transport into subsurface tile drains of different spacing. *Journal of Environmental Quality* 28, 997–1004.

Kladivko, E.J., Brown, L.C. and Baker, J.L. (2001) Pesticide transport to subsurface tile drains in humid regions of North America. *Critical Reviews in Environmental Science and Technology* 31, 1–62.

Kladivko, E.J., Frankenberger, J.R., Jaynes, D.B., Meek, D.W., Jenkinson, B.J. and Fausey, N.R. (2004) Nitrate leaching to subsurface drains as affected by drain spacing and changes in crop production systems. *Journal of Environmental Quality* 33, 1803–1813.

Klaus, J. and Zehe, E. (2010) Modelling rapid flow response of a tile-drained field site using a 2D physically based model: assessment of 'equifinal' model setups. *Hydrological Processes* 24, 1595–1609.

Köhne, J.M. and Gerke, H.H. (2005) Spatial and temporal dynamics of preferential bromide movement towards a tile drain. *Vadose Zone Journal* 4, 79–88.

Köhne, J.M., Júnior, J.A., Köhne, S. and Lennartz, B. (2006) Double-ring and tension infiltrometer measurements of saturated hydraulic conductivity and dye tracer movement. *Geophysical Research Abstracts* 8, 01863.

Köhne, S., Lennartz, B., Köhne, J.M. and Šimunek, J. (2006) Bromide transport at a tile-drained field site: experiment, and one- and two-dimensional equilibrium and non-equilibrium numerical modeling. *Journal of Hydrology* 321, 390–408.

Korsaeth, A., Bakken, L. and Riley, H. (2003) Nitrogen dynamics of grass as affected by N input regimes, soil texture and climate: lysimeter measurements and simulations. *Nutrient Cycling in Agroecosystems* 66, 181–199.

Kung, K.-J.S. (1990) Preferential flow in a sandy vadose zone: 1. Field observation. *Geoderma* 46, 51–58.

Kung, K.-J.S., Steenhuis, T.S., Kladivko, E.J., Gish, T.J., Bubenzer, G. and Helling, C.S. (2000a) Impact of preferential flow on the transport of adsorbing and non-adsorbing tracers. *Soil Science Society of America Journal* 64, 1290–1296.

Kung, K.-J.S., Kladivko, E.J., Gish, T.J., Steenhuis, T.S., Bubenzer, G. and Helling, C.S. (2000b) Quantifying preferential flow by breakthrough of silt loam soil. *Soil Science Society of America Journal* 64, 1296–1304.

Laubel, A., Jacobsen, O.H., Kronvang, B., Grant, R. and Andersen, H.E. (1999) Subsurface drainage loss of particles and phosphorus from field plot experiments and a tile-drained catchment. *Journal of Environmental Quality* 28, 576–584.

LAWA (Länderarbeitsgemeinschaft Wasser) (ed.) (1998) *Bewertung der Wasserbeschaffenheit von Fließgewässern in der Bundesrepublik Deutschland – chemische Gewässergüteklassifikation.* Kulturbuchverlag Berlin, Berlin.

Lennartz, B. and Kamra, S.K. (1998) Temporal variability of solute transport under vadose zone conditions. *Hydrological Processes* 12, 1939–1949.

Lennartz, B., Michaelsen, J., Wichtmann, W. and Widmoser, P. (1999) Time variance analysis of preferential solute movement at a tile-drained field site. *Soil Science Society of America Journal* 63, 39–47.

Magesan, G.N., White, R.E. and Scotter, D.R. (1995) The influence of flow rate on the concentrations of indigenous and applied solutes in mole-pipe drain effluent. *Journal of Hydrology* 172, 23–30.

Meyer-Windel, S., Lennartz, B. and Widmoser, P. (1999) Bromide and herbicide transport under steady-state and transient flow conditions. *European Journal of Soil Science* 50, 23–33.

Moravec, B.G., Keller, C.K., Smith, J.L., Allen-King, R.M., Goodwin, A.J., Fairley, J.P. and Larson, P.B. (2010) Oxygen-18 dynamics in precipitation and streamflow in a semi-arid agricultural watershed, Eastern Washington, USA. *Hydrological Processes* 24, 446–460.

Nash, J.E. and Sutcliffe, J.V. (1970). River flow forecasting through conceptual models. Part I: a discussion of principles. *Journal of Hydrology* 10, 282–290.

Nathan, R.J. and McMahon, T.A. (1990) Evaluation of automated techniques for base flow and recession analysis. *Water Resources Research* 26, 1465–1473.

Nieber, J.L. and Feddes, R.A. (1999) Solutions for combined saturated and unsaturated flow. In: Skaggs, R.W. and Schilfgaarde, J. (eds) Agricultural Drainage. *Agronomy* 38, 145–212.

Oygarden, L., Kvaerner, J. and Jenssen, P.D. (1997) Soil erosion via preferential flow to drainage systems in clay soils. *Geoderma* 76, 65–86.

Petersen, C.T., Hansen, S. and Jensen, H.E. (1997) Tillage-induced horizontal periodicity of preferential flow in the root zone. *Soil Science Society of America Journal* 61, 586–594.

Porro, I. and Wierenga, P.J. (1993) Transient and steady-state solute transport through large unsaturated soil columns. *Groundwater* 31, 193–199.

Rambow, J. and Lennartz, B. (1993) Laboratory method for studying pesticide dissipation in the vadose zone. *Soil Science Society of America Journal* 57, 1476–1479.

Richard T.L. and Steenhuis, T.S. (1988) Tile drain sampling of preferential flow on a field scale. *Journal of Contaminant Hydrology* 3, 307–325.

Robinson, M. (ed.) (1990) *Impact of Improved Land Drainage on River Flows.* Report No. 113, Institute of Hydrology, Wallingford, UK.

Robinson, M. and Rycroft, D.W. (1999) The impact of drainage on streamflow. In: Skaggs, R.W. and Schilfgaarde, J. (eds) Agricultural Drainage. *Agronomy* 38, 764–800.

Roth, K., Jury, W.A., Flühler, H. and Attinger, W. (1991) Transport of chloride through an unsaturated field soil. *Water Resources Research* 27, 2533–2541.

Russo, D., Jury, W.A. and Butters, G.L. (1989) Numerical analysis of solute transport during transient irrigation, 1. The effect of hysteresis and profile heterogeneity. *Water Resources Research* 25, 2109–2118.

Sassner, M., Jensen, K.H. and Destouni, G. (1994) Chloride migration in heterogeneous soil, 1. Experimental methodology and results. *Water Resources Research* 30, 735–745.

Skaggs, R.W. and Chescheir, G.M. (2003) Effects of subsurface drain depth on nitrogen losses from drained lands. *Transactions of the ASAE* 46, 237–244.

Skaggs, R.W., Brevé, M.A. and Gilliam, J.W. (1994) Hydrologic and water quality impacts of agricultural drainage. *Critical Reviews in Environmental Science and Technology* 24, 1–32.

Soracco, C.G., Lozano, L.A., Sarli, G.O., Gelati, P.R. and Filgueira, R.R. (2010) Anisotropy of saturated hydraulic conductivity in a soil under conservation and no-till treatments. *Soil and Tillage Research* 109, 18–22.

Spaling, H. (1994) Analyzing cumulative environmental effects of agricultural land drainage in southern Ontario, Canada. *Agriculture, Ecosystems and Environment* 53, 279–292.

Stamm, C., Flühler, H., Gächter, R., Leuenberger, J. and Wunderli, H. (1998) Preferential transport of phosphorus in drained grassland soils. *Journal of Environmental Quality* 27, 515–522.

Tan, C.S., Drury, C.F., Reynolds, W.D., Groenevelt, P.H. and Dadfar, H. (2002) Water and nitrate loss through tiles under clay loam soil in Ontario after 42 years of consistent fertilization and crop rotation. *Agriculture, Ecosystems and Environment* 93, 121–130.

Tiemeyer, B., Kahle, P. and Lennartz, B. (2006) Nutrient losses from artificially drained catchments in North-Eastern Germany at different scales. *Agricultural Water Management* 85, 47–57.

Tiemeyer, B., Lennartz, B. and Kahle, P. (2008) Analysing nitrate losses from an artificially drained lowland catchment (North-Eastern Germany) with a mixing model. *Agriculture, Ecosystems and Environment* 123, 125–136.

Tiemeyer, B., Kahle, P. and Lennartz, B. (2009) Phosphorus losses from an artificially drained rural lowland catchment in North-Eastern Germany. *Agricultural Water Management* 96, 677–690.

Tiemeyer, B., Kahle, P. and Lennartz, B. (2010) Designing monitoring programs for artificially drained catchments. *Vadose Zone Journal* 9, 14–24.

Tomer, M.D., Meek, D.W., Jaynes, D.B. and Hatfield, J.L. (2003) Evaluation of nitrate nitrogen fluxes from a tile-drained watershed in Central Iowa. *Journal of Environmental Quality* 32, 642–653.

Toride, N. and Leij, F.J. (1996) Convective-dispersive stream tube model for field-scale solute transport: I. Moment analysis. *Soil Science Society of America Journal* 60, 352–361.

Turtola, E. and Paajanen, A. (1995) Influence of improved subsurface drainage on phosphorus losses and nitrogen leaching from a heavy clay soil. *Agricultural Water Management* 28, 295–310.

Utermann, J., Kladivko, J.E. and Jury, W.A. (1990) Evaluating pesticide migration in tile-drained soils with a transfer function model. *Journal of Environmental Quality* 19, 707–714.

Van der Velde, Y., Rozemeijer, J.C., de Rooij, G.H., van Geer, F.C. and Broers, H.P. (2010) Field-scale measurement for separation of catchment discharge into flow route contributions. *Vadose Zone Journal* 9, 25–35.

Van Herpe, Y. and Troch, P.A. (2000) Spatial and temporal variations in surface water nitrate concentrations in a mixed land use catchment under humid temperate climatic conditions. *Hydrological Processes* 14, 2439–2455.

Van Wesenbeeck, I.J. and Kachanoski, R.G. (1991) Spatial scale dependence of in situ solute transport. *Soil Science Society of America Journal* 55, 3–7.

Villholth, K.G., Jensen, K.H. and Fredericia, J. (1998) Flow and transport processes in a macroporous subsurface-drained glacial till soil. I. Field investigations. *Journal of Hydrology* 207, 98–120.

Vinten, A.J.A., Vivian, B.J., Wright, F. and Howard, R.S. (1994) A comparative study of nitrate leaching from soils of different textures under similar climatic and cropping conditions. *Journal of Hydrology* 159, 197–213.

WHO (World Health Organization) (2006) *Guidelines for Drinking-water Quality. Volume I – Recommendations*, 3rd edn., incorporating first and second addenda . . . Available at: www.who.int/water_sanitation_health/dwq/gdwq3rev/en/ (accessed 20 July 2010).

Wichtmann, W. (1994) Stoffeintrag aus landwirtschaftlichen Dränflächen in Fließgewässer (Messung und Simulation). PhD thesis, University of Kiel, Germany.

Zehe, E. and Flühler, H. (2001) Preferential transport of isoproturon at a plot scale and field scale tile-drained site. *Journal of Hydrology* 247, 100–115.

Zhao, H., Zhang, X., Xu, S., Zhao, X., Xie, Z. and Wang, Q. (2010) Effect of freezing on soil nitrogen mineralization under different plant communities in a semi-arid area during a non-growing season. *Applied Soil Ecology* 45, 187–192.

13 Effect of Land Use and Soil Management on Soil Properties and Processes

Sjoerd W. Duiker*

Introduction

Global land use has changed dramatically in the last three centuries with an increasing impact of human beings on soil properties and processes. Some natural ecosystems have been replaced by crops, pastures, buildings or roads, while the remaining natural areas are often highly affected by human activities. Agriculture occupies a larger share of the world's lands. Cropland has increased to 1.8 billion hectares, and pasture to 2.8 billion hectares (Ramankutty *et al.*, 2008); cropland now covers 12% and pastures 22% of the world's land surface (Ramankutty *et al.*, 2008). Despite the increase in acreage, cropland per capita has declined from 0.76 to 0.35 ha per person between 1900 and 1990 owing to exponential population growth (Scholes *et al.*, 2005).

It is likely that both the expansion and intensification of cropland will continue to produce food, fibre and fuel. There is still land available to take into production, but this land will be of progressively poorer quality for crop production. Using available soil and climate data, Eswaran *et al.* (1999) analysed the available land for rainfed crop production in the world (Table 13.1). Only 13% of the land was classified as prime or good, with 42% marginal or very marginal and 45% unsuitable. Eswaran *et al.* (1999) noted that most prime and good land was already in use, while large tracts of marginal land were under forest in the Amazon basin, Central Africa and South-east Asia. These lands are highly sensitive to degradation and nutrient depletion, and it is desirable to maintain them under natural vegetation. Hence the need to first focus on increased production on existing farmland to meet current and future human needs. If marginal land is used for crop production, special soil management practices are needed to make and keep this land productive without negative environmental effects.

Unfortunately, sustainable farming methods are not widely used and yet the need for higher production remains. The Millennium Ecosystem Assessment (Hassan *et al.*, 2005) reported that: (i) many people still lack access to food, and between 1997 and 2002 undernourishment actually increased after a period of decrease; (ii) water scarcity is worsening – currently more than 40% of accessible continental runoff is withdrawn for human use, 70% of it used for irrigation; (iii) soil erosion and salinization as a result of poor grazing practices and deforestation have led to the desertification of 10–20% of drylands; (iv) globally, an estimated 9.4 million ha of forest are lost annually, with reforestation occurring in the northern hemisphere, but deforestation continuing

* Corresponding author: sduiker@psu.edu

Table 13.1. Land quality distribution in the world (Source: Eswaran *et al.*, 1999).

Land quality class	Properties	Area (million km²)	Per cent of global land surface
Prime	Soils are highly productive with few management-related constraints.	4.09	3
Good	Productive soils with few management constraints for production. Soils are typically more susceptible to degradation than prime soils or need more inputs to become productive.	12.42	10
Marginal	Owing to marginal productivity or susceptibility to degradation, these soils should only be used for crop production with continuous monitoring of soil degradation. Soil conservation and nutrient management plans are very important to sustain productivity.	43.68	33
Very marginal	These soils should preferably not be used for crop production owing to inherent constraints.	11.65	9
Unsuitable	These soils belong to very fragile ecosystems such as deserts and boreal forests that are not suited for crop production.	58.74	45

in the tropics; and (v) the intrinsic capacity of cultivated systems to support crop production is being undermined as a result of soil erosion, salinization and loss of agricultural biodiversity. To sustain a growing world population and to improve the sustainable use of land and water, therefore, better methods of soil management need to be developed and applied. The objective of this chapter is to review the effects of soil management on soil properties and processes to allow sustainable, productive agriculture to meet the needs of future generations.

Materials and Methods

The research results reviewed are primarily from North America and cover processes from the field to watershed scale. Results are predominantly from regions characterized by a humid temperate climate and an undulating landscape. Some important crops grown in this region are maize (*Zea mays* L.), soybeans (*Glycine max* (L.) Merr.), wheat (*Triticum aestivum* L.) and cotton (*Gossypium hirsutum* L.). The principles underpinning the analysis, however, apply globally. The chapter emphasizes effects of soil tillage on soil properties and processes, contrasting intensive inversion tillage with continuous no-tillage, while also discussing intermediary reduced-tillage systems. Besides tillage, a discussion is also provided on the importance of the return of crop residues to the soil, diverse crop rotations and continuous occupancy of the soil with living root systems to maintain environmental function and soil productivity.

A distinction is made between conservation tillage and conservation agriculture. Conservation tillage is a term that emphasizes only the type of tillage used and its effect on crop residue cover. No-tillage is one type of conservation tillage that was defined in the past as no tillage between harvest and planting (CTIC, 2004). The definition of no-tillage may have changed as in-crop cultivation became uncommon in the USA, but the emphasis of the term is still on not tilling. The terms no-tillage and conservation tillage lack two elements that are now considered fundamental for sustainable agriculture. First, they don't include the permanency of (no-) tillage. Secondly, crop diversity and the elimination of empty fallow periods is not part of the definitions. The modern concept of Conservation Agriculture includes both of these elements (Kassam *et al.*, 2009). The three components of conservation agriculture are (Kassam *et al.*, 2009):

(i) maintaining year-round organic cover over the soil, including cover crops and intercrops and/or the mulch provided by retaining residues from the previous crop; (ii) minimizing soil disturbance by tillage and thus seeding directly into untilled soil, eliminating tillage altogether once the soil has been brought to good condition, and minimizing soil disturbance from cultural operations; and (iii) diversifying crop rotations, sequences and associations adapted to local environmental conditions, and including appropriate legumes. Conservation agriculture emphasizes management practices that utilize ecological principles. It strives to build a soil that resembles that of a resilient forest floor, resulting in an optimum environment in the root zone to maximize water infiltration and root penetration, to increase biological activity and diversity to build soil architecture, to suppress plant pathogens, insect pests and weeds, to increase organic matter content, and to improve nutrient recycling. Higher infiltration and lower evaporation improve water use efficiency. Because of high diversity and low disturbance, pesticide use and the threat of herbicide resistance are reduced (Anderson, 2008). In contrast to many other conservation approaches, conservation agriculture allows productive but sustainable agriculture without taking land out of production; it also enables sustainable crop production on marginal lands, such as highly erodible lands, as will be illustrated in this chapter.

Results and Discussion

Soil management impacts on soil properties and processes

Soil is thoroughly modified to enable crop production, and this modification is more profound as soil is used more intensively. Removing natural vegetation, usually by fire, was necessary to start crop production in the early days of agriculture, and is still practised in many parts of the world today. No-tillage planting with a stick was common after natural forest was burned. This system was only possible at very low population pressure to accommodate fallow periods of 10–25 years to allow regeneration of natural vegetation and soil (Nye and Greenland, 1965). Long fallow periods became impossible as population pressure increased. New methods were needed to produce crops continuously on the same land. The Chinese, using very intensive methods of recycling and land care, were able to farm land continuously for 40 centuries (King, 1911). Similarly, the Egyptians farmed the borders of the Nile continuously for 50 centuries (James, 1979). Usually the land was prime land, with a regular supply of water, not excessively aggressive precipitation, and deep, fertile and level soils.

As land conversion continued, however, increasingly marginal land had to be reclaimed, and without proper management methods millions of acres were permanently converted into wastelands through salinization and erosion. Soil tillage was the major reason for soil degradation, resulting in continent-wide catastrophes such as the US Dust Bowl of the 1930s. This event was caused primarily by fence-row to fence-row tillage in the fragile prairie landscape (Bennett, 1947). The Dust Bowl led to the largest soil conservation effort in history. Conservation practices were designed to avoid losing soil productivity, but mouldboard ploughing was still considered a necessity for annual crop production so all conservation practices accommodated the use of this tillage tool (Bennett, 1947). During the same time, however, Faulkner (1943) started to question the rationale behind inversion tillage. Shallow tillage was still considered necessary, probably for weed control. The discovery and development of herbicides after World War II offered a new method of weed control which has revolutionized soil management. The adoption of conservation tillage, which includes no-tillage as well as reduced tillage systems that leave more than 30% of the crop residue at the soil surface after planting, increased dramatically. In 2004, conservation tillage was used on 40% of planted acres in the USA (CTIC, 2004). No-tillage, the most extreme form of conservation tillage, was used on 23% of planted acres. In 2005, no-tillage was used on almost 100 million ha worldwide, primarily in the Americas and Australia (Derpsch, 2008).

Management impacts on water erosion

Soil erosion by wind and water is recognized as the major form of soil degradation in the world, affecting 1.6 billion ha to varying degrees (Fu, 1989; Dregne, 1990, 1992; Bridges and Oldeman, 1999). Pimentel *et al.* (1995) estimated that one-third of the world's arable land had been lost as a result of erosion in the past 40 years, and that continued loss of productive farmland threatened future food production. It is now universally recognized that mulch cover is essential to control erosion (Shelton *et al.*, 2000). A threshold of 30% crop residue cover immediately after planting has been used in the USA as a minimum cover for controlling erosion (CTIC, 2004). This threshold is based on research studies showing that 30% residue cover can reduce soil erosion by approximately 70% compared with no cover (Fig. 13.1). Despite their importance, the levels of crop residues are usually insufficient for erosion control. Shelton *et al.* (1995) evaluated residue cover after planting maize into maize residue in 69 different tillage systems in Nebraska and found that only 24 resulted in more than 30% crop residue cover (Table 13.2). Most crops do not produce as much crop residue as maize. Most maize in the USA, for example, is rotated with the low-residue crop soybeans. In many other parts of the world, crop residue is customarily removed or burned. Therefore, to achieve or maintain at least 30% residue cover the use of no-tillage or some very low residue-disturbing tillage operation is necessary for erosion control.

Continuous no-tillage is needed to experience soil improvement. Long-term no-tillage increases near-surface organic matter

Table 13.2. Per cent crop residue cover for maize planted into maize residue in Nebraska after different tillage/planting combinations (without stalk chopping or knife fertilizer injection; averages of 2 years) (Source: Shelton *et al.*, 1995).

Tillage system	Crop residue cover (%)[a]
No-till, plant	56.0a
Blade plough, plant	41.3b
Field cultivate, plant	32.7bc
Till, plant	30.1c
Disc, plant	28.2cd
Disc, field cultivate, plant	20.1de
Blade plough, till, plant	19.7e
Chisel plough, disc, plant	19.1ef
Chisel plough (autumn), disc, plant	19.3ef
Disc, disc, plant	17.5fg
Disc (autumn), disc, plant	17.1g

[a]Treatments followed by the same letter are not significantly different by the *t*-test ($p < 0\ 1$).

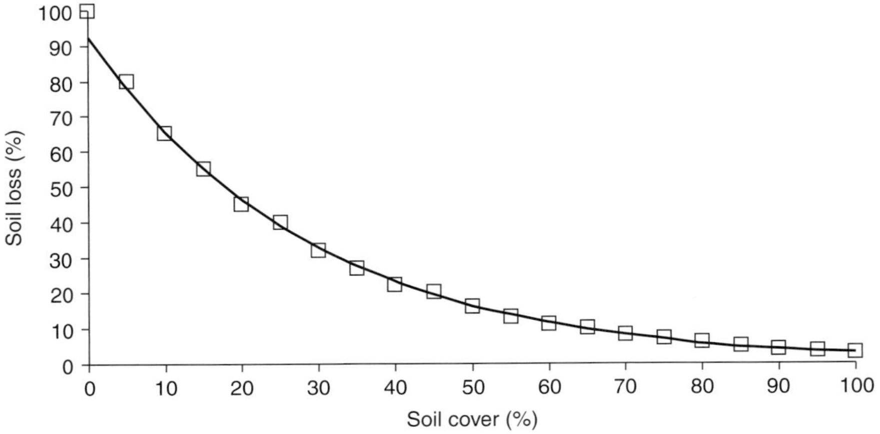

Fig. 13.1. Soil cover–soil loss relationship (data from Lyon *et al.*, 2000).

content and porosity, lowers bulk density and improves aggregate stability, leading to reduced water-dispersible clay, slaking and crusting, increased infiltration, and reduced runoff and erosion, even on highly erodible soils (Shipitalo and Edwards, 1998; Rhoton *et al.*, 2002). Besides increased resistance to inter-rill (sheet) erosion, no-tillage soil also becomes more resistant to rill erosion owing to increased consolidation and modification of surface properties (West *et al.*, 1992). It is important to avoid mixing of the shallow surface soil with lower layers (Franzluebbers, 2002). Even non-inversion tillage disturbs the surface layer and results in lower soil quality and increased susceptibility to degradation. In a study in the south-eastern USA, soil loss was three times higher from chisel/disced fields than from fields that had been under continuous no-tillage (Dabney *et al.*, 2004). Even after residue removal, soil loss was still three times greater in the tilled fields than in the no-till fields. The soil improvement with long-term no-tillage apparently increased the resiliency of the soil to erosion. One year after residue removal, however, soil erosion from the no-tillage field was equal to that from the tilled field. In other words, the improvement of surface soil properties disappeared in one year of fallow after residue removal. The combination of permanent no-tillage, sufficient crop residue return and occupation of the soil by living root systems is therefore important to achieve soil improvement and soil protection with conservation agriculture (see also Gilley and Doran, 1997; Rhoton *et al.*, 2002). Inversion tillage or lack of residue return can quickly reverse the soil improvement achieved by years of continuous no-till. In a no-tillage system where all or part of the crop residue is removed, including maize silage, a cover crop is essential.

The effect of conservation agriculture on soil loss varies among soil types. In a study in Pennsylvania, the effect of tillage system on soil erosion was determined at different times during the growing season on a well-drained and on a somewhat poorly drained soil (Verbree, 2008). On average, no-tillage reduced soil erosion by 93% on the well-drained soil, and by 78% on somewhat poorly drained soil (Table 13.3). The reduction in soil loss was notable throughout the season on the well-drained soil, but did not become significant until after harvest on the somewhat poorly drained soil. The benefits of conservation agriculture for erosion control can therefore be expected to be greatest on well-drained soils.

Surface structural stability and compacted subsoils affect the optimal type of conservation agriculture. Truman *et al.* (2009) determined the effects of tillage and cover crops on runoff and erosion of two sandy south-eastern

Table 13.3. Tillage effect on total sediment loss during a 1 h, 6 cm h^{-1} intensity rainfall simulation on a well-drained and a somewhat poorly drained soil in central Pennsylvania. Spring chisel/discing was compared with no-tillage in a maize silage system. A rye cover crop was used only in the no-tillage system. Both fields (slopes 7–8%) had been in continuous no-tillage production (predominantly maize) for at least 10 years before the study (Source: Verbree, 2008).

	Sediment loss (g m^{-2} h^{-1})			
	Well-drained Hagerstown silt loam		Somewhat poorly drained Buchanan silt loam	
Time of sediment loss	Long-term no-tillage	1-year-old chisel/ disc tillage	Long-term no-tillage	1-year-old chisel/ disc tillage
Immediately after planting	3.8[a]	24.6	4.6[a]	0.8
Mid-season	1.6[b]	28.2	0.3[a]	2.7
After silage harvest	1.0[b]	30.1	0.6[b]	22.3

[a]No significant effect of tillage (*F*-test, $p < 0.05$).
[b]Significant effect of tillage (*F*-test, $p < 0.05$).

coastal plain soils with naturally compacted subsoils (Compass and Tifton loamy sands), and on a Tennessee Valley (Decatur) silt loam soil degraded by years of intensive tillage and low residue return (Table 13.4). On the Compass loamy sand with natural hardpan, steady-state soil loss rates were highest with chisel/disc tillage. No-tillage led to a reduction of 65% in soil loss rates, and the inclusion of a black oat (*Avena strigosa* Schreb.) cover crop in the no-tillage system to an additional 25% reduction in soil loss. Autumn subsoiling with a paratill unit to 40 cm depth improved soil loss reduction only marginally. Clearly on this soil no-tillage and use of a cover crop were more important for soil conservation than subsoiling. On the Tifton loamy sand with natural hardpan, spring strip tillage in a killed rye cover crop resulted in steady-state soil loss reduction of 55% compared with conventional tillage. Strip tillage leaves a narrow strip of about 15 cm of bare soil into which cotton is planted, and the lower residue cover results in increased soil erosion potential compared with no-tillage. Strip tillage is therefore not recommended where rows go up and down slopes greater than 3%. Runoff was similarly affected to soil loss on these coastal plain soils (Table 13.4). Although subsoiling may not result in soil conservation benefits on these coastal plain soils, it is regularly performed to improve soil conditions for root growth and crop development.

The Decatur silt loam reacted differently to management (Table 13.4). This soil resembles an Oxisol with high content of iron oxides and stable soil structure under natural condition. Years of tillage for continuous cotton production, however, have resulted in weak surface structure. On this soil type, no-tillage alone resulted in similar final runoff rates to conventional disc tillage, and a cover crop did not improve infiltration. The greatest reduction of runoff and soil loss was obtained where autumn paratillage preceded a cover crop and cotton was planted using no-tillage in the spring. Apparently, the surface soil structure was largely destroyed owing to past management resulting in poor infiltration without tillage. Paratillage may become unnecessary on the Decatur silt loam soil once the surface soil structure has been restored through high residue return, use of cover crops and diverse crop rotations. This may not be possible on the coastal plain soils, which have inherently weak surface structure and hardpans at shallow depth that tend to re-form quickly after tillage (Fritton *et al.*, 1982).

In Ohio, measurements on small watershed scale confirmed that soil erosion on sloping land can be minimized if conservation tillage and cover crops are combined. At the Coshocton research station, soil loss was measured from small watersheds with slopes ranging from 7 to 13%. Soil loss decreased from 5.8 Mg ha^{-1} $year^{-1}$ in the

Table 13.4. Steady-state soil loss in 120 min rainfall simulation on three highly weathered Ultisols in the south-eastern USA (From Truman *et al.*, 2009).

	Steady-state soil loss rates (kg m^{-2} h^{-1})			Runoff rate after 2 h (mm h^{-1})		
Time of runoff	Compass loamy sand (Plinthic Paleudult)	Decatur silt loam (Rhodic Paleudult)	Tifton loamy sand (Plinthic Kandiudult)	Compass loamy sand (Plinthic Paleudult)	Decatur silt loam (Rhodic Paleudult)	Tifton loamy sand (Plinthic Kandiudult)
No-tillage	0.07	0.07		26	27	–
Conventional tillage	0.20	0.28	0.18	32	38	28
No-tillage + paratill + cover crop	0.01	0.02	–	2	4	–
No-tillage + cover crop	0.02	0.07	–	3	34	–
Strip tillage + cover crop	–	–	0.08	–	–	7

maize years in a maize-wheat-meadow rotation with conventional mouldboard plough tillage to 0.5 Mg ha^{-1} year^{-1} in a maize-soybean/cover crop rotation managed with no-tillage, para-tillage or chisel ploughing (Edwards *et al.*, 1993). Surface runoff also decreased when conservation tillage practices were used. On another watershed, also located in Coshocton, Ohio, soil loss during 4 years of no-tillage maize averaged 7 kg ha^{-1} year^{-1}, whereas it averaged 5335 kg ha^{-1} year^{-1} when the same watershed was managed with mouldboard ploughing under continuous maize (Shipitalo and Edwards, 1998).

If adopted on a large scale, conservation agriculture can dramatically affect sediment discharges from river watersheds. The impact of conservation tillage on sediment loss and runoff from large watersheds was studied in the Maumee River Basin in Ohio, Michigan and Indiana during a 26-year period (Myers *et al.*, 2000). Suspended sediment discharges in the Auglaize River decreased almost 50% during the period 1970–1998, which was attributed to a 65% increase in conservation tillage during that period. Similarly, in the Maumee River sediment discharge decreased by 11% while conservation tillage increased to 55%. Total water discharge was similar during this period, so natural climatic variations were not the likely explanation of the reduction in sediment discharge. The authors noted that the use of conservation tillage on relatively level, poorly drained, fine-textured soils with high runoff potential would be more beneficial for the reduction of sediment discharge at the outlet of large watersheds than the adoption of conservation tillage on coarser textured soils. The reason is that fine soil particles stay in suspension longer and can be transported large distances, whereas larger soil particles are deposited in farm fields and streams in the watershed and may not reach the watershed outlet. Similarly, crop residue cover and conservation tillage were found to be essential for soil conservation on heavy clay flatlands in the Mississippi watershed (Murphree and Mutchler, 1981). While medium and coarse soil particles were deposited in the watershed, suspended clay particles left a 260 ha watershed (Murphree *et al.*, 1985; Triplett and Dabney, 1999). These large watershed studies show the importance of the adoption of conservation tillage on heavy clay flatlands which might not normally be considered to have a soil conservation problem. It should also be noted that greater water quality benefits can be expected with conservation agriculture than with the use of conservation tillage practices such as chisel ploughing or discing.

Management impacts on tillage erosion

Soil properties are influenced by tillage itself owing to the mere operation of gravity in sloping landscapes, which moves the loosened soil downslope. Tillage erosion is defined as 'the loss and accumulation of soil resulting from the variable translocation of soil by tillage' (Lobb *et al.*, 1995). Tillage erosion only occurs on sloping land and leads to profile truncation on back slopes and profile burial on toe slopes. Tillage erosion rates are commonly as high as or higher than water erosion rates (range 3–600 Mg ha^{-1} year^{-1}; Lindstrom *et al.*, 1992; Van Oost *et al.*, 2006). The Canadian Agri-Environmental Indicator Project estimated that 50% of cropland in Canada was affected by unsustainable levels of tillage erosion (> 6 Mg ha^{-1} year^{-1}), whereas only 15% was subject to unsustainable water erosion rates (McRae *et al.*, 2000). Tillage erosion takes place: (i) because of a change in slope (topography-based tillage erosion; De Alba, 2001), and (ii) because of the effect of field boundaries (field boundary erosion). The former is important on mechanized farms with large fields without contour strips, whereas the latter is common on mechanized or animal-powered farms with contour strips or small fields. If tillage occurs on the contour, soil will be removed from the top and accumulate at the bottom of the cultivated strip. Over time, this may lead to the formation of terraces. However, undesirable soil variability will be introduced as a result of topsoil removal on the upslope side of the terrace, and accumulation on the downslope side. On convex slopes, tillage erosion is highest but water erosion is lowest, whereas on lower lying, concave parts of the landscape, water erosion is highest but tillage erosion is lowest. Tillage erosion therefore reinforces water erosion, by moving loose soil to parts of the field where rill and gully erosion are most intense (Kleinman *et al.*, 2006).

Tillage erosion is primarily a function of implement characteristics (tool shape, width, length), operation (depth, speed, direction), field shape and size, topography (slope gradient, curvature) and soil physical resistance to displacement (Van Muysen *et al.*, 2000; Van Oost *et al.*, 2006). The mouldboard plough is the most renowned for causing tillage erosion. Chisel ploughing, now more common in North America than mouldboard ploughing (in contrast to Europe), can result in similar tillage erosion, depending on the chisel point characteristics (Van Muysen *et al.*, 2000). Wide, twisted points or shovels will lead to higher tillage erosion than narrow, straight points. As the depth of chisel ploughing is typically shallower than that of mouldboard ploughing, it will usually result in less tillage erosion.

Secondary tillage operations cause tillage erosion as well, but, because of shallower depth of operation, soil loss rates are usually low compared with primary tillage. Animal-powered and manual tillage also cause tillage erosion, although rates are much lower than with tractor farming owing to shallower operation, slower speed and different implement characteristics (Van Oost *et al.*, 2006). However, manual or animal-powered tillage can also operate on very steep slopes, which leads to very high tillage erosion rates. Downslope mouldboard ploughing results in the highest tillage erosion rates, whereas upslope ploughing can move some soil upslope. Although there is usually a net movement of soil downslope, upslope-and-downslope tillage results in lower tillage erosion rates than downslope tillage alone. The depth of tillage has a large impact on tillage erosion, while speed is a less important factor. For example, increasing the tillage depth from 0.2 to 0.3 m can double tillage erosion (Van Oost *et al.*, 2006).

Tillage erosion is a major contributor to increased within-field variability, and leads to a lowering of productivity and nutrient use efficiency (Siebert and Scott, 1990; Bakker *et al.*, 2005). Tillage erosion causes a loss of humus-rich topsoil on convex slopes, and exposure of subsoils with undesirable characteristics on concave slopes and crests. For example, many Ultisols and Alfisols have silt-loam A horizons, but clay-loam or clay B horizons. When these subsoils are exposed, the productivity of the soil decreases. Calcitic subsoil horizons are common in areas with low rainfall. Soils with shallow fragipans or other hardpans are common in many parts of North America as well. When these subsoils are exposed, productivity decreases. Decreased depth to bedrock is another reason for decreased productivity. Soil accumulation in concavities results in richer soil in these locations, but this is unlikely to compensate for the loss of productivity on convexities. If tillage erosion is severe, unproductive subsoil may accumulate on top of deep topsoils in concave parts of the landscape, leading to further soil degradation. Although limited attention has been given to the effect of tillage erosion on hydraulic properties of the soil, it is likely that runoff increases where topsoil is removed owing to lower aggregate stability as a result of organic matter depletion and tillage disturbance.

Management impacts on soil organic carbon

Batjes (1999), using the FAO-UNESCO Soil Map of the World, estimated that the soils of the world contain 1468–1548 Pg organic carbon to a depth of 1 m. Soils therefore are a significant carbon pool besides the atmosphere (720 Pg), fossil deposits (5000 Pg), land plants (560 Pg) and oceans (38,000 Pg) (Schlesinger, 1995). Land-use change has contributed to the increase of carbon dioxide in the atmosphere, including deforestation, biomass burning, the conversion of natural to agricultural ecosystems, drainage of wetlands and tillage (Lal, 2004). Soil tillage is widely considered to be a major reason for the loss of organic matter in many of the world's soils (Reicosky *et al.*, 1995, 1997; Paul *et al.*, 1997; Lal *et al.*, 1998). A typical prairie soil lost almost 50% of its organic matter after 30 years of annual tillage (Donigian *et al.*, 1994). West and Post (2002), in a review of global trials, estimated that a change from conventional tillage to no-tillage can sequester 57 ± 14 g C m^{-2} $year^{-1}$ (excluding wheat-fallow systems), and that enhancing crop rotation complexity can sequester an average 20 ± 12 g C m^{-2} $year^{-1}$, excluding a change from continuous maize to a maize-soybean rotation. Crop rotation of cereals with legumes and forages and optimum fertilization practices were found to increase soil organic carbon (Halvorson *et al.*, 2002; Gregorich *et al.*, 2005). Eve *et al.* (2002) estimated that the increased adoption of conservation tillage, enrolment of cropland in the Conservation Reserve Program, cropping intensification (reduction of bare fallow) and improved grazing land management resulted in a net increase of 21.1 Tg soil carbon per year between 1982 and 1997.

The reasons for loss of organic matter after tillage include the physical release of CO_2 from soil pores and solution (Reicosky *et al.*, 1997), the exposure of previously protected organic matter to decomposition when aggregates are broken (Six *et al.*, 1999), the mixing of crop residues with soil (Reicosky *et al.*, 1995) and erosion losses of the topsoil (Lal *et al.*, 1998). Carbon dioxide losses after tillage can be extremely rapid. Reicosky *et al.* (1995) measured CO_2 losses exceeding the carbon content of the wheat crop residue in 19 days following autumn mouldboard ploughing. The type of tillage determines how much soil organic matter will be lost. The key determinant of carbon losses immediately after tillage is the volume of soil that is disturbed and connectivity of the soil pores with the atmosphere. Reicosky and Archer (2007), for example, measured two times higher carbon dioxide losses when ploughing at 20 cm depth instead of at 10 cm depth. If the soil is left rough and open to the atmosphere, carbon dioxide losses are much higher than if the soil is immediately smoothed with secondary tillage and the large pore cavities filled in (Fig. 13.2). Reduced tillage tools such as chisel and disc ploughs cause smaller losses of carbon dioxide shortly after tillage than does mouldboard ploughing.

Soil organic carbon content does not always increase with a change from conventional to no-tillage. In eastern Canada, greater organic carbon at the bottom of the plough layer of mouldboard ploughed soil compensated for lower organic carbon near the surface. The result was no difference in organic carbon content between soil managed with no-tillage and mouldboard ploughing (Gregorich *et al.*, 2005). Often, no significant difference in soil organic carbon content is observed when comparing reduced tillage with no-tillage (Franzluebbers and Arshad, 1996). Furthermore, shallow sampling may have caused overestimation of soil organic carbon gains with no-tillage (Baker *et al.*, 2007; Gal *et al.*, 2007). There is ample confirmation, however, of greater carbon accumulation near the soil surface with continuous no-tillage, which may be more critical to the soil's environmental function than the organic carbon content of the entire soil profile. Near-surface organic matter increases aggregate stability, improves water infiltration, favours the formation of continuous macropores, and acts as an organically bound slow-release fertilizer and a diverse food source for beneficial soil organisms (Franzluebbers, 2002). A soil organic matter stratification ratio of near-surface (0–5 cm) to below-surface (10–15 cm) organic matter may therefore be a better indicator to evaluate soil management practices than total organic matter content in the whole 0–15 cm depth (Franzluebbers, 2002; Sa and Lal, 2009).

Crop residue return is important to maintain soil organic carbon contents. In many integrated crop-livestock systems around the world, crop residues are used for feed or bedding. In dairy states such as Pennsylvania and New York, for example, approximately 30% of maize is harvested as silage, leading to almost complete removal of crop residues (NASS, 2005). Wheat, barley and oat straw is harvested for bedding or used on construction sites. In addition, there is now interest in using crop residues for fuel production. The US Departments of Agriculture and Energy evaluated the use of crop residue for biofuel production and estimated that almost one billion dry tons of biomass could be available from agricultural resources, potentially supplying 30% of current petroleum consumption (Perlack *et al.*, 2005). The study projected that up to 75% of crop residue could be removed sustainably if no-tillage was used on all cropland in the USA to protect soil from

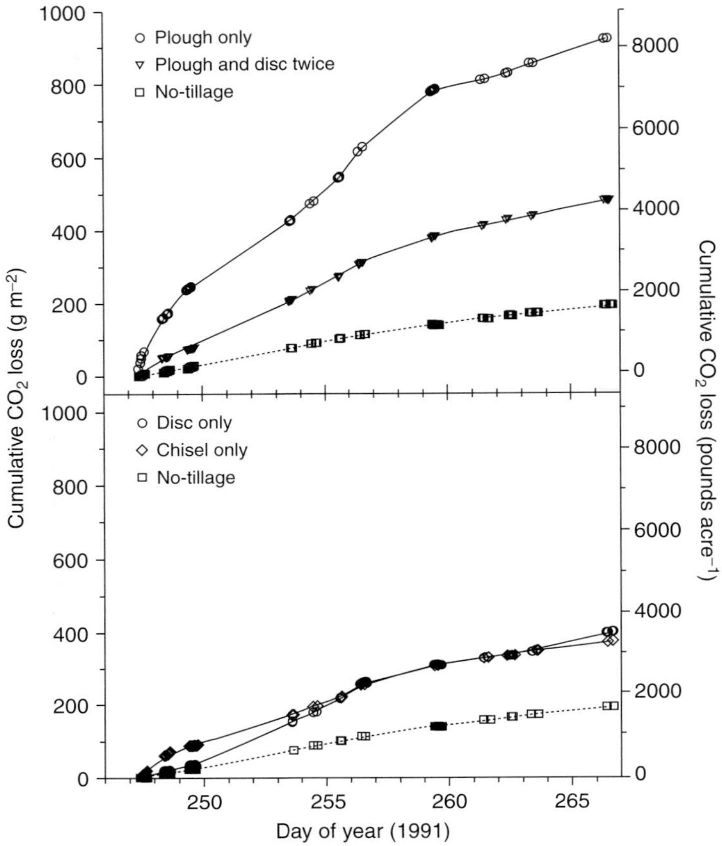

Fig. 13.2. Cumulative carbon dioxide loss after autumn tillage as affected by tillage type. Rapid loss occurs after tillage (Reicosky *et al.*, 1995).

erosion. Most of the residue to produce biofuel would come from maize. The study did not consider the need to return crop residue to soil to maintain carbon content. Johnson *et al.* (2006) estimated that minimum quantities of above-ground carbon input would be 2.5 ± 1.0 Mg C ha^{-1} year^{-1} if mouldboard ploughing was used and 1.8 ± 0.44 Mg C ha^{-1} year^{-1} if no-tillage or chisel ploughing was used. Based on this study, Wilhelm *et al.* (2007) compared the minimum amount of maize stover needed to maintain soil organic carbon content with that needed to provide water or wind erosion protection. In continuous maize cropping systems, a minimum of 7.58 Mg maize stover ha^{-1} year^{-1} needed to be retained if mouldboard ploughing was used, in contrast to 5.25 Mg ha^{-1} year^{-1} if conservation tillage practices were used. In maize-soybean cropping systems, 12.50 Mg maize stover ha^{-1} year^{-1} needed to be retained in the maize years of the rotation if mouldboard ploughing was used and 7.90 Mg ha^{-1} year^{-1} if conservation tillage was used. The average US maize yield was 8.75 Mg ha^{-1} (dry matter basis) in 2009, an all-time record. At a harvest index of 0.53, crop residue production was 7.76 Mg ha^{-1}. These results suggest that, if soil organic carbon content needs to be maintained or increased, no maize residue should be harvested. Environmental and agronomic functions of the soil might be negatively affected if organic carbon content decreases, especially in non-prime soils, as was shown in a study in Ohio. Two years of stover removal in 8-year-old continuous no-tillage fields reduced soil organic carbon content in a silt loam soil and reduced crop yield on a steeply sloping silt loam soil (Blanco-Canqui and Lal, 2007). In conclusion, crop residue is vital to maintain organic carbon content and most of it should be left in the field to maintain soil function.

Management impacts on infiltration and percolation

Crop residues are the first level of defence to water runoff. Raindrops may carry high kinetic energy, breaking down soil aggregates and causing seal formation on a bare soil surface. The seal, though thin, has very low hydraulic conductivity, causing water to run off instead of infiltrating. Residue mulch will break the force of raindrops and slow the formation of a surface seal. The importance of crop residue for infiltration was studied on an Oxisol in Parana, Brazil (Roth *et al.*, 1988). From 0 to 6 Mg ha^{-1} soybean mulch was applied to 7-year-old tillage systems of double-cropped soybean/wheat. Simulated rainfall was applied at 6 cm h^{-1} for one hour. Infiltration was highly correlated with mulch rate and not affected by tillage system (Fig. 13.3). On this Oxisol, which had high hydraulic conductivity, infiltration was primarily controlled by the formation of a surface seal. To accommodate 100% infiltration, 100% crop residue cover was needed in this study. The desired mulch levels were therefore 4 Mg ha^{-1} for wheat and 6 Mg ha^{-1} for soybeans. However, soybean or wheat residue production was only 2.5 Mg ha^{-1} and 1.5 Mg ha^{-1}, respectively. Roth *et al.* (1988) therefore stressed the importance of using permanent no-tillage to leave all crop residue at the soil surface, but also recommended a change in crop rotation to grow more high residue-producing crops such as maize, and cover crops such as black oats or cereal rye (*Secale cereale* L.) during fallow periods. The infiltration benefit of crop residue cover on this soil is characteristic of well-drained soils with good internal drainage that are sensitive to sealing and crusting (Andraski *et al.*, 2003; Bosch *et al.*, 2005; Leys, *et al.*, 2007; Zhang *et al.*, 2007). The fact that Oxisols, which are known for their high structural stability due to high concentrations of iron oxides, can become sensitive to crusting after years of intensive tillage suggests that most well-drained soils will show this response to mulching.

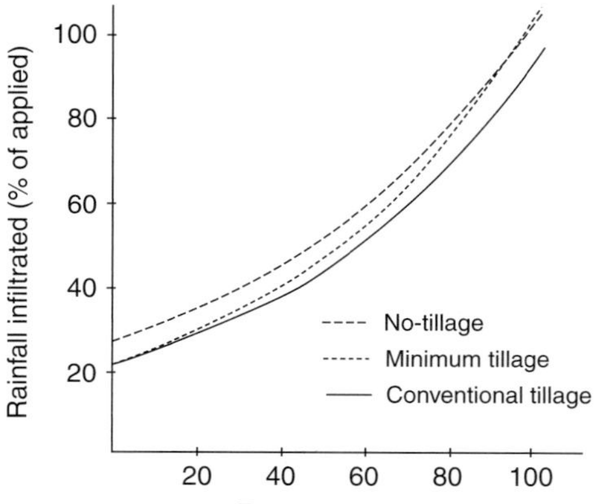

Fig. 13.3. Total infiltration as a percentage of applied rainfall versus degree of soil cover for conventional tillage, minimum tillage and no-tillage. Rainfall applied was 60 mm in one hour (reproduced with permission from Roth *et al.*, 1988).

Some studies show no or limited infiltration benefits of no-tillage. In many cases, low infiltration is caused by low levels of crop residue (e.g. Gomez *et al.*, 1999). In other cases, the research method used may explain the lack of response. For example, the use of infiltration rings, which exclude the impact of raindrops (e.g. Lal *et al.*, 1989; Lal and VanDoren., 1990), has little relevance to infiltration under natural rainfall. Also, rainfall simulation studies shortly after tillage do not reflect infiltration dynamics over the entire growing season. Sealing and reconsolidation usually occur after a few rainstorms and tillage effects may be reversed (e.g. Dabney *et al.*, 2004; Wilson *et al.*, 2004). If infiltration is governed by limited water transmission to the subsoil, however, situations can arise where the impact of residue on surface properties would have little consequence for infiltration because no further infiltration is possible. These conditions have been termed 'saturation excess' infiltration, which is common when soil is saturated as a result of excess precipitation, poor drainage or slow percolation (Dunne and Black, 1970).

Because of the balance between 'saturation excess' and 'infiltration excess' conditions, differences in infiltration due to tillage are commonly larger on well-drained soils than on poorly drained soils (Table 13.5).

The decrease of infiltration after tillage is generally slower in well-structured clay or clay loam soils than in structureless soils (Stroosnijder and Hoogmoed, 1984; Karunatilake and Van Es, 2002; Tang *et al.*, 2006). Structureless soils may be single-grained sandy soils, sodic soils or soils subject to many years of intensive tillage. The effect of tillage on infiltration also depends on climatic conditions and rainfall characteristics (Hoogmoed and Stroosnijder, 1984; Truman *et al.*, 2009). In periods with water deficit (evapotranspiration > precipitation), no-tillage tends to increase infiltration over tillage, whereas during periods of water excess infiltration differences are usually small or absent. No-tillage increases infiltration more over tillage at high rainfall intensity than at low rainfall intensity. The increase in infiltration with use of continuous no-tillage may be absent in level soils with poor internal drainage, such as clay soils with high swell/shrink capacity (Lal *et al.*, 1989). However, over time, earthworm populations may increase in no-tillage soil, which may improve the internal drainage of clay soils with low permeability (Shipitalo *et al.*, 2000).

At the watershed scale, effects of no-tillage on runoff may be even more pronounced than those observed in small-plot simulated rainfall studies. Runoff represented only 0.2% of precipitation from a no-tillage watershed in Ohio during 18 years of data collection (never more than 1% during any particular year), but it represented 17% of precipitation from a tilled watershed over a 4-year period (Shipitalo and Edwards, 1998). Runoff differences due to tillage vary during the year depending on local climatic conditions (Fig. 13.4). During the winter and autumn in the US Midwest, when evapotranspiration is less than precipitation and low-intensity cyclonic rainfall or snowfall is common, no difference in runoff between tillage systems is expected. In the late spring and early summer, when high-intensity convectional storms are common and evapotranspiration is greater than precipitation, runoff remains low with no-till, but increases with conventional tillage because of soil sealing and crusting. A similar response can be expected on well-drained soils and on soils with a seasonally high water table, because seasonally high water tables typically dissipate in late spring. Percolation may be affected more by tillage than by runoff or infiltration owing to a reduction in evaporation in no-tillage soil besides an increase in infiltration. Deep percolation, however, may not be affected by tillage because infiltration differences due to tillage occur during water deficit periods when little deep percolation occurs. The danger of increased groundwater contamination from surface-applied materials in no-tillage is therefore considered

Table 13.5. Long-term tillage effect on runoff during a 1 h, 6 cm h^{-1} intensity rainfall simulation on a well-drained and a somewhat poorly drained soil in central Pennsylvania (Source: Verbree, 2008).

	Runoff (cm)			
	Well-drained Hagerstown silt loam		Somewhat poorly drained Buchanan silt loam	
Tillage type	Long-term no-tillage	Chisel/disc	Long-term no-tillage	Chisel/disc
Immediately after planting	1.48[a]	1.89	2.2[b]	0.12
Mid-season	1.42[b]	5.34	1.23[a]	1.85
After silage harvest	0.33[b]	3.28	0.26[b]	2.9

[a]No significant effect of tillage (*F*-test, $p < 0.05$).
[b]Significant effect of tillage (*F*-test, $p < 0.05$).

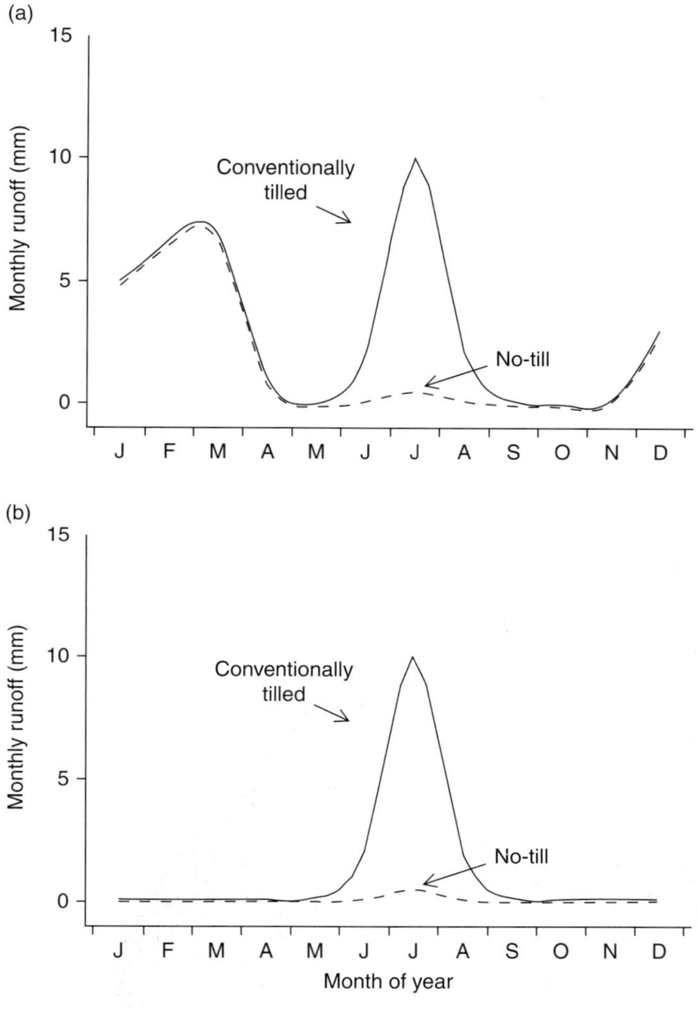

Fig. 13.4. Conceptualized effect of tillage on monthly surface runoff on well-drained soils (a) and on soils with restricted drainage (b) (reproduced with permission from Shipitalo *et al.*, 2000).

to be limited in humid temperate regions (Shipitalo *et al.*, 2000), except where earthworm burrows are immediately connected to drainage tile (Shipitalo and Gibbs, 2000).

Management impacts on soil structure

Long-term use of conservation agriculture leads to soil improvement. The protection of the surface by residue reduces sealing and crusting. The surface impact of high-intensity storms is greatly moderated in comparison with the bare tilled soil. The higher biological activity and concentration of organic matter in the surface soil under continuous no-tillage leads to improved aggregation and macroporosity. Fungal biomass has been found to be much greater in the surface of no-tillage soils than in conventionally tilled soils (Frey *et al.*, 1999), and biological activity in continuous no-tillage soil is commonly higher (Andrade *et al.*, 2003). Fine root mass is concentrated near the surface of no-tillage soils (Ball-Coelho *et al.*, 1998). Earthworm populations are typically higher in permanent no-tillage than in conventionally

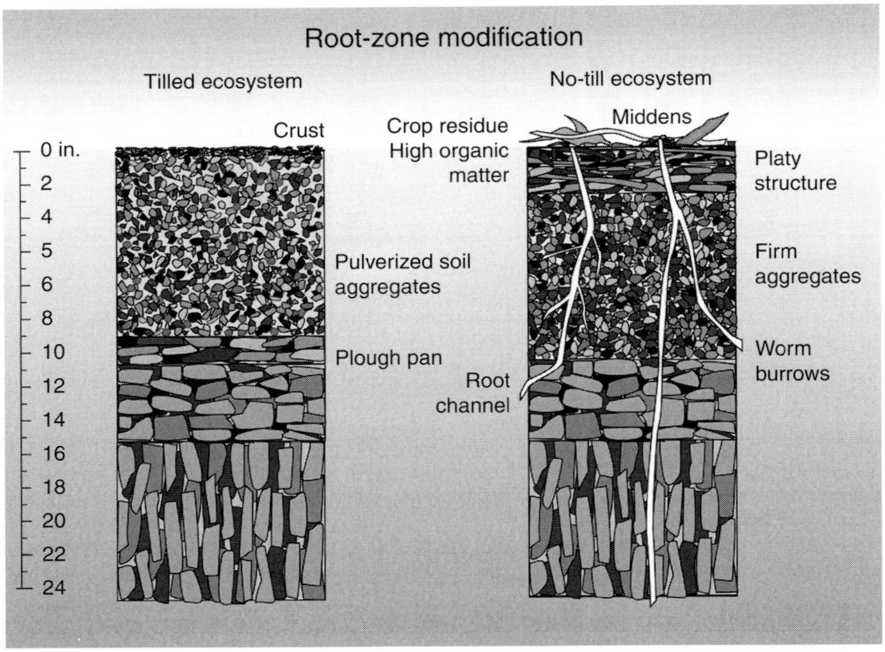

Fig. 13.5. Root-zone modification as affected by long-term usage of tillage systems (Duiker and Myers, 2005).

tilled soils (Kladivko *et al.*, 1997). All of these factors lead to greater surface aggregate stability and better soil tilth in long-term no-tillage soil (Tisdall and Oades, 1982). Macropores connected to the soil surface are more numerous in long-term no-tillage than in tilled soil, especially as a result of the effect of deep-burrowing earthworms. The night crawler, *Lumbricus terrestris*, creates 50–100 cm deep vertical burrows in the soil profile, which are capped by a mixture of casts, stable aggregates and crop residue that night crawlers accumulate above their burrows. Because the night crawler relies on surface residue for its livelihood, it may be absent in clean-tilled soil (Edwards and Lofty, 1982; Kladivko *et al.*, 1997). The root zone and soil profile, therefore, would completely change with long-term no-tillage compared with continuous tillage (Fig. 13.5).

Soil management for sustainable crop production

This review has shown many benefits of high crop-residue cover, continuous no-tillage, diverse crop rotations and continuous occupation of the soil with living crops. Soil erosion by water can be reduced, tillage erosion minimized, organic carbon losses reversed, soil biological activity restored and water infiltration increased. There are other factors, however, that will determine the adoption of these practices by farmers. A major concern would typically be how crop yields are affected by a new production practice. In particular, there continues to be a concern that crop yields may be reduced with no-tillage. This concern was addressed in a review of tillage effects on crop yields in 61 experiments that compared maize yields and 43 full-season soybean trials in the USA and Canada. On average, the review showed that crop yields did not vary between conventional tillage and no-tillage (Defelice *et al.*, 2006; Fig. 13.6 and Table 13.6).

Closer investigation showed that no-tillage soybean yields were higher or similar to those achieved with conventional tillage in all regions, soil types and crop rotations, except in the northern part of the USA and Canada, where a 4% yield reduction was observed (Defelice *et al.*, 2006). Similar results were

(a)

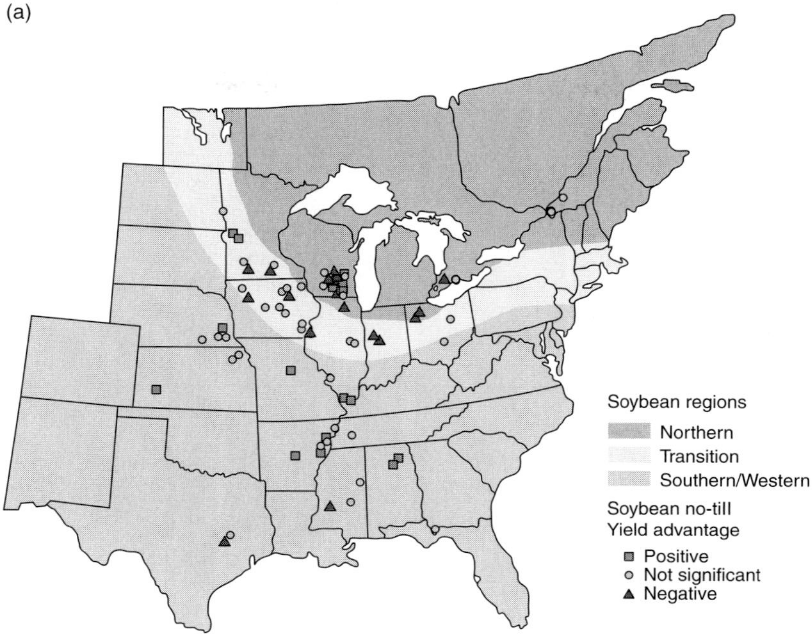

(b)

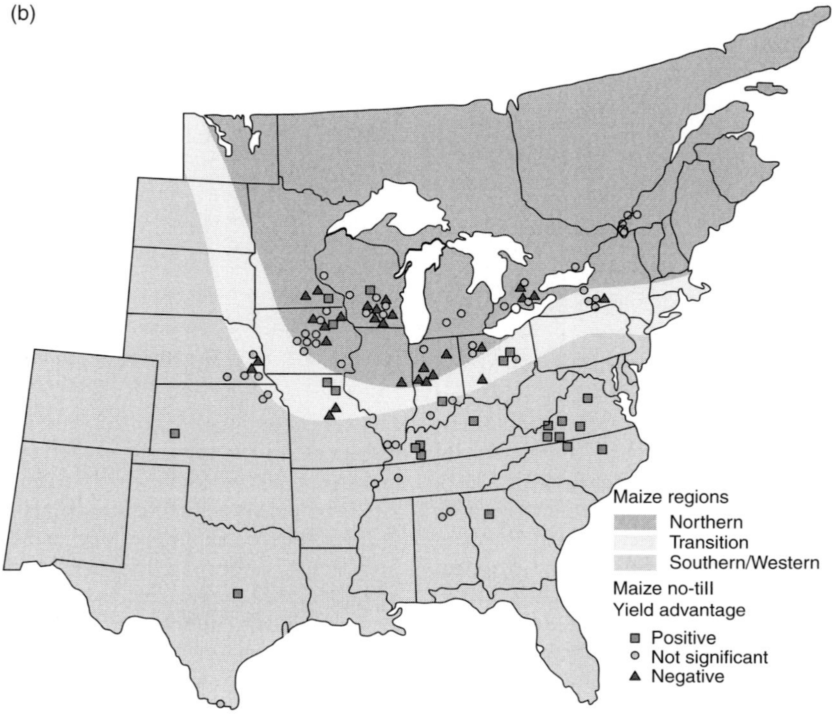

Fig. 13.6. No-tillage soybean (a) and maize (b) yield advantage in tillage trials in North America (reproduced with permission from Defelice *et al.*, 2006).

Table 13.6. Maize and soybean yield advantage of no-tillage over conventional tillage (Reproduced with permission from Defelice *et al.*, 2006).

		% Yield advantage of no-till (no. of comparisons)	
		Maize	Soybean
All experiments		−0.5 (104)	0.7 (67)
Experiments by geographical location	Southern/Western	12.2 (26)	5.0 (26)
	Transition	−1.8 (16)	−0.6 (22)
	Northern	−5.5 (62)	−3.9 (19)
Experiments by soil drainage	Moderate/well drained soils	2.0 (64)	2.0 (41)
	Poorly drained soils	−4.5 (40)	−1.4 (26)
Experiments by crop rotation	Maize-soybean rotation	1.9 (38)	0.7 (40)
	Continuous	−1.5 (60)	1.1 (23)

obtained with maize, which also showed yield reductions with no-tillage on poorly drained soil. Closer inspection of maize yields showed average yield increases of 7–13% with no-tillage in southern/western regions (where moisture stress is common), and maize yield reductions of 2.6% on poorly drained soils, 4% in continuous maize and 4–8% in no-tillage under different conditions in the northern regions (Defelice *et al.*, 2006; Fig. 13.6). The authors noted that in many studies no-tillage had been implemented for only a short time, and often under less than optimal management. They also noted that farmers consider many factors besides yield in a decision to practice no-tillage: management costs, farm size, labour and time constraints, and soil conservation. In some cases, government programmes make the switch to no-tillage a minor economic issue.

If needed, reduced no-tillage yields in northern regions or on poorly drained soils can be addressed with some form of strip tillage in which no more than 25% of the surface of soil is disturbed (Morrison and Sandabria, 2002; Archer and Reicosky, 2009). In other cases, specific soil structural problems caused by years of intensive tillage and reduced residue return may necessitate some shallow vertical tillage which fractures soil with minimal disturbance of the soil surface (Raper, 2000a,b; Truman *et al.*, 2009). These practices have to be combined with intensive and diverse crop rotations and/or cover cropping that return crop residue and root matter to the soil (Motta *et al.*, 2007; Truman *et al.*, 2009). Once soil structure has improved as a result of the use of these practices, it is probably unnecessary to continue the use of vertical tillage on these soils.

Other soils may suffer from naturally occurring hardpans that readily re-form, such as sandy coastal plain soils in the south-eastern USA (Truman *et al.*, 2009). Reconsolidation of the hardpans within one season may necessitate occasional use of vertical tillage on these soils (Torbert and Reeves, 1991; Reeves *et al.*, 1992; Raper *et al.*, 1994; Raper and Reeves, 2007). Again, these practices need to be combined with continuous crop residue cover, and diverse crop rotations and/or cover crops to return large quantities of crop residue to the soil.

Summary

The consequences of limited land and water resources and a growing world population make it necessary to develop and apply high-yielding yet sustainable cropping systems. Because nearly all prime land is already in production, cropping systems need to be developed for good and marginal lands as well. These lands often suffer from constraints that make them more susceptible to degradation, such as soil erosion by wind and water, tillage erosion, organic matter decline, soil structural degradation, and decline in biological activity and diversity leading to increased runoff and reduced percolation.

Unfortunately, unsustainable land use is widespread, leading to degradation of soil, water and air, and the disappearance of natural areas necessary for maintaining biological diversity.

The development and application of cropping systems with diverse crop rotations, in which year-round crop residue covers the soil, and soil disturbance by tillage is eliminated, or limited to below-surface fracturing, will be essential to our future. These concepts are embedded in conservation agriculture systems. Research studies have clearly shown the benefits of these systems, including their ability to produce crops economically on all soil types and under all climatic conditions. However, many factors preclude rapid adoption of conservation agriculture. These include the lack of a comprehensive vision for this practice among agronomists, engineers and soil scientists. This translates into a totally inadequate effort to improve on current practices for weed, disease and insect management, nutrient management (including manure), and the design of no-tillage planters, drills and harvesting machinery.

Field-wide tillage continues to be promoted in extension programmes, causing farmers to question the viability of conservation agriculture. Policies are in place favouring the status quo of tillage and monoculture cropping systems. Research in conservation agriculture systems (especially continuous no-till) still receives inadequate support. Hopefully, this chapter will strengthen support for conservation agriculture among leaders in science and technology, for it is essential to meeting the demands on agriculture for production and environmental conservation. This vision should translate in investments in agricultural research to improve conservation agriculture technology and in extension systems that help farmers to implement them. Current policies that discourage farmers from adopting conservation agriculture should be replaced by policies that encourage it. If these factors are addressed, rapid progress is possible, leading to a positive turnaround in environmental performance of current crop production systems.

References

Anderson, R. (2008) Diversity and no-till: keys for pest management in the U.S. Great Plains. *Weed Science* 56, 141–145.

Andrade, D.S., Colizzi-Filho, A. and Giller, K.E. (2003) The soil microbial community and soil tillage. In: El Titi, A. (ed.) *Soil Tillage in Agroecosystems*. CRC Press, Boca Raton, Florida, pp. 51–81.

Andraski, T.W., Bundy, L.G. and Kilian, K.C. (2003) Manure history and long-term tillage effects on soil properties and phosphorus losses in runoff. *Journal of Environmental Quality* 32, 1782–1789.

Archer, D.W. and Reicosky, D.C. (2009) Economic performance of alternative tillage systems in the northern corn belt. *Agronomy Journal* 101, 296–304.

Baker, J.M., Ochsner, T.E., Venterea, R.T. and Griffis, T.J. (2007) Tillage and soil carbon sequestration: what do we really know? *Agriculture, Ecosystems and Environment* 118, 1–5.

Bakker, M.M., Govers, G., Kosmas, C., Vanacker, V., Van Oost, K. and Rounsevell, M. (2005) Soil erosion as a driver of land-use change. *Agriculture Ecosystems and Environment* 105, 467–481.

Ball-Coelho, B.R., Roy, R.C. and Swanton, C.J. (1998) Tillage alters corn root distribution in coarse-textured soil. *Soil and Tillage Research* 45, 237–249.

Batjes, N.H. (1999) *Management Options for Reducing CO_2 Concentrations in the Atmosphere by Increasing Carbon Sequestration in the Soil*. ISRIC (World Soil Information), Wageningen, The Netherlands.

Bennett, H.H. (1947) *Elements of Soil Conservation*, 1st edn. McGraw-Hill, New York.

Blanco-Canqui, H. and Lal, R. (2007) Soil and crop response to harvesting corn residues for biofuel production. *Geoderma* 141, 355–362.

Bosch, D.D., Potter, T.L., Truman, C.C., Bednarz, C.W. and Strickland, T.C. (2005) Surface runoff and lateral subsurface flow as a response to conservation tillage and soil-water conditions. *Transactions of the ASAE* 48, 2137–2144.

Bridges, E.M. and Oldeman, L.R. (1999) Global assessment of human-induced soil degradation. *Arid Soil Research and Rehabilitation* 13, 319–325.

CTIC (2004) *Crop Residue Management Survey.* Conservation Technology Information Center, West Lafayette, Indiana.

Dabney, S.M., Wilson, G.V., McGregor, K.C. and Foster, G.R. (2004) History, residue, and tillage effects on erosion of loessal soil. *Transactions of the ASAE* 47, 767–775.

De Alba, S. (2001) Modelling the effects of complex topography and patterns of tillage and soil translocation by tillage with mouldboard plough. *Journal of Soil and Water Conservation* 56, 335–345.

Defelice, M.S., Carter, P.R. and Mitchell, S.B. (2006) Influence of tillage on corn and soybean yield in the United States and Canada. *Crop Management* (online) doi:10.1094/CM-2006-0626-01-RS.

Derpsch, R. (2008) No-tillage and conservation agriculture: a progress report. In: Goddard, T., Zoebisch, M.A., Gan, Y., Ellis, W. Watson, A. and Sombatpanit, S. (eds) *No-till Farming Systems.* Special Publication No. 3, World Association of Soil and Water Conservation, Bangkok, pp. 7–39.

Donigian, A.S. Jr, Barnwell, T.O., Jackson, R.B., Patwardhan, A.S., Weinreich, K.B., Rowell, A.L., Chinnaswamy, R.V. and Cole, C.V. (1994) *Assessment of Alternative Management Practices and Policies Affecting Soil Carbon in Agroecosystems of the Central United States.* Report EPA/600/4-94/067, US Environmental Protection Agency, Athens, Georgia.

Dregne, H.E. (1990) Erosion and soil productivity in Africa. *Journal of Soil and Water Conservation* 43, 1–436.

Dregne, H.E. (1992) Erosion and soil productivity in Asia. *Journal of Soil and Water Conservation* 47, 8–13.

Duiker, S.W. and Myers, J. (2005) *Steps Toward a Successful Transition to No-till.* Agricultural Research and Cooperative Extension, College of Agricultural Sciences, The Pennsylvania State University, University Park, Pennsylvania.

Dunne, T. and Black, R.D. (1970) An experimental investigation of runoff production in permeable soils. *Water Resources Research* 6, 478–490.

Edwards, C.A. and Lofty, J.R. (1982) The effect of direct drilling and minimal cultivation on earthworm populations. *Journal of Applied Ecology* 19, 723–734.

Edwards, W.M., Triplett, G.B., Van Doren, D.M., Owens, L.B., Redmond, C.E. and Dick, W.A. (1993) Tillage studies with a corn-soybean rotation: hydrology and sediment loss. *Soil Science Society of America Journal* 57, 1051–1055.

Eswaran, H., Beinroth, F. and Reich, P. (1999) Global land resources and population supporting capacity. *American Journal of Alternative Agriculture* 14, 129–136.

Eve, M.D., Sperow, M., Paustian, K., Follett, R.F., Mickler, R.A. and McNutty, S.G. (2002) National-scale estimation of changes in soil carbon stocks on agricultural lands. *Environmental Pollution* 116, 431–438.

Faulkner, E.H. (1943) *Plowman's Folly.* Grosset and Dunlap, New York.

Franzluebbers, A.J. (2002) Soil organic matter stratification ratio as an indicator of soil quality. *Soil and Tillage Research* 66, 95–106.

Franzluebbers, A.J. and Arshad, M.A. (1996) Soil organic matter pools with conventional and zero tillage in a cold, semiarid climate. *Soil and Tillage Research* 39, 1–11.

Frey, S.D., Elliott, E.T. and Paustian, K. (1999) Bacterial and fungal abundance and biomass in conventional and no-tillage agroecosystems along two climatic gradients. *Soil Biology and Biochemistry* 31, 573–585.

Fritton, D.D., Swader, F.N. and Hoddinott, K. (1982) Profile modification persistence in a fragipan soil. *Soil Science* 136, 124–130.

Fu, B. (1989) Soil erosion and its control in the loess plateau of China. *Soil Use and Management* 5, 76–82.

Gal, A., Vyn, T.J., Micheli, E., Kladivko, E.J. and McFee, W.W. (2007) Soil carbon and nitrogen accumulation with long-term no-till versus moldboard plowing overestimated with tilled-zone sampling depths. *Soil and Tillage Research* 96, 42–51.

Gilley, J.E. and Doran, J.W. (1997) Tillage effects on soil erosion potential and soil quality of a former Conservation Reserve Program site. *Journal of Soil and Water Conservation* 52, 184–188.

Gomez, J.A., Giraldez, J.V., Pastor, M. and Fereres, E. (1999) Effects of tillage method on soil physical properties, infiltration and yield in an olive orchard. *Soil and Tillage Research* 52, 167–175.

Gregorich, E.G., Rochetter, P., VandenBygaart, A.J. and Angers, D.A. (2005) Greenhouse contributions of agricultural soils and potential mitigation practices in Eastern Canada. *Soil and Tillage Research* 83, 53–72.

Halvorson, A.D., Wienhold, B.J. and Black, A. (2002) Tillage, nitrogen and cropping system effects on soil carbon sequestration. *Soil Science Society of America Journal* 66, 906–912.

Hassan, R., Scholes, R. and Ash, N. (eds) (2005). *Ecosystems and Human Well-being. Volume I: Current Status and Trends. Findings of the Condition and Trends Working Group, Millennium Ecosystem Assessment.* Island Press, Washington, DC.

Hoogmoed, W.B. and Stroosnijder, L. (1984). Crust formation on sandy soils in the Sahel. I. Rainfall and infiltration. *Soil and Tillage Research* 4, 5–23.

James, T.G.H. (1979) *An Introduction to Ancient Egypt.* Farrar Staus Giroux, New York.

Johnson, J.M.F., Allmaras, R.R. and Reicosky, D.C. (2006) Estimating source carbon from crop residues, roots and rhizode position using the national grain-yield database. *Agronomy Journal* 98, 622–636.

Karunatilake, U.P. and Van Es, H.M. (2002) Rainfall and tillage effects on soil structure after alfalfa conversion on a clay loam soil in New York. *Soil and Tillage Research* 67, 135–146.

Kassam, A., Friedrich, T., Shaxson, F. and Pretty, J. (2009) The spread of conservation agriculture: justification, sustainability and uptake. *International Journal of Agricultural Sustainability* 7, 292–320.

King, F.H. (1911) *Farmers of Forty Centuries or Permanent Agriculture in China, Korea and Japan.* Rodale Press, Emmaus, Pennsylvania.

Kladivko, E.J., Akhouri, N.M. and Weesies, G. (1997) Earthworm populations and species distributions under no-till and conventional tillage in Indiana and Illinois. *Soil Biology and Biochemistry* 29, 613–615.

Kleinman, P.J.A., Srinivasan, M.S., Dell, C.J., Schmidt, J.P., Sharpley, A.N. and Bryant, R.B. (2006) Role of rainfall intensity and hydrology in nutrient transport via surface runoff. *Journal of Environmental Quality* 35, 1248–1259.

Lal, R. (2004) Soil carbon sequestration impacts on global climate change and food security. *Science* 304, 1623–1627.

Lal, R., and VanDoren, D.M. Jr (1990) Influence of 25 years of continuous corn production by three tillage methods on water infiltration for two soils in Ohio. *Soil and Tillage Research* 16, 71–84.

Lal, R., Logan, T.J. and Fausey, N.R. (1989) Long-term tillage and wheel traffic effects on poorly drained mollic Ochraqualf in northwest Ohio. 2. Infiltrability, surface runoff, sub-surface flow and sediment transport. *Soil and Tillage Research* 14, 359–373.

Lal., R., Kimble, J.M., Follett, R.F. and Cole, C.V. (1998) *The Potential of U.S. Cropland to Sequester Carbon and Mitigate the Greenhouse Effect.* Sleeping Bear Press, Chelsea, Michigan.

Leys, A., Govers, G., Gillijns, K. and Poesen, J. (2007) Conservation tillage on loamy soils: explaining the variability in interrill runoff and erosion reduction. *European Journal of Soil Science* 58, 1425–1436.

Lindstrom, M.J., Nelson, W.W. and Schumacher, T.E. (1992) Quantifying tillage erosion rates due to moldboard plowing. *Soil and Tillage Research* 24, 243–255.

Lobb, D.A., Kachanoski, R.G. and Miller, M.H. (1995) Tillage translocation and tillage erosion on shoulder slope landscape positions measured using Cx-137 as a tracer. *Canadian Journal of Soil Science* 75, 211–218.

Lyon, D., Smith, J. and Fryrear, D. (2000) Wind erosion. In: Reeder, R. (ed.) *Conservation Tillage Systems and Management*, MWPS-45. MidWest Plan Service, Ames, Iowa, pp. 11–15.

McRae, T., Smith, C.A.S. and Gregorich, L.J. (2000) *Environmental Sustainability of Canadian Agriculture*: Report of the Agri-Environmental Indicator Project. Agriculture and Agri-Food Canada, Ottawa, Ontario.

Morrison J.E. Jr and Sanabria, J. (2002) One-pass and two-pass spring strip tillage for conservation row-cropping in adhesive clay soils. *Transactions of the ASAE* 45, 1263–1270.

Motta, A.C.V., Reeves, D.W., Burmester, C. and Feng, Y. (2007) Conservation tillage, rotations, and cover crop affecting soil quality in the Tennessee Valley: particulate organic matter, organic matter, and microbial biomass. *Communications in Soil Science and Plant Analysis* 38, 2831–2847.

Murphree, C.E. and Mutchler, C.K. (1981) Sediment yield from a flatland watershed. *Transactions of the ASAE* 24, 966–969.

Murphree, C.E., Mutchler, C.K. and McGregor, K.C. (1985) Sediment yield from a 259-ha flatlands watershed. *Transactions of the ASAE* 28, 1120–1123.

Myers, D.N., Metzker, K.D. and Davis, S. (2000) *Status and Trends in Suspended-sediment Discharges, Soil Erosion, and Conservation Tillage in the Maumee River Basin – Ohio, Michigan, and Indiana.* Water-Resources Investigations Report 00-4091, US Department of the Interior and US Geological Survey, Denver, Colorado.

NASS (National Agricultural Statistics Services) (2005) *Pennsylvania Agricultural Statistics 2004–2005.* USDA National Agricultural Statistics Services, Pennsylvania Field Office, Harrisburg, Pennsylvania.

Nye, P.H. and Greenland, D.J. (1965) *The Soil under Shifting Cultivation.* Technical Communication No. 51, Commonwealth Bureau of Soils, Harpenden. Commonwealth Agricultural Bureaux, Farnham Royal, UK [now CAB International, Wallingford, UK].

Paul, E.A., Paustian, K., Elliott, E.T. and Cole, C.V. (eds) (1997) *Soil Organic Matter in Temperate Agroecosystems: Long-term Experiments in North America.* CRC Press, Boca Raton, Florida.

Perlack, R.D., Wright, L.L., Turhollow, A.F., Graham, R.L., Stokes, B.J. and Erbach, D.C. (2005) Biomass as feedstock for a bioenergy and bioproducts industry: The technical feasibility of a billion-ton annual supply. US Department of Energy/US Department of Agriculture. DOE/GO-102005-2135/ORNL/TM-2005/66. Available at: http://www1.eere.energy.gov/biomass/pdfs/final_billionton_vision_report2.pdf (accessed 25 February 2011).

Pimentel, D., Harvey, C., Resodudarmo, P., Sinclair, K., Kurz, D., McNari, M., Crist, S., Sphritz, L., Fitton, L., Saffouri, R. and Blair, R. (1995) Environmental and economic costs of soil erosion and conservation benefits. *Science* 267, 1117–1123.

Ramankutty, N., Evan, A.T., Monfreda, C. and Foley, J.A. (2008) Farming the planet: 1. Geographic distribution of global agricultural lands in the year 2000. *Global Biogeochemical Cycles* 2(1), GB1003.

Raper, R.L. and Reeves, D.W. (2007) In-row subsoiling and controlled traffic effects on coastal plain soils. *Transactions of the ASABE* 50, 1109–1115.

Raper, R.L., Reeves, D.W., Burt, E.C. and Torbert, H.A. (1994) Conservation tillage and traffic effects on soil condition. *Transactions of the ASAE* 37, 763–768.

Raper, R.L., Reeves, D.W., Schwab, E.B. and Burmester, C.H. (2000a) Reducing soil compaction of Tennessee Valley soils in conservation tillage systems. *Journal of Cotton Science* 4, 84–90.

Raper, R.L., Reeves, D.W., Burmester, C.H. and Schwab, E.B. (2000b) Tillage depth, tillage timing, and cover crop effects on cotton yield, soil strength, and tillage energy requirements. *Applied Engineering in Agriculture* 16, 379–385.

Reeves, D.W., Rogers, H.H., Droppers, J.A., Prior, S.A. and Powell, J.B. (1992) Wheel-traffic effects on corn as influenced by tillage system. *Soil and Tillage Research* 23, 177–192.

Reicosky, D.C. and Archer, D.W. (2007) Moldboard plow tillage depth and short-term carbon dioxide releases. *Soil and Tillage Research* 94, 109–121.

Reicosky, D.C., Kemper, W.D., Langdale, G.W., Douglas C.L. Jr and Rasmussen, P.E. (1995) Soil organic matter changes resulting from tillage and biomass production. *Journal of Soil and Water Conservation* 50, 253–261.

Reicosky, D.C., Dugas, W.A. and Torbert, H.A. (1997) Tillage-induced soil carbon dioxide loss from different cropping systems. *Soil and Tillage Research* 41, 105–118.

Rhoton, F.E., Shipitalo, M.J. and Lindbo, D.L. (2002) Runoff and soil loss from Midwestern and southeastern US silt loam soils as affected by tillage practice and soil organic matter content. *Soil and Tillage Research* 66, 1–11.

Roth, C.H., Meyer, B., Frede, H.-G. and Derpsch, R. (1988) Effect of mulch rates and tillage systems on infiltrability and other soil physical properties of an Oxisol in Parana, Brazil. *Soil and Tillage Research* 11, 81–91.

Sa, J.C. de M. and Lal, R. (2009) Stratification ratio of soil organic matter pools as an indicator of carbon sequestration in a tillage chronosequence on a Brazilian Oxisol. *Soil and Tillage Research* 103, 46–56.

Schlesinger, M.E. (1995) An overview of the global carbon cycle. In: Lal, R., Kimble, J., Levine, E. and Stewart, B.A. (eds) *Soils and Global Change.* CRC/Lewis Publishers, Boca Raton, Florida.

Scholes, R., Hassan, R. and Ash, N.J. (2005) Summary: ecosystems and their services around the year 2000. In: Hassan, R., Scholes, R. and Ash, N. (eds) *Ecosystems and Human Well-being. Volume I: Current Status and Trends. Findings of the Condition and Trends Working Group, Millennium Ecosystem Assessment.* Island Press, Washington DC, pp. 1–23.

Shelton, D.P., Dickey, E.C., Kachman, S.D. and Fairbanks, K.T. (1995) Corn residue cover on the soil surface after planting for various tillage and planting systems. *Journal of Soil and Water Conservation* 50, 399–404.

Shelton, D., Jasa, P., Brown, L. and Hirschi, M. (2000) Water erosion. In: Reeder, R. (ed.) *Conservation Tillage Systems and Management.* MWPS-45. MidWest Plan Service, Ames, Iowa, pp. 17–22.

Shipitalo, M.J. and Edwards, W.M. (1998) Runoff and erosion control with conservation tillage and reduced-input practices on cropped watersheds. *Soil and Tillage Research* 46, 1–12.

Shipitalo, M.J. and Gibbs, F. (2000) Potential of earthworm burrows to transmit injected animal wastes to tile drains. *Soil Science Society of America Journal* 64, 2103–2109.

Shipitalo, M.J., Dick, W.A., Edwards, W.M. and Addiscott, T.M. (2000) Conservation tillage and macropore factors that affect water movement and the fate of chemicals. *Soil and Tillage Research* 53, 167–183.

Siebert, S.F. and Scott, T.W. (1990) Influence of topsoil removal and fertilizer application on peanut yields from an Indonesian Ultisol. *Agriculture, Ecosystem and Environment* 32, 213–221.

Six, J., Elliott, E.T. and Paustian, K. (1999) Aggregate and soil organic matter dynamics under conventional and no-tillage systems. *Soil Science Society of America Journal* 63, 1350–1358.

Stroosnijder, L. and Hoogmoed, W.B. (1984) Crust formation on sandy soil in the Sahel. II. Tillage and its effect on the water balance. *Soil and Tillage Research* 4, 321–337.

Tang, Z., Lei, T., Yu, J., Shainberg, I., Mamedov, A.I., Ben-Hur, M. and Levy, G.J. (2006) Runoff and interrill erosion in sodic soils treated with dry PAM and phosphogypsum. *Soil Science Society of America Journal* 70, 679–690.

Tisdall, J.M. and Oades, J.M. (1982) Organic matter and water-stable aggregates in soil. *Journal of Soil Science* 33, 141–163.

Torbert, H.A. and Reeves, D.W. (1991) Tillage and traffic effects on cotton yield and N requirement. In: *Proceedings of the Beltwide Cotton Producers Research Conference*, San Antonio, Texas. National Cotton Council, Memphis, Tennessee, pp. 931–935.

Triplett, G.B. and Dabney, S.M. (1999) Soil erosion and soybean production. In: Heatherly, L.G. and Hodges, H.F. (eds) *Soybean Production in the Midsouth.* CRC Press, Boca Raton, Florida, pp. 19–39.

Truman, C.C., Shaw, J.N., Flanagan, D.C., Reeves, D.W. and Ascough, J.C. III (2009) Conservation tillage to effectively reduce interrill erodibility of highly-weathered Ultisols. *Journal of Soil and Water Conservation* 64, 265–275.

Van Muysen, W., Govers, G., Van Oost, K. and Van Rompaey, A. (2000) The effect of tillage depth, tillage speed, and soil condition on chisel tillage erosivity. *Journal of Soil and Water Conservation* 55, 355–364.

Van Oost, K., Govers, G., De Alba, S. and Quine, T.A. (2006) Tillage erosion: a review of controlling factors and implications for soil quality. *Progress in Physical Geography* 30, 443–466.

Verbree, D.A. (2008) Sediment and nutrient losses in reduced tillage systems on dairy farms. MS thesis, The Pennsylvania State University, University Park, Pennsylvania.

West, L.T., Miller, W.P., Bruce, R.R., Langdale, G.W., Laflen, J.M. and Thomas, A.W. (1992) Cropping system and consolidation effects on rill erosion in the Georgia Piedmont. *Soil Science Society of America Journal* 56, 1238–1243.

West, T.O. and Post, W.M. (2002) Soil organic carbon sequestration rates by tillage and crop rotation: a global data analysis. *Soil Science Society of America Journal* 66, 1930–1946.

Wilhelm, W.W., Johnson, J.M.F., Karlen, D.L. and Lightle, D.T. (2007) Corn stover to sustain soil organic carbon further constrains biomass supply. *Agronomy Journal* 99, 1665–1667.

Wilson, G.V., Dabney, S.M., McGregor, K.C. and Barkoll, B.D. (2004) Tillage and residue effects on runoff and erosion dynamics. *Transactions of the ASAE* 47, 119–128.

Zhang, G.S., Chan, K.Y., Oates, A., Heenan, D.P. and Huang, G.B. (2007) Relationship between soil structure and runoff/soil loss after 24 years of conservation tillage. *Soil and Tillage Research* 92, 122–128.

14 Land Use and Agricultural Management Systems: Effects on Subsurface Drain Water Quality and Crop Yields

Allah Bakhsh and Ramesh S. Kanwar*

Introduction

The global population will increase to about 9 billion by the year 2050 (Kanwar, 2010). This will increase the stress on available land and water resources enormously, and the preservation of the quality of water bodies will require well thought out and sound policies. The impact of this increased population will also be severe on the environmental quality of land and water resources. Science-based information will be needed to understand the processes involved in land and water resource degradation that results from intensive agriculture, especially in areas of the world where major water bodies are showing signs of hypoxia (Alexander *et al.*, 1995). The development of sustainable agricultural production systems will be necessary to restore already damaged ecosystems, and sound principles of science and technology will need to be applied to minimize further environmental degradation.

Land use and localized management systems in agricultural watersheds have affected both surface and subsurface hydrological processes. These changes in hydrological processes have profound effects on the quality of the environment at local, regional and global scales. Hydrological processes at the landscape scale control the movement of agricultural chemicals that cause non-point source pollution of rivers and groundwater systems. Land use also affects the physical, biological and chemical properties of soils, and evapotranspiration regimes, which eventually affect the hydrological response of a watershed. Therefore, it is very important to understand local land-use practices in watersheds before determining the response of these watersheds as regards water balance and non-point source pollution.

Iowa's landscape has changed significantly over the past 150 plus years, and the effect of these changes on hydrological regimes is an excellent example to study. Therefore, this chapter focuses on various studies conducted in Iowa on the effects of land-use practices on subsurface hydrology and non-point source pollution (Bakhsh *et al.*, 2010). About 150 years ago, less than 3% of Iowa's land was under production agriculture. With the clearing of wooded areas and prairies, the installation of an extensive network of subsurface drains and open ditches, and the straightening of many perennial streams from 1910 to the 1930s, much of Iowa's landscape was altered and converted to agricultural use. Although this brought the benefits of agricultural production and an enhanced local economy, modification of the local and

* Corresponding author: rskanwar@iastate.edu

regional hydrology contributed to increased occurrences of peak flows from agricultural lands, resulting in increased soil erosion and direct transfer of non-point source pollutants to the Missouri and Mississippi Rivers In addition, an impact of intensive agricultural and animal production systems on the quality of surface and groundwater resources in Iowa has been observed (Baker *et al.*, 2005).

Non-point source nutrient pollution has been recognized as one of the major environmental and social issues in Iowa and the US Midwest for several reasons. First, excessive use of nutrients in agricultural watersheds can have serious impacts on the quality of surface and groundwater resources. Secondly, several states, including Iowa, are in the process of creating laws to reduce nitrate-nitrogen (NO_3-N) and phosphorus (P) loading from mineral fertilizers and manure to soil and water resources. Thirdly, the pollution of water by nutrients from manure supplied to croplands will determine environmental indicators for developing public policies on the management of manure.

During the past 30 plus years, agricultural and livestock production systems in Iowa have changed significantly (Bakhsh *et al.*, 2005). Today's animal production systems have become larger, and the public is concerned about the effects of animal production facilities on surface and groundwater quality. Of particular concern are surface runoff losses of N in the forms of NH_4-N, NO_3-N and organic-N, runoff losses of P, phosphate-phosphorus (PO_4-P) and organic-P, and leaching losses of NO_3-N, PO_4-P, silica and bacteria to groundwater. NH_4-N at concentrations of >2.0 mg l^{-1} can result in fish kills, and PO_4-P at levels as low as 0.05 mg l^{-1} can promote the growth of algae and speed up the process of eutrophication in lakes and reservoirs. In addition, several other water quality problems are emerging in Iowa and the rest of the Midwest; in particular, the presence of endocrine and pharmaceutically active chemicals in drinking water supplies and waste water and the implications of this pose new challenges to drinking water supplies.

Developing and implementing management strategies in watersheds to control water pollution efforts require two things: (i) the development and evaluation of on-the-farm management systems that have the potential to maintain sustainable economic yields and reduce water pollution, and (ii) large-scale implementation of best management practices in watersheds/landscapes, or the adoption of integrated farming systems that reflect significant reductions in N and P loadings in the water bodies. Previous strategies for addressing water quality problems from agricultural and livestock production systems have focused primarily on preventing point discharges of animal waste pollutants from confinement areas to water bodies (Gupta *et al.*, 2004), but more evidence is coming to light that animal manure applied to land has much broader water quality implications than just point source discharges (Power *et al.*, 2001; Bakhsh *et al.*, 2010).

Tillage disturbs the land surface and has been reported to affect the hydrological cycle, N use efficiency and the leaching of NO_3-N to subsurface drainage water (Randall and Iragavarapu, 1995; Kanwar *et al.*, 1997; Power *et al.*, 2001). Crop residues left under a no-tillage system affect evaporation, runoff, infiltration rates and tile flows through preferential flow paths such as cracks and wormholes (Gupta *et al.*, 2004). Compared with no-tillage, a conventional tillage system, such as mouldboard and chisel ploughing, incorporates crop residues, disturbs the soil structure and may impede water movements through the soil profile. The chisel plough system has also been reported to increase oxidation and N-mineralization processes in the soil profile (Green and Blackmer, 1995; Green *et al.*, 1995; Power *et al.*, 2001; Dinnes *et al.*, 2002). The effects of tillage on NO_3-N leaching losses for soils underlain by subsurface drainage systems become more complex when combined with a maize(corn)-soybean rotation and the application of organic/inorganic fertilizer.

In addition to land use and local management practices, weather and rainfall patterns (especially wet and dry weather cycles) affect the leaching of nutrients/agricultural chemicals through the soil profile and to water bodies. These effects have temporal variability due to varying climatic conditions, and become more complicated when they are combined with the effects of spatial variability of land and topography of the area. Bakhsh and Kanwar (2004) reported that land-use management practices based on spatial zones of the landscape and on topographic attributes help reduce NO_3-N leaching losses to subsurface drainage water. Changing climatic conditions also necessitate study of the effects of management on a long-term basis to understand and mitigate NO_3-N leaching losses to subsurface drainage water.

Hence, the objectives of this chapter are to evaluate the effects of long-term tillage, crop rotation, and manure and commercial fertilizer management systems on non-point source pollution (especially NO_3-N leaching losses to subsurface drain water) and crop yields. The evaluation is based on the results of a three-phase, long-term experimental study in 1979–2005. In the first phase of the experiments, a study was conducted between 1979 and 1992 to determine the long-term effects of tillage and crop rotation on crop yields, subsurface hydrology and water quality. In this study, all tillage and crop rotation (continuous maize and maize-soybean rotation) practices were established in 1979, but data on water quality were collected only from 1990 to 1992. The second phase of the experiments, from 1993 to 1998, was a 6-year study that investigated the effects of tillage, N management and crop rotation on crop yields and subsurface water hydrology. The third phase of the experiments was conducted from 2001 to 2005, and aimed to determine the effects of N-management practices (with a major focus on the application of swine manure) on NO_3-N leaching losses to subsurface drain water in the maize-soybean rotation systems. This chapter presents a summary of these long-term studies conducted in Iowa to evaluate the effects of several land-use practices and N-management systems on subsurface hydrology (specifically, subsurface drain flow) and non-point source pollution.

Materials and Methods

Study area

Field experiments were conducted at Iowa State University's north-eastern research centre near Nashua, Iowa. The soils at the site are Floyd loam (fine-loamy, mixed, mesic Aquic Hapludolls), Kenyon loam (fine-loamy, mixed, mesic Typic Hapludolls) and Readlyn loam (fine-loamy, mixed, mesic Aquic Hapludolls) (Kanwar *et al.*, 1997). These soils have 3 to 4% organic matter and belong to the Kenyon-Clyde-Floyd soil association; they are moderately well to poorly drained and lie over loamy glacial till. The soils have a seasonally high water table and benefit from an improved subsurface drainage system (Voy, 1995).

The research site has 36 plots of 0.4 ha (58.5 × 67 m in size each), with fully documented tillage and cropping records for the past 30 years (since 1979). The subsurface drainage system was installed in 1979 on all 36 plots. Experimental treatments on tillage (four tillage systems – mouldboard ploughing, chisel ploughing, ridge tillage and no-till), crop rotation and chemical management practices were also established in 1979 on each plot, with three replications of each treatment. Surface and subsurface water quality monitoring equipment was installed on the plots in 1988 (Kanwar *et al.*, 1999). No tile flow was observed during 1988 and 1989 because of dry conditions. Therefore, field data on tile flow, NO_3-N concentrations and leaching losses, and maize-soybean grain yields were available for analysis from 1990 through to 2005. The field experiments were conducted in three phases (1990 to 1992, 1993 to 1998 and 2001 to 2005) using a randomized complete block design, with three replications (as already stated). The years 1999 and 2000 were excluded from the analysis because they covered a transition period in which the continuous maize treatments were eliminated, the number of no-tillage plots was reduced and new manure treatments were added at the site.

Experimental treatments from 1990 to 2005

In the first phase of the study, from 1990 to 1992, two tillage systems (chisel plough versus no-till) and three crop rotations (continuous maize, maize after soybean, and soybean after maize) were arranged to evaluate their effects on tile flow, NO_3-N concentrations in drainage water, NO_3-N leaching losses in drainage water and grain yields. Anhydrous ammonia was pre-plant knife injected into the soil at a rate of 202 and 168 kg N ha^{-1} in the spring for the continuous maize and maize-soybean rotation treatments, respectively, for both tillage systems. Plots were chisel ploughed in the autumn after harvest and the field was cultivated in the spring. In both tillage systems there were either one or two cultivations during the growing season for weed control. The maize variety Golden Harvest H2343[1] and the soybean variety Sands of Iowa 237[1] were grown from 1990 to 1992 (Kanwar *et al.*, 1997).

In the second phase, from 1993 to 1998, the major focus was on the N-management system, tillage system (no-till and chisel plough) and crop rotation (continuous maize, maize after soybean, soybean after maize). Continuous maize under a no-till system was eliminated. The continuous maize treatment with chisel ploughing received 135 kg N ha^{-1} of urea ammonium nitrate (UAN) in the spring before planting. Similarly, the maize after soybean plots with both chisel plough and no-till systems received UAN at 110 kg N ha^{-1} before planting. The single pre-plant spring application of UAN was made using a spoke injector, which injected at 200 mm intervals and 250 mm from the maize rows (Baker *et al.*, 1989). Maize was planted in rows spaced 750 mm apart into a seedbed prepared by autumn chisel ploughing and field cultivating in the spring. Soybeans were drilled, no-till seeded, in 200 mm rows into maize stover from the previous year without any additional fertilizer application. The same varieties of maize (Golden Harvest 2343[1]) and soybeans (Sands of Iowa 237[1]) were planted during the 6-year experiment (Bakhsh *et al.*, 2000).

In the third phase of the study, from 2001 to 2005, the continuous maize treatments were eliminated altogether and manure treatments were included in the maize-soybean rotation system. This resulted in six treatments with the maize-soybean rotation system. Two of six treatments, with chisel ploughing, received spoke-injected UAN at 168 kg N ha^{-1} in the spring before planting during the maize phase of production. Liquid swine manure was obtained from a growing/finishing building. The application of manure to achieve the required N application rate of 168 kg N ha^{-1} was difficult owing to the non-uniform quality of the manure, which resulted in variable volatilization rates. The other problems associated with manure application at the site have been discussed by Karlen *et al.* (2004). Liquid swine manure was injected in the autumn for chisel ploughed plots and in the spring for plots with no-tillage. Spring manure application to no-tillage plots was preferred over autumn application to minimize nutrient losses. Maize plots with chisel ploughing and no-tillage received N application at an average rate of 178 and 177 kg N ha^{-1}, respectively. Fertilizer application rates are summarized in Table 14.1. Maize was planted in rows spaced 750 mm apart after preparing the seedbed with a field cultivator. Soybeans were drilled in 200 mm rows directly into maize stover from the previous year. Maize variety of NK 45-T5[1] (Monsanto Corp., St Louis, Missouri) at the rate of 80,160 seeds per ha^{-1} and soybean variety Asgrow 2105[1] or Kruger 2525[1] at the rate of 494,000 seeds ha^{-1} were planted from 2001 to 2005. Primary tillage was performed using a chisel plough in the autumn after the maize harvest. Secondary tillage was field cultivation before planting and during the plant growth period to control weeds. Weeds were controlled satisfactorily with herbicides and field cultivation. Maize and soybean grain yields were collected from each plot, tested for moisture content, and adjusted to a constant water content of 155 g kg^{-1} (15.5%) for maize and 130 g kg^{-1} (13%) for soybeans. Grain yield for each plot was measured using a commercial combine with all stover left in the field (Bakhsh *et al.*, 2005).

Subsurface drainage system and sampling procedure

Each plot at the site is drained separately and has subsurface drainage lines installed in the centre of the plot at a depth of 1.2 m below the ground surface, with a drain spacing of 28.5 m.

Table 14.1. Fertilizer application rates (kg ha^{-1}) to maize-soybean rotation plots for manure and UAN treatments, 2001–2005.

	Year					
Fertilizer[a]	2001	2002	2003	2004	2005	Average
Autumn manure application						
N	171	220	141	N/A	182	178
P_2O_5	169	162	71	N/A	75	134
K_2O	180	175	111	N/A	118	146
Spring manure						
N	123	224	213	N/A	148	177
P_2O_5	101	95	129	N/A	147	118
K_2O	146	154	136	N/A	110	137
Urea ammonium nitrate solution (UAN)						
Maize						
N	168	168	168	168	169	168
P_2O_5	67	67	67	67	67	67
K_2O				120		
Soybean						
P_2O_5		49	45	49	49	48
K_2O		159		134	134	

[a]Potentially available N from manure during first cropping season = 50% (Total Kjeldahl Nitrogen – NH_3-N) + NH_3-N; 202 and 168 kg N ha^{-1}, pre-plant anhydrous ammonia were applied to continuous maize and rotated maize plots, respectively, from 1990 to 1992; 135 and 110 kg N ha^{-1}, pre-plant urea ammonium nitrate (UAN) solution were applied to continuous maize and rotated maize plots, respectively, from 1993 to 1998.

Cross-contamination of plots was avoided by installing subsurface drainage lines on the northern and southern borders of each plot and isolating the eastern and western borders with berms (Kanwar *et al.*, 1999). The central subsurface drainage lines are intercepted at the end of the plots and are connected to individual sumps for measuring drainage effluents and collecting water samples for chemical analysis. The sumps are equipped with a 110 volt effluent pump, a water flow meter and an orifice tube to collect water samples. Data-loggers, connected to the water flow meters, record subsurface drainage flow continuously as a function of time. Composite water samples were collected for NO_3-N analysis using an orifice tube located on the discharge pipe of the sump pump. Approximately 0.2% of the water pumped from the sump flowed through a 5 mm diameter polyethylene tube to a water sampling bottle located in the collection sump each time the pump operated. Cumulative subsurface drain flows were recorded, and sampling bottles were removed two times per week from mid-March to the beginning of December during the study period. A more detailed description of the automated subsurface drainage system installed can be found in Kanwar *et al.* (1999).

The water samples collected for NO_3-N analysis were analysed spectrophotometrically using a Lachat Model AE ion analyser (Lachat Instruments, Milwaukee, Wisconsin[1]). The cumulative NO_3-N leaching loss with subsurface drainage water was calculated in kg N ha^{-1} by multiplying the NO_3-N concentrations in mg l^{-1} with the drainage effluent in mm; dividing this cumulative NO_3-N loss by the amount of drainage effluent for the entire monitoring season gave the flow-weighted NO_3-N concentrations for each year.

Statistical analysis

Subsurface drainage flow, NO_3-N leaching loss, flow-weighted NO_3-N concentrations, and maize and soybean yields were statistically analysed using data from the randomized complete block design. The PROC analysis of variance (ANOVA) procedure in SAS version 9.1 for windows (SAS Institute, 2003) was used, and ANOVA tables are reported for maize and soybean yields. A least significant difference (LSD) test at $p \leq 0.05$ was used to compare treatment means and evaluate treatment effects on subsurface water quality and crop yields.

Results and Discussion

Subsurface hydrology (tile flow) and nitrate nitrogen leaching losses from 1990 to 1992

The amount and distribution of precipitation during the growing season from March to November had a major impact on subsurface drainage 'tile' volume and NO_3-N leaching losses to tile water. The years 1990 and 1991 were wet, with growing season rainfall amounts of 1048 and 967 mm, respectively, compared with the 30-year average growing seasonal rainfall of 771 mm (Table 14.2) for the study area (Voy, 1995). The rainfall of 752 mm in 1992 was less than normal. Although the rainfall in 1990 was more than that in 1991 and 1992, a higher tile flow volume (subsurface drainage flow) of 261 mm was observed in 1991 (compared with 189 mm for 1990 and 113 mm for 1992) (Table 14.3). The year 1990 had heavy rainfall of 340 mm (236 + 104 mm) in the month of July; this was 227% more than the normal monthly rate. The year 1991, however, had more rainfall in the spring season (April and May) when crops were not growing actively and crop water requirements were low. This showed that not only the amount of rainfall but also its temporal distribution during the growing season had a pronounced effect on tile flow volume.

Treatment effects on tile flow volume were mostly non-significant for 1990 except for rotation effects with no-tillage; no-till continuous maize (CC) had a maximum tile flow volume of 274 mm, which was significantly ($p \leq 0.05$) greater than those from no-till rotated maize (158 mm) and rotated soybean with a chisel plough tillage system (157 mm) (Table 14.3). Continuous maize with no-tillage had significantly ($p \leq 0.05$) greater tile flow volumes than those from rotated maize with the no-till system from 1990 to 1992. When averaged across years (1990–1992), though, tillage effects on tile flow volume were not significant. Although no-till CC plots gave 35% higher tile flow volume than CC plots with chisel ploughing (263 versus 195 mm), the difference was not statistically significant ($p \leq 0.05$). Rotation effects with a no-till system were significant. However, a no-till rotation produced less tile flow volume from rotated maize plots than from CC plots (132 versus 263 mm). This showed that differences may not only be associated with crop but may also be due to rainfall and spatial variability effects, which might have resulted in higher tile flow volume for CC plots. No-till plots had higher tile flow because of the development of macropores and preferential flow paths in the soil profile and because of less evaporation owing to crop residue left on the soil surface (Kanwar *et al.*, 1997; Bakhsh *et al.*, 2005).

No-till continuous maize increased leaching loss 50 to 75% when compared with rotated no-till maize-soybean from 1990 to 1992 (Table 14.4). Similarly, rotation effects for a chisel plough system were also significant. Chisel ploughing with CC plots had twofold higher NO_3-N leaching losses than those from rotated maize and soybean plots. Tillage effects on NO_3-N leaching losses to tile flow with CC plots were not significant and similar for all years except the maize-soybean rotation in 1992. Overall, CC plots had almost twofold higher NO_3-N leaching losses than did rotated maize and soybean under both tillage systems because of receiving N applications each year and because of a higher N application rate. This difference was more pronounced for the no-tillage system owing to higher tile flows.

Flow-weighted NO_3-N concentrations (FWNC) are considered to be a better indicator for assessing the contamination level in the event that tile flow water joins water bodies (Bjorneberg *et al.*, 1998; Jaynes *et al.*, 1999). Tillage and crop rotation effects on FWNC were highly significant ($p \leq 0.01$) for all 3 years (1990–1992). On average, chisel ploughing with CC had significantly higher

Table 14.2. Growing season rainfall data for above and below (–) normal monthly rainfall (mm) from 1990 to 2005.

	Year																
Month[a]	1990	1991	1992	1993	1994	1995	1996	1997	1998	1999	2000	2001	2002	2003	2004	2005	Normal[a]
March		–13	–20	14	–48	–2	–13	–3	9	–44	–42	–13	–40	–23	–56	–42	54
April	15	85	4	1	–29	14	–59	–21	–19	51	–35	–24	22	11	–43	–28	87
May	0	66	–56	–12	–39	–48	1	–32	20	181	–1	39	–34	–10	176	1	109
June	54	36	–67	56	46	50	10	43	89	46	37	–57	–46	34	–47	81	121
July	236	–45	79	128	71	36	–71	14	–71	227	–20	–34	75	–28	51	–6	104
August	92	22	–36	118	–48	43	–13	19	118	–2	–5	–30	52	–91	–29	49	103
September	–37	–37	17	–15	–32	–32	–5	–46	–41	–48	–29	59	–39	–41	–33	78	90
October	–8	41	–40	–26	44	–7	11	42	122	–44	–1	–22	–8	–46	–12	–55	62
November	–21	41	99	–7	19	–23	51	–41	–17	–16	29	–15	–34	27	–5	–10	41
Total	1048	967	752	1026	756	802	683	747	979	1122	704	674	719	604	885	839	771

[a]Rainfall figures are interpreted as follows, for example: in July 1990 rainfall was 236 + 104 (normal monthly) = 340 mm; Normal (1951–1984) = average monthly rainfall data recorded at Charles city, Iowa, 20 km from the study area.

Table 14.3. The effects of tillage and cropping systems on subsurface drainage water flow (mm) from 1990 to 1992.

Tillage system	Cropping system[a]	1990	1991	1992	Average
Chisel plough	Continuous maize	182.7ab	273.3abc	128.0abc	194.7ab
	C-S-C	193.3ab	176.3c	165.3ab	178.3b
	S-C-S	156.7b	316.7a	69.3bc	180.9b
No-till	Continuous maize	274.3a	336.7a	178.0a	263.0a
	C-S-C	158.0b	180.0bc	59.3c	132.4b
	S-C-S	169.0ab	286.7ab	76.7abc	177.4b
Average		189.0	261.0	112.8	187.8
CV (%)		31.9	22.8	50.3	26.3
$LSD_{(0.05)}$		109.6	108.5	103.3	74.4

[a]C, maize; S, soybean.

Table 14.4. The effects of tillage and cropping systems on NO_3-N leaching losses (kg N ha^{-1}) to subsurface drainage water from 1990 to 1992.

Tillage system	Cropping system[a]	1990	1991	1992	Average
Chisel plough	Continuous maize	99.9a	76.1a	18.9a	64.9a
	C-S-C	52.4b	36.5c	16.6ab	35.2b
	S-C-S	51.2b	45.5bc	8.3bc	35.0b
No-till	Continuous maize	107.2a	62.5ab	19.9a	63.2a
	C-S-C	36.5b	30.7c	5.0c	24.1b
	S-C-S	31.6b	31.7c	7.3bc	23.6b
Average		63.1	47.2	12.7	41.0
CV (%)		31.6	22.0	44.2	30.8
$LSD_{(0.05)}$		36.2	18.9	10.2	14.7

[a]C, maize; S, soybean.

NO_3-N concentrations (32.6 mg l^{-1}) than CC plots with a no-tillage system (23 mg l^{-1}); values were also higher compared with rotated maize plots with the same tillage (Table 14.5). Rotation effects were also significant for plots with no-tillage practices. No-tillage with CC plots had significantly higher NO_3-N concentrations (23 mg l^{-1}) than rotated maize and soybean plots (16 and 13 mg l^{-1}) with no-tillage. Although no data are available, low NO_3-N concentrations with a no-tillage system compared with a chisel ploughed system can be associated with enhanced mineralization effects with chisel ploughing and the denitrification process with no-tillage practices (Gupta *et al.*, 2004; Thoma *et al.*, 2005). Flow-weighted NO_3-N concentrations are the integrated outcome of the complex interaction between precipitation patterns of dry and wet years. The year 1990 had the highest average NO_3-N concentrations (32.8 mg l^{-1}), and the highest rainfall (Table 14.3).

The accumulated N in the preceding years and more rainfall in July of 1990 gave maximum average maize grain yields of 10.75 Mg ha^{-1} for 1990. A minimum average maize grain yield of 8.72 Mg ha^{-1} was observed in 1991 after the wet year of 1990 (Table 14.6). Based on data averaged over the 3 years (1990–1992), chisel ploughing gave significantly higher maize grain yields than the no-till system with continuous maize plots (9.69 versus 8.46 Mg ha^{-1}). Rotated maize (CSC) plots had significantly higher maize grain yields than CC plots with chisel ploughing (10.45 versus 9.69 Mg ha^{-1}) and no-till plots (10.08 versus 8.46 Mg ha^{-1}). Higher maize grain yields with chisel ploughing versus no-tillage can be associated with better N availability in this tillage system (Levanon *et al.*, 1993).

Table 14.5. The effects of tillage and cropping system on flow-weighted NO_3-N concentrations (mg l^{-1}) in subsurface drainage water from 1990 to 1992.

Tillage system	Cropping system[a]	1990	1991	1992	Average
Chisel plough	Continuous maize	54.5a	28.0a	15.2a	32.6a
	C-S-C	28.3cd	20.8b	10.2bcd	19.7c
	S-C-S	32.8bc	14.4d	11.7b	19.6c
No-till	Continuous maize	39.1b	18.7bc	11.3bc	23.0b
	C-S-C	23.0de	16.8cd	8.4d	16.2d
	S-C-S	19.2e	11.0e	9.0cd	13.1e
Average		32.8	18.3	10.7	20.7
CV (%)		11.8	8.5	12.3	11.9
$LSD_{(0.05)}$		7.0	2.8	2.4	2.4

[a]C, maize; S, soybean.

Table 14.6. The effects of tillage and cropping systems on maize and soybean yields (Mg ha^{-1}) from 1990 to 1992.

Tillage system	Cropping system[a]	1990	1991	1992	Average
Chisel plough	Continuous maize	11.16a	8.75b	9.17b	9.69b
	C-S-C	11.25a	9.89a	10.21a	10.45a
No-till	Continuous maize	9.34b	7.32c	8.71b	8.46c
	C-S-C	11.24a	8.94b	10.05a	10.08ab
Average		10.75	8.72	9.54	9.67
CV (%)		1.75	3.04	4.24	2.93
$LSD_{(0.05)}$		0.37	0.53	0.81	0.41
Chisel plough	S-C-S	3.61a	3.34a	3.61a	3.52a
No-till	S-C-S	3.60a	3.24a	3.31b	3.38a
Average		3.60	3.29	3.46	3.45
CV (%)		2.86	2.07	0.54	2.12
$LSD_{(0.05)}$		0.36	0.24	0.07	0.16

[a]C, maize; S, soybean.

On average, about 31% of the applied fertilizer was leached to tile flow with CC plots under both tillage systems, while only 21 and 14% were leached under maize-soybean rotation using a chisel plough and no-till, respectively. Similar amounts of N were leached to tile flow from rotated soybean plots for both tillage systems although no N fertilizer was applied to the soybeans. This showed that soybean may not utilize a lot of the N left by the previous crop, and there was a greater flow-weighted NO_3-N concentration when soybean was under chisel-ploughed management than under no-till. This would indicate that no-till soybean in rotation would reduce NO_3-N leaching loss more than chisel ploughed soybean. Jaynes *et al.* (1999) reported NO_3-N drainage loss of 10 to 50% of the applied N and cautioned that care must be taken to interpret the nitrate drainage loss in terms of the applied fertilizer because it is not the only source contributing to the loss.

Effects of tillage, nitrogen management and crop rotation from 1993 to 1998

Growing season rainfall affected average tile flow volume, which ranged from a minimum of 70 mm in 1996 to 418 mm in 1993 (Table 14.7). Average minimum and maximum volumes of tile flow corresponded with the minimum and maximum amounts of growing season rainfall, respectively (Table 14.2). Rainfall was above normal in 1993, 1995 and 1998,

Table 14.7. The effects of tillage and cropping systems on subsurface drainage water flow (mm) from 1993 to 1998.

Tillage system	Cropping system[a]	1993	1994	1995	1996	1997	1998	Average
Chisel plough	Continuous maize	392.0bc	71.3b	112.0bc	42.8bc	61.0b	185.3b	144.1bc
	C-S-C	352.0bc	29.0b	67.3c	48.7bc	50.3b	186.7b	122.3c
	S-C-S	282.5c	56.0b	94.7bc	38.0c	55.0b	205.7ab	121.9c
No-till	C-S-C	492.3ab	164.7a	201.0ab	113.7a	133.3ab	372.3a	246.2ab
	S-C-S	572.3a	109.3ab	249.7a	106.0ab	211.3a	264.3ab	252.2a
Average		418.2	86.1	144.9	69.8	102.2	242.9	177.3
CV (%)		22.5	55.2	48.4	48.6	66.7	40.2	25.9
$LSD_{(0.05)}$		176.9	89.5	132.1	63.9	128.3	183.9	107.8

[a]C, maize; S, soybean.

which were the years of highest average tile volume, and below normal in 1994, 1996 and 1997, which were the years of lowest average tile volume.

No-till treatments had about 92% higher tile flow volumes than those from chisel ploughed systems from 1993 to 1998. In 1993, rotated soybean treatments with no-tillage had significantly higher tile flow volume (572 versus 283 mm) than rotated soybean treatments with chisel ploughing (Table 14.7). It is interesting to note that the plots with no-tillage always had the highest tile flow volume among all treatments in all years whether they were planted to maize or soybeans. Averaged across years (1993–1998), no-tillage treatments with soybeans had a twofold higher tile flow volume (252 versus 122 mm) than soybean with chisel ploughed tillage. Rotation effects on tile flow were not significant for either tillage system. No-tillage resulted in almost twofold (120 mm) higher tile flow than the chisel ploughing system because of increased macropore flow processes, less evaporation and greater infiltration due to the crop residues left with this system (Randall and Iragavarapu, 1995; Bjorneberg *et al.*, 1996; Kanwar *et al.*, 1997; Bakhsh *et al.*, 2000, 2002).

The average NO_3-N leaching loss to tile flow ranged from 6.6 kg N ha^{-1} in 1994 to 38.9 kg N ha^{-1} in 1993 (Table 14.8), which showed the effect of rainfall as well as that of low plant N uptake in 1994. Overall treatment effects on NO_3-N leaching loss to tile flow were significant ($p \leq 0.05$) only for 1996. This can be associated with minimum rainfall in 1996 and the effect of the previous year's low crop N uptake due to hail damage in 1995. Continuous maize plots with chisel ploughing had higher NO_3-N leaching loss in 1993 only, which may be due to higher N application rates in the years before 1993 and the higher relative N application rates compared with the maize-soybean rotation. Rotated no-till maize plots had the highest NO_3-N leaching loss during all years compared with all other treatments, which may be due to higher tile flow volumes. Averaged across 6 years (1993–1998), rotated maize plots with no-tillage had significantly higher NO_3-N leaching loss (25.7 versus 13.7 kg N ha^{-1}) than rotated maize plots with chisel ploughing (Table 14.8). This can be associated with higher tile flow rates and lower plant N uptake from these plots.

Growing season rainfall and hail damage in 1995 elevated flow-weighted NO_3-N concentrations to the highest level of 12.5 mg l^{-1} (Table 14.9), along with a decrease in yield (Table 14.10). The no-tillage system resulted in lower flow-weighted NO_3-N concentrations than the chisel ploughed system (Table 14.9) owing to dilution and possible denitrification. Bakhsh *et al.* (2000) reported that higher NO_3-N leaching loss with no-tillage can be due to higher tile flows and low crop N uptake compared with a chisel ploughed system. CC plots with chisel ploughing had slightly higher NO_3-N leaching loss than rotated maize plots (16.9 versus 13.7 kg N ha^{-1}; Table 14.8) because of the low average maize grain yield of 6.37 Mg ha^{-1} for CC plots (Table 14.10).

Table 14.8. The effects of tillage and cropping systems on NO_3-N leaching losses (kg N ha^{-1}) to subsurface drainage water from 1993 to 1998.

Tillage system	Cropping system[a]	1993	1994	1995	1996	1997	1998	Average
Chisel plough	Continuous maize	46.7a	7.8ab	16.0a	3.7b	3.8b	23.3a	16.9ab
	C-S-C	32.8b	2.7b	10.5a	6.3b	6.3ab	23.6a	13.7b
	S-C-S	32.3b	3.4b	10.2a	5.7b	3.7b	24.5a	13.3b
No-till	C-S-C	45.8ab	13.5a	25.2a	13.7a	16.6a	39.7a	25.7a
	S-C-S	37.1ab	5.6b	23.1a	12.9a	15.7ab	25.9a	20.0ab
Average		38.9	6.6	17.0	8.5	9.2	27.4	17.9
CV (%)		18.8	60.6	49.2	39.8	71.4	35.9	23.9
$LSD_{(0.05)}$		13.8	7.5	15.7	6.4	12.4	18.5	10.4

[a]C, maize; S, soybean.

Table 14.9. The effects of tillage and cropping systems on flow-weighted NO_3-N concentrations (mg l^{-1}) in subsurface drainage water from 1993 to 1998.

Tillage system	Cropping system[a]	1993	1994	1995	1996	1997	1998	Average
Chisel plough	Continuous maize	12.2a	11.0a	14.3ab	7.8b	7.2b	12.8a	10.9ab
	C-S-C	9.3c	9.3ab	15.5a	13.0a	12.4a	12.7a	12.0a
	S-C-S	11.5ab	6.2cd	10.9cd	15.1a	6.8b	11.9ab	10.4b
No-till	C-S-C	9.3bc	8.2bc	12.7bc	12.8a	12.3a	10.9ab	11.0ab
	S-C-S	6.5d	4.8d	9.0d	12.4a	7.3b	9.7b	8.3c
Average		9.7	7.9	12.5	12.2	9.2	11.6	10.5
CV (%)		11.8	15.6	10.2	17.2	13.1	10.0	12.2
$LSD_{(0.05)}$		2.2	2.3	2.4	4.0	2.3	2.2	1.2

[a]C, maize; S, soybean.

Table 14.10. The effects of tillage and cropping systems on maize-soybean yields (Mg ha^{-1}) from 1993 to 1998.

Tillage system	Cropping system[a]	1993	1994	1995	1996	1997	1998	Average
Chisel plough	Continuous maize	4.57a	5.79b	4.55c	6.99b	8.56b	7.80b	6.37c
	C-S-C	5.09a	7.93a	6.03a	8.81a	9.76a	9.74a	7.89a
No-till	C-S-C	4.34a	6.34b	5.15b	8.38a	9.51a	8.14b	6.97b
Average		4.67	6.68	5.24	8.06	9.8	8.56	7.08
CV (%)		8.84	6.59	4.17	2.73	1.89	2.19	7.25
$LSD_{(0.05)}$		0.93	1.00	0.49	0.49	0.39	0.42	0.37
Chisel plough	S-C-S	2.65a	3.60a	3.25a	4.14a	3.65a	4.03a	3.55a
No till	S-C-S	2.66a	3.47a	3.18a	4.16a	3.76a	4.23a	3.58a
Average		2.66	3.54	3.21	4.15	3.70	4.13	3.56
CV (%)		3.95	4.01	1.11	1.33	2.77	2.23	2.68
$LSD_{(0.05)}$		0.37	0.49	0.12	0.19	0.36	0.32	0.11

[a]C, maize; S, soybean.

Rotated maize treatments with chisel ploughing had the highest maize grain yield of 7.89 Mg ha^{-1}, followed by rotated maize plots with no-tillage (6.97 Mg ha^{-1}) and CC treatments with chisel ploughing (6.37 Mg ha^{-1}). Low maize grain yield also contributed to the increased NO_3-N leaching loss for the corresponding treatments because of low crop N uptake. On average, about 13% of the applied N fertilizer was leached to tile water from CC plots with chisel ploughing. Similarly, rotated maize treatments with chisel ploughing and the no-tillage system lost an average amount of 12% and 23% of the applied N fertilizer to tile water, respectively. This difference can be associated with higher tile flows, low maize grain yields and low N uptake with the no-tillage system (Randall and Iragavarapu, 1995; Kanwar *et al.*, 1997; Bakhsh *et al.*, 2000).

Effects of manure nitrogen management, tillage system and crop rotation from 2001 to 2005

Rainfall from March to November varied from 604 mm in 2003 to 885 mm in 2004 (Table 14.2). There was above normal rainfall in 2005 and below normal rainfall in 2001 and 2002. Treatment effects on tile flow were mostly non-significant, but seasonal effects were found to be highly significant ($p \leq 0.01$). Tile flow volume was 134 mm in 2001 and 30 mm in 2002 (Table 14.11). About 99% of the seasonal tile flow exited the fields before May in 2001 (data not shown here). The temporal distribution of rainfall during the growing season affected tile flow, especially with reference to evapotranspiration requirements. On average (2001–2005), no-till soybean treatments with manure application had 79 mm higher tile flow volume than soybean with chisel ploughing and UAN applications. Soybean treatments with manure and chisel ploughing gave about twofold higher tile flow volume (108 versus 57 mm) than soybean with UAN and chisel ploughed systems. The manure application probably increased tile flow volume because of better water retention effects.

Treatment effects on NO_3-N leaching losses to tile flow were mostly non-significant except in 2004 owing to above-normal rainfall in 2004. Average NO_3-N leaching loss to tile flow varied from 3.6 kg N ha^{-1} in 2002 to 27.1 kg N ha^{-1} in 2004 (Table 14.12) as a result of varying rainfall patterns during these years. Soybean treatments with manure application to maize in the previous year and chisel ploughing had the highest NO_3-N leaching loss of 28.3 kg N ha^{-1} in 2001. On average (2001–2005), maize treatments with autumn manure application and chisel ploughing had 82% higher NO_3-N leaching loss (20.6 versus 11.3 kg N ha^{-1}) than maize with UAN application under chisel ploughing, and 23% higher (20.6 versus 16.7 kg N ha^{-1}) than the maize spring-manure-treated with no-tillage

Table 14.11. The effects of tillage and cropping system on subsurface drainage water flow (mm) from 2001 to 2005.

Tillage system	Cropping system[a]	2001	2002	2003	2004	2005	Average
Chisel plough	C-S-C-FM	75.6a	46.1ab	55.4bc	142.6ab	39.9ab	71.9b
	C-S-C-UAN	73.2a	3.0c	50.4c	87.5ab	30.4b	48.9b
	S-C-S-UAN	91.3a	10.2bc	65.9abc	85.6ab	32.9b	57.2b
	S-C-S-FM	190.9a	10.7bc	140.9ab	98.6ab	99.3a	108.1ab
No-till	C-S-C-SM	169.9a	49.0ab	124.7abc	79.4b	80.3ab	100.7ab
	S-C-S-SM	205.6a	63.6a	145.6a	171.8a	92.3a	135.8a
Average		134.4	30.4	97.1	110.0	62.5	87.0
CV (%)		54.4	74.1	49.3	45.6	57.8	41.9
$LSD_{(0.05)}$		133.1	41.0	87.1	91.9	65.7	63.6

[a]C, maize; S, soybean; FM, autumn manure applied to maize only; SM, spring manure applied to maize only; UAN, urea ammonium nitrate solution fertilizer applied to maize only.

system. No-tillage with manure application resulted in similar NO_3-N leaching losses for both the maize and soybean phase of production (16.7 versus 16.4 kg N ha^{-1}). A similar response was observed for manure-treated plots with chisel ploughing, which also had similar NO_3-N leaching losses for maize and soybean rotation plots (20.6 versus 17.1 kg N ha^{-1}), respectively. The highest NO_3-N leaching loss of 20.6 kg N ha^{-1} with autumn manuring and chisel ploughing can be associated with the greater time that is available for mineralization processes in this system (Gupta *et al.*, 2004; Bakhsh *et al.*, 2005).

Average flow-weighted NO_3-N concentrations varied from 14.2 mg l^{-1} in 2002 to 24.5 mg l^{-1} in 2004 (Table 14.13). Treatment effects on flow-weighted NO_3-N concentrations were highly significant ($p \leq 0.01$) from 2001 to 2005. When averaged over the 5 years (2001–2005), maize treatments with manure in the autumn had significantly higher flow-weighted NO_3-N concentrations (27 mg l^{-1}) than maize plots with UAN application and chisel ploughing (20.1 mg l^{-1}) and also than spring manure applied no-tillage plots (17.7 mg l^{-1}). Higher flow-weighted NO_3-N concentrations with chisel ploughed and manure-treated plots showed the possible enhanced mineralization effects with the chisel ploughing system, dilution effects with no-tillage plots, and also differences in maize

Table 14.12. The effects of tillage and cropping system on NO_3-N leaching losses (kg N ha^{-1}) to subsurface drainage water from 2001 to 2005.

Tillage system	Cropping system[a]	2001	2002	2003	2004	2005	Average
Chisel plough	C-S-C-FM	19.4a	6.4a	15.3a	50.1a	12.0a	20.6a
	C-S-C-UAN	10.3a	0.4b	11.2a	27.5b	7.0b	11.3ab
	S-C-S-UAN	17.2a	1.8ab	12.0a	15.8b	6.1b	10.6b
	S-C-S-FM	28.3a	1.9ab	21.3a	19.4b	14.3ab	17.1ab
No-till	C-S-C-SM	20.8a	4.6ab	21.3a	17.2b	19.5a	16.7ab
	S-C-S-SM	17.2a	6.2a	15.6a	32.3b	10.6ab	16.4ab
Average		18.9	3.6	16.1	27.1	11.6	15.4
CV (%)		53.9	77.0	42.8	36.0	48.2	37.1
$LSD_{(0.05)}$		18.2	5.0	12.6	17.7	10.2	9.6

[a]C, maize; S, soybean; FM, autumn manure applied to maize only; SM, spring manure applied to maize only; UAN, urea ammonium nitrate solution fertilizer applied to maize only.

Table 14.13. The effects of tillage and cropping system on flow-weighted NO_3-N concentrations (mg l^{-1}) in subsurface drainage water from 2001 to 2005.

Tillage system	Cropping system[a]	2001	2002	2003	2004	2005	Average
Chisel plough	C-S-C-FM	24.9a	16.9a	26.8a	36.5a	30.1a	27.0a
	C-S-C-UAN	14.2c	11.4b	21.7b	30.2ab	23.1ab	20.1b
	S-C-S-UAN	18.8b	18.8a	18.2bc	18.6c	18.1bc	18.5b
	S-C-S-FM	15.8bc	19.3a	16.1c	19.9c	15.0c	17.0b
No-till	C-S-C-SM	12.4c	9.6b	18.1bc	23.1bc	25.2ab	17.7b
	S-C-S-SM	8.3d	9.6b	11.1d	18.8c	11.9c	11.9c
Average		15.7	14.2	18.6	24.5	20.6	18.7
CV (%)		13.2	20.5	12.8	16.4	20.4	13.6
$LSD_{(0.05)}$		3.8	5.3	4.3	7.3	7.6	4.2

[a]C, maize; S, soybean; FM, autumn manure applied to maize only; SM, spring manure applied to maize only; UAN, urea ammonium nitrate solution fertilizer applied to maize only.

grain yields due to varied crop N uptake. Maize treatments with autumn manure application and chisel ploughing had slightly higher maize grain yields (11.55 Mg ha^{-1}) than with plots receiving UAN and chisel ploughing (11.37 Mg ha^{-1}) and plots with spring manure with no-tillage practices (11.23 Mg ha^{-1}) (Table 14.14). The amount of N leached to tile water (for both the rotation plot under maize and that under soybean) as a percentage of applied fertilizer was 13% from UAN with a chisel ploughing system, 21% from autumn-applied manure with chisel ploughing and 19% from spring-applied manure with no-tillage practices. Randall and Iragavarapu (1995) reported similar results of about 20% of the applied N lost to tile water based on their 11-year study in Minnesota.

In another study conducted on small watersheds where seven individual sub-watersheds drain into Cedar River tributaries which eventually drain into the Mississippi River, it was found that the average NO_3-N concentrations in river tributaries (for sub-watersheds) are directly correlated to the percentage sub-watershed area under a maize-soybean row crop system (Table 14.15). Higher average NO_3-N concentrations are associated with a larger percentage of watershed under row crops, indicating the effect of nitrogen fertilizers or animal manure application on stream water quality.

The overall results of these studies indicated that nitrate concentrations were strongly related to land-use practices in Iowa and that the average NO_3-N concentrations in stream water exceeded the drinking water quality standard of 10 mg l^{-1} at times during the summer months. Therefore, nutrient management practices in the watersheds are extremely

Table 14.14. The effects of tillage and cropping system on maize-soybean yields (Mg ha^{-1}) for 2001 to 2005.

Tillage system	Cropping system[a]	2001	2002	2003	2004	2005	Average
Chisel plough	C-S-C-FM	11.08a	12.20a	10.20a	12.27b	11.98a	11.55a
	C-S-C-UAN	10.22b	12.01b	9.77a	12.85a	12.02a	11.37a
No-till	C-S-C-SM	10.58ab	12.03b	9.85a	11.60c	12.11a	11.23a
Average		10.62	12.08	9.94	12.24	12.04	11.38
CV (%)		3.01	0.46	7.13	1.82	1.67	2.77
$LSD_{(0.05)}$		0.73	0.12	1.61	0.51	0.46	0.47
Chisel plough	S-C-S-UAN	3.08b	3.62b	2.07a	4.00a	4.45b	3.44b
	S-C-S-FM	3.45a	3.75a	1.92a	3.99a	4.62a	3.55a
No-till	S-C-S-SM	2.97b	3.55b	1.91a	3.74b	4.65a	3.36c
Average		3.17	3.64	1.96	3.91	4.57	3.45
CV (%)		3.02	1.46	7.59	2.37	1.27	3.25
$LSD_{(0.05)}$		0.22	0.12	0.34	0.21	0.13	0.05

[a]C, maize; S, soybean; FM, autumn manure applied to maize only; SM, spring manure applied to maize only; UAN, urea ammonium nitrate solution fertilizer applied to maize only.

Table 14.15. Average NO_3-N concentrations (mg l^{-1}) for seven sub-watershed tributaries of the Cedar River in Iowa for the period of 2000–2003 (IDNR, 2004).

Sub-watershed	% Under row crop	2000	2001	2002	2003
Bear	74	9.1	11.1	9.5	6.2
Blue	61	8.9	8.9	7.7	5.7
Lime	80	13.1	14.6	12.0	11.1
Morgan	67	8.9	9.5	8.9	7.7
Mud	79	11.4	13.5	11.4	8.9
N. Bear	79	12.7	13.8	11.7	12.2
Otter	66	9.5	10.1	9.4	5.5

important for controlling the nutrient concentrations in streams and river waters. Several management systems (combinations of tillage, crop rotations and nutrient practices) presented in this chapter can be adopted by farmers within a watershed to reduce the transport of NO_3-N to surface or groundwater.

Summary

This long-term field study compared the effects of two tillage systems (chisel versus no-tillage), and N application rates of 202 and 168 kg N ha^{-1} to continuous maize and rotated maize, respectively, for (i) the 1990–1992 phase, (ii) 135 and 110 kg N ha^{-1} to continuous maize and rotated maize, respectively, for the 1993–1998 phase, and (iii) 168 kg N ha^{-1} to rotated maize for the 2001–2005 phase. Growing season precipitation patterns and the cycles of wet and dry years affected tile flow, flow-weighted NO_3-N concentrations, NO_3-N leaching losses to tile water and plant N uptake processes significantly ($p \leq 0.05$).

On average, chisel ploughing resulted in higher NO_3-N leaching losses of about 8 kg N ha^{-1} compared with the no-tillage system during the 1990–1992 phase, owing to the effects of higher N application rates. At the same time, no-tillage resulted in significantly higher NO_3-N leaching losses of 9 kg N ha^{-1} compared with chisel ploughing beneath the maize-soybean rotation during the 1993 to 1998 phase because of lower N application rates. Continuous maize treatments with chisel ploughing lost about 31% of the applied N fertilizer (202 kg N ha^{-1}) during 1990–1992, and 12% of the applied N fertilizer (135 kg N ha^{-1}) in 1993–1998, to tile water. Rotated maize plots with chisel ploughing lost 20% of the applied N fertilizer (168 kg N ha^{-1}) in 1990–1992, 12% of the applied fertilizer (110 kg N ha^{-1}) in 1993–1998, and 7% of the applied fertilizer (168 kg N ha^{-1}) in 2001–2005.

Tillage effects on flow-weighted NO_3-N concentrations were significant ($p = 0.05$) during the 1990–1992 period. Chisel ploughing with continuous maize and rotated maize plots resulted in significantly higher flow-weighted NO_3-N concentrations when compared with the no-till system (32.6 versus 23 mg l^{-1} and 19.7 versus 16.1 mg l^{-1}, respectively). For 2001–2005, rotated maize plots with chisel ploughing and autumn manuring resulted in significantly higher flow-weighted NO_3-N concentrations in comparison with no-till and spring manuring (27.0 versus 17.7 mg l^{-1}) and with chisel ploughing with spring UAN applications (27.0 versus 20.1 mg l^{-1}). Rotated soybean plots receiving no N application with chisel ploughing and no-tillage lost 35.2 and 24.1 kg N ha^{-1} during 1990–1992, and 13.3 and 20.0 kg N ha^{-1} in 1993–1998, to tile water, respectively. Leaching loss of NO_3-N from soybean plots during the 2001–2005 period for UAN application and autumn manuring with chisel ploughing were 10.6 and 17.1 kg N ha^{-1}, respectively; for spring manuring with no-tillage the loss was 16.4 kg N ha^{-1}. Rotated maize treatments had significantly higher grain yields than continuous maize treatments. Chisel ploughing with continuous maize showed higher grain yield than no-tillage in the 1990–1992 phase (9.69 versus 8.46 Mg ha^{-1}), and also increased yield with rotated maize for the 1993–1998 phase (7.89 versus 6.97 Mg ha^{-1}).

These results suggest that higher N application rates coupled with dry years have higher build-up of N pools as a result of mineralization processes and are more likely to have elevated NO_3-N concentrations in tile water. Higher N application rates to maize in the preceding year increased NO_3-N concentrations in tile water for the following soybean years receiving no N application. Therefore, proper N consideration needs to be given to soybean as well as to the effects of a preceding dry year to minimize the off-site transport of NO_3-N leaching losses and reduce the NO_3-N concentrations in tile water.

The next question is: where should we go from here in developing new knowledge on the use of best management practices? The focus of future research should concentrate on the use of watershed models in predicting the effects of complex agricultural systems (including tillage, crop rotations, nutrient application rates to match with spatial variable soil properties, and water use for different crops) on the profitability of the farmer and in reducing chemical losses to water bodies. Some scientists are calling this new approach 'predictive agriculture' in contrast to the currently practised 'prescriptive agriculture'. With today's computing power, we should be able to simulate highly complex soil–water–plant–animal systems and make predictions on farmer profitability in response to climate change, carbon sequestration and the ecological health of our planet.

Note

[1] Use of trade names is for reader information and does not imply any endorsement by Iowa State University or the US Department of Agriculture Agricultural Research Service.

References

Alexander, R.B., Smith, R.A. and Schwarz, G.E. (1995) The regional transport of point and nonpoint source nitrogen to the Gulf of Mexico. In: Kenner, LA. (ed.) *Proceedings of First Gulf of Mexico Hypoxia Conference*, 5–6 December 1995. US EPA Publication No. 855R97001, pp, 127–132. US Environmental Protection Agency National Center for Environment; Publication and Information, Washington, DC.

Baker, J.L., Colvin, T.S., Marley, S.J. and Dawelbeit, M. (1989) A point-injector applicator to improve fertilizer management. *Applied Engineering in Agriculture* 5, 334–338.

Baker, J.L., David, M.B. and Lemke, D.W. (2005) Understanding nutrient fate and transport, including the importance of hydrology in determining losses, and potential implications on management systems to reduce those losses. In: *Proceedings of Gulf Hypoxia and Local Water Quality Concerns Workshop*, held at Iowa State University on 26–28 September 2005, pp. 11–25.

Bakhsh, A. and Kanwar, R.S. (2004) Using discriminant analysis and GIS to delineate subsurface drainage patterns. *Transactions of the ASAE* 47, 689–699.

Bakhsh, A., Kanwar, R.S., Jaynes, D.B., Colvin, T.S and Ahuja, L.R. (2000) Prediction of NO_3-N losses with subsurface drainage from manured and UAN-fertilized plots using GLEAMS. *Transactions of the ASAE* 43, 69–77.

Bakhsh, A., Kanwar, R.S., Bailey, T.B., Cambardella, C.A., Karlen, D.L. and Colvin, T.S. (2002) Cropping system effects on NO_3-N losses with subsurface drainage water. *Transactions of the ASAE* 45, 1789–1797.

Bakhsh, A., Kanwar, R.S. and Karlen, D.L. (2005) Effects of liquid swine manure applications on NO_3-N leaching losses to subsurface drainage water. *Agriculture, Ecosystems and Environment* 109, 118–128.

Bakhsh, A., Kanwar, R.S. and Baker, J.L. (2010) N-application methods and precipitation pattern effects on subsurface drainage nitrate losses and crop yields. *Water, Air, and Soil Pollution* 212, 65–70.

Bjorneberg, D.L., Kanwar, R.S. and Melvin, S.W. (1996) Seasonal changes in flow and nitrate-N loss from subsurface drains. *Transactions of the ASAE* 39, 961–976.

Dinnes, D.L., Karlen, D.L., Jaynes, D.B., Kaspar, T.C., Hatfield, J.L., Colvin, T.S. and Cambardella, C.A. (2002) Nitrogen management strategies to reduce nitrate leaching in tile-drained Midwestern soils. *Agronomy Journal* 94, 153–171.

Green, C.J. and Blackmer, A.M. (1995) Residue decomposition effects on nitrogen availability to corn following after corn or soybean. *Soil Science Society of America Journal* 59, 1065–1070.

Green, C.J., Blackmer, A.M. and Horton, R. (1995) Nitrogen effects on conservation of carbon during corn residue decomposition in soil. *Soil Science Society of America Journal* 59, 453–459.

Gupta, S., Munyankusi, E., Moncrief, J., Zvomuya, F. and Hanewall, M. (2004) Tillage and manure application effects on mineral nitrogen leaching from seasonally frozen soils. *Journal of Environmental Quality* 33, 1238–1246.

IDNR (Iowa Department of Natural Resources) (2004) *Cedar River Assessment Survey Project.* IDNR/US Geological Survey, Iowa City, Iowa.

Jaynes, D.B., Hatfield, J.L. and Meek, D.W. (1999) Water quality in Walnut Creek watershed: herbicides and nitrate in surface waters. *Journal of Environmental Quality* 28, 45–59.

Kanwar, R.S. (2010) Sustainable water systems for agriculture and 21st century challenges. *Journal of Crop Improvement* 24, 1–19.

Kanwar, R.S., Colvin, T.S. and Karlen, D.L. (1997) Ridge, moldboard, chisel, and no-till effects on subsurface drainage water quality beneath two cropping system. *Journal of Production Agriculture* 10, 227–234.

Kanwar, R.S., Bjorneberg, D. and Baker, D. (1999) An automated system for monitoring the quality and quantity of subsurface drain flow. *Journal of Agricultural Engineering Research* 73, 123–129.

Karlen, D.L., Cambardella, C.A. and Kanwar, R.S. (2004) Challenges of managing swine manure. *Applied Engineering in Agriculture* 20, 693–699.

Levanon, D., Codling, E.E., Meisinger, J.J. and Starr, J.L. (1993) Mobility of agrochemicals through soil from two tillage systems. *Journal of Environmental Quality* 22, 155–161.

Power, J.F., Wiese, R. and Flowerday, D. (2001) Managing farming systems for nitrate control: a research review from Management System Evaluation Areas. *Journal of Environmental Quality* 30, 1866–1880.

Randall, G.W. and Iragavarapu, T.K. (1995) Impact of long-term tillage systems for continuous corn on nitrate leaching to subsurface drainage. *Journal of Environmental Quality* 24, 360–366.

SAS Institute (2003) *The SAS system for Windows, Release 9.1.* Cary, North Carolina.

Thoma, D.P., Gupta, S.C., Strock, J.S. and Moncrief, J.F. (2005) Tillage and nutrient source effects on water quality and corn grain yield from a flat landscape. *Journal of Environmental Quality* 34, 1102–1111.

Voy, K.D. (1995) *Soil Survey of Floyd County, Iowa.* US Department of Agriculture Soil Conservation Service in cooperation with Iowa Agricultural and Home Economics Experiment Station, Cooperative Extension Service, Iowa State University, Ames, Iowa.

15 Climate Data for Hydrological and Agronomic Modelling

Max P. Bleiweiss* and A. Salim Bawazir

Introduction

Meteorological data such as temperature, humidity, wind speed, precipitation, solar radiation and cloud cover are crucial to hydrological and agronomic modelling. These models are increasingly being used in addressing problems in water resources and in agriculture. They simulate natural processes such as flow of water, transport of sediments and contaminants, snow-melt runoff and groundwater recharge within a watershed. They are also used widely in agriculture to simulate crop production, nutrient transport, air pollution and the prediction of insect emergence. Simulations of these processes are vital to planning and management of water resources, environmental impact problems, and the impact on social and economic development. The available models that will be mentioned in this chapter are diverse in computational sophistication, mathematical algorithms used and data requirements. However, in general, they all require some sort of meteorological data as input. The intention of this chapter is not to discuss the model details – which are already detailed by authors of some other chapters of this book – but to discuss the different types of climate data available, give a general overview of climate station data and analysed or reanalysed data sets, describe how to retrieve these data, compare different data sets for accuracy or performance and state their importance. In this framework this chapter discusses the following:

- Climate data requirements for use in hydrological and agronomic models
- Procedures for comparing different meteorological data sets, for example, reanalysis model output to climate station measurement
- Procedures for comparing hydrological or agronomic model outputs with measured values, for example, hydrological model output compared with stream gauge measurements
- Procedures for creating homogeneous climate data sets for input into hydrological and agronomic models that are spatially distributed at the required time step. An example of climatic data input to the Soil and Water Assessment Tool (SWAT) model (Arnold *et al.*, 1998) on the Mimbres watershed in south-west New Mexico is presented (El-Sadek *et al.*, 2010).

The types of models that are generally used depend on the time scales and spatial scales being studied; ranging from hydrological

* Corresponding author: mbleiwei@taipan.nmsu.edu

modelling of large basins to small-scale modelling of crop growth, which then dictate the spatial and temporal scale of the climate data necessary to drive the models. The location of the basin or crops being modelled often dictates which data should be obtained.

Singh and Frevert (2006) grouped a variety of models under the label of 'watershed models'. They defined watershed models as models that 'simulate natural processes of the flow of water, sediment, chemicals, nutrients, and microbial organisms within a watershed, as well as quantifying the impact of human activities on these processes'. They further divided this category into large watershed models, streamflow models, streamflow and water quality models, urban watershed models, agricultural watershed models, and planning and management models. Another way in which to categorize watershed models would be according to the process that is being emphasized in the modelling efforts, such as rainfall runoff or hydraulic models (e.g. KINematic Runoff and EROSion (KINEROS2); Semmens *et al.*, 2008), snow-melt runoff models (e.g. the Snow Runoff Model (SRM); Martinec *et al.*, 2008), or more complete models that take into account the whole hydrological process, including groundwater recharge (e.g. the Semi-distributed Land Use Runoff Process (SLURP); Kite, 1995) and the Soil and Water Assessment Tool (SWAT); Arnold *et al.*, 1998). Watershed models can also be delineated based on their attention to spatial and temporal detail, for example the lumped or tank models (e.g. the Sacramento Catchment Model; Burnash, 1995), semi-distributed models (e.g. SLURP or SWAT) and fully distributed hydrological models (e.g. TOPographic Kinematic Approximation and Integration (TOPKAPI); Todini and Ciarapica, 2001) or GEOtop (Rigon *et al.*, 2006).

Besides the watershed models, soil-crop models and insect models are of interest. The European COST action 718-Meteorological Applications in Agriculture recently considered several soil-crop models and compared their performance and their data requirements (Kersebaum *et al.*, 2007). COST is the acronym for European Cooperation in the field of Scientific and Technical Research (http://www.cost.esf.org/). The several models considered in that study and their data (climate data) requirements are listed in Table 15.1.

Table 15.1. Models that participated in the COST (European Cooperation in the field of Scientific and Technical Research) 718 action (Kersebaum *et al.*, 2007).

			Climate data requirements[c]								
Model name	Model type[a]	Time step[b]	Tavg	Tmax	Tmin	CC	RH	Insol	Prec	W/S	ET_0
AMBAV	ET/SM	h	X			X	X	X	X	X	
SIMWASER	WB/CG	d		X	X		X	X	X	X	
THESEUS	SM/CG	d		X	X			X		X	
OPUS	SM/CG	d		X	X		X			X	
STAMINA	MM[d]/CG	–	–	–	–	–	–	–	–	–	–
AGROSIM	AE	d	X	X	X			X	X	X	
AGROTOOL	CM	d	X	X	X		X	X	X		
NDICEA	ND	w	X						X		X
SWAP/ANIMO	WB/ND	d		X	X		X	X	X	X	
SWIM	SM/WB	d	X					X	X		
HERMES	ND	d	X				X	X	X		
CERES	CG/ND	d		X	X			X	X		
FASSET	AE	d	X					X	X		
CANDY	AE	d	X					X	X		

[a]AE, Agro-Ecosystem; CG, Crop Growth; CM, Crop Model; ET, Evapotranspiration; MM, Micrometeorology; ND, Nitrogen Dynamics; SM, Soil Moisture; SPA, Soil-Plant-Atmosphere; WB, Water Balance.
[b]d, daily; h, hourly; w, weekly.
[c]CC, cloud cover; ET_0, reference ET; Insol, insolation; Prec, precipitation; RH, relative humidity; Tavg, average air temperature; Tmax, maximum air temperature; Tmin, minimum air temperature; W/S, wind speed.
[d]Simulates required meteorological parameters.

Climate varies as the earth's terrain and land cover vary and is influenced by the proximity of large bodies of water. Guyot (1998) defines three different regimes within which climate is considered to be uniform. These regimes, as defined by Guyot (1998), include 'the regional-climate scale', which characteristically involves distances of the order of 100 km for plain and 10 km for mountainous areas, 'the topoclimatic scale', which involves distances of the order of 10 km for plain and 1 km for mountainous areas, and 'the microclimatic scale', which involves distances of the order of 100 m for plain and 10 m for mountainous areas. Depending on the scale of the process being modelled, it is required that the input data be representative of that scale. For example, a bulk hydrological model approach may only require climate data collected from a single climate station within or nearby the watershed, while a semi-distributed or distributed model will require, increasingly, data from several stations that are close to one another in order to adequately capture the dynamics of the surrounding region.

The chapter has three main sections after this Introduction. Climate data types and how they are obtained are discussed in the next section (Climate Data Types). In the following section, 'Techniques for Comparing Data Sets/Model Results', the more common model efficiency measures are presented along with some suggested modifications to existing measures. Following the discussion of the efficiency measures, 'rules' are presented to ensure that underlying assumptions in their use are not being ignored. Before the discussion of model measures, a comparison between two different meteorological data sets is presented to show, by example, a suggested approach for the analysis of such data. The last section provides an example of the creation of a homogeneous data set for use in hydrological modelling, and is entitled: 'Data Set Development for the Mimbres Basin of the Rio Grande in New Mexico: An Example of Creating a Regularly Gridded Homogeneous Data Set from Disparate Sources'. This is demonstrated by obtaining a regularly gridded data set from several different sources whose temporal and spatial details are not the same.

Climate Data Types

Climate data are obtained from point sources such as climate stations. Figure 15.1 shows an example of a research climate station that is used to acquire a substantial amount of information as it also measures surface energy fluxes (net radiation, sensible heat, latent heat and soil heat fluxes). In some instances, these types of data from irregularly spaced stations may be interpolated to a regular grid (Daley, 1996). In other cases, the measuring system – such as weather radar – can produce gridded data output that can be processed to directly

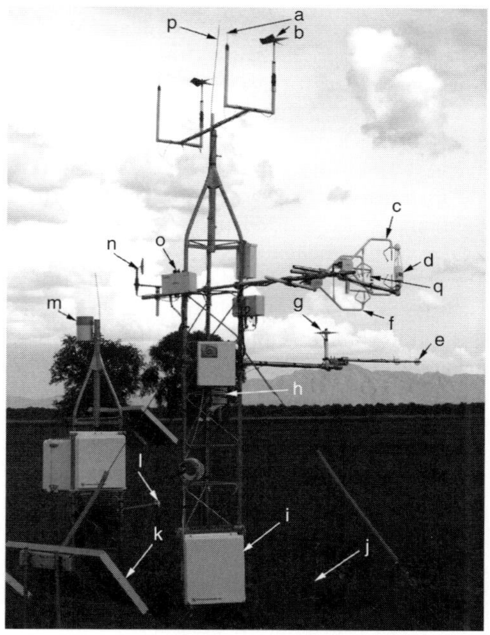

Fig. 15.1. Research climate station in a lucerne field used for measuring surface energy fluxes and weather parameters. The various instruments are: (a) a fine wire thermocouple, and (b) a propeller which form the two parts of the one-propeller eddy covariance (OPEC) system, (c) a 3D sonic anemometer (sensible heat sensor), (d) a LI-COR latent heat sensor, (e) a net radiometer, (f) a second 3D sonic anemometer, (g) a pyranometer, (h) a relative humidity and temperature sensor, (i) an enclosure for the data-logger, (j) the location of soil heat flux plates, and soil temperature and soil moisture sensors, (k) a solar panel for power, (l) an infrared sensor for measuring surface temperature, (m) a rain gauge, (n) a wind direction sensor, (o) a wind speed sensor, (p) a lightning rod, and (q) a krypton hygrometer.

yield a gridded climate product (precipitation) without intermediary interpolation. Still other sources of climate 'data' are provided by simulations obtained from atmospheric models that produce information on a regular grid according to some map projection or geographic coordinate system. An example of this would be output from the Weather Research and Forecasting Model (WRF) that is used by the US National Weather Service (of the US National Oceanic and Atmospheric Administration, NOAA) for weather forecasting. Model output may also be merged with a variety of observations to produce what are known as analysis or reanalysis data sets. An example of a reanalysis data set is the North American Regional Reanalysis (NARR) (Mesinger *et al.*, 2006) produced by the US National Centers for Environmental Prediction (NCEP, part of the US National Weather Service); this is discussed in more detail later in the chapter. In summary, the available data are in the form of point data, gridded point data, and data gridded because of either the nature of the measuring system or a modelling system (i.e. the sensor output, for example weather radar, is sampled on a regular grid or the model outputs information on a regular grid, for example the NARR output). These data sources can also be differentiated depending on whether the output is available directly from the measuring system or is derived from a modelled data set. In some situations, several different gridded data sets are merged to create a single output. For example, the PERSIANN (Precipitation Estimation from Remotely Sensed Information using Artificial Neural Networks) precipitation data set (Sorooshian *et al.*, 2000) is the result of merging several different gridded products to create a new estimate of precipitation. Climate data from point sources will be discussed next, followed by a description of gridded climate data – either interpolated point data, or weather radar and model-derived results.

Climate data from weather stations or point data

Weather stations are distributed irregularly in space and time. The particular distribution depends on several factors. Among these are the willingness of volunteers to operate stations, funding from governmental and/or private agencies, and population density and access to remote areas. In the USA, a volunteer network called COOP (cooperative) has been in existence since 1890 and currently consists of approximately 11,000 volunteers operating about 8000 stations. There are some volunteer observations from before this time but they are outside the formal COOP observer programme. Besides the COOP network, there are several other observation programmes in the USA that support a variety of goals:

- Remote Automated Weather Stations (RAWS) (http://raws.fam.nwcg.gov/) for the purpose of 'monitoring air quality, rating fire danger, and providing information for research applications'. There are approximately 2200 RAWS stations, primarily in the western USA. This is a multiagency network including, for example, the US Department of Agriculture (USDA) Forest Service, the US National Park Service, the US Bureau of Land Management, etc.
- The SNOwpack TELemetry (SNOTEL) (http://www.wcc.nrcs.usda.gov/snow/) network of the National Water and Climate Center of the USDA Natural Resources Conservation Service (NRCS), which collects 'snowpack and related climatic data in the Western United States'. There are over 730 SNOTEL sites in the western USA and Alaska.
- The Soil Climate Analysis Network (SCAN), which is similar to the SNOTEL network but is, instead, located where snow accumulation is not significant. The purpose of the SCAN network is to provide nationwide soil moisture and climate data. Like the SNOTEL data, these data are also available from the National Water and Climate Center of the USDA NRCS (http://www.wcc.nrcs.usda.gov/scan/).
- Automated Surface Observing Systems (ASOS) (http://www.nws.noaa.gov/asos/), which support 'weather forecast activities and aviation operations and, at the same time, support the needs of the meteorological, hydrological, and

climatological research communities'. This is the primary surface weather observing network in the USA and is supported by three US agencies: the National Weather Service, the Federal Aviation Agency and the Department of Defense. There are approximately 850 ASOS stations, primarily at airports.
- Local/Regional Networks, which include: Oklahoma Mesonet (http://www. mesonet.org/); MesoWest (Utah) (http://mesowest.utah.edu/index.html); and Texas Mesonet (http://mesonet.tamu.edu/).

The World Meteorological Organization (WMO) maintains a website listing its member states (http://www.wmo.int/pages/members/index_en.html). From that site, the member meteorological or hydrometeorological services can be reached, of which there are about 190. The services and information content offered by these member states vary considerably depending on agency resources. Besides accessing the individual agency, the World Data Center for Meteorology housed within the US National Climatic Data Center (NCDC) (http://www.ncdc.noaa.gov/oa/wdc/index.php) holds much of the world's data records for meteorology. These data can be accessed and used for hydrological modelling. Depending on the particular data set, it may be easier to obtain the data from another source, as there may be value-added from that other source. For example, data available from the NCDC may also be available from NOAA Regional Climate Centers (in the USA) in formats easier to ingest and manipulate (see, for example, http://www.ncdc.noaa.gov/oa/climate/regionalclimatecenters.html). Still another database that covers much of Africa is the Système d'Informations Environnementales sur les Ressources en Eau et leur Modélisation (Environmental Information System on Water Resources and their Modelling) (SIEREM) (Boyer *et al.*, 2006; available at http://www.hydrosciences.fr/sierem/) and this is sponsored by HydroSciences Montpellier, a Joint Research Unit of CNRS (Centre National de la Recherche Scientifique), IRD (Institut de Recherche pour le Développement) and the University of Montpellier 1 and University of Montpellier 2. Additional guidance for acquiring data sets for use in hydrological modelling is presented in Lacroix *et al.* (2000). The above list is not meant to be all-inclusive; but, instead, to indicate the diverse sources of climate data.

Gridded climate data from point data

There are several sources of gridded climate data that have been produced from point data. For example, in Australia, the Bureau of Meteorology (BOM) provides daily gridded data of rainfall, maximum temperature (Tmax), minimum temperature (Tmin) and 'solar exposure', which is defined as 'the total amount of solar energy falling on a horizontal surface'. The periods of record and spatial resolution are shown in Table 15.2. In the USA, gridded Tmax, Tmin and precipitation data are available from the NOAA NCEP Climate Prediction Center (CPC) (Table 15.3). These data are derived from analysis of COOP data. In the European Union, the European Climate Assessment & Dataset project (ECA&D) (http://eca.knmi.nl/) has produced and continues to produce a daily gridded data set named 'E-OBS' that consists of Tmin, Tmax and precipitation for the years 1950 to the present time on four different grids: a 0.25° and 0.5° regular geographic coordinate system as well as a 0.22° and 0.44° rotated pole grid, with the north pole at 39° 25" N, 162° W. The region covered is from 25° N to 75° N and 40° W to 75° E (Haylock *et al.*, 2008). The period of record and spatial resolution are shown in Table 15.4. Gridded data can relate the values

Table 15.2. Australian gridded data sets from the Bureau of Meteorology.

Parameter	Units	Period of record	Spatial resolution
Rainfall	mm	1900 onwards	0.05°/5 km
Max. temperature	°C	1911 onwards	0.05°/5 km
Min. temperature	°C	1911 onwards	0.05°/5 km
Solar exposure	MJ	1990 onwards	0.05°/5 km

Table 15.3. US gridded data sets derived from climate station data.[a]

Data set name	Period of record	Spatial resolution	Temporal resolution
US Unified Daily Precipitation Analysis	1948–1998	0.25° × 0.25°	Daily
Real-time gridded precipitation	1996–present	0.25° × 0.25°	Daily
Tmin	1948–present	0.5° × 0.5°	Daily
Tmax	1948–present	0.5° × 0.5°	Daily

[a]URLs for data download: ftp://ftp.cpc.ncep.noaa.gov/precip/daily_grids/tmax/; ftp://ftp.cpc.ncep.noaa.gov/precip/daily_grids/tmin/;ftp://ftp.cpc.ncep.noaa.gov/precip/wd52ws/us_daily/

Table 15.4. European gridded data sets (Version 3.0).[a]

Data set grid	Parameter[b]	Period of record
0.25° regular grid	TG TN TX RR	1950–2010
0.50° regular grid	TG TN TX RR	1950–2010
0.22° rotated grid	TG TN TX RR	1950–2010
0.44° rotated grid	TG TN TX RR	1950–2010

[a]URL for data download: http://eca.knmi.nl/download/ensembles/ensembles.php
[b]TG, daily mean temperature; TN, daily minimum temperature; TX, daily maximum temperature; and RR, daily precipitation sum.

presented as either being fixed to a corner of the grid cell or being representative in some manner to the centre of the cell (e.g. average over the cell, maximum in the cell, etc.). The assignment is made by the person who created the data set and then that information is carried over to the software used for further use of those data. It is really a matter of where the grid cell is located as, in general, the value is not considered to be only for some infinitely small point but to be in some way representative of the cell.

Besides the agency-produced gridded data, there are some tools available for producing customized gridded data sets. For example, a thin plate spline interpolation scheme can be applied to data arranged on an irregular grid to produce a regular grid (e.g. Press *et al.*, 1992). This has been done with COOP data in the Rio Grande Basin, USA to allow for a continuous time series of observations. Otherwise, any of the available climate stations within or near the basin contained missing data of sufficient magnitude for the model requirements for continuous data to be allowed only two stations to be used for a 2-year period of modelling. By using the irregularly spaced COOP network, on any given day, sufficient stations were reporting to allow the interpolation to proceed. In this case, each of the grid locations of the interpolated grid became the location of a pseudo-climate station with a continuous period of record over a long period (in this case, 25 years, though it could have been longer). Daley (1996) discussed several more robust techniques for interpolation between data that are on an irregular grid.

Gridded climate data whose native format is a grid

Some data are obtained from instruments that make measurements that are tied to a grid; for example, the Multi-sensor Precipitation Estimate (MPE) Stage IV precipitation data (Lin and Mitchell, 2005) were derived from NEXRAD (Next-Generation Radar, a radar network operated by the US National Weather Service) weather radar measurements which are sampled on a grid called the Hydrologic Rainfall Analysis Project (HRAP) grid. MPE Stage IV is used by the US River Forecast Centers and is an analysis obtained from consideration of NEXRAD weather radar data and precipitation gauges to produce a precipitation estimate at hourly intervals and with a spatial resolution of 4.7 km. This analysis uses the NEXRAD grid as the base grid and, as described in Lin and Mitchell (2005), a precipitation gauge analysis that uses hourly precipitation data to create a gauge-only grid that is compatible with the NEXRAD grid. The MPE Stage IV process then merges these and other data sets. An example of MPE Stage IV data is shown as a greyscale-shaded image in Fig. 15.2. Data derived from a satellite imager are naturally gridded; for example, the PERSIANN

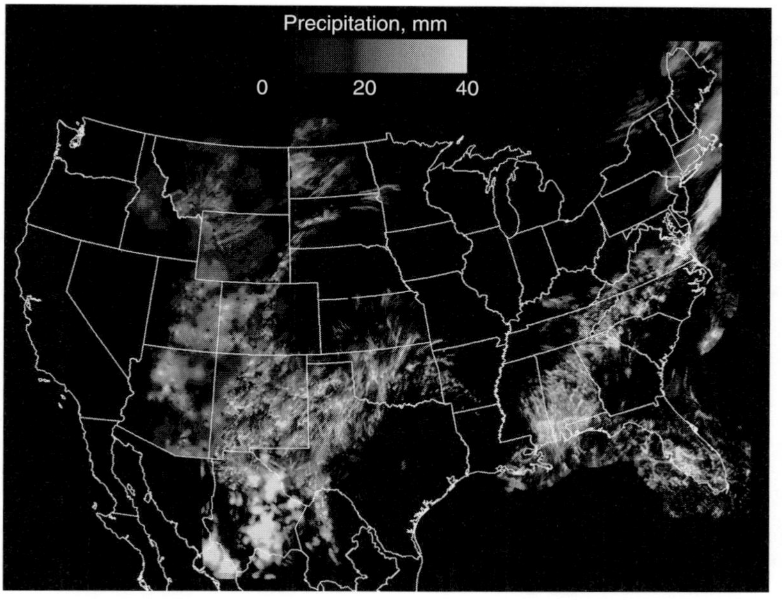

Fig. 15.2. Greyscale-shaded image of MPE (Multi-sensor Precipitation Estimate) Stage IV precipitation for 16 August 2006, a day that saw several floods in New Mexico. The scale in the upper portion of the image is in units of mm (kg m^{-2}). (Data provided by NCAR/EOL (US National Center for Atmospheric Research Earth Observing Laboratory) under the sponsorship of the National Science Foundation; http://data.eol.ucar.edu/).

data set, which is based on geosynchronous operational environmental satellite long-wave infrared (GOES IR) imagery, and PERSIANN-GT, which is based on GOES as well as on the satellite-based Tropical Rainfall Measurement Mission (TRMM) (Sorooshian *et al.*, 2000). In these examples, one may wish to distinguish between (i) 'real' data, i.e. data obtained using a sensor to measure a physical quantity (e.g. temperature with a thermometer, amount of rainfall with a rain gauge), (ii) 'data' that are the result of submitting the measured quantity (e.g. back-scattered signal strength from weather radar) to some algorithm that relates the measured quantity (back-scatter signal strength) to a physical parameter (rainfall amount), and (iii) a result obtained by 'analysing' different 'measurements' to a consistent product (MPE Stage IV).

Gridded climate data derived directly from other model/reanalysis data sets

A retrospective analysis or, as it is called, a reanalysis is a single data assimilation system used to merge some or all of the archived historical observations (ground observations, balloons, satellite data) and convert them into consistent gridded states of the atmosphere for climate applications. Several of these kinds of data sets are listed in Table 15.5. These can be used directly as they are or they can be downscaled using several different modelling techniques. Downscaling is the process of 'interpolating' a coarse-resolution data set to a finer resolution data set in space and/or time. This can be accomplished either by using statistical techniques (such as kriging using thin plate splines) or through a dynamic process (such as using a regional climate model to create a high-resolution data set from coarse-resolution global circulation model output). For example, the file for the parameter of interest is sampled at a location of interest (the grid cell located nearest to the location) over the time interval of interest to create a time series of that parameter. Similarly, additional parameters can be sampled until all of the information needed by the model is acquired. The location of the grid

Table 15.5. Available reanalysis data sets.

Data set name	Domain	POR	Spatial resolution	Temporal resolution	More information[a]
ERA-15	Global	1979 to 1993	2.5° × 2.5°	Four times daily and monthly	ERA-15
ERA-40	Global	Mid-1957 to mid-2002	2.5° × 2.5°	Four times daily and monthly	ERA-40
NCEP/NCAR Reanalysis 1	Global	1948/01/01 to present	2.5° × 2.5°	Four times daily, daily and monthly	NCEP 1
NCEP/NCAR Reanalysis 2	Global	1979/01 to 2008/12	2.5° × 2.5°	Four times, daily and monthly	NCEP 2
NARR	North America	1979/01/01 to present	32 km	Eight times daily and monthly	NARR
JRA-25	Global	1979 to December 2004	1.125° × 1.125°	Four times daily	JRA-25

[a]URLs for more information:
ERA-15: http://www.ecmwf.int/products/data/archive/descriptions/er/full.html
ERA-40: http://www.ecmwf.int/products/data/archive/descriptions/e4/index.html
NCEP 1: http://www.esrl.noaa.gov/psd/data/gridded/data.ncep.reanalysis.html
NCEP 2: http://www.esrl.noaa.gov/psd/data/gridded/data.ncep.reanalysis2.html
NARR: http://www.esrl.noaa.gov/psd/data/gridded/data.narr.html
JRA-25: http://jra.kishou.go.jp/JRA-25/AboutJRA25_en.html and http://www.cisl.ucar.edu/news/08/0714.jra-25.jsp

cell becomes the location of a pseudo-climate station in that there is no real climate station located there; instead, a climate data set for that location has been created and ascribed to that location. In some instances, some additional processing may be required to arrive at a daily average or total from the temporal sampling of the data set (namely, solar insolation at 3 h intervals needs to be summed to arrive at a total value).

Gridded climate data downscaled from model/reanalysis data sets

The reanalysis data sets presented in Table 15.5 can be used without modification or they can be downscaled using several different modelling techniques: the Regional Spectral Model (RSM; Juang and Kanamitsu, 1994), an atmospheric model such as the Fifth-Generation NCAR (US National Center for Atmospheric Research)/Penn State Mesoscale Model (MM5; Grell *et al.*, 1994), or the Advanced Research WRF (WRF-ARW; Skamarock *et al.*, 2008). Downscaling provided by the atmospheric models is very computer intensive and is not practical for application in developing long-term climate data sets for hydrological modelling. For example, using the NARR data in a nested version of WRF-ARW to downscale to 1 km spatial resolution over a domain of 125 × 125 km with hourly output and a simulation period of 30 hours takes about 30 hours on a four-processor desktop PC; however, on a high-performance computer it takes only about 3 h. This downscaling allows not only the acquisition of data with a finer spatial resolution, but the increased temporal resolution also allows for the development of some parameters that are not readily available from the 3 h data (viz. the 2 m air temperature every 3 h does not allow precise determination of Tmin or Tmax – generally, such a data set would have Tmin too high and Tmax too low because the actual minima and maxima are not captured in the 3 h data).

Techniques for Comparing Data Sets/Model Results

There are two cases to consider in this section on comparing data sets: the first is comparing model input parameters from different sources, and the second is obtaining a measure of success with the models by comparing model output with observations. In the first case, examples of comparing model input would include a comparison between precipitation data sets to better understand the model output (e.g. high discharge with little attendant precipitation, etc.) or to justify using a modelled parameter in place of a measured parameter (e.g. potential evapotranspiration (PET) from the NARR data set as opposed to a tower-based measurement or estimates of PET, etc.). In the second case of comparing model output with observations, determining how well the simulation has performed is of interest. Perhaps, both cases are really very similar, but they have just been treated differently in the literature. In any case, there are several measures of 'model efficiency' in use, and these will be described below after discussion of the comparison of input parameters. It should also be noted that there are other ways in which time series may be analysed than those presented below, such as spectral analysis in the frequency domain, and the building of stochastic models that attempt to represent the time series (Box and Jenkins, 1976). In the case of stochastic models, there exist additional categories depending on the purpose of the analysis, for example forecasting models, transfer function models and control models (Box and Jenkins, 1976). A discussion of these later forms of time series analyses is beyond the scope of this chapter because it is not appropriate for the analysis considered here.

Comparison of model input parameters (e.g. two different series of Tmax or PET, etc.)

An example of comparison of model input parameters is presented in this section using NARR-generated PET and comparing that with standardized reference evapotranspiration (ETsz) as derived from parameters measured with meteorological tower sensors. There are two problems that need to be addressed in this example: serial correlation in the data and errors in both coordinates (or uncertainty in both variables). The problem of errors in both variables (both the 'dependent' and 'independent' variables are observed with error) has not received much attention in the literature, either from the standpoint of theory or in actual use. However, it is quite usual for results to be presented where this error problem is ignored, this in spite of sufficient studies to allow for this problem to be addressed properly. We will first discuss the issue of errors in both variables and then (in the next subsection) address the issue of serial correlation.

In a series of papers by Isobe *et al.* (1990) and Feigelson and Babu (1992) considerable effort has been expended on understanding several least-squares linear regression procedures. Their focus was on a particular problem in astronomy with some similarity to our problem; namely, there are errors in both variables under consideration and these errors are either unknown or insignificant. The six methods that these researchers considered were: ordinary least-squares (OLS) regression of Y on X; OLS regression of X on Y; the bisector of the two OLS lines; orthogonal 'reduced major-axis' regression; 'reduced major-axis' regression; and mean ordinary least squares.

Isobe *et al.* (1990) and Feigelson and Babu (1992) emphasized the difference between 'least-squares' regression and other techniques that are 'non-least squares' linear regression methods (such as those that attempt to incorporate known measurement errors), as well as the choice between which variable would be the dependent variable and which would be the independent variable. Their definition of OLS linear regression was: 'ordinary least-squares regression of the dependent variable Y against the independent variable X, or OLS($Y \mid X$). In OLS($Y \mid X$), the regression line is defined to be that which minimizes the sum of the squares of the Y residuals'.

Isobe *et al.* (1990) and Feigelson and Babu (1992) also stated that the latter three methods, which are 'invariant to switching the variables X and Y variables, lead to completely different regression lines, both mathematically and in real applications'. Other important aspects related to the assumptions surrounding the use of OLS($Y \mid X$) are:

1. 'the true relation between the variables is linear;

2. the values of the independent variable are measured without error;

3. the observed values of the dependent variable are subject to errors which have zero mean, finite common variance, and are independent from point to point; and
4. the errors do not depend on the independent variable'.

In the example presented here, it is assumed that (**1**) is valid; for (**2**), that both the dependent and independent variables have error (or uncertainty); in the case of (**3**), we know that serial correlation exists; and, for point (**4**), the nature of the errors is unknown. Therefore, the use of ordinary linear regression in this case is inappropriate. In their articles, Isobe *et al.* (1990) and Feigelson and Babu (1992) provide formulae for all six regressions as well as an error analysis appropriate for each. Xu (2001) discussed both the use of regression analysis and model efficiency measures and demonstrated his methodology with a case study using a conceptual water balance model. He quite succinctly discussed and reminded us of these assumptions inherent in the use of regression analysis (as stated above) by saying that, if any of these conditions were not met, then the further use of OLS was inappropriate and any interpretation of results based on OLS would be 'fallacious'. However, it should be noted that Xu (2001) did not address the problem of errors in both the 'dependent' and 'independent' variables.

The implementation of the equations in Isobe *et al.* (1990) was accomplished in the commercial software language IDL® (Interactive Data Language) and made available as part of the IDLASTRO library (http://idlastro.gsfc.nasa.gov/) of IDL® routines for use in astronomical applications. The results of this regression analysis of ETsz versus NARR-generated PET are shown in Table 15.6. Figure 15.3 shows the results as a scatter plot of ETsz versus NARR-generated PET overlaid with three of the regression results: OLS($Y|X$), OLS($X|Y$) and orthogonal regression of ($Y|X$). The residuals from the orthogonal regression are presented in Fig. 15.4, where there is a small seasonal signature. Based on these results, different results may be obtained depending on whether one performs OLS($Y|X$) (Case (0)) or OLS($X|Y$) (Case (1)) or orthogonal regression (Case (4)). These first two cases also agree with findings using other programs, such as Microsoft Excel™ or Originlab ORIGIN®.

Serial correlation in data sets

Wilks (2006) discussed the case of serial correlation in data but did not address the problem of errors in both variables. He discussed how such correlation has impacts on the effective sample size and how this then changes the degrees of freedom that allow for statistical confidence levels to be established. The effective sample size (n') could

Table 15.6. Results of the regression analysis of the standardized reference evapotranspiration (ETsz) versus the North America Regional Reanalysis (NARR) potential evapotranspiration (PET).[a]

Case[b]	Intercept (*a*)	Slope (*b*)	Standard deviation of intercept (sigma_*a*)	Standard deviation of the slope (sigma_*b*)
(0)	0.1035	0.7221	0.07340	0.01085
(1)	−0.3817	0.7942	0.07278	0.01067
(2)	−0.1349	0.7575	0.06723	0.009957
(3)	−0.06658	0.7474	0.07061	0.01053
(4)	−0.1333	0.7573	0.06732	0.009774
(5)	−0.1391	0.7582	0.06721	0.009955

[a]The *Y*-axis is ETsz and the *X*-axis is NARR PET.
[b]Case (0), ordinary least squares (OLS) *Y* versus *X*; Case (1) OLS *X* versus *Y*; Case (2), OLS bisector; Case (3), orthogonal reduced major axis; Case (4), reduced major axis; Case (5) mean OLS.

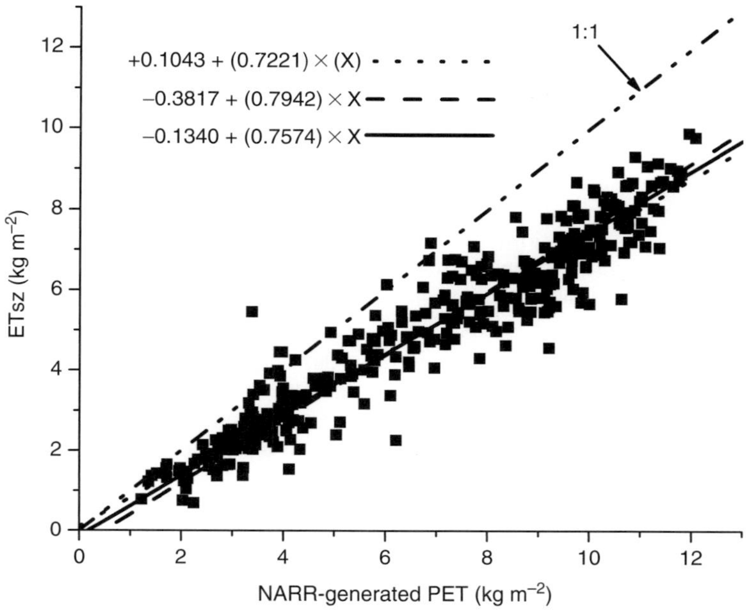

Fig. 15.3. Scatter plot of daily ETsz (standardized reference evapotranspiration) versus NARR (North America Regional Reanalysis)-generated daily PET (potential evapotranspiration) values for 2002, along with the results of OLS(*Y*/*X*) (dotted line), OLS(*X*/*Y*) (dashed line), and orthogonal regression of (*Y*/*X*) (solid line).

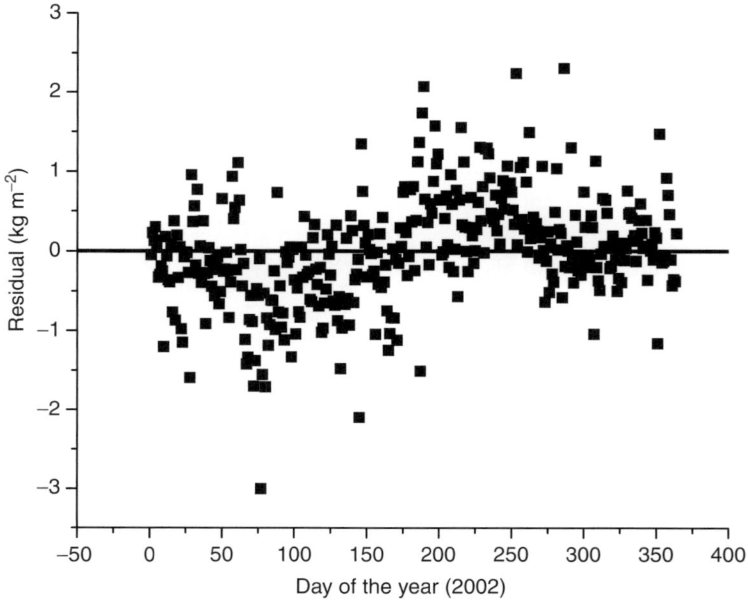

Fig. 15.4. The residuals from the orthogonal linear regression of ETsz (standardized reference evapotranspiration) versus NARR (North America Regional Reanalysis)-generated PET (potential evapotranspiration) values for 2002 plotted by day of year. Note the slight seasonal variation of the signal; perhaps, a better fit could be obtained by adjusting for this before the regression analysis.

be estimated approximately following Wilks (2006; his equation 5.12):

$$n' \cong n\frac{1-\rho_1}{1+\rho_1} \tag{15.1}$$

where ρ_1 is the lag − 1 autocorrelation coefficient and n is the number of samples. Also from Wilks (2006; his equation 5.13), the variance of a time average (of individual x values) over a sufficiently large sample could be determined as:

$$\text{Variance}\left[\bar{x}\right] \cong \frac{s^2}{n'} = \frac{s^2}{n}\left(\frac{1+\rho_1}{1-\rho_1}\right) \tag{15.2}$$

where s is the sample variance. The ratio $[(1 + \rho_1)/(1 - \rho_1)]$ acts as the 'inflation factor' which adjusts the variance of the sampling distribution of the time average to reflect the influence of the serial correlation. Wilks continued to discuss the impact of serial correlation on linear regression where the presence of this correlation invalidates the assumptions underlying the regression; that is to say, independence of the residuals. For the examination of the residuals for serial correlation, he suggested using the Durbin–Watson test statistic, d (Durbin and Watson, 1951):

$$d = \frac{\sum_{i=2}^{n}(e_i - e_{i-1})^2}{\sum_{i-1}^{n} e_i^2} \tag{15.3}$$

where the e_i values are the residuals from the linear regression. Although a rigorous statistical analysis is indicated, it can be stated that, in general, low values of d (≤1) indicate the presence of positive serial correlation while high values of d (≥2) indicate the presence of negative serial correlation. The range of d is from 0 to 4. Specifics for applying the Durbin–Watson test are given in Gujarati (2004). When serial correlation does exist, the variance used to determine the standard errors (s_e) in the estimate of the slope and intercept from the linear regression must be adjusted accordingly:

$$s_{e,new} = \left[s_e^2\frac{1+\rho_1}{1-\rho_1}\right]^{1/2} \tag{15.4}$$

where ρ_1 is as above. The impact of this is that the error estimates for slope and intercept from the regression analysis described above are adjusted upwards.

For the PET data set, the value ρ_1 at lag −1 is 0.35654 and the plot of the autocorrelation function is shown in Fig. 15.5. Given that sigma_a (the standard deviation of the intercept) and sigma_b (the standard deviation of the slope) are of the order 0.067 and 0. 010, respectively, and the correction factor is the square root of 2.11, the new sigma_a and sigma_b are 0.097 and 0.015, respectively. Even with these 'inflated' errors, the ratio between the slope and the estimated error in the slope is ~50, which indicates that its value is well established.

Comparing hydrological modelling simulations with observations

Assessing the performance of a model requires measures that may be used to compare the simulated output with the observations. An extensive literature exists that discusses various measures and also makes recommendations for using specific measures, depending on the particular author's perspective; it is becoming clear that many of the measures used in the past were insufficient to capture all aspects of comparability between two different time series, and that the use of some of the measures appears not to take into account the assumptions inherent in their use. Taking a somewhat different approach that certainly needs further investigation is recent work by Reusser *et al.* (2009), in which they attempted to compare the temporal dynamics between the simulated and observed model results. They analysed the effectiveness of over 40 different measures and assessed their suitability at different time scales.

Statistical measures used as indicators of model performance include summary statistics such as mean deviation and standard deviation (SD), test statistics and an analysis of residual errors (Loague and Green, 1991) such as maximum error (ME), root mean square error (RMSE), modelling efficiency (EF) and coefficient of residual mass (CRM). Other measures of model performance are average absolute deviations (AAD), sometimes referred to as mean absolute error (MAE), Wilmot's index of

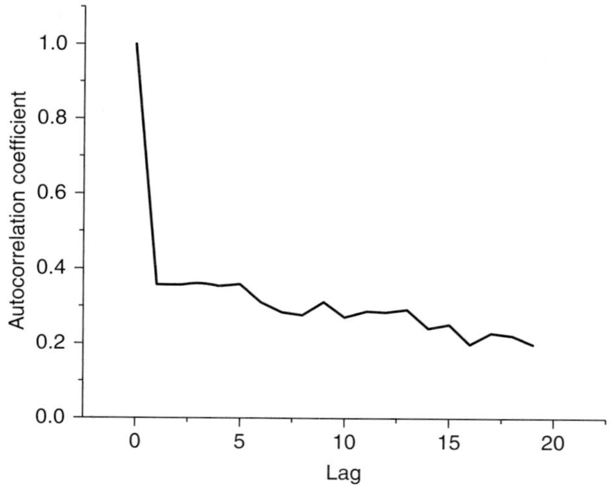

Fig. 15.5. The PET (potential evapotranspiration) data set autocorrelation coefficient plotted versus lag for lags from 0 to 19.

agreement (W) (Willmott *et al.*, 1985; Loague and Green, 1991) and the Nash–Sutcliffe efficiency (NSE) (Nash and Sutcliffe, 1970). The equations for the measures are given in Table 15. 7 along with their range of values.

Regarding test statistics such as Pearson's correlation coefficient (r) or the coefficient of determination (r^2), Willmott (1982) stated: 'Since the relationships between r and r^2 and model performance are not well-defined, and not consistent, r or r^2 should not be part of an array of model performance measures.' However, according to Willmott (1982), the intercept and slope derived from a least-squares regression are appropriate. It would seem as though this could be proper, assuming that the guidance in the previous section on comparing model input parameters is followed and that the assumptions implicit in the analysis are met. Legates and McCabe (1999) have examined several 'goodness-of-fit' measures in their article on hydrological and hydroclimatic model validation. In particular, they consider r^2, NSE and W. Besides NSE and W, they also discuss modified versions of both of these measures, which are defined by replacing the power of 2 in the exponents by a variable that is initially set to 1, as well as replacing the parens with absolute value brackets:

Modified modelling efficiency (EF) (NSE):

(See Equation 15.5 at the bottom of the page.)
where P_i are the predicted values and O_i are the observed values.

Modified W:

(See Equation 15.6 at the bottom of the page.)

$$\mathrm{E_j(Modified\ EF)} = \frac{\sum_{i=1}^{n}\left|O_i - \overline{O}\right|^j - \sum_{i=1}^{n}\left|P_i - O_i\right|^j}{\sum_{i=1}^{n}\left|O_i - \overline{O}\right|^j} \tag{15.5}$$

$$\mathrm{W_j(Modified\ W)} = 1 - \left[\frac{\sum_{i=1}^{n}\left|P_i - O_i\right|^j}{\sum\left(\left|O_i - \overline{O}\right| + \left|P_i - \overline{O}\right|\right)^j}\right] \tag{15.6}$$

Table 15.7. Recommended 'goodness-of-fit' measures for model validation (After Legates and McCabe, 1999).

Name of statistic	Equation[a]	[Range]/Ideal value
Maximum error (ME)	$\mathrm{ME} = Max\left\|P_i - O_i\right\|_{i-1}^{n}$	[0: + ∝]/0.0
Average absolute deviation or mean absolute error (AAD)	$\mathrm{AAD} = \frac{1}{n} \times \sum_{i=1}^{n}\left\|P_i - O_i\right\|$	[0: + ∝]/0.0
Root mean square error (RMSE)	$\mathrm{RMSE} = \left[\frac{\sum_{i=1}^{n}(P_i - O_i)^2}{n}\right]^{0.5} \times \frac{100}{\bar{O}}$	[0:+ ∝]/0.0
Modelling efficiency (Nash–Sutcliffe) (EF)	$\mathrm{EF} = \frac{\sum_{i=1}^{n}(O_i - \bar{O})^2 - \sum_{i=1}^{n}(P_i - O_i)^2}{\sum_{i=1}^{n}(O_i - \bar{O})^2}$	[−∝:1]/1.0
Modified EF (E_j)	$E_j = \frac{\sum_{i=1}^{n}\left\|O_i - \bar{O}\right\|^j - \sum_{i=1}^{n}\left\|P_i - O_i\right\|^j}{\sum_{i=1}^{n}\left\|O_i - \bar{O}\right\|^j}$	[−∝:1]/1.0
Wilmot's Index of Agreement (W)	$W = 1 - \left[\frac{\sum_{i=1}^{n}(P_i - O_i)^2}{\sum\left(\left\|O_i - \bar{O}\right\| + \left\|P_i - \bar{O}\right\|\right)^2}\right]$	[0:1]/1.0
Modified W (W_j)	$W_j = 1 - \left[\frac{\sum_{i=1}^{n}\left\|P_i - O_i\right\|^j}{\sum\left(\left\|O_i - \bar{O}\right\| + \left\|P_i - \bar{O}\right\|\right)^j}\right]$	[0:1]/1.0

[a] P_i are the predicted values, O_i are the observed values, n is the total number of observations and $\bar{O}$ is the mean of the observed data.

Their reason for changing the exponent from 2 to 1 is to remove the 'inflated influence' that squaring the errors and differences has on the outcome. Further, Legates and McCabe (1999) suggested that comparing with the mean value (which is what the NSE does) is not always appropriate in hydrological or climatic variables and that it might be better to compare with some 'baseline' value instead. Examples of these 'baselines' are temporal segments obtained from the observations such as climatology or persistence. These temporal segments could be obtained through monthly means, for example. The conclusion of Legates and McCabe (1999) provides some advice that would allow for better interpretation of model results:

> In addition to E_1 or E'_1, it is strongly recommended that the observed and modelled means and standard deviations, as well

as MAE or RMSE (and probably both), be reported. Scatter plots and residual and outlier analyses also are essential to an appropriate model assessment.

A paper produced by the American Society of Civil Engineers in 1993 (ASCE, 1993), based on recommendations by the ASCE Task Committee on Definition of Criteria for Evaluation of Watershed Models of the Watershed Management Committee, Irrigation and Drainage Division, concluded that three measures of efficiency were appropriate for continuous-hydrograph modelling and four measures were appropriate for event-based modelling. The measures for continuous-hydrograph modelling are:

1. Deviation of runoff volumes, Dv
2. Nash–Sutcliffe coefficient, R^2
3. Coefficient of gain from the daily mean, DG

and for single-event hydrograph modelling:

1. Simple per cent error in peak, PEP
2. Sum of squared residuals, G
3. Total sum of squared residuals, TSSR
4. Total sum of absolute residuals, TSAR

The symbols (Dv, R^2, DG, PEP, G, TSSR, TSAR) in the above listed measures are different from most of the ones used in our discussion.

There are several tools available as an aid to determine the value of some of these parameters. The first is an Internet-based tool called the 'Web-based Hydrograph Analysis Tool' (WHAT) (http://cobweb.ecn.purdue.edu/~what/). The second is a software package that may be downloaded to a Microsoft Windows® computer; it is called Integrated Resources for Evaluating Numerical Estimates (IRENE) (Fila *et al.*, 2003) (http://www.sipeaa.it/ASP/ASP2/IRENE.asp). Also, the software ANNIE (Interactive hydrologic analyses and data management) is part of the US Geological Survey Water Resources Applications Software suite (http://water.usgs.gov/software/ANNIE/). A description of ANNIE is found in their summary document: 'ANNIE is a program designed to help users interactively store, retrieve, list, plot, check, and update spatial, parametric, and time-series data for hydrologic models and analyses.'

Data Set Development for the Mimbres Basin of the Rio Grande in New Mexico: An Example of Creating a Regularly Gridded Homogeneous Data Set from Disparate Sources

As part of an effort to model several sub-basins in the Rio Grande Basin for the purpose of understanding the hydrological cycle of those sub-basins, a distributed climate data set was developed for use by the SWAT model (Arnold *et al.*, 1998). The model can be set up to use many different climate stations; it was felt that a semi-distributed model might perform better with a distributed data set. In addition, there are no climate stations in the Mimbres basin with the desired period of record or necessary suite of instruments. Actually, there are only three stations in the basin; two of those are at the top of a mountain at the basin boundary with a limited period of recordings and the third does not measure precipitation.

In order that all climate variables were available, the data had to be retrieved from several different data sets. The variables used were precipitation, Tmin, Tmax, average wind speed, average relative humidity (RH), daily insolation and PET. All of these are available from the NARR data set, although the spatial resolution for some of the parameters (precipitation, Tmin and Tmax) is not ideal (32 km), nor is the temporal resolution (3 h) adequate for precise determination of Tmin and Tmax. However, as the remaining variables (average wind speed, average RH, daily insolation and PET) could be obtained more satisfactorily from NARR, and because of the concept of air mass whereby atmospheric variables change slowly in the horizontal direction (Barry and Chorley, 2003), it was felt that these would be adequate for modelling in this case. An example of the spatial coverage and detail for one parameter of the NARR data set (PET) is shown in Fig. 15.6. Data for precipitation was obtained from the MPE Stage IV.

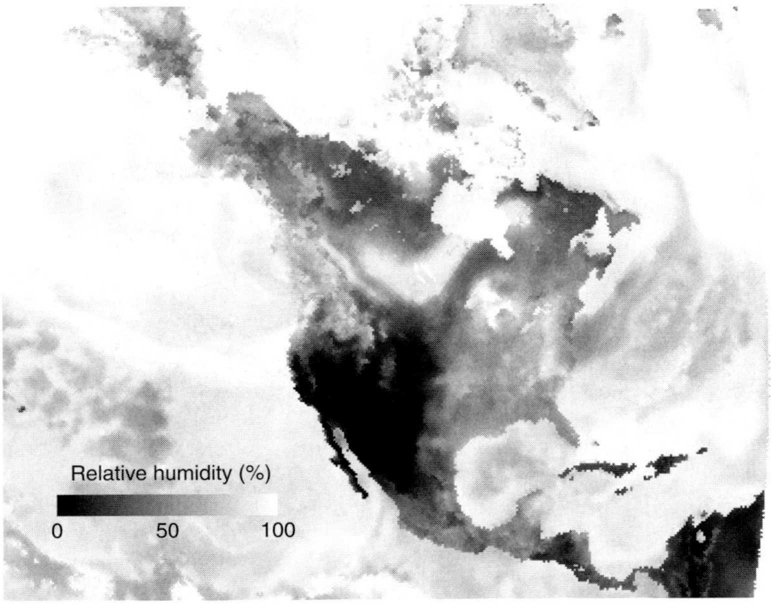

Fig. 15.6. Greyscale-shaded image of the NARR (North America Regional Reanalysis) RH (relative humidity) 3 h value from day 165 (2005) and averaged over the period 15.00 to 18.00 h GMT showing the domain of the NARR data set. The NARR data were also used as the source of average wind speed, potential evapotranspiration (PET) and daily insolation. All parameters have the same spatial coverage and resolution.

Data for Tmin and Tmax were obtained from a disaggregation of the Parameter-elevation Regressions on Independent Slopes Model (PRISM) data set (Daly *et al.*, 1994), which contains monthly average Tmin and Tmax at approximately 4 km spatial resolution for the continental USA. An example of a PRISM data set (for Tmin) is shown as a greyscale-shaded image in Fig. 15.7.

The data that are already in a gridded format are available in a variety of grid spacings and map projections or coordinate systems. In order to establish pseudo-climate stations for input to the models, these disparate data sets must be re-sampled and re-projected to a common grid spacing and map projection. For this, the software package 'Environment for Visualizing Information' (ENVI®) published by ITT Visual Information Solutions was used. This software package is built on IDL®, also by ITT Visual Information Solutions, and enables batch processing of data sets so that they can be ingested into one file format, grid spacing, and map projection or coordinate system, with output via re-sampling and re-projecting to a new grid spacing and map projection. The map projection and grid spacing that were used were the same as those used by the MPE Stage IV precipitation data set: the 'HRAP projection' (a type of polar stereographic map projection) and 4.7 × 4.7 km. The original file formats and grid details are given in Table 15.8 for the data sets that were used.

For the precipitation data set, the MPE Stage IV data were first output to the ENVI® file format (a flat binary file accompanied by a text header file) from the input 'GRIB' (GRIdded Binary) format. The locations of the grid cells that were within and just outside the Mimbres basin were identified by their sample/line (*i,j*) location. That location in each image was sampled in time and saved to a text file to form a time series of precipitation for that location. (An example of the MPE Stage IV is shown in Fig. 15.2 as a greyscale-shaded image that shows the spatial distribution of precipitation for one day.)

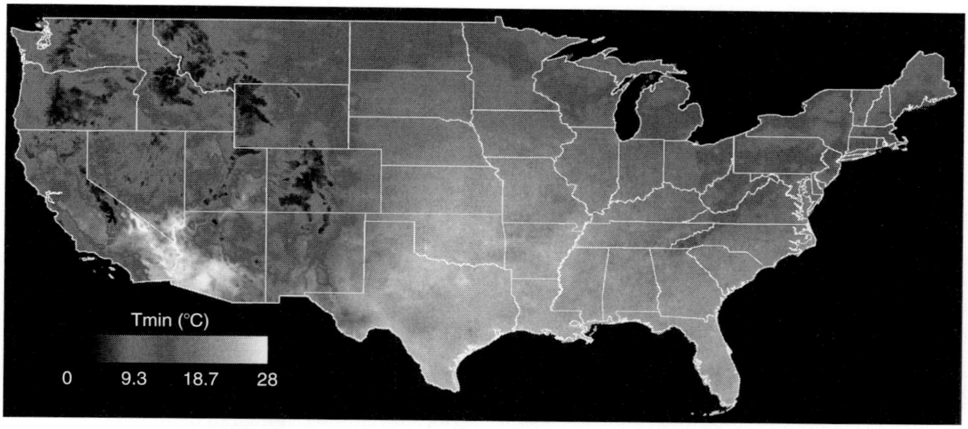

Fig. 15.7. Greyscale-shaded image of monthly average Tmin for July 2007 from the PRISM (Parameter-elevation Regressions on Independent Slopes Model) data set. These kinds of data were increased/decreased according to daily climate station data to create daily Tmin maps used for regional Tmin estimation; similarly, for Tmax (Data courtesy PRISM Climate Group, Oregon State University; http://www.prismclimate.org).

Table 15.8. Data sets used for creation of a distributed climate data set for use in hydrological modelling.

Parameter[a]	Data set[b]		Data format	Map projection
Tmax, Tmin	PRISM/Climate Station		ASCII grid/ASCII	Geographic coordinate system (2.50 arcmin)/ point data
Precipitation	Gridded	MPE Stage IV	GRIB (GRIdded Binary) file	4 km polar-stereographic NWS HRAP grid (4.7625 km at 60N)
		PERSIANN	Flat binary	Geographic coordinate system
		NARR	NetCDF (network Common Data Form) file	32 km Lambert Conformal
	Weather stations	COOP	Text	N/A
		RAWS	Text	N/A
		SNOTEL	Text	N/A
PET, SOLRAD, RH, Wind speed	NARR		NetCDF file	32 km Lambert Conformal

[a]PET, potential evapotranspiration; RH, relative humidity; SOLRAD, solar radiation (daily insolation); Tmax, maximum temperature; Tmin, minimum temperature.
[b]COOP, a US volunteer network; MPE, Multi-sensor Precipitation Estimate; NARR, North American Regional Reanalysis; PERSIANN, Precipitation Estimation from Remotely Sensed Information using Artificial Neural Networks; PRISM, Parameter-elevation Regressions on Independent Slopes Model; RAWS, Remote Automated Weather Stations; SNOTEL, SNOwpack TELemetry.

The Tmin and Tmax data sets resulted from disaggregation of the PRISM data. This process proceeded by making a regional additive/subtractive adjustment to the monthly average map so that it agreed with a climate station at a particular grid-point location on a daily basis. In particular, the PRISM data were re-projected, re-sampled and subset to the same region as the MPE Stage IV data set. Then, a climate station nearby the Mimbres basin (in this case a RAWS station) was selected as the station to provide the information for the disaggregation. The climate station's Tmax for a particular day in the month was compared with the PRISM monthly average Tmax for that month at the grid location closest to the climate station location. If the climate station's value was greater than (less than) the PRISM value, the whole prism map was adjusted upwards (downwards) by that amount, resulting in the daily Tmax map for that date. Similarly, this process was used for creating the daily Tmin maps. Obviously, the further away the climate station that provides the adjustment factor, the less reliable this technique becomes. Although not yet investigated, it may be possible to create a daily map that is more representative over a wider area by considering more adjustment stations and a more sophisticated adjustment technique (Daley, 1996). In addition, a recent paper by DiLuzio *et al.* (2008) showed how the PRISM data can be downscaled for Tmin, Tmax and precipitation using similar techniques in a more rigorous and complete manner. A subset of a thus derived daily Tmax file overlaid with the outline of the Mimbres basin is shown in Fig. 15.8, where one can easily see the gridded nature of the data – each cell represents the location of a pseudo-climate station. Figure 15.9 shows a greyscale-shaded DEM (digital elevation map) of the region in the vicinity of the Mimbres basin overlaid with an outline of the basin as well as the locations of the pseudo-climate stations.

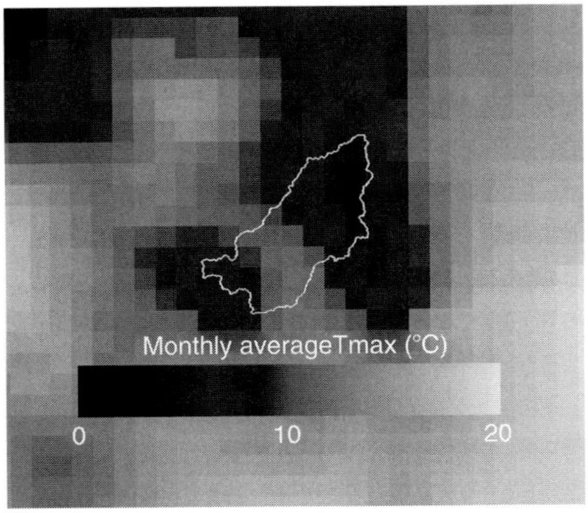

Fig. 15.8. A subset of a daily Tmax (maximum temperature) file overlaid with the outline of the Mimbres basin. Each cell represents the location of a pseudo-climate station. Each cell is 4.7 km on a side.

The remaining parameters (daily average RH, daily average wind speed, daily total potential evapotranspiration, and daily insolation (solar radiation)) were obtained from the NARR data sets by re-projecting, re-sampling, accumulating as necessary, and subsetting the original files that are arranged by parameter and year. These were approximately 500 MB of files in NetCDF (network Common Data Form) format. An example of a subset portion of NARR-accumulated daily PET is shown in Fig. 15.10.

Summary

Hydrological and agronomic models are increasingly being used to address problems in water resource management and to allow more efficient management of our agricultural resources. These models allow us to obtain a better understanding of the underlying processes that control the flow of water in our rivers and the development of crops. In order to use these models, we must find the necessary input information on which the models are based. Besides such information as soil type, terrain relief and land cover type, climate data are required at a variety of spatial and temporal scales. Generally, the agronomic models are at a spatial scale that requires climate information only for the site of the experiment; however, in those instances where the agronomic models are in a spatially distributed form, they, along with hydrological models, can benefit from climate data that are spatially distributed. The reason for this is that hydrological models are increasingly better able to take advantage of distributed processes owing to advances in computing technology.

In this chapter, we have described the variety of climate data sets that are available and given some details as to where these data may be obtained. In particular, we have described 'point' data obtained from climate stations, gridded data that have been created from processing point data on an irregular grid, and data that are, in their native format, gridded, i.e. weather radar precipitation data that are acquired on a regular grid, or analysis or reanalysis climate model data that are generated in

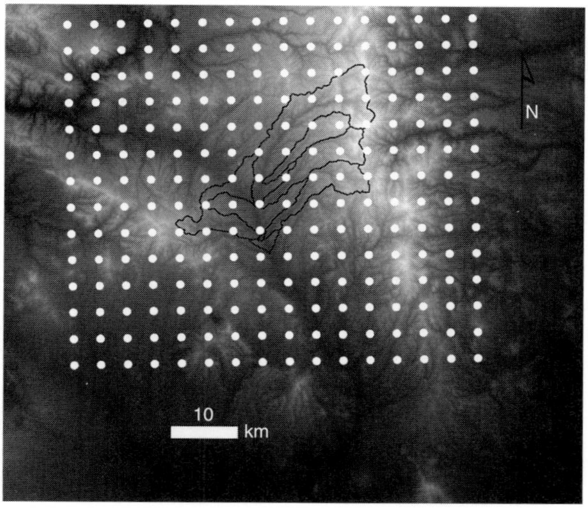

Fig. 15.9. Greyscale-shaded DEM (digital elevation model) overlaid with the outline of the Mimbres basin and 'dots' representing the location of the pseudo-climate stations. The separation of the stations is 4.7 km in each direction.

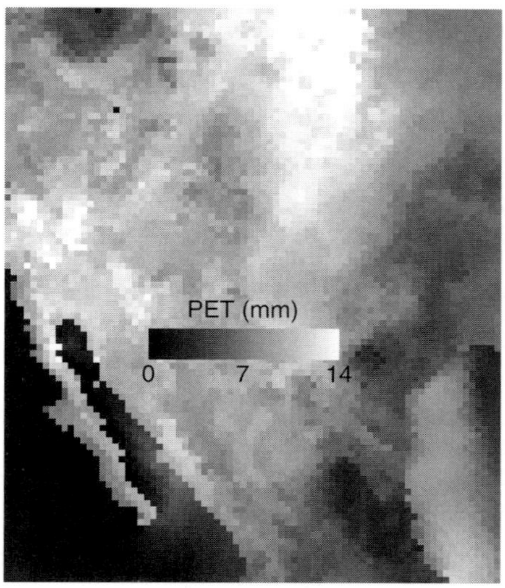

Fig. 15.10. Greyscale-shaded image of the accumulated NARR PET (North America Regional Reanalysis potential evapotranspiration) for day 180 (2002) and subset to the region from 20° N to 40° N and 95° W to 120° W. The scale is in kg m^{-2} (mm) water. In this scene, you can see the pixelation due to the 32 km spatial resolution. These subsets were further processed to the 4.7 km resolution of the MPE (Multi-sensor Precipitation Estimate) Stage IV precipitation data as well as re-projected to the same projection as the MPE Stage IV precipitation data for use in SWAT (Soil and Water Assessment Tool) modelling of the Mimbres basin of the Rio Grande in New Mexico.

a gridded format. In some instances, it may be required, or desired, to compare a particular climate parameter obtained from one data set with another. In order that this can be done in a rigorous manner, we have outlined the steps involved and presented a particular instance in which a reanalysis climate model for PET is compared with PET derived from climate-station-measured parameters. Also, where it is desired to build a distributed climate data set from disparate sources on to a homogeneous grid (spatially and temporally uniform between data types), we have described one such instance that was performed for the Mimbres basin in Southern New Mexico.

Because of the problems inherent in taking measurements of precipitation (for example) it may be preferable to create an analysis or reanalysis data set so that all of the parameters in a particular data set are physically meaningful – air temperature, relative humidity, solar radiation, precipitation, etc. Recent work by El-Sadek *et al.* (2010) has shown that there are large discrepancies between one precipitation data set (MPE Stage IV) and several different rain gauge measurements that are not consistent within a station or from year to year or between stations. In this particular instance, the discrepancy may be due to lack of rain gauge stations available to the MPE

Stage IV process, and a further analysis, or reanalysis, could be implemented to improve the MPE Stage IV. Nevertheless, if the atmospheric processes and land surface processes are brought into harmony, it would seem as though the hydrological and agronomic processes would be better described.

In addition to the above, we have also presented a review of the techniques and measures used for comparing hydrological and agronomic model outputs with observations. Based on work by Legates and McCabe (1999) only a limited number of measures provide the necessary information; these include a modified Nash–Sutcliffe efficiency and a modified Wilmot's index of agreement. Besides these measures, it is recommended that scatter plots along with residual and outlier analysis be presented.

Acknowledgement

Support for this work was provided by the National Geospatial Intelligence Agency (NGA), NGA University Research Initiative (NURI) grant HM1582-07-1-2030, the National Science Foundation EPSCoR grant EPS-081449 and the New Mexico State University (NMSU), College of Agriculture and Consumer Sciences (ACES) Agricultural Experiment Station.

References

Arnold, J.G., Srinivasan, R., Muttiah, R.S. and Williams, J.R. (1998) Large area hydrologic modeling and assessment. Part I: model development. *Journal of American Water Resources Association* 34, 73–89.

ASCE (American Society of Civil Engineers) (1993) Criteria for evaluation of watershed models. *Journal of Irrigation and Drainage Engineering* 119, 429–442.

Barry, R.G. and Chorley, R. (2003) *Atmosphere, Weather, and Climate*, 8th edn. Routledge, London.

Box, G.E.P. and Jenkins, G.M. (1976) *Time Series Analysis, Forecasting and Control*, revised edn. Holden-Day, Oakland, California.

Boyer, J.-F., Dieulin, C., Rouche, N., Cres, A., Servat, E., Paturel, J.E. and Mahé, G. (2006) SIEREM: an environmental information system for water resources. In: Demuth, S., Gustard, A., Planos, E., Scatena, F. and Servat, E. (eds) *Climate Variability and Change – Hydrological Impacts*. IAHS Publication No. 308, pp. 19–25.

Burnash, R.J.C. (1995) Chapter 10: The NWS river forecast system – catchment modeling. In: Singh, V.P. (ed.) *Computer Models of Watershed Hydrology*. Water Resources Publications, Littleton, Colorado, pp. 311–366.

Daley, R. (1996) *Atmospheric Data Analysis*. Cambridge University Press, Cambridge, UK.

Daly, C., Neilson, R.P. and Phillips, D.L. (1994) A statistical-topographic model for mapping climatological precipitation over mountainous terrain. *Journal of Applied Meteorology* 33, 140–158.

DiLuzio, M., Johnson, G.L., Daly, C., Eischeid, J.K. and Arnold, J.G. (2008) Constructing retrospective gridded daily precipitation and temperature datasets for the conterminous United States. *Journal of Applied Meteorology and Climatology* 47, 475–497.

Durbin, J. and Watson, G.S. (1951) Testing for serial correlation in least squares regression, II. *Biometrika* 38, 159–179.

El-Sadek, A.N., Bleiweiss, M.P., Shukla, M.K., Guldan, S.J. and Fernald, A. (2010) Alternative climate data sources for distributed hydrologic modeling on a daily time step. *Journal of Hydrological Processes* DOI: 10.1002/hyp.7917.

Feigelson, D.D. and Babu, G.J. (1992) Linear regression in astronomy. II. *Astrophysical Journal* 397, 55–67.

Fila, G., Bellocchi, G., Acutis, M. and Donatelli, M. (2003) IRENE: a software to evaluate model performance. *European Journal of Agronomy* 18, 369–372.

Grell, G., Dudia, J. and Stauffer, D. (1994) *A Description of the Fifth-generation Penn-State/NCAR Mesoscale Model*. NCAR Technical Report Note TN-398, National Center for Atmospheric Research, Boulder, Colorado.

Gujarati, D.N. (2003) *Basic Econometrics*, 4th edn. McGraw-Hill, Boston, Massachusetts.

Guyot, G. (1998) *Physics of the Environment and Climate*. Praxis Publishing, Worthing, UK.

Haylock, M.R., Hofstra, N., Klein Tank, A.M.G., Klok, E.J., Jones, P.D. and New, M. (2008) A European daily high-resolution gridded dataset of surface temperature and precipitation. *Journal of Geophysical Research (Atmospheres)* 113, D20119, doi:10.1029/2008JD10201.

Isobe, T., Feigelson, E.D., Akritas, M.G. and Babu, G.J. (1990) Linear regression in astronomy. I. *Astrophysical Journal* 364, 104–113.

Juang, H.-M. and Kanamitsu, M. (1994) The NMC nested regional spectral model. *Monthly Weather Review* 122, 3–26.

Kersebaum, K.C., Hecker, J.-M., Mirschel, W. and Wegehenkel, M. (eds) (2007) *Modelling Water and Nutrient Dynamics in Soil-Crop Systems*. Springer, Dordrecht, The Netherlands.

Kite, G.W. (1995) Chapter 15: The SLURP model. In: Singh, V.P. (ed.) *Computer Models of Watershed Hydrology*. Water Resources Publications, Littleton, Colorado, pp. 521–562.

Lacroix, M., Kite, G. and Droogers, P. (2000) *Using Datasets from the Internet for Hydrological Modeling: An Example from the Küçük Menderes Basin, Turkey*. IWMI Research Report 40, International Water Management Institute, Colombo, Sri Lanka.

Legates, D.R. and McCabe, G.J. Jr (1999) Evaluating the use of "goodness-of-fit" measures in hydrologic and hydroclimatic model validation. *Water Resources Research* 35, 233–241.

Lin, Y. and Mitchell, K.E. (2005) The NCEP Stage II/IV hourly precipitation analyses: development and applications. *Preprints, 19th Conference on Hydrology, American Meteorological Society*, 9–13 January 2005, San Diego, California, Paper 1.2.

Loague, K. and Green, R.E. (1991) Statistical and graphical methods for evaluating solute transport models: overview and application. *Journal of Contaminant Hydrology* 7, 51–73.

Martinec, J., Rango, A. and Roberts, R. (2008) *Snowmelt Runoff Model (SRM) User's Manual*. Special Report 100, Agricultural Experiment Station, New Mexico State University, Las Cruces, New Mexico.

Mesinger, F., DiMego, G., Kalnay, E., Mitchell, K., Shafran, P.C., Ebisuzaki, W., Jović, D., Woollen, J., Rogers, E., Berbery, E.H., Ek, M.B., Fan, Y., Grumbine, R., Higgins, W., Li, H., Lin, Y., Manikin, G., Parrish, D. and Shi, W. (2006) North America Regional Reanalysis. *Bulletin of the American Meteorological Society* 87, 343–360.

Nash, J.E. and Sutcliffe, J.V. (1970) River flow forecasting through conceptual models, part I – A discussion of principles, *Journal of Hydrology* 10, 282–290.

Press, W.H., Flannery, B.P., Teukolsky, S.A. and Vetterling, W.T. (1992) *Numerical Recipes in FORTRAN: The Art of Scientific Computing*, 2nd edn. Cambridge University Press, Cambridge, UK.

Reusser, D.E., Blume, T., Schaefli, B. and Zehe, E. (2009) Analysing the temporal dynamics of model performance for hydrological models. *Hydrology and Earth System Science* 13, 999–1018.

Rigon, R., Bertoldi, G. and Over, T.M. (2006) GEOtop: a distributed hydrological model with coupled water and energy budgets. *Journal of Hydrometeorology* 7, 371–388.

Semmens, D.J., Goodrich, D.C., Unkrich, C.L., Smith, R.E., Woolhiser, D.A. and Miller, S.N. (2008) KINEROS2 and the AGWA modeling framework. In: Wheater, H.S., Sorooshian, S. and Sharma, K.D. (eds) *Hydrological Modelling in Arid and Semi-Arid Areas*. Cambridge University Press, Cambridge, UK, pp. 41–48.

Singh, V.P. and Frevert, D.K. (eds) (2006) *Watershed Models*. CRC Press, Boca Raton, Florida.

Skamarock, W.C., Klemp, J.P., Dudhia, J., Gill, D.O., Barker, D.M., Duda, M.G., Huang, X., Wang, W. and Powers, J.G. (2008) *A Description of the Advanced Research WRF Version 3*. NCAR Technical Report Note TN-475, National Center for Atmospheric Research, Boulder, Colorado.

Sorooshian, S., Hsu, K.-L., Gao, X., Gupta, H.V., Imam, B. and Braithwaite, D. (2000) Evaluation of PERSIANN system satellite-based estimates of tropical rainfall. *Bulletin of the American Meteorological Society* 81, 2035–2046.

Todini, E. and Ciarapica, L. (2001) Chapter 12: The TOPKAPI model. In: Singh, V.P. and Frevert, D.K. (eds) *Mathematical Models of Large Watershed Hydrology*. Water Resources Publications, Littleton, Colorado.

Wilks, D.S. (2006) *Statistical Methods in the Atmospheric Sciences*, 2nd edn. Academic Press, Burlington, Massachusetts.

Willmott, C.J. (1982) Some comments on the evaluation of model performance. *Bulletin of the American Meteorological Society* 63, 1309–1313.

Willmott, C.J., Ackleson, S.G., Davis, R.E., Feddema, J.J., Klink, K.M., Legates, D.R., O'Donnell, J. and Rowe, C.M. (1985) Statistics for the evaluation and comparison of models. *Journal of Geophysical Research* 90(C5), 8995–9005.

Xu, C.-Y. (2001) Statistical analysis of parameters and residuals of a conceptual water balance model – methodology and case study. *Water Resource Management* 15, 75–92.

16 Climate Change and Soil Hydrology: European Perspectives

Satish Bastola, Suresh Kumar, Conor Murphy and John Sweeney*

Introduction

The climate of Europe has been changing along similar lines to that of the globe as a whole, with warming of 0.9 °C being experienced over the period 1901–2005 (Alcamo *et al.*, 2007). Though future changes in rainfall to be expected as a result of continued warming are subject to large uncertainties, mean rainfall intensity is expected to increase significantly across the continent, irrespective of local changes in annual or seasonal receipt (Giorgi *et al.*, 2004). Of more significance for soil processes and soil hydrology is the substantially increased likelihood of extreme precipitation events.

Soils represent a short- to long-term carbon storage medium. Increased temperatures can be expected to accelerate biological decomposition activity in the upper horizons and this will decrease carbon storage. These changes in soil organic matter also have implications for soil moisture storage, and changes in soil moisture driven by future changes in climate are likely to have significant implications for the structure and dynamic of ecosystems, agricultural production and water availability (Frederick and Major, 1997). Therefore, soils, and consequently soil hydrology, are likely to be highly sensitive to climate change. Soil hydrological processes determine how precipitation is partitioned into infiltration, runoff, evapotranspiration and groundwater recharge. Consequently, both surface water resources and the soil moisture store react to the changes in temperature and precipitation.

Precipitation and runoff are direct driving forces of soil erosion and sediment transport. Variation of precipitation will most likely lead to changes in surface runoff, soil erosion and sediment dynamics. In addition to the direct effect on soil arising from higher temperature and greater variability in rainfall, the effect from climate-induced increases or decreases of vegetative cover, alteration in land-use practices and water-use efficiency can also have a significant impact on soil. Increasing air temperatures also affect soil erosion indirectly by changing: (i) growing days for crop maturity, (ii) microbial activity levels, and (iii) crop management practices. Favis-Mortlock and Boardman (1995) suggest that a 7% increase in precipitation could lead to a 26% increase in erosion in the UK. Similarly, Pruski and Nearing (2002) show, using the HadCM3 (Hadley Centre Global Climate Model from the UK Met Office Hadley Centre for Climate Prediction and Research) prediction coupled with the Water Erosion Prediction Project-Carbon Dioxide

* Corresponding author: John.Sweeney@nuim.ie

(WEPP-CO2) model for eight locations in the USA, that erosion increased substantially where precipitation was predicted to increase significantly. Erosion rates are also sensitive to land use and existing conservation practices at the field or farm level. Soil erosion has been identified as the most severe hazard threatening the protection of soil in Europe. Kosmas *et al.* (1997) studied the effect of land use and precipitation on annual runoff and sediment loss at eight different sites along the northern Mediterranean region and the Atlantic coastline located in Portugal, Spain, France, Italy and Greece. They found that land use can greatly affect runoff and soil erosion in this region. Several studies have been conducted on the effects of climate change on soil erosion using computer simulation models such as the Universal Soil Loss Equation (USLE), SWAT (Soil and Water Assessment Tool) (Neitsch *et al.*, 2002), WEPP (Water Erosion Prediction Project) (Flanagan and Livingston, 1995), EUROSEM (European Soil Erosion Model) (Morgan *et al.*, 1998), KINEROS2 (Kinematic Runoff and Erosion Model) (Smith *et al.*, 1995) and LISEM (Limburg Soil Erosion Model) (De Roo *et al.*, 1996). These models have the potential to respond explicitly and rationally to changes in climate or land use, and have a potential for developing scenarios of change, thus helping in the assessment of policy or economic options.

Improved understanding of how climate change could influence soil hydrology has also been an important research topic. Hydrological simulation models are often used together with climate scenarios generated from Global Climate Models (GCMs) to evaluate the impacts of potential climate change on water resources and soil hydrological process. This approach is subject to a range of uncertainties associated with future emissions of greenhouse gases, the response of the climate system to these changes at global and local scales, and uncertainties associated with the impact models themselves. These uncertainties then cascade through the climate change impact assessment methodology, resulting in potentially large uncertainties being associated with critical future impacts at the local scale where key decisions are required in order to increase the resilience of water supply management and infrastructure to future changes. In an attempt to quantify major sources of uncertainties associated with climate change impact assessment, New and Hulme (2000) presented an approach to quantifying the uncertainties associated with the estimation of future greenhouse gas emissions, the climate sensitivity, and limitations and unpredictability in GCMs. Most early studies utilized a single hydrological model and ignored the modelling uncertainties associated with the structure of such models. Because hydrological models are inherently imperfect because they abstract and simplify real patterns and processes that are themselves imperfectly known and understood, and experiences with the calibration of hydrological models suggest that their parameters are inherently uncertain, it is essential to address them in the context of climate change assessments.

Although climate change will certainly affect soil hydrology, impacts will not be distributed uniformly over the globe. Therefore, this chapter aims at reviewing recent research developments in climate change and its consequence for soils, soil moisture and soil erosion from a European perspective and, additionally, aims at highlighting the importance of uncertainty characterization in the response of models at the river basin scale through a case study of Irish catchments.

Materials and Methods

A Pan-European Soil Erosion Risk Assessment model (PESERA) has been employed to model soil erosion in a changing climate at a European scale (Kirkby *et al.*, 2004). PESERA is a physically based and spatially distributed model developed for quantifying soil erosion in environmentally sensitive areas at a regional or European scale, and to assist with defining soil conservation strategies. It estimates erosion rates in individual storms using a sediment transport equation that has explicit terms for topography, overland flow and soil erodibility. Other soil characteristics and soil/land cover are implicitly incorporated as a soil runoff threshold. The model's robustness and flexibility has been demonstrated through its performance at different resolutions and across different agro-ecological zones. In order to simplify the spatial scale, two subregions of Europe, namely the northern region and the Mediterranean region were used to provide an insight into future erosion risks.

Hydrological simulation models are used together with climate change scenarios generated from GCMs to evaluate the impacts of potential climate change on water resources and soil hydrological processes. This approach is subject to a range of uncertainties. Therefore, to account for different sources of uncertainties in the hydrological impacts of climate change, a multi-model approach that combines multiple emission scenarios, multiple GCMs and multiple conceptual rainfall runoff models was implemented through the application of the methodology to selected Irish catchments. The six available regional climate scenarios derived from three GCM and two SRES (Special Report on Emissions Scenarios, prepared by the Intergovernmental Panel on Climate Change (IPCC) for the Third Assessment Report (TAR) in 2001) greenhouse gas emissions scenarios, namely A2 and B2, and downscaled for Ireland by Fealy and Sweeney (2007), were used to characterize future climate evolutions. The GCMs considered included: HadCM3; CGCM2, from the Canadian Centre for Climate Modelling and Analysis (CCCMA, Canada) and CSIRO-Mk2 from the Commonwealth Science and Industrial Research Organisation (CSIRO, Australia). The A2 and B2 scenarios represent future emissions levels that could be considered 'medium-high' (A2 emission) and 'medium-low' (B2 emission). From among the large number of models that can be used for the purpose of modelling flow in catchments, four rainfall runoff models were selected: HyMOD (a conceptual hydrological model; see Wagener *et al.*, 2001, for details), NAM (Nedbor-Afstromnings-Model – precipitation-runoff model; see Madsen, 2000, for detailed information), Tank (Sugawara, 1995) and TOPMODEL (Beven *et al.*, 1995). Each of these models varies in the way they conceptualize the key hydrological processes and in complexity, primarily related to the number of parameters requiring calibration. Among the four selected models, NAM and Tank describe the behaviour of each component of the hydrological cycle at the catchment level by using a group of conceptual elements. Conversely, TOPMODEL and HyMOD are both variable contributing area models. In TOPMODEL, the spatial variability is taken into account through indices derived from topography, whereas, in HyMOD, the model spatial variability within basin is modelled using a probability distribution function. All four models employ a single linear reservoir to model groundwater. The impact of climate change on water resources at the catchment scale is investigated using four Irish catchments: the River Blackwater at Ballyduff (2302 km^2), the River Suck at Bellagill (1219 km^2), the River Moy at Rahans (1803 km^2) and the River Boyne at Slane (2452 km^2). These four catchments were selected to represent the diverse hydrological responses of different catchments located throughout the Republic of Ireland.

Results and Discussion

Climate change scenarios for Europe

As already stated, the climate of Europe has been changing along similar lines to that of the globe as a whole. Unlike North America, Europe is exposed much more to the ingress of maritime air masses and its climate trends are therefore strongly influenced by sea surface temperature (SST) trends in the North Atlantic. Ting *et al.* (2009) have attributed recent trends in Atlantic SSTs to a combination of anthropogenic and natural influences which sometimes combine to amplify or mitigate climate trends in the region. A pronounced warming in SSTs and also in air temperatures has occurred especially since the 1990s, which were the warmest decade in the long instrumental records available from Europe. Not surprisingly, given the dominant maritime influence, European winters have tended to warm to a greater extent than summers. Precipitation changes have shown a marked geographical variation with an increase of 10–40% being observed in northern Europe and a decrease of up to 20% being observed in the southern parts of the continent (Alcamo *et al.*, 2007). Indications that mean rainfall intensity is increasing are also widely apparent across the continent (Alexander *et al.*, 2006).

A continuation of many of these trends is projected from contemporary modelling exercises. Warming is expected to be greatest in winter in the northern and eastern parts of Europe, while more continental parts are expected to show greatest warming during the summer. A major temperature increase of the order of 6 °C in summer temperatures is projected for the Iberian Peninsula and southern France (Good *et al.*, 2006). Indeed, the intensity and frequency of summer heatwaves are likely to increase across the continent; one such major event occurring in 2003 has been linked to over 35,000 excess deaths (Johnson *et al.*, 2005). Models also project a continuation of recent trends in precipitation, with mean annual precipitation projected to increase in the north and decrease in the south (Alcamo *et al.*, 2007). In winter, this is driven by increased cyclonic activity as more vigorous depressions from a warmer Atlantic bring maritime conditions further north and east than at present. In summer, the blocking effect of the extension of the Azores anticyclone deflects disturbances towards the Barents Sea, producing a widespread reduction in precipitation. Projections of wind speed changes are less confident and often highly influenced by the choice of models used (Räisänen *et al.*, 2004).

Mean rainfall intensity is expected to increase significantly across the continent, irrespective of local changes in annual or seasonal receipt (Giorgi *et al.*, 2004). Of more significance for soil processes and soil hydrology is the substantially increased likelihood of extreme precipitation events. Though these are most likely to occur in winter, enhanced convective activity during summer is also likely to produce short-duration high-intensity downpours with a substantially increased frequency. Increased summer heatwaves and drought frequency will also have important ramifications for soil processes as the century proceeds.

Climate change and soils

The sensitivity of soil hydrology to temperature is not straightforward. Kirschbaum (1995) suggested that a 1 °C increase in temperature could ultimately lead to a loss of over 10% of soil organic carbon in middle-to-high latitude locations, whereas the same temperature increase would lead to a loss of only 3% of soil organic carbon in the tropics. Much of northern Europe falls into the former category, suggesting that the north European soil organic carbon reservoir, particularly in peatland soils, may decrease significantly with global warming.

Changes in soil organic matter also have implications for soil moisture storage because soil organic matter can absorb up to 20 times its weight in water. However, an increased incidence of hot and dry summers and mild and wet winters will also have implications for soil moisture and erosion vulnerability. As well as being a major influence on soil processes, soil moisture is an important regulator of surface water flows. Fine-textured soils such as clays and peats can retain large volumes of water and can contribute to increased flood risk in high precipitation events if they are already close to field capacity. Projected increases both in precipitation intensity and in winter rainfall for upland areas of northern and north-western Europe relate to many areas with blanket bog, or to other wet soils such as gleys or peaty podzols derived from glacial deposits. These soils with poor infiltration thus render areas at lower levels more vulnerable to increased flooding. In contrast, coarser soils, derived from fluvioglacial deposits, especially in central and eastern Europe, facilitate infiltration and groundwater recharge better and can therefore mitigate the impact of both flood and drought events.

A further consideration of the relationship between climate change and soil centres on the fact that approximately half of Europe's soil carbon stocks of about 75 Gt are located in Scandinavia, the UK and Ireland, with the single biggest contributor to the total being peatlands. When such soils are drained, they release 20–40 t of CO_2 per annum per hectare. The first priority, therefore, in terms of using soils to combat climate change is to preserve as much as possible of existing peatlands and organic soils. The susceptibility of such soils to drying out and, in the case of peatlands, to cracking and disintegrating, poses a risk that such areas in Europe may become carbon sources rather than sinks as warming proceeds. In the case of the Irish peatlands, uncertainty exists as to whether disturbance has already rendered them sources rather than sinks (Ward *et al.*, 2007). Grassland and forest soils are also probably delivering net sequestration of carbon, with cultivated soils performing a similar role to a lesser extent. Land use and land-use change adversely affect soil carbon stocks, particularly when conversion to arable uses occurs.

Soil moisture in a changing climate

Changes in soil moisture driven by future changes in climate are likely to have significant implications for the structure and dynamics of ecosystems, agricultural production and water availability. In relation to ecosystem functioning and agriculture, a large number of studies have demonstrated the importance of soil moisture deficits as a key indicator of stress for vegetation and shown that a detailed knowledge of climate-induced changes in soil moisture patterns is critical in understanding the magnitude and spatial distribution of future changes in vegetation and crop yield (e.g. Porporato *et al.*, 2001). In terms of water resources, soil moisture is a key component of the hydrological cycle, controlling the partitioning of rainfall between runoff, evapotranspiration and deep infiltration (Daly and Porporato, 2005), thereby determining the water yield of a catchment. Various feedbacks exist between soil moisture and the biological and hydrological cycles. For example vegetation can influence the soil water regime by offsetting drier conditions through decreased transpiration (Etchevers *et al.*, 2002; Seneviratne *et al.*, 2002).

In contrast, dry soils can also cause a negative feedback by amplifying the impact and duration of heatwaves and prolonging the effects of drought (Nicholson, 2000). Brabson *et al.* (2005) using output from the HadCM3 global climate model showed that longer spells of extreme temperature are seen to arise both from the statistical increase in the frequency of extremes and from the extended periods of low soil moisture. Additionally, Fischer *et al.* (2007) showed that a lack of soil moisture during the record-breaking European heatwave of summer 2003 was considerably amplified by reductions in soil moisture beginning in spring. The authors highlighted that a lack of soil moisture strongly reduced latent heat cooling, thereby amplifying surface temperature anomalies. Simulations conducted by Fischer *et al.* (2007) indicated that, without negative soil moisture anomalies, the summer heat anomalies could have been reduced by around 40% in some regions.

It is obvious then that there is a need to understand how soil moisture is likely to respond to climate change. In meeting this need, the dominant approach to assessing climate change impacts is to use the output from global climate models to assess future changes in soil moisture content. This has not been an easy task to date, and future projections of soil moisture have been considered only by a few studies, not least because of a lack of observations, large ranges in the natural variability of soil moisture conditions and the complexity involved in assessing the temporal and spatial impacts for this variable. Jasper *et al.* (2006) highlighted that the impact of climate change on soil moisture depends on the interplay of highly complex and nonlinear processes such as infiltration, drainage, capillary rise, evapotranspiration and lateral subsurface flows, with each of these processes being influenced by local area characteristics such as soil type and texture, vegetation characteristics and slope. Therefore, even with a uniform change in temperature and precipitation, changes in soil moisture due to climate change are likely to show a high degree of spatial variability and significant levels of uncertainty.

Of the limited number of studies that have been conducted, the general findings to date indicate a likely increase in the moisture content of soils in winter and a substantial decrease in summer, particularly in southern Europe. In global-scale studies, a number of authors have used output from GCMs directly to simulate changes in soil moisture. Gerten *et al.* (2007) explored the effects of atmospheric CO_2 enrichment and climate change on soil moisture using a dynamic global vegetation and water balance model forced by five different scenarios of change in temperature, precipitation, radiation and atmospheric

CO_2 concentrations. Outputs from this large-scale assessment projected a decline in soil moisture for many regions for the period 2071–2100 compared with 1961–1990, with ecosystems in northern temperate latitudes at greatest risk. At the European scale, Gregory *et al.* (1997) reported that soil moisture in summer is likely to decrease by 10–30% in southern Europe, driven predominantly by reductions in precipitation and changes in evaporative demands. Following a process investigation, these authors suggested that drying of the soil during summertime is derived from an increase in evaporation in winter and spring due to higher temperatures and reduced snow cover, and a decrease in the net input of rainfall in summer. Reductions in soil moisture are suggested to be sufficiently large to produce a limiting effect on evaporation losses during the summer months in southern Europe.

The majority of GCM-based studies rely on very simple representations of the land-surface processes involved in determining soil moisture levels (Seneviratne *et al.*, 2002). In addition, Srinivasan *et al.* (2000) emphasized that, while the general trend is towards drier soil conditions with projected climate change, uncertainties are large. Many GCMs currently show poor skill at reproducing the observed seasonality in rainfall at regional scales, with obvious implications for their robustness in modelling soil moisture conditions. Unrealistic summer drying is suggested by several models. Recent efforts have focused on estimating future soil moisture conditions by employing downscaled output from GCMs to run more detailed models at scales more appropriate to understanding the complex response of soil moisture conditions.

Naden and Watts (2001) found that climate change could generally lead to decreased soil water content in the UK, with very marked decreases in the south of the country, in line with simulated changes in runoff. At the landscape scale, using a single vegetation type (grassland) and a limited number of soil types, the authors found that soil type is critical in determining future changes in moisture content, with the drying effect greatest for clay soils, while sandy soils were found to be least prone to severe reductions. However, climate change scenarios within this study are based on current values of stomatal resistance and leaf area index for grassland, and do not account for species response to enhanced carbon dioxide levels.

Etchevers *et al.* (2002) estimated the impacts of climate change on the hydrological budget of the Rhone catchment in Europe and found strong contrasts in the hydrological response of the catchment under future climate scenarios. When considering soil water content, northern parts of the catchment remained quite wet, whereas in the southern part of the catchment enhanced drying of soils was suggested, consistent with the findings of larger scale work discussed above. In Germany, Holsten *et al.* (2009) examined past trends and future effects of climate change on soil moisture dynamics in the Brandenburg region with an emphasis on Special Areas of Conservation (SACs). A decreasing trend in soil water content was shown by analysing simulation results for the period 1951–2003. Regionally downscaled climate scenarios, representing the range between wetter and drier realizations, were used to evaluate the future evolution of moisture conditions and available soil water. Results indicated a continuation of present drying trends with further decreases in average soil water ranging from −4% to −15% up to the middle of the current century. A high level of risk for wetlands was also identified (Holsten *et al.*, 2009).

In terms of wetlands, blanket peatlands are important and rare natural ecosystems throughout Europe, with many being located in SACs. Evans *et al.* (1999) used analogue conditions of the dry summer of 1995 to assess hydrological changes in peat hydrology for an area of upland blanket bog in the UK. Hulme (1997) suggested that extreme summers like that experienced in the UK in 1995 and the associated soil moisture deficits would be increasingly common by the mid-century. Reductions in summer rainfall and increases in evaporation will significantly affect the timing and quality of runoff from upland blanket peat, with the potential to trigger phases of peat erosion and alter the carbon flux from the peatland system (Evans *et al.*, 1999).

In perhaps the most detailed assessments to date, Jasper *et al.* (2006) examined future changes in summertime soil water patterns by considering climate change projections, soil and terrain characteristics, and slope and groundwater depth in the determination of future soil water evolution in the Thur river basin in Switzerland. The results of this work suggest that a warmer future climate with less precipitation in summer may significantly lower soil water content, leading to an increase in the frequency of water stress conditions. Critically, the increased level of detail incorporated into the modelling process suggests that the magnitude of future changes is closely related to land use, soil textural characteristics and slope. In relative terms, reductions in soil water were largest for soils in sloping areas with low water storage capacity, and also larger for forested areas than for cropland and grassland (Jasper *et al.*, 2006). In volumetric terms, the largest reductions were likely for flat areas with good water supply, mostly dominated by cropland. The role of groundwater was found to be important in these areas, particularly where the rooting zone of vegetation is connected to groundwater, and capillary rise can counteract soil water depletion. This offsetting effect is lost under drying conditions in the future through the disconnection of the rooting zone with groundwater. In steeper areas, where groundwater is not a factor, the driving processes determining soil moisture changes were found to be decreased precipitation and increased evapotranspiration. A key conclusion from Jasper *et al.* (2006) was that the degree of soil water depletion varies with climate scenario, land use, soil texture and topographic conditions, and that in order to fully understand future changes in soil moisture due to climate change reliable precipitation scenarios are required, along with the full representation of biophysical processes that control evapotranspiration, including vertical and lateral subsurface flows.

It is clear that soil hydrological responses to climate change are multifaceted and crucially important determinants of environmental productivity and stability in many parts of Europe, as well as important determinants of the pace of future climate change itself. Two key aspects will now be dealt with in greater detail. First, desiccated soils can be expected to be particularly vulnerable to wind and water erosion and this risk has been recognized as potentially severe, especially in southern Europe (Grimm *et al.*, 2002). Secondly, it is clear that better understanding of soil hydrology is an essential component for reducing uncertainty in modelling water resource changes and management.

Soil erosion in Europe in a changing climate

Erosion is a widespread form of soil degradation globally and poses severe limitations to sustainable agricultural land use as it reduces productivity of agricultural soils by removing the most fertile topsoil, where soils are shallow (Stone *et al.*, 1985; Verity and Anderson, 1990; Bakker *et al.*, 2005). Furthermore, severe erosion is commonly associated with the development of temporary or permanently eroded channels or gullies that can fragment farmland and cause deposition of sediments and accumulation of agrochemicals in water bodies (Steegen *et al.*, 2001). The soil removed by runoff from the land during a large storm accumulates below the eroded areas, in severe cases blocking roadways or drainage channels and inundating buildings. Erosion rates are very sensitive to both climate and land use, as well as to existing conservation practices at the field or farm level.

Based on work of Valentin (1998) and Nearing *et al.* (2004), global change is expected to exacerbate erosion problems through changes in rainfall conditions and land use. Therefore, it is very important to be able to assess the state of soil erosion at a European level using an objective methodology that allows the assessment to be repeatable as conditions change, and also to explore the broad-scale implications of prospective global environmental changes. Erosion by running water has been identified as the most severe hazard threatening the protection of soil in Europe. The revised Common Agricultural Policy (CAP) of the European Union (EU) set-aside programme is having a positive effects on the soil erosion risk (Van Rompaey *et al.*, 2001).

However, in order to study any risk under current and future climatic conditions, a correct assessment of erosion risk at local, national and European scales using an objective, spatially explicit methodology is important (Gobin *et al.*, 2003). Soil erosion indicators developed from a physically based model have the potential not only to provide information on the state of soil erosion at any given time, but also to assist in understanding the links between different factors causing erosion. This not only provides a basis for estimating the overall costs attributable to erosion under present and future conditions, but, equally, helps in identifying areas that could be severely affected and potential remedial actions that might be undertaken.

Modelling European soil erosion in a changing climate

USLE has been the most widely applied model in Europe (e.g. Van der Knijff *et al.*, 2000, 2002). This is now considered to be conceptually flawed, and other models are emerging. These are based on runoff thresholds (e.g. Kirkby *et al.*, 2000) or the MIR (Minimum Information Requirement) approach (Brazier *et al.*, 2001) applied to the more complex USDA WEPP model (Nearing *et al.*, 1989). A large number of physical process-based erosion models have also been developed over recent decades, such as SWAT, WEPP, EUROSEM, KINEROS2 and LISEM. Moreover, these models have the potential to respond explicitly and rationally to changes in climate or land use, and have a potential for developing scenarios of change, and thus helping in the assessment of policy or economic options. One shortcoming is the spatial application of these relatively complex models to large areas (i.e. > 100 km^2). The quality of the necessary input data is also frequently not sufficient.

PESERA was developed specifically to be applied to larger areas using the description of physical processes controlling sheet and rill erosion as a basis, in order to allow application using data available at the European scale. Cautiously, soil erosion and sediment transport estimated in PESERA represent only

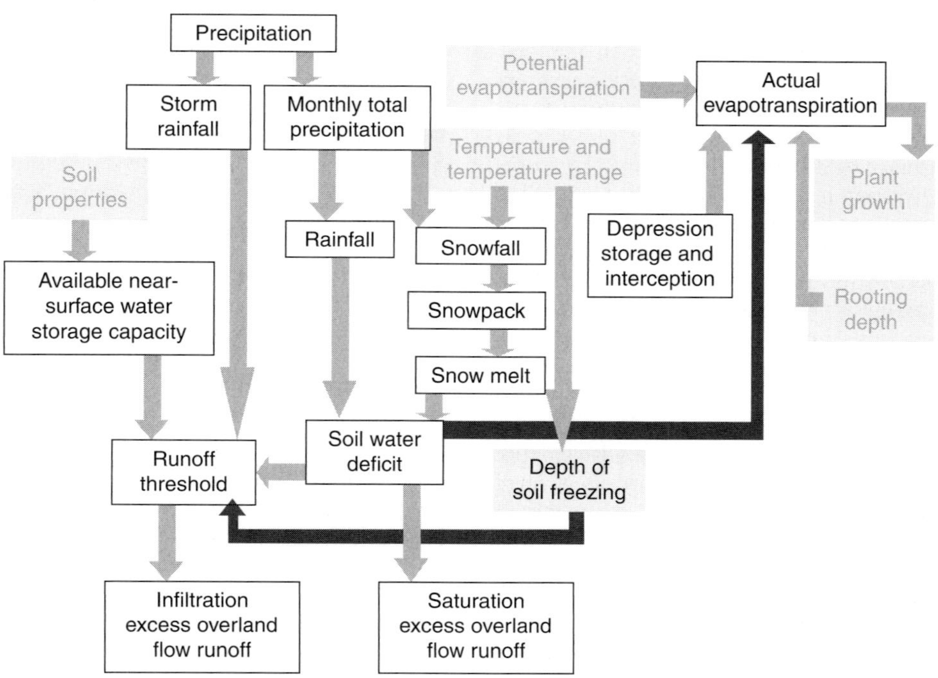

Fig. 16.1. Organization of the PESERA (Pan-European Soil Erosion Risk Assessment) model.

the dominant sheet and rill erosion processes (Kirkby *et al.*, 2004) important for agriculture. The PESERA model combines the effect of topography, climate and soil into a single integrated forecast of runoff and soil erosion (Fig. 16.1) that the USLE and RUSLE (Revised USLE) lack.

The PESERA model has been calibrated and validated at high and low resolutions (e.g. Gobin *et al.*, 2003). The 1 km^2 grid PESERA map (Plate 18) was validated through a visual, numerical and category comparison between measured and observed erosion rates (Gobin *et al.*, 2003). Although these results are promising, they are not sufficient to justify the application of the model at the European scale as a prediction and scenario analysis tool. Generally, parameters have been optimized for a local situation only and with a specific temporal resolution. According to the European Environment Agency (Turner *et al.*, 2001), there are three broad zones in Europe where soil erosion occurs: a southern (Mediterranean zone), a northern loess zone and an eastern zone. In the southern zone, severe soil erosion occurs because of intense seasonal rainfall. This is often associated with overgrazing, deforestation and a move away from traditional crops. Erosion in this zone may be long established. Many areas undergoing active tectonic uplift, such as parts of Spain, have a much higher rate of erosion than in the southern (Mediterranean) zone. The principal impact is that on-site soil productivity decreases as a result of thinning. The northern zone has moderate rates of water erosion. This mostly results from less intense rainfall falling on saturated, easily erodible soils. Impacts here are mainly off site, as the agricultural chemicals from the north's more intensive farming systems are moved into water bodies along with eroded sediments. Partially overlapping these two zones is the eastern zone, where former large state-controlled farms produced considerable erosion problems. For scenarios of alternative erosion futures, PESERA was run in conjunction with regional climate models (RCMs) and economic forecasts. The Hadley Centre's regional climate change prediction model (HadRM3) was used to drive the PESERA 1 km^2 grid model. Although the interrelations between climate change and changes in soil quality are complex and not fully understood (EEA, 2000), it is obvious that changes in climate will have significant impacts on the occurrence of erosion.

In order to simplify the spatial scale, two subregions were used to provide an insight into the future erosion risks. The first covers Belgium and parts of the southern UK, northern France, Germany, the Netherlands and Luxembourg, where there are good baseline data. The second, the Mediterranean region, covers the southern part of Spain and Portugal, which is considered to be particularly prone to erosion. This is because of long dry periods followed by heavy bursts of intense rainfall, falling on steep slopes with fragile soils and low vegetation cover (Gobin *et al.*, 2003). For the first region, the predicted soil erosion for the period 2071–2080 shows that the spread of erosion estimates is highest in late autumn and winter (October–February). The enhanced difference in erosion risk between summer and winter follows the pattern of the expected change in rainfall, with an increase in autumn and winter, and a clear decrease in the summer months. For the Mediterranean, the projected increase in winter rains in the HadRM3-A2B (A2B is one of the A2 scenarios) scenario for 2071–2080 does not lead to an increase in predicted erosion. On the contrary, erosion rates decline spectacularly. Only summer (July–August) shows increases in erosion rates due to a change of erosion pattern, or more an increase in the area for which erosion (risk) is predicted. Almost all Mediterranean environments are dominated by overland flow in the summer, but in wetter environments there are substantial subsurface flow and groundwater recharge in winter (up to 75% of the total rainfall). Drier environments are dominated by overland flow almost year-round (Gobin *et al.*, 2000). Given the spatial complexity of the model, erosion risk assessment at a local scale becomes problematic even under future climatic scenarios. Still, the overall picture at national or regional level does relate strongly to the land use. Experiments in the UK have shown that land use has a much more pronounced effect on erosion potential than climate change. However, soil vulnerability information, such as that provided by the PESARA study, is fundamental for determining appropriate land-use policies and responses to climate change (Hulme *et al.*, 1999).

Modelling the impacts of climate change on soil hydrology: addressing uncertainty

Soil hydrological processes determine how precipitation is partitioned into infiltration, runoff, evapotranspiration and groundwater recharge. Both surface water resources and the soil moisture store react to changes in temperature and precipitation. Hydrological models can be used to model these changes in both water stores. Conceptual hydrological models use relatively simple mathematical equations to conceptualize and aggregate the complex, spatially distributed and highly interrelated water, energy and vegetation processes in a watershed. Owing to the randomness in nature and the lack of complete knowledge of the hydrological system, uncertainty is an unavoidable element in any hydrological modelling study (Beven, 2000). In climate change impact studies, conceptual hydrological models forced with regional climate change scenarios analysed from GCMs are widely used, although this approach is subject to a range of uncertainties associated with output from both the climate models and the impact models.

A wide range of GCMs developed by various climate centres is available for simulating the earth's climate. These climate models differ in the way they simplify the climate system and aggregate the process in a space and time domain. Therefore the projections of water resources are likely to vary depending upon the choice of GCMs (e.g. Prudhomme *et al.*, 2003). Secondly, there exist large sources of uncertainty intrinsic to climate models. These can be classified as: initial conditions, boundary conditions, parameterizations and model structure. The uncertainty in climate projection arising from initial conditions is often neglected for results that are averaged over decades. The third source of uncertainty is mostly related to the computational aspects of modelling and to poor understanding of some of the important components of the climate system, and can be referred to as parameter and structural uncertainty. Tebaldi and Knutti (2007) argued that the quantification of all aspects of model uncertainty requires multi-model ensembles. Giorgi and Mearns (2002) introduced the Reliability Ensemble Averaging (REA) method for calculating the uncertainty range from ensembles of different atmosphere-ocean general circulation models. Similarly, Tebaldi *et al.* (2005) extended the REA method and proposed a Bayesian statistical model that combines information from a multimodal ensemble of atmosphere-ocean GCMs and observations to determine probability distributions of future temperature change on a regional scale. These assessments have been greatly facilitated by the Coupled Model Inter-comparison Project phase 3 (CMIP3) climate model output archive. Several studies used the output archived in Coupled Model inter-comparison projects to account for uncertainty in the GCM, and several others used the output from perturbed physics ensembles to evaluate the uncertainties arising from GCM model formulation (e.g. Murphy *et al.*, 2007). Furthermore, while output from GCMs reproduces the global- and continental-scale climate fairly well, GCMs are inadequate in impact studies owing to the differences in the spatial scales of GCMs and those needed for impact studies (Wilby and Wigley, 1997). This limitation has been widely addressed through downscaling. Methods of downscaling differ in the way that they reproduce various statistical characteristics of observed data in its downscaled results (Wilby and Wigley, 1997; Khan and Iqbal, 2009). New and Hulme (2000) presented an approach to quantifying uncertainties associated with the estimation of future greenhouse gas emissions, the climate sensitivity, and limitations and unpredictability in GCMs.

As with GCMs, uncertainty occurs in hydrological models from a variety of sources. These include: data uncertainty, parameter uncertainty, model structural uncertainty and state uncertainty. An extensive review of the causes of uncertainty in hydrological models and of various methods for assessing the uncertainty can be found in Melching (1995). Among various methods for assessing the uncertainty of hydrological models, the Generalized Likelihood Uncertainty

Estimation (GLUE) method has been extensively used (e.g. Freer *et al.*, 1996). Though the uncertainties that result from dependence on a single conceptual-mathematical model are typically much larger than those introduced through an inadequate choice of model parameter values, most hydrological uncertainty analyses ignore the former and focus exclusively on the latter. A review of the range of strategies for assessing structural uncertainties in environmental modelling are available in Refsgaard *et al.* (2006). These strategies can be broadly grouped into two, depending upon whether or not target data are available. In the application of hydrological models in climate change impact assessment, the structure of the hydrological model cannot be assessed directly using observation. Therefore the main strategy to account for modelling uncertainties is to do the extrapolation with multiple conceptual models. This approach has been used by Butts *et al.* (2002).

Though many impact studies have introduced methods to account for uncertainties associated with the estimation of future greenhouse gas emissions, climate sensitivity, and limitations and unpredictability in GCMs (e.g. New and Hulme, 2000), very few have looked at the uncertainties related to hydrological models. The use of hydrological response to the climatic scenarios resulting from the use of different hydrological models with different climate models will also allow more knowledge concerning the vital role of soil hydrological processes to be obtained.

Accounting for modelling uncertainties: a case study using basins located in Ireland

As already outlined in the Material and Methods section, the impact of climate change on water resources at the catchment scale was investigated using four Irish catchments: Blackwater at Ballyduff (2302 km^2), Suck at Bellagill (1219 km^2), Moy at Rahans (1803 km^2) and Boyne at Slane (2452 km^2). These four catchments were selected so that they represent the diverse hydrological responses of different catchments located throughout Ireland. Four hydrological models (HyMOD, NAM, TOPMODEL and Tank) were used with six sets of statistically downscaled climate scenarios (temperature and precipitation) derived from three GCMs (HadCM3, CCCMA and CSIRO-Mk2) and two emission scenarios (A2 and B2) (Fealy and Sweeney, 2007).

In order to examine the role of model uncertainty (parameter and structural uncertainty) in climate change impact studies and include a full consideration of impact model uncertainty, the GLUE method was used. The sets of behavioural predictions from different models were ranked and likelihood weighted to characterize the parameter as well as structural uncertainty propagated through hydrological models. The informal likelihood measure assumed is the Nash–Sutcliffe efficiency measure, widely used as a performance measure in hydrological modelling. To construct the total uncertainty, envelope inputs from the six climate change scenarios were used together with four hydrological models, each with a number of plausible model parameter sets. Predictions from the three GCMs were weighted based on the ability of the individual GCM to reproduce the properties of the observations.

The results show that the average width of the prediction interval arising from uncertainties associated with parameterization of hydrological modelling is nearly 40% of the average flow, increasing to nearly 70% of the average flow when different model structures are included. This indicates that the uncertainty arising from the uncertainty in the structure of hydrological model can have significant impact on the reliability of the future projection of water resources at catchment scale. Plate 19 illustrates the hydrological response (streamflow) simulated by the behavioural set of model parameters of four different hydrological models of the River Boyne catchment to HadCM3 forced with the A2 scenario. Plate 20 shows a seasonal prediction interval for the control (observed) period (1970–1990) and the future period (2020–2079). These were constructed based on the behavioural predictions obtained from the entire suite of models, which were ranked and likelihood weighted to produce upper 95%, lower 5% and median 50% quantiles of the simulation results. The overall uncertainty envelope increased to nearly 40% of the average streamflow for the three periods simulated, namely the 2020s, 2050s and 2070s. Similarly, Plate 21 shows the observed seasonal streamflow for the period 1970–1990 and the median seasonal streamflow estimated for the future period. There is a progressive increase of winter discharge when moving from the 2020s to the 2070s, and a progressive decrease in summer discharge from the 2020s to the 2070s. The case study shows that the role of hydrological model uncertainty is remarkably high and therefore it should be routinely considered in impact studies.

Summary and Conclusions

Detailed understanding of soil hydrology is needed for future protection of soils in Europe. For much of the continent, the EU Water Framework Directive will indirectly act as a vehicle to facilitate this by providing a regulatory influence on land use, together with an 'as-yet-to-be agreed' Soils Directive. However, it is clear that soil hydrological responses to climate change are multifaceted and crucially important determinants of environmental productivity and stability in many parts of Europe, as well as important determinants of the pace of future climate change itself. Therefore, they should be considered explicitly in long-term integrated river basin management. Changes in soil moisture due to climate change are likely to show a high degree of spatial variability and significant levels of uncertainty. To date, studies have indicated an increase in the moisture content of soils in winter and a substantial decrease in summer, particularly in southern Europe. Furthermore, the soil erosion modelled at national or regional levels relates strongly to the land use, indicating that erosion problems are also expected to be exacerbated through changes in land use.

Improving our understanding of how climate change could influence soil hydrology has been mostly hindered by scores of uncertainties. Many impact studies have introduced methods to account for uncertainties associated with the estimation of future greenhouse gas emissions, climate sensitivity, and limitations and unpredictability in GCMs; very few have looked at the uncertainties related to hydrological models. Hydrological modelling exhibits considerable uncertainties in both parameterization and structural characteristics. This has been addressed here by using a multi-model approach that combines multiple emission scenarios, GCMs and conceptual rainfall runoff models, and demonstrated through the modelling of four Irish catchments. The results indicate that the uncertainty arising from the uncertainty in the structure of hydrological models can have a significant impact on the reliability of the future projection of water resources at catchment scale.

Roughly 90% of the land in Ireland is under one or other varying forms of land use for agricultural production. The impacts of climate change on soil hydrological processes have a direct relevance in terms of biomass production and nutrient management as the agro-ecosystems are predominantly rainfed. The preparedness of agriculture to adopt irrigation regimes is entirely dependent on the water balance of the respective river basins. Hence, projecting the demands of water availability through hydrological modelling is a valuable tool for sustainable water management. Changing land-use patterns in Ireland will reflect water availability rather than temperatures, as this seems to be the most limiting factor during the crop growth period.

It may be concluded that climate change will have significant impacts on soil hydrological processes throughout Europe. While temperature increases will decrease carbon storage and reduce soil organic matter, with consequent implications for soil moisture storage, it is likely to be the projected precipitation changes that will have most impact. Soil moisture responses will be crucially important for regulating surface and groundwater resources, and ultimately for influencing agricultural potential. Soil erosion is likely to become a problem in southern Europe as a result of intense convective downpours in summer occurring after prolonged dry periods.

The majority of GCM-based studies rely on very simple representations of the land-surface processes involved in determining soil moisture levels. Many GCMs currently show poor skill at reproducing the observed seasonality in rainfall at regional scales, with obvious implications for their robustness in modelling soil moisture conditions. Future efforts are required to focus on estimating future soil moisture conditions by employing downscaled output from GCMs to run more detailed models at scales more appropriate to understanding the complex response of soil moisture conditions. Furthermore, hydrological modelling still exhibits considerable uncertainties in both parameterization and structural characteristics. These are further compounded by intrinsic uncertainties in the driving climate models, and this requires recognition in impact studies as well as further research designed to minimize them.

References

Alcamo, J., Floerke, M. and Maerker, M. (2007) Future long-term changes in global water resources driven by socio-economic and climatic changes. *Hydrological Sciences* 52, 247–275.

Alexander, L.V., Zhang, X., Peterson, T.C., Caesar, J., Gleason, B., Klein Tank, A.M.G., Haylock, M., Collins, D., Trewin, B., Rahimzadeh, F., Tagipour, A., Rupa Kumar, K., Revadekar, J., Griffiths, G., Vincent, L., Stephenson, D.B., Burn, J., Aguilar, E., Brunet, M., Taylor, M., New, M., Zhai, P., Rusticucci, M. and Vazquez-Aguirre, J.L. (2006) Global observed changes in daily climate extremes of temperature and precipitation. *Journal of Geophysical Research* 111, D05109, doi:10.1029/2005JD006290.

Bakker, M.M., Govers, G., Ewert, F., Rounsevell, M. and Jones, R. (2005) Variability in regional wheat yields as a function of climate, soil and economic variables: assessing the risk of confounding. *Agriculture, Ecosystems and Environment* 110, 195–209.

Beven, K., Lamb, R., Quinn, P., Romanowicz, R. and Freer, J. (1995) TOPMODEL. In: Singh, V.P. (ed.) *Computer Models of Watershed Hydrology*. Water Resources Publications, Littleton, Colorado, pp. 627–668.

Beven, K.J. (2000) *Rainfall-runoff Modelling – The Primer*. John Wiley, Chichester, UK.

Brabson, B.B., Lister, D.H., Jones, P.D. and Palutikof, J.P. (2005) Soil moisture and predictive spells of extreme temperatures in Britain. *Journal of Geophysical Research: Atmospheres* 110, D05104, doi:10.1029/2004JD005156.

Brazier, R.E., Rowan, J.S., Quinn, P. and Anthony, S. (2001) Towards an MIR (Minimum Information Requirement) approach to modelling on-site soil loss at the national scale. *Catena* 42, 59–79.

Butts, M.B., Hoest-Madsen, J. and Refsgaard, J.C. (2002) Hydrologic forecasting. In: Meyers, R.A. (ed.) *Encyclopaedia of Physical Science and Technology*. Academic Press, Oxford, UK, pp. 547–566.

Daly, E. and Porporato, A. (2005) A review of soil moisture dynamics: from rainfall infiltration to ecosystem response. *Environmental Engineering Science* 22, 9–24.

De Roo, A.P.J., Wesseling, C.G. and Ritsema, C.J. (1996) LISEM: a single-event physically based hydrological and soil erosion model for drainage basins. I. Theory, input and output. *Hydrological Processes* 10, 1107–1117.

EEA (2000) *Down to Earth: Soil Degradation and Sustainable Development in Europe. A Challenge for the 21st Century*. Environmental Issue Series No. 16, European Environment Agency (EEA), Copenhagen.

Etchevers, P., Golaz, C., Habets, F. and Noilhan, J. (2002) Impact of climate change on the Rhone river catchment hydrology. *Journal of Geophysical Research: Atmospheres* 107 (D16), 4293, doi:10.1029/2001JD000490.

Evans, M.G., Burt, T.P., Holden, J. and Adamson, J.K. (1999) Runoff generation and water table fluctuations in blanket peat: evidence from UK data spanning the dry summer of 1995. *Journal of Hydrology* 221, 141–160.

Favis-Mortlock, D.T. and Boardman, J. (1995) Nonlinear responses of soil erosion to climate change: a modelling study on the UK South Downs. *Catena* 25, 365–387.

Fealy, R. and Sweeney, J. (2007) Statistical downscaling of precipitation for a selection of sites in Ireland employing a generalised linear modelling approach. *International Journal of Climatology* 27, 2083–2094.

Fischer, E.M., Seneviratne, S.I., Vidale, P.L., Lüthi, D. and Schär, C. (2007) Soil moisture–atmosphere interactions during the 2003 European summer heat wave. *Journal of Climate* 20, 5081–5099.

Flanagan, D.C. and Livingston, S.J. (eds) (1995) *Water Erosion Prediction Project (WEPP) Version 95.7: User Summary*. NSERL Report No. 11, National Soil Erosion Research Laboratory, USDA Agricultural Research Service, West Lafayette, Indiana.

Frederick, K.D. and Major, D.C. (1997) Climate change and water resources. *Climatic Change* 37, 7–23.

Freer, J., Beven, K.J. and Ambroise, B. (1996) Bayesian uncertainty in runoff prediction and the value of data: an application of the GLUE approach. *Water Resources Research* 32, 2163–2173.

Gerten, D., Schaphoff, S. and Lucht, W. (2007) Potential future changes in water limitations of the terrestrial biosphere. *Climatic Change* 80, 277–299.

Giorgi, F. and Mearns, L.O. (2002) Calculation of average, uncertainty range, and reliability of regional climate changes from AOGCM simulations via the "reliability ensemble averaging" (REA) method. *Journal of Climate* 15, 1141–1158.

Giorgi, F., Bi, X. and Pal, J. (2004) Mean interannual and trends in a regional climate change experiment over Europe. II: Climate Change scenarios (2071–2100). *Climate Dynamics* 23, 839–858.

Gobin, A., Cerdan, O. and Govers, G. (eds) (2003) *Pan European Soil Erosion Risk Assessment: Third Annual Report*. EU 5th Framework Programme. Project No. QLK5-CT-1999-01323. Available at: http://eusoils.jrc.it (accessed 16 March 2010).

Good, P., Barring, L., Giannakopoulos, C., Holt, T. and Palutikof, J.P. (2006) Non-linear regional relationships between climate extremes and annual mean temperatures in model projections for 1961–2099 over Europe. *Climate Research* 13, 19–34.

Gregory, J.M., Mitchell, J.F.B. and Brady, A.J. (1997) Summer drought in northern midlatitudes in a time-dependent CO_2 climate experiment. *Journal of Climate* 10, 662–686.

Grimm, M., Jones, R.J.A. and Montanarella, L. (2002) *Soil Erosion Risk in Europe*. EUR 19939 EN, European Soil Bureau, Institute for Environment and Sustainability, European Commission Joint Research Centre, Ispra, Italy,.

Holsten, A., Vetter, T., Vohland, K. and Krysanova, V. (2009) Impact of climate change on soil moisture dynamics in Brandenburg with a focus on nature conservation areas. *Ecological Modelling* 220, 2076–2087.

Hulme, M. (1997) The climate in the UK from November 1994 to October 1995. *Weather* 52, 242–257.

Hulme, M., Barrow, E.M., Arnell, N.W., Harrison, P.A., Johns, T.C. and Downing, T.E. (1999) Relative impacts of human-induced climate change and natural climate variability. *Nature* 397, 688–691.

Jasper, K., Calanca, P. and Fuhrer, J. (2006) Changes in summertime soil water patterns in complex terrain due to climatic change. *Journal of Hydrology* 27, 550–563.

Johnson, H., Kovats, R.S., McGregor, G., Stedman, J., Gibbs, M. and Walton, H. (2005) The impact of the 2003 heatwave on daily mortality in England and Wales and the use of rapid weekly mortality estimates. *Euro Surveillance* 10, 168–171.

Khan, M.J. and Iqbal, M.T. (2009) Analysis of a small wind-hydrogen stand-alone hybrid energy system. *Applied Energy* 86, 2429–2442.

Kirkby, M.J., Le Bissonais, Y., Coulthard, T.J., Daroussin, J. and McMahon, M.D. (2000) The development of land quality indicators for soil degradation by water erosion. *Agriculture, Ecosystems and Environment* 81, 125–136.

Kirkby, M.J., Jones, R.J.A., Irvine, B., Gobin, A, Govers, G., Cerdan, O., Van Rompaey, A.J.J., Le Bissonnais, Y., Daroussin, J., King, D., Montanarella, L., Grimm, M., Vieillefont, V., Puigdefabregas, J., Boer, M., Kosmas, C., Yassoglou, N., Tsara, M., Mantel, S., Van Lynden, G.J. and Huting, J. (2004) Pan-European Soil Erosion Risk Assessment: The PESERA Map, Version 1 October 2003. Explanation of Special Publication Ispra 2004 No. 73 (S.P.I.04.73). European Soil Bureau Research Report No.16, EUR 21176 and map in ISO B1 format, European Commission Joint Research Centre, Ispra/Institute for Environment and Sustainability. Available from: Office for Official Publications of the European Communities, Luxembourg, and at: http://eusoils.jrc.ec.europa.eu/esdb_archive/pesera/ThePeseraMapBkLet52.pdf (accessed 4 March 2011).

Kirschbaum, M.U.F. (1995) The temperature dependence of soil organic matter decomposition and the effect of global warming on soil organic carbon storage. *Soil Biology and Biochemistry* 27, 753–760.

Kosmas, C., Danalatos, N.G., Cammeraat, L.H., Chabart, M., Diamantopoulos, J., Farand, R., Gutierrez, L., Jacob, A., Marques, H., Martinez-Fernandez, J., Mizara, A., Moustakas, N., Nicolau, J.M., Oliveros, C., Pinna, G., Puddu, R., Puigdefabregas, J., Roxo, M., Simao, A., Stamou, G., Tomasi, N., Usai, D. and Vacca, A. (1997) The effect of land use on runoff and soil erosion rates under Mediterranean conditions. *Catena* 29, 45–59.

Madsen, H. (2000) Automatic calibration of a conceptual rainfall-runoff model using multiple objectives. *Journal of Hydrology,* 235, 276–288.
Melching, C.S. (1995) Reliability estimation. In: Singh, V. P. (ed.) *Computer Models of Watershed Hydrology.* Water Resources Publications, Littleton, Colorado, pp. 69–118.
Morgan, R.P.C., Quinton, J.N. and Smith, R.E. (1998) The European Soil Erosion Model (EUROSEM): a dynamic approach for predicting sediment transport from fields and small catchments. *Earth Surface Processes and Landforms* 23, 527–544.
Murphy, J.M., Booth, B.B.B., Collins, M., Harris, G.R., Sexton, D.M.H. and Webb, M.J. (2007) A methodology for probabilistic predictions of regional climate change from perturbed physics ensembles. *Philosophical Transactions of the Royal Society A* 365, 1993–2028.
Naden, P.S. and Watts, C.D. (2001) Estimating climate-induced change in soil moisture at the landscape scale: an application to five areas of ecological interest in the UK. *Climatic Change* 49, 411–440.
Nearing, M.A., Foster, G.R., Lane, L.J. and Finkner, S.C. (1989) A process-based soil-erosion model for USDA-water erosion prediction project technology. *Transactions of the ASAE* 32, 1587–1593.
Nearing, M.A., Pruski, F.F. and O'Neal, M.R. (2004) Expected climate change impacts on soil erosion rates: a review. *Journal of Soil and Water Conservation* 59, 43–50.
Neitsch, S.L., Arnold, J.G., Kiniry, J.R., Williams, J.R. and King, K.W. (2002) *Soil and Water Assessment Tool Theoretical Documentation, Version 2000.* GSWRL Report 02-01, Grassland, Soil and Water Research Laboratory, USDA ARS (Agricultural Research Service)/BRC Report 2-05, Blackland Research and Extension Center, Texas Agricultural Experiment Station, Temple, Texas/TWRI Report TR-191, Texas Water Resources Institute, College Station, Texas.. Available at: http://swatmodel.tamu.edu/media/1290/swat2000theory.pdf (accessed 17 February 2011).
New, M. and Hulme, M. (2000) Representing uncertainty in climate change scenarios: a Monte-Carlo approach. *Integrated Assessment* 1, 203–213.
Nicholson, S. (2000) Land surface processes and Sahel climate. *Reviews of Geophysics* 38, 117–139.
Porporato, A., Laio, F., Ridolfi, L. and Rodriguez-Iurbe, I. (2001) Plants in water-controlled ecosystems: active role in hydrologic processes and response to water stress. III. Vegetation water stress. *Advances in Water Resources* 24, 725–744.
Prudhomme, C., Jakob, D. and Svensson, C. (2003) Uncertainty and climate change impact on the flood regime of small UK catchments. *Journal of Hydrology* 277, 1–23.
Pruski, F.F. and Nearing, M.A. (2002) Climate-induced changes in erosion during the 21st century for eight U.S. locations. *Water Resources Research* 38, 1298.
Räisänen, J., Hansson, U., Ullerstig, A., Döscher, R., Graham, L.P., Jones, C., Meier, M., Samuelsson, P. and Willén, U. (2004) European climate in the late 21st century: regional simulations with two driving global models and two forcing scenarios. *Climate Dynamics* 22, 13–31.
Refsgaard, J.C., Van der Sluijs, J.P., Brown, J. and Van der Keur, P. (2006) A framework for dealing with uncertainty due to model structure error. *Advances in Water Resources* 29, 1586–1597.
Seneviratne, S.I., Pal, J.S., Eltahir, E.A.B. and Schar, C. (2002) Summer dryness in a warmer climate: a process study with a regional climate model. *Climate Dynamics* 20, 69–85.
Smith, R.E., Goodrich, D.C. and Quinton, J.N. (1995) Dynamic, distributed simulation of watershed erosion: the KINEROS2 and EUROSEM models. *Journal of Soil and Water Conservation* 50, 517–520.
Srinivasan, G., Robock, A., Entin, J.K., Luo, L., Vinnikov, K.Y. and Viterbo, P. (2000) Soil moisture simulations in revised AMIP models. *Journal of Geophysical Research* 105, 26635–26644.
Steegen, A., Govers, G., Takken, I., Nachtergaele, J., Poesen, J. and Merckx, R. (2001) Factors controlling sediment and phosphorus export from two Belgian agricultural catchments. *Journal of Environmental Quality* 30, 1249–1258.
Stone, J.R., Gilliam, J.W., Cassel, D.K., Daniels, R.B., Nelson, L.A. and Kleiss, H.J. (1985) Effect of erosion and landscape position on the productivity of Piedmont soils. *Soil Science Society of America Journal* 49, 987–991.
Sugawara, M. (1995) Tank model. In: Singh, V.P. (ed.) *Computer Models of Watershed Hydrology.* Water Resources Publications, Littleton, Colorado, pp. 165–214.
Tebaldi, C. and Knutti, R. (2007) The use of the multimodel ensemble in probabilistic climate projections. *Philosophical Transactions of the Royal Society A* 365, 2053–2075.
Tebaldi, C., Smith, R.L., Nychka, D. and Mearns, L.O. (2005) Quantifying uncertainty in projections of regional climate change: a Bayesian approach to the analysis of multimodel ensembles. *Journal of Climate* 18, 1524–1540.

Ting, M., Kushnir, Y., Seager, R. and Li, C. (2009) Forced and internal 20th century SST trends in the North Atlantic. *Journal of Climate* 22, 1469–1481.

Turner, S., Lyons, H. and Favis-Mortlock, D. (2001) Analysis and mapping of soil problem areas (hot spots) in Europe. Final Report to the European Environment Agency. In: *Where Are the 'Hot Spots' of Soil Degradation in Europe?* CD-ROM distributed to EIONET for review. European Environment Agency, Copenhagen.

Valentin, C. (1998) Towards an improved predictive capability for soil erosion under global change. In: Boardman, J. and Favis-Mortlock, D. (eds) *Modelling Soil Erosion by Water.* Springer-Verlag, Berlin, pp., 7–17.

Van der Knijff, J., Jones, R.J.A. and Montanarella, L. (2000) *Soil Erosion Risk Assessment in Europe.* Office for Official Publications of the European Communities, Luxembourg.

Van der Knijff, J., Jones, R.J.A. and Montanarella, L. (2002) Soil erosion risk assessment in Italy. In: Rubio, J.L., Morgan, R.P.C., Asins, S. and Andreu, V. (eds) *Proceedings of the Third International Congress: Man and Soil at the Third Millennium.* Geoforma Ediciones, Logrono, Spain, pp. 1903–1913.

Van Rompaey, A., Govers, G., Van Hecke, E. and Jacobs, K. (2001) The impacts of land use policy on the soil erosion risk: a case study in the Belgian Loam Belt. *Agriculture, Ecosystems and Environment* 83, 83–94.

Verity, G.E. and Anderson, D.W. (1990) Soil erosion effects on soil quality and yield. *Canadian Journal of Soil Science* 70, 471–484.

Wagener, T., Boyle, D.P., Lees, M.J., Wheater, H.S., Gupta, H.V. and Sorooshian, S. (2001) A framework for development and application of hydrological models. *Hydrological and Earth Sciences* 5, 13–26.

Ward, S., Connolly, J., Walsh, J., Dahlman, L. and Holden, N.M. (2007) *Climate Change: Modelling Carbon Fluxes from Irish Peatlands: Towards the Development of a National Carbon Fluxes Inventory for Irish Peatland. Environmental Research, Technological Development and Innovation (ERTDI) Programme (2000–2006), Synthesis Report.* Environmental Research Report 53, Environmental Protection Agency (EPA), Johnstown, Ireland.

Wilby, R.L. and Wigley, T.M.L. (1997) Downscaling general circulation model output: a review of methods and limitations. *Progress in Physical Geography* 21, 530–548.

17 Modelling the Impacts of Climate Change on Water Balance and Agricultural and Forestry Productivity in Southern Portugal Using SWAT

João Pedro Nunes* and Júlia Seixas

Introduction

The western Mediterranean region is characterized by dry and sub-humid climates and drought-adapted agricultural and forestry practices capable of maintaining vegetation productivity in winter, when more water is available, while restricting water use during the summer drought (e.g. Puigdefábregas and Mendizabal, 1998). However, conditions for agriculture and forestry are in many cases marginal, and in recent years the adoption of intensive cultivation practices and the extension of agriculture to unsuitable lands have led to an increase in the extension of desertification phenomena (e.g. Puigdefábregas and Mendizabal, 1998; Vogiatzakis *et al.*, 2006). This is one of the regions in the globe expected to be most affected by climate change, with future climate scenarios pointing to an increase in climatic aridity, due to lower rainfall and a temperature-driven increase in evapotranspiration demands (Räisänen *et al.*, 2004; Giorgi, 2006). Therefore, there is a concern that this could decrease the climatic support capacity for present-day agricultural and forestry practices, potentially leading to land-use changes but also to land abandonment and desertification (Nunes *et al.*, 2008).

The impacts of climate change on the hydrological cycle and vegetation growth have received some attention in recent years, focusing on studies simulating the Mediterranean as a whole or as a subset of a global-scale application. Model results for this region (Wetherald and Manabe, 2002; Nohara *et al.*, 2006) point to a shift in the allocation of available rainwater, with a larger part used to replenish soil water storage and evapotranspirative demand at the expense of surface runoff. Therefore, surface runoff and river flow are expected to suffer the greatest impacts, although evapotranspiration and soil moisture are still expected to decrease, particularly during spring and summer. This change has also been theoretically demonstrated by Arora (2002); this author reports that, for dry climates (i.e. where evapotranspiration is limited by water availability rather than solar radiation), an increase in climatic aridity is expected to shift the allocation of rainwater towards evapotranspiration. Model results for vegetation productivity are more variable, pointing to a balance between positive impacts of climate change – increased atmospheric CO_2 and temperature – and negative impacts – mostly a deteriorating water balance (Cheddadi *et al.*, 2001; Olesen and Bindi, 2002; Morales

* Corresponding author: jpcn@ua.pt

et al., 2007); in the Mediterranean, the predicted outcome from these balances ranges from slightly positive to negative.

Regional studies, however, have been less common. In Portugal, Cunha *et al.* (2006) have simulated the hydrological impacts of climate change in a range of catchments, ranging from humid to semi-arid climatic conditions. Their results are consistent with the ones pointed out above, with decreases in runoff occurring mostly from spring to autumn; in winter, results point to differences between catchments, with an increase in runoff in humid climates for some scenarios. Groundwater recharge is also expected to decrease. As for impacts on vegetation productivity, Pinto *et al.* (2006) point to an overall decrease in wheat productivity (the most important rainfed culture in Portuguese drylands), with impact severity increasing with present-day climatic aridity, i.e. the most arid regions suffering the greater impacts. For forests, however, results by Pereira *et al.* (2006b) indicate an increase in productivity for the humid regions and a decrease for the dry regions; impacts for cork oak forestry (the most important cultivated forest in Portuguese drylands) are expected to be less severe, as these trees are more adapted to low water availability environments.

The results from these studies point to the potential of climate change to decrease soil moisture and effective evapotranspiration, although they agree on higher negative impacts for runoff. However, results for vegetation productivity are less clear, as the positive and negative impacts of climate change appear to be balanced, albeit with a high spatial variability. This issue was addressed by Nunes *et al.* (2008), who tested hypothetical climate change scenarios for southern Portugal with varying degrees of magnitude using an ecohydrological model. The scenarios were all within the range of predictions by climate models for different CO_2 emission scenarios. Their results indicate that increases in temperature and CO_2 could lead to higher vegetation productivity when combined with small decreases in rainfall (down to −10%); but, when combined with large decreases in rainfall (down to −40%), model results showed a productivity decrease for most vegetation types. In other words, the uncertainty in climate change scenarios, either in CO_2 emission levels or their consequences for temperature and rainfall, is large enough to include decreases and increases in vegetation growth.

This chapter describes an ecohydrological modelling exercise to estimate the impacts of drier climates on water and vegetation in southern Portugal. This included18 medium- and large-scale watersheds in two study areas: one humid, occupied by vineyards, croplands and commercial pine/eucalypt forests; and one semi-arid, occupied by croplands and cork oak stands. Two climate change scenarios were tested, with different magnitudes of changes to mean annual climate as well as to seasonal climate patterns. The work described builds on that by Nunes *et al.* (2008) by focusing on the seasonal pattern of changes to water balance, soil moisture and leaf phenology, and the consequences for productivity, water stress and temperature stress, including their spatial and inter-annual variability.

Materials and Methods

Study areas

As already stated, this work focused on two study areas in southern Portugal. The Alentejo study area includes eight watersheds in the south-western end of the Guadiana river basin, with a total area of 2778 km^2. The watersheds are mostly located over a plain, but the south-western part of the area includes a mountain range. Rocky and shallow lithosols occupy 82% of this study area, with the remainder occupied mostly by luvisols, both having developed over a schist and greywacke bedrock. The area is characterized by a dry (sub-humid to semi-arid) climate which has led to the development of extensive agriculture, mostly comprising winter wheat croplands (48% of the study area), with annual average yields of 1.4 Mg ha^{-1} (INE, 2006). The mountain areas have a wetter, humid climate, which also allows the establishment of semi-natural cork oak forestry (32%). This area is suffering an ongoing process of biophysical and human desertification, and many former agricultural fields are now covered by Mediterranean shrublands (18%), especially in mountainous regions.

The Ribatejo study area comprises ten watersheds with a total area of 1252 km^2, to the west of the Tejo river basin. The area is located on the slopes of several mountain ranges, with relatively deeper cambisols (75% of the study area), luvisols (17%) and vertisols (7%) developed over sedimentary bedrock. It is characterized by a humid climate, allowing for the development of intensive agriculture, especially winter wheat croplands (34% of the study area) and vineyards (21%); annual average yields are respectively 2.0 and 1.4 Mg ha^{-1} (INE, 2006). Commercial eucalypt and maritime pine forestry is also present (21%), as well as important areas of shrublands (10%), urban development (6%) and olive groves (4%).

Both study areas present a Mediterranean climate; the wet season (rainfall excess) extends approximately from October to March, while the rest of the year constitutes the dry season (rainfall deficit). The Alentejo area shows a warmer and drier climate than the Ribatejo area, with mean annual temperatures of 16.2 versus 15.0 °C, and mean annual rainfall of 586 versus 872 mm $year^{-1}$ (data for the cities of Beja and Rio Maior, respectively). The selection of humid and dry catchments for this study allows for an analysis of the impacts of a transition towards a drier climate, which could be a consequence of climate change. The areas are also described in detail by Nunes *et al.* (2008, 2009).

Ecohydrological modelling

The ecohydrological model applied in this study is SWAT (Soil and Water Assessment Tool, version 2000; Neitsch *et al.*, 2002). It is a process-based, spatially distributed and continuous model capable of simulating:

- Water balance and runoff, taking into account rainfall, evapotranspiration, surface and subsurface runoff, and deep aquifer recharge; runoff partitioning into surface and subsurface runoff is based on the Curve Number (CN) method and a kinematic percolation model.
- Vegetation growth, taking into account plant phenological development using a degree-day approach, and leaf area development, light interception and conversion into biomass based on the Monteith approach; the model includes the impacts of atmospheric CO_2 concentration on vegetation light-use efficiency, and allows for the scheduling of agricultural operations.

These processes are simulated with a daily time step at the Hydrological Response Unit (HRU) scale; HRUs are specific combinations of land use and soil type inside a given watershed. The watersheds themselves are linked through the river network.

SWAT was applied to eight watersheds in the Alentejo study area, and 12 in the Ribatejo study area, subdivided into sub-basins with an average area of around 10 km^2, using the NASA SRTM 90 × 90 m topographic maps (Jarvis *et al.*, 2006). Each sub-basin was divided into HRUs with an average area of 3.8 and 1.4 km^2 in Alentejo and Ribatejo, respectively, the difference in size being due to the more dispersed land-use patterns in the latter study area. Data sources for the HRUs included the CORINE land cover map for 1990 (EEA, 1995), selected to coincide with the calibration and validation period (see below); and the 1973 Portuguese soil map (Cardoso *et al.*, 1973). Model parameters for vegetation cover came from a bibliographic survey (see Nunes *et al.*, 2008, for details), while those for soil type were taken from the International Soil Reference and Information Centre (ISRIC) data set (Batjes, 2002).

The calibration and validation of the hydrological component used a split-sample approach, with 10 years of daily measurements (1980–1989). The length of the sample is below that of a typical 'climate normal' (30 years) and could therefore limit the validity of model calibration; this issue is further discussed below. Half of the watersheds were used for validation only and, in the remaining watersheds, alternatively the first (dry) half of the decade was used for calibration and the second (wet) half for validation, and vice versa. All meteorological and hydrological data were taken from the Portuguese water resources information system (INAG, 2006). Soil water balance results were evaluated from the surface-flow and base-flow fractions, as observations for evapotranspiration and soil moisture were not available. The predicted growth of agricultural species was evaluated using agricultural productivity statistics. The prediction of leaf area index (LAI) and harvest date for wheat were compared with published values for phenology, assuming that model predictions for peak LAI correspond roughly to the start of heading. Further details on the model application are given by Nunes *et al.* (2008).

Climate change scenarios

The selected climate change scenarios consider the SRES (Special Report on Emissions Scenarios, prepared by the Intergovernmental Panel on Climate Change (IPCC) for the Third Assessment Report (TAR) in 2001) B2 and A2 CO_2 emission scenarios, corresponding to atmospheric concentrations of 615 and 845 ppm in 2071–2100, an increase of 66 and 128% over the level of 370 ppm assumed by Nakicenovic and Swart (2000) to represent present-day conditions. Regional Climate Model (RCM) simulations for Europe using these scenarios were performed with a 50 × 50 km resolution and a daily time step within the PRUDENCE project (Christensen and Christensen, 2007). For these studies, results from the PROMES RCM (Gallardo *et al.*, 2001) were selected; four grid cells were used for the Alentejo study area and two grid cells in the Ribatejo study area. Biases between the PROMES control run for 1961–1990 and long-term averages of monthly measured rainfall and temperature were corrected using regression methods according to Wilby and Wigley (1997). Observed data included a 1 × 1 km grid of monthly average rainfall for Portugal estimated using geostatistical methods (Nicolau, 2002), and monthly average temperature for two stations in each study site.

The SWAT model was applied using these scenarios for three runs in each study area: one control run for 1961–1990, and two perturbed runs for 2071–2100 (for the B2 and A2 scenarios, respectively). Average annual and monthly results for each scenario were compared in order to estimate climate change impacts on actual evapotranspiration, runoff and river flow, soil water storage, vegetation phenology and annual productivity, and water and temperature stress days.

Results and Discussion

Model evaluation

The model performance statistics for monthly river flows in the Alentejo and Ribatejo study areas are presented in Table 17.1. They showed good results, and, although the model overestimated flow in both cases, this bias was small when compared with the average unsigned error. Table 17.1 shows results for the Nash–Sutcliffe model efficiency index (Nash and Sutcliffe, 1970), which compares

Table 17.1. SWAT calibration and validation statistics for mean monthly river flow in the two study areas in Portugal.

	Alentejo			Ribatejo		
Parameter	Calibration	Validation	All data	Calibration	Validation	All data
r^2	0.74 ($p < 0.01$)	0.77 ($p < 0.01$)	0.76 ($p < 0.01$)	0.82 ($p < 0.01$)	0.81 ($p < 0.01$)	0.81 ($p < 0.01$)
Bias ($m^3 s^{-1}$)	−0.01	0.22	0.13	0.05	0.14	0.11
Average unsigned error ($m^3 s^{-1}$)	0.78	0.84	0.82	0.27	0.50	0.41
Observed average and range ($m^3 s^{-1}$)	1.3 (0–44.6)	1.5 (0–60.4)	1.4 (0–60.4)	0.6 (0–9.4)	0.9 (0–19.4)	0.8 (0–19.4)
Model efficiency[a]	0.74	0.77	0.76	0.79	0.62	0.66

[a]Nash–Sutcliffe model efficiency index (Nash and Sutcliffe, 1970).

the variance of the error with the variance of the observations; the maximum result is 1, and values over 0.5 are usually taken as an indication of acceptable model performance.

Figure 17.1 compares observed with simulated values, normalized through the square root owing to the large range between the averaged and maximum observations, as seen in Table 17.1 (a similar approach was used by Jetten *et al.*, 2003). The dispersion of observed versus simulated values around the 1:1 agreement line showed a good performance in both cases when predicting extreme monthly flows, but was less satisfactory for low flow prediction. In fact, the average unsigned error is 58.6 % of the average flow for Alentejo and 51.2% for Ribatejo. However, this error is small compared with the range of observed values in both systems (Table 17.1). Because the wet season months dominate surface water balance in Mediterranean watersheds (Palutikof *et al.*, 1996), the fact that SWAT was capable of predicting extreme flows adds further confidence to model results. Furthermore, the differences of model performance between calibration and validation were small in terms of r^2, and in terms of model efficiency, particularly in the case of Alentejo, indicating a small level of over-calibration. The differences in model efficiency for Ribatejo could indicate some degree of over-calibration, perhaps as a result of the larger range of observed values present for validation. However, the model efficiency still indicated good model performance.

Further evaluation results for annual flow are given in Nunes *et al.* (2008); in general terms, the model performed better at the annual scale than at the monthly scale. The results shown above indicate that the model adequately simulated water balance in the study areas, although it was not possible to independently verify evapotranspiration results. As for soil moisture, the adequate simulation of monthly water balance implies that the seasonal variability of soil water content was also reasonably simulated. This assertion was also supported by the good model results for base-flow fraction; model efficiency values for base-flow simulation (0.74 in the Alentejo and 0.68 in the Ribatejo study areas) were similar to those shown in Table 17.1 for total flow. However, it was also not possible to independently verify soil moisture results.

SWAT results for vegetation growth were also evaluated. As detailed by Nunes *et al.* (2008), simulated average annual productivity for agricultural species (wheat, grapes and olives) were compared with published average annual statistics (INE, 2006); the prediction of crop calendar dates for winter wheat was also evaluated. The results are shown in Table 17.2. The model was able to

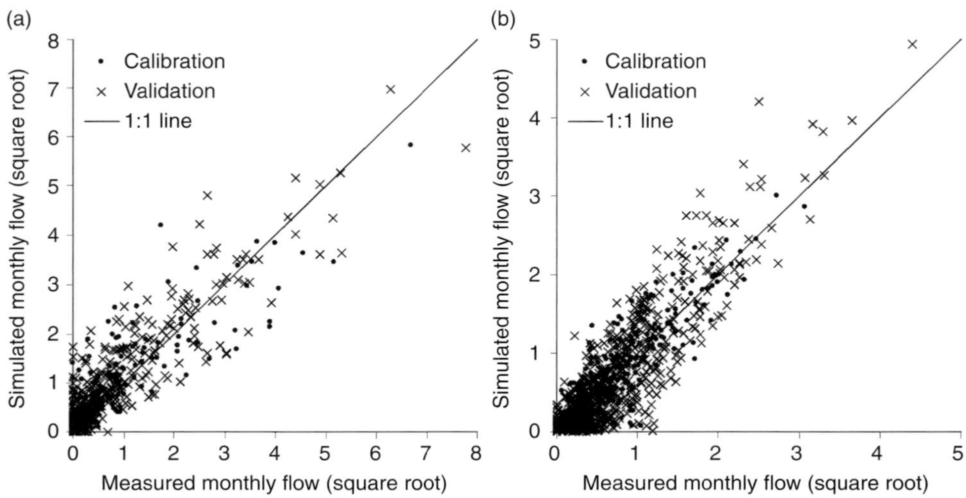

Fig. 17.1. Observed and SWAT-simulated average monthly river flow in the (a) Alentejo ($r^2 = 0.76$, $p < 0.01$) and (b) Ribatejo ($r^2 = 0.81$, $p < 0.01$) catchments given as square root of values ($m^{1.5}\ s^{-0.5}$).

Table 17.2. SWAT-simulated peak leaf area index (LAI) and observed flowering dates for winter wheat in Portugal; model results for the Alentejo and Ribatejo study areas are equal.

Crop	Study site		Harvest (Mg ha^{-1} year^{-1})[a]	Heading date[b]	Harvest date[b]
Olives	Alentejo	Predicted	0.2	–	–
		Observed	0.3	–	–
Grapes	Ribatejo	Predicted	2.0	–	–
		Observed	1.4	–	–
Winter wheat	Alentejo	Predicted	1.6	April[c]	July
		Observed	1.4	April to May[d]	June to July
	Ribatejo	Predicted	2.1	April[c]	July
		Observed	2.0	April to May[d]	June to July

[a]Observed values taken from INE (2006).
[b]Observed values taken from INIA-LQARS (2000).
[c]Simulated date for peak LAI (assumed to correspond roughly to the start of heading).
[d]Observed date for heading.

reasonably simulate olive production in the Alentejo area, but overestimated grape production in the Ribatejo area. Model results for winter wheat were reasonable in both study areas and, furthermore, the model was able to simulate the difference in wheat productivity between regions. As for the calendar dates, and despite this being necessarily a coarse comparison, the data do indicate that the model is capable of adequately simulating growth stages for winter wheat. Taken together, the evaluation for vegetation growth indicates that the model is reasonably simulating both seasonal growth patterns (in wheat) and annual biomass growth, although a comparison with measured LAI values was not performed owing to the lack of adequate data for the study period.

Table 17.3. Average annual change in temperature, rainfall and potential evapotranspiration (PET) predicted by the PROMES regional climate model for the B2 and A2 emissions scenarios (2071–2100) for the Alentejo and Ribatejo study areas; averages were computed in space and time.

Parameter	Scenario	Alentejo	Ribatejo
Temperature	Control	16.7°C	16.3°C
	B2	+2.7°C	+2.2°C
	A2	+3.6°C	+3.2°C
Rainfall	Control	618 mm	823 mm
	B2	−25.1%	−20.2%
	A2	−29.3%	−24.2%
PET	Control	1310 mm	1169 mm
	B2	+14.2%	+15.7%
	A2	+18.8%	+20.9%

Climate change scenarios

PROMES average annual climate change scenarios are given in Table 17.3. In both scenarios, the model indicated an increase in temperature (2.2 to 3.6 °C) combined with a decrease in rainfall (−20.2 to −29.3%). These scenarios also indicated a higher PET (potential evapotranspiration; 14.2 to 20.9%), with the net result of an increase in climatic aridity. Using the UNEP index (Middleton and Thomas, 1997), i.e. the ratio of rainfall to PET, the model showed an increase in climatic aridity in the Alentejo area for both scenarios; the climate remained within the semi-arid classification, but moved from the threshold for dry/sub-humid (0.47) to the threshold for arid (0.31 and 0.28 for the B2 and A2 scenarios, respectively). In the Ribatejo area, the model indicated a shift from humid conditions (0.70) to semi-arid conditions in both scenarios, although close to the threshold with dry/sub-humid (UNEP index 0.49 and 0.44, respectively for B2 and A2). Both scenarios can be said to have shown similar trends towards increasing aridity, with A2 leading to more severe changes than B2. Also, the PROMES scenarios showed more severe changes (in terms of temperature and rainfall) in the Alentejo area than in the Ribatejo area.

The climate changes were unevenly distributed throughout the year, as shown in Fig. 17.2. In both study areas, rainfall decreases are concentrated in the onset of the wet season (September and October) and in the onset of the dry season (May). The increase in PET occurred mostly during the summer dry season. Overall, the projected climate change scenarios pointed to an increase in climatic aridity, resulting in a shortening of the wet season by a few months. It should also be noted that in the Ribatejo area, and despite the downward trends for rainfall, the model predicted a 17.3% rainfall increase during January for the A2 scenario (Fig. 17.2).

Impacts on hydrological processes

Table 17.4 shows SWAT results for hydrological parameters when applied using the PROMES climate change scenarios. In both scenarios, SWAT predicted a decrease in the most important water balance components, which would be expected from the lower rainfall and higher evapotranspiration demand (Table 17.3). However, the model also predicted that the increase in climate aridity will be reflected mostly by lower runoff (–36.2 to –44.0%); effective evapotranspiration (EET) showed a much lower decrease (–10.2 to –21.6%). This was also reflected in

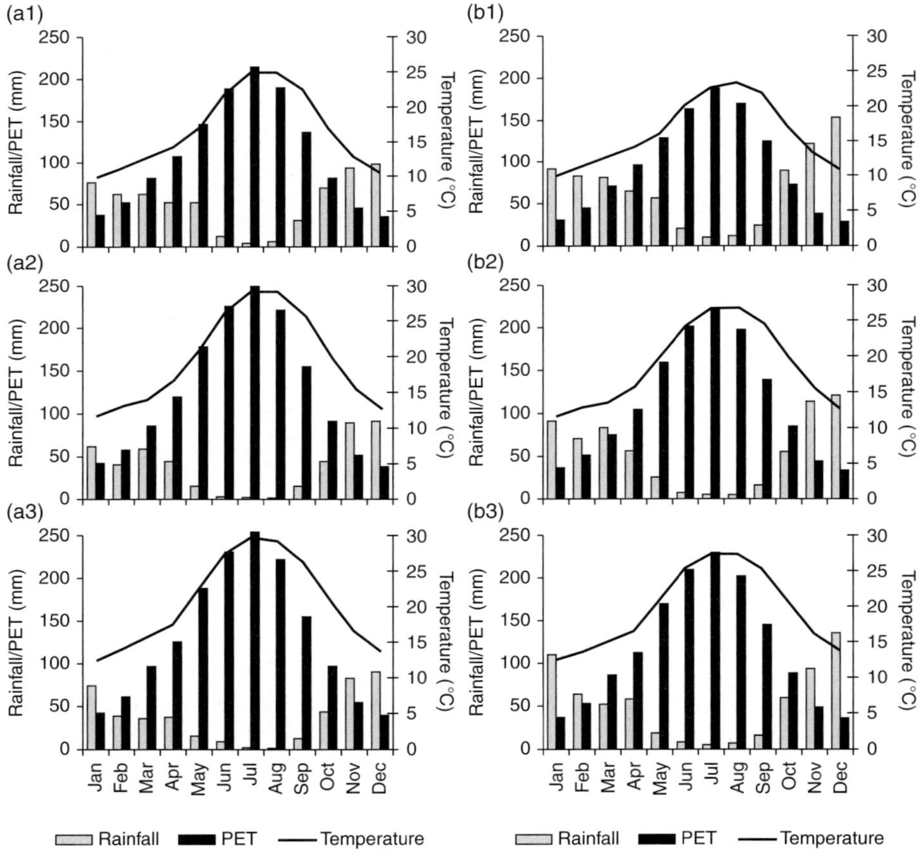

Fig. 17.2. Mean monthly rainfall, potential evapotranspiration (PET) and temperature: (a1 and b1, top) for 1961–1990; and as predicted by the PROMES regional climate model for the B2 (a2 and b2, middle) and A2 (a3 and b3, bottom) emissions scenarios in 2071–2100. The left column (a1 to 3) is for Alentejo and the right column (b1 to 3) for Ribatejo. Averages were computed in space and time.

results for soil moisture, which showed a relatively modest decrease (–10.0 to –21.7%) when compared with the predicted changes in rainfall (Table 17.3). Finally, more severe changes were predicted for the Alentejo area, accompanying the climate change scenarios shown earlier.

Changes in EET and soil moisture were also unevenly distributed throughout the year (Fig. 17.3), following the seasonal pattern of climate changes described earlier. Changes in EET were concentrated at the start of and during the dry season; the greater decreases occurred from April to June in the Alentejo area, and from May to July in the Ribatejo area. This followed, approximately, the periods when rainfall was predicted to decrease the most. In contrast, EET was predicted to slightly increase during the latter part of the wet season, when rainfall still exceeds PET. The seasonal patterns of soil moisture dynamics were driven by the changes to both rainfall and EET; the net result was that the most important changes occurred mostly during the dry season. In the B2 scenario, important decreases occurred during summer and autumn (from May to November) in both study areas. In the A2 scenario, these decreases occurred during spring, summer and autumn (from March to November), and in both scenarios the winter (December to February) had negligible soil moisture changes. In contrast, runoff changes were distributed throughout the year.

Figure 17.4 shows the inter-annual variability of model results for water balance (a) and soil moisture (b). Rainfall in both study areas was predicted to show a high variability, with similar patterns between control and climate change conditions; in the Alentejo area, the present-day trend for exceptionally high rainfall years was predicted to continue in both climate change scenarios. However, this variability was predicted to be reflected mostly in runoff rates; EET was not predicted to vary by a large degree between years. The same can be said about the inter-annual variability of soil moisture, which was predicted to remain similar to present-day conditions in both climate change scenarios. In this case, however, the trend for years with exceptionally low soil moisture rates (mostly between the minimum and 1st quartile) were predicted to also occur under climate change. Overall, model results did not indicate a trend for an increase in the inter-annual variability of climate, water balance and soil moisture, and therefore did not indicate changes to the frequency of severe agricultural droughts.

Table 17.4. Average annual change in effective evapotranspiration (EET), runoff and soil moisture predicted by SWAT for the B2 and A2 emissions scenarios (2071–2100) for the Alentejo and Ribatejo study areas; an area-weighted average was used to aggregate model results for the different watersheds.

Parameter	Scenario	Alentejo	Ribatejo
EET	Control	275 mm	394 mm
	B2	–17.4%	–10.2%
	A2	–21.6%	–13.5%
Runoff	Control	192 mm	300 mm
	B2	–40.5%	–36.2%
	A2	–44.0%	–41.0%
Soil moisture	Control	6.8% (v/v)	11.9% (v/v)
	B2	–17.0%	–10.0%
	A2	–21.7%	–13.5%

Impacts on agricultural and forestry productivity

Table 17.5 and Table 17.6 show model results for annual biomass productivity, and for water and temperature stress days for the Alentejo and Ribatejo study areas, respectively. Results are shown for the most important cultivated species (agriculture and forestry) for each area. For comparison purposes, results are also shown when considering only CO_2 increases; these were taken from simulations by Nunes *et al.* (2008). In most cases, the model pointed to a small increase in productivity, which could be attributed to a net positive effect of climate change. The simulated impacts of climate change on vegetation growth included negative and positive effects, as follows:

- faster phenological development
- increase in water and temperature stress days during summer
- decrease in temperature stress days during winter

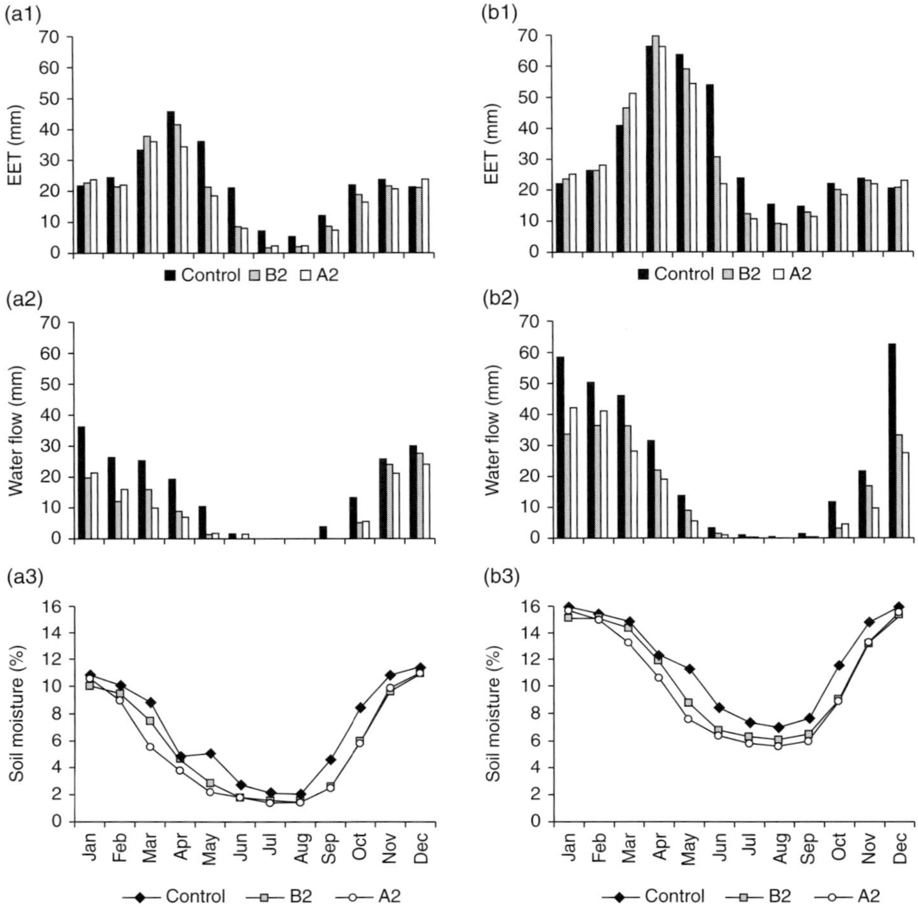

Fig. 17.3. (a1 and b1, top) Mean monthly effective evapotranspiration (EET), (a2 and b2, middle) runoff (water flow) and (a3 and b3, bottom) soil moisture for 1961–1990 (control) and as predicted by SWAT using data from the PROMES regional climate model for the B2 and A2 emissions scenarios in 2071–2100. The left column (a1 to 3) is for Alentejo and the right column (b1 to 3) for Ribatejo.

- higher solar radiation (6.8 to 8.8%) due to lower rainfall
- CO_2 fertilization increasing light-use efficiency.

The combination between negative and positive effects of climate change led to different results, according to vegetation type (Tables 17.5 and 17.6). Winter wheat responded more positively to scenario B2 (11.7 to 15.5%) than to scenario A2 (5.3 to 9.9%); woody vegetation types showed the opposite response, with both negative and positive impacts on growth included in the B2 scenario (−5.5 to 2.7%), in contrast to all positive impacts in A2 (2.8 to 34.8%).

In the case of wheat, the faster phenological development allowed for leaf development to occur during the wet season (i.e. with little changes to soil moisture), resulting in no changes to water stress days, and for harvesting to occur before the drier months. Considering a planting date of October, the onset of wheat heading and harvest were predicted to be anticipated by 1 month in the B2 scenario, and 2 months in the A2 scenario, when compared with current dates as given in Table 17.2. In the latter case, the agricultural

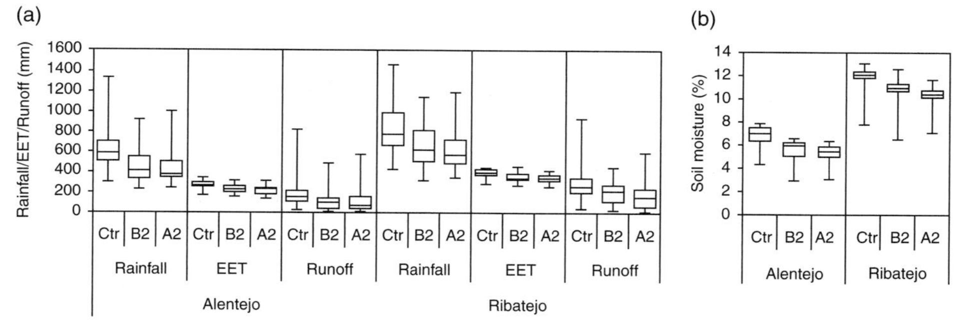

Fig. 17.4. (a) Inter-annual variability of mean annual rainfall, effective evapotranspiration (EET) and runoff, and (b) soil moisture in the Alentejo and Ribatejo study areas, as simulated by SWAT using data from the PROMES regional climate model for control (Ctr) conditions in 1961–1990, and for the B2 and A2 emissions scenarios in 2071–2100); an area-weighted average was used to aggregate model results for the different watersheds. Boxes represent the 1st, 2nd and 3rd quartiles, while the vertical bars represent the minimum and maximum values.

Table 17.5. Average annual change in biomass production and water and temperature stress days in the Alentejo study area for its major vegetation types, as predicted by SWAT B2 and A2 emissions scenarios (2071–2100). The biomass results include both the full climate change scenario (Full), and that which considers only the CO_2 increase element of the A2 and B2 scenarios (CO_2 only).

		Biomass production			
Vegetation type	Scenario	Full	CO_2 only[a]	Water stress (days)	Temperature stress (days)
Winter wheat	Control	3.6 Mg ha^{-1} year^{-1}		15.2	1.0
	B2	+15.5%	+14.6%	−0.1	−0.3
	A2	+5.3%	+21.9%	+0.4	−0.5
Cork oak	Control	1.5 Mg ha^{-1} year^{-1}		28.6	38.4
	B2	−3.2%	+13.5%	+0.3	−8.8
	A2	+34.8%	+19.3%	+4.4	−14.1

[a]Nunes *et al.* (2008)

calendar would approach the one currently observed on the Mediterranean coast of Africa (Frenken and Faurès, 1997), with wheat harvest occurring in May. The faster phenological development minimized changes to water stress, while reducing the availability of solar radiation during growth; this would be compensated by CO_2 fertilization, allowing for the more efficient use of available sunlight. The relationship between these factors could explain the differences in productivity predictions between the A2 and B2 scenarios. As for woody vegetation species, vegetation growth occurred during the entire year, and therefore the faster phenological development did not avoid the decrease in dry season soil moisture. The result pointed to an increase in the number of water stress days, particularly for the A2 scenario (Tables 17.5 and 17.6). Furthermore, the earlier onset of flowering and summer leaf fall would also decrease vegetation productivity. However, the results also point to a large decrease in temperature stress days, due especially to warmer winters, which was more important than the increase in water stress days. Combined with the response to increased CO_2 concentrations, the changes in climate combined to cause an increase in biomass productivity for maritime pine and vine in the B2 scenario, and for all species in the A2

Table 17.6. Average annual change in biomass production and water and temperature stress days in the Ribatejo study area for its major vegetation types, predicted by SWAT for the B2 and A2 emissions scenarios (2071–2100). The biomass results include both the full climate change scenario (Full), and that which considers only the CO_2 increase element of the A2 and B2 scenarios (CO_2 only).

		Biomass production			
Vegetation type	Scenario	Full	CO_2 only[a]	Water stress (days)	Temperature stress (days)
Winter wheat	Control	7.0 Mg ha^{-1} year^{-1}		0.0	1.1
	B2	+11.7%	+14.5%	+0.0	–0.3
	A2	+9.9%	+21.8%	+0.5	–0.6
Eucalypt	Control	5.4 Mg ha^{-1} year^{-1}		11.0	22.3
	B2	–5.5%	+1.2%	+2.8	–5.7
	A2	+2.8%	+1.8%	+6.6	–10.0
Maritime pine	Control	3.3 Mg ha^{-1} year^{-1}		1.0	9.8
	B2	+2.7%	+1.8%	+0.1	–1.8
	A2	+4.9%	+4.6%	+2.5	–3.7
Vine	Control	4.7 Mg ha^{-1} year^{-1}		1.4	30.7
	B2	+1.2%	+6.4%	-0.1	–4.9
	A2	+9.4%	+8.6%	+7.0	–10.7

[a]Nunes *et al.* (2008).

scenario (with cork oaks showing the highest response). Variability between vegetation responses could be explained by different phenological cycles, water-use efficiency during summer droughts, optimal temperatures and response to CO_2.

Figure 17.5 shows the spatial distribution of changes to average annual soil moisture and biomass productivity inside the study areas as predicted by the SWAT model, according to the most important vegetation and soil types. In both study areas, the variability of soil moisture changes was not large, except for the most important changes (values under the 1st quartile) which showed a large variability. There were several reasons behind this variability, including differences in present day climate, spatial variability of climate change scenarios (taken from the PROMES model with a 50 × 50 km resolution) and, in the Ribatejo area, a small tendency for greater decreases in soil moisture for forests (eucalypt and maritime pine) when compared with agriculture (winter wheat and vine). However, the only important explanatory variable for spatial variability was found in the Alentejo area, where soil moisture showed a larger decrease for luvisols than for lithosols. This could be explained by the much lower soil water holding capacity in lithosols when compared with that of luvisols; the available water content for luvisols is 31 mm, whereas for lithosols it is 100 mm (estimated from Batjes, 2002). This prevents lithosols from storing a large part of wet season rainfall both in present-day conditions and climate change scenarios (Nunes *et al.*, 2008) and therefore a decrease in rainfall would not affect soil moisture as much as in deeper soils.

In contrast, the spatial variability for biomass productivity changes was large for several vegetation types and scenarios, as can be seen in Fig. 17.5. This is more noticeable, in the Alentejo area, for wheat and cork oak in the B2 scenario, and, in the Ribatejo area, for wheat and maritime pine (both scenarios), and eucalypt (B2 scenario). Especially for maritime pine, this resulted in a spatial distribution ranging from increases to decreases in productivity, even if average annual trends were mildly positive (Tables 17.5 and 17.6). As in the previous case, there were several reasons behind this variability, including those mentioned earlier for soil moisture changes, as well as changes to soil moisture itself. Also similar to the previous case, the only important explanatory variable was the difference between lithosols and luvisols in the Alentejo area, and only in the

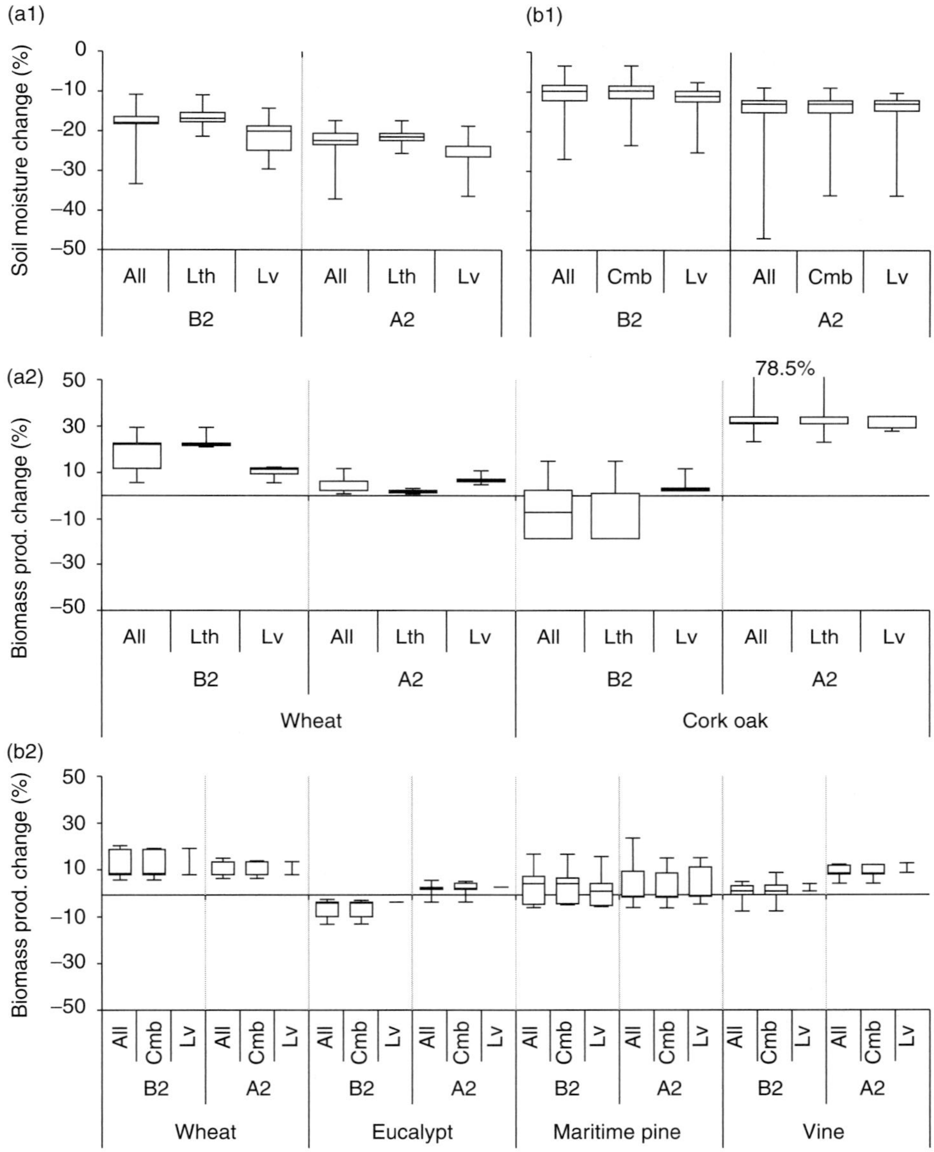

Fig. 17.5. Spatial variability of the mean annual soil moisture change (a1 and b1, top) and biomass productivity change (a2 and b2, middle and bottom), according to soil and vegetation type, as predicted by SWAT using data from the PROMES regional climate model for the B2 and A2 emissions scenarios in 2071–2100, for Alentejo (a1 and a2) and Ribatejo (b1 and b2); soils include lithosols (Lth), luvisols (Lv) and cambisols (Cmb). Boxes represent the 1st, 2nd and 3rd quartiles, while the vertical bars represent the minimum and maximum values.

B2 scenario. In this case, wheat experienced greater productivity increases in lithosols; the dependence on soil moisture in the topsoil would allow wheat to take advantage of the relatively lower decreases in soil moisture found for these soils, especially as most of the growth occurs during the wet season. For cork oaks, there were greater productivity increases for luvisols but a large spatial variability of productivity increases in lithosols; one explanation could be the ability of cork oaks to utilize water stored in the entire soil profile, allowing them to access more water in luvisols than in lithosols during the dry season.

Figure 17.6 shows SWAT modelled results for the inter-annual variability of biomass growth (top), water stress days (middle) and temperature stress days (bottom) for the most important vegetation types. In most cases, inter-annual variability was similar for present-day and climate change conditions, supporting the conclusion (detailed earlier) that model results did not indicate changes to the frequency of severe agricultural droughts. Also, in most cases, inter-annual variability was greater than the climate change impact on both biomass growth and water stress days. This indicates that the impacts of climate change would only be noticeable for long-term trends. The exception was the number of water stress days for maritime pine and especially for vine, which would show a greater variability under the A2 climate change scenario than for present-day conditions.

Discussion

As mentioned earlier, the water balance results from this work concur with previous modelling results for the Mediterranean, especially on the shift in the allocation of available

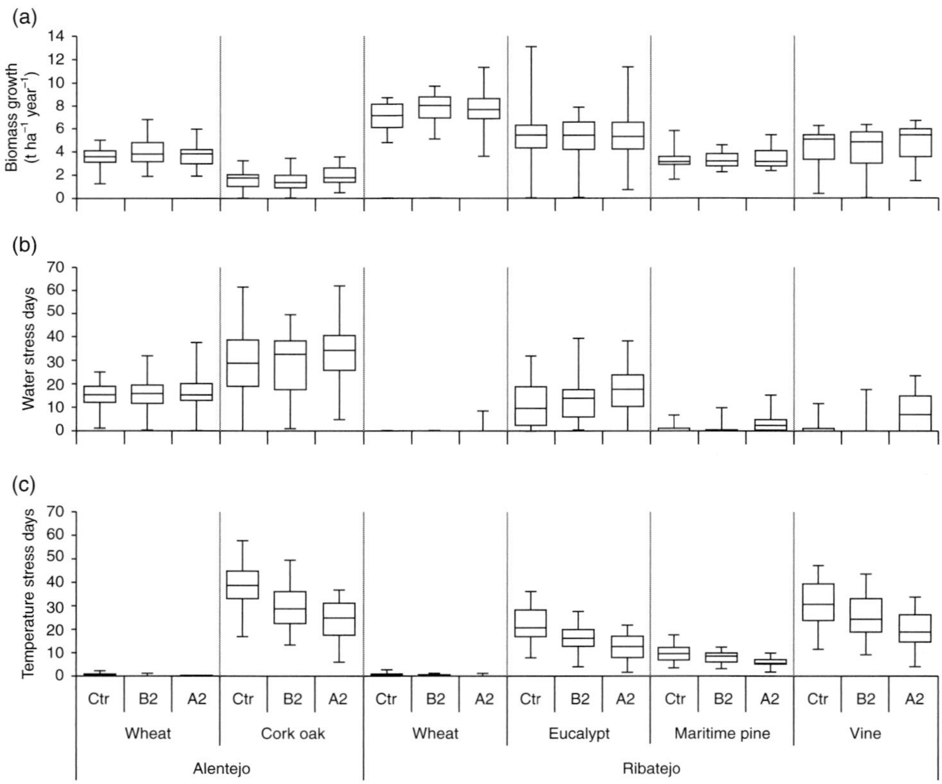

Fig. 17.6. Inter-annual variability of mean annual biomass growth (a, top), water stress days (b, middle) and temperature stress days (c, bottom) in the study areas, as simulated by SWAT using data from the PROMES for control (Ctr) conditions in 1961–1990, and for the B2 and A2 emissions scenarios in 2071–2100; an area-weighted average was used to aggregate model results for the different watersheds. Boxes represent the 1st, 2nd and 3rd quartiles, while the vertical bars represent the minimum and maximum values.

rainwater to replenish soil water storage at the expense of surface runoff. Estimates for changes in evapotranspiration, soil moisture and surface runoff shown in Table 17.4 for the A2 scenario were similar to those obtained by Wetherald and Manabe (2002) and Nohara *et al.* (2006) for the western Mediterranean (climate change scenarios IS92a and A1B, respectively), and by Cunha *et al.* (2006) for Portugal in scenario A2. The results for scenario B2 did not agree with those by Cunha *et al.* (2006), who predicted a slight increase in runoff; this is probably because the authors used model results predicting an increase in winter rainfall while the model results used in this work predict a decrease (Fig. 17.2). This highlights the uncertainty associated with climate model predictions and its effects on impact scenarios, as already highlighted, for example, by Giorgi (2005). However, in general terms the results in this work for hydrology confirm previous impact studies, as well as the theoretical predictions for drying climates by Arora (2002).

It should be noted that the results of this work did not predict an increase in the frequency of severe drought years, both in hydrological (Fig. 17.4) and agricultural (Fig. 17.6) terms, when compared with future average years. Nevertheless, when considering present-day water consumption, this would result in an increase in the water withdrawal ratio (i.e. consumption versus available resources), and therefore lead to an increase in severe water stress conditions (as defined by, for example, Alcamo *et al.*, 2003). In this sense, it can be said that the results from this work indicated the potential for an increase in the number of years with severe water stress over water resources, but further work is needed at the larger basin scale, including impacts on existing water management and collection systems.

The model results for winter wheat growth (Tables 17.5 and 17.6) did not agree well with trends expected by Olesen and Bindi (2002) for the Mediterranean and by Pinto *et al.* (2006) for Portugal, pointing to a small (10–15%) decrease in wheat productivity in the A2 scenario. For the remaining land covers, the predictions of productivity agreed better with other results, such as the trends for small increases for Mediterranean ecosystems predicted by Morales *et al.* (2007), the increase in productivity for sclerophyllous woodlands expected by Cheddadi *et al.* (2001), or the increase in forest productivity for the present-day humid regions of Portugal estimated by Pereira *et al.* (2006b). The spatial variability of vegetation productivity, including regions of negative and positive trends as shown in Fig. 17.5, also agreed with results by Morales *et al.* (2007). In contrast with this work, Pereira *et al.* (2006b) expected a decrease in cork oak productivity for southern Portugal.

As discussed earlier, different estimates of future CO_2 concentrations and their impact on vegetation productivity can have an important impact on vegetation productivity estimates, changing trends from negative to positive (see Nunes *et al.*, 2008). In this work, model results for vegetation growth depended on CO_2 fertilization to some extent; changes to the parameterization of CO_2 impacts on light-use efficiency could lead to important changes in the results from this study, which should be taken into account when evaluating the conclusions. It can be argued, however, that the same problem affects the other modelling studies discussed earlier, especially as the magnitude of the CO_2 fertilization effect is still a key uncertainty in assessing the impacts of climate change on ecosystems (Fischlin *et al.*, 2007). The results of this work therefore highlight how uncertainty in this issue can affect vegetation productivity predictions. Nevertheless, the impact of earlier phenological development on avoiding summer drought in Mediterranean climates is a result which merits further study. Also, it should be stressed that irrigation is the dominant water use in southern Portugal (about 80% of total), and therefore the runoff decreases predicted in this work would probably have a more severe consequence for the productivity of irrigated agriculture; this would be particularly important in drier years (Berry *et al.*, 2006). Further research focusing specifically on irrigated agriculture in the Mediterranean is still necessary.

This work also suffers from a number of limitations which should be addressed. First, all model results depend on the quality of the climate change scenarios provided by the

PROMES model; the uncertainty of climate model runs in the PRUDENCE project was addressed by Jacob *et al.* (2007).

Secondly, the conclusions are limited by the capacities of the SWAT model to adequately represent reality. The validity of the model calibration presented in this work for scenarios of climate change is difficult to assess. An effort has been made to address this issue by calibrating and validating the SWAT model for different watersheds and time periods with significantly different climate conditions, thus providing a calibrated parameter set which is valid under a large range of annual temperatures and rainfall amounts. However, the time period of available measurements did not include the full range of temperature conditions simulated in the climate change scenarios, and so the validity of the calibration under these scenarios cannot be fully assessed, which constitutes a limitation of this methodology. Moreover, the effects of CO_2 concentration changes cannot be assessed with the data currently available except through a literature comparison. Nevertheless, the statistical indicators of model performance showed that the application of SWAT to the Alentejo and Ribatejo study areas was robust for seasonal water balance predictions, meaning that model results were either good or satisfactory in a variety of different climatic conditions. The model is also capable of differentiating streamflow regimes in terms of base flow and surface flow. Therefore, it can be used to assess the impacts of climate change on water balance with a reasonable degree of confidence, particularly when considering relative results, at both the annual and seasonal scales. Furthermore, the results indicated that the model is capable of simulating the impact of different climate conditions on vegetation productivity, and therefore changes to this parameter can be assessed with a reasonable degree of confidence. However, the robustness of the simulation could potentially be increased by an analysis of the sensitivity of model results to changes in model parameters, especially those which affect model performance the most, or where there are more doubts about their value (Wilby, 2005; Dessai and Hulme, 2007). The simulation could also be improved with currently available satellite imagery for vegetation phenology as well as by more information on the impacts of CO_2 concentrations on productivity, as mentioned earlier.

Finally, the conclusions were also limited by potential effects of climate change not included in the SWAT model. In particular, the model does not simulate impacts of crossing aridity thresholds on vegetation growth and land use, such as:

- the increased mortality of woody species during drought years (e.g. Martinez-Vilalta *et al.*, 2002)
- the higher variability in wheat yields and economic returns, forcing farmers to abandon croplands (e.g. Puigdefábregas and Mendizabal, 1998; Berry *et al.*, 2006)
- changes in vegetation density and spatial patterns, especially important for undergrowth in cork oak woodlands (e.g. Puigdefábregas, 1998)
- changes in soil structural properties (e.g. Lavee *et al.*, 1998)
- increase in wildfire frequency (e.g. Mouillot *et al.*, 2002) – especially important for eucalypt and maritime pine forests.

Perturbations such as severe droughts and wildfires could trigger vegetation mortality and land-use changes, with recovery to the previous status prevented by new climatic conditions, the normal climatic variability of dryland climates and self-reinforcing mechanisms of land degradation (e.g. Puigdefábregas, 1998; Pereira *et al.*, 2006a); these pressures and mechanisms were not taken into account by the model. Overall, a number of authors (e.g. Kosmas *et al.*, 1999; Barboni *et al.*, 2004; Vicente-Serrano *et al.*, 2006) have observed that Mediterranean vegetation types occur above climatic aridity thresholds, and that these thresholds could take the above-mentioned effects into account. One typical observation is that forests and permanent cultures do not tolerate semi-arid conditions well unless irrigated, although sclerophyllous forests have a slightly greater tolerance; indicated UNEP aridity index thresholds ranged between 0.6 and 0.3. This would indicate that climate changes could severely decrease the suitability of the study areas for forestry and vineyards despite model results, owing to the aforementioned impacts of perturbations.

However, it must be noted that the high spatial variability of present-day climatic aridity in the study areas (as referred to earlier, and also by Nunes *et al.*, 2008) would limit these effects to only part of the watersheds. These studies also did not take into account possible mitigating effects of the increase in atmospheric CO_2 concentration on both vegetation biomass productivity and drought resistance, which could increase the tolerance of vegetation to climatic aridity (Cheddadi *et al.*, 2001). The subject of the impacts of increased aridity on vegetation cover through perturbations and changes in vegetation structure is still a key uncertainty when assessing the impacts of climate change on vegetation productivity (Fischlin *et al.*, 2007).

Summary

Overall, the results from this work indicated that a warmer and drier climate, caused by global climate change, would lead to a shift in rainfall from runoff to evapotranspiration in southern Portugal. Within the range of climate change scenarios tested in this work, runoff is expected to suffer larger impacts than evapotranspiration and soil moisture, although all are expected to decrease. Results also indicated that changes to soil moisture would concentrate in spring, summer and autumn, being less evident in the winter wet season. The model also predicted an increase in vegetation productivity for most species, and especially in the A2 scenario, due essentially to CO_2 fertilization and lower temperature stress in winter; for wheat, this was coupled with an earlier phenological development and harvest, which allowed some avoidance of summer water stresses.

The simulations showed a degree of spatial variability in these trends and, in the Alentejo study area, the importance of shallow lithosols and their reduced capacity for soil moisture retention. Also, the inter-annual variability in biomass growth and water stress was expected to remain more important than the climate change signal, with a potential to mask long-term trends. It should be stressed that irrigated agriculture could suffer more important impacts owing to lower available water resources, and further research is needed to address this issue.

Finally, the limitations of this work also point to potential research areas on the subject of the impacts of climate change on water balance and vegetation productivity. Future research should focus on estimating the impacts of CO_2 concentrations on vegetation productivity, including severe vegetation disturbances (such as drought-induced mortality and wildfires), on the simulations, addressing stresses to water use in irrigation-dependent agricultural systems, and taking into account socio-economic issues such as farmers' responses to changes in agricultural productivity.

Ongoing research on this subject, for the Portuguese case study, includes:

- the impacts of climate change on irrigation requirements and groundwater recharge in southern Portugal
- scenarios of future agricultural practices, including adaptation to climate change
- the impacts of climate change on drought and wildfire frequency and severity, and the consequences for forest cover and water balance
- comparison with the impacts of climate change in other climatic regions, namely study areas with continental climates in central Europe (Germany and Ukraine).

References

Alcamo, J., Doll, P., Henrichs, T., Kaspar, F., Lehner, B., Rosch, T. and Siebert, S. (2003) Global estimates of water withdrawals and availability under current and future "business-as-usual" conditions. *Hydrological Sciences Journal – Journal des Sciences Hydrologiques* 48, 339–348.

Arora, V.K. (2002) The use of the aridity index to assess climate change effect on annual runoff. *Journal of Hydrology* 265, 164–177.

Barboni, D., Harrison, S.P., Bartlein, P.J., Jalut, G., New, M., Prentice, I.C., Sanchez-Goñi, M.-F., Spessa, A., Davis, B. and Stevenson, A.C. (2004) Relationships between plant traits and climate in the Mediterranean region: a pollen data analysis. *Journal of Vegetation Science* 15, 635–646.

Batjes, N.H. (2002) *Soil Parameter Estimates for the Soil Types of the World for Use in Global and Regional Modelling (Version 2D1; July 2002).* ISRIC Report 2002/02c, International Food Policy Research Institute (IFPRI)/International Soil Reference and Information Centre (ISRIC), Wageningen, The Netherlands.

Berry, P.M., Rounsevell, M.D.A., Harrison, P.A. and Audsley, E. (2006) Assessing the vulnerability of agricultural land use and species to climate change and the role of policy in facilitating adaptation. *Environmental Science and Policy* 9, 189–204.

Cardoso, J.C., Bessa, M.T. and Marado, M.B. (1973) Carta dos solos de Portugal – 1:1.000.000. *Agronomia Lusitana* 33, 481–602.

Cheddadi, R., Guiot, J. and Jolly, D. (2001) The Mediterranean vegetation: what if the atmospheric CO_2 increased? *Landscape Ecology* 16, 667–675.

Christensen, J.H. and Christensen, O.B. (2007) A summary of the PRUDENCE model projections of changes in European climate by the end of this century. *Climatic Change* 81, 7–30.

Cunha, L.V., Ribeiro, L., Oliveira, R.P. and Nascimento, J. (2006) Recursos hídricos. In: Santos, F.D. and Miranda, P. (eds) *Alterações Climáticas em Portugal: Cenários, Impactos e Medidas de Adaptação - Projecto SIAM II.* Gradiva Publicações, Lisbon, pp. 115–168.

Dessai, S. and Hulme, M. (2007) Assessing the robustness of adaptation decisions to climate change uncertainties: a case study on water resources management in the East of England. *Global Environmental Change* 17, 59–72.

EEA (1995) *CORINE Land Cover.* European Environmental Agency (EEA), Commission of the European Communities, Copenhagen.

Fischlin, A., Midgley, G.F., Price, J.T., Leemans, R., Gopal, B., Turley, C., Rounsevell, M.D.A., Dube, O.P., Tarazona, J. and Velichko, A.A. (2007) Ecosystems, their properties, goods, and services. In: Parry, M.L., Canziani, O.F., Palutikof, J.P., van der Linden, P.J. and Hanson, C.E. (eds) *Climate Change 2007: Impacts, Adaptation and Vulnerability. Contribution of Working Group II to the Fourth Assessment Report of the Intergovernmental Panel on Climate Change.* Cambridge University Press, Cambridge, UK, pp. 211–272.

Frenken, K. and Faurès, J.-M. (1997) *Irrigation Potential in Africa: A Basin Approach.* FAO Land and Water Bulletin No. 4, United Nations Food and Agriculture Organization (FAO), Rome.

Gallardo, C., Arribas, A., Prego, J.A., Gaertner, M.A. and De Castro, M. (2001) Multi-year simulations using a regional-climate model over the Iberian Peninsula: Current climate and doubled CO_2 scenario. *Quarterly Journal of the Royal Meteorological Society* 127, 1659–1681.

Giorgi, F. (2005) Climate change prediction. *Climatic Change* 73, 239–265.

Giorgi, F. (2006) Climate change hot-spots. *Geophysical Research Letters* 33(8), L08707.

INAG (2006) *SNIRH: Sistema Nacional de Informação de Recursos Hídricos.* Instituto da Água (INAG), Lisbon.

INE (2006) *INFOLINE – Serviço de informação Online do INE.* Instituto Nacional de Estatística (INE), Lisbon.

INIA-LQARS (2000) *Manual de Fertilização das Culturas.* Instituto Nacional de Investigação Agrária – Laboratório Químico Agrícola Rebelo da Silva (INIA-LQARS), Lisbon.

Jacob, D., Barring, L., Christensen, O.B., Christensen, J.H., de Castro, M., Deque, M., Giorgi, F., Hagemann, S., Lenderink, G., Rockel, B., Sanchez, E., Schar, C., Seneviratne, S.I., Somot, S., van Ulden, A. and van den Hurk, B. (2007) An inter-comparison of regional climate models for Europe: model performance in present-day climate. *Climatic Change* 81, 31–52.

Jarvis, A., Reuter, H.I., Nelson, A. and Guevara, E. (2006) *Hole-Filled Seamless Srtm Data V3.* International Centre for Tropical Agriculture (CIAT), Cali, Colombia.

Jetten, V., Govers, G. and Hessel, R. (2003) Erosion models: quality of spatial predictions. *Hydrological Processes* 17, 887–900.

Kosmas, C., Kirkby, M. and Geeson, N. (1999) *Manual on Key Indicators of Desertification and Mapping Environmentally Sensitive Areas to Desertification.* MEDALUS Project Report EUR 18882, European Commission, Brussels.

Lavee, H., Imeson, A.C. and Sarah, P. (1998) The impact of climate change on geomorphology and desertification along a Mediterranean-arid transect. *Land Degradation and Development* 9, 407–422.

Martinez-Vilalta, J., Pinol, J. and Beven, K. (2002) A hydraulic model to predict drought-induced mortality in woody plants: an application to climate change in the Mediterranean. *Ecological Modelling* 155, 127–147.

Middleton, N. and Thomas, D. (1997) *World Atlas of Desertification.* United Nations Environment Program (UNEP), London.

Morales, P., Hickler, T., Rowell, D.P., Smith, B. and Sykes, M.T. (2007) Changes in European ecosystem productivity and carbon balance driven by regional climate model output. *Global Change Biology* 13, 108–122.

Mouillot, F., Rambal, S. and Joffre, R. (2002) Simulating climate change impacts on fire frequency and vegetation dynamics in a Mediterranean-type ecosystem. *Global Change Biology* 8, 423–437.

Nakicenovic, N. and Swart, R. (2000) *Special Report on Emissions Scenarios. A Special Report of Working Group III of the Intergovernmental Panel on Climate Change.* Cambridge University Press, Cambridge, UK.

Nash, J.E. and Sutcliffe, J.V. (1970) River flow forecasting through conceptual models. Part 1: discussion of principles. *Journal of Hydrolology* 10, 282–290.

Neitsch, S.L., Arnold, J.G., Kiniry, J.R., Williams, J.R. and King, K.W. (2002) *Soil and Water Assessment Tool Theoretical Documentation, Version 2000.* GSWRL Report 02-01, Grassland, Soil and Water Research Laboratory, USDA ARS (Agricultural Research Service)/BRC Report 2-05, Blackland Research and Extension Center, Texas Agricultural Experiment Station, Temple, Texas/TWRI Report TR-191, Texas Water Resources Institute, College Station, Texas.. Available at: http://swatmodel.tamu.edu/media/1290/swat2000theory.pdf (accessed 17 February 2011).

Nicolau, M.R.R.C. (2002) *Modelação e Mapeamento da Distribuição Espacial de Precipitação. Uma Aplicação a Portugal Continental.* New University of Lisbon, Lisbon.

Nohara, D., Kitoh, A., Hosaka, M. and Oki, T. (2006) Impact of climate change on river discharge projected by multimodel ensemble. *Journal of Hydrometeorology* 7, 1076–1089.

Nunes, J.P., Seixas, J. and Pacheco, N.R. (2008) Vulnerability of water resources, vegetation productivity and soil erosion to climate change in Mediterranean watersheds. *Hydrological Processes* 22, 311–3134.

Nunes, J.P., Seixas, J., Keizer, J.J. and Ferreira, A.J.D. (2009) Sensitivity of runoff and soil erosion to climate change in two Mediterranean watersheds. Part I: model parameterization and evaluation. *Hydrological Processes* 23, 1202–1211.

Olesen, J.E. and Bindi, M. (2002) Consequences of climate change for European agricultural productivity, land use and policy. *European Journal of Agronomy* 16, 239–262.

Palutikof, J.P., Conte, M., Casimiro Mendes, J., Goodess, C.M. and Espirito Santo, F. (1996) Climate and climate change. In: Brandt, C.J. and Thornes, J.B. (eds) *Mediterranean Desertification and Land Use.* John Wiley, Chichester, UK, pp. 43–86.

Pereira, J.S., Chaves, M.M., Caldeira, M.C. and Correia, A.V. (2006a) Water availability and productivity. In: Morison, J.I.L. and Morecroft, M.D. (eds) *Plant Growth and Climate Change.* Blackwell, Oxford, pp. 118–145.

Pereira, J.S., Correia, A.V., Correia, A.C., Ferreira, M.T., Onofre, N., Freitas, H. and Godinho, F. (2006b) Florestas e biodiversidade. In: Santos, F.D. and Miranda, P. (eds) *Alterações Climáticas em Portugal: Cenários, Impactos e Medidas de Adaptação – Projecto SIAM II.* Gradiva Publicações, Lisbon, pp. 301–344.

Pinto, P.A., Braga, R. and Brandão, A.P. (2006) Agricultura. In: Santos, F.D. and Miranda, P. (eds) *Alterações Climáticas em Portugal: Cenários, Impactos e Medidas de Adaptação – Projecto SIAM II.* Gradiva Publicações, Lisbon, pp. 209–232.

Puigdefábregas, J. (1998) Ecological impacts of global change on drylands and their implications for desertification. *Land Degradation and Development* 9, 393–406.

Puigdefábregas, J. and Mendizabal, T. (1998) Perspectives on desertification: western Mediterranean. *Journal of Arid Environments* 39, 209–224.

Räisänen, J., Hansson, U., Ullerstig, A., Döscher, R., Graham, L.P., Jones, C., Meier, H.E.M., Samuelsson, P. and Willén, U. (2004) European climate in the late twenty-first century: regional simulations with two driving global models and two forcing scenarios. *Climate Dynamics* 22, 13–31.

Vicente-Serrano, S.M., Cuadrat-Prats, J.M. and Romo, A. (2006) Aridity influence on vegetation patterns in the middle Ebro Valley (Spain): evaluation by means of AVHRR images and climate interpolation techniques. *Journal of Arid Environments* 66, 353–375.

Vogiatzakis, I.N., Mannion, A.M. and Griffiths, G.H. (2006) Mediterranean ecosystems: problems and tools for conservation. *Progress in Physical Geography* 30, 175–200.

Wetherald, R.T. and Manabe, S. (2002) Simulation of hydrologic changes associated with global warming. *Journal of Geophysical Research – Atmospheres* 108 (D19), 4379.

Wilby, R.L. (2005) Uncertainty in water resource model parameters used for climate change impact assessment. *Hydrological Processes* 19, 3201–3219.

Wilby, R.L. and Wigley, T.M.L. (1997) Downscaling general circulation model output: a review of methods and limitations. *Progress in Physical Geography* 21, 530–548.

18 Soil Erosion by Water Under Future Climate Change

David Favis-Mortlock* and Donal Mullan

Introduction

'Mud': this humble word is one of the shorter and older[1] words in the English language. It is also the main focus of this chapter, as that part of the hydrological cycle which involves flow over the sloping parts of the earth's surface ('runoff') very commonly involves interaction with soil. This interaction results in mud and/or muddy sediment.

Runoff may well have mud below it or in it, but it always has weather above it. The second main theme of this chapter is the interaction of surficial hydrology and sediment transport with weather and climate (the longer term expression of weather): in particular, the interaction of soil erosion by water with future climate. Spatially, the emphasis is on hillslopes, the sloping portions of catchments, rather than streams or rivers. We also focus, as much as is possible, on the global situation, although our case study is set in temperate western Europe.

Hillslope hydrology, sediment transport and climate change

Since the 1960s, many hydrologists and geomorphologists have found it useful to think of catchments in terms of 'systems theory' (e.g. von Bertalanffy, 1968). The systems view conceptualizes catchments in terms of inputs, outputs and storage, and so provides a simplified, big-picture perspective of a complicated system. Thus, the many processes and factors which determine (for example) the rate at which runoff or sediment leaves a hillslope catchment may be lumped together into 'those factors which affect input' and 'those factors which affect output', with storage depending on the dynamic interplay between these two. This systems perspective is distinct from the so-called process-based (or physically based, or physics-based) view (e.g. Burt *et al.*, 2008), which emphasizes the physical (and chemical and biological) processes, such as infiltration, or detachment of sediment by runoff, that occur within the catchment.

We start by considering those hydrological and erosional processes that provide inputs for the hillslope hydrology and hillslope sediment transport systems, and consider their sensitivity to future climate change (Table 18.1).

Inputs to hillslope runoff and sediment transport

In most parts of the world, the main input to hillslope runoff (also known as 'overland flow' – the two terms are used interchangeably here) is rainfall. In colder regions snow melt is

* Corresponding author: david.favis-mortlock@ouce.ox.ac.uk

Table 18.1. Inputs to the hillslope hydrology and sediment transport systems, also showing key processes and 'best-guess' responses to future climate change.

Water or sediment?	Process	Sensitivity to future climate change
Water	Rainfall	Runoff amount affected by changes in future rainfall amount Saturation excess overland flow affected by the temporal clustering of future rainfall events Infiltration excess overland flow affected by changes in rainfall intensity and clustering; crusting/capping affected by changes in rainfall energy
	Snow melt	Snow-melt-generated runoff affected mainly by changes in temperature
	Return flow	Not affected by future climate change
Sediment (see Fig. 18.1)	RD–ST (raindrop detachment–splash transport)	Affected by changes in future rainfall energy
	RD–RIFT (raindrop detachment–raindrop-induced flow transport)	Affected by changes in future rainfall energy
	RD–FT (raindrop detachment–flow transport)	Affected by changes in future rainfall energy, and changes in runoff depth/flow velocity
	FD–FT (flow detachment–flow transport)	Affected by changes in runoff depth/flow velocity

also important; this may result from the melting of snow, ice or glaciers. A third input can occur in certain circumstances. Topographically lower areas within a catchment may experience subsurface return flow; this is water which has previously infiltrated in some upslope location, moved laterally downslope, and then exfiltrated to once again become (or join) hillslope runoff.

RAINFALL. Rainfall is, in general, the most important contributor to hillslope runoff, but a given precipitation event (i.e. a 'storm') may or may not give rise to runoff. Whether it does so depends on many factors, including the antecedent water content of the soil, the intensity with which the rain falls, and the physical/chemical state of the soil surface. The two mechanisms that produce hillslope runoff are saturation excess and infiltration excess. They are described separately here, but more often than not they interact.

Saturation excess overland flow occurs when the soil is saturated and depression storage is filled; if rainfall continues, then the rainfall will immediately produce surface runoff. Pre-storm soil moisture content is one factor that affects the time until the soil becomes saturated. This, in its turn, depends on a variety of factors, including pedology (especially soil texture, and any crusting/capping), vegetation type and extent, and meteorological factors such as the quantity of earlier rainfall, evaporation rate and the length of time since previous rainfall. Thus the temporal clustering of rainfall events is one factor that must be considered when we consider soil erosion by water under future changed climates, because it will play a part in determining the likelihood of saturation excess hillslope runoff.

Note too that the soil in the more low-lying portions of catchment hillslopes tends to become saturated more quickly than the soil at higher elevations, in part as a result of lateral movement of subsurface moisture. There is, therefore, a spatial dimension to the generation of saturation excess overland flow: it is not generated uniformly over the whole catchment. Thus the proportion of the catchment area that generates saturation excess runoff depends on the spatial distribution of pre-storm moisture content.

As well as depending on the pedological, vegetational and meteorological factors discussed in the previous paragraph, the spatial distribution of pre-storm moisture content will also depend on the rate of lateral flow of subsurface moisture, and on the time available between storms for this lateral flow to take place. As inter-storm time is also a factor in determining the quantity of pre-storm soil moisture, its role in also determining the spatial distribution of pre-storm soil moisture suggests that the temporal clustering of future storms will be especially important, in a far-from-simple way, with regard to the future generation of saturation-excess overland flow.

Infiltration-excess overland flow (also known as 'Hortonian overland flow') occurs when the rate at which rain arrives exceeds the rate at which water can infiltrate the soil, assuming that any depression storage has already been filled. Important factors here are the rainfall rate, and the rate at which water can infiltrate through the soil surface. This type of runoff is most common in summer in temperate regions, when rainfall intensities are high and the soil infiltration capacity has been reduced because of surface sealing or crusting. It is also important in arid and semi-arid regions, which experience infrequent but high-intensity rainfalls. The intensity of future rainfall will therefore be important for the future occurrence of infiltration-excess overland flow. In many parts of the world, future rainfall intensities are projected to increase (e.g. Solomon *et al.*, 2007), although the extent and severity of such increases are a matter of active research. None the less, increases in rainfall intensity have already been noted in the USA (Karl and Knight, 1998) and Europe (Osborn *et al.*, 1999; note that these have tended to occur mostly in winter). However, it must be emphasized that knowledge is currently still rather limited regarding the extent and spatial occurrence of future increases in rainfall intensity.

In contrast to saturation-excess overland flow, infiltration-excess hillslope runoff tends to be relatively uniform over a whole catchment (or at least that portion of it which is receiving rain from a single storm). However, a major factor that influences the occurrence of infiltration-excess overland flow is the state of the soil surface, in particular any crusting or capping. Crusting/capping can be the result of physical, chemical or biological agencies (Le Bissonnais and Singer, 1992). Rainfall energy is also an important factor in crusting/capping, thus any future increases in rainfall kinetic energy (as a result of greater rainfall intensities or larger raindrop sizes, for example) may well exacerbate soil sealing by crusting and capping. This, in turn, would lead to increases in infiltration-excess runoff, although these increases would not necessarily be uniform over the whole catchment, rather depending on the spatial variation of soil properties and land uses. Thus, future rainfall intensity and energy are likely to be important for the occurrence of infiltration-excess runoff, just as the temporal clustering of future rainfall is likely to be important for the occurrence of saturation-excess runoff.

SNOW MELT. Future increases in temperature during the colder half of the year and/or nearer the poles will of course affect the occurrence of snow-melt-generated runoff. Because greater warming is forecast – indeed, is already occurring – in polar regions (Arctic Climate Impact Assessment, 2005) compared with other regions of the planet, this disruption to existing temporal and spatial patterns of snow-melt runoff on hillslopes is likely to be severe, although relatively little research has been carried out so far (cf. Day, 2009).

RETURN FLOW. This is unlikely to be affected by future climate change.

SEDIMENT INPUTS TO HILLSLOPE RUNOFF. Hillslope runoff may acquire sediment load by means of a variety of processes (Fig. 18.1). Of these, raindrop detachment–splash transport (RD–ST), raindrop detachment–raindrop-induced flow transport (RD–RIFT) and raindrop detachment–flow transport (RD–FT) (Kinnell, 2006) are all driven by raindrop impact and so will be influenced mainly by changes in raindrop energy.

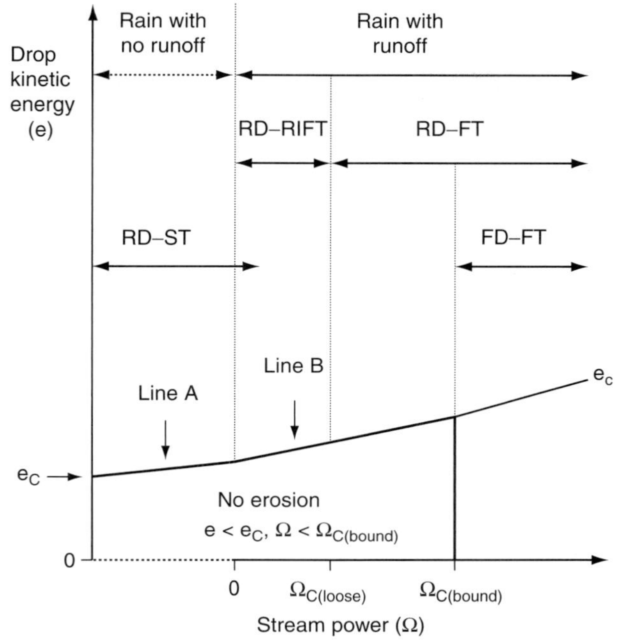

Fig. 18.1. Detachment and transport processes associated with variations in raindrop and flow energies: e_c, critical raindrop energy to cause detachment – raindrop-induced erosion occurs when drop energy exceeds e_c; line A, line for e_c when raindrops are detaching soil particles from the soil surface before flow development – the slope on this line is used to indicate increasing resistance to detachment caused by, for example, crust development; line B, line for e_c when raindrops are detaching soil particles from the soil surface when flow has developed – the slope on this line is used to indicate increasing utilization of raindrop energy in penetrating the flow when flow depth increases as stream power increases; $\Omega_{C(loose)}$, critical stream power required to transport loose (previously detached) soil particles; $\Omega_{C(bound)}$, critical stream power required to detach particles bound within the soil surface (held by cohesion and inter-particle friction); RD–ST, raindrop detachment–splash transport; RD–RIFT, raindrop detachment–raindrop-induced flow transport; RD–FT, raindrop detachment–flow transport; FD–FT, flow detachment–flow transport (from Kinnell, 2006).

More important than any of these, in terms of its environmental impact, is flow detachment–flow transport (FD–FT). This gives rise to erosion channels of all sizes, from micro-rills to gullies. FD–FT is driven by runoff velocity, so any increases in the speed and/or depth of runoff will increase FD–FT. This means that increases in runoff resulting from any of the mechanisms discussed previously might be expected to, all else being equal, result in increased erosion by water.

Climatically driven changes in soil erosion by water

There are thus many processes – almost all of which interact – that influence the movement of water and sediment over catchment hillslopes. As discussed above, there is a clear potential for future climate change to influence these processes in ways which will increase the future risk of soil erosion by water. However, to determine exactly how these processes will respond to future climate change is far from easy. This is both because of the number of processes that must be considered, and because of the complexity of their interactions.

None the less, an obvious practical question is: can we quantify this increased risk, even if only approximately? One strategy is to work with simplified landscape situations, such as a single hillslope under a single crop, with spatially uniform soil properties. This is the approach that we follow in the second, case study, half of this chapter. We might then consider the effects, on such a simplifed landscape situation, of changes in such climatic factors as rainfall amounts and intensities, number of precipitation days, ratio of rain to snow, evapotranspiration rates, etc. But even then our calculations would rapidly become more difficult because of various secondary impacts of climate change, some of which are biological (e.g. changes in plant biomass production, plant residue decomposition rates, soil microbial activity, etc.) and others which are under the control of humans (e.g. shifts in land use which are necessary – or desirable – in accommodating a new climatic regime) (Nearing *et al.*, 2004).

Because of this escalating complexity, even with initially simplified landscape situations, it is necessary to further simplify the problem so that we can consider only the most important climatic factors. We may then classify the impacts of future climate change on runoff and soil erosion by water under two headings: those impacts arising directly from future climate change, and the more indirect impacts largely arising from future land-use change (which may, however, be driven, in part, by future climate change).

Direct effects

The most obvious reason for soil erosion by water to change in response to changes in climate is a change in the erosive power of rainfall, i.e. a change in erosivity (e.g. Favis-Mortlock and Savabi, 1996; Williams *et al.*, 1996; Favis-Mortlock and Guerra, 1999; Nearing, 2001; Pruski and Nearing, 2002; Soil and Water Conservation Society, 2003). Increases in global temperature lead to increases in the moisture-holding capacity of the atmosphere at a rate of about 7% per 1 °C. This results in increased water vapour in the atmosphere, and ultimately in a more vigorous hydrological cycle (Nearing *et al.*, 2005), promoting a trend towards more intense precipitation events (Trenberth *et al.*, 2003). Climate models are predicting a continued increase in intense precipitation events during the 21st century (IPCC, 2007). The timing of precipitation will also change in some cases, but this is less amenable to generalization.

Indirect effects

Indirect consequences of climate change for soil erosion generally relate to changes in plant biomass, and hence changes in protection of the soil from erosion, i.e. erodibility. Complex changes in plant biomass resulting from changes in climate have the potential to both increase erosion rates through faster residue decomposition from increased microbial activity (Nearing *et al.*, 2005), and decrease erosion rates through an increase in soil surface canopy cover and biological ground cover (Rosenzweig and Hillel, 1998). More significant changes in plant biomass will also occur owing to shifting agricultural practice and hence land use to accommodate the new climatic regime (Williams *et al.*, 1996). Reacting to changes in climate could range from changing planting dates to the implementation of new crops or complete land-use changes, which have the potential to significantly alter soil erosion rates and patterns (Nearing *et al.*, 2005). For example, the introduction of new crops suited to warmer conditions, such as maize and sunflowers, increases the risk of erosion as both take a significant amount of time to provide adequate crop cover in early summer (Boardman and Favis-Mortlock, 1993). A shift in the spatial distribution of crops is also a land-use factor that may have an impact on future rates of erosion. The thermal altitudinal limit for agriculture is projected to rise by approximately 200 m per 1 °C increase in air temperature (Hulme *et al.*, 1993). This will enable farmers to replace permanent grasslands with grass leys or cereals, practices which have the potential to significantly increase erosion rates, particularly in high rainfall areas with sloping land (Boardman and Favis-Mortlock, 1993). For example, the conversion of grass to barley in west Derbyshire, England, in the 1980s led to severe erosion and the flooding of a village (Boardman and Spivey, 1987).

Background to the case study: future soil erosion by water in Ireland

The following case study focuses upon soil erosion by water under a future, changed climate in temperate western Europe: more specifically an agricultural hillslope site in the north of Ireland (Loughmuck hillslope, outside Omagh, in County Tyrone). Soil erosion by water is currently not common in Ireland (Favis-Mortlock, 2006); hence, the aim of this study was to determine whether future changes in climate and land use might render erosion more likely.

The study site is a pasture field which is ploughed approximately once every 10 years to reseed with a new sward of grass. It was chosen on the basis of observed erosion, with measured rates of deposition providing an opportunity for model validation under present-day conditions. Favis-Mortlock and Guerra (1999) note that two conditions must be met at any location for there to be problematic water erosion: there must be sufficient rainfall, and this must occur at a time when the soil surface is insufficiently protected. Erosion at Loughmuck hillslope occurred (see Figs 18.2 and 18.3) as a result of the coincident occurrence of these two factors following heavy rainfall in October 2008, when the field had recently been ploughed and reseeded with grass.

Evidence for past erosion in Ireland

Observation of present-day erosion illustrates that soil erosion by water can occur in Ireland under certain conditions, although it is not currently a major issue in Ireland owing to the predominance of permanent grassland, and hence protection of the soil from rainfall and runoff (Favis-Mortlock, 2006). Gardiner and Burke (1982) point out, however, that erosion does occur and was common in the past when cultivation was more frequent. Evans (1963) indicates that tillage was more important than pasture in Irish agriculture between 1750 and 1850, with population pressure resulting in Ulster farming undergoing a shift from the long tradition of livestock husbandry towards intensive arable farming based on 'the trinity of oats, flax and the potato'. In addition, Ó Danachair (1970) reveals that between 1780 and 1840, there was more land under cultivation in Ireland than ever before or since. A high vulnerability to soil erosion at this time has therefore been postulated by McEntee (1994), because year-after-year cropping, increased acreage of tilled land, and even

Fig. 18.2. Rills on Loughmuck hillslope outside Omagh, County Tyrone, Ireland.

Fig. 18.3. Measuring deposition at the foot of Loughmuck hillslope outside Omagh, County Tyrone, Ireland.

reclamation of marginal land on higher hills and bog margins were all prevalent. McEntee (1994) notes that evidence for past erosion is provided from various sources, including landscape features, documentation on traditional farming practices, and sedimentary and palynological evidence.

Irish landscape features include increased topsoil depth downslope (McCabe and Collins, 1977), typical of the drumlin belt of north-east Ireland, where topsoil depth varies as a result of erosion of the steep drumlin sides. Other landscape features include differences in level from one field to another, referred to as a 'colluvial step' (McEntee, 1994), formed by the accumulation of colluvium against a field bank running across a slope. Farming practices such as 'sliping' (Evans, 1963), whereby soil was restored by drawing it upslope in horse-drawn slide carts, also provide evidence for past erosion in the north of Ireland (McEntee, 1994). This method was further illustrated in an account of land use in the Warrenpoint lowlands and the western Mournes of County Down, in which Boal (1963) commented: 'Where soil is washed down from the top of fields, exposing the underlying rock, horse-drawn sledges are used to cart the soil back up the hill.' Erosion has been linked to cultivation through sedimentary and palynological evidence from lakes and peat (Thompson and Edwards, 1982; Hall, 1990). Cultivation of erosion-prone crops such as flax (Souchère *et al.*, 2003) was prevalent in the 17th- and 18th-century intensification of agriculture in Ireland (Hall, 1989). The discovery of flax pollen in lake sediments from County Down around this time may therefore attest to soil erosion by water in this part of Ireland. Thus, soil erosion by water certainly occurred in the past in Ireland, but it is a rather minor problem at present: could it become a future problem?

The potential for increased future soil erosion in Ireland: direct impacts

Warming of the climate system is unequivocal, as is now evident from observations of increases in global average air and ocean temperatures, widespread melting of snow and ice, and rising global average sea level (IPCC, 2007). Global mean surface temperatures have risen by 0.74 °C ± 0.18 °C in the 100-year period from 1906 to 2005, with 11 of the 12 years between 1995 and 2006 ranking among the warmest years in the instrumental record of global surface temperature since 1850 (IPCC, 2007). Climate change projections for Northern Ireland indicate increasing annual mean temperatures accompanied by more frequent and extreme hot days. Summers will become more reliably dry while more intense rainfall days are expected in winter and spring (Arkell *et al.*, 2007). The range of future projections for temperature and precipitation in Northern Ireland are provided in Table 18.2, based on UKCIP02 (UK Climate Impacts Programme 2002) scenarios (Hulme *et al.*, 2002) derived from a UK Met Office general circulation model (GCM) coupled to a regional circulation model (RCM).

Potential for indirect impacts

As discussed above, one of the indirect impacts of climate change on soil erosion may be a change in land use to accommodate new crops suited to the new climatic conditions. In Northern Ireland, the indirect potential for increased erosion is highlighted in a recent research project undertaken by the Agri-Food and Biosciences Institute (AFBI) for the UK Department for Environment, Food and Rural Affairs (DEFRA) to project changes in land use in Northern Ireland until the year 2025 (DEFRA, 2007). Previous reviews, stakeholder discussions and expert judgement were used as the basis for collating the projections, in terms of percentage change against the baseline year of 2004. The study revealed expected declines in livestock for all animal categories, and an increase in the areas of wheat and oilseed rape, driven by the demand for biofuels. An outline summary of the main land-use changes for Northern Ireland for the period from 2004 to 2025 is given in Fig. 18.4.

Table 18.2. Future temperature and precipitation scenarios for Northern Ireland based on four UKCIP02 (UK Climate Impacts Programme 2002) scenarios for three future time slices.

Climate variable		2020s	2050s	2080s
Temperature	Annual mean	+0.5 to 1.0 °C	+0.5 to 2.0 °C	+1.0 to 3.5 °C
	Spring mean	+0.5 to 1.0 °C	+0.5 to 2.0 °C	+1.0 to 3.0 °C
	Summer mean	+0.5 to 1.0 °C	+1.0 to 2.5 °C	+1.0 to 3.5 °C
	Autumn mean	+0.5 to 1.0 °C	+1.0 to 2.5 °C	+1.5 to 4.0 °C
	Winter mean	0 to +1.0 °C	+0.5 to 1.5 °C	+1.0 to 2.5 °C
	Inter-annual variability	Winter and particularly spring will become more reliably warm. Autumn and especially summer will vary more widely from year to year.		
	Diurnal range	Winter nights will warm more than winter days; summer days will warm more than summer nights, although summer evenings will be warmer.		
	Extremes	Extremely warm days will become more frequent and hotter. Heat waves will become more likely. The number of cold days will also decline.		
	Sea surface	0 to 1.0 °C	+0.5 to 1.5 °C	+1.0 to 2.5 °C
Precipitation	Annual mean	WNV[a]	Up to −10%	Up to −10%
	Spring mean	WNV	Up to +10%	Up to +10%
	Summer mean	Up to −20%	−10 to −30%	−20 to −50%
	Autumn mean	WNV	WNV	Up to −10%
	Winter mean	Up to +10%	Up to +15%	Up to +25%
	Inter-annual variability	Summer will become more reliably dry. Precipitation in autumn, winter and spring will become more variable.		
	Extremes	More intense rain days in winter and spring. Greater probability of extreme rainfall on any given winter day. Intense summer storms may occur. Seasonally, very dry summers and very wet winters are likely.		

[a]WNV, will not vary.

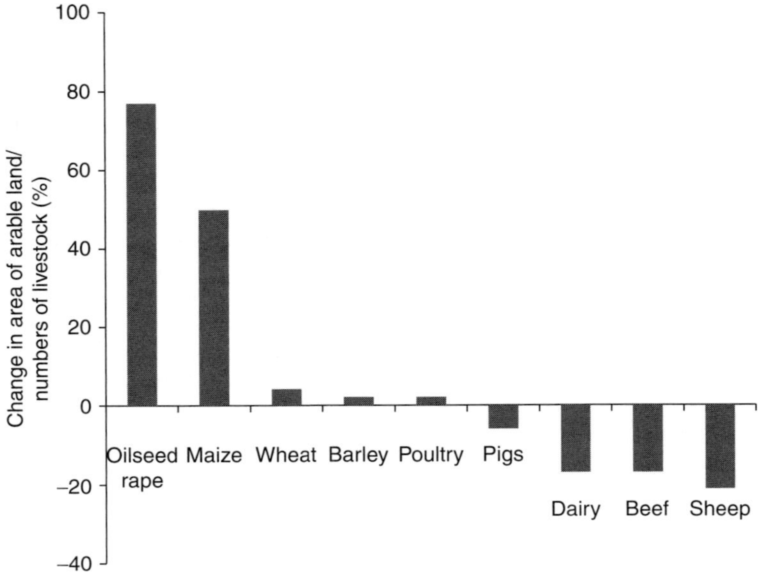

Fig. 18.4. Projected changes in land use in Northern Ireland by 2025 (source: UK Department of Environment, Food and Rural Affairs (DEFRA)).

Materials and Methods

Modelling soil erosion by water

Soil erosion models are the mathematical descriptors used to represent the erosion process (Tiwari *et al.*, 2000), based upon our understanding of the physical laws and landscape processes that occur in the natural world. Such modelling translates these components into mathematical relationships, either empirical or physics-based, describing the fundamental erosional processes of detachment, transport and deposition (Doe and Harmon, 2001). They are typically simplifications of reality that describe the processes of erosion and the factors controlling them, resulting in some output of soil loss dependent on the model objectives (Morgan and Quinton, 2001). Methods for predicting soil loss under a wide range of conditions are necessary both to estimate 'acceptable' levels of erosion and to determine the effects of different conservation strategies (Morgan, 2005). Soil erosion models have made a significant contribution to these causes through playing a vital role in the understanding of erosional processes and in advancing applications for the prediction and design of conservation strategies (Favis-Mortlock *et al.*, 2001). Prediction models have become increasingly important tools in the assessment of soil erosion, as adequate and reliable models can be used to evaluate a range of management scenarios without expensive and time-consuming field tests (Lal, 1998; Toy *et al.*, 2002). Computer models can simulate catchments and erosion processes and may be able to account for many of the complex interactions that affect erosion rates (Zhang, L. *et al.*, 1996). Models range from very simplistic expressions to very complex mathematical representations of functions and processes based upon the laws of nature (Doe and Harmon, 2001).

Selected model for Loughmuck hillslope

The Water Erosion Prediction Project (WEPP) erosion model is used in this case study to model future erosion rates for Loughmuck hillslope. WEPP is a physically based continuous simulation model that predicts soil loss and deposition using a spatially and temporally distributed approach (Foster and Lane, 1987; Nearing *et al.*, 1989; Laflen *et al.*, 1991, 1997; Flanagan and Nearing, 1995). The WEPP model was developed to herald a new generation of erosion prediction technology (Nearing *et al.*, 1989) and to serve as an improved tool for soil and water conservation planning and assessment (Zhang, X.C. *et al.*, 1996) based on the fundamentals of stochastic weather generation, infiltration theory, hydrology, soil physics, plant science, hydraulics and erosion mechanics (Flanagan and Nearing, 1995). It is one of the most widely used tools for simulating water erosion and sediment yield (Pieri *et al.*, 2007). The model has been tested and applied in various geographical locations around the world, including the USA (Savabi, 1993; Savabi *et al.*, 1995; Huang *et al.*, 1996; Pruski and Nearing, 2002; Laflen *et al.*, 2004; Zhang *et al.*, 2004; O'Neal *et al.*, 2005; Zhang, 2005; Zhang and Nearing, 2005), Australia (Rosewell, 2001), the UK (Favis-Mortlock, 1994; Favis-Mortlock and Savabi, 1996; Brazier *et al.*, 2000), Brazil (Favis-Mortlock and Guerra, 1999), China (Zhang and Liu, 2005), South Korea (Kim *et al.*, 2009), Switzerland (Hebel and Siegrist, 1998) and Italy (Spadaro *et al.*, 2004; Pieri *et al.*, 2007). There are four input files required to drive WEPP, and the parameterization of WEPP using these input files is detailed in Table 18.3.

Predicting the effects of climate change on erosion: early modelling approaches

Some of the first studies of the effect of climate change on soil erosion (Boardman *et al.*, 1990; Favis-Mortlock *et al.*, 1991; Boardman and Favis-Mortlock, 1993; Lee *et al.*, 1993; Nicks, 1993; Nicks and Williams, 1993) developed a relatively 'what-if' scenario manner towards projecting future erosion rates through gradational increases or decreases in precipitation and/or temperature. Such changes to meteorological variables were constructed by considering a number of plausible climate scenarios, the parameters of which lay within the range of expected extreme values. These plausible climate scenarios were typically gathered from the best future climate information available at the time, which ranged from qualitative information on analogues drawn from formerly existing climates to some early results from climate models (e.g. Department of the Environment, 1988). For example, Boardman *et al.* (1990) utilized such an approach to study the impacts of climate change upon soil erosion on agricultural land in England and Wales through increasing mean annual temperature by 2 °C, 3 °C and 4 °C, and by increasing mean annual precipitation by +5%, +10% and +15%. These climate change scenarios were selected on the basis of information from the (UK) Department of the Environment's review of climate change scenarios (Department of the Environment, 1988). Both quantitative and qualitative information on future climate change for the southern UK was considered in this review, including modelled predictions of temperature and rainfall increases, and qualitative statements that by 2050 the southern UK climate would be similar to that of south-west France (Department of the Environment, 1988). These climate change scenarios were used to model future erosion

Table 18.3. Description of the data sources used to parameterize the four input files required to drive the Water Erosion Prediction Project (WEPP) erosion model for Loughmuck hillslope.

Input file	Source of data
Slope	Differential global positioning system (GPS)
Soil	Soil sampling undertaken by Department of Agriculture and Rural Development of Northern Ireland
Management	Plant files from existing WEPP database
	Adjusted local plant parameters taken from Cruickshank (1997)
	Date of operations from interviewing farmer
Climate	Observed climate data from British Atmospheric Data Centre (BADC)
	Changed climate data generated using the Statistical DownScaling Model (SDSM) and input to the WEPP weather generator CLIGEN

rates using the Erosion Productivity Impact Calculator (EPIC) model. Similarly, the EPIC model was employed by Lee *et al.* (1993) to simulate the sensitivity of the US maize belt to climate change and elevated CO_2. Soil erosion rates were modelled under 36 possible future climates, including temperature increases of +2 °C, precipitation changes of ±10% and ±20%, wind speed changes of ±10% and ±20%, and atmospheric CO_2 concentrations of 350 and 625 ppmv. These scenarios were not chosen on the basis of results from any specific climate models, but rather were considered representative of the types of changes forecast by these models for the maize belt.

Limitations of early modelling approaches

A common theme in the two examples given above and of the other aforementioned studies is the relatively general 'what-if' scenarios approach to projecting future climatic changes for input to erosion models. Adjusted climate parameters in such studies tend to reflect stepped increases or decreases in meteorological variables, based around the 'best guess' possible climate futures available at the time rather than on any robust or objective climate modelling framework. Almost certainly, this approach to changed climate data is a product of the lack of availability and imprecision of climate model output at the time these studies were conducted. Boardman *et al.* (1990) note that 'the lack of precision in current climatic models and especially predictions of future distribution and intensity of rainfall make assessing changes in erosion amounts and rates a hazardous operation'. In this regard, the examination of possible future scenarios that point the way to what may happen in the future was preferred to utilizing scenarios output by specific climate models. The spatial scale of any set of climate change projections must also be considered in a climate impacts study. For example, in the Boardman *et al.* (1990) study, the spatial range of climate projections was denoted as Southern UK. As meteorological variables such as precipitation vary markedly over relatively small spatial scales, such 'broad-brush' projections for an entire region cannot be expected to accurately represent changes at the much finer spatial scales in which point-scale processes such as soil erosion operate.

Developing climate scenarios from general circulation models

With the advancement of climate models and increased publication of material surrounding climate change projections in the early-to-mid 1990s (e.g. IPCC, 1990; Houghton *et al.*, 1992; Hulme *et al.*, 1993, 1995), climate model output from GCMs became more rapidly utilized as the basis for future climate change data in a wide range of impact sectors. Subsequently (since the mid-late 1990s), GCMs have emerged as the key basis for representing changed climate conditions in studies of future soil erosion (Favis-Mortlock and Boardman, 1995). GCMs are numerical models representing physical processes in the atmosphere, ocean, cryosphere and land surface. They are currently the most suitable tools for predicting future climate change (IPCC, 2007).

Issues of spatial scale in GCMs

Despite the widespread use of GCMs in assessing the impacts of climate change on soil erosion and other impact sectors, they are fundamentally restricted in their ability to provide detailed climate impact assessments owing to their coarse spatial resolution (typically 50,000 km^2) and inability to resolve important sub-grid features such as clouds and topography (Wilby *et al.*, 2002). The representation of, for example, orography and land surface characteristics is greatly simplified in GCMs compared with reality, consequently resulting in a loss of characteristics which may have a significant impact on regional climate. Finer resolution projections of climate change are increasingly needed for impact assessments on point-scale processes such as soil erosion.

Towards site-specific climate change projections for soil erosion studies

Downscaling techniques are utilized to bridge the spatial and temporal resolution gap between what is currently provided by GCMs and the requirements of impact assessors (Wilby and Wigley, 1997). There are several different approaches to downscaling, with the choice of downscaling technique largely determined by data availability for model calibration and by the necessary variables for impact assessment (Wilby *et al.*, 2002). Downscaling can broadly be grouped into two main types: dynamic or empirical (Crawford, 2007). Within these broad groups, various downscaling methods exist. Figure 18.5 indicates some of the most popular downscaling techniques and Table 18.4 highlights the key advantages and limitations associated with these techniques. The two key downscaling techniques used in soil erosion studies are now considered in greater detail: the change-factor approach, and the more recent adoption of statistical downscaling methods.

Change-factor downscaling of future climate scenarios

Change-factor downscaling is an established and conceptually straightforward technique for downscaling future climate change projections for impact assessments (Arnell and Reynard, 1996; Arnell, 1998; Pilling and Jones, 1999; Arnell *et al.*, 2003; Diaz-Nieto and Wilby, 2005). This method involves perturbing present-day baseline observational data (commonly time series of monthly or daily data) with grid-box scale changes inferred from climate model and GCM experiments (Crawford, 2007). Changes in climate variables are often expressed as absolute or percentage changes from the baseline climate. One of the first studies on the impacts of future climate change on soil erosion making direct use of GCM data was that of Favis-Mortlock and Boardman (1995). They used the EPIC model to study the impact of climate change on erosion in the South Downs, southern England. In this study, the EPIC weather generator (WXGEN) was perturbed with IPCC (Intergovernmental Panel on Climate Change) climate model projections for southern England (Hulme *et al.*, 1993). Absolute temperature changes and percentage changes in precipitation from these projections were simply added to the WXGEN-generated

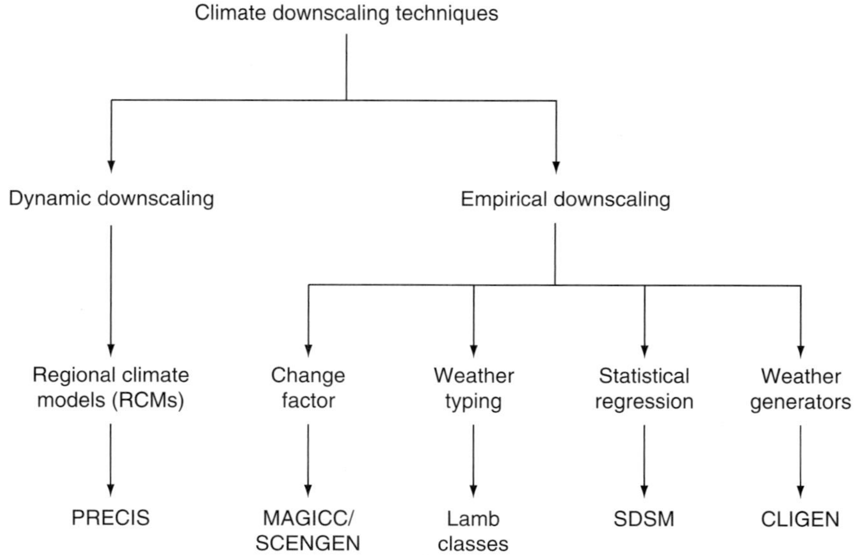

Fig. 18.5. Climate downscaling techniques (after Crawford, 2007). CLIGEN, the WEPP weather generator; Lamb classes, Lamb weather classes (Lamb, 1972); MAGICC, Model for the Assessment of Greenhouse-gas Induced Climate Change; PRECIS, a regional climate model; SCENGEN, Climate Change Scenario Generator; SDSM, Statistical DownScaling Model.

Table 18.4. The basic concepts and main advantages/disadvantages associated with downscaling techniques.

Method	Basic concepts	Main advantages	Main disadvantages
Change factor	Perturbing baseline climate observational data with grid-box scale changes inferred from GCMs[a]	Simple Inexpensive Low data requirements and quick to process	Dependent on realism of host GCM Resolution of changes too coarse Fails to capture spatial changes in variance
Dynamic downscaling	Nesting of a high resolution RCM within a coarse resolution GCM to generate regional-scale climate change projection	Consistent with host GCM High resolution (c. 50 km) Addresses full climate signal Resolves small-scale atmospheric features better than host GCM Ability to simulate different external forcings, e.g. ecosystem changes	Computationally demanding Expensive Require high level of technical expertise Dependent on realism of host GCM Unlikely to produce ensemble projections No feedback to GCM
Weather typing	Grouping local meteorological variables in relation to difference classes of atmospheric circulation	Based on sensible linkages between large-scale climate and local-scale weather Applicable to wide range of environmental variables as well as multi-site applications	Subjectivity apparent in classification of each weather 'type' Time-consuming Entirely dependent on stationary circulation to surface climate relationships
Stochastic weather generation	Statistical models of observed weather, with climate change scenarios generated stochastically using revised parameter sets scaled in direct proportion to corresponding variable changes in a GCM	Ability to reproduce many observed climate statistics Operates at a user-specified temporal resolution Production of large ensembles of scenarios for risk analysis	Uncertainty with method for adjusting future parameters in a physically meaningful way May produce unexpected or inconsistent results
Statistical downscaling	Develops empirical relationships between observed large-scale atmospheric predictors and local-scale predictands and perturbed for future climate conditions with input from GCM predictor variables	Creates site-specific climate projections Operates at daily time step Inexpensive Creates ensemble projections Access to full suite of atmospheric predictor variables	Predictor–predictand relationships assumed to be time invariant Data intensive Choice of GCM grid box and predictors significantly affects results Often requires data transformation Requires climatological expertise

[a]GCM, general circulation model.

daily data for each month of the observed climatology. Favis-Mortlock and Savabi (1996) perturbed the WEPP weather generator (CLIGEN) with climate model output from the Model for the Assessment of Greenhouse-gas Induced Climate Change (MAGICC) (Viner and Hulme, 1992). MAGICC is not a GCM, but rather it uses a set of reduced-form models to emulate the behaviour of fully three-dimensional, dynamic GCMs. The key advantage to the change-factor approach is the simplicity and speed of calculation, and direct scaling of projections in line with changes predicted by the host GCM (Crawford, 2007). In addition, the approach can easily be used to generate spatially correlated climate data for multiple stations for watershed modelling.

In modifying baseline climate with direct GCM and climate model output using the 'change factor' approach directly to simulate the impact of climate change upon soil erosion, these studies used a form of 'implicit downscaling' (Zhang, 2007). This method incorporates climatic characteristics of the target station as well as the areal-averaged relative climate changes projected at the GCM grid level, but fails to consider differences between climate variability at the GCM grid scale and at the target station (Zhang, 2007). Such studies clearly fail to represent realistic conditions of precipitation changes under global climate change, because changes to the number of wet days, precipitation duration and peak intensity of rainfall events are likely to accompany changes in precipitation amount and yet are only partially considered in the implicit method (Pruski and Nearing, 2002). Studies of climate change impacts on soil erosion in the USA by Pruski and Nearing (2002) indicate that, for every 1% increase in total rainfall, erosion rates would increase only by 0.85% if there were no corresponding increase in rainfall intensity. This study also revealed, however, that, if both rainfall amount and intensity were to increase together, predicted erosion rates would increase by 1.7% for every 1% increase in total rainfall. This illustrates that intense precipitation events are of great concern in assessing the potential impacts of climate change on soil erosion and surface hydrology, because severe soil erosion is often caused by infrequent heavy storms. Climate models are predicting a continued increase in intense precipitation events during the 21st century (IPCC, 2007). Therefore, treatment of future precipitation variability, and not just amount, is imperative for the assessment of climatic impacts on soil erosion and hydrology (Zhang *et al.*, 2009).

Statistical downscaling of future climate scenarios

In contrast to the implicit method outlined by Zhang (2007), the explicit method of downscaling uses mathematical transfer functions to downscale GCM or RCM native grid-box projections to the target station, with stochastic weather generators subsequently used to generate daily weather series using climate parameters modified by the spatially downscaled climate scenarios (e.g. Semenov and Barrow, 1997; Wilby, 1997; Kilsby *et al.*, 1998; Diaz-Nieto and Wilby, 2005; Zhang, 2005). Statistical downscaling methods rely on identifying and developing mathematical transfer functions or empirical relationships between observed large-scale predictors and the surface environmental variable of interest (local-scale predictands) (Fealy and Sweeney, 2007; Wilby and Dawson, 2007). This approach is based on the concept that regional climate is governed by the relationship between the synoptic climate state and local physiographic features, represented by a statistical model (Crawford, 2007). The output from a GCM is then fed into the statistical model for estimation of corresponding local and regional climate variables (Wilby *et al.*, 2004). The predictor–predictand relationship should explain a large amount of the observed variability and the expected changes in the mean climate should lie within the range of its natural variability (von Storch *et al.*, 1993). A common trait among statistical downscaling techniques is their sensitivity to choice of atmospheric predictor variables (Wilby and Dawson, 2001). Statistical downscaling has become a very popular method of creating downscaling climate scenarios, based around its creation of ensemble forecasts, which provide uncertainty analysis of future projections (Crawford, 2007). Their ability to provide site-specific information is fundamental to climate change impact studies and is often the only practicable means of generating climate scenarios for point-scale processes such as soil erosion (e.g. Favis-Mortlock and Boardman, 1995).

Zhang (2005) developed a method for statistically downscaling GCM monthly output at grid scale to daily output at local station scale using univariate transfer functions by calibrating probability distributions of GCM-projected monthly precipitation and temperature to match those of local climatology for the period 1950–1999. These methods were used to model soil erosion under future climate change in Oklahoma (Zhang, 2005), and in the southern Loess Plateau of China (Zhang and Liu, 2005; Zhang *et al.*, 2009). This method involves first spatially downscaling GCM estimates of climate change at the native GCM grid scale to station-scale climate data using transfer functions, as outlined previously, with a subsequent temporal downscaling of monthly values to daily values at the station scale. This method allows the site-specific assessment of soil erosion under future climate change.

Generating changed climate data for Loughmuck hillslope

In order to provide changed climate data with the appropriate spatial resolution as input to WEPP for the simulation of future soil erosion rates, statistical downscaling methods are used to downscale three climate variables for a climate station called Carrigans, located 10 km from the field site. The three climate variables downscaled are daily maximum temperature, daily minimum temperature and daily precipitation. The Statistical DownScaling Model (SDSM) is used to downscale GCM output to the station scale. SDSM is a decision support tool for assessing local climate change impacts using a robust statistical downscaling technique (Crawford, 2007). The model facilitates the rapid development of multiple, low-cost, single-site scenarios of daily surface weather variables under current and regional future climate forcing (Wilby and Dawson, 2001; Wilby *et al.*, 2002). SDSM is frequently described as a hybrid between a regression-based approach and a weather generator, because large-scale daily circulation patterns and atmospheric moisture variables are used to condition local-scale weather generator parameters at individual sites (Wilby and Harris, 2006). The underlying philosophy of SDSM relies on the establishment of multiple regressions between local-scale predictands (such as daily rainfall and temperature) and regional-scale predictors (such as mean sea level pressure and surface vorticity) (Crawford, 2007). The established relationships are then applied to the circulation simulated by a GCM in order to generate predictions of local climate. Here, output by three GCMs and two emissions scenarios, as detailed in Table 18.5, are used to downscale future climate projections at the station scale from the period 1961–2099. These future scenarios were split into three future time slices centred on the 2020s, 2050s and 2080s. The use of multiple GCMs and emissions scenarios helps to address the uncertainties inherent in climate models (Fealy and Sweeney, 2008).

In contrast to the method employed by Zhang (2005), SDSM provides daily data, which removes the requirement for temporal downscaling, and is motivated by the assumption that GCMs simulate large-scale atmospheric circulation better than they simulate surface climate variables (Murphy, 2000). The parameters relating to precipitation in the WEPP weather generator CLIGEN are: mean precipitation on wet days for each month; standard deviation and skewness of daily precipitation for each month; probability of a wet day following a wet day; probability of a wet day following a dry day; average maximum half-hourly precipitation; and a

Table 18.5. The general circulation models (GCMs) and emissions scenarios used for downscaling climate change projections in this case study.

GCM[a]	Origin	Emissions scenario considered[b]
HadCM3	UK	A2 and B2
CGCM2	Canada	A2 and B2
CSIROMk2	Australia	A2 and B2

[a]CGCM2, from the Canadian Centre for Climate Modelling and Analysis (CCCMA); CSIROMk2, from the Commonwealth Science and Industrial Research Organization; HadCN3, from the Met Office Hadley Centre for Climate Prediction and Research.
[b]A2 and B2 scenarios from SRES (Special Report on Emissions Scenarios) prepared by the Intergovernmental Panel on Climate Change (IPCC) for the Third Assessment Report (TAR) in 2001.

cumulative distribution of time to peak as a fraction of the storm duration. Because of the daily temporal resolution of SDSM output, each of these statistics can be calculated from the daily downscaled output for each of the three future time slices and then input into the WEPP weather generator CLIGEN in the form of monthly changes. In this respect, an index of simple daily intensity is produced. The method fails, however, to capture sub-daily future climate information and provide a handle on rainfall intensity at finer temporal scales – it cannot provide perturbed climate data for WEPP intensity parameters such as maximum half-hourly precipitation. In the absence of any robust method, sensitivity analysis is used here to represent sub-daily future rainfall intensity characteristics.

Considering-future land-use change

Land-use and land-cover change has been recognized as an important driver of environmental change on all spatial and temporal scales (Turner *et al.*, 1994). Patterns of land use, land-cover change and land management are shaped by the interaction of economic, environmental, social, political and technological forces on local to global scales, with most authors highlighting the case of policies as of significant importance in driving land-use changes (Lambin *et al.*, 2001). The complexity of land-use systems therefore requires multidisciplinary analyses to determine how land-use patterns will alter in a changing climate (Clayton and Radcliffe, 1996).

Initial efforts aimed at modelling land-use change primarily focused on biophysical attributes, given the availability of such data. However, given the need for multidisciplinary analyses in projecting future land-use change, more recent models have considered a broader range of factors, including the impact of future climate change. For example, models such as the Climate Land Use Allocation Model (CLUAM) (Hossell *et al.*, 1995) have been used to examine the likely effects of climate change on the extent and pattern of agricultural production and land use in England and Wales, integrating a number of components such as the world food market, population growth, and economic growth and technology. This illustrates the need for the incorporation of data on a wide range of socio-economic drivers of change. In order to construct future land-use scenarios, however, such models invariably make multiple assumptions regarding future cultural, socio-economic and environmental conditions (Turner *et al.*, 1995; Wilbanks and Kates, 1999). This illustrates the difficulty in objectively projecting future land-use change, leaving expert extrapolation from current practices the only feasible approach in many cases (Favis-Mortlock and Guerra, 1999). It is, therefore, perhaps unsurprising that most studies of soil erosion under future climate change have adopted the simplest extrapolation of all, with the approach that present land use will continue unchanged into the future (Boardman *et al.*, 1990; Favis-Mortlock *et al.*, 1991; Boardman and Favis-Mortlock, 1993; Phillips *et al.*, 1993; Botterweg, 1994; Favis-Mortlock and Boardman, 1995; Favis-Mortlock and Savabi, 1996; Lee *et al.*, 1996; Kallio *et al.*, 1997; Lee, 1998; Favis-Mortlock and Guerra, 1999). In this study, a simple wheat-wheat-barley rotation is simulated under the future climate change scenarios. This change of land use is selected because it is common across parts of Ireland in the present day, and because it allows an examination of the erosion response at Loughmuck hillslope to a significant land-use conversion. Planting and harvesting dates were left unchanged from the present-day simulations as no information was available on how such dates would change in response to the climate scenarios developed here.

Description of WEPP simulations

Running present-day simulations

In this study, WEPP version 2006.5 was first run under present-day conditions to provide a baseline with which simulations under future climate change can be compared. Approximate deposition measurements also allow WEPP to be validated for present-day simulations. The length of the simulation was 500 years, allowing a sufficient number of tillage–no-tillage cycles to be modelled (50 tillage years, 450 no-tillage years). This represents the long-term average annual erosion rate, accounting for erosion during the one tillage year occurring once every 10 years, as well as the remaining nine no-tillage years of the cycle. In order to validate WEPP under present-day conditions, the 50 tillage years from the same 500 year simulation described above are extracted and averaged, giving an average annual erosion rate during tillage years only. Because measured deposition data for this hillslope were obtained during a year in which tillage occurred, the modelled average annual erosion rate during tillage years should be comparable with the measured rate.

Future simulations under climate change only

Simulations under future climate change are then carried out for the hillslope. Daily maximum and minimum temperature and daily precipitation, including their variance and skewness, are perturbed in CLIGEN using SDSM output. Because SDSM output is daily in resolution, this simply involved calculating these statistics separately for each of the three future time slices, and averaging them by month for input into CLIGEN. The number of wet days is also modified using SDSM output. Future rates of soil erosion are modelled for three future time slices centred on the 2020s, 2050s and 2080s, using downscaled results from three GCMs and two emissions scenarios. In this respect, results are presented for three time slices, three GCMs and two emissions scenarios (3 × 3 × 2), generating a total of 18 scenarios of future soil erosion for the hillslope. The mean erosion rate for each time slice is also calculated. The percentage change from the baseline erosion rate is also indicated. Rainfall intensity changes are then considered by performing a sensitivity analysis on CLIGEN's maximum half-hourly precipitation parameter. In this study, a 50% increase is applied evenly to all months of the year, in order to examine erosion response to large increases in rainfall intensity.

WEPP simulations with land-use change

As already discussed, most previous studies of soil erosion under future climate change have not considered changes in land use, despite the potential for changes in land use, which will almost certainly be affected by future climate change, to have an impact on soil erosion rates. Given the difficulty in objectively projecting future changes in land use, this study considers a random change from pasture to arable land use, with a change from grass to a wheat-wheat-barley rotation, which is common across parts of Ireland in the present day. This particular land-use scenario is selected in order to represent the erosion response of the hillslope to a significant land-use conversion.

Results and Discussion

Both increases and decreases in soil erosion rates are projected for each of the three future time slices at Loughmuck hillslope. These changes reflect differences in the broad range of scenarios employed in the study. The results of the WEPP simulations for each of these future scenarios are now presented, along with discussion to account for changes in future erosion rates at Loughmuck hillslope. First, however, the present-day rates of erosion are displayed and discussed.

Present-day erosion rates

Table 18.6 displays the long-term average annual erosion rate, and also compares the short-term average annual erosion rate (tillage years only) with measured rates of

Table 18.6. Modelled and measured present-day erosion rates (Mg ha^{-1}) for Loughmuck hillslope.

Modelled	Long-term average annual erosion rate	0.35
	Short-term average annual erosion rate for tillage years only	3.54
Measured	Measured deposition rate	5.1 ± 1.7

deposition. The measured rate compares reasonably with the short-term modelled erosion rate, at least for the mean, with the standard deviation around the mean deposition rate introducing a band of uncertainty that decreases confidence in the validation. The uncertainty around the measured deposition rate arises from the relative subjectivity in allocating the contributing area of the field to erosion, where hillslope length is easily delineated, but hillslope width is rather more subjective. The modelled erosion rate during tillage years lies within the uncertainty bounds around the measured deposition rate, and, as the model was not calibrated in any way, this may be considered a satisfactory result (cf. Favis-Mortlock, 1994).

Future erosion rates

Direct impacts of climate change

Table 18.7 illustrates the impacts of climate change (changes in daily maximum and minimum temperature and rainfall) on future rates of soil erosion at Loughmuck hillslope. The mean of each GCM-emissions scenario combination illustrates a decrease in the average annual erosion rate for all three future time slices, from −37% for the 2080s to −53% for the 2050s. The magnitude of change varies depending on the GCM (HadCM3, CGCM2 or CSIROMk2, see Table 18.5) and emissions scenario (A2 or B2, see Table 18.5), with the largest decrease (−77%) occurring under the HadCM3-A2 scenario for the 2050s, and only one of the 18 combinations resulting in an increase in erosion. Precipitation decreases of between 5% (HadCM3-B2 scenario for the 2020s) and 22% (CGCM2-A2 for the 2080s) illustrate the predominant factor responsible for reducing soil erosion rates. Critically, however, it is the timing of precipitation and the amount of precipitation per wet day (daily precipitation intensity) rather than merely the average annual precipitation amount that generally controls the response of erosion to climate change. The 'window of opportunity' when tillage occurs once every 10 years is 'opened' in the autumn months. Therefore, it is the precipitation amount during these months and daily precipitation intensity during these months that affect erosion rates most. The HadCM3 and CGCM2 models particularly illustrate this trend (see Fig. 18.6), with an increase in precipitation amount in the autumn and winter months, compared with the rest of the year, consequently resulting in generally higher soil erosion rates compared with the results from the CSIROMk2 model. In addition, daily precipitation intensity increases during these key autumn months when compared with the observed baseline daily intensity rate. This explains why increases in runoff rates can occur even when average annual precipitation decreases. Indeed, the maximum increase in runoff from any of the future scenarios (161% under HadCM3-B2 for the 2020s) occurs at a time when average annual precipitation decreases by 5%. This highlights the role that seasonal variation in precipitation and daily precipitation intensity play in affecting future rates of soil erosion, illustrating the need to look beyond average annual precipitation amount. Table 18.8 exhibits the response of the average annual erosion rate at Loughmuck hillslope when rainfall intensity changes are added to the climate change simulations. The mean GCM-emissions scenario combination indicates an increase in average annual erosion rates for each future time slice of between 19% increase for the 2050s and 60% for the 2020s. Considerable inter-GCM differences are apparent in this trend, however, with large increases (up to 109% under the CSIROMk2-B2 scenario for the 2020s) and decreases (up to 37% under the CSIROMk-B2 scenario in the 2050s) from single GCM-emission scenario combinations. As the sub-daily rainfall intensity changes are constant for all scenarios, variations in erosion rates under these future scenarios reflect the same set of factors responsible for variations under the downscaled climate scenarios only.

Indirect impacts of climate change

Future rates of soil erosion for Loughmuck hillslope under a changed land use for the same climate change simulations as indicated in Tables 18.7 and 18.8 are displayed in Table 18.9. The present-day baseline erosion rate under the new land use is 3.34 Mg ha^{-1},

Table 18.7. Changes in average annual precipitation, runoff and soil erosion under climate change projections from combinations of three general circulation models (HadCM3, CGCM2 and CSIROMk2) with two emissions scenarios (A2 and B2) for three future time slices.

Scenario[a]	Precipitation depth (mm)	Change (%)	Runoff depth (mm)	Change (%)	Soil loss rate (Mg ha^{-1})	Change (%)
Baseline	1201	0	0.84	0	0.35	0
2020s						
HadCM3-A2	1108	−8	1.47	75	0.17	−51
CGCM2-A2	1079	−10	0.32	−62	0.12	−66
CSIROMk2-A2	1104	−8	0.28	−67	0.15	−57
HadCM3-B2	1140	−5	2.22	161	0.16	−54
CGCM2-B2	1053	−12	1.22	45	0.34	−3
CSIROMk2-B2	1097	−9	0.69	−18	0.36	3
Mean	1097	−9	1.03	22	0.22	−38
2050s						
HadCM3-A2	1045	−13	0.36	−57	0.08	−77
CGCM2-A2	1048	−13	1.38	64	0.24	−31
CSIROMk2-A2	1127	−6	0.61	−27	0.19	−46
HadCM3-B2	1104	−8	0.83	−1	0.24	−31
CGCM2-B2	1036	−14	0.73	−13	0.13	−63
CSIROMk2-B2	1096	−9	0.17	−80	0.10	−71
Mean	1076	−10.5	0.68	−19	0.16	−53
2080s						
HadCM3-A2	1076	−10	1.83	118	0.32	−9
CGCM2-A2	937	−22	0.05	−94	0.09	−74
CSIROMk2-A2	1002	−17	0.80	−5	0.25	−29
HadCM3-B2	1133	−6	1.27	51	0.33	−6
CGCM2-B2	1027	−14	0.56	−33	0.22	−37
CSIROMk2-B2	1032	−14	0.35	−58	0.12	−66
Mean	1034	−14	0.81	−4	0.22	−37

[a]See Table 18.5 for details.

an 854% increase from the baseline erosion rate under grass. Under this new arable rotation, such large increases in erosion can be attributed to the extension of the one-in-ten-year 'window of opportunity' to an annual window with tillage occurring every autumn. When WEPP is run under changed land use for the future time slices, erosion rates generally decrease from the present-day baseline rate under the new rotation, reflecting the same factors responsible for decreasing erosion rates under climate change projections only, i.e. decreasing precipitation as output by each of the GCMs. The erosion rates still represent considerable increases from the original present-day baseline erosion rate under grass, however, owing to the influence of the changed land use, with an increase in tillage increasing the exposure of the soil to the forces of rainfall and flowing water.

The full range of impacts

Table 18.10 illustrates the response of soil erosion at Loughmuck hillslope under the full range of climate change impacts employed in this study: the direct impacts (downscaled climate change projections and added rainfall intensity changes) and the indirect impacts (change in land use). Astronomical increases in erosion are evident under all scenarios, with the lowest increase of 1106% under the CSIROMk2-A2 scenario for the 2020s up to the highest increase of 2871% under the HadCM3-B2 scenario for the 2020s. Such high percentage changes in average annual erosion rates translate to absolute values of up to 10.4 Mg ha^{-1}. The dramatic increases in erosion rates reflect a combination of all the factors discussed with respect to the aforementioned scenarios. The key two factors are the land-use conversion towards increased tillage (indirect impact) and the increased sub-daily rainfall intensity (direct impact), which combine to result in large increases in soil erosion at Loughmuck hillslope. As the rainfall intensity changes and land-use changes are identical for all 18 GCM-emission scenario combinations, the magnitude of increase is controlled entirely by the downscaled climate

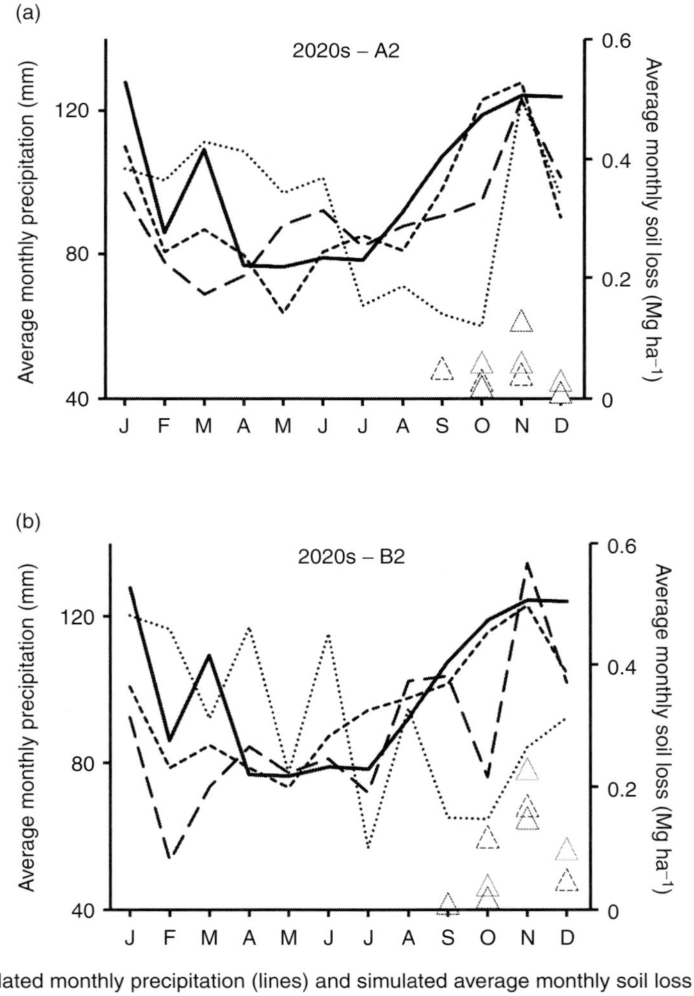

Fig. 18.6. Simulated monthly precipitation (lines) and simulated average monthly soil loss (triangles) from combinations of three general circulation models (GCMs: HadCM3, CGCM2 and CSIROMk2) with two emissions scenarios (A2 and B2) for three future time slices (2020s, 2050s and 2080s). See Table 18.5 for details of models and scenarios.
Key:
Baseline precipitation ——— CGCM2 precipitation — —
HadCM3 precipitation ------- CSIROMk2 precipitation ....
Soil loss under HadCM3 △ Soil loss under CGCM2 △ Soil loss under CSIROMk2 △

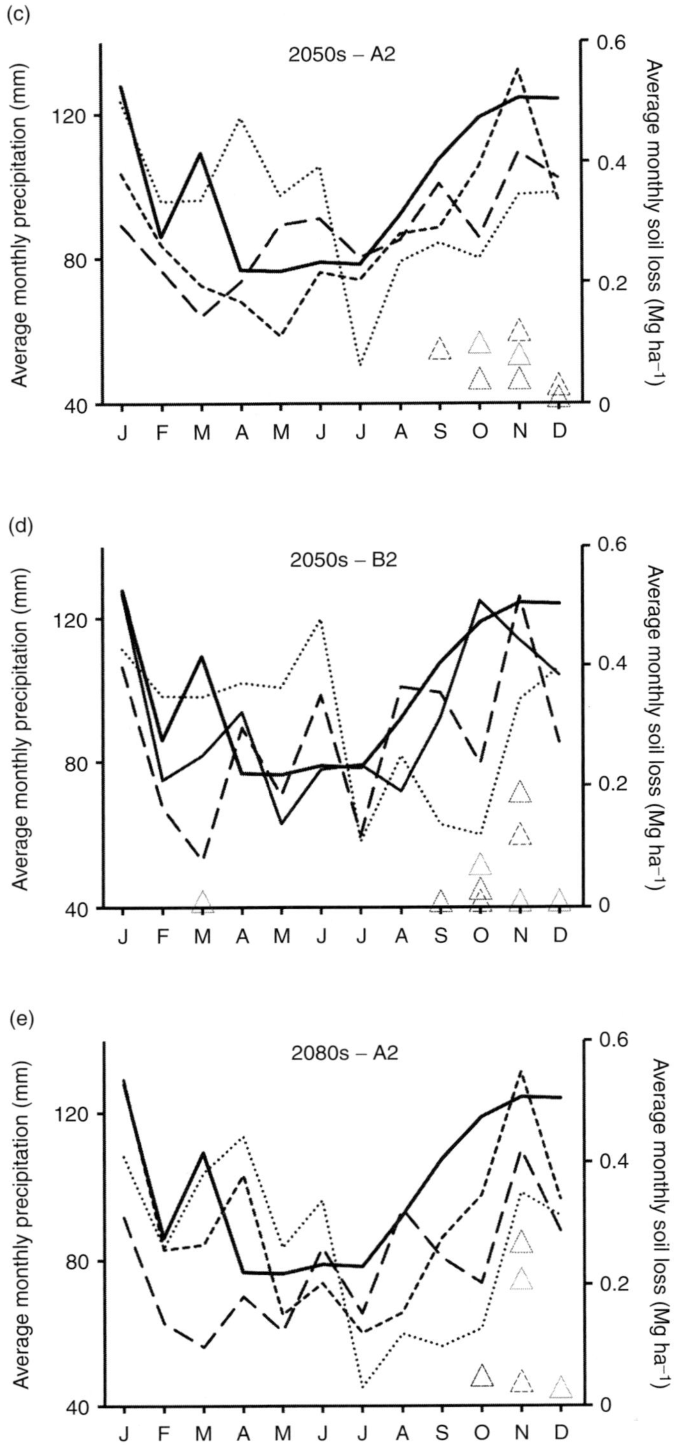

Fig. 18.6. Continued.

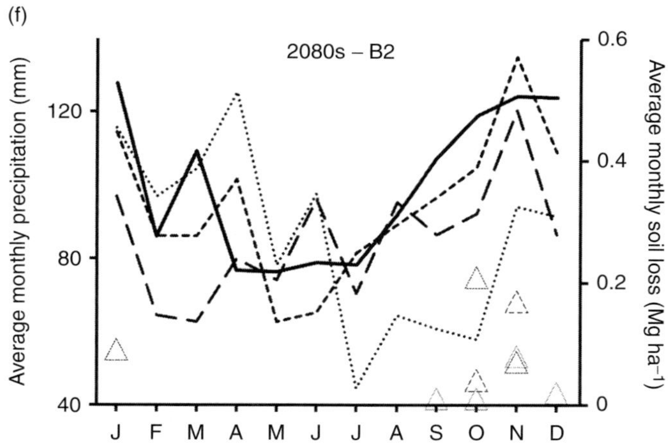

Fig. 18.6. Continued.

Table 18.8. Changes in precipitation, runoff and soil erosion with 50% increases in the average maximum 0.5 h rainfall intensity added to climate change projections from combinations of three general circulation models (HadCM3, CGCM2 and CSIROMk2) with two emissions scenarios (A2 and B2) for three future time slices.

Scenario[a]	Precipitation depth (mm)	Change (%)	Runoff depth (mm)	Change (%)	Soil loss rate (Mg ha^{-1})	Change (%)
Baseline	1201	0	0.84	0	0.35	0
2020s						
HadCM3-A2	1108	−8	2.00	138	0.69	97
CGCM2-A2	1079	−10	0.73	−13	0.46	31
CSIROMk2-A2	1104	−8	0.46	−45	0.32	-9
HadCM3-B2	1140	−5	2.76	229	0.54	54
CGCM2-B2	1053	−12	1.73	106	0.62	77
CSIROMk2-B2	1097	−9	1.07	27	0.73	109
Mean	1097	−9	1.45	74	0.56	60
2050s						
HadCM3-A2	1045	−13	0.60	−29	0.26	−26
CGCM2-A2	1048	−13	1.70	102	0.47	34
CSIROMk2-A2	1127	−6	0.99	18	0.64	83
HadCM3-B2	1104	−8	1.16	38	0.57	63
CGCM2-B2	1036	−14	1.15	37	0.33	−6
CSIROMk2-B2	1096	−9	0.30	−64	0.22	−37
Mean	1076	−10.5	0.98	17	0.42	19
2080s						
HadCM3-A2	1076	−10	2.26	169	0.69	97
CGCM2-A2	937	−22	1.04	24	0.41	17
CSIROMk2-A2	1002	−17	1.08	29	0.37	6
HadCM3-B2	1133	−6	1.57	87	0.62	77
CGCM2-B2	1027	−14	0.77	−8	0.36	3
CSIROMk2-B2	1032	−14	0.44	−47	0.27	−23
Mean	1034	−14	1.19	42	0.45	30

[a]See Table 18.5 for details.

Table 18.9. Changes in precipitation, runoff and soil erosion with a changing land use under climate change projections from combinations of three general circulation models (HadCM3, CGCM2 and CSIROMk2) with two emissions scenarios (A2 and B2) for three future time slices.

Scenario[a]	Precipitation depth (mm)	Change (%)	Runoff depth (mm)	Change (%)	Soil loss rate (Mg ha^{-1})	Change (%)
Baseline	1201	0	0.84	0	0.35	0
Land use change	1201	0	3.67	337	3.34	854
2020s						
HadCM3-A2	1108	−8	3.81	354	3.96	1031
CGCM2-A2	1079	−10	2.64	214	1.78	409
CSIROMk2-A2	1104	−8	1.37	63	1.46	317
HadCM3-B2	1140	−5	4.83	475	5.16	1374
CGCM2-B2	1053	−12	3.59	327	3.32	849
CSIROMk2-B2	1097	−9	3.38	302	3.64	940
Mean	1097	−9	3.27	289	3.22	820
2050s						
HadCM3-A2	1045	−13	2.12	152	1.69	383
CGCM2-A2	1048	−13	3.64	333	3.11	789
CSIROMk2-A2	1127	−6	2.46	193	2.07	491
HadCM3-B2	1104	−8	2.45	192	2.4	586
CGCM2-B2	1036	−14	2.61	211	2.65	657
CSIROMk2-B2	1096	−9	1.65	96	1.37	291
Mean	1076	−10.5	2.49	196	2.22	533
2080s						
HadCM3-A2	1076	−10	3.59	327	3.3	843
CGCM2-A2	937	−22	3.11	270	2.88	723
CSIROMk2-A2	1002	−17	2.5	198	2.53	623
HadCM3-B2	1133	−6	4.09	387	3.66	946
CGCM2-B2	1027	−14	2.71	223	2.81	703
CSIROMk2-B2	1032	−14	1.86	121	1.62	363
Mean	1034	−14	2.98	254	2.80	700

[a]See Table 18.5 for details.

change scenarios. The highest increases in erosion under the full range of impacts occur when either precipitation amount or simple daily intensity increases during the key tillage window in autumn.

Uncertainties

When dealing with a range of possible future scenarios of climate and land-use change, it is clear that a number of uncertainties exist, and that these uncertainties feed into the WEPP model, which itself possesses a number of limitations. The uncertainties in this study can broadly be grouped into four categories:

1. Uncertainty surrounding future climate change projections (emissions and GCMs)
2. Limitations associated with the downscaling technique
3. Assumptions regarding future conditions of land use and rainfall intensity changes
4. Shortcomings of the WEPP model for modelling future erosion rates.

Uncertainty surrounding future climate change projections

Future projections of climate are inherently uncertain. Such uncertainty becomes immediately apparent with respect to future emissions scenarios, which represent the first step

Table 18.10. Changes in average annual precipitation, runoff and soil erosion under a changed land use and with rainfall intensity changes added for climate change projections from combinations of three general circulation models (HadCM3, CGCM2 and CSIROMk2) with two emissions scenarios (A2 and B2) for three future time slices.

Scenario[a]	Precipitation depth (mm)	Change (%)	Runoff depth (mm)	Change (%)	Soil loss rate (Mg ha^{-1})	Change (%)
Baseline	1201	0	0.84	0	0.35	0
Land use change	1201	0	3.67	337	3.34	854
2020s						
HadCM3-A2	1108	−8	8.8	948	9.91	2731
CGCM2-A2	1079	−10	7.72	819	5.85	1571
CSIROMk2-A2	1104	−8	4.25	406	4.22	1106
HadCM3-B2	1140	−5	10.09	1101	10.4	2871
CGCM2-B2	1053	−12	8.67	932	8.07	2206
CSIROMk2-B2	1097	−9	7.84	833	7.89	2154
Mean	1097	−9	7.90	840	7.72	2107
2050s						
HadCM3-A2	1045	−13	6.27	646	5.03	1337
CGCM2-A2	1048	−13	8.73	939	7.8	2129
CSIROMk2-A2	1127	−6	7.26	764	6.52	1763
HadCM3-B2	1104	−8	6.46	669	6.22	1677
CGCM2-B2	1036	−14	7.55	799	7.25	1971
CSIROMk2-B2	1096	−9	5.3	531	4.38	1151
Mean	1076	−10.5	6.93	725	6.20	1671
2080s						
HadCM3-A2	1076	−10	8.31	889	8.14	2226
CGCM2-A2	937	−22	8.91	961	8.03	2194
CSIROMk2-A2	1002	−17	6.61	687	6.39	1726
HadCM3-B2	1133	−6	10.24	1119	9.05	2486
CGCM2-B2	1027	−14	7.76	824	7.57	2063
CSIROMk2-B2	1032	−14	5.7	579	4.86	1289
Mean	1034	−14	7.92	843	7.34	1997

[a]See Table 18.5 for details.

in the climate modelling process. The future impacts of climate change will depend on the future political, social, demographic and economic characteristics of the world. Therefore, given the uncertainty in such a broad range of future conditions, clearly assumptions and simplifications must be made about the driving agents for GCMs. In addition, the GCMs themselves are subject to major conceptual sources of uncertainty, centred on atmospheric particles, atmospheric water vapour and cloud parameterization (Crawford, 2007). In dealing with such a broad range of uncertainties, 'scenarios' becomes a key word in climate impact studies. A scenario differs from a forecast in that it is a plausible future, whereas a forecast is the most likely future. A set of scenarios, therefore, span the range of likely future developments (Parry, 2002; Crawford, 2007). In this study, use of multiple GCMs and emissions scenarios (two emissions scenarios and three GCMs for three future time slices) helps to address the uncertainties inherent in global climate modelling by providing a range of equally plausible future conditions.

Limitations associated with the downscaling technique

In addition, the downscaling technique also represents a significant source of uncertainty.

Wilby *et al.* (2004) note that the major theoretical weakness of statistical downscaling techniques is that their basic assumption is not verifiable, i.e. that statistical relationships derived for the present-day climate also hold under future climate forcing. Predictor–predictand relationships are thus assumed to be time invariant, yet it is well recognized that transfer functions may become invalid or weights attached to different predictors may change under the future climate regime (Wilby *et al.*, 2004). Hewitson (1999) illustrates the danger of the time-invariance assumption, noting that atmospheric moisture content exerts some influence over present-day precipitation occurrence and amounts, but is expected to assume much greater significance in the future. Relationships therefore must be critically and carefully assessed as it is impossible to check future climate conditions with observational records (Arnell, 2002). However, tests of stationarity of statistical transfer functions using comparable relationships in RCMs suggest that the stationarity or time-invariance assumption may be robust provided that the choice of predictors is judicious (Charles *et al.*, 1999).

Problems due to assumptions

In the absence of any methodology to more robustly predict future changes in land use and sub-daily rainfall intensity, assumptions must be made regarding such changes. In this study, one single scenario of each is employed. The single scenario of land use (change from grass to a wheat-wheat-barley rotation) and the single scenario of rainfall intensity (+50% maximum half-hourly precipitation) were chosen for modelling future erosion rates as they both represent considerable changes from present-day conditions, and could model future erosion rates given the possibility of considerable environmental changes. In order to reduce the uncertainty associated with these assumptions, a range of intermediate (and indeed more extreme) scenarios could also be modelled to examine the response of erosion rates to each. For example, maximum half-hourly precipitation could be increased in 10% increments, and a wider range of land-use change scenarios could be examined. This study was designed to pick out an intentionally considerable change in each of these scenarios in order to examine the impacts on future rates of soil erosion given such changes. In addition, it is recognized that changes to the growing season resulting from climate change could shift planting and harvesting dates, and these could be critical to the timing of erosion (e.g. O'Neal *et al.*, 2005). These dates were left unchanged in the current study, but could have been changed in a 'what if' manner to examine likely changes in the response of erosion to shifting management dates. The scenarios employed here are almost certainly unrealistic, but they point the way to potential future conditions – unlike those previous studies that leave such parameters unchanged from the present day.

WEPP model limitations

WEPP suffers from many of the generic limitations associated with erosion models. With process-based models such as WEPP, the equations governing processes tend to be derived at small scales and under very specific physical conditions (Beven, 1989). However, models such as WEPP use some of these equations at larger spatial scales and under different physical conditions. This issue of 'lumping up' small-scale physics to larger spatial scales is certainly questionable (Beven, 1989). As a process-based, distributed model, the computational and data requirements of WEPP are vast, thus limiting its applicability to areas where sufficient data are available. The nature of this type of model also presents the problem of over-parameterization, with a large number of processes considered and the associated parameters running the risk of having a high degree of uncertainty associated with model inputs, which translates itself through to the outputs in the model (Merritt *et al.*, 2003). Despite its process-based nature, however, there is still a large degree of empiricism in the WEPP model. For example, WEPP's routing of inter-rill runoff into rills is derived from an empirical relationship based on extensive rainfall simulation studies on a number of different soils (Meyer, 1981). However, WEPP is well suited for studying the complex interactions involved in climate change impact assessments on soil erosion as it accounts for most of the processes and interactions whereby climate change affects runoff and erosion (Nearing *et al.*, 2004).

As discussed by Favis-Mortlock and Guerra (1999), there should be decreased danger of invalid extrapolations of relationships on which the model is based when a physically based model is used instead of an empirically based model. Confidence in the model may be further enhanced if it validates reasonably well for the present day without calibration, as it did in this study. However, Favis-Mortlock (1994) warns against the possibility of producing 'the right answer for the wrong reason,' if, for example, two or more model shortcomings cancel each other out by biasing results in opposite directions. One of the specific problems with WEPP that has direct relevance for this study is that the model tends to overpredict erosion on slope lengths longer than 100 m. Therefore, applying it to Loughmuck hillslope, at 180 m long, may have resulted in over-predicted rates of soil loss.

In order to improve the current ability of WEPP to predict the impact of climate change on future rates of soil erosion, one of the key issues that could be addressed is the temporal resolution at which rainfall intensity changes can be modified. It is well documented that severe soil erosion is linked to high-magnitude low-frequency rainfall events, yet WEPP rainfall intensity can only be directly modified by changing the maximum half-hourly precipitation parameter (although internal changes to sub-daily intensity do occur through modification of the daily rainfall amount). If the maximum half-hourly precipitation parameter was changed to a maximum of 15 min precipitation, for example, better estimates of runoff and soil erosion may be made possible. However, such modifications to WEPP would then be required to wait until the temporal resolution at which GCM-output-changed climate data could match such high resolution.

Summary

Future rates of soil erosion may increase or decrease at Loughmuck hillslope, depending on the mix of factors considered. If downscaled climate change projections are considered in isolation, i.e. without representing sub-daily rainfall intensity changes and land-use change, then future rates of soil erosion are projected to decrease as a result of decrease in precipitation. However, when either or both these additional factors are included in the modelling process, erosion rates are projected to increase dramatically. Thus, this study indicates that, for this site at least, it is not just rainfall amount and timing that has an impact on soil erosion, but also the sub-daily intensity with which this rainfall arrives. However, the two factors found to affect future erosion rates most in this study – land-use change and sub-daily rainfall intensity – are also the factors that are currently the most difficult to predict with any confidence. Hence, in the absence of any robust prediction method, a 'what if' scenario-based approach has been employed here, based on assumptions about future conditions.

Statistical downscaling of future climate change scenarios, as described in this study, represents a notable improvement over those previous studies which considered only relative climate changes at the native GCM grid-box scale. The downscaling methodology employed here permits a robust representation of expected site-specific future climate change scenarios, with an associated representation of climatic variability. However, the development of downscaling methodologies that also include rainfall changes at the sub-daily scale is now a pressing research requirement if we wish to robustly model future rates of soil erosion. An additional pressing research need is a framework for assessing likely changes in future land use.

Finally, this particular present-day case study illustrates the potential for future climate change to create a soil erosion problem in a location where present-day erosion is of rather minor significance. This can occur because of increases in the erosivity of future rainfall (in particular, because of increases in sub-daily rainfall intensity) and/or changes in land use resulting in greater erodibility. Of the two, changes in land use may give rise to the greatest increase in erosion by water. This is hardly surprising: the importance of land use in determining rates of erosion has been known for 50 years or more.

The extent to which this case study, set in temperate western Europe, provides an analogue for other locations, is unknown. If, however, this case study is even a crude approximation for what will happen at other currently non-eroding sites, then there is cause for concern regarding the extent and severity of soil erosion by water under future, changed climates. We must not be complacent regarding this problem.

Note

[1] From the Middle English *mudde*, *c.* 1300; probably via Middle Low German and Middle Dutch *modde*. http://dictionary.reference.com/browse/mud

References

Arctic Climate Impact Assessment (2005) *Arctic Climate Impact Assessment – Scientific Report.* Cambridge University Press, Cambridge, UK.

Arkell, B., Darch, G. and McEntee, P. (eds) (2007) *SNIFFER UKCC13: Preparing for a Changing Climate in Northern Ireland.* SNIFFER (Scotland and Northern Ireland Forum for Environmental Research), Edinburgh, UK.

Arnell, N.W. (1998) Climate change and water resources in Britain. *Climatic Change* 39, 83–110.

Arnell, N.W. (2002) *Hydrology and Global Environmental Change.* Prentice Hall, London.

Arnell, N.W. and Reynard, N.S. (1996) The effects of climate change due to global warming on river flows in Great Britain. *Journal of Hydrology* 183, 397–424.

Arnell, N.W., Hudson, D.A. and Jones, R.G. (2003) Climate change scenarios from a regional climate model: estimating change in runoff in southern Africa. *Journal of Geophysical Research* 108, 1–17.

Beven, K. (1989) Changing ideas in hydrology – the case of physically-based models. *Journal of Hydrology* 105, 157–172.

Boal, F.W. (1963) County Down. In: Symons, L. (ed.) *Land Use in N. Ireland.* University of London Press, London, pp. 218–233.

Boardman, J. and Favis-Mortlock, D.T. (1993) Climate change and soil erosion in Britain. *The Geographical Journal* 159, 179–183.

Boardman, J. and Spivey, D. (1987) Flooding and erosion in west Derbyshire, April 1983. *East Midland Geographer* 10, 36–44.

Boardman, J., Evans, R., Favis-Mortlock, D.T. and Harris, T.M. (1990) Climate change and soil erosion on agricultural land in England and Wales. *Land Degradation and Rehabilitation* 2, 95–106.

Botterweg, P. (1994) Modelling the effects of climate change on runoff and erosion in central southern Norway. In: Rickson, R.J. (ed.) *Conserving Soil Resources- European Perspectives.* CAB International, Wallingford, UK, pp. 273–285.

Brazier, R.E., Beven, K.J., Freer, J. and Rowan, J.S. (2000) Equifinality and uncertainty in physically based soil erosion models: applications of the GLUE methodology to WEPP – The Water Erosion Prediction Project – for sites in the UK and USA. *Earth Surface Processes and Landforms* 25, 825–845.

Burt, T.P., Chorley, R.J., Brunsden, D., Cox, N.J. and Goudie, A.S. (eds) (2008) *The History of the Study of Landforms or the Development of Geomorphology,* Volume 4: *Quaternary and Recent Processes and Forms (1890–1965) and the Mid-Century Revolutions.* Geological Society of London, London.

Charles, S.P., Bates, B.C., Whetton, P.H. and Hughes, J.P. (1999) Validation of downscaling models for changed climate conditions: case study of southwestern Australia. *Climate Research* 12, 1–14.

Clayton, A. and Radcliffe, N.J. (1996) *Sustainability: A Systems Approach.* Earthscan, London.

Crawford, T. (2007) Future climate change: modelling the implications of shifts in rainfall characteristics for runoff in Northern Ireland. PhD thesis, Queen's University Belfast, Belfast, Northern Ireland.

Cruickshank, J.G. (ed.) (1997) *Soil and Environment: Northern Ireland.* Agricultural and Environmental Science Department, Queen's University Belfast, Belfast, Northern Ireland.

Day, C.A. (2009) Modelling impacts of climate change on snowmelt runoff generation and streamflow across western US mountain basins: a review of techniques and applications for water resource management. *Progress in Physical Geography* 33, 614–633.

DEFRA (2007) *Baseline Projections for Agriculture and Implications for Emissions to Air and Water: Final Report.* Department for Environment, Food and Rural Affairs (DEFRA), London.
Department of the Environment (1988) *Possible Impacts of Climate Change on the Natural Environment in the United Kingdom.* Department of the Environment (DETR, now DEFRA), London.
Diaz-Nieto, J. and Wilby, R.L. (2005) A comparison of statistical downscaling and climate change factor methods: implications on low flows in the River Thames, United Kingdom. *Climatic Change* 69, 245–268.
Doe, W.W. and Harmon, R.S. (2001) Introduction to soil erosion and landscape evolution modeling. In: Harmon, R.S. and Doe, W.W. (eds) *Landscape Erosion and Evolution Modeling.* Kluwer Academic/Plenum Publishers, New York.
Evans, E.E. (1963) The prehistoric and historic background. In: Symons, L. (ed.) *Land Use in N. Ireland.* University of London Press, London, pp. 24–44.
Favis-Mortlock, D.T. (1994) Use and abuse of soil erosion models in Southern England. PhD thesis, University of Brighton, UK.
Favis-Mortlock, D.T. (2006) Ireland. In: Boardman, J. and Poesen, J. (eds) *Soil Erosion in Europe.* Wiley, Chichester, UK, pp. 455–462.
Favis-Mortlock, D.T. and Boardman, J. (1995) Nonlinear responses of soil erosion to climate change: a modelling study on the UK South Downs. *Catena* 25, 365–387.
Favis-Mortlock, D.T. and Guerra, A.J.T. (1999) The implications of general circulation model estimates of rainfall for future erosion: a case study from Brazil. *Catena* 37, 329–354.
Favis-Mortlock, D.T. and Savabi, M.R. (1996) Shifts in rates and spatial distributions of soil erosion and deposition under climate change. In: Anderson, M.G. and Brooks, S.M. (eds) *Advances in Hillslope Processes. Vol. 1.* Wiley, Chichester, UK.
Favis-Mortlock, D.T., Evans, R., Boardman, J. and Harris, T.M. (1991) Climate change, winter wheat yield and soil erosion on the English South Downs. *Agricultural Systems* 37, 415–433.
Favis-Mortlock, D.T., Boardman, J. and MacMillan, V. (2001) The limits of erosion modelling: why we should proceed with care. In: Harmon, R.S. and Doe, W.W. III (eds) *Landscape Erosion and Evolution Modeling.* Kluwer Academic/Plenum Publishers, New York, pp. 477–516.
Fealy, R. and Sweeney, J. (2007) Statistical downscaling of precipitation for a selection of sites in Ireland employing a generalised linear modelling approach. *International Journal of Climatology* 27, 2083–2094.
Fealy, R. and Sweeney, J. (2008) Statistical downscaling of temperature, radiation and potential evapotranspiration to produce a multiple GCM ensemble mean for a selection of sites in Ireland. *International Journal of Climatology* 41, 1–27.
Flanagan, D.C. and Nearing, M.A. (eds) (1995) *USDA Water Erosion Prediction Project Hillslope Profile and Watershed Model Documentation.* NSERL Report No. 10, National Soil Erosion Research Laboratory, USDA Agricultural Research Service, West Lafayette, Indiana.
Foster, G.R. and Lane, L.J. (1987) *User Requirements USDA Water Erosion Prediction Project (WEPP)*, NSERL Report No. 1, National Soil Erosion Research Laboratory, USDA Agricultural Research Service, West Lafayette, Indiana.
Gardiner, M.J. and Burke, W. (1982) Soil erosion and conservation measures in Ireland. In: Prendergast, A.G. (ed.) *Soil Erosion.* Abridged Proceedings of the Workshop on 'Soil Erosion and Conservation: Assessment of the Problems and the State of the Art in EEC Countries'. CEC Report EUR 8427 EN, Commission of the European Communities, Florence, pp. 53–57.
Hall, V.A. (1989) The historical and palynological evidence for flax cultivation in mid Co. Down. *Ulster Journal of Archaeology* 52, 5–10.
Hall, V.A. (1990) Recent landscape history from a Co. Down lake deposit. *New Phytologist* 115, 377–383.
Hebel, B. and Siegrist, S. (1998) Plot size application and validation of WEPP for Swiss conditions. In: *Proceedings of the International Workshop 'Experiences with Soil Erosion Models'*, 6–8 October 1998, Prague. *Wiener Mitteilungen: Wasser, Abwasser, Gewässer* 151, S, 149–160.
Hewitson, B.C. (1999) *Deriving Regional Precipitation Scenarios from General Circulation Models.* Water Research Commission Report 751/1/99, Pretoria, South Africa.
Hossell, J.E., Jones, P.J., Marsh, J.S., Parry, M.L., Rehman, T. and Tranter, R.B. (1995) The likely effects of climate change on agricultural land use in England and Wales. *Geoforum* 27, 149–157.
Houghton, J.T., Callander, B.A. and Varney, S.K. (eds) (1992) *The Supplementary Report to the IPCC Scientific Assessment.* Cambridge University Press, Cambridge, UK.
Huang, C.H., Bradford, J.M. and Laflen, J.M. (1996) Evaluation of the detachment transport coupling concept in the WEPP rill erosion equation. *Soil Science Society of America Journal* 60, 734–739.
Hulme, M., Hossell, J.E. and Parry, M.L. (1993) Climate change and land use in the UK. *The Geographical Journal* 159, 131–147.

Hulme, M., Jiang, T. and Wigley, T.M.L. (1995) *SCENGEN, a Climate Change Scenario Generator: User Manual.* Climatic Research Unit, University of East Anglia, Norwich, UK.

Hulme, M., Jenkins, G.J., Lu, X., Turnpenny, J.R., Michell, T.D., Jones, R.G., Lowe, J., Murphy, J.M., Hassell, D., Boorman, P., MacDonald, R. and Hill, S. (2002) *Climate Change Scenarios for the United Kingdom: The UKCIP02 Scientific Report.* Tyndall Centre for Climate Change Research, School of Environmental Sciences, University of East Anglia, Norwich, UK.

IPCC (Intergovernmental Panel on Climate Change) (1990) *Climate Change: The IPCC Scientific Assessment.* Cambridge University Press, Cambridge, UK.

IPCC (2007) *Climate Change 2007: IPCC Fourth Assessment Report.* Available at: http://www.ipcc.ch/ (accessed 17 October 2008).

Kallio, K., Rekolainen, S., Ekholm, P., Granlud, K., Laine, Y., Johnsson, H. and Hoffman, M. (1997) Impacts of climate change on agricultural nutrient losses in Finland. *Boreal Environmental Research* 2, 1040–1042.

Karl, T.R. and Knight, R.W. (1998) Secular trends of precipitation amount, frequency, and intensity in the United States. *Bulletin of the American Meteorological Society* 79, 1107–1119.

Kilsby, C.G., Cowpertwait, P.S.P., O'Connell, P.E. and Jones, P.D. (1998) Predicting rainfall statistics in England and Wales using atmospheric circulation variables. *International Journal of Climatology* 18, 523–539.

Kim, M.K., Flanagan, D.C., Frankenberger, J.R. and Meyer, C.R. (2009) Impact of precipitation changes on runoff and soil erosion in Korea using CLIGEN and WEPP. *Journal of Soil and Water Conservation* 64, 154–162.

Kinnell, P.I.A. (2006) Simulations demonstrating interaction between coarse and fine sediment loads in rain-impacted flow. *Earth Surface Processes and Landforms* 31, 355–367.

Laflen, J.M., Elliot, W.J., Simanton, R., Holzhey, S. and Kohl, K.D. (1991) WEPP soil erodibility experiments for rangeland and cropland soils. *Journal of Soil and Water Conservation* 46, 39–44.

Laflen, J.M., Elliot, W.J., Flanagan, D.C., Meyer, C.R. and Nearing, M.A. (1997) WEPP – Predicting water erosion using a process-based model. *Journal of Soil and Water Conservation* 52, 96–102.

Laflen, J.M., Flanagan, D.C. and Engel, B.A. (2004) Soil erosion and sediment yield prediction accuracy using WEPP. *Journal of the American Water Resources Association* 40, 289–297.

Lal, R. (ed.) (1998) *Soil Quality and Soil Erosion.* CRC Press, Boca Raton, Florida.

Lamb, H.H. (1972) British Isles weather types and a register of daily sequence of circulation patterns, 1861–1971. *Geophysical Memoir* 116, Her Majesty's Stationery Office (HMSO), London.

Lambin, E.F., Turner, B.L., Geist, H., Agbola, S., Angelsen, A., Bruce, J.W., Coomes, O.T., Dirzo, R., Fischer, G., Folke, C., George, P.S., Homewood, K., Imbernon, J., Leemans, R., Li, X.-B., Moran, E.F., Mortimore, M., Ramakrishnan, P.S., Richards, J.F., Skånes, H., Steffen, W., Stone, G.D., Svedin, U., Veldkamp, T.A., Vogel, C. and Xu, J.-C. (2001) The causes of land-use and land-cover change: moving beyond the myths. *Global Environmental Change* 11, 261–269.

Le Bissonnais, Y. and Singer, M.J. (1992) Crusting, runoff and erosion response to soil water content and successive rainfalls. *Soil Science Society of America Journal* 56, 1898–1903.

Lee, J.J. (1998) Cross-scale aspects of EPA erosion studies. In: Boardman, J. and Favis-Mortlock, D.T. (eds) *Modelling Soil Erosion by Water.* NATO-ASI Global Change Series I-55, Springer Verlag, Heidelberg, Germany, pp. 191–200.

Lee, J.J., Phillips, D.L. and Liu, R. (1993) The effects of trends in tillage practices on erosion and carbon content of soils in the US corn belt. *Water, Air and Soil Pollution* 70, 389–401.

Lee, J.J., Phillips, D.L. and Dobson, R.F. (1996) Sensitivity of the US Corn belt to climate change and elevated CO_2: II. Soil erosion and organic carbon. *Agricultural Systems* 52, 503–521.

McCabe, F. and Collins, J.F. (1977) Soil type, soil slope and topsoil depth relationships on a Co. Cavan drumlin. *Irish Geographer* 10, 19–27.

McEntee, M.A. (1994) Cultivation-induced topsoil movement and its effects on soil properties in North East Ireland. PhD thesis, Queen's University Belfast, Belfast, Northern Ireland.

Merritt, W.S., Letcher, R.A. and Jakeman, A.J. (2003) A review of erosion and sediment transport models. *Environmental Modelling and Software* 18, 761–799.

Meyer, L.D. (1981) How rain intensity affects interrill erosion. *Transactions of the ASAE* 24, 1472–1475.

Morgan, R.P.C. (2005) *Soil Erosion and Conservation,* 3rd edn. Blackwell, Malden, Massachusetts/Oxford, UK/Carlton, Victoria, Australia.

Morgan, R.P.C. and Quinton, J.N. (2001) Erosion modeling. In: Harmon, R.S. and Doe, W.W. III (eds) *Landscape Erosion and Evolution Modeling.* Kluwer Academic/Plenum Publishers, New York, pp. 117–143.

Murphy, J. (2000) Predictions of climate change over Europe using statistical and dynamical downscaling techniques. *International Journal of Climatology* 20, 489–501.

Nearing, M.A. (2001) Potential changes in rainfall erosivity in the U.S. with climate change during the 21st century. *Journal of Soil and Water Conservation* 56, 229–232.

Nearing, M.A., Foster, G.R., Lane, L.J. and Finckner, S.C. (1989) A process-based soil erosion model for USDA-Water Erosion Prediction Project technology. *Transactions of the ASAE* 32, 1587–1593.

Nearing, M.A., Pruski, F.F. and O'Neal, M.R. (2004) Expected climate change impacts on soil erosion rates: a review. *Journal of Soil and Water Conservation* 59, 43–50.

Nearing, M.A., Jetten, V., Baffaut, C., Cerdan, O., Couturier, A., Hernandez, M., Le Bissonnais, Y., Nichols, M.H., Nunes, J.P., Renschler, C.S., Souchère, V. and van Oost, K. (2005) Modeling response of soil erosion and runoff to changes in precipitation and cover. *Catena* 61, 131–154.

Nicks, A.D. (1993) Modeling hydrologic impacts of global change using stochastically generated climate change scenarios. In: Eckstein, Y. and Zaporozec, A. (eds) *Industrial and Agricultural Impacts on the Hydrologic Environment, Proceedings of the Second USA/CIS Joint Conference on Environment Hydrology and Hydrogeology.* Water Environment Federation, Alexandria, Virginia.

Nicks, A.D. and Williams, R.D. (1993) Estimating water resources of a mixed land use basin under changing climate and environment. In: *Proceedings of International Conference on Environmentally Sound Water Resources Utilization, Bangkok*, pp. II-8–II-17.

Ó Danachair, C. (1970) The use of the spade in Ireland, In: Gailey, A. and Fenton, A. (eds) *The Spade in Northern and Atlantic Europe.* Ulster Folk Museum and Institute of Irish Studies, Queen's University Belfast, pp. 49–56.

O'Neal, M.R., Nearing, M.A., Vining, R.C., Southworth, J. and Pfeifer, R.A. (2005) Climate change impacts on soil erosion in Midwest United States with changes in crop management. *Catena* 61, 165–184.

Osborn, T.J., Hulme, M., Jones, P.D. and Basnett, T.A. (1999) Observed trends in the daily intensity of United Kingdom precipitation. *International Journal of Climatology* 20, 347–364.

Parry, M.L. (2002) Scenarios for climate impact and adaptation assessment. *Global Environmental Change* 12, 149–153.

Phillips, D.L., White, D. and Johnson, C.B. (1993) Implications of climate change scenarios for soil erosion potential in the USA. *Land Degradation and Rehabilitation* 4, 61–72.

Pieri, L., Bittelli, M., Wu, J.Q., Dun, S., Flanagan, D.C., Rossi Pisa, P., Ventura, F. and Salvatorelli, F. (2007) Using the Water Erosion Prediction Project (WEPP) model to simulate field-observed runoff and erosion in the Apennines mountain range, Italy. *Journal of Hydrology* 336, 84–97.

Pilling, C. and Jones, J.A.A. (1999) High resolution climate change scenarios: implications for British runoff. *Hydrological Processes* 13, 2877–2895.

Pruski, F.F. and Nearing, M.A. (2002) Climate-induced changes in erosion during the 21st century for eight U.S. locations. *Water Resources Research* 38 (12), 1298, doi:10.1029/2001WR000493.

Rosenzweig, C. and Hillel, D. (1998) *Climate Change and the Global Harvest. Potential Impacts of the Greenhouse Effect on Agriculture.* Oxford University Press, New York.

Rosewell, C.J. (2001) Evaluation of WEPP for runoff and soil loss prediction in Gunnedah, NSW, Australia. *Australian Journal of Soil Resources* 9, 230–243.

Savabi, M.R. (1993) Modeling subsurface drainage and surface runoff with WEPP. *Journal of Irrigation and Drainage Engineering* 119, 801–813.

Savabi, M.R., Flanagan, D.C., Hebel, B. and Engel, B.A. (1995) Application of WEPP and GIS-GRASS to a small watershed in Indiana. *Journal of Soil and Water Conservation* 50, 477–483.

Semenov, M.A. and Barrow, E.M. (1997) Use of a stochastic weather generator in the development of climate change scenarios. *Climatic Change* 35, 397–414.

Soil and Water Conservation Society (2003) Conservation Implications of Climate Change: Soil Erosion and Runoff from Cropland: A Report from the Soil and Water Conservation Society, January 2003. SWCS, Ankeny, Iowa. Available at: http://www.swcs.org/documents/filelibrary/advocacy_publications_before_2005/Climate_changefinal_112904154622.pdf (accessed 6 March 2011).

Solomon, S., Qin, D., Manning, M., Chen, Z., Marquis, M., Averyt, K.B., Tignor, M. and Miller, H.L. (eds) (2007) *Contribution of Working Group I to the Fourth Assessment Report of the Intergovernmental Panel on Climate Change, 2007.* Cambridge University Press, Cambridge, UK.

Souchère, V., King, C., Dubreuil, N., Lecomte-Morel, V., Le Bissonais, Y. and Chalat, M. (2003) Grassland and crop trends: role of the European Union Common Agricultural Policy and consequences for runoff and soil erosion. *Environmental Science and Policy* 6, 7–16.

Spadaro, G., Flanagan, D.C., Cosentino, S. and Mantineo, M. (2004) Estimation of soil erosion using the WEPP model for Mediterranean conditions. *European Society of Agronomy* 2, 2–4.

Thompson, R. and Edwards, K.J. (1982) A Holocene palaeomagnetic record and a geomagnetic master curve from Ireland. *Boreas* 11, 335–349.

Tiwari, A.K., Risse, L.M. and Nearing, M.A. (2000) Evaluation of WEPP and its comparison with USLE and RUSLE. *Transactions of the ASAE* 43, 1129–1135.

Toy, T.J., Foster, G.R. and Renard, K.G. (2002) *Soil Erosion: Processes, Prediction, Measurement and Control.* Wiley, New York.

Trenberth, K.E., Dai, A., Rasmussen, R.M. and Parsons, D.B. (2003) The changing character of precipitation. *Bulletin of the American Meteorological Society* 84, 1205–1217.

Turner, B.L., Meyer, W.B. and Skole, D.L. (1994) Global land-use/land cover change: towards an integrated study. *Ambio* 23, 91–95.

Turner, B.L., Skole, D., Sanderson, S., Fischer, G., Fresco, L. and Leemans, R. (1995) *Land-Use and Land-Cover Change: Science/Research Plan.* IGBP Report No. 35/HDP Report N. 7, IGBP (International Geosphere-Biosphere Programme) Secretariat, Stockholm, Sweden.

Viner, D. and Hulme, M. (1992) *Climate Change Scenarios for Impact Studies in the U.K.: General Circulation Models, Scenario Construction Methods and Applications for Impact Assessment.* Report Prepared for the UK Department of the Environment. Climatic Research Unit, University of East Anglia, Norwich, UK.

von Bertalanffy, L. (1968). *General System Theory: Foundations, Development, Applications.* George Braziller, New York.

von Storch, H., Zorita, E. and Cubasch, U. (1993) Downscaling of global climate change estimates to regional scales: an application to Iberian rainfall in wintertime. *Journal of Climate* 6, 1161–1171.

Wilbanks, T.J. and Kates, R.W. (1999) Global change in local places. *Climatic Change* 43, 601–628.

Wilby, R.L. (1997) Non-stationarity in daily precipitation series: implications for GCM down-scaling using atmospheric circulation indices. *International Journal of Climatology* 17, 439–454.

Wilby, R.L. and Dawson, C.W. (2001) *Using SDSM Version 2.2 – A Decision Support Tool for the Assessment of Regional Climate Change Impacts, Version 2.2 User Manual.* King's College London, London.

Wilby, R.L. and Dawson, C.W. (2007) *SDSM 4.2 – A Decision Support Tool for the Assessment of Regional Climate Change Impacts.* Lancaster University, Lancaster, UK.

Wilby, R.L. and Harris, I. (2006) A framework for assessing uncertainties in climate change impacts: low flow scenarios for the River Thames, UK. *Water Resources Research* 42, W02419.1–W02419.10.

Wilby, R.L. and Wigley, T.M.L. (1997) Downscaling general circulation model output: a review of methods and limitations. *Progress in Physical Geography* 21, 530–548.

Wilby, R.L., Dawson, C.W. and Barrows, E.M. (2002) SDSM – a decision support tool for the assessment of regional climate change impacts. *Environmental Modelling and Software* 17, 147–159.

Wilby, R.L., Charles, S.P., Zorita, E., Timbal, B., Whetton, P. and Mearns, O.L. (2004) *Guidelines for the Use of Climate Scenarios developed from Statistical Downscaling Methods.* Available at: http://www.ipcc-data.org/guidelines/dgm_no2_v1_09_2004.pdf (accessed 6 March 2011).

Williams, J., Nearing, M.A., Nicks, A., Skidmore, E., Valentine, C., King, K. and Savabi, R. (1996) Using soil erosion models for global change studies. *Journal of Soil and Water Conservation* 51, 381–385.

Zhang, L., O'Neill, A.L. and Lacey, S. (1996) Modelling approaches to the prediction of soil erosion in catchments. *Environmental Software* 11, 123–133.

Zhang, X.C. (2005) Spatial downscaling of global climate model output for site-specific assessment of crop production and soil erosion. *Agricultural and Forest Meteorology* 135, 215–229.

Zhang, X.C. (2007) A comparison of explicit and implicit spatial downscaling of GCM output for soil erosion and crop production assessments. *Climatic Change* 84, 337–363.

Zhang, X.C. and Liu, W.Z. (2005) Simulating potential response of hydrology, soil erosion, and crop productivity to climate change in Changwu tableland region on the Loess Plateau of China. *Agricultural and Forest Meteorology* 131, 127–142.

Zhang, X.C. and Nearing, M.A. (2005) Impact of climate change on soil erosion, runoff and wheat productivity in central Oklahoma. *Catena* 61, 185–195.

Zhang, X.C., Nearing, M.A., Risse, L.M. and McGregor, K.C. (1996) Evaluation of WEPP runoff and soil loss predictions using natural runoff plot data. *Transactions of the ASAE* 39, 855–863.

Zhang, X.C., Nearing, M.A., Garbrecht, J.D. and Steiner, J.L. (2004) Downscaling monthly forecasts to simulate impacts of climate change on soil erosion and wheat production. *Soil Science Society of America Journal* 68, 1376–1385.

Zhang, X.C., Liu, W.Z., Li, Z. and Zheng, F.L. (2009) Simulating site-specific impacts of climate change on soil erosion and surface hydrology in southern Loess Plateau of China. *Catena* 79, 237–242.

19 Microwave Remote Sensing of Soil Hydraulic Properties

Thomas Schmugge*

Introduction

The hydraulic properties of soil are important for determining the rates of water and energy flows in the soil. While point measurements of these properties can be made, their numbers are limited and their extrapolation to large areas is questionable owing to problems of spatial heterogeneity. Sharma *et al.* (1980) showed that measurements of infiltration parameters across a 24-acre field in Chickasha, Oklahoma varied by an order of magnitude. Nielsen *et al.* (1973) reported variations of a similar magnitude for a 150 ha field in Davis, California. These results imply that an infiltration curve based on average parameters from a limited number of point samples is not sufficient to describe the infiltration rate for a watershed. This problem is more acute for climate models, where the areas considered are on the order of tens of kilometres. In this chapter, we will provide some examples of how remotely sensed data can alleviate this problem. Remote sensing makes use of electromagnetic energy that is reflected or emitted from the land surface to make observations of that surface. These include reflected solar energy, e.g. with Landsat, and reflected microwave energy, i.e. imaging radars, emitted thermal radiation in the infrared part of the spectrum for surface temperatures and in the microwave part for surface soil moisture. Thus, these observations are primarily of the surface layer. Long wavelength microwave sensors (>10 cm) can give information about a thicker surface layer of at most 5 cm thick for soil moisture sensing. As a result, remote sensing techniques require use of some sort of model in conjunction with the surface observations to obtain subsurface properties.

The moisture status of the surface soil layer is a function of many factors, including the soil hydraulic properties. Blanchard and O'Neill (1983) were able to distinguish between three soil types at the Beaver Dam site of the US Department of Agriculture (USDA) Agricultural Research Service (ARS) Beltsville Agricultural Research Center (BARC) by observing the change of the microwave emission or brightness temperature, T_B, as the soils dried after an irrigation. Figure 19.1 shows the increase in microwave emissivity as the three soils dried from a saturated condition, along with the measured infiltration rates. It is seen that for the sandy soil (Beaver Dam), with the highest infiltration rate, the emissivity increased the quickest as the surface layer dried most rapidly. The next step in this process is to use

* Corresponding author: tschmugge@psl.nmsu.edu

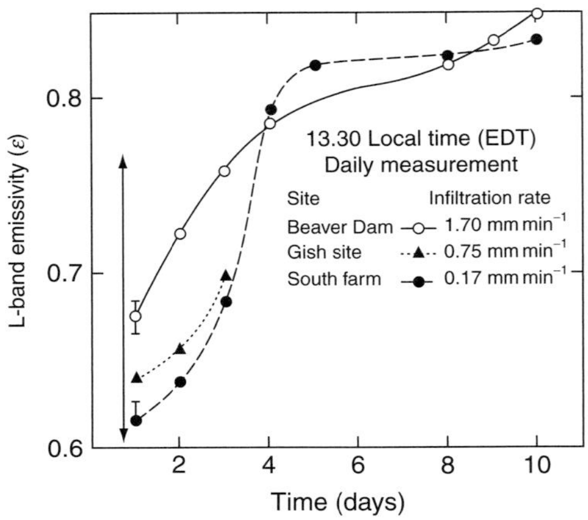

Fig. 19.1. Observations of the change of microwave emissivity as a soil dries for three sample soils at the Beltsville Agricultural Research Center in Maryland. The measured infiltration rates for the three soils are also given. The sand contents of the three soils were: Beaver Dam, 67%; Gish, 24%; and South Farm, 31%. Note: the L-band implies a 21 cm wavelength radiometer (From Blanchard and O'Neill, 1983).

this observed change in the soil's microwave emission to estimate quantitatively soil properties.

To obtain information about soil properties we must use models to relate the remotely sensed surface observations to subsurface parameters. This is best done with models that couple the surface layer observations to the subsurface properties. In general, this will require repetitive measurements so that the model predictions of temporal variation can be compared with the data. The model parameters can then be adjusted to yield better agreement with observations. We will describe the capabilities of microwave radiometers for soil moisture sensing and present a discussion of the application of this combined technique to both a field experiment and aircraft observations to estimate soil properties. Examples of doing this with a time series of microwave observations of surface soil moisture are included. At the present time, this process has been demonstrated only with microwave radiometric observations of surface soil moisture and so only this approach for the remote sensing of soil moisture will be discussed.

Materials and Methods

Microwave radiometry and soil moisture

After more than 40 years of research on the use of passive microwave radiometers for soil moisture sensing (Schmugge *et al.*, 1974) the basic capabilities are well understood and are the basis for the recently launched (November 2009) Soil Moisture and Ocean Salinity Mission (SMOS) by the European Space Agency (ESA) (Kerr *et al.*, 2010).

At microwave frequencies, the most striking feature of the emission from the earth's surface is the large contrast between water and land. This emissivity contrast is due to the large difference between the dielectric constant of water (~80) and that of dry soils (~3.5). This arises from the ability of the electric dipole of the water molecule to align itself in response to the electric field at microwave frequencies. The dielectric constant is of importance here because it describes the propagation characteristics of an electromagnetic wave in the medium. These characteristics include the velocity of propagation, the wavelength in the medium and the absorption of energy in the medium. Figure 19.2 presents the change in the dielectric constant for three soils as a function of their moisture content. The soils range from a sand to a heavy clay, and all three display a wide range in their dielectric constants as the moisture content changes. It is the dielectric contrast at the boundary between two media that determines the reflection and transmission coefficients of electromagnetic waves at such a boundary and thus the surface emissivity. The soil emissivity at the longer microwave wavelengths, about 10 to 30 cm, is a strong function of its moisture content because of the large dielectric contrast between dry soil and water.

The resulting emissivity for soil changes from about 0.95 for dry soil to about 0.6 for wet soils at a rate of approximately 0.01 per 1% moisture content. The rate is a function of the soil texture, being greater for lighter, sandy soils and smaller for heavier, clayey soils as indicated by the dielectric constants in Fig. 19.2. The rate will also be reduced by surface features such as roughness and

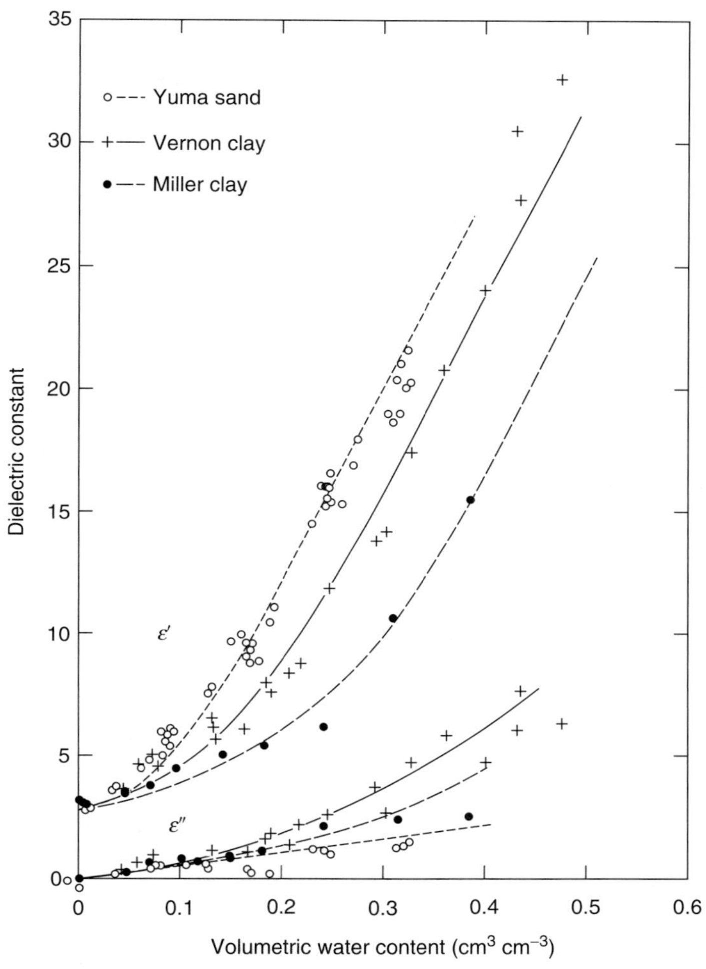

Fig. 19.2. Laboratory measurements of the real and imaginary parts of the dielectric constant for three soils as a function of soil moisture. The curves are from the Wang and Schmugge Model (1980). The upper three curves are for the real part (ε') of the dielectric constant and the lower three are for the imaginary part (ε'').

vegetation cover. The observed emissivity is determined by the moisture content in a thin layer at the surface of the order of a few tenths of a wavelength thick. Therefore, it has been found that the longer wavelengths (>10 cm) are more effective for moisture sensing, both from sampling depth considerations and because of their greater ability to penetrate a vegetation canopy (Jackson and Schmugge, 1989; Schmugge and Jackson, 1994). Knowledge of this surface layer moisture content is very important for determining the energy balance at the soil surface, and its rate of change can be used to infer information about the hydraulic properties of the soil.

A microwave radiometer measures the thermal emission from the surface, and at these wavelengths the intensity of the observed radiation is proportional to the product of the thermodynamic temperature of the soil and the surface emissivity (the Rayleigh–Jeans approximation to the Planck radiation law). This product is commonly called the brightness temperature (T_B). The intensity of the radiation observed by a radiometer is given by:

$$T_B = \tau[RT_{sky} + (1-R)T_{surf}] + T_{atm} \quad (19.1)$$

where R is the surface reflectivity, τ is the atmospheric transmission, T_{sky} is the sky brightness, T_{surf} is the soil surface temperature and T_{atm} is the direct atmospheric contribution. This process is represented schematically in Fig. 19.3. The first term is the reflected sky brightness, RT_{sky}, which depends on the atmospheric conditions and frequency. For the frequencies of interest, T_{sky} = 5 to 10K for the normal range of atmospheric conditions, with 3K of it being the constant cosmic background radiation; for a surface reflectivity of at most 0.4, for a wet smooth soil, its contribution to the observed T_B will generally be less than 5K. The last term in Eqn 19.1 is the direct atmospheric contribution, T_{atm}, and, as noted for T_{sky}, will be about 5K. At these longer microwave wavelengths, i.e. >10 cm, the atmospheric transmission typically will be about 99%, so we are left with the emission from the soil, i.e. the second term, as the main contributor to the observed T_B. So the dominant term, the second, is the radiation emitted by the soil surface which is the product of the surface emissivity, (1 – R), and the temperature of the soil, T_{surf}.

It is this variation of surface emissivity with soil moisture that is exhibited through observations of T_B made with microwave radiometers on remote platforms. The technique has been demonstrated over large areas from both aircraft and satellite platforms. A 21 cm radiometer was on board the Skylab satellite in the mid-1970s. The instrument had 100 km resolution and was non-imaging; however, the T_B observations made were well correlated with ground moisture indicators such as antecedent precipitation and modelled soil moisture (see, for example: Eagleman and Lin, 1976; Wang, 1985; Jackson *et al.*, 2004). Imaging 21 cm radiometers were successfully flown on aircraft in several field experiments conducted since the mid-1980s, e.g. FIFE (Wang *et al.*, 1990), Monsoon '90 (Schmugge *et al.*, 1994), HAPEX-Sahel (Chanzy *et al.*, 1997), Washita '92 (Jackson *et al.*, 1995) and the Southern Great Plains Experiment (SGP97, Jackson *et al.*, 1999). In all of these experiments, the microwave observations were well correlated with ground measurements of soil moisture for a wide range of vegetation conditions. (See Monerris and Schmugge, 2009, for a recent survey of the current status.)

Unfortunately, it is only a relatively thin soil layer at the surface whose moisture content determines the surface emissivity, typically about 5 cm for a 21 cm wavelength

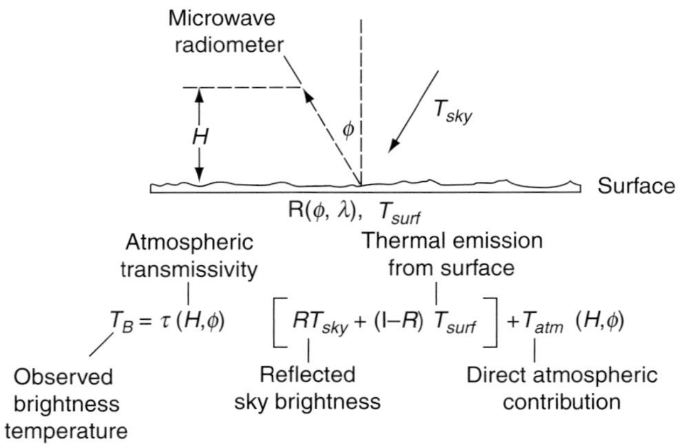

Fig. 19.3. Schematic diagram of the sources contributing to the microwave radiometer signal, T_B, the brightness temperature. H, altitude of sensor; R, surface reflectivity; T_{atm}, direct atmospheric contribution; T_{sky}, sky brightness; T_{surf}, soil surface temperature; ϕ, incidence of radiometer angle; λ, wavelength of microwave radiation; τ, atmospheric transmission.

measurement. This layer is too thin to be of direct use for many applications. However, if we make repeated measurements of this surface layer, then inferences of subsurface properties can be made. In this chapter we shall give examples of the use of this technique.

Estimation of soil properties

The moisture status of the surface soil layer is a function of many factors, among which are the soil's hydraulic properties. As noted earlier, Blanchard and O'Neill (1983) were able to distinguish between three soil types by observing the change of T_B as the soils dried after an irrigation; Fig. 19.1 shows the increase in microwave emissivity as these three soils dried from a saturated condition, along with the measured infiltration rates. It is seen that for the sandy soil with the highest infiltration rate (Beaver Dam) emissivity increased the quickest. The next step in this process is to use repeated observations of this observed change in the soil's microwave emission to quantitatively estimate soil properties. One of the first groups to do this was Camillo *et al.* (1986), who used a coupled soil water and energy budget model with a microwave emission model to predict T_B. The observed and predicted values of T_B were compared and the soil properties were adjusted to minimize the differences. Burke *et al.* (1997, 1998) carried this analysis further with a different soil model and we will summarize this work here. The block diagram of the approach is shown in Fig. 19.4. The adjustable soil model coefficients are those that relate the soil water content (θ) and hydraulic conductivity ($K(h)$) to the soil matric potential (ψ) (Campbell, 1974):

$$\theta = \theta_{sat} \times \left(\frac{\psi_e}{\psi}\right)^{1/b} \tag{19.2}$$

$$K(h) = K_S \times \left(\frac{\psi_e}{\psi}\right)^{1/b} \tag{19.3}$$

where θ_{sat} and K_s are the moisture content and conductivity at saturation, ψ_e is the air entry potential and b is a soil parameter that describes the shape of the water release characteristic. The energy and water balance model for the soil was used to obtain temperature and moisture profiles for the soil. These were then used as inputs for the microwave emission model. This is a coherent radiative transfer model developed by Wilheit (1978) which describes the propagation of electromagnetic energy in a stratified medium, in this case the soil. The observed brightness temperature T_B can be represented as:

$$T_B = \Sigma f_i T_i + RT_{sky} \tag{19.4}$$

where $\Sigma f_i + R = 1$. Here T_B is the emerging energy in temperature units (K), T_i is the temperature of the ith layer, R is the surface reflectivity and RT_{sky} is the reflected sky brightness. The f_i values are related to the electromagnetic absorption properties of the soil layers and are functions of the soil's dielectric properties that depend on the moisture content. The dielectric constant for soil as a function of its moisture content is obtained with the semi-empirical model developed by Wang and Schmugge (1980). This model requires knowledge of the soil texture and density.

Results and Discussion

Field measurements

The experimental data were acquired with a truck-mounted radiometer operating at the 21 cm wavelength over a loam soil field at the Beltsville Agricultural Research Center (BARC) in Maryland. The radiometer and truck were provided by the NASA (National Aeronautics and Space Administration) Goddard Space Flight Center. The radiometer has a 3 dB beam width of ~13° resulting in a ground footprint of ~1.5 m from a nominal height of 6 m and a calibration accuracy of ±3K. To provide a range of soils, three controlled plots of 10 m by 10 m were constructed with raised borders and surrounded by plastic to a depth of 35 cm. The native soil was a loamy sand (75% sand, 5% clay). Other soils

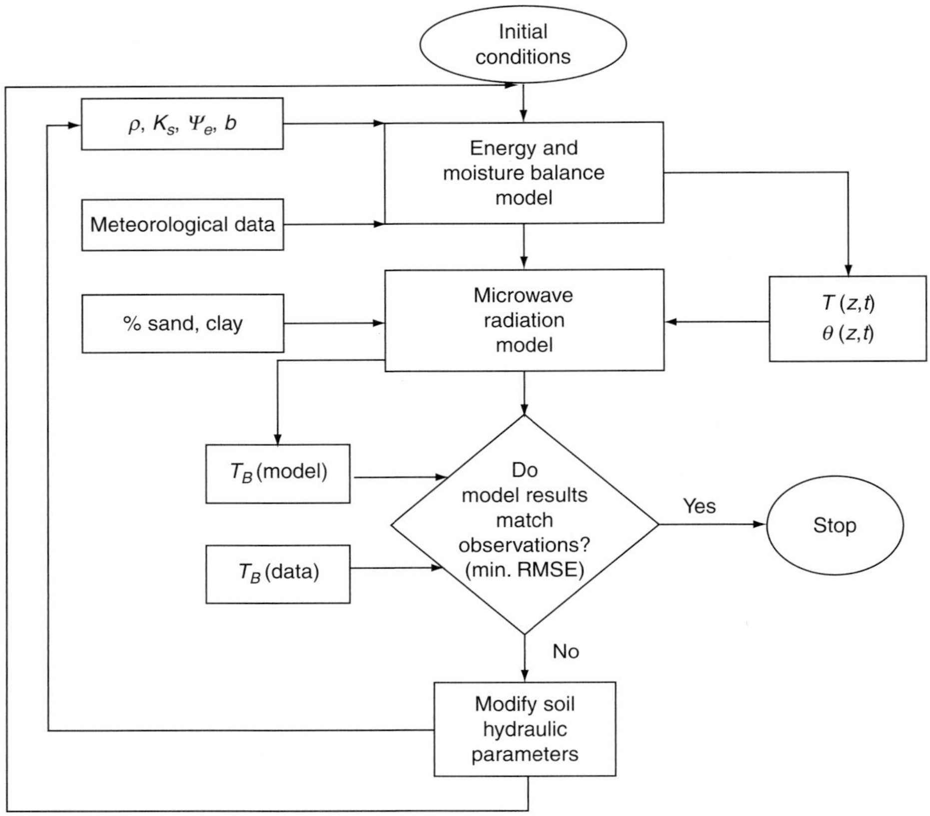

Fig. 19.4. Flow chart of the procedure for estimating the soil hydraulic properties from repeated microwave observations. Note that the soil texture information is used to determine the dielectric properties of the soil used in the microwave radiation model. *b*, a soil parameter that describes the shape of the water release characteristics; K_s, soil saturated hydraulic conductivity; RMSE, root mean square error; *T*, soil temperature as a function of depth (*z*) and time (*t*); Ψ_e, air entry potential; ρ, soil bulk density, θ, soil water content as a function of depth (*z*) and time (*t*).

(a loam and a sandy loam) were imported and laid to a depth of 25 cm over the underlying native soil. Plot 1 contained the native soil, whereas Plot 2 had loam (45% sand, 22% clay) down to 25 cm, then the native soil; likewise, Plot 3 had sandy loam (65% sand, 10% clay) over the native soil. The characteristics of the soils are given in Table 19.1, along with measured values of K_s, ψ_e, ρ (soil bulk density) and *b*.

During 24–26 July 1984, a dry-down situation was observed after 10–12 cm of water was applied to each plot. The microwave brightness temperatures were observed every half-hour during the day when surface conditions were changing most rapidly, and every 2 hours at night. Soil moisture samples of the 0–5 cm layer were taken every 2 hours. An automated weather station was set up to provide the local meteorological conditions. These measurements continued for 72 hours after the irrigation, when a rain event interrupted the dry-down. These data provided the basis for the Camillo *et al.* (1986) analysis. This experiment was essentially repeated on days 211–220 and 252–268 in 1985, i.e. in early August and mid-September, and provided the basis for the Burke *et al.* (1998) analysis.

The results of one such dry-down are presented in Fig. 19.5, where the calculated values of T_B, the curve, are compared with the

Table 19.1. Soil characteristics for three experimental plots at the Beltsville Agricultural Research Center in Maryland.

Soil type	Sand (%)	Clay (%)	ρ (g cm^{-3})[a]	b[a]	Ψ_e (J kg^{-1})[a]	K_s (kg·s m^{-3})[a]
Loam	45	22	1.31 ± 0.05	5.7 ± 0.4	−0.42 ± 0.15	26 ± 8
Sandy loam	65	10	1.35 ± 0.04	3.5 ± 0.2	−1.11 ± 0.23	71 ± 25
Loamy sand	75	5	1.37 ± 0.07	3.4 ± 0.2	−0.82 ± 0.14	112 ± 19

b, a soil parameter that describes the shape of the water release characteristics; K_s, soil saturated hydraulic conductivity; Ψ_e, air entry potential; ρ, soil bulk density.

observed values for the loam soil. The calculated values were obtained using the best-fit parameters derived from the process described in Fig. 19.4. A similar comparison with the measured values of the 0–5 cm soil moisture is given in Fig. 19.6. Here the agreement is not as good, with the radiometer results frequently being lower than the measurements. This lack of agreement indicates that the radiometer is responding to the moisture in a different layer (probably drier and thinner) than the top 0–5 cm of the soil.

This fitting procedure consists of varying the hydraulic parameters K_s, ψ_e, ρ and b until the root mean square error (RMSE) between the predicted and observed time series of T_B values are minimized for each parameter. The results of the process are presented in Fig. 19.7, which summarizes the sensitivity of the estimated hydraulic parameters to T_B. The RMSE of the predicted and observed T_B values is plotted against the parameter value varied in the simulation. To produce each graph, one parameter was varied while the other three were kept constant. The results for two separate dry-downs are shown with good agreement. The horizontal line at RMSE = 5K is an estimate of the combined uncertainty of the measured and modelled T_B values, and its intersection with the curves yields an indication of the range of uncertainty in the derived values. The best-fit values, averaged for the two dry-downs, along with the combined ranges and the derived results from Cosby *et al.* (1984), are given in Table 19.2 for the three soils. It is clear from Fig. 19.7 that there is a well-defined minimum for all the parameters except K_s, which appears to be ill defined by this procedure. In Table 19.2 the values of K_s for both dry-downs are given and these show that for the three soils the results are similar in that there are large differences between the two tries. However, measured values of K_s also show considerable variation. The values derived from the T_B model agree reasonably well with those predicted by the Cosby *et al.* (1984) formulation based on texture and density.

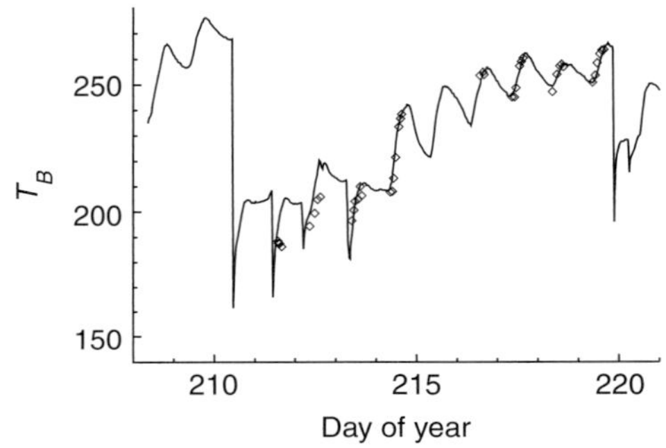

Fig. 19.5. Observed (diamonds) and modelled (the curve) values of T_B (the microwave brightness temperature) for a loam soil at the Beltsville Agricultural Research Center in Maryland. The modelled values were calculated with the best-fit values for the soil parameters.

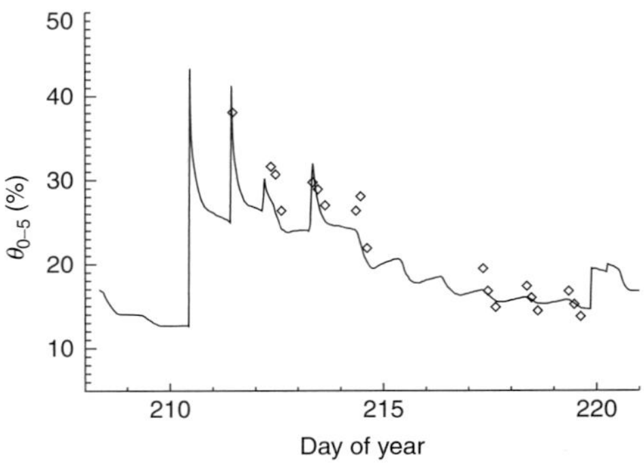

Fig. 19.6. Observed (diamonds) and modelled (the curve) values of θ, the water content of the top 0–5 cm soil layer soil moisture for a loam soil at the Beltsville Agricultural Research Center in Maryland. The modelled values were calculated with the best-fit values for the soil parameters.

Table 19.2. Results for parameter extraction from T_B observations (microwave brightness temperature) for three experimental dry-down plots at the Beltsville Agricultural Research Center in Maryland. Within each soil type, for ρ, b and Ψ_e, the first line of data is the average result of two dry-downs and the second line is the range of values for RMSE < 5K for the two dry-downs; for K_s, the results from two dry-downs are shown instead of the average and no range is given. For all the entries, the third line is the values derived from the Cosby *et al.* (1984) model (Adapted from Burke *et al.*, 1998).

Soil type	ρ (g cm^{-3})	b	Ψ_e (J kg^{-1})	K_s (kg.s m^{-3})
Loam	1.32	4.3	−0.38	52, 36
	1.21–1.43	3.6–4.9	−0.20 to −0.55	
		6.4 ± 2.4	−2.09 ± 0.44	41
Sandy loam	1.39	3.6	−1.1	99, 49
	1.32–1.50	3.0–4.0	−0.9 to −1.45	
		4.5 ± 1.8	1.20 ± 0.45	87
Loamy sand	1.36	3.3	−0.72	112, 181
	1.27–1.48	3.0–3.7	−0.4 to −1.05	
		3.7 ± 1.5		125

b, a soil parameter that describes the shape of the water release characteristics; K_s, soil saturated hydraulic conductivity; Ψ_e, air entry potential; ρ, soil bulk density.

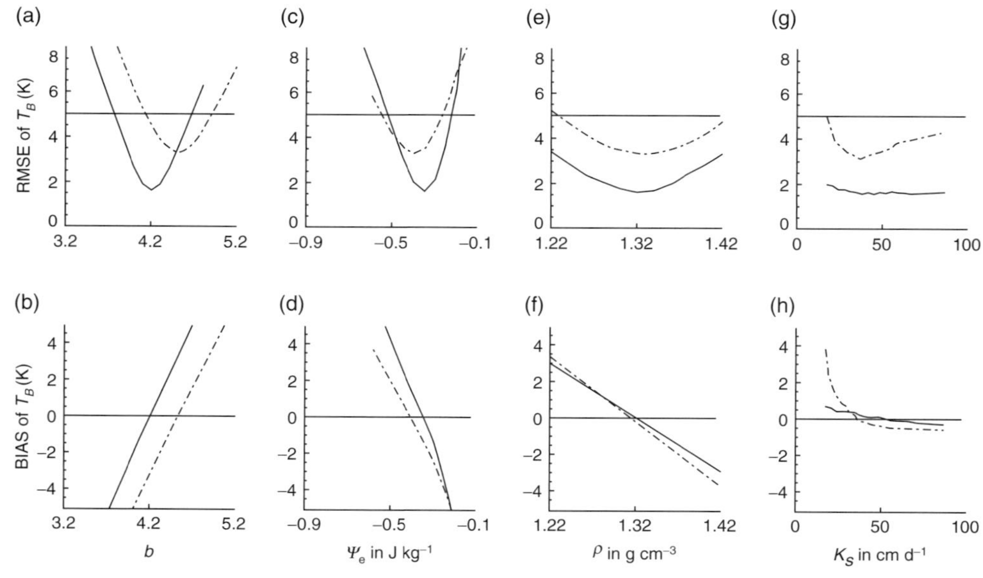

Fig. 19.7. The root mean square error (RMSE) (top) and bias (bottom) of the modelled microwave brightness temperature T_B as functions of different soil hydraulic properties: (a, b) soil b value (a soil parameter that describes the shape of the water release characteristics); (c, d) air entry potential (Ψ_e); (e, f) soil bulk density (ρ); and (g, h) saturated hydraulic conductivity (K_s). The horizontal lines represent an acceptable RMSE of 5K and a bias of 0K. The values for two successive dry-downs are shown, indicating the robustness of the technique. The solid curve is for the first dry-down, i.e. the case shown in Fig. 19.5.

Aircraft measurements

In the early 1990s there were a number of large-scale experiments which studied the contribution that remotely sensed data can make to improving our understanding of the land surface–atmosphere interaction with the goal of increasing our understanding of the role that the land surface plays in atmospheric models. In several of these experiments, remotely sensed soil moisture data were acquired from an aircraft platform. These include the Monsoon '90 experiment conducted in Walnut Gulch in south-eastern Arizona in July and August of 1990, and the HAPEX-Sahel experiment conducted in Niger in West Africa in 1991 and 1992. The instrument in both cases was the PushBroom Microwave Radiometer (PBMR) operating at the 21 cm wavelength. The PBMR receives horizontally polarized radiation in four beams at ±8° and ±24° from nadir with an angular field of 16°. This yields a spatial resolution of about 0.3 H, where H is the aircraft altitude, and a swath width for the four beams of 1.2 H. Another experiment was the Washita'92 large-scale study of remotely sensed soil moisture and hydrology conducted on the Little Washita watershed in south-west Oklahoma in June 1992 (Jackson *et al.*, 1995). The soil moisture sensing for this experiment used the Electrically Scanned Thinned Array Radiometer (ESTAR), also operating at the 21 cm wavelength. This radiometer was a test for the aperture synthesis technique used with the SMOS radiometer.

The Hydrologic Atmospheric Pilot EXperiment in the Sahel (HAPEX-Sahel) in 1991 and 1992 had an intensive observation period (IOP) in August–October 1992. The aim of the experiment was to improve the parameterization of land surface–atmosphere interactions at the global circulation model (GCM) grid-box scale, i.e. at about 1° by 1° (Goutorbe *et al.*, 1994). The experiment combined remote sensing and ground-based measurements with hydrological and meteorological modelling to develop aggregation techniques for use in estimating the land surface–atmosphere interaction for large areas such as the Sahel. One of the remote sensing components was the PBMR for surface soil moisture mapping. During the IOP in August 1992, a 30 × 8 km area was mapped with the PBMR on nine occasions. Comparisons of the microwave observations with ground measurements indicated that the PBMR was quantitatively responding to the soil moisture variations (Chanzy *et al.*, 1997).

Hollenbeck *et al.* (1996) used the PBMR data from HAPEX-Sahel to obtain soil hydraulic information by change detection between a series of images. They compared the microwave brightness temperature (T_B) images separated by 2 days as the soil dried after a significant rain event. The relative change in brightness temperature/emissivity for these two images displayed a pattern similar to that of the soils map for the region. A test of soil heterogeneity based on detection of the deviations of relative change from an average rate in a series of images by more than the instrument was developed. This test is accurate enough to demonstrate the existence of soil heterogeneity between a pair of images in a dry-down. Areas of relatively fast and slow drying can be mapped coinciding with geomorphological feature soil patterns that can be expected to show the observed moisture dynamics. The advantage of this approach is there is no need for an exact knowledge of the calibration relationship. This was only a qualitative comparison; to be useful a more quantitative analysis is needed.

Mattikalli *et al.* (1995) used data collected during the Washita'92 study (Jackson *et al.*, 1995) to develop relationships that quantified the ratio of per cent sand to per cent clay and the saturated hydraulic conductivity in terms of the change components of the microwave brightness temperature.

The Monsoon '90 study was a field campaign conducted over the USDA ARS Walnut Gulch experimental watershed in south-eastern Arizona during June–September 1990 (Kustas and Goodrich, 1994). The objective was to assess the feasibility of using remotely sensed data coupled with water and energy balance models to obtain large-area estimates of hydrological and energy fluxes for semi-arid rangelands. The PBMR soil moisture data from the Monsoon '90 experiment provided an opportunity to assess the feasibility of obtaining soil hydraulic information from repetitive observations of soil moisture over the area. In this case, six flights were flown over the site after a rather large rain event (>5 cm) so that the spatial nature of the dry-down could be observed. With these data, Santanello *et al.* (2007) used a different approach from that of Camillo *et al.* (1986) and Burke *et al.* (1998) to obtain soil properties from the repetitive observations. They assumed that the soil moisture estimates of Schmugge *et al.* (1994) were valid and compared them with those estimated from a Land Surface Model NOAH (Chen *et al.*, 1996; Ek *et al.*, 2003). In the model, the hydraulic parameters were estimated from the soil textures, which were varied to get the best fit with the observed soil moisture values. The resulting estimates of soil texture corresponded well with the soils of the watershed.

Summary

The upper few centimetres of the vadose zone serve as a continuum between the soil and the atmosphere and thus control the moisture and energy fluxes at the interface. The hydraulic properties of the soil in the vadose zone are important for these exchanges. To estimate the hydraulic properties of soils it is necessary to have some information on the flow or temporal change of water content in the soil. This requires repetitive measurements of soil moisture in the surface, which can be done remotely. This chapter is a review of the basic principles of the remote sensing of soil moisture with microwave radiometers, and it describes how repetitive observations can be used to estimate soil hydraulic properties. Examples of how repetitive observations of the microwave emission from the soil, coupled with a soil model, can be used to estimate soil properties are given using data from field and aircraft platforms. The results show good agreement with classical measurements of such properties. The analyses by Camillo *et al.* (1986) and Burke *et al.* (1998) have shown that it is possible to infer soil hydraulic properties from repeated microwave brightness temperature (T_B) observations with a reasonable degree of accuracy. What is uncertain at the present time is the required frequency of these observations in order to make valid inferences. Working with daily microwave observations of T_B, Mattikali *et al.* (1995) have shown that there is a statistically significant relationship between K_s and changes of T_B. The data they worked with were image data over a large area in Oklahoma, and indicate the possibility of obtaining these estimates on an areal basis. This result is reinforced by the results of Hollenbeck *et al.* (1996), who were able to relate the temporal change of T_B on an areal basis to soil differences. Santanello *et al.* (2007) obtained quantitative soil results over a large area using remotely sensed soil moisture data from the Monsoon '90 experiment. The results presented here are encouraging but do require further validation in a wider range of conditions.

With the soil moisture data for large areas coming available from the SMOS satellite (Kerr *et al.*, 2010) and NASA's future SMAP (Soil Moisture Active Passive) mission (Entakhabi *et al.*, 2010) it will be very interesting to see if any of these techniques can be applied to 40 km estimates of surface soil moisture. At this scale, the soil data would be very useful for incorporation into the atmospheric models used for climate studies and weather forecasting.

Acknowledgements

A portion of this chapter was published in the American Geophysical Union Monograph: *Assessment of Non-Point Source Pollution in the Vadose Zone* (Schmugge, 1999). I would also like to acknowledge Dr Eleanor Burke for providing Figs 19.5, 19.6 and 19.7 from her thesis on this topic.

References

Blanchard, B.J. and O'Neill, P.E. (1983) Estimation of the hydraulic character of soils with passive microwave systems, In: *Proceedings of National Conference on Advances in Infiltration*. American Society of Agricultural Engineering, St Joseph, Michigan, pp. 215–225.

Burke, E.J., Gurney, R.J., Simmonds, L.P. and Jackson, T.J. (1997) Calibrating s soil water and energy budget model with remotely sensed data to obtain quantitative information about the soil. *Water Resources Research* 33, 1689–1697.

Burke, E.J., Gurney, R.J., Simmonds, L.P. and O'Neill, P.E. (1998) Using a modeling approach to predict soil hydraulic properties from passive microwave measurements. *IEEE Transactions on Geoscience and Remote Sensing* 36, 454–462.

Camillo, P.J., O'Neill, P.E. and Gurney, R.J. (1986) Estimating soil hydraulic parameters using passive microwave data. *IEEE Transactions on Geoscience and Remote Sensing* 24, 930–936.

Campbell, G.S. (1974) A simple method for determining unsaturated conductivity from moisture retention data. *Soil Science* 117, 311–314.

Chanzy, A., Schmugge, T.J., Calvet, J.-C., Kerr, Y., van Oevelen, P., Grosjean, O. and Wang, J.R. (1997) Airborne microwave radiometry on a semi-arid area during HAPEX-Sahel. *Journal of Hydrology* 188, 287–311.

Chen, F., Mitchell, K., Schaake, J., Xue, Y.-K., Pan, H.-L., Koren, V., Duan, Q.Y., Ek, M. and Betts, A. (1996) Modeling of land-surface evaporation by four schemes and comparison with FIFE observations. *Journal of Geophysical Research* 101, 7251–7268.

Cosby, B.J., Hornberger, G.M., Clapp, R.B. and Ginn, T.R. (1984) A statistical exploration of the relationship of soil moisture characteristics to the physical properties of soils. *Water Resources Research* 20, 682–690.

Eagleman, J.R. and Lin, W.C. (1976) Remote sensing of soil moisture by a 21-cm passive radiometer. *Journal of Geophysical Research* 81, 3660–3666.

Ek, M.B. and Holtslan, A.A.M. (2003) Influence of soil moisture on boundary layer cloud development. *Journal of Hydrometeorology* 70, 86–99.

Entekhabi, D, Njoku, E.G., O'Neill, P.E., Kellogg, K.H., Crow, W.T., Edelstein, W.N., Entin, J.K., Goodman, S.D., Jackson, T.J., Johnson, J., Kimball, J., Piepmeier, J.R., Koster, R.D., Martin, N., McDonald, K.C., Moghaddam, M., Moran, S., Reichle, R., Shi, J.C., Spencer, M.W., Thurman, S.W., Tsang, L. and Van Zyl, J. (2010) The Soil Moisture Active Passive (SMAP) Mission. *Proceedings of the IEEE* 98, 704–716.

Goutorbe, J.P., Lebel, T., Tinga, A., Bessemoulin, P., Brouwer, J., Dolman, H., Engman, E.T., Gash, J.G.C., Hoepffner, M., Kabat, P., Kerr, Y.H., Monteny, B., Prince, S.D., Said, F., Sellers, P. and Wallace, J. (1994) HAPEX-SAHEL: a large scale study of land atmosphere interactions in the semi-arid tropics. *Annales Geophysicae* 12, 53–64.

Hollenbeck, K.J., Schmugge, T.J., Hornberger, G.M. and Wang, J.R. (1996) Identifying soil hydraulic heterogeneity by detection of relative change in passive microwave remote sensing observations. *Water Resources Research* 32, 139–148.

Jackson, T.J. and Schmugge, T.J. (1989) Passive microwave remote sensing system for soil moisture: some supporting research. *IEEE Transactions on Geoscience and Remote Sensing* 27, 225–235.

Jackson, T.J., Le Vine, D.M., Swift, C.T., Schmugge, T. and Schiebe, F.R. (1995) Large area mapping of soil moisture using the ESTAR passive microwave radiometer in Washita'92. *Remote Sensing of Environment* 54, 27–37.

Jackson, T.J., Le Vine, D.M., Hsu, A.Y., Oldak, A., Starks, P.J., Swift, C.T., Isham, J.D. and Haken, M. (1999) Soil moisture mapping at regional scales using microwave radiometry: the Southern Great Plains hydrology experiment. *IEEE Transactions on Geoscience and Remote Sensing* 37, 2136–2151.

Jackson, T.J., Hsu, A.Y., van de Grienc, A. and Eaglemon, J.R. (2004) Skylab L-band microwave radiometer observations of soil moisture. *International Journal of Remote Sensing* 24, 2585–2606.

Kerr, Y.H., Waldteufel, P., Wigneron, J.-P., Delwart, S., Cabot, F., Boutin, J., Escorihuela, M.-J., Font, J., Reul, N., Gruhier, C., Juglea, S.E., Drinkwater, M.R., Hahne, A., Martín-Neira, M. and Mecklenburg, S. (2010) The SMOS mission: new tool for monitoring key elements of the global water cycle. *Proceedings of the IEEE* 98, 666-687.

Kustas, W.P. and Goodrich, D. (1994) Preface to Special Issue: MONSOON '90 Multidisciplinary Experiment. *Water Resources Research* 30, 1211–1225.

Mattikalli, N.M., Engman, E.T., Ahuja, L.R. and Jackson, T.J. (1995) Estimation of soil properties from microwave remote sensing of soil moisture. In: Engman, E.T., Guyot, G. and Marino, C.M. (eds) *Remote Sensing for Agriculture, Forestry and Natural Resources. Proceedings SPIE* 2585, 89–101.

Monerris, A. and Schmugge, T. (2009) Soil moisture estimation using L-band radiometry. In: Jedlovec, G. (ed.) *Advances in Geoscience and Remote Sensing.* InTech – Open Access Publisher, Rijeka, Croatia. Available at: http://www.intechopen.com/articles/show/title/soil-moisture-estimation-using-l-band-radiometry (accessed 7 March 2011).

Nielsen, D.R., Biggar, J.W. and Erh, K.T. (1973) Spatial variability of field-measured soil-water properties. *Hilgardia* 42, 215–260.

Santanello, J.A., Peters-Lidard, C.D., Garcia, M.E., Mocko, D.M., Tischler, M.A., Moran, M.S. and Thoma, D.P. (2007) Using remotely-sensed estimates of soil moisture to infer soil texture and hydraulic properties across a semi-arid watershed. *Remote Sensing of Environment* 110, 79–97.

Schmugge, T.J. (1999) Estimating near-surface soil hydraulic properties with microwave remote sensing. In: Corwin, D.L., Loague, K. and Ellsworth, T.R. (eds) *Assessment of Non-Point Source Pollution in the Vadose Zone*. American Geophysical Union, Washington, DC, pp. 217–222.

Schmugge, T.J. and Jackson, T.J. (1994) Mapping surface soil moisture with microwave radiometers. *Meteorology and Atmospheric Physics* 54, 213–223.

Schmugge, T., Gloersen, P., Wilheit, T.T. and Geiger, F. (1974) Remote sensing of soil moisture with microwave radiometers. *Journal of Geophysical Research* 79, 317–323.

Schmugge, T.J., Jackson, T.J., Kustas, W.P., Roberts, R., Parry, R., Goodrich, D., Amer, S. and Weltz, M.A. (1994) PBMR observations of surface soil moisture in Monsoon '90. *Water Resources Research* 30, 1321–1327.

Sharma, M.L., Gander, G.A. and Hunt, C.G. (1980) Spatial variability of infiltration in a watershed. *Journal of Hydrology* 45, 101–122.

Wang, J.R. (1985) Effect of vegetation on soil moisture sensing observed from orbiting microwave radiometers. *Remote Sensing of Environment* 17, 141–151.

Wang, J.R. and Schmugge, T.J. (1980) An empirical model for the complex dielectric permittivity of soils as a function of water content. *IEEE Transactions on Geoscience and Remote Sensing* 18, 288–295.

Wang, J.R., Schmugge, T.J., Shiue, J.C. and Engman, E.T. (1990) The L-band PBMR measurements of surface soil moisture in FIFE. *IEEE Transactions on Geoscience and Remote Sensing* 28, 906–914.

Wilheit, T.T. (1978) Radiative transfer in a plane stratified dielectric. *IEEE Transactions on Geoscience and Electronics* 16, 138–143.

Index